Basic Curves

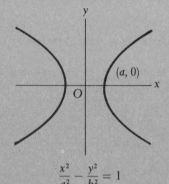

$$\frac{x^2}{a^2} - \frac{y^2}{b^2} = 1$$

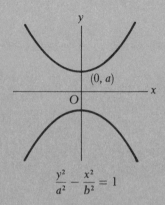

$$\frac{y^2}{a^2} - \frac{x^2}{b^2} = 1$$

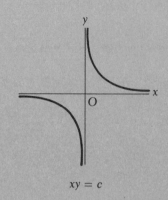

$$xy = c$$

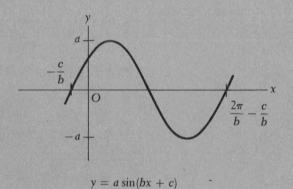

$$y = a \sin(bx + c)$$

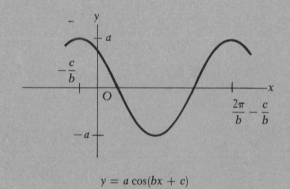

$$y = a \cos(bx + c)$$

Geometric Formulas

Triangle	Circle	Cylinder	Cone	Sphere

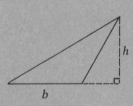 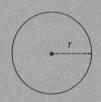 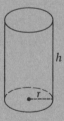 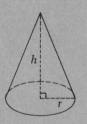

Triangle
$$A = \tfrac{1}{2}bh$$

Circle
$$A = \pi r^2$$
$$c = 2\pi r$$

Cylinder
$$V = \pi r^2 h$$
$$A = 2\pi rh + 2\pi r^2$$

Cone
$$V = \tfrac{1}{3}\pi r^2 h$$

Sphere
$$V = \tfrac{4}{3}\pi r^3$$
$$A = 4\pi r^2$$

Basic
Technical
Mathematics

Fourth Edition

Other Benjamin/Cummings Titles of Related Interest

Basic Technical Mathematics with Calculus, Fourth Edition, by Allyn J. Washington

Basic Technical Mathematics with Calculus, Metric Version, Fourth Edition, by Allyn J. Washington

Student Solutions Manual for the Washington Series in Basic Technical Mathematics, Fourth Edition, by Anne Zeigler and Frances Bazen

Introduction to Technical Mathematics, Third Edition, by Allyn J. Washington and Mario F. Triola

Technical Calculus with Analytic Geometry, Second Edition, by Allyn J. Washington

Introduction to Geometry (module), by Allyn J. Washington

Problems Supplement for Technical Mathematics, by Larry Pucke and Tom Stark

Problems in Technical Mathematics for Electricity/Electronics, by Forrest Barker

Basic Mathematics for the Trades, by E. J. Mowbray and J. C. Reeder, in consultation with Allyn J. Washington

Arithmetic and Beginning Algebra, by Allyn J. Washington

Essentials of Basic Mathematics, Third Edition, by Allyn J. Washington, Samuel H. Plotkin, and Carolyn E. Edmond

Basic Technical Mathematics

Fourth Edition

Allyn J. Washington

Dutchess Community College

The Benjamin/Cummings Publishing Company
Menlo Park, California ● Reading, Massachusetts
Wokingham, Berkshire, U.K. ● Amsterdam
Don Mills, Ontario ● Sydney

Sponsoring editors: Craig S. Bartholomew and Sally D. Elliott

Production coordinator: Charles Hibbard

Cover designer: Gary Head

Book designer: Gary Head

Copyeditor: Linda Thompson

Library of Congress Cataloging in Publication Data

Washington, Allyn J.
  Basic technical mathematics.

  Includes index.
  1. Mathematics—1861–   I. Title.
QA39.2.W37 1984   510   84-14534
ISBN 0-8053-9550-4

CDEFGHIJ-DO-89876

The Benjamin/Cummings Publishing Company, Inc.
2727 Sand Hill Road
Menlo Park, California 94025

Preface

Scope of the Book

This book is intended primarily for students in technical and pre-engineering technology programs or other programs for which a coverage of basic mathematics is required.

Chapters 1 through 19 provide the necessary background in algebra and trigonometry for analytic geometry and calculus courses, and Chapters 1 through 20 provide the background necessary for calculus courses. There is an integrated treatment of mathematical topics, primarily algebra and trigonometry, which are necessary for a sound mathematical background for the technician. Numerous applications from many fields of technology are included, primarily to indicate where and how mathematical techniques are used. However, it is not necessary that the student have a specific knowledge of the technical area from which any given problem is taken.

It is assumed that students using this text will have a background including algebra and geometry. However, the material is presented in sufficient detail for use by those whose background is possibly deficient to some extent in these areas. The material presented here is sufficient for two or three semesters.

One of the primary reasons for the arrangement of topics in this text is to present material in an order that allows a student to take courses in allied technical areas, such as physics and electricity, concurrently. These allied courses normally require a student to know certain mathematical topics by certain definite times; yet the traditional order of topics in mathematics makes it difficult to attain this coverage without loss of continuity. However, the material in this book can be rearranged to fit any appropriate sequence of topics if this is deemed necessary. Another feature of this text is that certain topics that are traditionally included primarily for mathematical completeness have been omitted.

The approach used here is basically an intuitive one. It is not unduly rigorous mathematically, although all appropriate terms and concepts are introduced as needed and given an intuitive or algebraic foundation. The book's aim is to help the student develop a feeling for mathematical methods, and not simply to provide a collection of formulas. The text emphasizes that it is essential for the student to be fluent in algebra and trigonometry in order to understand and succeed in any subsequent work in mathematics.

New Features

This fourth edition of *Basic Technical Mathematics* includes all the basic features of the first three editions. However, **most sections have been rewritten** to some degree to include additional explanatory material, examples, and exercises. Some sections have been extensively revised. Among the new features of this edition are the following:

- **Calculator material** is integrated throughout the text. Specific operations are introduced where appropriate, and proper calculator use is shown in examples.

- New sections are Section 1–4, which introduces the **calculator,** Section 2–4, which covers graphs of functions defined by tables of data, and interpolation (as a general concept), Section 4–2, which introduces **slope** and the **graph of the straight line,** and Section 6–4, which covers the **graph of the quadratic function.**

- There is a new appendix section giving 14 selected computer programs in **BASIC.** These programs are keyed to the text material by use of a computer symbol (as shown to the left).

- Certain sections of the third edition have been divided into two sections to provide more detailed coverage. Thus Section 1–13 is a section devoted completely to **applied verbal problems,** Section 8–1 is devoted only to the **graphical introduction of vectors,** and Section 21–2 is devoted only to **measures of central tendency.** Also, the coverage of **geometric topics** in Appendix C has been significantly expanded.

- Additional significant mathematical notation is used. In Section 9–3, the displacement of a sine curve is now shown directly as $-c/b$; in Section 11–4, the polar form $r \angle \theta$ of a complex number is introduced and used extensively; in Section 16–2, open and solid circles are now used to show graphical boundaries of inequalities.

- Other changes in topic coverage are in Section 3–3, where the use of the calculator is stressed for finding values of the trigonometric functions and tables are only lightly covered, Section 10–3, where rationalization of denominators is shown as an appropriate procedure but not as a required part of the simplification of a radical, and in Section 12–5, where the properties of logarithms are now the primary emphasis in calculations with logarithms.

- There are a number of new exercises related to **mechanics and modern technical areas.**

- There is a special margin symbol (shown at the left) which is used to clearly indicate points with which students commonly make errors or which they tend to have more difficulty in handling.

- There is a much **greater use of figures** in this edition: more than 410 (an increase of more than 35%) in the text and more than 450 in the answer section. Many of the new figures assist the student in the examples and exercises.

- There are now more than **6,450 exercises,** an increase of about 15%. The exercises are grouped so that there generally is an even-numbered exercise equivalent to each odd-numbered exercise.

- Many new examples have been added, and more detail has been included with others. There are now more than **850 examples,** an increase of about 10%.

Other features are:

- The many **examples** included in this text are often used advantageously to **introduce concepts,** as well as to **clarify and illustrate points** made in the text.

- **Graphical methods** are used extensively.

- **Stated problems are included a few at a time** to allow the student to develop techniques of solution.

- Those topics that experience has shown to be more difficult for the student have been developed in more detail, with many examples.

- **The order of coverage can be changed** in several places without loss of continuity, and certain sections may be omitted without loss of continuity. Any omissions or changes in order will, of course, depend on the type of course and the completeness required.

- The **chapter on statistics** is included in order to introduce the student to statistical and empirical methods.

- **Review exercises** are included after each chapter. These may be used either for additional problems or for review assignments.

- The answers to all the odd-numbered exercises are given at the back of the book. Included are answers to **graphical problems** and other types that are not always included in textbooks.

Acknowledgments

The author gratefully acknowledges the contributions of the following people, all of whom read all or parts of the Fourth Edition manuscript. Their detailed comments and valuable suggestions were of great assistance.

Elizabeth L. Bliss
Trident Technical College

Glenn R. Boston
Catawba Valley Technical College

Cheryl Cleaves
State Technical Institute at
  Memphis

Ronald C. Curry
Brevard Community College

Margaret Hobbs
State Technical Institute at
  Memphis

Wendell A. Johnson
The University of Akron

Jon C. Luke
Indiana University–Purdue
  University at Indianapolis

S. Paul Maini
Suffolk County Community
    College

Frank Mannasmith
Missouri Institute of Technology

Harold Oxsen
Diablo Valley College

David A. Petrie
Cypress College

Larry Rubin
Nashville State Technical Institute

John A. Seaton
Prince George's Community
    College

Robert Seaver
Lorain County Community
    College

Tom Stark
Cincinnati Technical College

Paul D. Williams
Pennsylvania State University,
    Fayette Campus

The author also sincerely thanks the following people, including those who reviewed earlier editions, for their valuable comments and contributions.

A. David Allen
Ricks College

Frances Bazen
Florence Darlington Technical
    College

Weston H. Beale
Camden County College

Frank Blou
Essex County College

Irwin P. Bramson
Daniel Webster College

Dale Bryson
Umpqua Community College

Frank Caldwell
York Technical College

Robert Campbell
North Shore Community College

Paul Carter
State University of New York,
    Morrisville

Al Coens
Lincoln Trail College

James E. Corbett, Jr.
J. S. Reynolds Community
    College

John Dennin
State University of New York,
    Morrisville

Tom Divver
Spartanburg Technical College

Alfred B. Fehlmann, Jr.
University of Dayton

Sanford Galat
Los Angeles Valley College

Judith Gersting
Indiana University–Purdue
    University at Indianapolis

R. Joanna Hampton
Southern Illinois University

John R. Hinton
Columbus Technical Institute

Mark Juliani
Community College of Allegheny
    County

Ellen Kowalczyk
Madison Area Technical College

Larry R. Lance
Columbus Technical Institute

Hollan H. Land
State Technical Institute at
    Memphis

Walter H. Lange
University of Toledo, Community
    and Technical College

Ellen Kowalczyk
Madison Area Technical College

Larry R. Lance
Columbus Technical Institute

Hollan H. Land
State Technical Institute at
  Memphis

Walter H. Lange
University of Toledo, Community
  and Technical College

M. Levine
Virginia Western Community
  College

Mark Manchester
State University of New York,
  Morrisville

Edwin P. McCravey
Midlands Technical Education
  Center

Owen McKinnie
New River Community College

Barbara Miller
Lorain County Community
  College

Frank Montgomery
Randolph Technical College

Randy Parker
Lenoir Community College

Marilyn L. Peacock
Tidewater Community College

John Pfetzing
Sinclair Community College

Maurice M. Roy
New Brunswick Community
  College

Robert Saluga
Clermont General-Technical
  College

B. Samimy
University of Cincinnati

Shirish Shah
Community College of Baltimore

Gary L. Thesing
Lake Superior State College

Al Vaughn
Ulster County Community College

William K. Viertel, formerly of
  State University of New York
  Agricultural and Technical
  College at Canton

Richard C. Wheeler
Wentworth Institute

Anne Zeigler
Florence Darlington Technical
  College

In addition, I wish to thank Gail Brittain, Stephen Lange, Michael Mayer, John Davenport, Mario Triola, and Jeffrey Clark of Dutchess Community College for their help and suggestions for this and earlier editions. In particular, I wish to thank Mario Triola for his very valuable assistance in checking proofs and many of the answers, and Jeffrey Clark for checking the computer programs.

The assistance, cooperation, and diligent efforts of Charles Hibbard and others on the Benjamin/Cummings staff during the production of this text are also greatly appreciated. Finally, special mention is due my wife, who helped with her patience during the preparation of this edition, and for checking text and answers of earlier editions.

A. J. W.

Contents

CHAPTER 1

Fundamental Concepts and Operations

Mathematics has played a most important role in the development and understanding of the various fields of technology and in the endless chain of technological and scientific advances of our time. With the mathematics we shall develop in this text, many kinds of applied problems can and will be solved. Of course, we cannot solve the more advanced types of problems which arise, but we can form a foundation for the more advanced mathematics which is used to solve such problems. Therefore, the development of a real understanding of the mathematics presented in this text will be of great value to you in your future work.

A thorough understanding of algebra is essential in the study of any of the fields of elementary mathematics. It is important for the reader to learn and understand the basic concepts and operations presented here, or the development and the applications of later topics will be difficult to comprehend. Unless the algebraic operations are understood well, the result will be a weak foundation for further work in mathematics and in many of the technical areas where mathematics may be applied.

We shall begin our study of mathematics by reviewing some of the basic concepts and operations that deal with numbers and symbols. With these we shall be able to develop the topics in algebra which are necessary for further progress into other fields of mathematics, such as trigonometry and calculus.

1–1 Numbers and Literal Symbols

The way we represent numbers today has been evolving for thousands of years. *The first numbers used were those which stand for whole quantities, and these we call the* **positive integers** *(or* **natural numbers**). The positive integers are represented by the symbols 1, 2, 3, 4, and so forth.

Of course, it is necessary to have numbers to represent parts of certain quantities, and for this purpose fractional quantities are introduced. *The name* **positive rational number** *is given to any number that we can represent by the division of one positive integer by another. Numbers that cannot be designated by the division of one integer by another are termed* **irrational.**

Example A

The numbers 5, 18, and 19 are positive integers. They are also rational numbers, since they may be written as $\frac{5}{1}$, $\frac{18}{1}$, and $\frac{19}{1}$. Normally we do not write the 1s in the denominators.

The numbers $\frac{1}{2}$, $\frac{5}{8}$, $\frac{11}{3}$, and $\frac{106}{17}$ are positive rational numbers, since the numerator and the denominator of each are integers.

The numbers $\sqrt{2}$, $\sqrt{3}$, and π are irrational. It is not possible to find two integers which represent these numbers when one of the integers is divided by the other. For example, $\frac{22}{7}$ is not an *exact* representation of π; it is only an *approximation*.

In addition to the positive numbers, it is necessary to introduce **negative numbers,** not only because we need to have a numerical answer to problems such as $5 - 8$, but also because the negative sign is used to designate direction. *Thus,* -1, -2, -3, *and so on are the* **negative integers.** *The number* **zero** *is an integer, but it is neither positive nor negative. This means that the* **integers** *are the numbers* . . . , -3, -2, -1, 0, 1, 2, 3, . . . , *and so on.*

The integers, the rational numbers, and the irrational numbers, which include all such numbers which are zero, positive, or negative, make up what we call the **real number system.** We shall use real numbers throughout this text, with one important exception. In Chapter 11 we shall be using **imaginary numbers,** *which is the name given to square roots of negative numbers.* (The symbol j is used to designate $\sqrt{-1}$, which is not part of the real number system.) However, until Chapter 11, when we discuss the operations on imaginary numbers in detail, it will be necessary only to recognize them if they occur.

Example B

The number 7 is an integer. It is also a rational number since $7 = \frac{7}{1}$, and it is a real number since the real numbers include all of the rational numbers.

The number 3π is irrational, and it is real since the real numbers include all of the irrational numbers.

The numbers $\sqrt{-10}$ and $-\sqrt{-7}$ are imaginary.

The number $\frac{1}{8}$ is rational and real. The number $\sqrt{5}$ is irrational and real.

The number $\frac{-3}{7}$ is rational and real. The number $-\sqrt{7}$ is irrational and real.

The number $\frac{\pi}{6}$ is irrational and real. The number $\frac{\sqrt{-3}}{2}$ is imaginary.

A **fraction** *may contain any number or symbol representing a number in its numerator or in its denominator.* Thus, a fraction may be rational, irrational, or imaginary.

Example C   The numbers $\frac{2}{7}$ and $\frac{-3}{2}$ are fractions, and they are also rational.

The numbers $\frac{\sqrt{2}}{9}$ and $\frac{6}{\pi}$ are fractions, but they are not rational numbers. It is not possible to express either as the ratio of one integer to another.

The number $\frac{\sqrt{-3}}{2}$ is a fraction, and it is also imaginary.

The real numbers may be represented as points on a line. We draw a horizontal line and designate some point on it by O, which we call the **origin** (see Fig. 1–1). The number zero, which is an integer, is located at this point. Then equal intervals are marked off from this point toward the right, and the positive integers are placed at these positions. The other positive rational numbers are located between the positions of the integers.

Now we can give the direction interpretation to negative numbers. By starting at the origin and proceeding to the left, in the **negative direction,** we locate all the negative numbers. *As shown in Fig. 1–1, the positive numbers are to the right of the origin and the negative numbers to the left.* Representing numbers in this way will be especially useful when we study graphical methods.

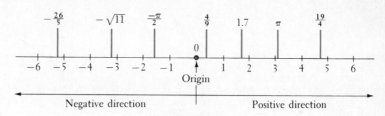

Figure 1-1

It will not be proved here, but the rational numbers do not take up all the positions on the line; the remaining points represent irrational numbers.

Another important mathematical concept we use in dealing with numbers is the **absolute value** of a number. By definition, *the absolute value of a positive number is the number itself, and the absolute value of a negative number is the corresponding positive number (obtained by changing its sign).* We may interpret the absolute value as being the number of units a given number is from the origin, regardless of direction. The absolute value is designated by $|\ |$ placed around the number.

Example D   The absolute value of 6 is 6, and the absolute value of -7 is 7. We designate these by $|6| = 6$ and $|-7| = 7$.

Other examples are $|-\pi| = \pi$, $|\frac{7}{5}| = \frac{7}{5}$, $|-\sqrt{2}| = \sqrt{2}$, and $|0| = 0$.

On the number scale, *if a first number is to the right of a second number, then the first number is said to be* **greater than** *the second. If the first number is to* the left of the second, it is **less than** *the second number*. The symbol $>$ is used to designate "is greater than," and the symbol $<$ is used to designate "is less than." These are called **signs of inequality.**

Example E

$$6 > 3, \quad 8 > -1, \quad 5 < 9, \quad 0 > -4, \quad -2 > -4, \quad -1 < 0$$

Every number, except zero, has a **reciprocal.** *The reciprocal of any number is 1 divided by that number.*

Example F

The reciprocal of 7 is $\frac{1}{7}$. The reciprocal of $\frac{2}{3}$ is

$$\frac{1}{\frac{2}{3}} = 1 \cdot \frac{3}{2} = \frac{3}{2}$$

The reciprocal of π is $1/\pi$. The reciprocal of -5 is $-\frac{1}{5}$. Notice that the negative sign is retained in the reciprocal.

In applications, numbers are often used to represent some type of measurement and therefore have certain units of measurement associated with them. *Such numbers are referred to as* **denominate numbers.** For a discussion of units of measurement and the symbols which are used, see Appendix B. The following example illustrates the use of units and the symbols which represent them.

Example G

To indicate that an object is 10 feet long, we represent the length as 10 ft.

To indicate that the speed of a projectile is 150 meters per second, we represent the speed as 150 m/s. (Note the use of s for second. We use s rather than sec, which is also common.)

To indicate that the volume of a container is 500 cubic centimeters, we represent the volume by 500 cm^3. (Also common are cc and cu cm.)

To this point we have been dealing with numbers in their explicit form. However, it is normally more convenient to state definitions and operations on numbers in a generalized form. *To do this we represent the numbers by letters, often referred to as* **literal numbers.**

For example, we can say, "if a is to the right of b on the number scale, then a is greater than b, or $a > b$." This is more convenient than saying, "if a first number is to the right of a second number, then the first number is greater than the second number." The statement "the reciprocal of a number a is $1/a$" is another example of using letters to stand for numbers in general.

In an algebraic discussion, certain letters are sometimes allowed to take on any value, while other letters represent the same number throughout the discussion. *Those literal numbers which may vary in a given problem are called* **variables,** *and those which are held fixed are called* **constants.**

Common usage today normally designates the letters near the end of the alphabet as variables and letters near the beginning of the alphabet as constants. There are exceptions, but these are specifically noted. Letters in the middle of the alphabet are also used, but their meaning in any problem is specified.

Example H

The electric resistance R of a wire may be related to the temperature T by the equation $R = aT + b$. R and T may take on various values, and a and b are fixed for any particular wire. However, a and b may change if a different wire is considered. Here, R and T are the variables and a and b are constants.

Exercises 1–1

In Exercises 1 through 4, designate each of the given numbers as being an integer, rational, irrational, real, or imaginary. (More than one designation may be correct.)

1. $3, -\pi$ 2. $\dfrac{5}{4}, \sqrt{-4}$ 3. $-\sqrt{-6}, \dfrac{\sqrt{7}}{3}$ 4. $-\dfrac{7}{3}, \dfrac{\pi}{6}$

In Exercises 5 through 8, find the absolute value of each of the given numbers.

5. $3, \dfrac{7}{2}$ 6. $-4, \sqrt{2}$ 7. $-\dfrac{6}{7}, -\sqrt{3}$ 8. $-\dfrac{\pi}{2}, -\dfrac{19}{4}$

In Exercises 9 through 16, insert the correct sign of inequality ($>$ or $<$) between the given pairs of numbers.

9. $6 \quad 8$ 10. $7 \quad 5$ 11. $\pi \quad -1$ 12. $-4 \quad 3$

13. $-4 \quad -3$ 14. $-\sqrt{2} \quad -9$ 15. $-\dfrac{1}{3} \quad -\dfrac{1}{2}$ 16. $0.2 \quad 0.6$

In Exercises 17 through 20, find the reciprocals of the given numbers.

17. $3, -2$ 18. $\dfrac{1}{6}, -\dfrac{7}{4}$ 19. $-\dfrac{5}{\pi}, x$ 20. $-\dfrac{8}{3}, \dfrac{y}{b}$

In Exercises 21 through 24, locate the given numbers on a number line as in Fig. 1–1.

21. $2.5, -\dfrac{1}{2}$ 22. $\sqrt{3}, -\dfrac{12}{5}$ 23. $-\dfrac{\sqrt{2}}{2}, 2\pi$ 24. $\dfrac{123}{19}, -\dfrac{\pi}{6}$

In Exercises 25 through 34, solve the given problems.

25. List the following numbers in numerical order, starting with the smallest: $-1, 9, \pi,$ $\sqrt{5}, |-8|, -|-3|, -18$.

26. List the following numbers in numerical order, starting with the smallest: $\frac{1}{5}, -\sqrt{10},$ $-|-6|, -4, 0.25, |-\pi|$.

27. If a and b are positive integers and $b > a$, what type of number is represented by (a) $b - a$, (b) $a - b$?

28. If a and b represent positive integers, what kind of number is represented by (a) $a + b$, (b) a/b, (c) $a \times b$?

29. Describe the location of a number x on the number line when (a) $x > 0$, (b) $x < -4$.

30. Describe the location of a number x on the number line when (a) $|x| < 1$, (b) $|x| > 2$.

31. The pressure P and volume V of a certain body of gas are related by the equation $P = c/V$ for certain conditions. Identify the symbols as variables or constants.

32. The velocity v and height h of an object are related by the equation $v = \sqrt{2ah}$. Identify the literal symbols as variables or constants.

33. In writing a laboratory report, a student wrote "$-20°C > -30°C$." Is this statement correct?

34. After 2 s, the current in a certain circuit is less than 3 A. Using t to represent time and i to represent current, this statement may be written, "for $t > 2$ s, $i < 3$ A." In this way write the statement, "less than three meters from the light source the illuminance is greater than eight lumens (lm) per square meter." (Let I represent illuminance and s represent distance.)

1–2 Fundamental Laws of Algebra and Order of Operations

In performing the basic operations with numbers, we know that certain basic laws are valid. *These basic statements are called the* **fundamental laws of algebra.**

For example, we know that if two numbers are to be added, it does not matter in which order they are added. Thus $5 + 3 = 8$, as well as $3 + 5 = 8$. For this case we can say that $5 + 3 = 3 + 5$. This statement, generalized and assumed correct for all possible combinations of numbers to be added, is called the **commutative law** for addition. The law states that *the sum of two numbers is the same, regardless of the order in which they are added*. We make no attempt to prove this in general, but accept its validity.

In the same way we have the **associative law** for addition, which states that *the sum of three or more numbers is the same, regardless of the manner in which they are grouped for addition*. For example,

$$3 + (5 + 6) = (3 + 5) + 6$$

The laws which we have just stated for addition are also true for multiplication. Therefore, *the product of two numbers is the same, regardless of the order in which they are multiplied*, and *the product of three or more numbers is the same, regardless of the manner in which they are grouped for multiplication*. For example, $2 \cdot 5 = 5 \cdot 2$ and $5 \cdot (4 \cdot 2) = (5 \cdot 4) \cdot 2$.

There is one more very important law, called the **distributive law.** It states that *the product of one number and the sum of two or more other numbers is equal to the sum of the products of the first number and each of the other numbers of the sum*. For example,

$$4(3 + 5) = 4 \cdot 3 + 4 \cdot 5$$

In practice these laws are used intuitively, except perhaps for the distributive law. However, it is necessary to state them and to accept their validity, so that we may build our later results with them.

Not all operations are associative and commutative. For example, division is not commutative, since the indicated order of division of two numbers does matter. For example, $\frac{6}{5} \neq \frac{5}{6}$ ($\neq$ is read "does not equal").

Using literal symbols, the fundamental laws of algebra are as follows:

Commutative law of addition: $a + b = b + a$

Associative law of addition: $a + (b + c) = (a + b) + c$

Commutative law of multiplication: $ab = ba$

Associative law of multiplication: $a(bc) = (ab)c$

Distributive law: $a(b + c) = ab + ac$

Having identified the fundamental laws of algebra, we shall state the laws which govern the operations of addition, subtraction, multiplication, and division of signed numbers. These laws will be of primary and direct use in all of our work.

1. *To add two real numbers with like signs, add their absolute values and affix their common sign to the result.*

Example A

$$(+2) + (+6) = +(2 + 6) = +8$$
$$(-2) + (-6) = -(2 + 6) = -8$$

2. *To add two real numbers with unlike signs, subtract the smaller absolute value from the larger absolute value and affix the sign of the number with the larger absolute value to the result.*

Example B

$$(+2) + (-6) = -(6 - 2) = -4$$
$$(+6) + (-2) = +(6 - 2) = +4$$

3. *To subtract one real number from another, change the sign of the number to be subtracted, and then proceed as in addition.*

Example C

$$(+2) - (+6) = +2 + (-6) = -(6 - 2) = -4$$
$$(-a) - (-a) = -a + a = 0$$

The second part of Example C shows that subtracting a number from itself results in zero, even if the number is negative. Therefore, we see that *subtracting a negative number is equivalent to adding a positive number of the same absolute value*. This reasoning is the basis of the rule which states, "the negative of a negative number is a positive number."

4. *The product (or quotient) of two real numbers of like signs is the product (or quotient) of their absolute values. The product (or quotient) of two real numbers of unlike signs is the negative of the product (or quotient) of their absolute values.*

Example D

$$(+3)(+5) = +(3 \cdot 5) = +15 \qquad \frac{+3}{+5} = +\left(\frac{3}{5}\right) = +\frac{3}{5}$$

$$(-3)(+5) = -(3 \cdot 5) = -15 \qquad \frac{-3}{+5} = -\left(\frac{3}{5}\right) = -\frac{3}{5}$$

$$(-3)(-5) = +(3 \cdot 5) = +15 \qquad \frac{-3}{-5} = +\left(\frac{3}{5}\right) = +\frac{3}{5}$$

When we have an expression in which there is a combination of the basic operations, we must be careful to perform them in the proper order. Generally it is clear by the grouping of numbers as to the proper order of performing these operations. Numbers are grouped by the use of symbols such as **parentheses, ( ),** and the **bar, ——,** between the numerator and the denominator of a fraction. However, if the order of at least some of the operations is not defined by specific groupings, we use the following **order of operations** related to additions, subtractions, multiplications, and divisions.

5. *Operations within specific groupings are done first. In completing the operations within specific groupings and otherwise where the order of operations is not indicated by specific grouping, multiplications and divisions are performed first, and then additions and subtractions are performed.*

In performing additions and subtractions, or multiplications and divisions, we work from left to right.

Example E

The expression $20 \div (2 + 3)$ is evaluated by first adding $2 + 3$ and then dividing 20 by 5 to obtain the result 4. Here, the grouping of $2 + 3$ is clearly shown by the parentheses.

The expression $20 \div 2 + 3$ is evaluated by first dividing 20 by 2 and adding this quotient of 10 to 3 in order to obtain the result of 13. Here no specific grouping is shown, and therefore the division is performed before the addition.

The expression $16 - 2 \times 3$ is evaluated by first multiplying 2 by 3 and subtracting the product from 16 in order to obtain the result of 10. We do *not* first subtract 2 from 16.

When evaluating expressions, it is generally more convenient to change the operations and numbers so that the result is obtained by addition and subtraction of unsigned numbers (equivalent to positive numbers). This is especially true when more than one addition or subtraction is involved.

Example F

$$(-6) - 2(-4) + \frac{25}{-5} = (-6) - (-8) + (-5) = -6 + 8 - 5 = -3$$

Here we see that the multiplication and the division were performed first, and then the addition and the subtraction were performed. We also see the change to unsigned numbers before the addition and subtraction were actually done.

Example G

$$\frac{40}{(+7) + (-3)(+5)} = \frac{40}{+7 + (-15)} = \frac{40}{7 - 15} = \frac{40}{-8} = -5$$

Here the multiplication in the denominator was performed first, and then the addition was performed. It was necessary to evaluate the denominator before the division could be performed. We also see again the use of unsigned numbers in the denominator.

1–3 Operations with Zero

Since the basic operations with zero tend to cause some difficulty, we shall demonstrate them separately in this section.

If a represents any real number, the various operations with zero are defined as follows:

$$a \pm 0 = a \quad \text{(the symbol } \pm \text{ means ''plus or minus'')}$$
$$a \cdot 0 = 0$$
$$\frac{0}{a} = 0 \quad \text{if } a \neq 0$$

 Note that there is no answer defined for division by zero. To understand the reason for this, consider the problem of 4/0. If there were an answer to this expression, it would mean that the answer, which we shall call b, should give 4 when multiplied by 0. That is, $0 \cdot b = 4$. However, no such number b exists, since we already know that $0 \cdot b = 0$. Also, the expression 0/0 has no meaning, since $0 \cdot b = 0$ for any value of b which may be chosen. Thus, *division by zero is not defined.* All other operations with zero are the same as for any other number.

Example A $5 + 0 = 5,$ $7 - 0 = 7,$ $0 - 4 = -4,$

$$\frac{0}{6} = 0, \qquad \frac{0}{-3} = 0, \qquad \frac{5 \cdot 0}{7} = 0,$$

$\frac{8}{0}$ is undefined, $\frac{7 \cdot 0}{0 \cdot 6}$ is undefined

There is no need for confusion in the operations with zero. They will not cause any difficulty if we remember that

division by zero is undefined

and that this is the only undefined operation.

Exercises 1–2, 1–3

In Exercises 1 through 36, evaluate each of the given expressions by performing the indicated operations.

1. $6 + 5$

2. $8 + (-4)$

3. $(-4) + (-7)$

4. $(-3) + (+9)$

5. $16 - 7$

6. $(+8) - (+11)$

7. $-9 - (-6)$

8. $8 - (-4)$

9. $(8)(-3)$

10. $(+9)(-3)$

11. $(-7)(-5)$

12. $(+5)(-8)$

13. $\frac{-9}{+3}$

14. $\frac{-18}{-6}$

15. $\frac{-60}{-3}$

16. $\frac{+28}{-7}$

17. $(-2)(+4)(-5)$

18. $(+3)(-4)(+6)$

19. $\frac{(+2)(-5)}{10}$

20. $\frac{-64}{(2)(-4)}$

21. $9 - 0$

22. $\frac{0}{-6}$

23. $\frac{+17}{0}$

24. $\frac{(+3)(0)}{0}$

25. $8 - 3(-4)$

26. $20 + 8 \div 4$

27. $3 - 2(6) + \frac{8}{2}$

28. $0 - (-6)(-8) + (-10)$

29. $\frac{(+3)(-6)(-2)}{0 - 4}$

30. $\frac{7 - (-5)}{(-1)(-2)}$

31. $\frac{24}{3 + (-5)} - 4(-9)$

32. $\frac{-18}{+3} - \frac{4 - 6}{-1}$

33. $(-7) - \frac{-14}{2} - 3(2)$

34. $-7(-3) + \frac{+6}{-3} - (-9)$

35. $\frac{(+3)(-9) - 2(-3)}{3 - 10}$

36. $\frac{(+2)(-7) - 4(-2)}{-9 - (-9)}$

In Exercises 37 through 44, determine which of the fundamental laws of algebra is demonstrated.

37. $(6)(7) = (7)(6)$

38. $6 + 8 = 8 + 6$

39. $6(3 + 1) = 6(3) + 6(1)$

40. $4(5 \cdot 7) = (4 \cdot 5)(7)$

41. $3 + (5 + 9) = (3 + 5) + 9$

42. $8(3 - 2) = 8(3) - 8(2)$

43. $(2 \times 3) \times 9 = 2 \times (3 \times 9)$

44. $(3 \cdot 6) \cdot 7 = 7 \cdot (3 \cdot 6)$

In Exercises 45 through 50, answer the given questions.

45. What is the sign of the product of an even number of negative numbers?

46. What is the sign of the product of an odd number of negative numbers?

47. Using the division of 4 by 2 and then 2 by 4, show that division is not commutative.

48. Is subtraction commutative? Illustrate.

49. In a month, a home computer dealer sold eight computers, which cost her $2000 each; on each, she made a profit of $1000. Set up the expression which would be used to determine the total amount the dealer received for these computers. What fundamental law is illustrated?

50. A boat travels at 6 mi/h in still water. A stream flows at 4 mi/h. If the boat travels downstream for 2 h, set up the expression which would be used to evaluate the distance traveled. What fundamental law is illustrated?

1–4 The Calculator

In the preceding sections, a number of basic calculations have been used. Throughout the remainder of the book, there will be numerous calculations to do. Many of them can be done mentally, but many others will require the use of a calculator.

A scientific calculator can be used for any of the calculations found in the book. Many of the more advanced models can also perform many types of calculations beyond the needs of this text. However, the calculator that performs only the basic arithmetic operations is in itself not sufficient for many of the calculations found in this text.

In Appendix D there is a discussion of the use of the basic keys on a scientific calculator. Using Appendix D as a reference, we shall show how the calculator can be used to do calculations wherever it may be necessary. In many examples in the remainder of the text, as basic types of calculation are encountered, we shall show the sequence of keys that is to be used.

If you are not sufficiently familiar with your calculator, you should review your calculator manual or Appendix D, for *you must be familiar with the order in which the keys must be used on your particular calculator*. The order in which the operations are performed by the calculator, which determines the order in which the keys are used, is called the **logic** of the calculator. The sequence of keys we show in the examples of this book is that of a calculator using *algebraic* *logic* and it is proper for many calculators, but it may require a change for use on other calculators.

The following examples show the use of a calculator in making calculations of the types shown in the previous sections.

Example A

Find the reciprocal of 25.

By using the definition of a reciprocal, we know that we can find the reciprocal of 25 by dividing 1 by 25. The sequence of calculator keys for this is

$$ 1 \quad \boxed{\div} \quad 25 \quad \boxed{=} $$

The result is 0.04.

However, most scientific calculators have a key for reciprocals, the $\boxed{1/x}$ key, as shown in Appendix D. By using it, the sequence of keys is simply

$$ 25 \quad \boxed{1/x} $$

Again, the result is 0.04.

Example B

Calculate the value of $38.3 - 12.9(-3.58)$.

Recalling the proper order of operations, we know that the product $12.9(-3.58)$ is to be done before the subtraction. However, if your calculator uses algebraic logic, the numbers can be entered as shown, and the calculator will do the multiplication first. Also, the sign of -3.58 can be entered by use of the $\boxed{+/-}$ key. Therefore, the sequence of keys is

$$ 38.3 \quad \boxed{-} \quad 12.9 \quad \boxed{\times} \quad 3.58 \quad \boxed{+/-} \quad \boxed{=} $$

and the display shows 84.482.

If your calculator does not use algebraic logic, it may be necessary to perform the multiplication first.

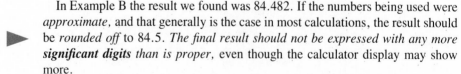

In Example B the result we found was 84.482. If the numbers being used were *approximate,* and that generally is the case in most calculations, the result should be *rounded off* to 84.5. *The final result should not be expressed with any more* **significant digits** *than is proper,* even though the calculator display may show more.

A discussion of the terms **rounding off, significant digits,** and **approximate numbers** can be found in Appendix B.

Example C

Calculate the value of $\dfrac{0.8695}{1.042 + 3.529} + (0.1563)(2.259)$.

The proper order of operations tells us first to find the value of the denominator of the first expression, then to do the division in the first expression and the multiplication in the second expression, and finally to add the results. A calculator with algebraic logic and parentheses will perform these operations in the proper order. Noting that the fraction in the first expression can be written as $0.8695 \div (1.042 + 3.529)$, we have the following sequence of keys.

.8695 $\boxed{\div}$ $\boxed{(}$ 1.042 $\boxed{+}$ 3.529 $\boxed{)}$

$\boxed{+}$ .1563 $\boxed{\times}$ 2.259 $\boxed{=}$

The calculator display shows the result as 0.5433027, which is 0.5433 when rounded off.

Exercises 1-4

In all of the following exercises, perform the required calculations on a scientific calculator.

In Exercises 1 through 8, perform the indicated operations. In Exercises 1 through 4, round off results to three significant digits and in Exercises 5 through 8, round off results to four significant digits.

1. $3.16 + 53.9 \div 117.8$

2. $807 - 245 \times 3.19$

3. $0.702 - 0.566 \div 0.108$

4. $46.7 \times 0.923 + 39.8 \times 0.362$

5. $\dfrac{23.962 \times 0.01537}{10.965 - 8.249}$

6. $\dfrac{0.69378 + 0.04997}{257.4 \times 3.216}$

7. $\dfrac{3872}{503.1} - \dfrac{2.056 \times 309.6}{395.2}$

8. $\dfrac{1}{0.5926} + \dfrac{3.6957}{2.935 - 1.054}$

In Exercises 9 through 12, verify the basic laws of algebra by calculating the value of each side of the equal sign. Do not round off results. Indicate which law is illustrated.

9. $37.962 + 5.049 = 5.049 + 37.962$

10. $0.3526 \times 2.4953 = 2.4953 \times 0.3526$

11. $68.572(5.0496 + 1.9256) = 68.572 \times 5.0496 + 68.572 \times 1.9256$

12. $0.0273(2.56 \times 8.99) = (0.0273 \times 2.56) \times 8.99$

In Exercises 13 through 16, find the products of the indicated signed numbers. Round off results to three significant digits.

13. -1.46×5.62

14. $(39.6) \times (-2.57)$

15. $(-0.3627) \times (-2.69)$

16. $(-7.064) \times (-0.0962)$

In Exercises 17 through 26, perform the indicated calculations.

17. Find the reciprocal of 2.8 by (a) using the definition of a reciprocal, (b) using the $\boxed{1/x}$ key. Round off the result, assuming that 2.8 is approximate.

18. Find the reciprocal of 0.48 by (a) using the definition of a reciprocal, (b) using the $\boxed{1/x}$ key. Round off the result, assuming that 0.48 is approximate.

19. Is $\pi > 3.1416$? Determine the answer by using the key sequence $\boxed{\pi}$ $\boxed{-}$ 3.1416.

20. Show that the value of π is not exactly the same as $\frac{22}{7}$ by using the key $\boxed{\pi}$ and by dividing 22 by 7.

21. If we find the decimal equivalent of a rational number, at some point some sequence of digits will start repeating endlessly. For an irrational number there is never such an endlessly repeating sequence of digits. Find the decimal equivalents of (a) $\frac{8}{33}$, (b) π. Note the repetition for $\frac{8}{33}$ and that no such repetition occurs for π.

22. Following Exercise 21, show that the decimal equivalent of the fraction $\frac{124}{990}$ indicates that it is rational.

23. Divide 2 by 0. What is the calculator display?

24. Divide 0 by 0. What is the calculator display?

25. In playing a particular video game, a player scored 17,845 points in 12 min 30 s. On the average, how many points did the player score each second?

26. In order to find the length of belting between two pulleys, it is necessary to calculate the value of

$$2 \times 136 + \frac{3.25(36.0 + 28.5)}{2}$$

Find the length of the belting (in inches) by performing the calculation and rounding off the result to three significant digits.

1–5 Exponents

We have introduced numbers and the fundamental laws which are used with them in the fundamental operations. Then we showed the use of the calculator in performing these operations. Also, we have shown the use of literal numbers to represent numbers. In this section we shall introduce some basic terminology and notation which are important to the basic algebraic operations developed in the following sections.

In multiplication we often encounter a number which is to be multiplied by itself several times. Rather than writing this number over and over repeatedly, we use the notation a^n, where a is the number being considered and n is the number of times it appears in the product. *In the expression a^n, the number a is called the* **base,** *the number n is called the* **exponent,** *and in words, the expression is read as the "nth power of a."*

Example A

$$4 \cdot 4 \cdot 4 \cdot 4 \cdot 4 = 4^5 \quad \text{(the fifth power of 4)}$$
$$(-2)(-2)(-2)(-2) = (-2)^4 \quad \text{(the fourth power of } -2)$$
$$a \cdot a = a^2 \quad \text{(the second power of } a, \text{ called ``}a \text{ squared'')}$$
$$(\tfrac{1}{5})(\tfrac{1}{5})(\tfrac{1}{5}) = (\tfrac{1}{5})^3 \quad \text{(the third power of } \tfrac{1}{5}, \text{ called ``}\tfrac{1}{5} \text{ cubed'')}$$
$$8 \cdot 8 \cdot 8 \cdot 8 \cdot 8 \cdot 8 \cdot 8 \cdot 8 \cdot 8 = 8^9 \quad \text{(the ninth power of 8)}$$

The basic operations with exponents will now be stated symbolically. We first state them for positive integers as exponents. Therefore, if m and n are positive integers, we have the following important operations for exponents.

$$a^m \cdot a^n = a^{m+n} \tag{1–1}$$

$$\frac{a^m}{a^n} = a^{m-n} \quad (m > n, \, a \neq 0) \qquad \frac{a^m}{a^n} = \frac{1}{a^{n-m}} \quad (m < n, \, a \neq 0) \tag{1–2}$$

$$(a^m)^n = a^{mn} \tag{1–3}$$

$$(ab)^n = a^n b^n, \qquad \left(\frac{a}{b}\right)^n = \frac{a^n}{b^n} \quad (b \neq 0) \tag{1–4}$$

Two forms are shown for Eq. (1–2) in order that the resulting exponent is a positive integer. This is generally, but not always, the best form of result. After the following three examples we will discuss zero and negative exponents. This will cover the case for $m = n$, and also show that the first form of Eq. (1–2) is its basic form.

In applying Eqs. (1–1) and (1–2), the base a must be the same for the exponents to be added or subtracted. When a problem involves a product of different bases, *only exponents of the same base may be combined.* In the following three examples, Eqs. (1–1) to (1–4) are verified and illustrated.

Example B

Applying Eq. (1–1), we have

$$a^3 \times a^5 = a^{3+5} = a^8$$

We see that this result is correct since we can also write

$$a^3 \times a^5 = (a \times a \times a)(a \times a \times a \times a \times a) = a^8.$$

Applying the first form of Eqs. (1–2), we have

$$\frac{a^5}{a^3} = a^{5-3} = a^2, \qquad \frac{a^5}{a^3} = \frac{\cancel{a} \times \cancel{a} \times \cancel{a} \times a \times a}{\cancel{a} \times \cancel{a} \times \cancel{a}} = a^2$$

Applying the second form of Eqs. (1–2), we have

$$\frac{a^3}{a^5} = \frac{1}{a^{5-3}} = \frac{1}{a^2}, \qquad \frac{a^3}{a^5} = \frac{\cancel{a} \times \cancel{a} \times \cancel{a}}{\cancel{a} \times \cancel{a} \times \cancel{a} \times a \times a} = \frac{1}{a^2}$$

Example C

Applying Eq. (1–3), we have

$$(a^5)^3 = a^{5(3)} = a^{15}, \qquad (a^5)^3 = (a^5)(a^5)(a^5) = a^{5+5+5} = a^{15}$$

Applying the first form of Eqs. (1–4), we have

$$(ab)^3 = a^3b^3, \qquad (ab)^3 = (ab)(ab)(ab) = a^3b^3$$

Applying the second form of Eqs. (1–4), we have

$$\left(\frac{a}{b}\right)^3 = \frac{a^3}{b^3}, \qquad \left(\frac{a}{b}\right)^3 = \left(\frac{a}{b}\right)\left(\frac{a}{b}\right)\left(\frac{a}{b}\right) = \frac{a^3}{b^3}$$

Example D

Other illustrations of the use of Eqs. (1–1) to (1–4) are as follows:

$$\frac{(3 \times 2)^4}{(3 \times 5)^3} = \frac{3^4 2^4}{3^3 5^3} = \frac{3 \times 2^4}{5^3}$$

$$(-x^2)^3 = [(-1)x^2]^3 = (-1)^3(x^2)^3 = -x^6$$

$$ax^2(ax)^3 = ax^2(a^3x^3) = a^4x^5$$

$$\frac{(ry^3)^2}{r(y^2)^4} = \frac{r^2y^6}{ry^8} = \frac{r}{y^2}$$

 In the third illustration, note that ax^2 means a times the square of x, and does *not* mean a^2x^2, whereas $(ax)^3$ *does* mean a^3x^3.

As we previously pointed out, Eqs. (1–1) to (1–4) were developed for use with positive integers as exponents. We shall now show how their use can be extended to include zero and negative integers as exponents.

In Eq. (1–2), if $n = m$, we would have $a^m/a^m = a^{m-m} = a^0$. Also $a^m/a^m = 1$, since any nonzero quantity divided by itself equals 1. Therefore, for Eqs. (1–2) to hold when $m = n$, we have

$$a^0 = 1, \quad (a \neq 0) \tag{1–5}$$

Equation (1–5) gives the definition of zero as an exponent. Since a has not been specified, this equation states that *any nonzero algebraic expression raised to the zero power is* 1. Also, the other laws of exponents are valid for this definition.

Example E

Equation (1–1) states that $a^m \cdot a^n = a^{m+n}$. If $n = 0$, we thus have $a^m \cdot a^0 = a^{m+0} = a^m$. Since $a^0 = 1$, this equation could be written as $a^m(1) = a^m$. This provides further verification for the validity of Eq. (1–5).

Example F

$$5^0 = 1, \qquad (2x)^0 = 1, \qquad (ax + b)^0 = 1$$
$$(a^2xb^4)^0 = 1, \qquad (a^2b^0c)^2 = a^4b^0c^2 = a^4c^2$$
$$2t^0 = 2(1) = 2$$

 We note in the last illustration that only t is raised to the zero power. If the quantity $2t$ were raised to the zero power, it would be written as $(2t)^0$.

If we apply the first form of Eqs. (1–2) to the case where $n > m$, the resulting exponent is negative. This leads us to the definition of a negative exponent.

Example G

Applying the first form of Eqs. (1–2) to a^2/a^7, we have

$$\frac{a^2}{a^7} = a^{2-7} = a^{-5}$$

Applying the second form of Eqs. (1–2) to the same fraction leads to

$$\frac{a^2}{a^7} = \frac{1}{a^{7-2}} = \frac{1}{a^5}$$

In order that these results can be consistent, it must be true that

$$a^{-5} = \frac{1}{a^5}$$

Following the reasoning in Example G, if we define

$$a^{-n} = \frac{1}{a^n}, \quad (a \neq 0) \tag{1–6}$$

then all of the laws of exponents will hold for negative integers.

Example H

$$3^{-1} = \frac{1}{3}, \qquad 4^{-2} = \frac{1}{16}, \qquad \frac{1}{a^{-3}} = a^3, \qquad a^4 = \frac{1}{a^{-4}}$$

Example I

$$(a^0b^2c)^{-2} = \frac{1}{(b^2c)^2} = \frac{1}{b^4c^2}$$
$$\left(\frac{a^3t}{b^2x}\right)^{-2} = \frac{(a^3t)^{-2}}{(b^2x)^{-2}} = \frac{(b^2x)^2}{(a^3t)^2} = \frac{b^4x^2}{a^6t^2}$$

When we discussed the operations with signed numbers in Section 1–2, we noted that it is necessary to follow a particular order when performing the basic operations. Since raising a number to a power is a form of multiplication, this operation is also done before additions and subtractions. In fact it is done before divisions and other multiplications. Therefore, *the order of operations is* (1) *specific groupings,* (2) *powers,* (3) *multiplications and divisions, and* (4) *additions and subtractions.*

Example J

$$8 - (-1)^2 - 2(-3)^2 = 8 - (+1) - 2(+9) = 8 - 1 - 18 = -11$$

 Note that we squared -1 and -3 as the first operation. The next operation was finding the product in the last term. Finally the subtractions were performed. We did *not* change the sign of -1 before we squared it, for this would have been incorrect.

Frequently an algebraic expression is known, and it is necessary to **evaluate** the expression for given values of the literal symbols. *The evaluation is done by* **substituting** *the numbers for the literal symbols and then calculating the value of the expression.*

If the evaluation is done on a calculator, we use the $\boxed{x^2}$ key to square numbers and the $\boxed{x^y}$ key for other powers. (See Appendix D for the use of the calculator keys.) The $\boxed{x^y}$ key can be used only if x is positive (or zero). For raising a negative number to a power, we must either determine the sign of the result by inspection or use successive multiplications.

Example K

In order to find the surface area of a ball bearing, we must evaluate the expression $4\pi r^2$, where r is the radius of the bearing.

For a ball bearing of radius 2.55 mm, we find its surface area by substituting the values of π and r into $4\pi r^2$, and we get

$$4(3.14)(2.55)^2$$

as the expression. On a calculator, it is not necessary to substitute 3.14 (or any other decimal equivalent), for π, but we simply use the $\boxed{\pi}$ key. Also, we use the $\boxed{x^2}$ key for 2.55^2. A calculator which uses algebraic logic will square the 2.55 before performing the multiplication. The sequence of keys is

$$4 \quad \boxed{\times} \quad \boxed{\pi} \quad \boxed{\times} \quad 2.55 \quad \boxed{x^2} \quad \boxed{=}$$

and the display will show 81.712825. Thus, rounding off the result to three significant digits (this is the accuracy of r) we find that the surface area is 81.7 mm^2. (See Appendix B for units of measurement.)

Exercises 1–5

In Exercises 1 through 44, simplify the given expressions. Express results with positive exponents only.

1. $x^3 x^4$ 2. $y^2 y^7$ 3. $2b^4 b^2$ 4. $3k(k^5)$

5. $\dfrac{m^5}{m^3}$ 6. $\dfrac{x^6}{x}$ 7. $\dfrac{n^5}{n^9}$ 8. $\dfrac{s}{s^4}$

9. $(a^2)^4$ 10. $(x^8)^3$ 11. $(t^5)^4$ 12. $(n^3)^7$

13. $(2n)^3$ 14. $(ax)^5$ 15. $(ax^4)^2$ 16. $(3a^2)^3$

17. $\left(\dfrac{2}{b}\right)^3$ 18. $\left(\dfrac{x}{y}\right)^7$ 19. $\left(\dfrac{x^2}{2}\right)^4$ 20. $\left(\dfrac{3}{n^3}\right)^3$

21. 7^0 22. $(8a)^0$ 23. $3x^0$ 24. $6v^0$

25. 6^{-1} 26. 10^{-3} 27. $\dfrac{1}{s^{-2}}$ 28. $\dfrac{1}{t^{-5}}$

29. $(-t^2)^7$ 30. $(-y^3)^5$ 31. $(2x^2)^6$ 32. $(-c^4)^4$

33. $(4xa^{-2})^0$ 34. $3(ab^{-1})^0$ 35. $b^5 b^{-3}$ 36. $2c^4 c^{-7}$

37. $(5^0 x^2 a^{-1})^{-1}$ 38. $(3m^{-2}n^4)^{-2}$ 39. $\left(\dfrac{4a}{x}\right)^{-3}$ 40. $\left(\dfrac{2b^2}{y^5}\right)^{-2}$

41. $(-8gs^3)^2$ 42. $ax^2(-a^2 x)^2$ 43. $\dfrac{15a^2 n^5}{3an^6}$ 44. $\dfrac{(ab^2)^3}{a^2 b^8}$

In Exercises 45 through 48, evaluate the given expressions.

45. $7(-4) - (-5)^2$ 46. $\dfrac{12}{-3} - (-1)^3$

47. $6 + (-2)^5 - (-2)(8)$ 48. $9 - 2(-3)^4 - (-7)$

In Exercises 49 through 52, perform the indicated operations on a calculator. In each case round off the results to three significant digits.

49. $2.38(60.7)^2 - 2540$ 50. $(0.513)(-2.778) - (-3.67)^3$

51. $\dfrac{3.07(-1.86)}{(-1.86)^4 + 1.596}$ 52. $\dfrac{(15.66)^2 - (14.07)^3}{-3.68}$

In Exercises 53 and 54, solve the given problems.

53. In analyzing the deformation of a certain beam, it might be necessary to simplify the expression

$$\frac{wx(-2Lx^2)}{24EI}$$

Perform this simplification.

54. In a certain electric circuit, it might be necessary to simplify the expression

$$\frac{gM}{j\omega C(\omega^2 M^2)} \qquad (\omega \text{ is the Greek omega})$$

Perform this simplification.

In Exercises 55 and 56, evaluate the given expressions.

55. In order to find the electric power (in watts) consumed by an electric light, the expression i^2R must be evaluated, where i is the current in amperes (A) and R is the resistance in ohms (Ω). Find the power consumed if $i = 0.525$ A and $R = 250\ \Omega$.

56. Due to the forces acting on a certain beam, the deflection of the beam, in inches, is found by evaluating the expression

$$\frac{x^3 - 10x^2}{168}$$

where x is the distance, in feet, from one end of the beam. Find the deflection if $x = 7.55$ ft.

1–6 Scientific Notation

In technical and scientific work we often encounter numbers which are either very large or very small in magnitude. Illustrations of such numbers are given in the following example.

Example A

Television signals travel at about 30,000,000,000 cm/s. The mass of the earth is about 6,600,000,000,000,000,000,000 tons. A typical protective coating used on aluminum is about 0.0005 in. thick. The wavelength of some X rays is about 0.000000095 cm.

Writing numbers such as these is inconvenient in ordinary notation, as shown in Example A, particularly when the numbers of zeros needed for the proper location of the decimal point are excessive. Therefore, a convenient notation, known as **scientific notation,** is normally used to represent such numbers.

A number written in scientific notation is expressed as the product of a number greater than or equal to 1 and less than 10, and a power of 10. Symbolically this can be written as

$$P \times 10^k$$

where $1 \leq P < 10$, and k is an integer. The following example illustrates how numbers are written in scientific notation.

Example B

$$340,000 = 3.4(100,000) = 3.4 \times 10^5$$

$$0.017 = \frac{1.7}{100} = \frac{1.7}{10^2} = 1.7 \times 10^{-2}$$

$$0.000503 = \frac{5.03}{10000} = \frac{5.03}{10^4} = 5.03 \times 10^{-4}$$

$$6.82 = 6.82(1) = 6.82 \times 10^0$$

From Example B we can establish a method for changing numbers from ordinary notation to scientific notation. *The decimal point is moved so that only one nonzero digit is to its left. The number of places moved is the value of k. It is positive if the decimal point is moved to the left, and it is negative if it is moved to the right.* Consider the illustrations in the following example.

Example C

$$340000 = 3.4 \times 10^5 \qquad\qquad 0.017 = 1.7 \times 10^{-2}$$
5 places   2 places

$$0.000503 = 5.03 \times 10^{-4} \qquad 6.82 = 6.82 \times 10^0$$
4 places   0 places

To change a number from scientific notation to ordinary notation, the procedure above is reversed. The following example illustrates the procedure.

Example D

To change 5.83×10^6 to ordinary notation, we must move the decimal point 6 places to the right. Therefore, additional zeros must be included for the proper location of the decimal point. This means we write $5.83 \times 10^6 = 5,830,000$.

To change 8.06×10^{-3} to ordinary notation, we must move the decimal point 3 places to the left. Again, additional zeros must be included. Thus, $8.06 \times 10^{-3} = 0.00806$.

An illustration of the importance of scientific notation is demonstrated by the metric system use of prefixes on units to denote certain powers of 10. These are illustrated in Appendix B.

As we see from the previous examples, scientific notation provides an important application of the use of exponents, positive and negative. Also, scientific notation provides a practical way to handle calculations involving numbers of very large or very small magnitudes.

The calculation can be made by first expressing all numbers in scientific notation. Then the actual calculation can be performed on numbers between 1 and 10, with the laws of exponents used to find the proper power of 10 for the result. It is proper to leave the result in scientific notation.

Example E

In determining the result of 95,600,000,000,000/0.0286, we may set up the calculation as

$$\frac{9.56 \times 10^{13}}{2.86 \times 10^{-2}} = \left(\frac{9.56}{2.86}\right) \times 10^{15} = 3.34 \times 10^{15}$$

The power of 10 here is sufficiently large that we would normally leave the result in this form.

Example F

In evaluating the product $(7.50 \times 10^9)(6.44 \times 10^{-3})$, we obtain

$$(7.50 \times 10^9)(6.44 \times 10^{-3}) = 48.3 \times 10^6$$

However, since the answer is to be given in scientific notation, we should express it as the product of a number between 1 and 10 and a power of 10, which means we should rewrite it as

$$48.3 \times 10^6 = (4.83 \times 10)(10^6)$$
$$= 4.83 \times 10^7$$

In Example E the quotient $9.56 \div 2.86$ can be found by use of a calculator, but it would not be possible even to enter the number 95,600,000,000,000 in that form on most calculators. However, most scientific calculators have the feature of scientific notation, and it is possible to enter numbers in scientific notation form as well as to have answers given automatically in scientific notation form.

To enter a number into a calculator in scientific notation form, we use the $\boxed{\text{EE}}$ key (see Appendix D). If scientific notation form has not been used and the display cannot show the answer in standard form, the calculator will automatically use scientific notation form to display the answer.

Example G

To perform the calculation of Example E by use of the scientific notation feature of a calculator, we first express each number in scientific notation, as shown in Example E.

$$\frac{95,600,000,000,000}{0.0286} = \frac{9.56 \times 10^{13}}{2.86 \times 10^{-2}}$$

The sequence of keys now used for the calculation is

9.56 $\boxed{\text{EE}}$ 13 $\boxed{\div}$ 2.86 $\boxed{\text{EE}}$ 2 $\boxed{+/-}$ $\boxed{=}$

and the display will be 3.3426573   15, which means that the quotient is equal to 3.34×10^{15}, rounded off.

Exercises 1–6

In Exercises 1 through 8, change the numbers from scientific notation to ordinary notation.

1. 4.5×10^4 2. 6.8×10^7 3. 2.01×10^{-3} 4. 9.61×10^{-5}

5. 3.23×10^0 6. 8.40×10^0 7. 1.86×10 8. 5.44×10^{-1}

In Exercises 9 through 16, change the given numbers from ordinary notation to scientific notation.

9. 40000   10. 5600000   11. 0.0087   12. 0.702

13. 6.89   14. 1.09   15. 0.063   16. 0.0000908

In Exercises 17 through 20, perform the indicated calculations without the use of a calculator and by first expressing all numbers in scientific notation. (See Example E.)

17. $(28,000)(2,000,000,000)$ 18. $(50,000)(0.006)$

19. $\dfrac{88,000}{0.0004}$ 20. $\dfrac{0.00003}{6,000,000}$

In Exercises 21 through 28, perform the indicated calculations by use of a scientific calculator which has scientific notation. In each case, round off results to the number of significant digits used in each of the numbers.

21. $(1280)(865,000)(43.8)$ 22. $(0.0000659)(0.00486)(3,190,000,000)$

23. $\dfrac{(0.0732)(6710)}{0.00134}$ 24. $\dfrac{(2430)(97,100)}{0.00452}$

25. $(3.642 \times 10^{-8})(2.736 \times 10^{5})$ 26. $\dfrac{9.368 \times 10^{-12}}{4.651 \times 10^{4}}$

27. $\dfrac{(3.1075 \times 10^{-5})(1.0772 \times 10^{14})}{6.6482 \times 10^{7}}$ 28. $\dfrac{7.3009 \times 10^{-2}}{(5.9843)(2.5036 \times 10^{-20})}$

In Exercises 29 through 40, change any numbers in ordinary notation to scientific notation or change any numbers in scientific notation to ordinary notation. (See Appendix B for an explanation of symbols used.)

29. The stress on a certain structure is 22,500 lb/in.^{2}.

30. The half-life of uranium 235 is 710,000,000 years.

31. The power of a radio signal received from a lunar probe is 1.6×10^{-12} W.

32. Some computers can perform an addition in 4.5×10^{-12} s.

33. A certain electrical resistor has a resistance of 4.5×10^{3} Ω.

34. To attain an energy density of that in some laser beams, an object would have to be heated to about 10^{30} °C.

35. One foot of steel pipe will increase about 0.000011 ft for a 1°C rise in temperature.

36. The wavelength of yellow light is about 0.00000059 m.

37. The mass of a proton is 1.67×10^{-24} g.

38. The average distance from the earth to the sun is 9.29×10^{7} mi.

39. The altitude of a communications satellite is 36,000 km.

40. The total acreage of national forests in the United States is approximately 187,000,000 acres.

In Exercises 41 through 44, perform the indicated calculations by first expressing all numbers in scientific notation.

41. There are about 161,000 cm in one mile. What is the area in square centimeters of 1 square mile.

42. A particular virus is a sphere 0.0000175 cm in diameter. What volume does the virus occupy?

43. The resistance R, in ohms, of a wire of length l and cross-sectional area A is given by $R = \rho l / A$, where ρ (the Greek rho) is known as the resistivity. Find R for a wire for which $\rho = 0.0000000175 \ \Omega \cdot m$, $l = 0.150$ m, and $A = 0.000000435 \ m^2$.

44. For a certain gas, the product of the volume and pressure is $16.5 \ Pa \cdot cm^3$. If the pressure is 0.00108 Pa, what is the volume?

1–7 Roots and Radicals

A problem that is often encountered is: What number multiplied by itself n times gives another specified number? For example, we may ask: What number squared is 9? We can see that either $+3$ or -3 is a proper answer to this question. *We therefore call either $+3$ or -3 a* **square root** *of* 9, since $(+3)^2 = 9$ or $(-3)^2 = 9$.

In order to have a general notation for the square root of a number, one which is not ambiguous and does not represent more than one number, *we define the* **principal square root** *of a to be positive if a is positive and represent it by* $\sqrt{a}$. This means that $\sqrt{9} = 3$ and not -3.

The general notation for **the principal nth root** of a is $\sqrt[n]{a}$. (When $n = 2$, it is common practice not to put the 2 where n appears.) *The* $\sqrt{}$ *sign is called a* **radical sign.** Unless we state otherwise, when we refer to the root of a number, we refer to the principal root. (In Chapter 11 we consider roots other than principal roots.)

Example A		
	$\sqrt{2}$	(the square root of two)
	$\sqrt[3]{2}$	(the cube root of two)
	$\sqrt[4]{2}$	(the fourth root of two)
	$\sqrt[7]{6}$	(the seventh root of six)
	$\sqrt[3]{8}$	(the cube root of eight, which also equals two)

In considering the question of what number squared is 9, we saw that either $+3$ or -3 gives a proper result. Therefore, in order to avoid ambiguity, we defined the principal square root of a positive number to be positive. (Again, $\sqrt{9} = 3$ and not -3.) In order to have a single defined value for other roots, and to consider only real number roots, *we define the principal nth root of a to be positive if a is positive, and the principal nth root of a to be negative if a is negative and n is odd. (If a is negative and n is even, the roots are not real.)*

Example B

$$\sqrt{4} = 2 \quad (\sqrt{4} \neq -2), \quad \sqrt{169} = 13 \quad (\sqrt{169} \neq -13),$$
$$-\sqrt{64} = -8, \quad -\sqrt{81} = -9, \quad -\sqrt[4]{256} = -4,$$
$$\sqrt[3]{27} = 3, \quad \sqrt[3]{-27} = -3, \quad -\sqrt[3]{27} = -(+3) = -3$$

Another important property of radicals is that *the square root of a product of positive numbers is the product of the square roots.* That is,

$$\sqrt{ab} = \sqrt{a}\sqrt{b} \tag{1–7}$$

This property is useful in simplifying radicals. It is most useful if either a or b is *a* **perfect square,** *which is the square of a rational number.* Consider the following example.

Example C

$$\sqrt{8} = \sqrt{(4)(2)} = \sqrt{4}\sqrt{2} = 2\sqrt{2}$$
$$\sqrt{75} = \sqrt{(25)(3)} = \sqrt{25}\sqrt{3} = 5\sqrt{3}$$
$$\sqrt{80} = \sqrt{(16)(5)} = \sqrt{16}\sqrt{5} = 4\sqrt{5}$$

When we simplify a radical and do not give its *approximate* decimal equivalent, we are generally interested in an *exact* value. In some cases the exact value is preferred, although a decimal value can easily be found on a calculator.

If a decimal value for a square root is acceptable, we can use the $\boxed{\sqrt{x}}$ key on a calculator. The method for finding other roots on a calculator is discussed in Chapter 10.

Example D

The calculator sequence for finding $\sqrt{29.1}$ is

$$29.1 \quad \boxed{\sqrt{x}}$$

and the display is 5.3944416, which means that

$$\sqrt{29.1} = 5.39 \quad \text{(rounded off)}$$

Another important point regarding the simplification of a radical is that *all operations under a radical must be done before finding the root or using Eq.* (1–7). That is, the horizontal bar of a radical sign is a grouping symbol which groups the numbers and symbols under it.

Example E

$$\sqrt{16 + 9} = \sqrt{25} = 5, \text{ but}$$
$$\sqrt{16 + 9} \text{ is } \textbf{\textit{not}} \sqrt{16} + \sqrt{9} = 4 + 3 = 7$$
$$\sqrt{2^2 + 6^2} = \sqrt{4 + 36} = \sqrt{40} = \sqrt{4}\sqrt{10} = 2\sqrt{10}, \text{ but}$$
$$\sqrt{2^2 + 6^2} \text{ is } \textbf{\textit{not}} \sqrt{2^2} + \sqrt{6^2} = 2 + 6 = 8$$

In defining the principal square root, we did not define the square root of a negative number. Earlier *we defined the square root of a negative number to be an* **imaginary number.** There are certain very important places where it is necessary to recognize when imaginary numbers occur, but until Chapter 11 we will not use them otherwise.

It should be emphasized that although the square root of a negative number gives an imaginary number, the cube root of a negative number gives a negative real number. More generally, *the even root of a negative number is imaginary, and the odd root of a negative number is real.* A more detailed discussion of exponents, radicals, and imaginary numbers is found in Chapters 10 and 11.

Example F

$$\sqrt{-64} \text{ is imaginary,} \quad \sqrt[3]{-64} = -4$$
$$\sqrt{-243} \text{ is imaginary,} \quad \sqrt[5]{-243} = -3$$

Exercises 1–7

In Exercises 1 through 32, simplify the given expressions. In each of 1 through 16, the result is an integer.

1. $\sqrt{25}$ 2. $\sqrt{81}$ 3. $-\sqrt{121}$ 4. $-\sqrt{36}$

5. $-\sqrt{49}$ 6. $\sqrt{100}$ 7. $\sqrt{400}$ 8. $-\sqrt{900}$

9. $\sqrt[3]{125}$ 10. $\sqrt[4]{16}$ 11. $\sqrt[3]{-216}$ 12. $\sqrt[5]{-32}$

13. $(\sqrt{5})^2$ 14. $(\sqrt{19})^2$ 15. $(\sqrt[3]{31})^3$ 16. $(\sqrt[4]{53})^4$

17. $\sqrt{18}$ 18. $\sqrt{32}$ 19. $\sqrt{12}$ 20. $\sqrt{50}$

21. $\sqrt{44}$ 22. $\sqrt{54}$ 23. $\sqrt{63}$ 24. $\sqrt{90}$

25. $2\sqrt{48}$ 26. $4\sqrt{108}$ 27. $\dfrac{7^2\sqrt{81}}{3^2\sqrt{49}}$ 28. $\dfrac{2^5\sqrt[5]{243}}{3\sqrt{144}}$

29. $\sqrt{36 + 64}$ 30. $\sqrt{25 + 144}$ 31. $\sqrt{3^2 + 9^2}$ 32. $\sqrt{8^2 - 4^2}$

In Exercises 33 through 36, find the value of each square root by use of a calculator. Round off the result to the accuracy of the given number.

33. $\sqrt{85.4}$ 34. $\sqrt{3762}$ 35. $\sqrt{0.4729}$ 36. $\sqrt{0.0627}$

In Exercises 37 through 42, solve the given problems.

37. The time, in seconds, required for an object to fall h feet due to gravity can be found by evaluating $0.25\sqrt{h}$. Find the time, to the nearest 0.1 s, for an object to fall 1500 ft.

38. The period, in seconds, of a pendulum whose length is L feet can be found by evaluating $1.11\sqrt{L}$. Find the period, to the nearest 0.01 s, of a pendulum for which $L = 1.75$ ft.

39. A square parcel of land has an area of 1600 m². What is the length of a side of the parcel?

40. A cubical water tank holds 512 ft³. What is the length of an edge of the tank?

41. Is it always true that $\sqrt{a^2} = a$?

42. If $0 < x < 1$ (x between 0 and 1), is $x > \sqrt{x}$?

1–8 Addition and Subtraction of Algebraic Expressions

It is the basic characteristic of algebra that letters are used to represent numbers. Since we have used literal symbols to represent numbers, even if in a general sense, we may conclude that all operations valid for numbers are valid for these literal symbols. In this section we shall discuss the terminology and methods for combining literal symbols.

Addition, subtraction, multiplication, division, and taking of roots are known as **algebraic operations.** *Any combination of numbers and literal symbols which results from algebraic operations is known as an* **algebraic expression.**

When an algebraic expression consists of several parts connected by plus signs and minus signs, each part (along with its sign) is known as a **term** *of the expression. If a given expression is made up of the product of a number of quantities, each of these quantities, or any product of them, is called a* **factor** *of the expression.*

> **It is important to distinguish clearly between terms and factors since some operations that are valid for terms are not valid for factors, and conversely.**

Example A

$3xy + 6x^2 - 7x\sqrt{y}$ is an algebraic expression with three terms. They are $3xy$, $6x^2$, and $-7x\sqrt{y}$.

The first term $3xy$ has individual factors of 3, x, and y. Also, any product of these factors is also a factor of $3xy$. Thus, additional expressions which are factors of $3xy$ are $3x$, $3y$, xy, and $3xy$ itself.

Example B

$7x(y^2 + x) - \dfrac{x + y}{6x}$ is an algebraic expression with terms $7x(y^2 + x)$ and $\dfrac{-(x + y)}{6x}$.

The term $7x(y^2 + x)$ has individual factors of 7, x, and $(y^2 + x)$, as well as products of these factors. The factor $y^2 + x$ has two terms, y^2 and x.

The numerator of the term $-\dfrac{x + y}{6x}$ has two terms, and the denominator has individual factors of 2, 3, and x. The minus sign can be treated as a factor of -1.

An algebraic expression containing only one term is called a **monomial.** *An expression containing two terms is a* **binomial,** *and one containing three terms is a* **trinomial.** *Any expression containing two or more terms is called a* **multinomial.** Thus, any binomial or trinomial expression can also be considered as a multinomial.

In any given term, the numbers and literal symbols multiplying any given factor constitute the **coefficient** *of that factor. The product of all the numbers in explicit form is known as the* **numerical coefficient** *of the term. All terms which differ at most in their numerical coefficients are known as* **similar** *or* **like** *terms.* That is, similar terms have the same variables with the same exponents.

Example C

$7x^3\sqrt{y}$ is a monomial. It has a numerical coefficient of 7. The coefficient of $\sqrt{y}$ is $7x^3$, and the coefficient of x^3 is $7\sqrt{y}$.

Example D

$4 \cdot 2b + 81b - 6ab$ is a multinomial of three terms. The first term has a numerical coefficient of 8, the second has a numerical coefficient of 81, and the third has a numerical coefficient of -6 (the sign of the term is attached to the numerical coefficient). The first and second terms are similar, since they differ only in their numerical coefficient. The third term is not similar to either of the others, for it has the factor a.

The commutative law tells us that $x^2y^3 = y^3x^2$. Therefore, the two terms of the expression $3x^2y^3 + 5y^3x^2$ are similar.

In adding and subtracting algebraic expressions, we combine similar terms. In doing so, we are combining quantities which are alike. All of the similar terms may be combined into a single term, and the final simplified expression will be made up entirely of terms which are not similar.

Example E

$3x + 2x - 5y = 5x - 5y$. Since there are two similar terms in the original expression, they are added together, so the simplified result has two unlike terms.

$6a^2 - 7a + 8ax$ cannot be simplified, since none of the terms are like terms.

Similarly, $6a + 5c + 2a - c = 6a + 2a + 5c - c = 8a + 4c$. (Here we use the commutative and associative laws.)

In writing algebraic expressions, it is often necessary to group certain terms together. For this purpose we use **symbols of grouping.** In this text we use **parentheses** ( ), **brackets** [ ], and **braces** { }. The **bar,** which is used with radicals and fractions, also groups terms. The bar attached to the radical sign groups the terms below it, and the bar separating the numerator and denominator of a fraction groups the terms above and below it. In earlier sections, when we discussed the order of operations, we used parentheses and the bar for grouping.

When adding and subtracting algebraic expressions, it is often necessary to remove symbols of grouping. To do so we must *change the sign of **every term** within the symbols if the grouping is preceded by a minus sign.* If the symbols of grouping are preceded by a plus sign, each term within the symbols retains its original sign. This is a result of the distributive law.

Example F

(1) $2a - (3a + 2b) - b = 2a - 3a - 2b - b = -a - 3b$

(2) $-(2x - 3c) + (c - x) = -2x + 3c + c - x = 4c - 3x$

Example G

(1) $3 - 2(m^2 - 2) = 3 - 2m^2 + 4 = 7 - 2m^2$

(2) $4(t - 3 - 2t^2) - (6t + t^2 - 4) = 4t - 12 - 8t^2 - 6t - t^2 + 4$
$$= -9t^2 - 2t - 8$$

It is fairly common to have expressions in which more than one symbol of grouping is to be removed in the simplification. Normally, *when several symbols of grouping are to be removed, it is more convenient to remove the innermost symbols first.* This is illustrated in the following example.

Example H

(1) $3ax - [ax - (5s - 2ax)] = 3ax - [ax - 5s + 2ax]$
$$= 3ax - ax + 5s - 2ax = 5s$$

(2) $[a^2b - ab + (ab - 2a^2b)] - \{[(3a^2b + b) - (4ab - 2a^2b)] - b\}$
$$= [a^2b - ab + ab - 2a^2b] - \{[3a^2b + b - 4ab + 2a^2b] - b\}$$
$$= a^2b - ab + ab - 2a^2b - \{3a^2b + b - 4ab + 2a^2b - b\}$$
$$= a^2b - ab + ab - 2a^2b - 3a^2b - b + 4ab - 2a^2b + b$$
$$= -6a^2b + 4ab$$

Calculators and computers use only parentheses for grouping symbols. Therefore, in writing expressions in which only parentheses are used to group terms it is frequently necessary to use one set of parentheses within another set to group more than one set of symbols. These are called **nested parentheses.** The following example illustrates the simplification of an algebraic expression with nested parentheses. It will be noted that the innermost parentheses are removed first.

Example I

$$2 - (3x - 2(5 - (7 - x))) = 2 - (3x - 2(5 - 7 + x))$$
$$= 2 - (3x - 10 + 14 - 2x)$$
$$= 2 - 3x + 10 - 14 + 2x$$
$$= -x - 2$$

One of the most common errors made by beginning students is changing the sign of only the first term when removing symbols of grouping preceded by a minus sign. *Remember, if the symbols are preceded by a minus sign, we must change the sign of **all** terms.*

Exercises 1–8

In the following exercises, simplify the given algebraic expressions.

1. $5x + 7x - 4x$
2. $6t - 3t - 4t$
3. $2y - y + 4x$
4. $4c + d - 6c$
5. $2a - 2c - e + 3c - a$
6. $x - 2y + 3x - y + z$
7. $a^2b - a^2b^2 - 2a^2b$
8. $xy^2 - 3x^2y^2 + 2xy^2$
9. $s + (4 + 3s)$
10. $5 + (3 - 4n + p)$
11. $v - (4 - 5x + 2v)$
12. $2a - (b - a)$
13. $2 - 3 - (4 - 5a)$
14. $\sqrt{x} + (y - 2\sqrt{x}) - 3\sqrt{x}$
15. $(a - 3) + (5 - 6a)$
16. $(4x - y) - (2x - 4y)$
17. $-(t - 2u) + (3u - t)$
18. $2(x - 2y) + (5x - y)$
19. $3(2r + s) - (5s - r)$
20. $3(a - b) - 2(a - 2b)$
21. $-7(6 - 3c) - 2(c + 4)$
22. $-(5t + a^2) - 2(3a^2 - 2st)$
23. $-[(6 - n) - (2n - 3)]$
24. $-[(a - b) - (b - a)]$
25. $2[4 - (t^2 - 5)]$
26. $3[3 - (a - 4)]$
27. $-2[-x - 2a - (a - x)]$
28. $-2[-3(x - 2y) + 4x]$
29. $a\sqrt{xy} - [3 - (a\sqrt{xy} + 4)]$
30. $9v - [6 - (v - 4) + 4v]$
31. $8c - \{5 - [2 - (3 + 4c)]\}$
32. $7y - \{y - [2y - (x - y)]\}$
33. $5p - (q - 2p) - [3q - (p - q)]$
34. $-(4 - x) - [(5x - 7) - (6x + 2)]$
35. $-2\{-(4 - x^2) - [3 + (4 - x^2)]\}$
36. $-\{-[-(x - 2a) - b] - a\}$
37. $3a - (6 - (a + 3))$
38. $-2x + 2((2x - 1) - 5)$
39. $-(3t - (7 + 2t - (5t - 6)))$
40. $a^2 - 2(x - 5 - (7 - 2(a^2 - 2x) - 3x))$

41. When discussing gear trains, the expression $-(-R - 1)$ is found. Simplify this expression.

42. In analyzing a certain electric circuit, the expression $I_1 + I_2 - (I_2 - I_3)$ is found. Simplify this expression. (The numbers below the I's are subscripts. Different subscripts denote different unknowns.)

43. Under certain conditions, the expression for finding the profit on given sales involves simplifying the expression $5(x + 1) - (400 + 2x)$. Simplify this expression.

44. In developing the theory for an elastic substance, we find the following expression:

$$[(B + \tfrac{4}{3}\alpha) + 2(B - \tfrac{2}{3}\alpha)] - [(B + \tfrac{4}{3}\alpha) - (B - \tfrac{2}{3}\alpha)]$$

Simplify this expression.

1–9 Multiplication of Algebraic Expressions

To find the product of two or more monomials, we use the laws of exponents as given in Section 1–5 and the laws for multiplying signed numbers as stated in Section 1–2. We first multiply the numerical coefficients to determine the numerical coefficient of the product. Then we multiply the literal factors, remembering that *the exponents may be combined only if the base is the same*. Consider the illustrations in the following example.

Example A

(1) $3ac^3(4sa^2c) = 12a^3c^4s$

(2) $(-2b^2y)(-9aby^5) = 18ab^3y^6$

(3) $2xy(-6cx^2)(3xcy^2) = -36c^2x^4y^3$

If a product contains a monomial which itself is raised to a power, *we must first raise it to the indicated power* before proceeding with the remainder of the multiplication.

Example B

(1) $3(2a^2x)^3(-ax) = 3(8a^6x^3)(-ax) = -24a^7x^4$

(2) $2s^3(-st^4)^2(4s^2t) = 2s^3(s^2t^8)(4s^2t) = 8s^7t^9$

We find the product of a monomial and a multinomial by using the distributive law, which states that we *multiply each term of the multinomial by the monomial*. We must be careful to assign the correct sign to each term of the result, using the rules for multiplication of signed numbers. Also, we must properly combine literal factors in each term of the result.

Example C

(1) $2ax(3ax^2 - 4yz) = 2ax(3ax^2) - (2ax)(4yz) = 6a^2x^3 - 8axyz$

(2) $5cy^2(-7cx - ac) = (5cy^2)(-7cx) + (5cy^2)(-ac)$
$$= -35c^2xy^2 - 5ac^2y^2$$

In practice, it is generally not necessary to write out the middle step as it appears in the example above. We can generally write the answer directly. For example, the first part of Example C would usually appear as

$$2ax(3ax^2 - 4yz) = 6a^2x^3 - 8axyz$$

We find the product of two or more multinomials by using the distributive law and the laws of exponents. The result is that we *multiply each term of one multinomial by each term of the other, and add the results*.

Example D

(1) $(x - 2)(x + 3) = x(x) + x(3) + (-2)(x) + (-2)(3)$
$$= x^2 + 3x - 2x - 6 = x^2 + x - 6$$

(2) $(x - 2y)(x^2 + 2xy + 4y^2) = x^3 + 2x^2y + 4xy^2 - 2x^2y - 4xy^2 - 8y^3$
$$= x^3 - 8y^3$$

Finding the power of an algebraic expression is equivalent to multiplying the expression by itself the number of times indicated by the exponent. In practice, it is often convenient to write the power of an algebraic expression in this form before multiplying. Consider the following example.

Example E

(1) $(x + 5)^2 = (x + 5)(x + 5) = x^2 + 5x + 5x + 25$
$$= x^2 + 10x + 25$$

(2) $2(3 - 2x)^2 = 2(3 - 2x)(3 - 2x)$
$$= 2(9 - 6x - 6x + 4x^2)$$
$$= 2(9 - 12x + 4x^2)$$
$$= 18 - 24x + 8x^2$$
$$= 8x^2 - 24x + 18$$

(3) $(2a - b)^3 = (2a - b)(2a - b)(2a - b)$
$$= (2a - b)(4a^2 - 4ab + b^2)$$
$$= 8a^3 - 8a^2b + 2ab^2 - 4a^2b + 4ab^2 - b^3$$
$$= 8a^3 - 12a^2b + 6ab^2 - b^3$$

We should note in the first illustration that

▶ $(x + 5)^2$ is **not** equal to $x^2 + 25$

since the term $10x$ is not included. We must follow the proper procedure for multiplication in squaring an expression and not simply square each of the terms within the parentheses.

Exercises 1–9

In the following exercises, perform the indicated multiplications.

1. $(a^2)(ax)$

2. $(2xy)(x^2y^3)$

3. $-ac^2(acx^3)$

4. $-2s^2(-4cs)^2$

5. $(2ax^2)^2(-2ax)$

6. $6pq^3(3pq^2)^2$

7. $a(-a^2x)^3(-2a)$

8. $-2m^2(-3mn)(m^2n)^2$

9. $a^2(x + y)$

10. $2x(p - q)$

11. $-3s(s^2 - 5t)$

12. $-3b(2b^2 - b)$

13. $5m(m^2n + 3mn)$

14. $a^2bc(2ac - 3a^2b)$

15. $3x(x - y + 2)$

16. $b^2x^2(x^2 - 2x + 1)$

17. $ab^2c^4(ac - bc - ab)$

18. $-4c^2(9gc - 2c + g^2)$

19. $ax(cx^2)(x + y^3)$ 20. $-2(-3st^3)(3s - 4t)$

21. $(x - 3)(x + 5)$ 22. $(a + 7)(a + 1)$

23. $(x + 5)(2x - 1)$ 24. $(4t + s)(2t - 3s)$

25. $(2a - b)(3a - 2b)$ 26. $(4x - 3)(3x - 1)$

27. $(2s + 7t)(3s - 5t)$ 28. $(5p - 2q)(p + 8q)$

29. $(x^2 - 1)(2x + 5)$ 30. $(3y^2 + 2)(2y - 9)$

31. $(x^2 - 2x)(x + 4)$ 32. $(2ab^2 - 5t)(-ab^2 - 6t)$

33. $(x + 1)(x^2 - 3x + 2)$ 34. $(2x + 3)(x^2 - x + 5)$

35. $(4x - x^3)(2 + x - x^2)$ 36. $(5a - 3c)(a^2 + ac - c^2)$

37. $2(a + 1)(a - 9)$ 38. $-5(y - 3)(y + 6)$

39. $2x(x - 1)(x + 4)$ 40. $ax(x + 4)(7 - x^2)$

41. $(2x - 5)^2$ 42. $(x - 3)^2$

43. $(x + 3a)^2$ 44. $(2m + 1)^2$

45. $(xyz - 2)^2$ 46. $(b - 2x^2)^2$

47. $2(x + 8)^2$ 48. $3(a + 4)^2$

49. $(2 + x)(3 - x)(x - 1)$ 50. $(3x - c^2)^3$

51. $3x(x + 2)^2(2x - 1)$ 52. $[(x - 2)^2(x + 2)]^2$

53. Under given conditions, when determining the income from a business enterprise, the expression $(40 - x)(200 + 5x)$ is found. Perform the indicated multiplication.

54. The expression $a + b(1 - X) + c(1 - X)^2$ is found in determining a certain chemical volume. Perform the indicated multiplications.

55. The analysis of the deflection of a certain concrete beam involves the expression $w(l^2 - x^2)^2$. Perform the indicated multiplication.

56. In finding the maximum power in a particular electric circuit, the expression $(R + r)^2 - 2r(R + r)$ is used. Multiply and simplify.

1–10 Division of Algebraic Expressions

To find the quotient of one monomial divided by another, we use the laws of exponents as given in Section 1–5 and the laws for dividing signed numbers as stated in Section 1–2. Again, the exponents may be combined only if the base is the same.

Example A

(1) $\dfrac{16x^3y^5}{4xy^2} = \dfrac{16}{4}(x^{3-1})(y^{5-2}) = 4x^2y^3$

(2) $\dfrac{-6a^2xy^2}{2axy^4} = -\left(\dfrac{6}{2}\right)\dfrac{a^{2-1}x^{1-1}}{y^{4-2}} = -\dfrac{3a}{y^2}$

As noted in the second illustration, we use only positive exponents in the final result, unless there are specific instructions otherwise.

The quotient of a multinomial divided by a monomial is found by dividing each term of the multinomial by the monomial and adding the results. This process is a result of the equivalent operation with arithmetic fractions, which can be shown as

$$\frac{a + b}{c} = \frac{a}{c} + \frac{b}{c}$$

$\left(\text{Be careful: Although } \dfrac{a + b}{c} = \dfrac{a}{c} + \dfrac{b}{c}, \text{ we must note that } \dfrac{c}{a + b} \text{ is not } \dfrac{c}{a} + \dfrac{c}{b}.\right)$

Example B

(1) $\dfrac{4x^3y - 8x^3y^2 + 6x^2y^4}{2x^2y} = \dfrac{4x^3y}{2x^2y} - \dfrac{8x^3y^2}{2x^2y} + \dfrac{6x^2y^4}{2x^2y}$

$\qquad\qquad\qquad\qquad\qquad = 2x - 4xy + 3y^3$

(2) $\dfrac{a^3bc^4 - 6abc + 9a^2b^3c - 3}{3ab^2c^3} = \dfrac{a^2c}{3b} - \dfrac{2}{bc^2} + \dfrac{3ab}{c^2} - \dfrac{1}{ab^2c^3}$

In practice we would usually not write the middle step shown in the first illustration of Example B. The divisions are done by inspection, and the expression would appear as shown in the second illustration. However, we must remember that each term in the numerator is divided by the monomial in the denominator.

If each term of an algebraic sum is a number or is of the form ax^n where n is a nonnegative integer, we call the expression a **polynomial** *in x.* The distinction between a multinomial and a polynomial is that a polynomial does not contain terms like $\sqrt{x}$ or $1/x^2$, whereas a multinomial may contain such terms. Also, a polynomial may consist of only one term, whereas a multinomial must have at least two terms. In a polynomial, *the greatest value of n which appears is the* **degree** *of the polynomial.*

Example C

$3 + 2x^2 - x^3$ is a polynomial of degree 3

$x^4 - 3x^2 - \sqrt{x}$ is not a polynomial

$4x^5$ is a polynomial of degree 5

The first two expressions are also multinomials since each contains more than one term. The second illustration is not a polynomial due to the presence of the $\sqrt{x}$ term. The third illustration is a polynomial since the exponent is a positive integer. It is also a monomial.

The problem often arises of dividing one polynomial by another. To solve this problem, we first arrange the dividend (the polynomial to be divided) and the divisor in descending powers of the variable. Then we divide the first term of the dividend by the first term of the divisor. The result gives the first term of the quotient. Next, we multiply the entire divisor by the first term of the quotient and subtract the product from the dividend. We divide the first term of this difference by the first term of the divisor. This gives the second term of the quotient. We multiply this term by each of the terms of the divisor and subtract this result from the first difference. We repeat this process until the remainder is either zero or a term which is of lower degree than the divisor. This process is similar to that of long division of numbers.

Example D

Divide $6x^2 + x - 2$ by $2x - 1$. (This division can also be indicated by $(6x^2 + x - 2) \div (2x - 1)$ or in the fractional form $\dfrac{6x^2 + x - 2}{2x - 1}$.)

We set up the division in the same way we would for long division in arithmetic. Then following the procedure outlined above, we have the following.

$$
\begin{array}{r}
3x + 2 \\
2x - 1 \,\overline{\big)\, 6x^2 + x - 2} \\
\underline{6x^2 - 3x} \qquad \textit{(subtract)} \\
4x - 2 \\
\underline{4x - 2} \qquad \text{(subtract)} \\
0
\end{array}
$$

The remainder is zero and the quotient is $3x + 2$. Note that when we subtracted $-3x$ from x, we obtained $4x$.

Example E

Divide $4x^3 + 6x^2 + 1$ by $2x - 1$. Since there is no x-term in the dividend, it is advisable to leave space for any x-terms which might arise.

$$
\begin{array}{r}
2x^2 + 4x + 2 \qquad \text{(quotient)} \\
\text{(divisor)} \quad 2x - 1 \,\overline{\big)\, 4x^3 + 6x^2 + 1} \qquad \text{(dividend)} \\
\underline{4x^3 - 2x^2} \\
8x^2 + 1 \\
\underline{8x^2 - 4x} \\
4x + 1 \\
\underline{4x - 2} \\
3 \qquad \text{(remainder)}
\end{array}
$$

Exercises 1–10

In the following exercises, perform the indicated divisions.

1. $\dfrac{8x^3 y^2}{-2xy}$
              2. $\dfrac{-18b^7 c^3}{bc^2}$

3. $\dfrac{-16r^3 t^5}{-4r^5 t}$
              4. $\dfrac{51mn^5}{17m^2 n^2}$

5. $\dfrac{(15x^2)(4bx)(2y)}{30bxy}$

6. $\dfrac{(5st)(8s^2t^3)}{10s^3t^2}$

7. $\dfrac{6(ax)^2}{-ax^2}$

8. $\dfrac{12a^2b}{(3ab^2)^2}$

9. $\dfrac{a^2x + 4xy}{x}$

10. $\dfrac{2m^2n - 6mn}{2m}$

11. $\dfrac{3rst - 6r^2st^2}{3rs}$

12. $\dfrac{-5a^2n - 10an^2}{5an}$

13. $\dfrac{4pq^3 + 8p^2q^2 - 16pq^5}{4pq^2}$

14. $\dfrac{a^2xy^2 + ax^3 - 4ax^2}{ax}$

15. $\dfrac{2\pi fL - \pi fR^2}{\pi fR}$

16. $\dfrac{2(ab)^4 - a^3b^4}{3(ab)^3}$

17. $\dfrac{3ab^2 - 6ab^3 + 9a^3b}{9a^2b^2}$

18. $\dfrac{4x^2y^3 + 8xy - 12x^2y^4}{2x^2y^2}$

19. $\dfrac{x^{n+2} + ax^n}{x^n}$

20. $\dfrac{3a(x + y)b^2 - (x + y)}{a(x + y)}$

21. $(2x^2 + 7x + 3) \div (x + 3)$

22. $(3x^2 - 11x - 4) \div (x - 4)$

23. $\dfrac{x^2 - 3x + 2}{x - 2}$

24. $\dfrac{2x^2 - 5x - 7}{x + 1}$

25. $\dfrac{x - 14x^2 + 8x^3}{2x - 3}$

26. $\dfrac{6x^2 + 6 + 7x}{2x + 1}$

27. $(4x^2 + 23x + 15) \div (4x + 3)$

28. $(6x^2 - 20x + 16) \div (3x - 4)$

29. $\dfrac{x^3 + 3x^2 - 4x - 12}{x + 2}$

30. $\dfrac{3x^3 + 19x^2 + 16x - 20}{3x - 2}$

31. $\dfrac{2x^4 + 4x^3 + 2}{x^2 - 1}$

32. $\dfrac{2x^3 - 3x^2 + 8x - 2}{x^2 - x + 2}$

33. $\dfrac{x^3 + 8}{x + 2}$

34. $\dfrac{x^3 - 1}{x - 1}$

35. $\dfrac{x^2 - 2xy + y^2}{x - y}$

36. $\dfrac{3a^2 - 5ab + 2b^2}{a - 3b}$

37. In determining the volume of a certain gas, the following expression is used:

$$\frac{RTV^2 - aV + ab}{RT}$$

Perform the indicated division.

38. In hydrodynamics the following expression is found:

$$\frac{2p + v^2d + 2ydg}{2dg}$$

Perform the indicated division.

39. The expression for the total resistance of three resistances in parallel in an electric circuit is

$$\frac{R_1R_2R_3}{R_2R_3 + R_1R_3 + R_1R_2}$$

Find the reciprocal of this expression, and then perform the indicated division.

40. The following expression is found when analyzing the motion of a certain object:

$$\frac{s + 6}{s^2 + 8s + 25}$$

Find the reciprocal of this expression, and perform the indicated division.

1–11 Equations

The basic operations for algebraic expressions that we have developed are used in the important process of solving equations. In this section we show how the basic algebraic operations are used in solving equations, and in the following sections we demonstrate some of the important applications°of equations.

An **equation** *is an algebraic statement that two algebraic expressions are equal.* It is possible that many values of the letter representing the **unknown** will **satisfy** the equation; that is, many values may produce equality when **substituted** in the equation. It is also possible that only one value for the unknown will satisfy the equation (and this will be true of nearly all of the equations we solve in this section). Or possibly there may be no values which satisfy the equation, although the statement is still an equation.

Example A

The equation $x^2 - 4 = (x - 2)(x + 2)$ is true for all values of x. For example, if we substitute $x = 3$ we have $9 - 4 = (3 - 2)(3 + 2)$ or $5 = 5$. If we let $x = -1$, we have $-3 = -3$. *An equation that is true for all values of the unknown is termed an* **identity.**

The equation $x^2 - 2 = x$ is true if $x = 2$ or if $x = -1$, but it is not true for any other values of x. If $x = 2$ we obtain $2 = 2$, and if $x = -1$ we obtain $-1 = -1$. However, if we let $x = 4$, we obtain $14 = 4$, which of course is not correct. *An equation valid only for certain values of the unknown is termed a* **conditional equation.** These equations are those which are generally encountered.

Example B

The equation $3x - 5 = x + 1$ is true only for $x = 3$. When $x = 3$ we obtain $4 = 4$; if we let $x = 2$, we obtain $1 = 3$, which is not correct.

The equation $x + 5 = x + 1$ is not true for any value of x. For any value of x we try, we will find that the left side is 4 greater than the right side. However, it is still an equation.

To **solve** *an equation we find the values of the unknown which satisfy it.* There is one basic rule to follow when solving an equation:

Perform the same operation on both sides of the equation.

We do this to isolate the unknown and thus to find its values.

By performing the same operation on both sides of an equation, the two sides remain equal. Thus,

we may add the same number to both sides, subtract the same number from both sides, multiply both sides by the same number, or divide both sides by the same number (not zero).

Example C

In solving the following equations, we note that we may isolate x, and thereby solve the equation, by performing the indicated operation.

$x - 3 = 12$	$x + 3 = 12$	$\dfrac{x}{3} = 12$	$3x = 12$
Add 3 to both sides.	Subtract 3 from both sides.	Multiply both sides by 3.	Divide both sides by 3.
$x - 3 + 3 = 12 + 3$	$x + 3 - 3 = 12 - 3$	$3\left(\dfrac{x}{3}\right) = 3(12)$	$\dfrac{3x}{3} = \dfrac{12}{3}$
$x = 15$	$x = 9$	$x = 36$	$x = 4$

Each can be checked by substitution in the original equation. (The term *transposing* is often used to denote the result of adding or subtracting a term from both sides of the equation. In transposing, a term is moved from one side of the equation to the other, and its sign is changed.)

The solution of an equation generally requires a combination of the basic operations. The following examples illustrate the solution of such equations.

Example D

Solve the equation $2t - 7 = 9$.

We are to perform basic operations to both sides of the equation to finally isolate t on one side. The steps to be followed are suggested by the form of the equation, and in this case are as follows.

$$2t - 7 = 9 \qquad \text{(add 7 to both sides)}$$
$$2t = 16 \qquad \text{(divide both sides by 2)}$$
$$t = 8$$

Therefore, we conclude that $t = 8$. Checking in the original equation, we see that we have $2(8) - 7 = 9$, $16 - 7 = 9$, or $9 = 9$. Therefore, the solution checks.

Example E

Solve the equation $3n + 4 = n - 6$.

$$2n + 4 = -6 \qquad \text{(n subtracted from both sides)}$$
$$2n = -10 \qquad \text{(4 subtracted from both sides)}$$
$$n = -5 \qquad \text{(both sides divided by 2)}$$

Checking in the original equation, we have $-11 = -11$.

Example F

Solve the equation $x - 7 = 3x - (6x - 8)$.

$$x - 7 = 3x - 6x + 8 \qquad \text{(parentheses removed)}$$
$$x - 7 = -3x + 8 \qquad \text{(x's combined on right)}$$
$$4x - 7 = 8 \qquad \text{($3x$ added to both sides)}$$
$$4x = 15 \qquad \text{(7 added to both sides)}$$
$$x = \tfrac{15}{4} \qquad \text{(both sides divided by 4)}$$

Checking in the original equation, we obtain (after simplifying) $-\tfrac{13}{4} = -\tfrac{13}{4}$.

Note that we always check in the **original** equation. This is done since errors may have been made in finding the later equations.

If an equation contains decimals, the best procedure is to set up the solution first, before actually performing the calculations. Once the unknown is isolated, a calculator can be used to perform the calculations.

Example G

Solve the equation $3.26 - 5.13(2.75 - 1.86x) = 9.42$.

Following the procedure outlined above, we have the following.

$$3.26 - (5.13)(2.75) + (5.13)(1.86x) = 9.42$$
$$(5.13)(1.86x) = 9.42 - 3.26 + (5.13)(2.75)$$
$$x = \frac{9.42 - 3.26 + (5.13)(2.75)}{(5.13)(1.86)}$$
$$= 2.12 \qquad \text{(rounded off)}$$

In checking we find that the left side of the original equation is 9.38. The difference is a result of rounding off. If we use the unrounded calculator value, we get 9.42.

Exercises 1–11

In Exercises 1 through 28, solve the given equations.

1. $x - 2 = 7$ 2. $x - 4 = 1$ 3. $x + 5 = 4$ 4. $s + 6 = 3$

5. $\dfrac{t}{2} = 5$ 6. $\dfrac{x}{4} = 2$ 7. $4x = 20$ 8. $2x = 12$

9. $3t + 5 = -4$ 10. $5x - 2 = 13$

11. $5 - 2y = 3$ 12. $8 - 5t = 18$

13. $3x + 7 = x$ 14. $6 + 8y = 5 - y$

15. $2(s - 4) = s$ 16. $3(n - 2) = -n$

17. $6 - (r - 4) = 2r$ 18. $5 - (x + 2) = 5x$

19. $2(x - 3) - 5x = 7$ 20. $4(x + 7) - x = 7$

21. $x - 5(x - 2) = 2$ 22. $5x - 2(x - 5) = 4x$

23. $7 - 3(1 - 2p) = 4 + 2p$ 24. $3 - 6(2 - 3t) = t - 5$

25. $5.8 - 0.3(x - 6.0) = 0.5x$ 26. $1.9t = 0.5(4.0 - t) - 0.8$

27. $0.15 - 0.24(y - 0.50) = 0.63$ 28. $27.5(5.17 - 1.44x) = 73.4$

In Exercises 29 through 32, solve the given problems.

29. What conclusion can be made about the equation $2(x - 3) + 1 = 2x - 5$?

30. What conclusion can be made about the equation $1 - (3 - x) = x - 2$?

31. Show that the equation $3(x + 2) = 3x + 4$ is not valid for any values of x.

32. Show that the equation $7 - (2 - x) = x + 2$ is not valid for any values of x.

1–12 Formulas and Literal Equations

Equations and their solutions are of great importance in most fields of technology and science. They are used to attain, study, and confirm information of all kinds. One of the most important applications occurs in the use of formulas in mathematics, physics, engineering, and other fields. *A* **formula** *is an equation which expresses a rule and uses letters to represent certain quantities.* For example, the formula for the area of a circle is $A = \pi r^2$. The symbol A stands for the area as does the expression πr^2, and the formula states that the area is found by multiplying the square of the radius by π.

Often it is necessary to solve a formula or any equation containing more than one literal symbol for a particular letter or symbol which appears in it. We do this in the same manner as we solve any equation: We isolate the letter or symbol desired by use of the basic algebraic operations.

Example A   Solve $A = \pi r^2$ for π.

$$\frac{A}{r^2} = \pi \qquad \text{(both sides divided by } r^2\text{)}$$

$$\pi = \frac{A}{r^2} \qquad \text{(since each side equals the other, it makes no difference which expression appears on the left)}$$

Since the symbol for which we are solving is usually shown on the left of the equal sign, we switched sides, as shown.

Example B   A formula relating acceleration a, velocity v, initial velocity v_0, and time t, is $v = v_0 + at$. Solve for t.

$$v - v_0 = at \qquad \text{(} v_0 \text{ subtracted from both sides)}$$

$$t = \frac{v - v_0}{a} \qquad \text{(both sides divided by } a \text{ and then sides switched)}$$

As we can see from Examples A and B, we can solve for the indicated literal number just as we solved for the unknown in the previous section. That is, we perform the basic algebraic operations on the various literal numbers which appear in the same way we perform them on explicit numbers. Other illustrations appear in the following examples.

Example C

In the study of the forces on a certain beam, the equation

$$M = \frac{L(wL + 2P)}{8}$$

is used. Solve for P.

$$8M = \frac{8L(wL + 2P)}{8} \qquad \text{(multiply both sides by 8)}$$

$$8M = L(wL + 2P) \qquad \text{(simplify right side)}$$

$$8M = wL^2 + 2LP \qquad \text{(remove parentheses)}$$

$$8M - wL^2 = 2LP \qquad \text{(subtract } wL^2 \text{ from each side)}$$

$$P = \frac{8M - wL^2}{2L} \qquad \text{(divide both sides by } 2L \text{ and switch sides)}$$

Example D

The effect of temperature is important when accurate instrumentation is required. The volume V of a precision container at temperature T in terms of the volume V_0 at temperature T_0 is given by

$$V = V_0[1 + b(T - T_0)]$$

where b depends on the material of which the container is made. Solve for T.

Since we are to solve for T, we must isolate the term containing T. This can be done by first removing the grouping symbols, and then isolate the term with T.

$$V = V_0[1 + b(T - T_0)] \qquad \text{(original equation)}$$

$$V = V_0[1 + bT - bT_0] \qquad \text{(remove parentheses)}$$

$$V = V_0 + bTV_0 - bT_0V_0 \qquad \text{(remove brackets)}$$

$$V - V_0 + bT_0V_0 = bTV_0 \qquad \text{(subtract } V_0 \text{ and add } bT_0V_0 \text{ to both sides)}$$

$$T = \frac{V - V_0 + bT_0V_0}{bV_0} \qquad \text{(divide both sides by } bV_0 \text{ and switch sides)}$$

If we wish to determine the value of any literal number in an expression for which we know values of the other literal numbers, we should *first solve for the required symbol and then substitute the given values*. This is illustrated in the following example.

Example E

The electric resistance R, in ohms, of a resistor changes with temperature T according to the formula $R = R_0(1 + \alpha T)$, where R_0 is the resistance at $0°C$. For a given resistor $R_0 = 712\ \Omega$ and $\alpha = 0.00455/°C$. Determine the value of T for $R = 825\ \Omega$.

Following the procedure given above, we first solve for T and then substitute the given values.

$$R = R_0 + R_0\alpha T$$
$$R - R_0 = R_0\alpha T$$
$$T = \frac{R - R_0}{\alpha R_0}$$

Now substituting, we have

$$T = \frac{825 - 712}{(0.00455)(712)}$$
$$= 34.9°C$$

Here a calculator is used to perform the calculation, and the result is rounded to three significant digits.

Exercises 1–12

In Exercises 1 through 8, solve for the indicated letter.

1. $ax = b$, for x
2. $cy + d = 0$, for y
3. $4n + 1 = m$, for n
4. $bt - 3 = a$, for t
5. $ax + 6 = 2ax - c$, for x
6. $s - 6n^2 = 3s + 4$, for s
7. $\frac{1}{2}t - (4 - a) = 2a$, for t
8. $7 - (p - \frac{1}{3}x) = 3p$, for x

In Exercises 9 through 28, each of the given formulas arises in the technical or scientific area of study listed. Solve for the indicated letter.

9. $E = IR$, for R (electricity)
10. $F = pDL$, for p (mechanics)
11. $P = 2\pi Tf$, for T (mechanics: torsion)
12. $PV = RT$, for T (chemistry: gas law)
13. $P = \frac{\pi^2 EI}{L^2}$, for E (mechanics)
14. $V = \frac{4}{3}\pi r^3$, for π (geometry)

15. $v = v_0 - gt$, for g (physics: motion)

16. $A_1 = A(M + 1)$, for M (photography)

17. $a = V(k - PV)$, for k (biology: muscle contractions)

18. $T = (F_1 - F_2)r$, for F_1 (machine design)

19. $Q_1 = P(Q_2 - Q_1)$, for Q_2 (refrigeration)

20. $s = s_0 + v_0t - 16t^2$, for v_0 (physics: motion)

21. $l = a + (n - 1)d$, for n (mathematics: progressions)

22. $T_2w = q(T_2 - T_1)$, for T_1 (chemistry: energy)

23. $L = \pi(r_1 + r_2) + 2d$, for r_1 (physics: pulleys)

24. $r_e + r_c(1 - a) = \dfrac{1}{h}$, for r_e (electricity: transistors)

25. $V_0 = \dfrac{V_r A}{1 + \beta A}$, for β (electricity: integrated circuit)

26. $R = \dfrac{2(E - E_p)}{m_0 g}$, for E (modern physics)

27. $R = \dfrac{wL}{H(w + L)}$, for H (architecture: interior lighting)

28. $PV^2 = RT(1 - e)(V + b) - A$, for T (thermodynamics)

In Exercises 29 through 32, determine the value of the indicated symbol for the given values of the other symbols.

29. A formula used in a particular architectural design is $p = \pi r + 2r + 2w$. Determine the value of w if $r = 22.5$ in. and $p = 172$ in.

30. The voltage in a certain circuit is given by the formula $V_2 = V_1 - IR$, where V_1 is another voltage, I is the current, and R is the resistance. Find I, in amperes, if $V_1 = 8.535$ V, $V_2 = 7.260$ V, and $R = 10.60\ \Omega$.

31. A formula relating the Fahrenheit temperature F and the Celsius temperature C is $F = \frac{9}{5}C + 32$. Find the Celsius temperature which corresponds to $90.2°F$.

32. In forestry, a formula used to determine the volume V of a log is $V = \frac{1}{2}L(B + b)$, where L is the length of the log and B and b are the areas of the ends. Find b, in square feet, if $V = 38.6\ \text{ft}^3$, $L = 16.1$ ft, and $B = 2.63\ \text{ft}^2$.

1–13 Applied Verbal Problems

Mathematics is very useful in most technical areas, because with it we can solve many kinds of applied problems. Some of these problems are in formula form and can therefore be solved directly. However, in practice it is often necessary to set up equations to be solved by using known formulas and given conditions. Such problems are first formulated as verbal problems, and it is necessary to translate them into mathematical terms for solution.

Usually the most difficult part in solving a stated problem is identifying the information that leads to the equation. Often this is due to the fact that some of the information is implied, but not explicitly stated, in the problem. A careful reading of the problem and an understanding of all terms and expressions are very important to being able to set up the equation properly for solution.

Since a careful reading and analysis are important to the solution of stated problems, it is possible only to give a general guideline to follow. Thus, (1) *read the statement of the problem carefully; (2) clearly identify the unknown quantities, assign an appropriate letter to represent one of them, and specify the others in terms of this unknown; (3) analyze the statement clearly to establish the necessary equation; and (4) solve the equation, checking the solution in the original statement of the problem.* Carefully read the following examples.

Example A

Two machine parts together weigh 17 lb. If one weighs 3 lb more than the other, what is the weight of each?

Since the weight of each part is required, we write

let w = the weight of the lighter part

as a way of establishing the unknown for the equation. Any appropriate letter could be used, and we could have let it represent the heavier part.

Also, since one weighs 3 lb more than the other, we can write

let $w + 3$ = the weight of the heavier part

Since the two parts together weigh 17 lb, we have the equation

$$w + (w + 3) = 17$$

This can now be solved.

$$2w + 3 = 17$$
$$2w = 14$$
$$w = 7$$

Thus, the lighter part weighs 7 lb and the heavier part weighs 10 lb. This checks with the original statement of the problem.

Example B

A certain microprocessor chip is rectangular, and its length is 2.0 mm more than its width. Find the dimensions of the chip if its perimeter is 26.4 mm.

Since the dimensions, the length and the width, are required, we

let w = the width of the chip

Since the length is 2.0 mm more than its width, we know that

$w + 2.0$ = the length of the chip

The perimeter (distance around) of a rectangle is twice the length plus twice the width. This gives us the equation

$$2(w + 2.0) + 2w = 26.4$$

since the perimeter is given as 26.4 mm. This is the equation we need.

Solving this equation, we have

$$2w + 4.0 + 2w = 26.4$$
$$4w = 22.4$$
$$w = 5.6 \text{ mm}$$

and

$$w + 2 = 7.6 \text{ mm}$$

Thus, the length is 7.6 mm and the width is 5.6 mm. We see that these values check with the statements in the original problem.

Example C

A man rowing x mi/h covers 8 mi in 1 h when going downstream. By rowing twice as fast while going upstream, he is able to cover only 7 mi in 1 h. Find the original rate of speed of his rowing, x, and the rate of flow of the stream.

In this problem we have let x equal the man's original rate of rowing. From the fact that the man was able to cover 8 mi in 1 h while going downstream, we conclude that his rate plus the rate of the stream equals 8 mi/h. Thus, $8 - x$ is the rate of the stream. When he is going upstream, the stream retards his progress, which means that the rate at which he actually proceeds upstream is $2x - (8 - x)$. He goes at this rate for 1 h, traveling 7 mi. Using the formula $d = rt$ (distance equals rate times time), we have

$$7 = [2x - (8 - x)](1)$$

Solving for x, we obtain

$$7 = 2x - (8 - x)$$
$$7 = 2x - 8 + x$$
$$15 = 3x$$
$$x = 5$$

Therefore, the original rate of rowing was 5 mi/h, and the stream flows at the rate of 3 mi/h. We see that this checks in that the distance covered in 1 h going downstream is $(5 + 3)(1) = 8$ mi, and the distance covered in 1 h going upstream is $[(2)(5) - 3] = 7$ mi.

 Note that we check the solution with the ***statement*** of the problem. This is done because the original equation may be in error.

Example D

A solution of alcohol and water contains 2 L of alcohol and 6 L of water. How much pure alcohol must be added to this solution so that the resulting solution with be $\frac{2}{5}$ alcohol?

First we let x equal the number of liters of alcohol to be added. The statement of the problem tells us that we want the volume of alcohol to be $\frac{2}{5}$ of the total volume of the final mixture. The final total volume of alcohol will be $2 + x$, and the final total volume of the mixture of water and alcohol will be $8 + x$. This means that

$$2 + x = \frac{2}{5}(8 + x)$$

Multiplying each side by 5, we have

$$5(2 + x) = 2(8 + x)$$
$$10 + 5x = 16 + 2x$$
$$3x = 6$$
$$x = 2$$

Therefore, 2 L of alcohol are to be added to the solution. Note that this result checks, since there would be 4 L of alcohol of a total volume of 10 L when 2 L of pure alcohol are added to the original solution.

Exercises 1–13

Solve the following problems by first setting up an appropriate equation.

1. Together, two computers cost $7800 per month to rent. If one costs twice as much as the other, what is the monthly cost of each?

2. The combined capacity of two oil tanks is 1375 gallons. If one tank holds 275 gal more than the other tank, what is the capacity of each?

3. The sum of three electric currents is 12.7 A. If the smallest is 2.2 A less than the next, which in turn is 2.3 A less than the largest, what are the values of the three currents?

4. In order to produce equilibrium on a particular beam, the sum of two forces must equal a third force. If the second of the two forces is 6.4 N more than the first and the third force is four times the first, what are the forces?

5. A developer purchased 70 acres of land for $900,000. If part of the land cost $20,000 per acre and the remainder cost $10,000 per acre, how much did the developer buy at each price?

6. A vial contains 2000 mg, which is to be used for two dosages. One patient is to be administered 660 mg more than the other. How much should be administered to each?

7. In the design of a bridge, an engineer determines that four fewer 18-m girders are needed for the span than 15-m girders. How many 18-m girders are needed?

8. A fuel oil storage depot had an 8-weeks supply on hand. However, cold weather caused the supply to be used in 6 weeks when 5000 gal extra were used each week. How many gallons were in the original supply?

9. The length of a spring increases 0.25 ft for each pound it supports. If the spring is 8 ft long when 6 lb is hung from it, what is the length of the spring when no weight is hung from it?

10. A temperature measured in degrees Fahrenheit is 32 more than $\frac{9}{5}$ the corresponding reading in degrees Celsius. If the Celsius reading of a certain room is 25°C, what is the Fahrenheit reading?

11. An architect designs a rectangular window such that the width of the window is 18 in. less than the height. If the perimeter of the window is 180 in., what are its dimensions?

12. It takes 300 cm of trim to go around a rectangular solar panel. The length of the panel is 90 cm more than the width. What are the dimensions of the panel?

13. A primary natural gas pipeline feeds into three smaller pipelines, each of which is 2.6 km longer than the main pipeline. If the total length of the four pipelines is 35.4 km, what is the length of each section of the line?

14. A square tract of land is enclosed with fencing and then divided in half by additional fencing parallel to two of the sides. If 75 m of fencing are used, what is the length of one side of the tract?

15. In a race, one race car stalls at the beginning and starts 30 s after a second car. However, the first car travels at 260 ft/s and the second car travels at 240 ft/s. How long will it take the first car to overtake the second car? If the race is 20 mi in length, who will come in first?

16. A supersonic jet made one trip by averaging 100 mi/h less than the speed of sound for 1 h and then averaging 400 mi/h more than the speed of sound for 3 h. If the trip covered 3980 mi, what is the speed of sound?

17. Two supersonic jets, originally 5400 mi apart, start at the same time and travel toward each other. Find the speed of each if one travels 400 mi/h faster than the other and they meet in 1.5 h.

18. A corporate executive leaves the manufacturing plant and travels on an interstate highway at 55 mi/h to the corporate headquarters. The executive later returns in a helicopter, which travels at 125 mi/h on a route parallel to the highway. If the total travel time is 1.8 h, how far is it from the manufacturing plant to the corporate headquarters?

19. A certain type of engine uses a fuel mixture of 15 parts of gasoline to 1 part of oil. How much gasoline must be mixed with a gasoline-oil mixture, which is 75% gasoline, to make 8 L of the required mixture for the engine?

20. An alloy weighing 20 lb is 30% copper. How many pounds of another alloy, which is 80% copper, must be added in order for the final alloy to be 50% copper?

1–14 Exercises for Chapter 1

In Exercises 1 through 12, simplify the given expressions.

1. $(-2) + (-5) - (+3)$

2. $(+6) - (+8) - (-4)$

3. $\dfrac{(-5)(+6)(-4)}{(-2)(+3)}$

4. $\dfrac{(-9)(-12)(-4)}{24}$

5. $-5 - 2(-6) + \dfrac{-15}{+3}$

6. $3 - 5(-2) - \dfrac{12}{-4}$

7. $\dfrac{18}{3 - 5} - (-4)^2$

8. $-(-3)^2 - \dfrac{-8}{(-2) - (-4)}$

9. $\sqrt{16} - \sqrt{64}$

10. $-\sqrt{144} + \sqrt{49}$

11. $(\sqrt{7})^2 - \sqrt[3]{8}$

12. $-\sqrt[4]{16} + (\sqrt{6})^2$

In Exercises 13 through 24, simplify the given expressions. Where appropriate, express results with positive exponents only.

13. $(-2rt^2)^2$

14. $(3x^4y)^3$

15. $\dfrac{18m^3n^4t}{3mn^5t^3}$

16. $\dfrac{15p^4q^2r}{5pq^5r}$

17. $(x^0y^{-1}z^3)^2$

18. $(3a^0b^{-2})^3$

19. $\dfrac{-16s^{-2}(st^2)}{-2st^{-1}}$

20. $\dfrac{-35x^{-1}y(x^2y)}{5xy^{-1}}$

21. $\sqrt{45}$

22. $\sqrt{48}$

23. $\sqrt{4 + 16}$

24. $\sqrt{9 + 36}$

In Exercises 25 through 28, perform the indicated operations on a calculator. In each case round off the results to four significant digits.

25. $37.38 - 16.92(1.067)^2$

26. $\dfrac{8.896 \times 10^{-12}}{3.5954 + 6.0449}$

27. $(0.6723)^3 - \sqrt{0.1958 + 2.844}$

28. $\dfrac{1}{0.03568} + \dfrac{37{,}466}{29.63^2}$

In Exercises 29 through 60, perform the indicated operations.

29. $a - 3ab - 2a + ab$

30. $xy - y - 5y - 4xy$

31. $6xy - (xy - 3)$

32. $-(2x - b) - 3(x - 5b)$

33. $(2x - 1)(x + 5)$

34. $(x - 4y)(2x + y)$

35. $(x + 8)^2$

36. $(2x + 3y)^2$

37. $\dfrac{2h^3k^2 - 6h^4k^5}{2h^2k}$

38. $\dfrac{4a^2x^3 - 8ax^4}{2ax^2}$

39. $4a - [2b - (3a - 4b)]$

40. $3b - [2b + 3a - (2a - 3b)] + 4a$

41. $2xy - \{3z - [5xy - (7z - 6xy)]\}$

42. $x^2 + 3b + [(b - y) - 3(2b - y + z)]$

43. $(2x + 1)(x^2 - x - 3)$

44. $(x - 3)(2x^2 - 3x + 1)$

45. $-3y(x - 4y)^2$

46. $-s(4s - 3t)^2$

47. $3p[(q - p) - 2p(1 - 3q)]$

48. $3x[2y - r - 4(s - 2r)]$

49. $\dfrac{12p^3q^2 - 4p^4q + 6pq^5}{2p^4q}$

50. $\dfrac{27s^3t^2 - 18s^4t + 9s^2t}{9s^2t}$

51. $(2x^2 + 7x - 30) \div (x + 6)$

52. $(4x^2 + 15x - 21) \div (2x + 7)$

53. $\dfrac{3x^3 - 7x^2 + 11x - 3}{3x - 1}$

54. $\dfrac{x^3 - 4x^2 + 7x - 12}{x - 3}$

55. $\dfrac{4x^4 + 10x^3 + 18x - 1}{x + 3}$

56. $\dfrac{8x^3 - 14x + 3}{2x + 3}$

57. $-3\{(r + s - t) - 2[(3r - 2s) - (t - 2s)]\}$

58. $(1 - 2x)(x - 3) - (x + 4)(4 - 3x)$

59. $\dfrac{2y^3 + 9y^2 - 7y + 5}{2y - 1}$

60. $\dfrac{6x^2 + 5xy - 4y^2}{2x - y}$

In Exercises 61 through 72, solve the given equations.

61. $3s + 8 = 5s$

62. $6n = 14 - n$

63. $3x + 1 = x - 9$

64. $4y - 3 = 5y + 7$

65. $6x - 5 = 3(x - 4)$

66. $-2(4 - y) = 3y$

67. $2s + 4(3 - s) = 6$

68. $-(4 - v) = 2(2v - 5)$

69. $3t - 2(7 - t) = 5(2t + 1)$

70. $6 - 3x - (8 - x) = x - 2(2 - x)$

71. $2.7 + 2.0(2.1x - 3.4) = 0.1$

72. $0.250(6.721 - 2.44x) = 2.08$

In Exercises 73 through 80, change any numbers in ordinary notation to scientific notation or change any numbers in scientific notation to ordinary notation. (See Appendix B for an explanation of symbols which are used.)

73. The escape velocity (the velocity required to leave the earth's gravitational field) of a rocket is in excess of 25,000 mi/h.

74. When the first pictures of the surface of Mars were transmitted to Earth, Mars was 213,000,000 mi from Earth.

75. The ratio of the charge to the mass of an electron is 1.76×10^{11} C/kg.

76. Atmospheric pressure is about 1.013×10^5 Pa.

77. An oil film on water is about 0.0000002 in. thick.

78. Under tension, a certain wire stretches 0.00025 m.

79. The viscosity of air is about 1.8×10^{-5} N·s/m².

80. The electric field intensity in a certain electromagnetic wave is 2.5×10^{-3} V/m.

In Exercises 81 through 96, solve for the indicated letter. Where noted the given formula arises in the technical or scientific area of study listed.

81. $3s + 2 = 5a$, for s

82. $5 - 7t = 6b$, for t

83. $3(4 - x) = 8 + 2n$, for x

84. $6 - 3b = 5(7 - 2v)$, for v

85. $B = \dfrac{\phi}{A}$, for A (ϕ is Greek phi) (electricity: magnetism)

86. $D = \dfrac{KI^2t}{A}$, for t (medicine: cell damage)

87. $I = P + Prt$, for t (business: interest)

88. $v^2 = v_0^2 + 2gh$, for h (physics: motion)

89. $I(r + nR) = nE$, for R (electricity)

90. $R(R_1 + R_2) = R_1R_2$, for R (electricity)

91. $D_p = \dfrac{ND_0}{N + 2}$, for D_0 (mechanics: gears)

92. $Y_{n+1} = \dfrac{R_D X_n + X_0}{R_D + 1}$, for X_n (chemistry: distillation)

93. $L = L_0[1 + \alpha(t_2 - t_1)]$, for α (the Greek alpha)   (physics: heat)

94. $P_a - P_b = \dfrac{32LV\mu}{g_c D^2}$, for μ (the Greek mu)   (fluid dynamics)

95. $S = \frac{1}{2}n[2a + (n - 1)d]$, for a (mathematics: progressions)

96. $M = Rx - P(x - a)$, for a (mechanics: beams)

In Exercises 97 through 100, evaluate the given expressions. Perform all calculations by use of a scientific calculator and round results to three significant digits.

97. The change in length of a steel girder can be found by evaluating the expression $aL(T_2 - T_1)$. Find the change in length, in feet, of a girder for which $a = 0.670 \times 10^{-5}/°F$, $L = 75.0$ ft, $T_2 = 75.6°F$, and $T_1 = 39.8°F$.

98. The combined resistance R of two electric resistors in parallel is found from the formula given in Exercise 90. Find R, in ohms, if $R_1 = 3.78 \times 10^3\ \Omega$ and $R_2 = 8.75 \times 10^3\ \Omega$.

99. Three forces on a certain beam are related by the equation $3.50F_1 = 1.80F_2 + 3.25F_3$. Find F_3 if $F_1 = 2.63$ N and $F_2 = 2.50$ N.

100. A formula which gives the height h of an object in terms of its velocity v, initial velocity v_0, and the acceleration due to gravity is given in Exercise 88. Find h, in feet, if $v_0 = 55.0$ ft/s, $v = 25.5$ ft/s, and $g = -32.2$ ft/s^2.

In Exercises 101 through 104, perform the indicated operations.

101. When determining the center of mass of a certain object, the expression $[(8x - x^2) - (x^2 - 4x)]$ is used. Simplify this expression.

102. In determining the value of sales in a particular business enterprise, the expression $(700 + 100n)(12 - n)$ is found. Multiply and simplify this expression.

103. Simplify the following expression, which arises in determining the final temperature of a certain mixture of objects originally at different temperatures:

$$500(0.22)(T_f - 20) + 120(T_f - 20) - 22(75 - T_f)$$

104. In studying the dispersion of light, the expression $(k + 2)[(n - jK)^2 - 1]$ is found. Expand this expression.

In Exercises 105 through 114, solve the stated problems by first setting up an appropriate equation.

105. The combustion of carbon usually results in the production of both carbon monoxide and carbon dioxide. If 500 kg of oxygen are available for combustion and it is desired that nine times as much oxygen be converted to carbon dioxide as is converted to carbon monoxide, how much oxygen would be converted to each of the compounds?

106. The electric current in one transistor is three times that in another transistor. If the sum of the currents is 0.012 A, what is the current in each?

107. The cost of producing a first type of pocket calculator is three times the cost of producing a second type. The total cost of producing two of the first type and three of the second type is $45. What is the cost of producing each type?

108. An air sample contains 4 ppm (parts per million) of two harmful pollutants. The concentration of one is four times the other. What is the concentration of each?

109. The rectangular front of a microwave oven is trimmed around the edge and there is an additional strip parallel to the base (the longer dimension). If the base is 10 cm more than the height and a total of 310 cm of trim is used, what are the dimensions of the oven?

110. A rectangular field is enclosed with a fence of four strands of barbed wire. The length is 20 m more than the width, and a total of 1360 m of barbed wire is used for the fencing. What are the dimensions of the field?

111. Two motorboats, 55.3 mi apart, start toward each other. One travels at the rate of 22.5 mi/h and the other at 17.0 mi/h. When will they meet?

112. A bicycle rider averaged 14.0 mi/h in going from a friend's home and 16.0 mi/h in returning. The total time of travel was 2.25 h. How long did the return trip take?

113. One grade of oil has 0.50% of a certain additive and a second grade has 0.75% of the same additive. How many liters of the first grade of oil must be added to the second grade in order to have 1000 L with 0.65% of the additive?

114. Fifty pounds of a cement-sand mixture is 40% sand. How many pounds of sand must be added for the resulting mixture to be 60% sand?

CHAPTER 2

Functions and Graphs

In Chapter 1 we established and reviewed many of the basic algebraic operations. These are important in the development of the topics to be taken up in the remainder of this book.

In mathematics and most areas of technology, the way in which various quantities are related is important. In this chapter we first present the concept of a function, which is a way of giving a relation between quantities. Then we develop the use of graphs, which are used extensively to show and study relationships in a convenient way.

2–1 Functions

In the later part of Chapter 1, we discussed the solution of equations, with applications to formulas. In most of the formulas, one quantity was given in terms of one or more other quantities. It is obvious, then, that the various quantities are related by means of the formula. We see that in the study of natural events, a relationship can be found to exist between certain quantities.

If we were to perform an experiment to determine whether or not a relationship exists between the distance an object drops and the time it falls, observation of the results would indicate (approximately, at least) that $s = 16t^2$, where s is the distance in feet and t is the time in seconds. We would therefore see that distance and time for a falling object are related.

A similar study of the pressure and the volume of a gas at constant temperature would show that as pressure increases, volume decreases according to the formula $PV = k$, where k is a constant. Electrical measurements of current and voltage with respect to a particular resistor would show that $V = kI$, where V is the voltage, I is the current, and k is a constant.

Considerations such as these lead us to one of the most important and basic concepts in mathematics. *Whenever a relationship exists between two variables such that for every permissible value of the first, there is only one corresponding value of the second, we say that the second variable is a* **function** *of the first variable.*

The first variable is called the **independent variable,** since values can be assigned to it arbitrarily. *The second variable is called the* **dependent variable,** since its value is determined by the choice of the independent variable. *Values of the independent variable and dependent variable are to be* **real numbers.** Therefore, it is possible that there are restrictions on the possible values of the variables.

There are many ways to express functions. Formulas such as those we have discussed define functions. Other ways to express functions are by means of tables, charts, and graphs.

Example A

In the equation $y = 2x$, we see that y is a function of x, since for each value of x there is only one value of y. For example, if $x = 3$, $y = 6$ and no other value. By arbitrarily assigning values to x, we make it the independent variable and y the dependent variable.

Example B

The volume of a cube of edge e is given by $V = e^3$. Here V is a function of e, since for each value of e there is one value of V. The dependent variable is V and the independent variable is e.

If the equation relating the volume and edge of a cube were written as $e = \sqrt[3]{V}$, that is, if the edge of a cube were expressed in terms of its volume, we would say that e is a function of V. In this case e would be the dependent variable and V the independent variable.

Example C

The power P developed in a certain resistor by a current I is given by $P = 4I^2$. Here P is a function of I. The dependent variable is P, and the independent variable is I.

Example D

If the formula in Example C is written as $I = \frac{1}{2}\sqrt{P}$, then I is a function of P. The independent variable is P, and the dependent variable is I. Even though P is the independent variable, it is restricted to values which are positive or zero. (This is written as $P \geq 0$.) Otherwise, the values of I would be imaginary. As noted earlier, we shall restrict ourselves to real numbers. Also, as defined, $I \geq 0$, since $\sqrt{P}$ cannot be negative.

Example E

For the equation $y = 2x^2 - 6x$, we say that y is a function of x. The dependent variable is y, and the independent variable is x. Some of the values of y corresponding to chosen values of x are given in the following table.

x	-2	-1	0	$\frac{1}{2}$	1	2	3	π	10
y	20	8	0	$-\frac{5}{2}$	-4	-4	0	$2\pi^2 - 6\pi$	140

For convenience of notation,

the phrase "function of x" is written as $f(x)$.

This, in turn, means that the statement "y is a function of x" may be written as $y = f(x)$. (Note that $f(x)$ does **not** mean f times x.)

Example F

If $y = 6x^3 - 5x$, we may say that y is a function of x, where this function $f(x)$ is $6x^3 - 5x$. We may also write $f(x) = 6x^3 - 5x$. It is common to write functions in this form, rather than in the form $y = 6x^3 - 5x$. However, y and $f(x)$ represent the same expression.

One of the most important uses of this notation is to designate the value of a function for a particular value of the independent variable. That is, *the value of the function $f(x)$ when $x = a$ is written as $f(a)$.*

Example G

For the function $f(x) = 3x^2 - 5$, the value of $f(x)$ for $x = 2$ may be represented as $f(2)$. Thus, substituting 2 for x, we have

$$f(2) = 3(2^2) - 5 = 7$$

In the same way, the value of $f(x)$ for $x = -1$ is

$$f(-1) = 3(-1)^2 - 5 = -2$$

The value of $f(x)$ for $x = -2.73$ is

$$f(-2.73) = 3(-2.73)^2 - 5 = 17.4$$

where a calculator was used and the result is rounded off.

In certain instances we need to define more than one function of x. Then we use different symbols to denote the functions. For example, $f(x)$ and $g(x)$ may represent different functions of x, such as $f(x) = 5x^2 - 3$ and $g(x) = 6x - 7$. Special functions are represented by particular symbols. For example, in trigonometry we shall come across the "sine of the angle ϕ," where the sine is a function of ϕ. This is designated by $\sin \phi$.

Example H

If $f(x) = \sqrt{3x} + x$ and $g(x) = ax^4 - 5x$, then

$$f(3) = \sqrt{3(3)} + 3 = 3 + 3 = 6$$

and

$$g(3) = a(3^4) - 5(3) = 81a - 15$$

There are occasions when we wish to evaluate a function in terms of a literal number rather than an explicit number. However, whatever number a represents in $f(a)$, we substitute a for x in $f(x)$.

Example I

If $g(t) = 4t^2 - 5t$, to find $g(a^3)$ we substitute a^3 for t in the function. Thus,

$$g(a^3) = 4(a^3)^2 - 5(a^3) = 4a^6 - 5a^3$$

For the same function,

$$
\begin{aligned}
g(b + 1) &= 4(b + 1)^2 - 5(b + 1) \\
&= 4(b^2 + 2b + 1) - 5(b + 1) \\
&= 4b^2 + 8b + 4 - 5b - 5 \\
&= 4b^2 + 3b - 1
\end{aligned}
$$

Example J

The resistance of a particular resistor as a function of temperature is $R = 10.0 + 0.10T + 0.001T^2$. If a given temperature T is increased by 10°C, what is the value of R for the increased temperature as a function of the temperature T?

We are to determine R for a temperature of $T + 10$. Since

$$f(T) = 10.0 + 0.10T + 0.001T^2$$

we know that

$$
\begin{aligned}
f(T + 10) &= 10.0 + 0.10(T + 10) + 0.001(T + 10)^2 \\
&= 10.0 + 0.10T + 1.0 + 0.001T^2 + 0.02T + 0.1 \\
&= 11.1 + 0.12T + 0.001T^2
\end{aligned}
$$

A function may be looked upon as a set of instructions. These instructions tell us how to obtain the value of the dependent variable for a particular value of the independent variable, even if the set of instructions is expressed in literal symbols.

Example K

The function $f(x) = x^2 - 3x$ tells us to "square the value of the independent variable, multiply the value of the independent variable by 3, and subtract the second result from the first." An analogy would be a computer which was programmed so that when a number was entered into the computer, it would square the number and then subtract 3 times the value of the number. This is represented in diagram form in Figure 2–1.

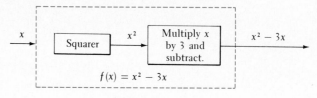

Figure 2–1

The functions $f(t) = t^2 - 3t$ and $f(n) = n^2 - 3n$ are the same as the function $f(x) = x^2 - 3x$, since the operations performed on the independent variable are the same. Although different literal symbols appear, this does not change the function.

As we mentioned earlier, the values of the independent variable and the dependent variable must be real numbers. This means that the values chosen for the independent variable must give real number values for the dependent variable. Therefore, *values of the independent variable that lead to division by zero or to imaginary values of the dependent variable may not be used.* Example D and the following example illustrate this point.

Example L

The function $f(u) = \dfrac{u}{u - 4}$ is not defined if $u = 4$, since this would require division by zero. Therefore, the values of u are restricted to values other than 4.

The function $g(s) = \sqrt{3 - s}$ is not defined for real numbers if s is greater than 3, since such values would result in imaginary values for $g(s)$. Thus, the values of s are restricted to values equal to or less than 3.

Practical physical considerations may also restrict the values of the independent variable. For example, the surface area of a cube, as a function of its edge, is given by $A = 6e^2$. However, since the edge of a cube cannot be negative and there is no cube if $e = 0$, the values of e must be positive.

Exercises 2-1

In Exercises 1 through 16, determine the appropriate functions.

1. Express the area A of a circle as a function of its radius r.

2. Express the area A of a circle as a function of its diameter d.

3. Express the circumference c of a circle as a function of its radius r.

4. Express the circumference c of a circle as a function of its diameter d.

5. Express the area A of a rectangle of width 5 as a function of its length l.

6. Express the volume V of a right circular cone of height 8 as a function of the radius r of the base.

7. Express the area A of a square as a function of its side s; express the side s of a square as a function of its area A.

8. Express the perimeter p of a square as a function of its side s; express the side s of a square as a function of its perimeter p.

9. Express the distance d traveled in t hours at 55 mi/h.

10. Express the cost C of fencing a rectangular field 100 m long and w meters wide if fencing costs $5 per meter.

11. A rocket weighs 3000 tons at liftoff. If the first-stage engines burn fuel at the rate of 10 tons/s, find the weight w of the rocket as a function of the time t while the first-stage engines operate.

12. A total of x feet is cut from a 24-ft board. Express the remaining length y as a function of x.

13. Express the simple interest I on $200 at 4% per year as a function of the number of years t.

14. A chemist adds x liters of a solution which is 50% alcohol to 100 L of a solution which is 70% alcohol. Express the number n of liters of alcohol in the final solution as a function of x.

15. A TV repairman charges $25 plus $20/h for labor costs for a house call. Express the labor cost C as a function of the length h, in hours, of a house call.

16. A taxi fare is 55¢ plus 10¢ for every $\frac{1}{5}$ mile traveled. Express the fare F as a function of the distance s traveled.

In Exercises 17 through 28, evaluate the given functions.

17. Given $f(x) = 2x + 1$, find $f(1)$ and $f(-1)$.

18. Given $f(x) = 5x - 9$, find $f(2)$ and $f(-2)$.

19. Given $f(x) = 5 - 3x$, find $f(-2)$ and $f(4)$.

20. Given $f(x) = 7 - 2x$, find $f(5)$ and $f(-4)$.

21. Given $f(n) = n^2 - 9n$, find $f(3)$ and $f(-5)$.

22. Given $f(v) = 2v^3 - 7v$, find $f(1)$ and $f(\frac{1}{2})$.

23. Given $\phi(x) = \dfrac{6 - x^2}{2x}$, find $\phi(1)$ and $\phi(-2)$.

24. Given $H(q) = \dfrac{8}{q} + 2\sqrt{q}$, find $H(4)$ and $H(16)$.

25. Given $g(t) = at^2 - a^2 t$, find $g(-\frac{1}{2})$ and $g(a)$.

26. Given $s(y) = 6\sqrt{y} - 3$, find $s(9)$ and $s(a^2)$.

27. Given $K(s) = 3s^2 - s + 6$, find $K(-s)$ and $K(2s)$.

28. Given $T(t) = 5t + 7$, find $T(-2t)$ and $T(t + 1)$.

In Exercises 29 through 32, evaluate the given functions by use of a calculator. Round off results to the accuracy of the value of the independent variable used.

29. Given $f(x) = 5x^2 - 3x$, find $f(3.86)$ and $f(-6.92)$.

30. Given $g(t) = \sqrt{t + 1.0604} - 6t^3$, find $g(0.9261)$ and $g(-0.3256)$.

31. Given $F(y) = \dfrac{2y^2}{y + 0.03685}$, find $F(0.02474)$ and $F(-0.08466)$.

32. Given $f(x) = \dfrac{x^4 - 2.0965}{6x}$, find $f(1.9654)$ and $f(-2.3865)$.

In Exercises 33 through 36, state the instructions of the function in words as in Example K.

33. $f(x) = x^2 + 2$ 34. $f(x) = 2x - 6$

35. $g(y) = 6y - y^3$ 36. $\phi(s) = 8 - 5s + s^5$

In Exercises 37 through 40, state any restrictions that might exist on the values of the independent variable.

37. $Y(y) = \dfrac{y + 1}{y - 1}$ 38. $G(z) = \dfrac{1}{(z - 4)(z + 2)}$

39. $F(y) = \sqrt{y - 1}$ 40. $X(x) = \dfrac{6}{\sqrt{1 - x}}$

In Exercises 41 through 48, solve the given problems.

41. The pressure p exerted on an area A by a force of 250 N, as a function of A, is given by $p = 250/A$. Find the pressure on an area of 0.50 m^2.

42. The distance s that a freely falling body travels as a function of the time t is given by $s = 16t^2$, where s is measured in feet and t is measured in seconds. How far does an object fall in 2 s?

43. If the temperature within a certain refrigerator is maintained at 273 K (0°C), its *coefficient of performance* p as a function of the external temperature T (in kelvins) is

$$p = \frac{273}{T - 273}$$

What is its coefficient of performance at 308 K (35°C)?

44. The vertical distance of a point on a suspension cable from its lowest point as a function of the horizontal distance from the lowest point is given by

$$f(x) = \frac{x^4 + 600x^2}{2,000,000}$$

where x is measured in feet. How far above the lowest point is a point on the cable at a horizontal distance of 50 ft from the lowest point?

45. A piece of wire 60 in. long is cut into two pieces, one of which is bent into a circle and the other into a square. Express the total area of the two figures as a function of the perimeter of the square.

46. The net profit P made on selling 20 items, if each costs \$15, as a function of the price p is $P = 20(p - 15)$. What is the profit if the price is \$28? \$12?

47. The voltage of a certain thermocouple as a function of the temperature is given by $E = 2.8T + 0.006T^2$. Find the voltage when the temperature is $T + h$.

48. The electrical resistance of a certain ammeter as a function of the resistance of the coil of the meter is

$$R = \frac{10R_c}{10 + R_c}$$

Find the function which represents the resistance of the meter if the resistance of the coil is doubled.

2–2 Rectangular Coordinates

One of the most valuable ways of representing functions is by graphical representation. By using graphs we are able to obtain a "picture" of the function, and by using this picture we can learn a great deal about the function.

To make a graphical representation, we recall that numbers can be represented by points on a line. For a function we have values of the independent variable and the corresponding values of the dependent variable. Therefore, it is necessary to use two lines to represent the values from these sets of numbers. We do this most conveniently by placing the lines perpendicular to each other.

*We place one line horizontally and label it the **x-axis.*** The numbers of the set for the independent variable are normally placed on this axis. *The other line we place vertically, and label the **y-axis.*** Normally the y-axis is used for values of the dependent variable. *The point of intersection is called the* **origin.** This is the **rectangular coordinate system.**

On the x-axis, positive values are to the right of the origin, and negative values are to the left of the origin. On the y-axis, positive values are above the origin, and negative values are below it. *The four parts into which the plane is divided are called* **quadrants,** which are numbered as in Fig. 2–2.

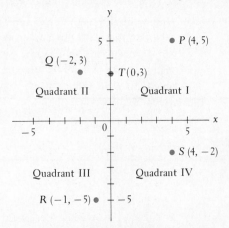

Figure 2–2

A point P on the plane is designated by the pair of numbers (x, y), where x is the value of the independent variable and y is the corresponding value of the dependent variable. *The x-value, called the* **abscissa,** *is the perpendicular distance of P from the y-axis. The y-value, called the* **ordinate,** *is the perpendicular distance of P from the x-axis.* The values x and y together, written as (x, y), are the **coordinates** of the point P.

Example A

The positions of points $P(4, 5)$, $Q(-2, 3)$, $R(-1, -5)$, $S(4, -2)$, and $T(0, 3)$ are shown in Fig. 2–2, which is located on the previous page. We see that this representation allows for *one point for any pair of values* (x, y). Also, we note that the point $T(0, 3)$ is on the y-axis. Any such point which is on either axis is not *in* any of the quadrants.

Example B

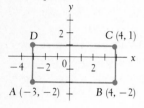

Figure 2–3

Three vertices of the rectangle in Fig. 2–3 are $A(-3, -2)$, $B(4, -2)$, and $C(4, 1)$. What is the fourth vertex?

We use the fact that opposite sides of a rectangle are equal and parallel to find the solution. Since both vertices of the base AB of the rectangle have a y-coordinate of -2, the base is parallel to the x-axis. Therefore, the top of the rectangle must also be parallel to the x-axis. Thus, the vertices of the top must both have a y-coordinate of 1, since one of them has a y-coordinate of 1. In the same way the x-coordinates of the left side must both be -3. Therefore, the fourth vertex is $D(-3, 1)$.

Example C

Where are all the points whose ordinates are 2?

Since the ordinate is the y-value, we can see that the question could be stated as: "Where are all points for which $y = 2$?" Since all such points are 2 units above the x-axis, the answer can be stated as "on a line 2 units above the x-axis."

Example D

Where are all points (x, y) for which $x < 0$ and $y < 0$?

Noting that $x < 0$ means "x is less than zero," or "x is negative," and that $y < 0$ means the same for y, we want to determine where both x and y are negative. Our answer is "in the third quadrant," since both coordinates are negative for all points in the third quadrant, and this is the only quadrant for which this is true.

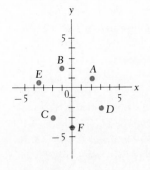

Figure 2-4

Exercises 2-2

In Exercises 1 and 2, determine (at least approximately) the coordinates of the points specified in Fig. 2-4.

1. A, B, C 2. D, E, F

In Exercises 3 and 4, plot (at least approximately) the given points.

3. $A(2, 7), B(-1, -2), C(-4, 2)$ 4. $A(3, \frac{1}{2}), B(-6, 0), C(-\frac{5}{2}, -5)$

In Exercises 5 through 8, plot the given points and then join these points, in the order given by straight-line segments. Name the geometric figure formed.

5. $A(-1, 4), B(3, 4), C(1, -2)$ 6. $A(0, 3), B(0, -1), C(4, -1)$

7. $A(-2, -1), B(3, -1), C(3, 5), D(-2, 5)$

8. $A(-5, -2), B(4, -2), C(6, 3), D(-3, 3)$

In Exercises 9 through 12, find the indicated coordinates.

9. Three vertices of a rectangle are $(5, 2), (-1, 2)$, and $(-1, 4)$. What are the coordinates of the fourth vertex?

10. Two vertices of an equilateral triangle are $(7, 1)$ and $(2, 1)$. What is the abscissa of the third vertex?

11. P is the point $(3, 2)$. Locate point Q such that the x-axis is the perpendicular bisector of the line segment joining P and Q.

12. P is the point $(-4, 1)$. Locate point Q such that the line segment joining P and Q is bisected by the origin.

In Exercises 13 through 28, answer the given questions.

13. Where are all the points whose abscissas are 1?

14. Where are all the points whose ordinates are -3?

15. Where are all points such that $y = 3$?

16. Where are all points such that $x = -2$?

17. Where are all the points whose abscissas equal their ordinates?

18. Where are all the points whose abscissas equal the negative of their ordinates?

19. What is the abscissa of all points on the y-axis?

20. What is the ordinate of all points on the x-axis?

21. Where are all the points for which $x > 0$?

22. Where are all the points for which $y < 0$?

23. Where are all the points for which $x < -1$?

24. Where are all the points for which $y > 4$?

25. Where are all points (x, y) for which $x > 0$ and $y < 0$?

26. Where are all points (x, y) for which $x < 0$ and $y > 1$?

27. In which quadrants is the ratio y/x positive?

28. In which quadrants is the ratio y/x negative?

2-3 The Graph of a Function

Now that we have introduced the concepts of a function and the rectangular coordinate system, we are in a position to determine the graph of a function. In this way we shall obtain a visual representation of a function.

The graph of a function is the set of all points whose coordinates (x, y) satisfy the functional relationship $y = f(x)$. Since $y = f(x)$, we can write the coordinates of the points on the graph as $[x, f(x)]$. Writing the coordinates in this manner tells us exactly how to find them. *We assume a certain value for x and then find the value of the function of x. These two numbers are the coordinates of the point.*

Since there is no limit to the possible number of points which can be chosen, we normally select a few values of x, obtain the corresponding values of the function, and plot these points. These points are then connected by a *smooth* curve (not short straight lines from one point to the next), and are connected from left to right. In this section we shall develop only the method of plotting points to obtain the graph of a function. We shall develop other methods for some functions in later chapters.

Example A

Graph the function $f(x) = 3x - 5$.

For purposes of graphing, we let $y = f(x)$, or $y = 3x - 5$. We then let x take on various values and determine the corresponding values of y. Note that once we choose a given value of x, we have no choice about the corresponding y-value, as it is determined by evaluating the function. If $x = 0$, we find that $y = -5$. This means that the point $(0, -5)$ is on the graph of the function $3x - 5$. Choosing another value of x, for example, 1, we find that $y = -2$. This means that the point $(1, -2)$ is on the graph of the function $3x - 5$. Continuing to choose a few other values of x, we tabulate the results, as shown in Fig. 2-5. It is best to arrange the table so that the values of x increase; then there is no doubt how they are to be connected, for they are then connected in the order shown. Finally, we connect the points as shown in Fig. 2-5, and see that the graph of the function $3x - 5$ is a straight line.

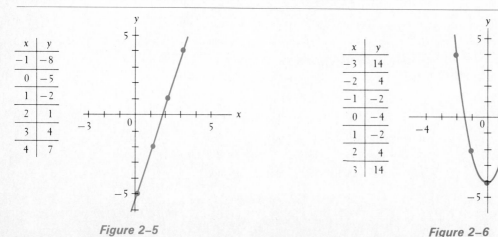

x	y
-1	-8
0	-5
1	-2
2	1
3	4
4	7

x	y
-3	14
-2	4
-1	-2
0	-4
1	-2
2	4
3	14

Figure 2-5   Figure 2-6

Example B

Graph the function $f(x) = 2x^2 - 4$.

First, we let $y = 2x^2 - 4$ and tabulate the values as shown in Fig. 2–6. In determining the values in the table, we must take particular care to obtain the correct values of y for negative values of x. Mistakes are relatively common when dealing with negative numbers. We must carefully use the laws for signed numbers. For example, for the case of $y = 2x^2 - 4$, if $x = -2$, we have $y = 2(-2)^2 - 4 = 2(4) - 4 = 8 - 4 = 4$. Once the values are obtained, we plot and connect the points with a smooth curve, as shown in Fig. 2–6.

There are some special points which should be noted. Since the graphs of most common functions are smooth, any irregularities in the graph should be carefully checked. In these cases it usually helps to take values of x between those values where the question arises. Also, if the function is not defined for some value of x (remember, *division by zero is not defined, and only real values of the variables are permissible*), the function does not exist for that value of x. Finally, in applications, we must be careful to plot values that are meaningful; often negative values for quantities, such as time, are not physically meaningful. The following examples illustrate these points.

Example C

Graph the function $y = x - x^2$.

First we determine the values in the table, as shown with Fig. 2–7. Again we must be careful with negative values of x. For $x = -1$, we have $y = (-1) - (-1)^2 = -1 - (+1) = -1 - 1 = -2$. Once all the values have been found and plotted, we note that $y = 0$ for both $x = 0$ and $x = 1$. The question arises—what happens between these values? Trying $x = \frac{1}{2}$, we find that $y = \frac{1}{4}$. Using this point completes the necessary information. Note that in plotting these graphs we do not stop the graph with the last point determined, but indicate that the curve continues.

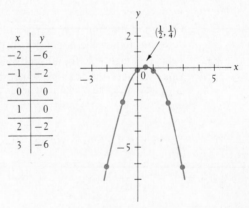

x	y
-2	-6
-1	-2
0	0
1	0
2	-2
3	-6

Figure 2–7

Example D

Graph the function $y = 1 + \dfrac{1}{x}$.

In finding the points on this graph, as shown in Fig. 2–8, we note that y is not defined for $x = 0$. Thus we must be careful not to have any part of the curve cross the y-axis ($x = 0$). Although we cannot let $x = 0$, we can choose other values of x between -1 and 1 which are close to zero. In doing so, we find that as x gets closer to zero, the points get closer and closer to the y-axis, although they do not reach or touch it. In this case the y-axis is called an **asymptote** of the curve.

x	y
-4	$3/4$
-3	$2/3$
-2	$1/2$
-1	0
$-1/2$	-1
$-1/3$	-2
$1/3$	4
$1/2$	3
1	2
2	$3/2$
3	$4/3$
4	$5/4$

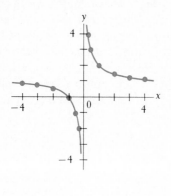

Figure 2–8

x	y
-1	0
0	1
1	1.4
2	1.7
3	2
4	2.2
5	2.4
6	2.6
7	2.8
8	3

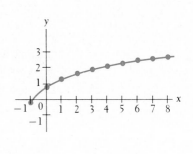

Figure 2–9

Example E

Graph the function $y = \sqrt{x + 1}$.

When finding the points for the graph, we may not let x equal any negative value less than -1, for all such values of x lead to imaginary values for y. Also, since we have the positive square root indicated, all values of y are positive or zero. See Fig. 2–9. Note that the graph starts at the point $(-1, 0)$.

Example F

The power of a certain voltage source as a function of the load resistance is given by

$$P = \frac{100R}{(0.5 + R)^2}$$

where P is measured in watts and R in ohms. Plot the power as a function of the resistance.

Since a negative value for the resistance has no physical significance, we need not plot P for negative values of R. The following table of values is obtained.

R(ohms)	0	0.25	0.50	1.0	2.0	3.0	4.0	5.0	10.0
P(watts)	0.0	44.4	50.0	44.4	32.0	24.5	19.8	16.5	9.1

The values 0.25 and 0.50 are used for R when it is found that P is less for $R = 2$ than for $R = 1$. In this way a smoother curve is obtained (see Fig. 2–10).

Also note that the scale on the P-axis is different from that on the R-axis. This reflects the different magnitudes and ranges of values used for each of the variables. Different scales are normally used in such cases.

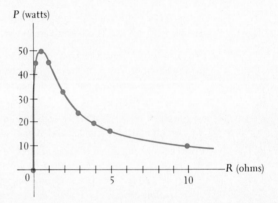

Figure 2–10

Functions of a particular type have graphs which are of a specific form, and many of them have been named. We noted that the graph of the function in Example A is a **straight line**. A **parabola** is illustrated in Example B and in Example C. The graph of the function in Example D is a **hyperbola**. Other types of graphs are found in the exercises and in later chapters. The straight line is considered again in Chapter 4 and the parabola in Chapter 6. A more detailed analysis of several of these curves is found in Chapter 20. The use of graphs is extensive in mathematics and in nearly all areas of application. For example, in Section 2–5 we see how graphs can be used to solve equations. Many other uses of graphical methods appear in the later chapters.

Exercises 2–3

See Appendix E for a computer program for plotting the graph of $y = x^2$.

In Exercises 1 through 32, graph the given functions.

1. $y = 3x$
2. $y = -2x$
3. $y = 2x - 4$
4. $y = 3x + 5$
5. $y = 7 - 2x$
6. $y = 5 - 3x$
7. $y = \frac{1}{2}x - 2$
8. $y = 6 - \frac{1}{3}x$
9. $y = x^2$

10. $y = -2x^2$ 11. $y = 3 - x^2$ 12. $y = x^2 - 3$

13. $y = \frac{1}{2}x^2 + 2$ 14. $y = 2x^2 + 1$ 15. $y = x^2 + 2x$

16. $y = 2x - x^2$ 17. $y = x^2 - 3x + 1$ 18. $y = 2 + 3x + x^2$

19. $y = x^3$ 20. $y = -2x^3$ 21. $y = x^3 - x^2$

22. $y = 3x - x^3$ 23. $y = x^4 - 4x^2$ 24. $y = x^3 - x^4$

25. $y = \frac{1}{x}$ 26. $y = \frac{1}{x + 2}$ 27. $y = \frac{1}{x^2}$

28. $y = \frac{1}{x^2 + 1}$ 29. $y = \sqrt{x}$ 30. $y = \sqrt{4 - x}$

31. $y = \sqrt{16 - x^2}$ 32. $y = \sqrt{x^2 - 16}$

In Exercises 33 through 46, graph the given functions. In each case plot the first variable as the ordinate and the second variable as the abscissa.

33. The velocity v (in feet per second) of an object under the influence of gravity, as a function of time t (in seconds), is given by $v = 100 - 32t$. If the object strikes the ground after 4 s, graph v as a function of t.

34. If \$1000 is placed in an account earning 6% simple interest, the amount A in the account after t years is given by the function $A = 1000(1 + 0.06t)$. If the money is withdrawn from the account after 6 years, plot A as a function of t.

35. The consumption c of fuel, in liters per hour (L/h), of a certain engine is determined as a function of the number r, in revolutions per minute (r/min) of the engine to be $c = 0.011r + 4.0$. This formula is valid from 500 r/min to 3000 r/min. Plot c as a function of r.

36. The heat capacity (in joules per kilogram) of an organic liquid is related to the temperature (in degrees Celsius) by the equation $c_p = 2320 + 4.73T$ for the temperature range of $-40°C$ to $120°C$. Graphically show that this equation does not satisfy experimental data above $120°C$. Plot the graph of the equation and the following data points.

c_p	3030	3220	3350
T	140	160	200

37. The maximum speed v, in miles per hour, at which a car can safely travel around a circular turn of radius r, in feet, is given by $r = 0.42v^2$. Plot r as a function of v.

38. The energy in the electric field around an inductor is given by $E = \frac{1}{2}LI^2$, where I is the current in the inductor and L is the inductance. Plot E as a function of I (a) if $L = 1$ unit, (b) if $L = 0.1$ unit, (c) if the E-axis is marked off in units of L.

39. If a rectangular tract of land has a perimeter of 600 m, its area as a function of its width is $A = 300w - w^2$. Plot A as a function of w.

40. A formula used to determine the number N of board feet of lumber which can be cut from a 4-ft section of a log of diameter d, in inches, is $N = 0.22d^2 - 0.71d$. Plot N as a function of d for values of d from 10 in. to 40 in.

41. The illuminance (in lumens per square meter) of a certain source of light as a function of the distance r (in meters) from the source is given by $I = 400/r^2$. Plot $I = f(r)$.

42. The value V, in dollars, of a certain model of a car depreciates according to its age t, in years, according to the formula $V = \frac{12,000}{t + 1}$. Plot $V = f(t)$ for $t < 9$.

43. The total profit P, in dollars, a manufacturer makes in producing x units of a certain product is given by $P = 4x^3 - 18x^2 - 120x - 200$. Plot P as a function of x for values to $x = 10$.

44. The deflection y of a beam at a horizontal distance x from one end is given by $y = -k(x^4 - 30x^2 + 1000x)$, where k is a constant. Plot the deflection (in terms of k) as a function of x (in feet) if there are 10 ft between the end supports of the beam.

45. Plot the graphs of $y = x$ and $y = |x|$ on the same coordinate system. Note how the graphs differ.

46. Plot the graphs of $y = 2 - x$ and $y = |2 - x|$ on the same coordinate system. Note how the graphs differ.

2–4 Graphs of Functions Defined by Tables of Data

In the previous section, we showed how to construct the graph of a function, where the function was defined by a formula or equation. However, as we stated in Section 2–1, there are other ways of expressing functions. One of the most important ways of showing the relationship between variables is by use of a table of values obtained by observation or experimentation.

Often the data from an experiment indicate that the variables could have a formula which relates them, although the formula may not be known. Data from experiments from the various fields of science and technology generally have variables which are related in such a way. In this case, when we are plotting the graph, the points should be connected by a smooth curve.

Statistical data in tabular form often give values which are taken only for certain intervals or are averaged over specified intervals. In such cases, no real meaning can be given to the intervals between the points on the graph. On these graphs, the points should be connected by straight-line segments, where this is done only to make the points stand out better and make the graph easier to read. Example A illustrates this type of graph.

Example A

The electric energy usage, in kilowatt-hours, for a particular all-electric house for each month of a certain year is given in the following table. Plot these data.

Month	Jan	Feb	Mar	Apr	May	Jun
Energy usage	2626	3090	2542	1875	1207	892

Month	July	Aug	Sep	Oct	Nov	Dec
Energy usage	637	722	825	1437	1825	2427

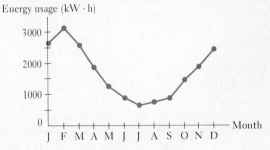

Figure 2–11

We know that there is no meaning to the intervals between the months, since we have the total number of kilowatt-hours for each month. Therefore, we use straight-line segments, but only to make the points stand out better.

In the next example we illustrate data that give a graph in which the points are connected by a smooth curve and not by straight-line segments.

Example B

An object was heated to 150.0°C. Its temperature was then recorded each minute, giving the values in the following table. Plot the graph.

Time (minutes)	0.0	1.0	2.0	3.0	4.0	5.0
Temperature (°C)	150.0	142.8	138.5	135.2	132.7	130.8

Since the temperature changed in a continuous way, there is meaning to the values in the intervals between points. Therefore these points are joined by a smooth curve, as in Fig. 2–12. Also note that most of the vertical scale was used for the required values, with the indicated break in the scale between 0 and 130.

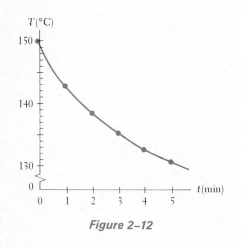

Figure 2–12

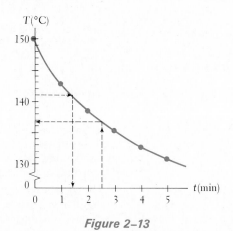

Figure 2–13

If the graph relating two variables is known, values can be obtained directly from the graph. In finding such a value through the inspection of the graph, we are *reading the graph*.

Example C

For the cooling object in Example B, we can estimate values of one variable for given values of the other.

If we want to know the temperature after 2.5 min, we estimate 0.5 of the interval between 2 and 3 on the t-axis and mark it. See Fig. 2–13. Then we draw a line vertically from this point to the graph. From the point where it intersects the graph, we draw a horizontal line to the T-axis. We now estimate that the line crosses at $T = 136.7$°C. Obviously, the number of tenths is a rough estimate.

In the same way, if we want to determine how long the object took to cool down to 141.0°, we go from 141.0 on the T-axis to the graph and then to the t-axis. This crosses at about $t = 1.4$ min.

We can see from Example C that we can estimate values from a graph. However, unless a very accurate graph is drawn with expanded scales for both variables, only quite approximate values can be found. There is a method, using the table of values itself, which generally gives reasonably accurate results. *The method is called* **linear interpolation.**

Linear interpolation assumes that if a particular value of one variable lies between two of those listed in the table, then the corresponding value of the other variable is at the same proportional distance between its listed values. From the point of view of the graph, linear interpolation assumes that the two points defined in the table are connected by a straight line. Although this is generally not correct, it is a very good approximation if the values in the table are sufficiently close together.

Example D

For the cooling object in Example B, we can use interpolation to find the temperature of the object after 1.4 min. Since 1.4 min is $\frac{4}{10}$ of the way from 1.0 min to 2.0 min, we shall assume that the value of T which we want is $\frac{4}{10}$ of the way between 142.8 and 138.5, the values of T for 1.0 min and 2.0 min, respectively. The difference between these values is 4.3, and $\frac{4}{10}$ of 4.3 is 1.7 (rounded off to tenths). Subtracting (the values of T are decreasing) 1.7 from 142.8, we obtain 141.1. Thus, the required value of T is about 141.1°C. (Note that this agrees well with the result in Example C.)

Another method of indicating the interpolation is shown in Fig. 2–14. From the figure we see that

$$\frac{x}{4.3} = \frac{0.4}{1.0}$$

$$x = 1.7 \qquad \text{(rounded off)}$$

Therefore,

$$142.8 - 1.7 = 141.1°C$$

is the required value of T. If the values of T had been increasing, we would have added 1.7 to the value of T for 1.0 min.

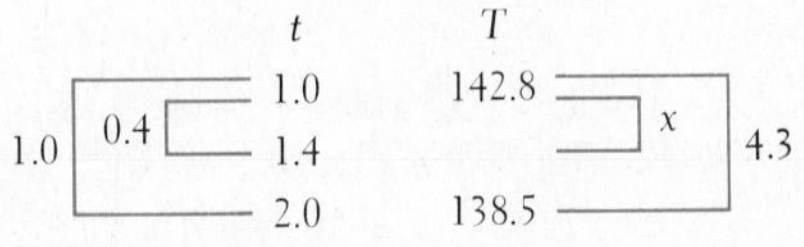

Figure 2–14

Exercises 2–4

In Exercises 1 through 8, represent the data graphically.

1. The rainfall (in inches) in a certain city was recorded as follows.

Year	1976	1977	1978	1979	1980	1981	1982	1983	1984
Rainfall	35.4	36.7	40.4	40.2	38.2	32.8	33.4	41.2	40.4

2. The hourly temperatures (in degrees Fahrenheit) on a certain day were recorded as follows.

Hour	6 A.M.	7	8	9	10	11	12	1 P.M.	2	3	4
Temperature	16	18	20	25	32	36	39	41	39	42	36

3. The diesel fuel production, in thousands of gallons, at a certain refinery during a 10-week period was as follows.

Week	1	2	3	4	5	6	7	8	9	10
Production	765	780	840	850	880	840	845	820	760	810

4. The number of isotopes of some of the heavier elements are in the following table.

Atomic number	82	84	86	88	90	92	94
Number of isotopes	16	19	12	11	12	14	11

These elements, by atomic number, are: 82, lead; 84, polonium; 86, radon; 88, radium; 90, thorium; 92, uranium; 94, plutonium.

5. The density (in kilograms per cubic meter) of water from 0° to 10°C is given in the following table.

Temperature	0	1	2	3	4	5
Density	999.85	999.90	999.94	999.96	999.97	999.96

Temperature	6	7	8	9	10
Density	999.94	999.90	999.85	999.78	999.69

6. The voltage (in volts) and current (in milliamperes) for a certain electrical experiment were measured as follows.

Voltage	10	20	30	40	50	60	70	80
Current	145	188	220	255	285	315	335	370

7. An experiment measuring the load (in kilograms) on a spring and the scale reading of the spring (in centimeters) produced the following results.

Load	0	1	2	3	4	5	6
Reading	7.0	7.6	8.2	8.8	9.4	10.1	12.8

8. The vapor pressure (in kilopascals) of ethane gas as a function of temperature (in degrees Celsius) is given by the following table.

Temperature	-70	-60	-50	-40	-30	-20	-10	0	15
Pressure	260	380	570	790	1060	1430	1840	2390	3270

In Exercises 9 and 10, use the graph in Fig. 2–13, which relates the temperature of the cooling object and time. Find the indicated values by reading the graph.

9. (a) For $t = 4.3$ min, find T. (b) For $T = 145.0°C$, find t.

10. (a) For $t = 1.8$ min, find T. (b) For $T = 133.5°C$, find t.

In Exercises 11 and 12, use the following table, which gives the voltage produced by a certain thermocouple as a function of the temperature of the thermocouple. Plot the graph. Find the indicated values by reading the graph.

Temperature (°C)	0	10	20	30	40	50
Voltage (volts)	0.0	2.9	5.9	9.0	12.3	15.8

11. (a) For $T = 26°C$, find V. (b) For $V = 13.5$ V, find T.

12. (a) For $T = 32°C$, find V. (b) For $V = 4.8$ V, find T.

In Exercises 13 through 16, find the indicated values by means of linear interpolation.

13. Using the table with Exercise 5, find the density for $T = 8.6°C$.

14. Using the table with Exercise 6, find the current for a voltage of 56 V.

15. Using the table with Exercise 7, find the load on the spring for a scale reading of 9.2.

16. Using the table with Exercise 8, find the temperature for a vapor pressure of 1220 kPa.

In Exercises 17 through 20, use the following table, which gives the pressure and volume of air as measured in an experiment. For Exercises 17 and 18, it is necessary to plot the graph and then find the indicated values by reading the graph. In Exercises 19 and 20, find the indicated values by means of linear interpolation.

Volume (cm³)	200	180	160	140	120	100
Pressure (kPa)	200	220	250	285	335	400

17. (a) For $P = 290$ kPa, find V. (b) For $V = 175$ cm³, find P.

18. (a) For $P = 215$ kPa, find V. (b) For $V = 128$ cm³, find P.

19. Find P for $V = 152$ cm³.   20. Find V for $P = 325$ kPa.

In Exercises 21 through 24, use the following table, which gives the fraction f (as a decimal) of the total heating load of a certain system that will be supplied by a solar collector of area A, in square meters. Find the indicated values by means of linear interpolation.

f	0.22	0.30	0.37	0.44	0.50	0.56	0.61
A (m²)	20	30	40	50	60	70	80

21. For $A = 36$ m², find f.   22. For $A = 52$ m², find f.

23. For $f = 0.59$, find A.   24. For $f = 0.27$, find A.

In Exercises 25 through 28, a method of finding values beyond those given is considered. By using a straight line segment to extend a graph beyond the last known point we can estimate values from the extension of the graph. The method is known as **linear extrapolation.** Use this method to estimate the required values from the given graphs.

25. Using Fig. 2–13, estimate T for $t = 5.3$ min.

26. Using the graph for Exercises 11 and 12, estimate V for $T = 55°C$.

27. Using the graph for Exercises 17 through 20, estimate P for $V = 94$ cm³.

28. Using the graph for Exercises 17 through 20, estimate P for $V = 210$ cm³.

2–5 Solving Equations Graphically

It is possible to solve equations by the use of graphs. This is particularly useful when algebraic methods cannot be applied conveniently to the equation. Before taking up graphical solutions, however, we shall briefly introduce the related concept of the zero of a function.

Those values of the independent variable for which the function equals zero are known as the **zeros** *of the function. To find the zeros of a given function, we must set the function equal to zero and solve the resulting equation.* Using functional notation, we are looking for the solutions of $f(x) = 0$.

Example A

Find any zeros of the function $f(x) = 3x - 9$.

To find the zeros we let $f(x) = 0$, which means that $3x - 9 = 0$. Thus we obtain $x = 3$, which means that 3 is a zero of the function $f(x) = 3x - 9$.

Graphically, the zeros of a function are found where the curve crosses the x-axis. *These points are called the* **x-intercepts** *of the curve.* The function is zero at these points since the x-axis represents all points for which $y = 0$, or $f(x) = 0$.

Example B

Graphically determine any zeros of the function $f(x) = x^2 - 2x - 1$.

First we set $y = x^2 - 2x - 1$ and then find the points for the graph. After plotting the graph in Fig. 2–15, we see that the curve crosses the x-axis at approximately $x = -0.4$ and $x = 2.4$ (estimating to the nearest tenth). Thus the zeros of this function are approximately -0.4 and 2.4. Checking these values in the function, we have $f(-0.4) = -0.04$ and $f(2.4) = -0.04$, which shows that -0.4 and 2.4 are very close to the exact values.

x	y
-2	7
-1	2
0	-1
1	-2
2	-1
3	2
4	7

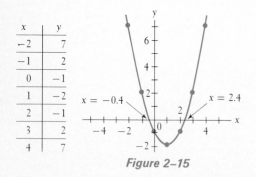

Figure 2–15

x	y
-4	0.3
-3	0.5
-2	1
-1	2
0	3
1	2
2	1
3	0.5
4	0.3

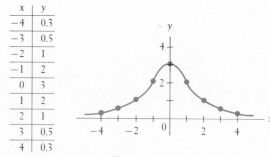

Figure 2–16

Example C

Graphically determine any zeros of the function $f(x) = \dfrac{6}{x^2 + 2}$.

Graphing the function $y = \dfrac{6}{x^2 + 2}$ in Fig. 2–16, we see that it does not cross the x-axis anywhere. Therefore, this function does not have any real zeros. We see, therefore, that not all functions have real zeros.

We can now see how to solve equations graphically and how this is related to the zeros of a function. To solve an equation, we collect all terms on one side of the equal sign, giving the equation $f(x) = 0$. This equation is solved graphically by first setting $y = f(x)$ and graphing this function. We then find those values of x for which $y = 0$. This means that we are finding the zeros of this function. The following examples illustrate the method.

Example D

Solve the equation $x^2(2x + 3) = 3x$ graphically.

We first collect all terms of the left side of the equal sign. This leads to the equation $2x^3 + 3x^2 - 3x = 0$. We then let $y = 2x^3 + 3x^2 - 3x$ and graph this function, as shown in Fig. 2–17. From the graph we see that $y = 0$ (which is equivalent to $2x^3 + 3x^2 - 3x = 0$) for the approximate values $x = -2.2$, $x = 0.0$, and $x = 0.7$. Therefore, these values are approximate solutions to the original equation. (Actually, $x = 0.0$ is exact, since we found this value when obtaining the table.) Checking these values in the original equation, we obtain $-6.776 = -6.6$ for $x = -2.2$, $0 = 0$ for $x = 0.0$, and $2.156 = 2.1$ for $x = 0.7$, which shows that the approximate solutions we obtained are very close to the exact solutions. We also note that we were able to find the solution $x = 0.7$ by being careful with values around $x = 0$, where we noticed that $f(x)$ was positive for both $x = -1$ and $x = 1$.

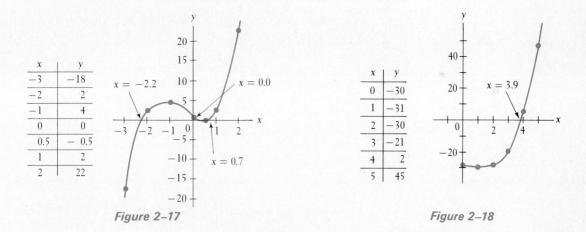

x	y
-3	-18
-2	2
-1	4
0	0
0.5	-0.5
1	2
2	22

Figure 2–17

x	y
0	-30
1	-31
2	-30
3	-21
4	2
5	45

Figure 2–18

Example E

A box, whose volume is 30 in.^{3}, is made with a square base and a height which is 2 in. less than the length of a side of the base. To find the dimensions of the box, we must solve the equation $x^2(x - 2) = 30$, where x is the length of a side of the base. (Verify the equation.) What are the dimensions of the box?

We are to solve the equation $x^2(x - 2) = 30$ graphically. First, we write the equation as $x^3 - 2x^2 - 30 = 0$. Now we set $y = x^3 - 2x^2 - 30$ and graph this equation as shown in Fig. 2–18. Only positive values of x are used, since negative values of x have no meaning to the problem. From the graph we see that $x = 3.9$ is the approximate solution. Therefore, the approximate dimensions are 3.9 in., 3.9 in., and 1.9 in.

Exercises 2–5

In Exercises 1 through 4, find the zeros of the given functions algebraically as in Example A.

1. $5x - 10$ 2. $7x - 4$ 3. $4x + 9$ 4. $2 - 3(x - 5)$

In Exercises 5 through 12, find the zeros of the given functions graphically. Check the solutions in Exercises 5 through 8 algebraically. Check the solutions in Exercises 9 through 12 by substituting in the function.

5. $2x - 7$ 6. $3x - 2$ 7. $5x - (3 - x)$ 8. $3 - 2(2x - 7)$

9. $x^2 + x$ 10. $2x^2 - x$ 11. $x^2 - x + 3$ 12. $x^2 + 3x - 5$

In Exercises 13 through 28, solve the given equations graphically.

13. $7x - 5 = 0$ 14. $8x + 3 = 0$ 15. $6x = 15$

16. $7x = -18$ 17. $x^2 + x - 5 = 0$ 18. $x^2 - 2x - 4 = 0$

19. $2x^2 - x = 7$ 20. $x(x - 4) = 9$ 21. $x^3 - 4x = 0$

22. $x^3 - 3x - 3 = 0$ 23. $x^4 - 2x = 0$ 24. $2x^5 - 5x = 0$

25. $\sqrt{2x + 2} = 3$ 26. $\sqrt{x} + 3x = 7$

27. $\dfrac{1}{x^2 + 1} = 0$ 28. $x - 2 = \dfrac{1}{x}$

In Exercises 29 through 36, solve the given problems graphically.

29. Under certain conditions the velocity (in feet per second) of an object as a function of the time (in seconds) is given by the equation $v = 60 - 32t$. When is the velocity zero?

30. For tax purposes a corporation assumes that one of its computers depreciates in value according to the equation $V = 90,000 - 12,000t$, where V is the value, in dollars, of the computer after t years. According to this formula, when will the computer be fully depreciated (no value)?

31. A box is suspended by two ropes, as shown in Fig. 2–19. In analyzing the tensions in the supporting ropes, it was found that they satisfied the following equation: $0.60T_1 + 0.87T_2 = 10.0$. Find T_1 if $T_2 = 8.0$ N.

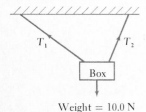

T_1 T_2

Box

Weight = 10.0 N

Figure 2–19

32. The perimeter of a field 50 m wide and l meters long is $p = 100 + 2l$. For what value of l is the perimeter 450 m?

33. Under given conditions a company finds that the profit p in producing x articles of a certain type is $p = 90x - x^2 - 1000$. For what values of x is the profit zero?

34. In the study of the velocities of nuclear fission fragments, it is found that under certain conditions the velocity would be zero if the expression $(1 + b)^2 - 3b$ were zero. For what values of b ($b > 0$), if any, is the velocity zero?

35. In order to find the distance x (in feet) from one end of a certain beam to the point where the deflection is zero, it is necessary to solve the equation $x^3 + x^2 - 5x = 0$. Determine where the deflection is zero.

36. If two electrical resistors, one of which is 1 Ω greater than the other, are placed in parallel, their combined resistance R_T as a function of the smaller resistance R is

$$R_T = \frac{R(R + 1)}{2R + 1}$$

What are the values of the resistors if $R_T = 10$ Ω?

2–6 Exercises for Chapter 2

In Exercises 1 through 4, determine the appropriate functions.

1. Find the volume of a right circular cylinder of height 8 ft as a function of the radius of the base.
2. Find the surface area A of a cube as a function of one of the edges e.
3. Find the temperature in degrees Fahrenheit as a function of degrees Celsius ($32°F = 0°C$ and $212°F = 100°C$).
4. Fencing 1000 m long is to be used to enclose three sides of a rectangular field. Express the area of the field as a function of its length (parallel to the fourth side).

In Exercises 5 through 12, evaluate the given functions.

5. Given $f(x) = 7x - 5$, find $f(3)$ and $f(-6)$.
6. Given $g(x) = 8 - 3x$, find $g(3)$ and $g(-4)$.
7. Given $F(u) = 3u - 2u^2$, find $F(-1)$ and $F(-3)$.
8. Given $h(y) = 2y^2 - y - 2$, find $h(5)$ and $h(-3)$.
9. Given $H(h) = \sqrt{1 - 2h}$, find $H(-4)$ and $H(2h)$.
10. Given $\phi(v) = \dfrac{3v - 2}{v + 1}$, find $\phi(-2)$ and $\phi(v + 1)$.
11. Given $f(x) = 3x^2 - 2x + 4$, find $f(x + h) - f(x)$.
12. Given $F(x) = x^3 + 2x^2 - 3x$, find $F(3 + h) - F(3)$.

In Exercises 13 through 16, evaluate the given functions by use of a calculator. Round off results to the accuracy of the value of the independent variable used.

13. Given $f(x) = 8.07 - 2x$, find $f(5.87)$ and $f(-4.29)$.
14. Given $g(x) = 7x - x^2$, find $g(45.81)$ and $g(-21.85)$.
15. Given $G(y) = \dfrac{y - 0.087629}{3y}$, find $G(0.17427)$ and $G(0.053206)$.
16. Given $h(t) = \dfrac{t^2 - 4t}{t^3 + 564}$, find $h(8.91)$ and $h(-4.91)$.

In Exercises 17 through 28, graph the given functions.

17. $y = 4x + 2$
18. $y = 5x - 10$
19. $y = 4x - x^2$
20. $y = x^2 - 8x - 5$
21. $y = 3 - x - 2x^2$
22. $y = 6 + 4x + x^2$

23. $y = x^3 - 6x$ 24. $y = 3 - x^3$

25. $y = 2 - x^4$ 26. $y = x^4 - 4x$

27. $y = \dfrac{x}{x + 1}$ 28. $y = \sqrt{25 - x^2}$

In Exercises 29 through 40, find any real zeros of the indicated functions by examining their graphs. Estimate the answer to the nearest tenth where necessary. Use the functions from Exercises 17 through 28, as indicated.

29. Exercise 17   30. Exercise 18   31. Exercise 19

32. Exercise 20   33. Exercise 21   34. Exercise 22

35. Exercise 23   36. Exercise 24   37. Exercise 25

38. Exercise 26   39. Exercise 27   40. Exercise 28

In Exercises 41 through 48, solve the given equations graphically.

41. $7x - 3 = 0$ 42. $2x + 11 = 0$ 43. $x^2 + 1 = 6x$

44. $3x - 2 = x^2$ 45. $x^3 - x^2 = 2 - x$ 46. $5 - x^3 = 2x^2$

47. $\dfrac{1}{x} = 2x$ 48. $\sqrt{x} = 2x - 1$

In Exercises 49 through 52, answer the given questions.

49. Where are all the points (x, y) whose abscissas are 1 and for which $y > 0$?

50. Three vertices of a rectangle are $(6, 5)$, $(-4, 5)$, and $(-4, 2)$. What are the coordinates of the fourth vertex?

51. An equation which is found in electronics is

$$h = \frac{\alpha}{1 - \alpha}$$

    Find h when $\alpha = 0.925$. That is, since $h = f(\alpha)$, find $f(0.925)$.

52. The percent p of wood lost in cutting it into boards 1.5 in. thick due to the thickness t, in inches, of the saw blade is $p = \dfrac{100t}{t + 1.5}$. Find p if $t = 0.4$ in. That is, since $p = f(t)$, find $f(0.4)$.

In Exercises 53 through 64, plot the indicated functions.

53. In determining the forces which act on a certain structure, it was found that two of them satisfied the equation $F_2 = 0.75F_1 + 45$, where the forces are in newtons. Plot the graph of F_2 as a function of F_1.

54. A company which makes digital clocks determines that in order to sell x clocks, the price must be $p = 50 - 0.25x$. Plot the graph of p as a function of x, for $x = 1$ to $x = 60$. (Although only the points for integral values of x have real meaning, join the points.)

55. A computer-leasing firm charges $150 plus $100 for every hour the computer is used during the month. What is the function relating the monthly charge C and the number of hours h that the computer is used? Plot a graph of the function for use up to 50 h.

56. There are 5000 L of oil in a tank which has a capacity of 100,000 L. It is filled at the rate of 7000 L/h. Determine the function relating the number of liters N and the time t while the tank is being filled. Plot N as a function of t.

57. A surveyor measuring the elevation of a point must consider the effect of the curvature of the earth. An approximate relation between the effect H (in feet) of this curvature, and the distance D (in miles) between the surveyor and the point of which he is measuring the elevation, is given by $H = 0.65D^2$. Plot H as a function of D.

58. It was determined that a good approximation for the cost C, in cents per mile, of operating a certain car at a constant speed v, in miles per hour, is given by $C = 0.025v^2 - 1.4v + 35$. Plot C as a function of v for $v = 10$ mi/h to $v = 60$ mi/h.

59. The resonant frequency f (in hertz) of a certain electric circuit is given by $f = 12{,}500/\sqrt{L}$, where L is the inductance (in henrys) in the circuit. Plot f as a function of L. (Take values of L from 0.25 H to 4.00 H.)

60. The pneumatic resistance (in lb-s/ft^5) of a certain type of tubing is given by $R = 0.05/A^2$, where A is the cross-sectional area (in square feet). Plot R as a function of A. (Take values of A from 0.001 ft^2 to 0.01 ft^2.)

61. The sales (in millions of dollars) of a certain corporation from 1976 through 1984 are shown in the following table.

Year	1976	1977	1978	1979	1980	1981	1982	1983	1984
Sales	12	14	17	18	20	24	32	35	39

62. The amplitude (in centimeters) of a certain pendulum, measured after each swing, as a function of time (in seconds), is given by the following table. Plot the graph of amplitude as a function of time.

Time	0.0	1.2	2.4	3.6	4.8
Amplitude	5.0	2.8	1.6	0.9	0.5

63. The length (in inches) of a brass rod is measured as a function of the temperature (in degrees Celsius). From the following table, plot the length as a function of temperature (make your scale meaningful).

Temperature	0	100	200	300	400	500
Length	100.0	100.2	100.4	100.7	101.0	101.2

64. The solubility (in kilograms of solute per cubic meter of water) of potassium nitrate as a function of temperature is given in the following table. Plot solubility as a function of temperature (in degrees Celsius).

Temperature	0	10	20	30	40	50	60	80
Solubility	133	209	316	458	639	855	1100	1690

In Exercises 65 and 66, solve the given equations graphically.

65. Under certain conditions, the distance s, in feet, that an object is above the ground is given by $s = 96t - 16t^2$, where t is the time in seconds. When is the object 100 ft above the ground?

66. A computer, using data from a refrigeration plant, estimated that in the event of a power failure, the temperature, in degrees Celsius, in the freezers would be given by $T = \dfrac{4t^2}{t + 2} - 20$, where t is the number of hours after the power failure. How long would it take for the temperature to reach 0°C?

CHAPTER 3

The Trigonometric Functions

The solutions to a great many kinds of applied problems involve the use of triangles, especially right triangles. Problems which can be solved by the use of triangles include the determination of distances which cannot readily be determined directly, such as the widths of rivers and distances between various points in the universe. Also, problems involving forces and velocities lend themselves readily to triangle solution. Even certain types of electric circuits are analyzed by the use of triangles.

The basic properties of triangles will allow us to set up some very basic and useful functions involving their sides and angles. These functions are very important to the development of mathematics and to its applications in most fields of technology.

Thus we come to the study of **trigonometry,** the literal meaning of which is ''triangle measurement.'' In this chapter we shall introduce the basic trigonometric functions and some of their elementary applications. Later chapters will consider additional topics in trigonometry. We shall begin our study by considering the concept of an angle.

3–1 Angles

An **angle** *is defined as being **generated** by rotating a **half-**line about its endpoint from an initial position to a terminal position. A half-line is that portion of a line to one side of a fixed point on the line. We call the initial position of the half-line the **initial side,** and the terminal position of the half-line the **terminal side.** The fixed point is the **vertex.*** The angle itself is the amount of rotation from the initial side to the terminal side.

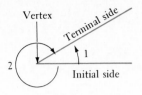

Vertex

Terminal side

Initial side

Figure 3–1

If the rotation of the terminal side from the initial side is counterclockwise, the angle is said to be ***positive***. *If the rotation is clockwise the angle is* ***negative***. In Fig. 3–1, angle 1 is positive and angle 2 is negative.

There are many symbols used to designate angles. Among the most widely used are certain Greek letters such as θ (theta), ϕ (phi), α (alpha), and β (beta). Capital letters representing the vertex (e.g. $\angle A$ or simply A) and other literal symbols, such as x and y, are also used commonly.

Two measurements of angles are widely used. These are the **degree** and the **radian.** In our work in this chapter, we will use degrees. However, since degrees and radians are both used on calculators (and computers), we will briefly show the relationship between them in this section. In Chapter 7 we will define and discuss the use of radians.

A degree is defined as $\frac{1}{360}$ *of one complete rotation.* The symbol $°$ is used to designate degrees.

Example A

The angles $\theta = 30°$, $\phi = 140°$, $\alpha = 240°$, and $\beta = -120°$ are shown in Fig. 3–2.

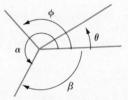

Figure 3–2

One method of subdividing a degree is to use decimal parts of a degree. Most calculators use degrees in decimal form. Another traditional way is to divide a degree into 60 equal parts called **minutes;** each minute is divided into 60 equal parts called **seconds.** The symbols $'$ and $''$ are used to designate minutes and seconds, respectively.

In Fig. 3–2 we note that angles α and β have the same initial and terminal sides. *Such angles are called* **coterminal angles.** An understanding of coterminal angles is important in certain concepts in trigonometry.

Example B

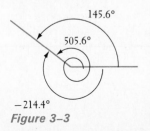

145.6°

505.6°

−214.4°

Figure 3–3

Determine the values of two angles which are coterminal with an angle of 145.6°.

Since there are 360° in a complete rotation, we can find one coterminal angle by considering the angle which is 360° larger than the given angle. This gives us an angle of 505.6°. Another method of finding a coterminal angle is to subtract 145.6° from 360°, and then consider the resulting angle to be negative. This means that the original angle and the derived angle would make up one complete rotation, when put together. This method leads us to the angle of −214.4° (see Fig. 3–3). These methods could be employed repeatedly to find other coterminal angles.

In using a calculator or computer it may be necessary to change an angle expressed in radians to one expressed in degrees. For this we use the fact that 1 radian is approximately 57.30°, or 1 rad = 57.30°.

Example C

Express 1.35 rad in degrees.

In order to do this we multiply 1.35 by 57.30. This gives 77.4. Thus,

$$1.35 \text{ rad} = 77.4°$$

Traditionally it has been very common to use degrees and minutes in tables, and most calculators use degrees and decimal parts. Therefore, it may also be necessary to change from one form to the other. The method of making these changes is illustrated in the following examples.

Example D

To change the angle 43°24′ to decimal form, we use the fact that $1' = (\frac{1}{60})°$. Therefore, $24° = (\frac{24}{60})° = 0.4°$. Thus,

$$43°24' = 43.4°$$

Also, 17°53′ is changed to decimal form as follows:

$$53' = \left(\frac{53}{60}\right)° = 0.88°$$

Thus, $17°53' = 17.88°$ (to the nearest hundredth), or $17°53' = 17.9°$ (to the nearest tenth).

Example E

To change 154.36° to an angle measured to the nearest minute, we use the fact that $1° = 60'$. Therefore,

$$0.36° = 0.36(60') = 21.6'$$

To the nearest minute, we have $154.36° = 154°22'$.

If the initial side of the angle is the positive x-axis and the vertex is at the origin, the angle is said to be in **standard position.** The angle is then determined by the position of the terminal side. If the terminal side is in the first quadrant, the angle is called a *first-quadrant angle.* Similar terms are used when the terminal side is in the other quadrants. *If the terminal side coincides with one of the axes, the angle is a* **quadrantal angle.** When an angle is in standard position, the terminal side can be determined if we know any point, other than the origin, on the terminal side.

Example F

To draw a third-quadrant angle of 205°, we simply measure 205° from the positive x-axis and draw the terminal side. See angle α in Fig. 3–4.

Also in Fig. 3–4, θ is in standard position and the terminal side is uniquely determined by knowing that it passes through the point (2, 1). The same terminal side passes through (4, 2) and $(\frac{11}{2}, \frac{11}{4})$, among other points. Knowing that the terminal side passes through any one of these points makes it possible to determine the terminal side.

Angle β in Fig. 3–4 is a quadrantal angle since its terminal side is the positive y-axis.

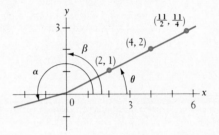

Figure 3–4

Exercises 3–1

In Exercises 1 through 4, draw the given angles.

1. 60°, 120°, −90°   2. 330°, −150°, 450°
3. 50°, −360°, −30°   4. 45°, 225°, −250°

In Exercises 5 through 12, determine one positive and one negative coterminal angle for each angle given.

5. 45°   6. 73°   7. 150°   8. 162°
9. 70°30′   10. 153°47′   11. 278.1°   12. 197.6°

In Exercises 13 through 16, change the given angles in radians to equal angles expressed in degrees. Round off results.

13. 0.265 rad   14. 0.838 rad   15. 1.447 rad   16. 3.642 rad

In Exercises 17 through 24, change the given angles to equal angles expressed in decimal form. In Exercises 21 through 24, round off results to hundredths.

17. 15°12′   18. 246°48′   19. 86°3′   20. 157°39′
21. 301°16′   22. 4°47′   23. 96°8′   24. 38°28′

In Exercises 25 through 32, change the given angles to equal angles expressed to the nearest minute.

25. 47.5°	26. 315.8°	27. 19.75°	28. 84.55°
29. 5.62°	30. 238.21°	31. 24.92°	32. 142.87°

In Exercises 33 through 40, draw angles in standard position such that the terminal side passes through the given point.

33. (4, 2)	34. (−3, 8)	35. (−3, −5)	36. (6, −1)
37. (−7, 5)	38. (−4, −2)	39. (2, −5)	40. (1, 6)

In Exercises 41 and 42, change the given angles to equal angles expressed in decimal form to the nearest thousandth. In Exercises 43 and 44 change the given angles to equal angles expressed to the nearest second. Use a calculator.

41. 21°42′36″	42. 7°16′23″	43. 86.274°	44. 57.019°

3–2 Defining the Trigonometric Functions

Let us place an angle θ in standard position and drop perpendicular lines from points on the terminal side to the x-axis, as shown in Fig. 3–5. In doing this we set up similar triangles, each with one vertex at the origin and one side along the x-axis. Using the basic fact from geometry that *corresponding sides of similar triangles are proportional*, we may set up equal ratios of corresponding sides.

Example A

Figure 3–5

In Fig. 3–5 we can see that triangles *ORP* and *OSQ* are similar triangles (they have exactly the same shape), since their corresponding angles are equal (each has the same angle at *O*, a right angle, and equal angles at *P* and *Q*). This means that the ratios of the lengths of corresponding sides are equal.

For example, sides *RP* and *SQ* are corresponding sides, and sides *OR* and *OS* are corresponding sides. The ratio of *RP* to *OR*, which is $\dfrac{RP}{OR}$, must equal the ratio of *SQ* to *OS*, which is $\dfrac{SQ}{OS}$. This means that

$$\frac{RP}{OR} = \frac{SQ}{OS}$$

This proportion holds for any positions (except at *O*) which *P* or *Q* may have on the terminal side of θ. (Here we have used the terms *ratio* and *proportion*. A detailed discussion of these terms is found in Chapter 17.)

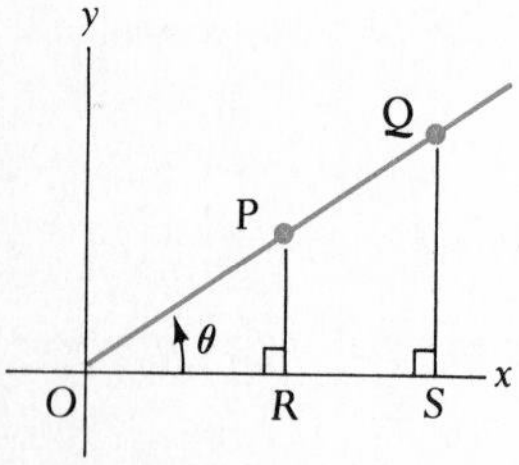

There are three important distances associated with a given point on the terminal side of an angle. *They are the* **abscissa** *(x-coordinate), the* **ordinate** *(y-coordinate), and the* **radius vector** *(the distance from the origin to the point).* The radius vector is denoted by *r*.

Example B

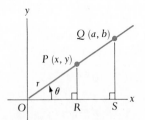

Figure 3–6

In Fig. 3–6 we show the same angle θ as in Fig. 3–5, except that we have given coordinates to points P and Q and have shown the radius vector r. The value of r would be the distance OP for point P, or the distance OQ for point Q.

In Example A we showed that triangles ORP and OSQ are similar. We also gave the proportion which showed the equality of the ratios of two pairs of corresponding sides. Since RP is the ordinate of P, OR is the abscissa of P, SQ is the ordinate of Q, and OS is the abscissa of Q,

$$\frac{RP}{OR} = \frac{SQ}{OS} \quad \text{is the same as} \quad \frac{y}{x} = \frac{b}{a}$$

For any position (except at O) of Q on the terminal side of θ, the ratio of its ordinate to its abscissa will be equal to $\frac{y}{x}$.

For any angle θ in standard position, there are six different ratios which may be set up. Because of the property of similar triangles, each of these ratios is the same, regardless of the point on the terminal side which is chosen. For a different angle, with a different terminal side, the ratios have different values. Thus, the ratios depend on the position of the terminal side, which means that the ratios depend on the size of the angle. In this way we see that *the ratios are functions of the* **angle**. *These functions are called the* **trigonometric functions** and are defined as follows (see Fig. 3–7):

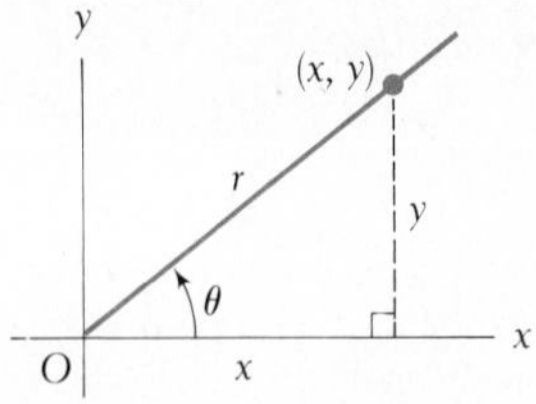

Figure 3–7

$$\text{sine } \theta = \frac{y}{r} \qquad \text{cosine } \theta = \frac{x}{r}$$

$$\text{tangent } \theta = \frac{y}{x} \qquad \text{cotangent } \theta = \frac{x}{y} \tag{3–1}$$

$$\text{secant } \theta = \frac{r}{x} \qquad \text{cosecant } \theta = \frac{r}{y}$$

In this chapter we shall restrict our attention to the trigonometric functions of acute angles. However, it should be emphasized that the definitions in Eqs. (3–1) are general and may be used for angles of any magnitude. Discussion of the trigonometric functions of angles in general, along with additional important properties, is found in Chapters 7 and 19.

For convenience, the names of the various functions are usually abbreviated to $\sin \theta$, $\cos \theta$, $\tan \theta$, $\cot \theta$, $\sec \theta$, and $\csc \theta$. Note that a given function is not defined if the denominator is zero. The denominator is zero in $\tan \theta$ and $\sec \theta$ for $x = 0$, and in $\cot \theta$ and $\csc \theta$ for $y = 0$. In all cases we will assume that $r > 0$. If $r = 0$ there would be no terminal side and therefore no angle.

In the following examples we will find specific values for the trigonometric functions. We will express values as $\sin \theta = 0.525$, for example. It is a common error to omit the angle and give the value as $\sin = 0.525$. This is a meaningless expression, for *we must show the angle* for which we have the value of the function.

Example C

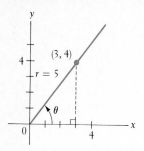

Figure 3–8

Determine the trigonometric functions of the angle with a terminal side passing through the point (3, 4).

By placing the angle in standard position, as shown in Fig. 3–8, and drawing the terminal side through (3, 4), we find by use of the Pythagorean theorem that

$$r = \sqrt{3^2 + 4^2} = \sqrt{25} = 5$$

Using the values $x = 3$, $y = 4$, and $r = 5$, we find that

$$\sin\theta = \frac{4}{5} \qquad \cos\theta = \frac{3}{5} \qquad \tan\theta = \frac{4}{3}$$

$$\cot\theta = \frac{3}{4} \qquad \sec\theta = \frac{5}{3} \qquad \csc\theta = \frac{5}{4}$$

Example D

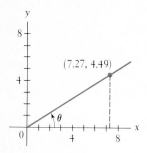

Figure 3–9

Find the trigonometric functions of the angle whose terminal side passes through (7.27, 4.49). Round off the results to three significant digits.

We show the angle and the given point in Fig. 3–9. From the Pythagorean theorem, we have

$$r = \sqrt{7.27^2 + 4.49^2} = 8.545$$

(Here we keep one extra significant digit since r is an intermediate result; see Appendix B.) Therefore, we have the following values for the trigonometric functions of θ.

$$\sin\theta = \frac{4.49}{8.545} = 0.525 \qquad \cos\theta = \frac{7.27}{8.545} = 0.851$$

$$\tan\theta = \frac{4.49}{7.27} = 0.618 \qquad \cot\theta = \frac{7.27}{4.49} = 1.62$$

$$\sec\theta = \frac{8.545}{7.27} = 1.18 \qquad \csc\theta = \frac{8.545}{4.49} = 1.90$$

In finding these values on a calculator, it would not actually be necessary to round off the value of r, as its value can be kept in memory. The reason for showing the value of r here is to show the values used in the calculation of each of the trigonometric functions. A sequence of steps on a calculator which can be used to find r and place its value in memory is

$$7.27 \quad \boxed{x^2} \quad \boxed{+} \quad 4.49 \quad \boxed{x^2} \quad \boxed{=} \quad \boxed{\sqrt{x}} \quad \boxed{\text{STO}}$$

At this point the necessary divisions can be performed using the $\boxed{\text{RCL}}$ key for r. For example, to find $\sin\theta$ we continue the calculator sequence with

$$4.49 \quad \boxed{\div} \quad \boxed{\text{RCL}} \quad \boxed{=}$$

If one of the trigonometric functions is known, we can determine the other functions of the angle using the Pythagorean theorem and the definitions of the functions. The following example illustrates the method.

Example E

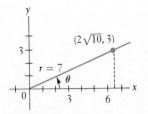

Figure 3–10

If we know that $\sin \theta = \frac{3}{7}$ and that θ is a first-quadrant angle, we know that the ratio of the ordinate to the radius vector (of y to r) is 3 to 7. Therefore, the point on the terminal side for which y is 3 can be determined by use of the Pythagorean theorem. The x-value for this point is

$$x = \sqrt{7^2 - 3^2} = \sqrt{49 - 9} = \sqrt{40} = 2\sqrt{10}$$

Therefore, the point $(2\sqrt{10}, 3)$ is on the terminal side, as shown in Fig. 3–10.

Therefore, using the values $x = 2\sqrt{10}$, $y = 3$, and $r = 7$ we have the other trigonometric functions of θ. They are

$$\cos \theta = \frac{2\sqrt{10}}{7}, \qquad \tan \theta = \frac{3}{2\sqrt{10}}, \qquad \cot \theta = \frac{2\sqrt{10}}{3},$$

$$\sec \theta = \frac{7}{2\sqrt{10}}, \qquad \csc \theta = \frac{7}{3}$$

The values shown here are *exact*. *Approximate* decimal values can be found by the use of a calculator.

Exercises 3–2

In Exercises 1 through 12, determine the values of the trigonometric functions of the angles whose terminal sides pass through the given points. Give answers in exact form, as in Examples C and E.

1. (6, 8)   2. (5, 12)   3. (15, 8)   4. (24, 7)

5. (9, 40)   6. (16, 30)   7. $(1, \sqrt{15})$ 8. $(\sqrt{3}, 2)$

9. (1, 1)   10. (6, 5)   11. (5, 2)   12. $\left(1, \frac{1}{2}\right)$

In Exercises 13 through 16, determine the values of the trigonometric functions of the angles whose terminal sides pass through the given points. Give answers in decimal form. Round off the results to the accuracy of the coordinates given.

13. (3.25, 5.15)   14. (0.687, 0.943)

15. (0.08623, 0.01327)   16. (37.65, 21.87)

In Exercises 17 through 24, use the given trigonometric functions to find the indicated trigonometric functions. In Exercises 17 through 20, give answers in exact form. In Exercises 21 through 24, give answers in decimal form, rounded off to three significant digits.

17. Given $\cos \theta = \frac{12}{13}$, find $\sin \theta$ and $\cot \theta$.

18. Given $\sin \theta = \frac{1}{2}$, find $\cos \theta$ and $\csc \theta$.

19. Given $\tan \theta = 1$, find $\sin \theta$ and $\sec \theta$.

20. Given $\sec \theta = \frac{4}{3}$, find $\tan \theta$ and $\cos \theta$.

21. Given sin θ = 0.750, find cot θ and csc θ.

22. Given cos θ = 0.300, find sin θ and tan θ.

23. Given cot θ = 0.254, find cos θ and tan θ.

24. Given csc θ = 1.20, find sec θ and cot θ.

In Exercises 25 through 28, each of the listed points is on the terminal side of an angle. Show that each of the indicated functions is the same for each of the points.

25. (3, 4), (6, 8), (4.5, 6), sin θ and tan θ

26. (5, 12), (15, 36), (7.5, 18), cos θ and cot θ

27. (2, 1), (4, 2), (8, 4), tan θ and sec θ

28. (3, 2), (6, 4), (9, 6), csc θ and cos θ

In Exercises 29 through 32, answer the given questions.

29. From the definitions of the trigonometric functions, it can be seen that csc θ is the reciprocal of sin θ. What function is the reciprocal of cos θ?

30. Following Exercise 29, what function is the reciprocal of tan θ?

31. Multiply the expression for cot θ by that for sin θ. Is the result the expression for any of the other functions?

32. Divide the expression for sin θ by that for cos θ. Is the result the expression for any of the other functions?

3–3 Values of the Trigonometric Functions

We have been able to calculate the trigonometric functions if we knew one point on the terminal side of the angle. However, in practice it is more common to know the angle in degrees, for example, and to be required to find the functions of this angle. Therefore, we must be able to determine the trigonometric functions of angles in degrees.

One way to determine the functions of a given angle is to make a scale drawing. That is, we draw the angle in standard position using a protractor and then measure the lengths of the values of x, y, and r for some point on the terminal side. By using the proper ratios we may determine the functions of this angle.

We may also use certain geometric facts to determine the functions of some particular angles. The following two examples illustrate this procedure.

Example A

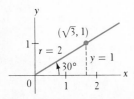

Figure 3–11

From geometry we know that the side opposite a 30° angle in a right triangle is one-half the hypotenuse. By using this fact and letting $y = 1$ and $r = 2$ (see Fig. 3–11), we determine that sin 30° $= \frac{1}{2}$. Also, by use of the Pythagorean theorem we determine that $x = \sqrt{3}$. Therefore, cos 30° $= \frac{\sqrt{3}}{2}$ and tan 30° $= \frac{1}{\sqrt{3}}$. In a similar way we may determine the values of the functions of 60° to be as follows: sin 60° $= \frac{\sqrt{3}}{2}$, cos 60° $= \frac{1}{2}$, and tan 60° $= \sqrt{3}$.

Example B

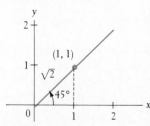

Figure 3–12

Determine sin 45°, cos 45°, and tan 45°.

From geometry we know that in an isosceles right triangle the angles are 45°, 45°, and 90°. We know that the sides are in proportion 1, 1, $\sqrt{2}$, respectively. Putting the 45° angle in standard position, we find $x = 1$, $y = 1$, and $r = \sqrt{2}$ (see Fig. 3–12). From this we determine

$$\sin 45° = \frac{1}{\sqrt{2}}, \qquad \cos 45° = \frac{1}{\sqrt{2}}, \quad \text{and} \quad \tan 45° = 1$$

Summarizing the results for 30°, 45°, and 60°, we have the following table.

θ	30°	45°	60°	(decimal approximations) 30°	45°	60°
$\sin \theta$	$\dfrac{1}{2}$	$\dfrac{1}{\sqrt{2}}$	$\dfrac{\sqrt{3}}{2}$	0.500	0.707	0.866
$\cos \theta$	$\dfrac{\sqrt{3}}{2}$	$\dfrac{1}{\sqrt{2}}$	$\dfrac{1}{2}$	0.866	0.707	0.500
$\tan \theta$	$\dfrac{1}{\sqrt{3}}$	1	$\sqrt{3}$	0.577	1.000	1.732

The scale-drawing method is only approximate, and the geometric methods work only for certain angles. However, the values of the functions may be determined through more advanced methods (using calculus and what are known as power series).

All the values of the trigonometric functions have been programmed into the scientific calculator. Therefore, we now show how the calculator is used to find these values.

To find the values of sin θ, cos θ, and tan θ, we use the $\boxed{\text{SIN}}$, $\boxed{\text{COS}}$, and $\boxed{\text{TAN}}$ keys, respectively. The angle is entered, and then the appropriate function key is used. *If your calculator has a degree-radian key or switch, be sure that it is set for degrees.*

Example C

Using a calculator to find sin 37°, we use the sequence 37 $\boxed{\text{SIN}}$. The display shows the value 0.6018150. Therefore,

$$\sin 37° = 0.6018150$$

To find tan 67.36°, we use the sequence 67.36 $\boxed{\text{TAN}}$. This gives us

$$\tan 67.36° = 2.3976264$$

Not only are we able to find values of the functions if we know the angle, but also we can find the angle if we know the value of a function. This requires that we reverse the procedures mentioned above. In doing this, we are actually using another important type of mathematical function, an **inverse trigonometric function.** These are discussed in some detail in Chapter 19. For calculator purposes at this point, it is sufficient to recognize the notation which is used. The notation for "the angle whose sine is x" is $\text{Sin}^{-1} x$, or Arcsin x. Similar meanings are given to $\text{Cos}^{-1} x$, Arccos x, $\text{Tan}^{-1} x$, and Arctan x. On a calculator the $\boxed{\text{ARCSIN}}$, $\boxed{\text{SIN}^{-1}}$, or the sequence of $\boxed{\text{INV}}$ $\boxed{\text{SIN}}$ keys are used to find Arcsin x.

Example D

If we know that the cosine of an angle is 0.3527, we can find the angle by using the calculator sequence .3527 $\boxed{\text{ARCCOS}}$. This gives us a display of 69.347452. Thus,

$$\text{if } \cos \theta = 0.3527, \text{ then } \theta = 69.35°$$

where the result has been rounded off.

It is generally possible to set up the solution of a problem in terms of the sine, cosine, or tangent. However, if a value of the cotangent, secant, or cosecant of an angle is needed, it can also be found on the calculator.

From the definitions of the trigonometric functions we see that csc $\theta = r/y$ and sin $\theta = y/r$. This means that *the value of csc θ is the reciprocal of the value of sin θ.* In the same way, *sec θ is the reciprocal of cos θ and cot θ is the reciprocal of tan θ.* Thus, by use of the reciprocal key, $\boxed{1/x}$, the values of these functions can be found.

Example E

To find the value of sec 27.82°, we first find cos 27.82° and then we find the reciprocal of this value. This leads to the calculator sequence 27.82 $\boxed{\text{COS}}$ $\boxed{1/x}$. The display is 1.1306869. Therefore,

$$\text{sec } 27.82° = 1.131$$

if the result is rounded off.

Before the extensive use of calculators, the values of the trigonometric functions were generally obtained from tables. Since it may be necessary to use such a table to obtain or check the value of a trigonometric function, we briefly outline its use here.

Since measuring angles in degrees and minutes has been common, a table of these values is given in Appendix F. Table 3 gives values of the trigonometric functions to each 10′.

To obtain a value of a function from Table 3, we note that the angles from 0° to 45° are listed in the left-hand column and are read down. The angles from 45° to 90° are listed on the right-hand side and are read up. The functions for angles from 0° to 45° are listed along the top, and those for the angles from 45° to 90° are listed along the bottom.

Example F

Using Table 3, find tan 42°20′ and sin 64°40′.
   Tan 42°20′ is found under tan θ (at top) and to the right of 20′ (which appears under 42°). Tan 42°20′ = 0.9110.
   Sin 64°40′ is found over sin θ (at bottom) and to the left of 40′ (which appears *over* 64°). Sin 64°40′ = 0.9038.

Example G

Given that tan θ = 2.378, find θ to the nearest 10′.
   We look for 2.378, or the nearest number to it, in the columns for tan θ in Table 3. Since the nearest number which appears is 2.375, and this appears over tan θ, we conclude that θ = 67°10′ to the nearest 10′.

We can see that the calculator is easier to use than a table, and it can also give values to a much greater degree of accuracy. However, these tables were used extensively until the late 1970s.

Exercises 3–3

In Exercises 1 through 4, use a protractor to draw the given angle. Measure off 10 units (centimeters are convenient) along the radius vector. Then measure the corresponding values of x and y. From these values determine the trigonometric functions of the angle.

1. 40°   2. 75°   3. 15°   4. 53°

In Exercises 5 through 16, find the values of the trigonometric functions by use of a calculator. Round off results to the accuracy of the given angle.

5. sin 22.4°   6. cos 72.5°   7. tan 57.6°   8. sin 36.0°

9. cos 15.71°   10. tan 8.653°   11. sin 83.792°   12. cos 46.725°

13. cot 67.78°   14. csc 22.81°   15. sec 50.4°   16. cot 41.8°

In Exercises 17 through 28, find θ for each of the given trigonometric functions by use of a calculator. Round off results to the accuracy of the given value.

17. cos θ = 0.3261   18. tan θ = 2.470

19. sin θ = 0.9114   20. cos θ = 0.0427

21. tan θ = 0.20741   22. sin θ = 0.10946

23. cos θ = 0.65007   24. tan θ = 5.7706

25. csc θ = 1.245   26. sec θ = 2.045

27. cot θ = 0.14434   28. csc θ = 1.0125

In Exercises 29 through 32, find the value of each of the trigonometric functions from Table 3 in Appendix F.

29. sin 19°0′   30. cos 43°0′   31. tan 67°20′   32. cot 76°50′

In Exercises 33 through 36, use Table 3 to find θ to the nearest 10′ for each of the given trigonometric functions.

33. tan θ = 0.8441   34. sin θ = 0.9175

35. cos θ = 0.1260   36. tan θ = 1.523

Linear interpolation, as described in Section 2–4, can be used to find values of the trigonometric functions for angles expressed to the nearest minute from Table 3. In Exercises 37 through 40, use interpolation to find the value of each of the trigonometric functions. In Exercises 41 through 44, use interpolation to find θ to the nearest 1′ for each of the given trigonometric functions.

37. tan 28°56′   38. cos 48°44′   39. sin 61°15′   40. tan 53°12′

41. cos θ = 0.2960   42. tan θ = 0.1086

43. sin θ = 0.5755   44. cot θ = 0.8070

In Exercises 45 through 48, solve the given problems.

45. When a projectile is fired into the air, its horizontal velocity v_x is related to the velocity v with which it is fired by the relation $v_x = v(\cos \theta)$, where θ is the angle between the horizontal and the direction in which it is fired. What is the horizontal velocity of a projectile fired with velocity 225 ft/s at an angle of 36.0° with respect to the horizontal?

46. The coefficient of friction between an object moving down an inclined plane with constant speed and the plane equals the tangent of the angle that the plane makes with the horizontal. If the coefficient of friction between a metal block and a wooden plane is 0.150, at what angle (to the nearest tenth of a degree) is the plane inclined?

47. The voltage at any instant in a coil of wire which is turning in a magnetic field is given by $E = E_{max} (\cos \alpha)$, where E_{max} is the maximum voltage and α is the angle the coil makes with the field. If the maximum voltage produced by a certain coil is 125.8 V, what voltage is generated when the coil makes an angle of 55.58° with the field?

48. When a light ray enters glass from the air it bends somewhat toward a line perpendicular to the surface. The *index of refraction* of the glass is defined as

$$n = \frac{\sin i}{\sin r}$$

where i is the angle between the perpendicular and the ray in the air and r is the angle between the perpendicular and the ray in the glass. A typical case for glass is $i = 59.0°$ and $r = 34.5°$. Find the index of refraction for this case. See Fig. 3–13.

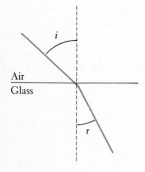

Air
Glass

Figure 3–13

3–4 The Right Triangle

From geometry we know that a triangle, by definition, consists of three sides and has three angles. If one side and any other two of these six parts of the triangle are known, it is possible to determine the other three parts. One of the three known parts must be a side, for if we know only the three angles, we can conclude only that an entire set of similar triangles has those particular angles.

Example A

Assume that one side and two angles are known. Then we may determine the third angle by the fact that the sum of the angles of a triangle is always 180°. Of all the possible similar triangles having these three angles, we have the one with the particular side which is known. Only one triangle with these parts is possible (in the sense that all triangles with the given parts are congruent and have equal corresponding angles and sides).

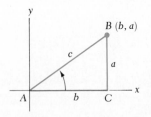

Figure 3–14

To **solve a triangle** *means that, when we are given three parts of a triangle (at least one a side), we are to find the other three parts.* In this section we are going to demonstrate the method of solving a right triangle. *Since one angle of the triangle will be 90°, it is necessary to know one side and one other part.* Also, we know that the sum of the three angles of a triangle is 180°, and this in turn tells us that *the sum of the other two angles, both acute, is 90°. Any two acute angles whose sum is 90° are said to be* **complementary.**

For consistency, when we are labeling the parts of the right triangle *we shall use the letters A and B to denote the acute angles, and C to denote the right angle. The letters a, b, and c will denote the sides opposite these angles, respectively. Thus, side c is the hypotenuse of the right triangle.* See Fig. 3–14.

In solving right triangles we shall find it convenient to express the trigonometric functions of the acute angles in terms of the sides. By placing the vertex of angle A at the origin and the vertex of right angle C on the positive x-axis, as shown in Fig. 3–15, we have the following ratios for angle A in terms of the sides of the triangle.

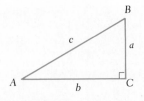

Figure 3–15

$$\sin A = \frac{a}{c} \qquad \cos A = \frac{b}{c} \qquad \tan A = \frac{a}{b}$$

$$\cot A = \frac{b}{a} \qquad \sec A = \frac{c}{b} \qquad \csc A = \frac{c}{a}$$

$$(3\text{–}2)$$

If we should place the vertex of B at the origin, instead of the vertex of angle A, we would obtain the following ratios for the functions of angle B (see Fig. 3–16):

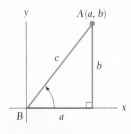

Figure 3–16

$$\sin B = \frac{b}{c} \qquad \cos B = \frac{a}{c} \qquad \tan B = \frac{b}{a}$$

$$\cot B = \frac{a}{b} \qquad \sec B = \frac{c}{a} \qquad \csc B = \frac{c}{b}$$

$$(3\text{–}3)$$

Inspecting these results, we may generalize our definitions of the trigonometric functions of any acute angle α of a right triangle to be as follows:

$$\sin \alpha = \frac{\text{side opposite } \alpha}{\text{hypotenuse}} \qquad \cos \alpha = \frac{\text{side adjacent } \alpha}{\text{hypotenuse}}$$

$$\tan \alpha = \frac{\text{side opposite } \alpha}{\text{side adjacent } \alpha} \qquad \cot \alpha = \frac{\text{side adjacent } \alpha}{\text{side opposite } \alpha} \qquad (3\text{–}4)$$

$$\sec \alpha = \frac{\text{hypotenuse}}{\text{side adjacent } \alpha} \qquad \csc \alpha = \frac{\text{hypotenuse}}{\text{side opposite } \alpha}$$

Using the definitions in this form, we can solve right triangles without placing the angle in standard position. The angle need only be a part of any right triangle.

We note from the above discussion that $\sin A = \cos B$, $\tan A = \cot B$, and $\sec A = \csc B$. From this we conclude that *cofunctions of acute complementary angles are equal.* The sine function and cosine function are cofunctions, the tangent function and cotangent function are cofunctions, and the secant function and cosecant function are cofunctions. From this we can see how the tables of trigonometric functions are constructed. Since $\sin A = \cos (90° - A)$, the number representing either of these need appear only once in the tables.

Example B

Given $a = 4$, $b = 7$, and $c = \sqrt{65}$ ($C = 90°$), find $\sin A$, $\cos A$, and $\tan A$ (see Fig. 3–17).

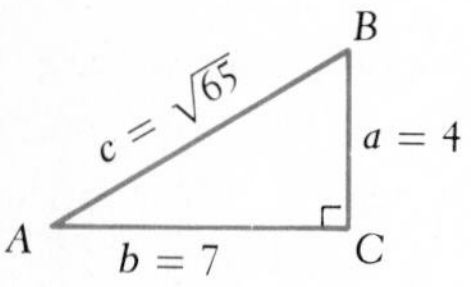

Figure 3–17

$$\sin A = \frac{\text{side opposite angle } A}{\text{hypotenuse}} = \frac{4}{\sqrt{65}} = 0.496$$

$$\cos A = \frac{\text{side adjacent angle } A}{\text{hypotenuse}} = \frac{7}{\sqrt{65}} = 0.868$$

$$\tan A = \frac{\text{side opposite angle } A}{\text{side adjacent angle } A} = \frac{4}{7} = 0.571$$

Example C

In Fig. 3–17, we have

$$\sin B = \frac{\text{side opposite angle } B}{\text{hypotenuse}} = \frac{7}{\sqrt{65}} = 0.868$$

$$\cos B = \frac{\text{side adjacent angle } B}{\text{hypotenuse}} = \frac{4}{\sqrt{65}} = 0.496$$

$$\tan B = \frac{\text{side opposite angle } B}{\text{side adjacent angle } B} = \frac{7}{4} = 1.75$$

We also note that $\sin A = \cos B$ and $\cos A = \sin B$.

We are now ready to solve right triangles. We do this by first expressing the unknown parts in terms of the known parts. Results should be rounded off to the accuracy of the given parts. (Again, discussions of rounding and significant digits are given in Appendix B.)

Example D

Given $A = 50.0°$ and $b = 6.70$, solve the right triangle of which these are parts (see Fig. 3–18).

Since $\dfrac{a}{b} = \tan A$, we have $a = b \tan A$. Thus,

$$a = 6.70(\tan 50.0°)$$
$$= 7.98$$

Here the calculator sequence is 6.7 $\boxed{\times}$ 50 $\boxed{\text{TAN}}$ $\boxed{=}$.

Since $A = 50.0°$, $B = 90.0° - 50.0° = 40.0°$. From $\dfrac{b}{c} = \cos A$, we have

$$c = \frac{b}{\cos A} = \frac{6.70}{\cos 50.0°} = 10.4$$

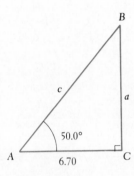

Figure 3–18

We have found that $a = 7.98$, $c = 10.4$, and $B = 40.0°$. It might be noted that we could have computed c by using an equation involving a. However, any error which may have been made in calculating a would also cause c to be in error. Thus, it is generally best to use given values in calculations.

Example E

Given $b = 56.82$ and $c = 79.55$, solve the right triangle of which these are parts (see Fig. 3–19).

Since $\cos A = \dfrac{b}{c}$, we have

$$\cos A = \frac{56.82}{79.55} = 0.7143$$

This means that $A = 44.42°$. We can find A directly on a calculator, without recording the value of $\cos A$, by the following sequence: 56.82 $\boxed{\div}$ 79.55 $\boxed{=}$ $\boxed{\text{ARCCOS}}$.

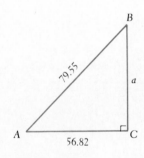

Figure 3–19

Since $A + B = 90°$, we have

$$B = 90.00° - 44.42° = 45.58°$$

We solve for a by use of the Pythagorean theorem, since in this way we can express a in terms of the given parts.

$$a^2 + b^2 = c^2 \quad \text{or} \quad a = \sqrt{c^2 - b^2}$$

(continued on next page)

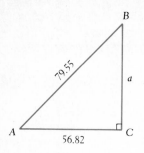

Figure 3–19

Thus,

$$a = \sqrt{79.55^2 - 56.82^2}$$
$$= 55.67$$

The above means we have determined that $a = 55.67$, $A = 44.42°$, and $B = 45.58°$.

In this example we expressed results to four significant digits because the given sides were expressed to this accuracy. In Example D the given sides and results were expressed to three significant digits. (Although this is consistent with the guidelines for significant digits in Appendix B and is reasonably accurate, an angle of *any* magnitude expressed to $0.1°$ has functions that are accurate to three significant digits.)

Example F

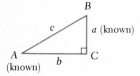

Figure 3–20

Assuming that A and a are known, express the unknown parts of the right triangle in terms of a and A (see Fig. 3–20).

Since $\dfrac{a}{b} = \tan A$, we have $b = \dfrac{a}{\tan A}$. Since A is known, $B = 90° - A$. Since $\dfrac{a}{c} = \sin A$, we have $c = \dfrac{a}{\sin A}$.

Exercises 3–4

In Exercises 1 through 4, draw appropriate figures and verify through observation that only one triangle may contain the given parts (that is, any others which may be drawn will be congruent—have equal corresponding sides and angles).

1. A $30°$ angle is included between sides of 3 in. and 6 in.

2. A side of 4 in. is included between angles of $40°$ and $70°$.

3. A right triangle with a hypotenuse of 5 cm and a leg of 3 cm.

4. A right triangle with a $70°$ angle between the hypotenuse and a leg of 5 cm.

In Exercises 5 through 24, solve the right triangles which have the given parts. Round off results. Refer to Fig. 3–21.

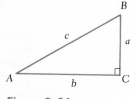

Figure 3–21

5. $A = 77.8°$, $a = 6700$

6. $A = 18.4°$, $c = 8.97$

7. $a = 150$, $c = 345$

8. $a = 93.2$, $c = 124$

9. $B = 32.1°$, $c = 23.8$

10. $B = 64.3°$, $b = 0.652$

11. $b = 82.1$, $c = 88.6$

12. $a = 5920$, $b = 4110$

13. $A = 32.10°$, $c = 56.85$

14. $B = 12.60°$, $c = 18.42$

15. $a = 56.73$, $b = 44.09$

16. $a = 9.908$, $c = 12.63$

17. $B = 37.5°$, $a = 0.862$

18. $A = 52.6°$, $b = 8.03$

19. $B = 74.18°$, $b = 1.849$

20. $A = 51.36°$, $a = 369.2$

21. $a = 591.87$, $b = 264.93$

22. $b = 2.9507$, $c = 5.0864$

23. $A = 12.975°$, $b = 14.592$

24. $B = 84.942°$, $a = 7413.5$

See Appendix E for a computer program for solving a right triangle.

In Exercises 25 through 28, find the part of the indicated triangle labeled either x or A.

25. The triangle in Fig. 3–22(a).   26. The triangle in Fig. 3–22(b).

27. The triangle in Fig. 3–22(c).   28. The triangle in Fig. 3–22(d).

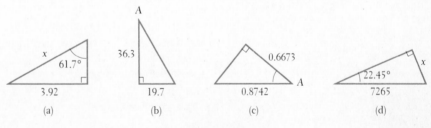

Figure 3–22

In Exercises 29 through 32, solve the right triangles which have the given parts. Express angles to the nearest 10' by use of Table 3. Refer to Fig. 3–21.

29. $B = 37°40'$, $a = 0.886$ 30. $A = 70°10'$, $a = 137$

31. $b = 86.7$, $c = 167$ 32. $a = 6.85$, $b = 2.12$

In Exercises 33 through 36, the parts listed refer to those in Fig. 3–21 and are assumed to be known. Express the other parts in terms of these known parts.

33. A, c 34. a, b 35. B, a 36. b, c

3–5 Applications of Right Triangles

Many applied problems can be solved by setting up the solutions in terms of right triangles. These applications are essentially the same as solving right triangles, although it is usually one specific part of the triangle that we wish to determine. The following examples illustrate some of the basic applications of right triangles.

Example A

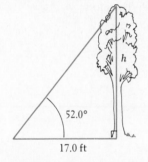

Figure 3–23

A tree has a shadow 17.0 ft long. From the point on the ground at the end of the shadow, the **angle of elevation** (the angle between the horizontal and the line of sight, when the object is above the horizontal) of the top of the tree is measured to be 52.0°. How tall is the tree?

By drawing an appropriate figure, as shown in Fig. 3–23, we note the information given and that which is required. Here we have let h be the height of the tree. Thus, we see that

$$\frac{h}{17.0} = \tan 52.0°$$

or

$$h = 17.0(\tan 52.0°)$$
$$= 21.8 \text{ ft}$$

Example B

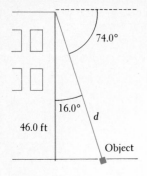

Figure 3–24

From the roof of a building 46.0 ft high, the **angle of depression** (the angle between the horizontal and the line of sight, when the object is below the horizontal) of an object in the street is 74.0°. What is the distance of the observer from the object?

Again we draw an appropriate figure (Fig. 3–24). Here we let d represent the required distance. From the figure we see that

$$\frac{46.0}{d} = \cos 16.0°$$

$$d = \frac{46.0}{\cos 16.0°}$$

$$= 47.9 \text{ ft}$$

Example C

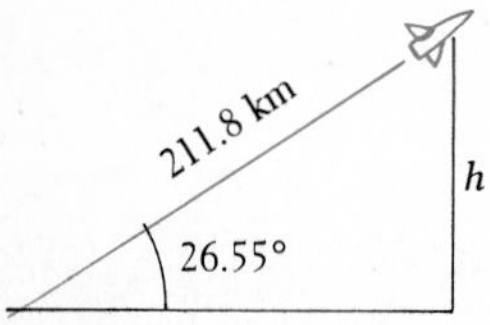

Figure 3–25

A missile is launched at an angle of 26.55° with respect to the horizontal. If it travels in a straight line over level terrain for 2.000 min and its average speed is 6355 km/h, what is its altitude at this time?

In Fig. 3–25 we have let h represent the altitude of the missile after 2.000 min (altitude is measured on a perpendicular). Also, we determine that the missile has flown 211.8 km in a direct line from the launching site. This is found from the fact that it travels at 6355 km/h for $\frac{1}{30.00}$ h (2.000 min) and from the fact that $(6355 \text{ km/h})\left(\frac{1}{30.00} \text{ h}\right) = 211.8$ km. Therefore,

$$\frac{h}{211.8} = \sin 26.55°$$

$$h = 211.8(\sin 26.55°)$$

$$= 94.67 \text{ km}$$

Example D

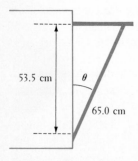

Figure 3–26

A shelf is supported by a straight 65.0-cm support attached to the wall at a point 53.5 cm below the bottom of the shelf. What angle does the support make with the wall? See Fig. 3–26.

In the figure we see that angle θ is to be determined. Therefore, we have

$$\cos \theta = \frac{53.5}{65.0}$$

or

$$\theta = 34.6°$$

to the nearest tenth of a degree.

Example E

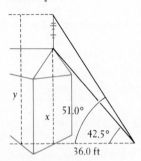

Figure 3–27

A television antenna is on the roof of a building. From a point on the ground 36.0 ft from the building, the angles of elevation of the top and the bottom of the antenna are 51.0° and 42.5°, respectively. How tall is the antenna?

In Fig. 3–27 we let x represent the distance from the top of the building to the ground, and y represent the distance from the top of the antenna to the ground. Therefore,

$$\frac{x}{36.0} = \tan 42.5°$$
$$x = 36.0(\tan 42.5°)$$
$$= 33.0 \text{ ft}$$

and

$$\frac{y}{36.0} = \tan 51.0°$$
$$y = 36.0(\tan 51.0°)$$
$$= 44.5 \text{ ft}$$

The length of the antenna is the difference of these distances, or 11.5 ft.

Exercises 3–5

In the following exercises, solve the given problems. Draw an appropriate figure unless the figure is given.

1. The angle of elevation of the sun is 48.0° at the time a television tower casts a shadow 346 ft long on level ground. Find the height of the tower.

2. A rope is stretched from the top of a vertical pole to a point 10.5 m from the foot of the pole. The rope makes an angle of 28.0° with the pole. How tall is the pole?

3. A loading platform is 3.25 ft above the ground. How long must a ramp be in order that it makes an angle of 20.0° with the ground?

4. A 20.0-ft ladder leans against the side of a house. The angle between the ground and ladder is 70.0°. How far from the house is the foot of the ladder?

5. The headlights of an automobile are set such that the beam drops 2.0 in. for each 25 ft in front of the car. What is the angle between the beam and the road?

6. A bullet was fired such that it just grazed the top of a table. It entered a wall, which was 12.60 ft from the graze in the table, at a point 4.63 ft above the table top. At what angle was the bullet fired above the horizontal? See Fig. 3–28.

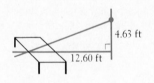

Figure 3–28

7. From the top of a cliff 112.5 m high, the angle of depression to a boat is 23.35°. How far is the boat from the foot of the cliff?

8. A robot is on the surface of Mars. The angle of depression from a camera in the robot to a rock on the surface of Mars is 13.33°. The camera is 196.0 cm above the surface. How far is the camera from the rock?

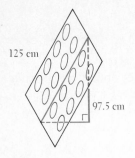

125 cm

97.5 cm

Figure 3–29

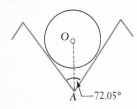

O

A —72.05°

Figure 3–30

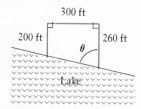

300 ft

200 ft   260 ft

θ

Lake

Figure 3–31

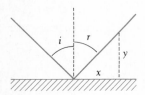

i r

y

x

Figure 3–32

9. In designing a new building, a doorway is 2.65 ft above the ground. A ramp for the disabled, at an angle of 6.0° with the ground, is to be built to the doorway. How long will the ramp be?

10. On a test flight, during the landing of the space shuttle *Columbia*, the ship was 325 ft above the end of the landing strip. It then came in on a constant angle of 7.5° with the landing strip. How far from the end of the landing strip did it first touch?

11. On a blueprint, the walls of a rectangular room are 5.65 cm and 3.85 cm long, respectively. What is the angle between the longer wall and a diagonal across the room?

12. A rectangular solar panel is 125 cm long. It is supported by a vertical rod 97.5 cm long (see Fig 3–29). What is the angle between the panel and the horizontal?

13. A jet cruising at 590 mi/h climbs at an angle of 15°30′. What is its gain in altitude in 2 min?

14. Ten rivets are equally spaced on the circumference of a circular plate. If the center-to-center distance between two adjacent rivets is 6.25 cm, what is the radius of the circle?

15. A guardrail is to be constructed around the top of a circular observation tower. The diameter of the observation area is 12.3 m. If the railing is constructed with 30 equal straight sections, what should be the length of each section?

16. In gauging screw threads a wire is placed in the thread. For the cross section of wire and V-thread shown in Fig. 3–30 determine the distance *OA* if the diameter of the wire is 0.05562 in. and the angle at *A* is 72.05°.

17. The level of a pond is 14.5 m above the level of a nearby stream. A straight 75.0-m drainage pipe extends from the pond surface to a point 3.0 m above the stream. What angle does the pipe make with the horizontal?

18. A lakefront property is shown in Fig. 3–31. What is the angle θ (to the nearest 10′) between the shoreline and the property line?

19. An astronaut circling the moon at an altitude of 100 mi notes that the angle of depression of the horizon is 23.8°. What is the radius of the moon?

20. A surveyor wishes to determine the width of a river. She sights a point *B* on the opposite side of the river from point *C*. She then measures 412.5 ft from *C* to *A* such that *C* is a right angle. She then finds that $\angle A = 56.67°$. How wide is the river?

21. If a light ray strikes a reflecting surface (Fig. 3–32), the angle of reflection *r* equals the angle of incidence *i*. If a light ray has an angle of incidence of 42.15°, what is the distance *y* from the plane of the surface of a point on the reflected ray, if the horizontal distance *x* from the point of incidence is 7.421 cm?

22. A woman considers building a dormer on her house. What are the dimensions *x* and *y* of the dormer in her plans (see Fig. 3–33)?

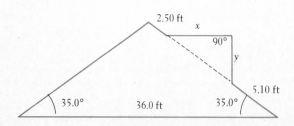

2.50 ft

x

90°

y

5.10 ft

35.0°   36.0 ft   35.0°

Figure 3–33

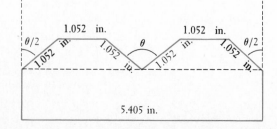

1.052 in.   1.052 in.

$\theta/2$ 1.052 in.   1.052 in. θ 1.052 in.   1.052 in. $\theta/2$

5.405 in.

Figure 3–34

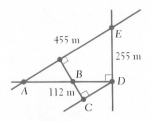

Figure 3–35

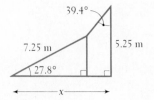

Figure 3–36

23. A machine part is shown in Fig. 3–34. What is the angle θ?

24. A way of representing the impedance and resistance in an alternating-current circuit is equivalent to letting the impedance be the hypotenuse of a right triangle and the resistance be the side adjacent to the phase angle. If the resistance in a given circuit is 25.40 Ω and the phase angle is 24.55°, what is the impedance?

25. A surveyor sights two points directly ahead. Both are at an elevation 18.525 m lower than the point from which he observes them. How far apart are the points if the angles of depression are 13.500° and 21.375°, respectively?

26. A flagpole is atop a building. From a point on the ground 720 ft from the building, the angles of elevation of the top and bottom of the flagpole are 33.5° and 31.2°, respectively. How tall is the flagpole?

27. Some of the streets in a certain city intersect as shown in Fig. 3–35. The distances between intersections A and E, D and E, and B and C are shown in the figure. How far is it between intersections B and D?

28. A supporting girder structure is shown in Fig. 3–36. Find the length x.

3–6 Exercises for Chapter 3

In Exercises 1 through 4, find the smallest positive angle and the smallest negative angle (numerically) coterminal with, but not equal to, the given angles.

  1. 17.0°   2. 248.3°   3. $-217.5°$ 4. $-7.6°$

In Exercises 5 through 8, change the given angles to equal angles expressed in decimal form.

  5. 31°54′   6. 174°45′   7. 38°6′   8. 321°27′

In Exercises 9 through 12, change the given angles to equal angles expressed to the nearest minute.

  9. 17.5°   10. 65.4°   11. 49.7°   12. 126.25°

In Exercises 13 through 16, determine the trigonometric functions of the angles whose terminal side passes through the given points. Give the answers in exact form.

  13. (24, 7)   14. (5, 4)   15. (4, 4)   16. (1.2, 0.5)

In Exercises 17 through 20, use the given trigonometric functions to find the indicated trigonometric functions. Give answers in decimal form, rounded off to three significant digits.

  17. Given $\sin \theta = \frac{5}{13}$, find $\cos \theta$ and $\cot \theta$.

  18. Given $\cos \theta = \frac{3}{8}$, find $\sin \theta$ and $\tan \theta$.

  19. Given $\tan \theta = 2$, find $\cos \theta$ and $\csc \theta$.

  20. Given $\cot \theta = 4$, find $\sin \theta$ and $\sec \theta$.

In Exercises 21 through 24, find the values of the trigonometric functions by use of a calculator. Round off the results to the accuracy of the given angle.

  21. $\sin 72.1°$ 22. $\cos 40.3°$ 23. $\tan 61.64°$ 24. $\sin 49.09°$

In Exercises 25 through 28, find θ for each of the given trigonometric functions by use of a calculator. Round off the results to the accuracy of the given value.

25. $\cos \theta = 0.950$ 26. $\sin \theta = 0.63052$

27. $\tan \theta = 1.574$ 28. $\cos \theta = 0.1345$

In Exercises 29 through 32, find the value of each of the trigonometric functions from Table 3.

29. $\tan 59°20'$ 30. $\sin 82°40'$ 31. $\cos 55°30'$ 32. $\tan 23°50'$

In Exercises 33 through 36, use Table 3 to find θ to the nearest $10'$ for each of the given trigonometric functions.

33. $\sin \theta = 0.5323$ 34. $\tan \theta = 1.264$

35. $\cos \theta = 0.4669$ 36. $\sin \theta = 0.1048$

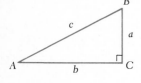

Figure 3–37

In Exercises 37 through 48, solve the right triangles which have the given parts. Round off the results. Refer to Fig. 3–37.

37. $A = 17.0°, b = 6.00$ 38. $B = 68.1°, a = 1080$

39. $a = 81.0, b = 64.5$ 40. $a = 1.06, c = 3.82$

41. $A = 37.5°, a = 12.0$ 42. $B = 15.7°, c = 12.6$

43. $b = 6.508, c = 7.642$ 44. $a = 72.14, b = 14.37$

45. $A = 49.67°, c = 0.8253$ 46. $B = 4.375°, b = 5.682$

47. $a = 11.652, c = 15.483$ 48. $a = 724.39, b = 852.44$

In Exercises 49 through 74, solve the given applied problems.

49. In analyzing the forces on a certain hinge, one of the vertical forces F_y was found from the equation $F_y = F \cos \theta$, where θ is the angle between one part of the hinge and the vertical. Find the value of F_y if $F = 56.0$ N and $\theta = 37.5°$.

50. The velocity v, in feet per second, of an object which slides s feet down an inclined plane is given by $v = \sqrt{2gs \sin \theta}$, where g is the acceleration due to gravity and θ is the angle which the plane makes with the horizontal. Find v if $g = 32.2$ ft/s², $s = 35.5$ ft, and $\theta = 25.2°$.

51. An approximate equation found in the diffraction of light through a narrow opening is $\sin \theta = \lambda/d$, where λ (the Greek lambda) is the wavelength of the light and d is the width of the opening. If $\theta = 1.167°$ and $d = 30.05$ μm, what is the wavelength of the light?

52. An equation used for the instantaneous value of electric current in an alternating-current circuit is $i = I_m \cos \theta$. Calculate i for $I_m = 56.00$ mA and $\theta = 10.55°$.

53. In determining the height h of a building which is 220 m distant, a surveyor may use the equation $h = 220 \tan \theta$, where θ is the angle of elevation to the top of the building. What should θ be if $h = 130$ m?

54. A formula used with a certain type of gear is $D = \frac{1}{4}N \sec \theta$, where D is the pitch diameter of the gear, N is the number of teeth on the gear, and θ is called the spiral angle. If $D = 6.75$ in. and $N = 20$, find θ.

55. A ship's captain, desiring to travel due south, discovers that, due to an improperly functioning instrument, he has gone 22.62 mi in a direction 4.050° east of south. How far from his course (to the east) is he?

56. A helicopter pilot notes that she is 150 m above a certain rooftop. If the angle of depression to the rooftop is 18.0°, how far on a direct line from the rooftop is the helicopter?

57. An observer 3500 ft from the launch pad of a rocket measures the angle of elevation to the rocket soon after liftoff to be 54.0°. How high is the rocket, assuming it has moved vertically?

58. The length of a kite string (assumed straight) is 560 ft. The angle of elevation of the kite is 64°. How high is the kite?

59. A roadway rises 120 ft for every 2200 ft along the road. Find the angle of inclination of the roadway.

60. A hemispherical bowl is standing level. Its inside radius is 6.50 in. and it is filled with water to a depth of 3.10 in. Through what angle may it be tilted before the water will spill?

61. A railroad embankment has a level top 22.0 ft wide, equal sloping sides of 14.5 ft, and a height of 7.20 ft. What is the width of the base of the embankment?

62. The horizontal distance between the extreme positions of a certain pendulum is 9.50 cm. If the length of the pendulum is 38.0 cm, through what angle does it swing?

63. A certain straight section of a natural gas pipeline is 3.25 km long, and it rises at an angle of 3.25°. How much higher is one end of the section than the other end?

64. A Coast Guard boat which is 2.75 km from a straight beach can travel at 37.5 km/h. By traveling along a line which is at 69.0° with the beach, how long will it take to reach the beach? See Fig. 3–38.

65. The span of a roof is 32.25 ft. Its rise is 7.500 ft at the center of the span. What angle does the roof make with the horizontal?

66. If the impedance in a certain alternating-current circuit is 56.50 Ω and the resistance in the circuit is 17.45 Ω, what is the phase angle? (See Exercise 24 of Section 3–5.)

67. A bridge is 12.50 m above the surrounding area. If the angle of elevation of the approach to the bridge is 4.625°, how long is the approach?

68. The windshield on an automobile is inclined 42.5° with respect to the horizontal. Assuming that the windshield is flat and rectangular, what is its area if it is 4.80 ft wide and the bottom is 1.50 ft in front of the top?

69. A person in a tall building observes an object drop from a window 20.0 ft away and directly opposite the observer. If the distance the object drops as a function of time is $s = 16t^2$, how far is the object from the observer (on a direct line) after 2.00 s?

70. The distance from ground level to the underside of a cloud is called the *ceiling*. One observer 1200 m from a searchlight aimed vertically notes that the angle of elevation of the spot of light on a cloud is 76°. What is the ceiling?

71. A laser beam is transmitted with a "width" of 0.00200°. What is the diameter of a spot of the beam on an object 52,500 km distant? See Fig. 3–39.

72. What angle is subtended at the eye of an observer 5.00 mi from an airplane which is 155 ft long and is perpendicular to the line of sight?

73. A hang glider is directly above the shore of a small lake. An observer is 375 m along a straight line from the shore. From the observer, the angle of elevation of the hang glider is 42.0°, and the angle of depression of the shore is 25.0°. How far above the shore is the hang glider?

74. Part of the supporting structure for a microwave transmitter is shown in Fig. 3–40. Find the length x of the indicated support piece.

Figure 3–38

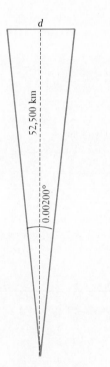

Figure 3–39

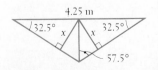

Figure 3–40

CHAPTER 4

Systems of Linear Equations; Determinants

In Chapter 1 we introduced the topic of solving equations and showed some of the types of technical problems which could be solved. Many of the equations we encountered at that time were examples of a very important type of equation, the **linear equation.** In this chapter we further consider the linear equation and its graph. Then we show how systems of two or three linear equations are solved.

4–1 Linear Equations

In general, *an equation is termed* **linear** *in a given set of variables if each term contains only one variable, to the first power, or is a constant.*

Example A

$5x - t + 6 = 0$ is linear in x and t, but $5x^2 - t + 6 = 0$ is not, due to the presence of x^2.

The equation $4x + y = 8$ is linear in x and y, but $4xy + y = 8$ is not, due to the presence of xy.

The equation $x - 6y + z - 4w = 7$ is linear in x, y, z, and w, but $x - \frac{6}{y} + z - 4w = 7$ is not, due to the presence of $\frac{6}{y}$ where y appears in the denominator.

An equation which can be written in the form

$$ax + b = 0 \qquad\qquad (4\text{--}1)$$

is known as a **linear equation in one unknown.** We have already discussed the solution to this type of equation in Section 1–11. In general, *the* **solution,** *or* **root,** *of the equation is* $x = -b/a$. Also, it will be noted that finding the solution is equivalent to finding the zero of the **linear function** $f(x) = ax + b$.

Example B

The equation $2x + 7 = 0$ is a linear equation of the form of Eq. (4–1) with $a = 2$ and $b = 7$.

From Chapter 2 we know that the linear function $2x + 7$ can be shown as $f(x) = 2x + 7$. To find the zero of this function, we set $f(x) = 0$ and get $2x + 7 = 0$.

The solution to the equation $2x + 7 = 0$, or the zero of the function $2x + 7$, is $-\frac{7}{2}$.

There are a great number of applied problems which involve more than one unknown. When forces are analyzed in physics, equations with two unknowns (the forces) often result. In electricity, equations relating several unknown currents arise. Numerous kinds of stated problems from various technical areas involve equations with more than one unknown. The use of equations involving more than one unknown is well established in technical applications.

Example C

A very basic law of direct-current electricity, known as Kirchhoff's first law, may be stated as "The algebraic sum of the currents entering any junction in a circuit is zero." If three wires are joined at a junction, this law leads to the linear equation

$$i_1 + i_2 + i_3 = 0$$

where i_1, i_2, and i_3 are the currents in each of the wires.

When determining two forces F_1 and F_2 acting on a beam, we might encounter an equation such as

$$2F_1 + 4F_2 = 200$$

An equation which can be written in the form

$$ax + by = c \qquad\qquad (4\text{--}2)$$

is known as a **linear equation in two unknowns.** In Chapter 2 we considered many equations which can be written in this form. We found that for each value of x, there is a corresponding value for y. Each of these pairs of numbers is a **solution** to the equation, although we did not call it that at the time. *A solution is any set of numbers, one for each variable, which satisfies the equation.* When we represent the solutions in the form of a graph, we see that the graph of any linear equation in two unknowns is a straight line. Also, graphs of linear equations in one unknown, those for which $a = 0$ or $b = 0$, are also straight lines. Thus we see the significance of the name *linear*. In the next section we shall further consider the graph of the linear equation.

Example D

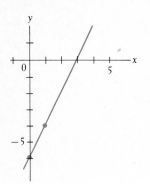

The equation $2x - y - 6 = 0$ is a linear equation in two unknowns, x and y, since it can be written in the form of Eq. (4–2) as $2x - y = 6$. Following the methods of Chapter 2, to graph this equation we would write it in the more convenient form of $y = 2x - 6$. We see from Fig. 4–1 that the graph is a straight line. The coordinates of any point on the line give us a solution of this equation. For example, the point $(1, -4)$ is a point on the line. This means that $x = 1$, $y = -4$ is a solution of the equation. In the same way, $x = 0$, $y = -6$ is a solution since the point $(0, -6)$ is on the line. This means that there is an unlimited number of solutions to the equation.

Figure 4–1

Two linear equations, each containing the same two unknowns,

$$a_1x + b_1y = c_1$$
$$a_2x + b_2y = c_2$$

(4–3)

are said to form a **system of simultaneous linear equations.** *A solution of the system is any pair of values (x, y) which satisfies both equations.* Methods of finding the solutions to such systems are the principal concern of this chapter.

Example E

The two linear equations

$$2x - y = 5$$
$$3x + 2y = 4$$

form a system of simultaneous linear equations. The solution of this system is $x = 2$, $y = -1$. These values satisfy both equations, since $2(2) - (-1) = 4 + 1 = 5$ and $3(2) + 2(-1) = 6 - 2 = 4$. This is the only pair of values which satisfies *both* equations. Methods for finding such solutions are taken up in later sections of this chapter.

Exercises 4–1

In Exercises 1 through 4, determine whether or not the given pairs of values are solutions of the given linear equations in two unknowns.

1. $2x + 3y = 9$; $(3, 1)$, $(5, \frac{1}{3})$ 2. $5x + 2y = 1$; $(2, -4)$, $(1, -2)$

3. $-3x + 5y = 13$; $(-1, 2)$, $(4, 5)$ 4. $x - 4y = 10$; $(2, -2)$, $(2, 2)$

In Exercises 5 through 8, for each given value of x, determine the value of y which gives a solution to the given linear equations in two unknowns.

5. $5x - y = 6$; $x = 1$, $x = -2$ 6. $2x + 7y = 8$; $x = -3$, $x = 2$

7. $x - 5y = 12$; $x = 3$, $x = -4$ 8. $3x - 2y = 9$; $x = \frac{2}{3}$, $x = -3$

In Exercises 9 through 16, determine whether or not the given pair of values is a solution of the given system of simultaneous linear equations.

9. $x - y = 5$ $x = 4$, $y = -1$ 10. $2x + y = 8$ $x = -1$, $y = 10$
    $2x + y = 7$ $3x - y = -13$

11. $x + 5y = -7$ $x = -2$, $y = 1$ 12. $-3x + y = 1$ $x = \frac{1}{3}$, $y = 2$
    $3x - 4y = -10$ $6x - 3y = -4$

13. $2x - 5y = 0$ $x = \frac{1}{2}$, $y = -\frac{1}{5}$ 14. $6x - y = 5$ $x = 1$, $y = -1$
    $4x + 10y = 4$ $3x - 4y = -1$

15. $3x - 2y = 2.2$ $x = 0.6$, $y = -0.2$
    $5x + y = 2.8$

16. $x - 7y = -3.2$ $x = -1.1$, $y = 0.3$
    $2x + y = 2.5$

In Exercises 17 through 20, answer the given questions.

17. If a board 84 in. long is cut into two pieces such that one piece is 6 in. longer than the other, the lengths x and y of the two pieces are found by solving the system of equations

$$x + y = 84$$
$$x - y = 6$$

Are the resulting lengths 45 in. and 39 in.?

18. If two forces F_1 and F_2 support a 98-lb weight, the forces can be found by solving the following system of equations. Are the forces 60 lb and 70 lb?

$$0.7F_1 - 0.6F_2 = 0$$
$$0.7F_1 + 0.8F_2 = 98$$

19. Under certain conditions two electric currents i_1 and i_2 can be found by solving the following system of equations. Are the currents $-\frac{1}{3}$ A and 1 A?

$$3i_1 + 4i_2 = 3$$
$$3i_1 - 5i_2 = -6$$

20. Using the data that fuel consumption for transportation contributes a percent p_1 of pollution which is 16% less than the percent p_2 of all other sources combined, the equations

$$p_1 + p_2 = 100$$
$$p_2 - p_1 = 16$$

can be set up. Are the percents $p_1 = 58$ and $p_2 = 42$?

4–2 Graphs of Linear Equations

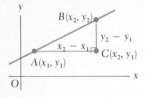

Figure 4–2

As we mentioned in Section 4–1, the main topic of this chapter is finding solutions of systems of linear equations. Before we take up the first method, which is a graphical method, we shall develop additional ways of graphing a linear equation.

Consider the line which passes through points A and B in Fig. 4–2. Point A has coordinates (x_1, y_1) and point B has coordinates (x_2, y_2). Point C is directly horizontal from point A and directly vertical from point B. Therefore, point C has coordinates (x_2, y_1), and there is a right angle at C.

One way of measuring the steepness of a line is to determine the vertical distance between two points in relation to the horizontal distance between them. Therefore, *we define the* **slope** *of a line through two points as the difference in the y-coordinates divided by the difference in the x-coordinates.* For points A and B, this means

$$m = \frac{y_2 - y_1}{x_2 - x_1} \tag{4–4}$$

where m is the slope of the line.

Example A

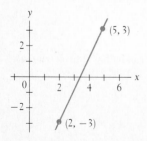

Figure 4–3

Find the slope of the line which passes through the points $(2, -3)$ and $(5, 3)$.

In Fig. 4–3, we draw the line through the two given points. By taking $(5, 3)$ as (x_2, y_2), then (x_1, y_1) is $(2, -3)$. Although we may choose either point as (x_2, y_2), *once the choice is made the order must be maintained.* Using Eq. (4–4), the slope is

$$m = \frac{3 - (-3)}{5 - 2} = \frac{6}{3} = 2$$

This means that the line rises 2 units for each unit it moves from left to right.

Example B

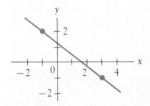

Figure 4–4

Find the slope of the line which passes through the points $(-1, 2)$ and $(3, -1)$.

In Fig. 4–4, we draw the line through these two points. Taking (x_2, y_2) as $(3, -1)$ and (x_1, y_1) as $(-1, 2)$, the slope is

$$m = \frac{-1 - 2}{3 - (-1)} = \frac{-3}{3 + 1} = -\frac{3}{4}$$

This means that the line *falls* 3 units for each 4 units it moves from left to right.

We note in Example A that *as x increases, y increases and that the slope is* **positive**. In Example B we see that *as x increases, y decreases and that the slope is* **negative**. We can also see that *the larger the numerical value of the slope, the more nearly vertical is the line*. Also, if the slope is numerically small, the line rises or falls slowly. This is further illustrated in Example C.

Example C

In Fig. 4–5(a), a line with a slope of $+5$ is illustrated. It rises sharply.

In Fig. 4–5(b), a line with a slope of $+\frac{1}{2}$ is illustrated. It rises slowly.

In Fig. 4–5(c), a line with a slope of -5 is shown. It falls sharply.

In Fig. 4–5(d), a line with a slope of $-\frac{1}{2}$ is shown. It falls slowly.

For each of these lines, we have shown the difference in the y-coordinates and the difference in the x-coordinates between two points. The ratio of these differences in each case gives the slope of each line.

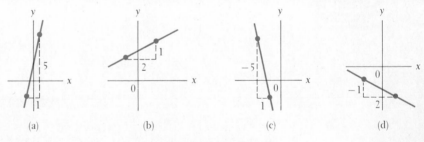

(a)   (b)   (c)   (d)

Figure 4–5

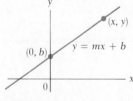

Figure 4–6

We shall now show how the slope is related to the equation of a straight line. In Fig. 4–6, if we have two points, $(0, b)$ and a general point (x, y), the slope of this line is

$$m = \frac{y - b}{x - 0}$$

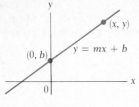

Figure 4-6

Simplifying this, we have $mx = y - b$, or

$$y = mx + b \qquad\qquad (4\text{–}5)$$

In Eq. (4–5), m is the slope of the line and b is the y-coordinate of the point where it crosses the y-axis. *This is called the **y-intercept** of the line,* and its coordinates are $(0, b)$. *Equation (4–5) is known as the* **slope-intercept** *form of the equation of a straight line.* If an equation is written in the form of Eq. (4–5), *the coefficient of x is the slope and the constant is the y-intercept of the line.* The point $(0, b)$ and simply the value of b are both referred to as the y-intercept. Since the x-coordinate of the y-intercept must be 0, this should cause no confusion.

Example D

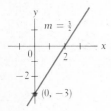

Figure 4-7

Find the slope and the y-intercept of the line $y = \frac{3}{2}x - 3$.

　　Since the equation is written as y as a function of x, it is in the form of Eq. (4–5). Therefore, since the coefficient of x is $\frac{3}{2}$, the slope of the line is $\frac{3}{2}$. Also, since we can write the equation as $y = \frac{3}{2}x + (-3)$, we see that the constant is -3, which means that the y-intercept is the point $(0, -3)$. The line is shown in Fig. 4–7.

Example E

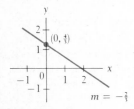

Figure 4-8

Find the slope and y-intercept of the line $2x + 3y = 4$.

　　Here, *we must first write the equation in the slope-intercept form.* Solving for y gives us

$$y = -\frac{2}{3}x + \frac{4}{3}$$

Therefore, the slope is $-\frac{2}{3}$ and the y-intercept is the point $(0, \frac{4}{3})$. The line is shown in Fig. 4–8.

Example F

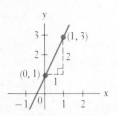

Figure 4-9

Graph $y = 2x + 1$ by determining the slope and the y-intercept of the line.

　　Since the equation is solved for y, the coefficient of x is the slope. Therefore, the slope of the line is 2. Also, since the b-term is the y-intercept, we see that the y-intercept of this line is $(0, 1)$.

　　We can use this information to sketch the line, as shown in Fig. 4–9. Since the slope is 2, we know that y increases 2 units for each unit of increase of x. Thus, starting at the y-intercept $(0, 1)$, if x increases by 1, y increases by 2, and we are at the point $(1, 3)$. The line must pass through $(1, 3)$, as well as through $(0, 1)$. Therefore, we draw the line through these points.

Another way of sketching the graph of a straight line is to find two points on the line and then draw the line through them. Two points which are easily determined are those where the line crosses the y-axis and the x-axis. We already know that the point where it crosses the y-axis is the y-intercept. In the same way, *the point where it crosses the x-axis is called the **x-intercept**,* and the coordinates of the x-intercept are $(a, 0)$. These points are easily found because in each case one of the coordinates is zero. By setting $x = 0$ and $y = 0$, in turn, and determining the corresponding value of the other unknown, we obtain the coordinates of the intercepts. A third point should be found as a check. This method is sufficient unless the line passes through the origin. Then both intercepts are at the origin and one more point must be determined, or we must use the slope-intercept method. Example G illustrates how a line is sketched by finding its intercepts.

Example G

Plot the graph of $2x - 3y = 6$ by finding its intercepts and one check point (see Fig. 4–10).

First let $x = 0$. This gives $-3y = 6$, or $y = -2$. Thus, the point $(0, -2)$ is on the graph. Next we let $y = 0$, and this gives $2x = 6$, or $x = 3$. Thus, the point $(3, 0)$ is on the graph. The point $(0, -2)$ is the y-intercept, and $(3, 0)$ is the x-intercept. These two points are sufficient to plot the line, but we should find another point as a check. Choosing $x = 1$, we find $y = -\frac{4}{3}$. Thus, the point $(1, -\frac{4}{3})$ should be on the line. From Fig. 4–10 we see that it is on the line.

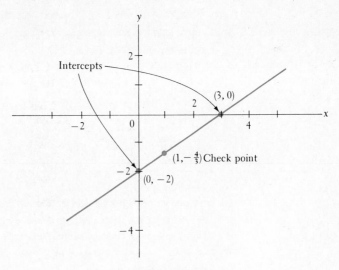

Figure 4–10

Further details and discussion of the concept of slope and of the graphing of linear equations are found in Chapter 20.

Exercises 4–2

In Exercises 1 through 8, find the slopes of the lines which pass through the given points.

1. $(1, 0), (3, 8)$ 2. $(3, 1), (2, 7)$

3. $(-1, 2), (-4, 17)$ 4. $(-1, -2), (2, 10)$

5. $(5, -3), (-2, -5)$ 6. $(3, -4), (-7, 1)$

7. $(4, 5), (-2, 2)$ 8. $(-5, 3), (1, 4)$

In Exercises 9 through 16, sketch the lines with the given slopes and y-intercepts.

9. $m = 2, b = -1$ 10. $m = 3, b = 1$

11. $m = -3, b = 2$ 12. $m = -4, b = -2$

13. $m = \dfrac{1}{2}, b = 0$ 14. $m = \dfrac{2}{3}, b = -1$

15. $m = -1, b = \dfrac{5}{2}$ 16. $m = -\dfrac{1}{3}, b = -\dfrac{3}{2}$

In Exercises 17 through 24, find the slopes and y-intercepts of the lines with the given equations and sketch the graphs.

17. $y = -2x + 1$ 18. $y = -4x$

19. $y = x + 4$ 20. $y = \dfrac{4}{5}x + 2$

21. $5x - 2y = 4$ 22. $6x - 2y = 7$

23. $x + 3y = 3$ 24. $2x + 6y = 3$

In Exercises 25 through 32, find the x-intercepts and the y-intercepts of the lines with the given equations and sketch the lines.

25. $x + 2y = 4$ 26. $3x + y = 3$

27. $4x - 3y = 12$ 28. $x - 5y = 5$

29. $y = 3x + 6$ 30. $y = -2x - 4$

31. $y = -x + 3$ 32. $y = 2x + 3$

In Exercises 33 through 36, sketch the indicated lines.

33. The diameter of the large end, d (in inches), of a certain type of machine tool can be found from the equation $d = 0.2l + 1.2$, where l is the length of the tool. Sketch d as a function of l, for values of l to 10 in.

34. The length l of a solar panel is related to its width w by the equation $l = -w + 150$, where the dimensions are measured in centimeters. Sketch l as a function of w.

35. Two forces, F_1 and F_2 (in newtons), act on a supporting brace. The equation relating the forces is $0.5F_1 + 0.6F_2 = 30$. Sketch the graph of F_2 as a function of F_1.

36. Two electric currents, I_1 and I_2 (in amperes), in part of a circuit in a microcomputer are related by the equation $4I_1 - 5I_2 = 2$. Sketch I_2 as a function of I_1. These currents can be considered to be negative.

4–3 Solving Systems of Two Linear Equations in Two Unknowns Graphically

We shall now take up the problem of solving for the unknowns when we have a system of two simultaneous linear equations in two unknowns. In this section we shall show how the solution may be found graphically. The sections which follow will discuss other basic methods of solution.

Since a solution of a system of simultaneous linear equations in two unknowns is any pair of values (x, y) which satisfies both equations, graphically, *the solution would be the coordinates of the point of intersection of the two lines.* This must be the case, for the coordinates of this point constitute the only pair of values to satisfy *both* equations. (In some special cases there may be no solution, in others there may be many solutions. See Examples D and E.)

Therefore, when we solve two simultaneous linear equations in two unknowns graphically, *we must plot the graph of each line and determine the point of intersection.* This may, of course, lead to approximate results if the lines cross at values between those chosen to determine the graph.

Example A

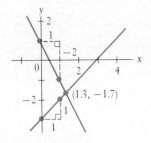

Figure 4–11

Solve the system of equations

$$y = x - 3$$
$$y = -2x + 1$$

Since each of the equations is in slope-intercept form, we see that $m = 1$ and $b = -3$ for the first line and that $m = -2$ and $b = 1$ for the second line. Using these values we sketch the lines, as shown in Fig. 4–11.

From the figure we see that the lines cross at about $(1.3, -1.7)$. This means that the solution of the system is approximately $x = 1.3$, $y = -1.7$. (The actual exact solution is $x = \frac{4}{3}$, $y = -\frac{5}{3}$.)

Example B

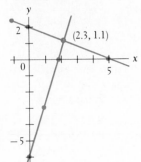

Figure 4–12

Solve the system of equations

$$2x + 5y = 10$$
$$3x - y = 6$$

We could write each equation in slope-intercept form in order to sketch the lines. Also, we could use the form in which they are written to find the intercepts. Choosing to find the intercepts and draw lines through them, we let $y = 0$; then $x = 0$. Therefore, we find that the intercepts of the first line are the points $(5, 0)$ and $(0, 2)$. A third point is $(-1, \frac{12}{5})$. The intercepts of the second line are $(2, 0)$ and $(0, -6)$. A third point is $(1, -3)$. Plotting these points and drawing the proper straight lines, we see that the lines cross at about $(2.3, 1.1)$. [The exact values are $(\frac{40}{17}, \frac{18}{17})$.] The solution of the system of equations is approximately $x = 2.3$, $y = 1.1$ (see Fig. 4–12).

Example C

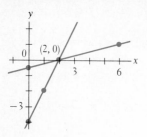

Figure 4–13

Solve the system of equations

$$x - 4y = 2$$
$$-2x + y = -4$$

The intercepts and a third point for the first line are $(2, 0)$, $(0, -0.5)$, and $(6, 1)$. For the second line they are $(2, 0)$, $(0, -4)$, and $(1, -2)$. Since they have one point in common, the point $(2, 0)$, we conclude that the exact solution to the system is $x = 2$, $y = 0$ (see Fig. 4–13).

Example D

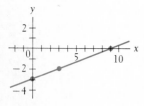

Figure 4–14

Solve the system of equations

$$x = 2y + 6$$
$$6y = 3x - 6$$

Writing each of these equations in slope-intercept form, we have

$$y = \frac{1}{2}x - 3$$

$$y = \frac{1}{2}x - 1$$

We note that each line has a slope of $\frac{1}{2}$ and that the y-intercepts are $b = -3$ and $b = -1$. This means that the y-intercepts are different, but the slopes are the same. Since the slope indicates that each line rises $\frac{1}{2}$ unit for y for each unit of x, *the lines are parallel and do not intersect* (see Fig. 4–14). Therefore, *there are no solutions. Such a system is called* **inconsistent.**

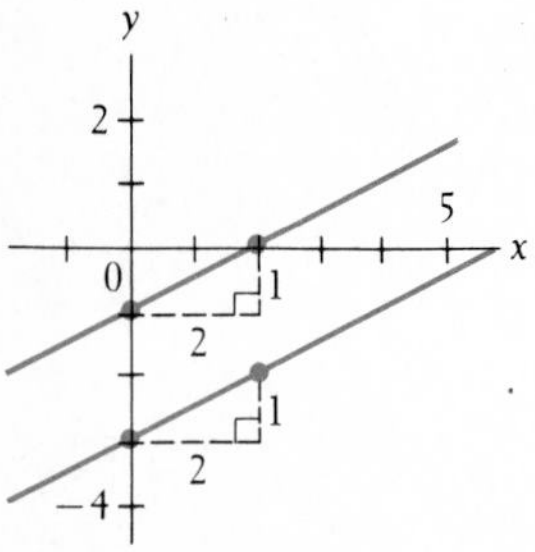

Example E

Solve the system of equations

$$x - 3y = 9$$
$$-2x + 6y = -18$$

The intercepts and a third point for the first line are $(9, 0)$, $(0, -3)$, and $(3, -2)$. In determining the intercepts for the second line, we find that they are $(9, 0)$ and $(0, -3)$, which are also the intercepts of the first line (see Fig. 4–15). As a check we note that $(3, -2)$ also satisfies the equation of the second line. This means the two lines are really the same line, and *the coordinates of any point on this common line constitute a solution of the system. Such a system is called* **dependent.**

Figure 4–15

Exercises 4–3

In the following exercises, solve each system of equations graphically. Estimate the answer to the nearest tenth of a unit if necessary.

1. $y = -x + 4$
   $y = x - 2$

2. $y = \frac{1}{2}x - 1$
   $y = -x + 8$

3. $y = 2x - 6$
   $y = -\frac{1}{3}x + 1$

4. $y = \frac{1}{2}x - 4$
   $y = 2x + 2$

5. $3x + 2y = 6$
   $x - 3y = 3$

6. $4x - 3y = -8$
   $6x + y = 6$

7. $2x - 5y = 10$
   $3x + 4y = -12$

8. $-5x + 3y = 15$
   $2x + 7y = 14$

9. $x - 4y = 8$
   $2x + 5y = 10$

10. $y = 4x - 6$
    $y = 2x + 4$

11. $y = -x + 3$
    $y = -2x + 3$

12. $x - 6 = 6y$
    $y = 3 - 3x$

13. $x - 4y = 6$
    $2y = x + 4$

14. $x + y = 3$
    $3x - 2y = 14$

15. $-2x + 2y = 7$
    $4x - 2y = 1$

16. $2x - 3y = -5$
    $3x + 2y = 12$

17. $x = 4y + 2$
    $3y = 2x + 3$

18. $x - 2y = 4$
    $3x = -2y + 7$

19. $8x - 7y = 3$
    $7y + x = 7$

20. $5x - 2y = 7$
    $3x + 4y = 8$

21. $x - 5y = 10$
    $2x - 10y = 20$

22. $18x - 3y = 7$
    $2y = 1 + 12x$

23. $y = 3x$
    $x - 2y = 6$

24. $4x - y = 3$
    $2x + 3y = 0$

25. $5x = y + 3$
    $4x = 2y - 3$

26. $5x + 7y = 5$
    $2x - 3y = 4$

27. $3x = 8y + 12$
    $-6x + 16y = 6$

28. $y = 6x + 2$
    $12x - 2y = -4$

29. The perimeter of a rectangular area is 24 km, and the length is 6 km longer than the width. The dimensions l and w can be found by solving the equations

$$2l + 2w = 24$$
$$l - w = 6$$

(Notice that the first equation represents the perimeter of this rectangle and the second equation represents the relationship of the length to the width.)

30. To assemble a particular piece of machinery, 18 bolts are used. There are two kinds of bolts, and 4 more of one kind are used than the other. The numbers of each kind, a and b, can be found by solving the equations

$$a + b = 18$$
$$a = b + 4$$

31. In electricity, applying Ohm's law to a particular circuit gives the equations needed to find a specified voltage E and current I (in amperes) as

$$E - 4I = 0$$
$$E + 6I = 9$$

32. A computer requires 12 s to do two series of calculations. The first series of calculations requires twice as much time as the second series. The times t_1 and t_2 can be found by solving the equations

$$t_1 + t_2 = 12$$
$$t_1 = 2t_2$$

4–4 Solving Systems of Two Linear Equations in Two Unknowns Algebraically

We have just seen how a system of two linear equations can be solved graphically. This technique is good for obtaining a "picture" of the solution. Finding the solution of systems of equations by graphical methods has one difficulty: The results are usually approximate. If exact solutions are required, we must turn to other methods. In this section we shall present two algebraic methods of solution.

The first method involves the *elimination of one variable by* **substitution.** To follow this method, *we first solve one of the equations for one of the unknowns. This solution is then substituted into the other equation,* resulting in one linear equation in one unknown. This linear equation can then be solved for the unknown it contains. By substituting this value into one of the original equations, we can find the corresponding value of the other unknown. The following two examples illustrate the method.

Example A

Use the method of elimination by substitution to solve the system of equations

$$x - 3y = 6$$
$$2x + 3y = 3$$

The first step is to solve one of the equations for one of the unknowns. The choice of which equation and which unknown depends on ease of algebraic manipulation. In this system, it is somewhat easier to solve the first equation for x. Therefore, performing this operation we have

$$x = 3y + 6$$

We then substitute this expression into the second equation in place of x, giving

$$2(3y + 6) + 3y = 3$$

Solving this equation for y, we obtain

$$6y + 12 + 3y = 3$$
$$9y = -9$$
$$y = -1$$

We now put the value $y = -1$ into the first of the original equations. Since we have already solved this equation for x in terms of y, we obtain

$$x = 3(-1) + 6 = 3$$

Therefore, the solution of the system is $x = 3, y = -1$. As a check, we substitute these values in the second equation. We obtain $2(3) + 3(-1) = 6 - 3 = 3$, which verifies the solution.

Example B

Use the method of elimination by substitution to solve the system of equations

$$-5x + 2y = -4$$
$$10x + 6y = 3$$

It makes little difference which equation or which unknown is chosen. Therefore, choosing to solve the first equation for y, we obtain

$$2y = 5x - 4$$
$$y = \frac{5x - 4}{2}$$

Substituting this expression into the second equation, we have

$$10x + 6\left(\frac{5x - 4}{2}\right) = 3$$

We now proceed to solve this equation for x.

$$10x + 3(5x - 4) = 3$$
$$10x + 15x - 12 = 3$$
$$25x = 15$$
$$x = \frac{3}{5}$$

Substituting this value into the expression for y, we obtain

$$y = \frac{5(3/5) - 4}{2} = \frac{3 - 4}{2} = -\frac{1}{2}$$

Therefore, the solution of this system is $x = \frac{3}{5}, y = -\frac{1}{2}$. Substituting these values in the second equation shows that the solution checks.

The method of elimination by substitution is useful if one equation can easily be solved for one of the unknowns. However, the numerical coefficients often make this method somewhat cumbersome. So we shall now develop another algebraic method of solving a system of linear equations.

The second method is the *elimination of a variable by means of* **addition or subtraction.** To use this method *we multiply each equation by a number chosen so that the coefficients for one of the unknowns will be numerically the same in* **both** *equations*. If these numerically equal coefficients are opposite in sign, we **add** the two equations. If the numerically equal coefficients have the same signs, we subtract one equation from the other. That is, we **subtract** the left side of one equation from the left side of the other equation, and also do the same to the right sides. After adding or subtracting, we have a simple linear equation in one unknown, which we then solve for the unknown. We then substitute this value into one of the original equations to obtain the value of the other unknown.

Example C

Use the method of elimination by addition or subtraction to solve the system of equations

$$x - 3y = 6$$
$$2x + 3y = 3$$

We look at the coefficients to determine the best way to eliminate one of the unknowns. In this case, since the coefficients of the y-terms are numerically the same and are opposite in sign, we may immediately add the two equations to eliminate y. Adding the left sides together and the right sides together, we obtain

$$x + 2x - 3y + 3y = 6 + 3$$
$$3x = 9$$
$$x = 3$$

Substituting this value into the first equation, we obtain

$$3 - 3y = 6$$
$$-3y = 3$$
$$y = -1$$

The solution $x = 3$, $y = -1$ agrees with the results obtained for the same problem illustrated in Example A of this section.

Example D

Use the method of elimination by addition or subtraction to solve the system of equations

$$3x - 2y = 4$$
$$x + 3y = 2$$

Looking at the coefficients of x and y, we see that it is necessary to multiply the second equation by 3 in order to make the coefficients of x the same. In order to make the coefficients of y numerically the same, we have to multiply the first equation by 3 and the second equation by 2. Therefore, the more convenient method is probably to multiply the second equation by 3 and eliminate x. Doing this (be careful to multiply the terms on *both* sides; a common error is to forget to multiply the value on the right) and then subtracting the second equation from the first, we have

$$3x - 2y = 4$$
$$\underline{3x + 9y = 6}$$
$$-11y = -2 \qquad \textbf{(\textit{subtracting})}$$
$$y = \frac{2}{11}$$

In order to find the value of x, we substitute $y = \frac{2}{11}$ into one of the original equations. Choosing the second equation (its form is somewhat simpler), we have

$$x + 3\left(\frac{2}{11}\right) = 2$$
$$11x + 6 = 22 \qquad \text{(multiplying each term by 11)}$$
$$x = \frac{16}{11}$$

We arrive at the solution $x = \frac{16}{11}$, $y = \frac{2}{11}$. Substituting these values into both of the original equations shows that the solution checks.

Example E

As we noted in Example D, we can solve the system of equations by first multiplying the first equation by 3 and the second equation by 2, thereby eliminating y. Doing this we have

$$9x - 6y = 12$$
$$\underline{2x + 6y = 4}$$
$$11x = 16 \qquad \text{(adding the equations)}$$
$$x = \frac{16}{11}$$

At this point we can find the value of y by substituting $x = \frac{16}{11}$ into one of the original equations, or we could eliminate x as is done in Example D. Substitution in the first of the original equations gives us

$$3\left(\frac{16}{11}\right) - 2y = 4$$
$$48 - 22y = 44 \qquad \text{(multiplying each term by 11)}$$
$$-22y = -4$$
$$y = \frac{2}{11}$$

Therefore, the solution is $x = \frac{16}{11}$, $y = \frac{2}{11}$, as we obtained in Example D.

The best form in which to have the equations for solution of the system by the addition or subtraction method is the form shown in Examples C and D. That is, the x-term and then the y-term are on the left (the reverse order in both equations would also be acceptable) and the constant is on the right. If the equations are not written in this form, they both should be written this way before proceeding with the solution.

Example F

In solving the system of equations

$$3x = 2y + 4$$
$$3y + x - 2 = 0$$

we should first rewrite the equations in the form noted above. Doing this, we have

$$3x - 2y = 4$$
$$x + 3y = 2$$

We then proceed with the solution. This system is the same as in Examples D and E.

Example G

Use the method of elimination by addition or subtraction to solve the system of equations

$$4x - 2y = 3$$
$$2x - y = 2$$

When we multiply the second equation by 2 and subtract, we obtain

$$
\begin{array}{rcr}
4x - 2y = & 3 \\
\underline{4x - 2y = } & \underline{4} \\
0 = & -1
\end{array}
$$

Since we know 0 does not equal -1, we conclude that there is no solution. When we obtain a result of $0 = a\ (a \neq 0)$, the system of equations is inconsistent. If we obtain the result $0 = 0$, the system is dependent.

After solving systems of equations by these methods, it is always a good policy to check the results by substituting the values of the two unknowns into the other original equation to see that the values satisfy the equation. Remember, we are finding the one pair of values which satisfies both equations.

Linear equations in two unknowns are often useful in solving stated problems. In such problems we must read the statement carefully in order to identify the unknowns and the method of setting up the proper equations. Exercises 29, 30, and 32 of Section 4–3 give statements and the resulting equations, which the reader should be able to derive. The following example gives a complete illustration of the method.

Example H

By weight, one alloy is 70% copper and 30% zinc. Another alloy is 40% copper and 60% zinc. How many grams of each of these would be required to make 300 g of an alloy which is 60% copper and 40% zinc?

Let A = required number of grams of first alloy, and B = required number of grams of second alloy. We know that the total weight of the final alloy is 300 g, which leads us to the equation $A + B = 300$. We also know that the final alloy will contain 180 g of copper (60% of 300). The weight of copper from the first alloy is $0.70A$ and that from the second is $0.40B$. This leads to the equation $0.70A + 0.40B = 180$. These two equations can now be solved simultaneously.

$$A + \quad B = 300$$
$$0.70A + 0.40B = 180$$

$$4A + 4B = 1200 \quad \text{(multiplying the first equation by 4)}$$
$$\underline{7A + 4B = 1800} \quad \text{(multiplying the second equation by 10)}$$
$$3A \qquad = 600 \quad \text{(subtracting the first equation from the second)}$$
$$A = 200 \text{ g}$$
$$B = 100 \text{ g} \quad \text{(by substituting into the first equation)}$$

Substitution shows that this solution checks with the given information.

Exercises 4–4

In Exercises 1 through 12, solve the given systems of equations by the method of elimination by substitution.

1. $x = y + 3$
   $x - 2y = 5$

2. $x = 2y + 1$
   $2x - 3y = 4$

3. $y = x - 4$
   $x + y = 10$

4. $y = 2x + 10$
   $2x + y = -2$

5. $x + y = -5$
   $2x - y = 2$

6. $3x + y = 1$
   $3x - 2y = 16$

7. $2x + 3y = 7$
   $6x - y = 1$

8. $2x + 2y = 1$
   $4x - 2y = 17$

9. $3x + 2y = 7$
   $2y = 9x + 11$

10. $3x + 3y = -1$
    $5x = -6y - 1$

11. $4x - 3y = 6$
    $2x + 4y = -5$

12. $5x + 4y = -7$
    $3x - 5y = -6$

In Exercises 13 through 24, solve the given systems of equations by the method of elimination by addition or subtraction.

13. $x + 2y = 5$
    $x - 2y = 1$

14. $x + 3y = 7$
    $2x + 3y = 5$

15. $2x - 3y = 4$
    $2x + y = -4$

16. $x - 4y = 17$
    $3x + 4y = 3$

17. $2x + 3y = 8$
    $x = 2y - 3$

18. $3x - y = 3$
    $4x = 3y + 14$

19. $x + 2y = 7$
    $2x + 4y = 9$

20. $3x - y = 5$
    $-9x + 3y = -15$

21. $2x - 3y - 4 = 0$
    $3x + 2 = 2y$

22. $3x + 5 = -4y$
    $3y = 5x - 2$

23. $0.3x = 0.7y + 0.4$
    $0.5y = 0.7 - 0.2x$

24. $2.50x + 2.25y = 4.00$
    $3.75x - 6.75y = 3.25$

In Exercises 25 through 32, solve the given systems of equations by either method of this section.

25. $2x - y = 5$
$6x + 2y = -5$

26. $3x + 2y = 4$
$6x - 6y = 13$

27. $6x + 3y + 4 = 0$
$5y = -9x - 6$

28. $1 + 6y = 5x$
$3x - 4y = 7$

29. $3x - 6y = 15$
$4x - 8y = 20$

30. $2x + 6y = -3$
$-6x - 18y = 5$

31. $1.2y + 10.8 = -8.4x$
$3.6x + 4.8y + 13.2 = 0$

32. $0.66x + 0.66y = -0.77$
$0.33x - 1.32y = 1.43$

In Exercises 33 through 36, solve the given systems of equations by an appropriate algebraic method.

33. An electrical experiment results in the following equations for currents (in amperes) i_1 and i_2 of a certain circuit:

$$i_1 = 3i_2$$
$$4i_1 - 2i_2 = 5$$

Find i_1 and i_2.

34. In determining the dimensions of a rectangular solar panel, the following equations are used:

$$2l + 2w = 460$$
$$l = w + 100$$

Find the length l and width w, in centimeters, of the panel.

35. Two grades of gasoline are mixed in order to get a mixture which has 1.5% of a certain special additive. By mixing x liters of a grade with 1.8% of the additive to y liters of a grade with 1.0% of the additive, 10,000 L of the mixture are produced. The equations used to find x and y are

$$x + y = 10{,}000$$
$$0.018x + 0.010y = 0.015(10{,}000)$$

Find x and y. (Verify the equations.)

36. In finding moments M_1 and M_2 (in foot-pounds) of certain forces acting on a given beam, the following equations are used:

$$10M_1 + 3M_2 = -140$$
$$4M_1 + 14M_2 = -564$$

Find M_1 and M_2.

In Exercises 37 through 44, set up appropriate systems of two linear equations in two unknowns, and solve the systems algebraically.

37. A piece of wire is 18 m long. Where must it be cut for one piece to be 3 m longer than the other piece?

38. The voltage across an electric resistor equals the current times the resistance. The sum of two resistances is 16 Ω. When a current of 4 A passes through the smaller resistance and 3 A through the larger resistance, the sum of the voltages is 52 V. What is the value of each resistance?

39. One microcomputer can perform a calculations per second, and a second microcomputer can perform b calculations per second. If the first calculates for 3 s and the second calculates for 2 s, 17.0 million calculations are performed. If the times are reversed, 15.5 million calculations are performed. Find the calculating rates a and b.

40. An airplane travels 1860 mi in 2 h with the wind and then returns in 2.5 h traveling against the wind. Find the speed of the airplane with respect to the air and the velocity of the wind.

41. A chemist has a 20% solution and an 8% solution of sulfuric acid. How many milliliters of each solution should he mix in order to obtain 180 mL of a 15% solution?

42. For proper dosage a drug must be a 10% solution. How many milliliters of a 5% solution and of a 25% solution should be mixed to obtain 1000 mL of the required solution?

43. The cost of booklets at a printing firm consists of a fixed charge and a charge for each booklet. The total cost of 1000 booklets is $550, and the total cost of 2000 booklets is $800. What is the fixed charge and the charge for each booklet?

44. Analyze the following statement from a student's laboratory report. ''The sum of the forces was 10 N, and twice the first force was 3 N less twice the second force.''

4–5 Solving Systems of Two Linear Equations in Two Unknowns by Determinants

Consider two linear equations in two unknowns, as given in Eq. (4–3):

$$a_1x + b_1y = c_1$$
$$a_2x + b_2y = c_2 \tag{4–3}$$

If we multiply the first of these equations by b_2 and the second by b_1, we obtain

$$a_1b_2x + b_1b_2y = c_1b_2$$
$$a_2b_1x + b_2b_1y = c_2b_1 \tag{4–6}$$

If we now subtract the second equation of (4–6) from the first, we obtain

$$a_1b_2x - a_2b_1x = c_1b_2 - c_2b_1$$

which by use of the distributive law may be written as

$$(a_1b_2 - a_2b_1)x = c_1b_2 - c_2b_1 \tag{4–7}$$

Solving Eq. (4–7) for x, we obtain

$$x = \frac{c_1 b_2 - c_2 b_1}{a_1 b_2 - a_2 b_1} \tag{4-8}$$

In the same manner, we may show that

$$y = \frac{a_1 c_2 - a_2 c_1}{a_1 b_2 - a_2 b_1} \tag{4-9}$$

If the denominator $a_1 b_2 - a_2 b_1 = 0$, there is no solution for Eqs. (4–8) and (4–9), since division by zero is not defined.

The expression $a_1 b_2 - a_2 b_1$, which appears in each of the denominators of Eqs. (4–8) and (4–9), is an example of a special kind of expression called a **determinant of the second order.** The determinant $a_1 b_2 - a_2 b_1$ is denoted by the symbol

$$\begin{vmatrix} a_1 & b_1 \\ a_2 & b_2 \end{vmatrix}$$

Thus, by definition, a determinant of the second order is given by

$$\begin{vmatrix} a_1 & b_1 \\ a_2 & b_2 \end{vmatrix} = a_1 b_2 - a_2 b_1 \tag{4-10}$$

The numbers a_1 and b_1 are called the **elements** *of the first* **row** *of the determinant. The numbers a_1 and a_2 are the elements of the first* **column** *of the determinant. In the same manner, the numbers a_2 and b_2 are the elements of the second row, and the numbers b_1 and b_2 are the elements of the second column. The numbers a_1 and b_2 are the elements of the* **principal diagonal,** *and the numbers a_2 and b_1 are the elements of the* **secondary diagonal.** *Thus, one way of stating the definition indicated in Eq. (4–10) is that the value of a determinant of the second order is found by taking the product of the elements of the principal diagonal and subtracting the product of the elements of the secondary diagonal.*

A diagram which is often helpful for remembering the expansion of a second-order determinant is shown in Fig. 4–16. The following examples illustrate how we carry out the evaluation of determinants.

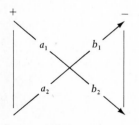

Figure 4–16

Example A

$$\begin{vmatrix} 1 & 4 \\ 3 & 2 \end{vmatrix} = 1(2) - 3(4) = 2 - 12 = -10$$

Example B

$$\begin{vmatrix} -5 & 8 \\ 3 & 7 \end{vmatrix} = (-5)(7) - 3(8) = -35 - 24 = -59$$

Example C

$$\begin{vmatrix} 4 & 6 \\ 3 & 17 \end{vmatrix} = 4(17) - (3)(6) = 68 - 18 = 50$$

$$\begin{vmatrix} 4 & 6 \\ -3 & 17 \end{vmatrix} = 4(17) - (-3)(6) = 68 + 18 = 86$$

$$\begin{vmatrix} 4 & 6 \\ -3 & -17 \end{vmatrix} = 4(-17) - (-3)(6) = -68 + 18 = -50$$

Note the signs of the terms being combined.

We note that the numerators of Eqs. (4–8) and (4–9) may also be written as determinants. The numerators of Eqs. (4–8) and (4–9) are

$$\begin{vmatrix} c_1 & b_1 \\ c_2 & b_2 \end{vmatrix} \quad \text{and} \quad \begin{vmatrix} a_1 & c_1 \\ a_2 & c_2 \end{vmatrix} \tag{4–11}$$

Therefore, the solutions for x and y of the system of equations (4–3) may be written directly in terms of determinants, without any algebraic operations, as

$$x = \frac{\begin{vmatrix} c_1 & b_1 \\ c_2 & b_2 \end{vmatrix}}{\begin{vmatrix} a_1 & b_1 \\ a_2 & b_2 \end{vmatrix}} \quad \text{and} \quad y = \frac{\begin{vmatrix} a_1 & c_1 \\ a_2 & c_2 \end{vmatrix}}{\begin{vmatrix} a_1 & b_1 \\ a_2 & b_2 \end{vmatrix}} \tag{4–12}$$

For this reason determinants provide a very quick and easy method of solution of systems of equations. Here again, the denominator of each of Eqs. (4–12) is the same. *The determinant of the denominator is made up of the coefficients of x and y. Also, we can see that the determinant of the numerator of the solution for x may be obtained by replacing the column of a's by the column of c's. The numerator of the solution for y may be obtained from the determinant of the denominator by replacing the column of b's by the column of c's.* This result is often referred to as **Cramer's rule.**

The following examples illustrate the method of solving systems of equations by determinants.

Example D

Solve the following system of equations by determinants:

$$2x + y = 1$$
$$5x - 2y = -11$$

First we set up the determinant for the denominator. This consists of the four coefficients in the system written as shown. Therefore, the determinant of the denominator is

$$\begin{vmatrix} 2 & 1 \\ 5 & -2 \end{vmatrix}$$

For finding x, the determinant in the numerator is obtained from this determinant by replacing the first column by the constants which appear on the right sides of the equations. Thus, the numerator for the solution for x is

$$\begin{vmatrix} 1 & 1 \\ -11 & -2 \end{vmatrix}$$

For finding y, the determinant in the numerator is obtained from the determinant of the denominator by replacing the second column by the constants which appear on the right sides of the equations. Thus, the determinant for the numerator for finding y is

$$\begin{vmatrix} 2 & 1 \\ 5 & -11 \end{vmatrix}$$

Now we set up the solutions for x and y using the determinants above.

$$x = \frac{\begin{vmatrix} 1 & 1 \\ -11 & -2 \end{vmatrix}}{\begin{vmatrix} 2 & 1 \\ 5 & -2 \end{vmatrix}} = \frac{1(-2) - (-11)(1)}{2(-2) - (5)(1)} = \frac{-2 + 11}{-4 - 5} = \frac{9}{-9} = -1$$

$$y = \frac{\begin{vmatrix} 2 & 1 \\ 5 & -11 \end{vmatrix}}{\begin{vmatrix} 2 & 1 \\ 5 & -2 \end{vmatrix}} = \frac{2(-11) - (5)(1)}{-9} = \frac{-22 - 5}{-9} = 3$$

Therefore, the solution to the system of equations is $x = -1$, $y = 3$.

    Since the determinant in the denominators is the same, it needs to be evaluated only once. This means that three determinants are to be evaluated in order to solve the system.

Example E

Solve the following system of equations by determinants:

$$x = y + 4$$
$$y + 2x - 11 = 0$$

Since these equations are not written in the form of Eqs. (4–3), we must first write them in that form so that we can properly determine the elements of the determinants. Doing this, we proceed with the solution.

$$x - y = 4$$
$$2x + y = 11$$

$$x = \frac{\begin{vmatrix} 4 & -1 \\ 11 & 1 \end{vmatrix}}{\begin{vmatrix} 1 & -1 \\ 2 & 1 \end{vmatrix}} = \frac{4 - (-11)}{1 - (-2)} = 5$$

$$y = \frac{\begin{vmatrix} 1 & 4 \\ 2 & 11 \end{vmatrix}}{\begin{vmatrix} 1 & -1 \\ 2 & 1 \end{vmatrix}} = \frac{11 - 8}{3} = 1$$

Therefore, the solution is $x = 5$, $y = 1$.

Example F

Solve the following system of equations by determinants:

$$5x + 7y = 4$$
$$3x - 6y = 5$$

$$x = \frac{\begin{vmatrix} 4 & 7 \\ 5 & -6 \end{vmatrix}}{\begin{vmatrix} 5 & 7 \\ 3 & -6 \end{vmatrix}} = \frac{-24 - 35}{-30 - 21} = \frac{59}{51}$$

$$y = \frac{\begin{vmatrix} 5 & 4 \\ 3 & 5 \end{vmatrix}}{\begin{vmatrix} 5 & 7 \\ 3 & -6 \end{vmatrix}} = \frac{25 - 12}{-51} = -\frac{13}{51}$$

Therefore, the solution is $x = \frac{59}{51}$, $y = -\frac{13}{51}$.

Example G

Two investments totaling \$18,000 yield an annual income of \$700. If the first investment has an interest rate of 5.5% and the second a rate of 3.0%, what is the value of each of the investments?

Let x = the value of the first investment, and y = the value of the second investment. We know that the total of the two investments is \$18,000. This leads to the equation $x + y = 18,000$. The first investment yields $0.055x$ dollars annually, and the second yields $0.03y$ dollars annually. This leads to the equation $0.055x + 0.03y = 700$. These two equations are then solved simultaneously.

$$x + y = 18,000$$
$$0.055x + 0.030y = 700$$

$$x = \frac{\begin{vmatrix} 18,000 & 1 \\ 700 & 0.03 \end{vmatrix}}{\begin{vmatrix} 1 & 1 \\ 0.055 & 0.03 \end{vmatrix}} = \frac{540 - 700}{0.03 - 0.055} = \frac{160}{0.025} = 6400$$

The value of y can be found most easily by substituting this value of x into the first equation.

$$y = 18,000 - x = 18,000 - 6400 = 11,600$$

Therefore, the values invested are \$6400 and \$11,600, respectively.

Certain points should be made here. As we noted in Example E, *the equations must be in the form of Eqs. (4–3) before the determinants are set up*. This is because the equations for the solutions in terms of determinants are based on that form of writing the system. Also, if either of the unknowns is missing from either equation, its coefficient is taken as zero, and zero is put in the appropriate position in that determinant. Finally, if the determinant of the denominator is zero and that of the numerator is not zero, the system is inconsistent. If determinants of both numerator and denominator are zero, the system is dependent.

Exercises 4–5

In Exercises 1 through 12, evaluate the given determinants.

1. $\begin{vmatrix} 2 & 4 \\ 3 & 1 \end{vmatrix}$ 2. $\begin{vmatrix} -1 & 3 \\ 2 & 6 \end{vmatrix}$ 3. $\begin{vmatrix} 3 & -5 \\ 7 & -2 \end{vmatrix}$ 4. $\begin{vmatrix} -4 & 7 \\ 1 & -3 \end{vmatrix}$

5. $\begin{vmatrix} 8 & -10 \\ 0 & 4 \end{vmatrix}$ 6. $\begin{vmatrix} -4 & -3 \\ 9 & -2 \end{vmatrix}$ 7. $\begin{vmatrix} -2 & 11 \\ -7 & -8 \end{vmatrix}$ 8. $\begin{vmatrix} -6 & 12 \\ -15 & 3 \end{vmatrix}$

9. $\begin{vmatrix} 0.7 & -1.3 \\ 0.1 & 1.1 \end{vmatrix}$ 10. $\begin{vmatrix} 0.20 & -0.05 \\ 0.28 & 0.09 \end{vmatrix}$ 11. $\begin{vmatrix} 16 & -8 \\ 42 & -15 \end{vmatrix}$ 12. $\begin{vmatrix} 43 & -7 \\ -81 & 16 \end{vmatrix}$

See Appendix E for a computer program for solving a system of equations by determinants.

In Exercises 13 through 32, solve the given systems of equations by use of determinants. (These systems are the same as those for Exercises 13 through 32 of Section 4-4.)

13. $x + 2y = 5$
    $x - 2y = 1$

14. $x + 3y = 7$
    $2x + 3y = 5$

15. $2x - 3y = 4$
    $2x + y = -4$

16. $x - 4y = 17$
    $3x + 4y = 3$

17. $2x + 3y = 8$
    $x = 2y - 3$

18. $3x - y = 3$
    $4x = 3y + 14$

19. $x + 2y = 7$
    $2x + 4y = 9$

20. $3x - y = 5$
    $-9x + 3y = -15$

21. $2x - 3y - 4 = 0$
    $3x + 2 = 2y$

22. $3x + 5 = -4y$
    $3y = 5x - 2$

23. $0.3x = 0.7y + 0.4$
    $0.5y = 0.7 - 0.2x$

24. $2.50x + 2.25y = 4.00$
    $3.75x - 6.75y = 3.25$

25. $2x - y = 5$
    $6x + 2y = -5$

26. $3x + 2y = 4$
    $6x - 6y = 13$

27. $6x + 3y + 4 = 0$
    $5y = -9x - 6$

28. $1 + 6y = 5x$
    $3x - 4y = 7$

29. $3x - 6y = 15$
    $4x - 8y = 20$

30. $2x + 6y = -3$
    $-6x - 18y = 5$

31. $1.2y + 10.8 = -8.4x$
    $3.6x + 4.8y + 13.2 = 0$

32. $0.66x + 0.66y = -0.77$
    $0.33x - 1.32y = 1.43$

In Exercises 33 through 36, solve the given systems of equations by use of determinants.

33. An object traveling at a constant velocity v (in feet per second) is 25 ft from a certain reference point after one second, and 35 ft from it after two seconds. Its initial distance s_0 from the reference point and its velocity can be found by solving the equations

$$s_0 + v = 25$$
$$s_0 + 2v = 35$$

Find s_0 and v.

34. An agricultural expert estimated that a certain organism spreads over an area A, in square kilometers, after t years according to the equation $A = at + b$, where a and b are constants. If $A = 6.0 \text{ km}^2$ when $t = 2$ years and $A = 7.5 \text{ km}^2$ when $t = 3$ years, then a and b can be determined by solving the system of equations

$$6.0 = 2a + b$$
$$7.5 = 3a + b$$

Solve for a and b, and find A as a function of t.

35. If $6000 is invested, part at 7% and part at 12%, with a total annual income of $560, the amounts A and B at these rates can be found by solving the system of equations

$$A + B = 6000$$
$$0.07A + 0.12B = 560$$

Find A and B.

36. When determining two forces F_1 and F_2 acting on a certain object, the following equations are obtained:

$$0.500F_1 + 0.600F_2 = 10$$
$$0.866F_1 - 0.800F_2 = 20$$

Find F_1 and F_2 (in newtons).

In Exercises 37 through 42, set up appropriate systems of two linear equations in two unknowns and then solve the system by use of determinants.

37. Two meshing gears together have 89 teeth. One of the gears has 4 less than twice the number of teeth of the other gear. How many teeth does each gear have?

38. A rocket is launched so that it averages 3000 mi/h. An hour later, another rocket is launched along the same path at an average speed of 4500 mi/h. Find the times of flight, t_1 and t_2, of the rockets when the second rocket overtakes the first.

39. A computer manufacturer received two types of computer components in a shipment of 5600 components. Type A components cost \$0.75 each and type B components cost \$0.45 each; the total cost of the shipment was \$3780. How many of each were in the shipment?

40. A roof truss is in the shape of an isosceles triangle. The perimeter of the truss is 41 m, and the base is 5 m longer than a rafter (neglecting overhang). Find the length of the base and the length of a rafter.

41. The resistance of a certain wire as a function of temperature can be found from the following equation: $R = \alpha T + \beta$. If the resistance is $0.4\ \Omega$ at $20°C$ and $0.5\ \Omega$ at $80°C$, find α and β, and then the resistance as a function of temperature.

42. The velocity of sound in steel is 15,900 ft/s faster than the velocity of sound in air. One end of a long steel bar is struck and an observer at the other end measures the time it takes for the sound to reach him. He finds that the sound through the bar takes 0.012 s to reach him and that the sound through the air takes 0.180 s. What are the velocities of sound in air and in steel?

4–6 Solving Systems of Three Linear Equations in Three Unknowns Algebraically

Many problems involve the solution of systems of linear equations which involve three, four, and occasionally even more unknowns. Solving such systems algebraically or by determinants is essentially the same as solving systems of two linear equations in two unknowns. Graphical solutions are not used, since a linear equation in three unknowns represents a plane in space. In this section we shall discuss the algebraic method of solving a system of three linear equations in three unknowns.

A system of three linear equations in three unknowns written in the form

$$
\begin{aligned}
a_1 x + b_1 y + c_1 z &= d_1 \\
a_2 x + b_2 y + c_2 z &= d_2 \\
a_3 x + b_3 y + c_3 z &= d_3
\end{aligned}
\qquad (4\text{–}13)
$$

has as its solution the set of values x, y, and z which satisfy all three equations simultaneously. The method of solution involves multiplying *two* of the equations by the proper numbers to eliminate *one* of the unknowns between these equations. We then repeat this process, using a *different pair* of the original equations, being sure that we eliminate the same unknown as we did between the first pair of equations. At this point we have two linear equations in two unknowns which can be solved by any of the methods previously discussed. The unknown originally eliminated may then be found by substitution into one of the original equations. It is wise to check these three values in one of the other original equations.

Example A

Solve the following system of equations:

$$(1) \qquad x + 2y - z = -5$$
$$(2) \qquad 2x - y + 2z = 8$$
$$(3) \qquad 3x + 3y + 4z = 5$$

$$
\begin{array}{lll}
(4) & 2x + 4y - 2z = -10 & \text{(1) multiplied by 2} \\
 & \underline{2x - y + 2z = 8} & \text{(2)} \\
(5) & 4x + 3y = -2 & \text{adding (4) and (2)} \\
\\
(6) & 4x - 2y + 4z = 16 & \text{(2) multiplied by 2} \\
 & \underline{3x + 3y + 4z = 5} & \text{(3)} \\
(7) & x - 5y = 11 & \text{subtracting} \\
\\
 & 4x + 3y = -2 & \text{(5)} \\
(8) & \underline{4x - 20y = 44} & \text{(7) multiplied by 4} \\
(9) & 23y = -46 & \text{subtracting} \\
(10) & y = -2 \\
\\
(11) & x - 5(-2) = 11 & \text{substituting (10) in (7)} \\
(12) & x = 1 \\
(13) & 1 + 2(-2) - z = -5 & \text{substituting (12) and (10) in (1)} \\
(14) & z = 2
\end{array}
$$

To check, we substitute the solution $x = 1$, $y = -2$, $z = 2$ in (2).

$$2(1) - (-2) + 2(2) \stackrel{?}{=} 8$$
$$8 = 8 \qquad \text{It checks.}$$

It should be noted that Eqs. (1), (2), and (3) could be solved just as well by eliminating y between (1) and (2), and then again between (1) and (3). We would then have two equations to solve in x and z. Also, z could have been eliminated between Eqs. (1) and (3) to obtain the second equation in x and y.

Example B

Solve the following system of equations:

(1) $$4x + y + 3z = 1$$
(2) $$2x - 2y + 6z = 11$$
(3) $$-6x + 3y + 12z = -4$$

(4) $8x + 2y + 6z = 2$ (1) multiplied by 2
       $2x - 2y + 6z = 11$ (2)
(5) $10x \quad\quad + 12z = 13$ adding

(6) $12x + 3y + 9z = 3$ (1) multiplied by 3
       $-6x + 3y + 12z = -4$ (3)
(7) $18x \quad\quad - 3z = 7$ subtracting

       $10x + 12z = 13$ (5)
(8) $72x - 12z = 28$ (7) multiplied by 4
(9) $82x \quad\quad = 41$ adding

(10) $$x = \frac{1}{2}$$

(11) $18\left(\frac{1}{2}\right) - 3z = 7$ substituting (10) in (7)

(12) $$-3z = -2$$

(13) $$z = \frac{2}{3}$$

(14) $4\left(\frac{1}{2}\right) + y + 3\left(\frac{2}{3}\right) = 1$ substituting (13) and (10) in (1)

(15) $$2 + y + 2 = 1$$
(16) $$y = -3$$

Therefore, the solution is $x = \frac{1}{2}$, $y = -3$, $z = \frac{2}{3}$. Checking the solution in (2), we have $2(\frac{1}{2}) - 2(-3) + 6(\frac{2}{3}) = 1 + 6 + 4 = 11$.

Example C

Three forces, F_1, F_2, and F_3, are acting on a beam. Find the forces (in pounds). The forces are determined by solving the following equations:

(1) $$F_1 + F_2 + F_3 = 25$$
(2) $$F_1 + 2F_2 + 3F_3 = 59$$
(3) $$2F_1 + 2F_2 - F_3 = 5$$

(4) $3F_1 + 3F_2 + 3F_3 = 75$ (1) multiplied by 3
       $F_1 + 2F_2 + 3F_3 = 59$ (2)
(5) $2F_1 + F_2 \quad\quad = 16$ subtracting

(6)	$3F_1 + 3F_2$	$= 30$	(1) added to (3)
	$2F_1 + F_2$	$= 16$	(5)
(7)	$F_1 + F_2$	$= 10$	(6) divided by 3
(8)	F_1	$= 6$	subtracting
(9)	$6 + F_2 = 10$		(8) substituted in (7)
(10)	$F_2 = 4$		
(11)	$6 + 4 + F_3 = 25$		substituting (8) and (10) in (1)
(12)	$F_3 = 15$		

Therefore, the three forces are 6 lb, 4 lb, and 15 lb, respectively. This solution can be checked by substituting in Eq. (2) or Eq. (3).

Example D

A triangle has a perimeter of 37 in. The longest side is 3 in. longer than the next longest, which in turn is 8 in. longer than the shortest side. Find the length of each side.

Let a = length of the longest side, b = length of the next-longest side, and c = length of the shortest side. Since the perimeter is 37 in., we have $a + b + c = 37$. The statement of the problem also leads to the equations $a = b + 3$ and $b = c + 8$. These equations are then put into standard form and solved simultaneously.

(1)	$a + b + c = 37$		
(2)	$a - b = 3$		rewriting the second equation
(3)	$b - c = 8$		rewriting the third equation
(4)	$a + 2b = 45$		adding (1) and (3)
	$a - b = 3$		(2)
(5)	$3b = 42$		subtracting
(6)	$b = 14$		
(7)	$a - 14 = 3$		substituting (6) in (2)
(8)	$a = 17$		
(9)	$14 - c = 8$		substituting (6) in (3)
(10)	$c = 6$		

Therefore, the three sides of the triangle are 17 in., 14 in., and 6 in.

Checking the solution, the sum of the lengths of the three sides is 17 in. + 14 in. + 6 in. = 37 in., and the perimeter was given to be 37 in.

For systems of equations with more than three unknowns, the solution is found in a manner similar to that used with three unknowns. For example, with four unknowns one of the unknowns is eliminated between three different pairs of equations. The result is three equations in the remaining three unknowns. The solution then follows the procedure used with three unknowns.

Exercises 4–6

In Exercises 1 through 16, solve the given systems of equations.

1. $x + y + z = 2$
   $x - z = 1$
   $x + y = 1$

2. $x + y - z = -3$
   $x + z = 2$
   $2x - y + 2z = 3$

3. $2x + 3y + z = 2$
   $-x + 2y + 3z = -1$
   $-3x - 3y + z = 0$

4. $2x + y - z = 4$
   $4x - 3y - 2z = -2$
   $8x - 2y - 3z = 3$

5. $5x + 6y - 3z = 6$
   $4x - 7y - 2z = -3$
   $3x + y - 7z = 1$

6. $3r + s - t = 2$
   $r - 2s + t = 0$
   $4r - s + t = 3$

7. $2x - 2y + 3z = 5$
   $2x + y - 2z = -1$
   $4x - y - 3z = 0$

8. $2u + 2v + 3w = 0$
   $3u + v + 4w = 21$
   $-u - 3v + 7w = 15$

9. $3x - 7y + 3z = 6$
   $3x + 3y + 6z = 1$
   $5x - 5y + 2z = 5$

10. $8x + y + z = 1$
    $7x - 2y + 9z = -3$
    $4x - 6y + 8z = -5$

11. $x + 2y + 2z = 0$
    $2x + 6y - 3z = -1$
    $4x - 3y + 6z = -8$

12. $3x + 3y + z = 6$
    $2x + 2y - z = 9$
    $4x + 2y - 3z = 16$

13. $2x + 3y - 5z = 7$
    $4x - 3y - 2z = 1$
    $8x - y + 4z = 3$

14. $2x - 4y - 4z = 3$
    $3x + 8y + 2z = -11$
    $4x + 6y - z = -8$

15. $r - s - 3t - u = 1$
    $2r + 4s - 2u = 2$
    $3r + 4s - 2t = 0$
    $r + 2t - 3u = 3$

16. $3x + 2y - 4z + 2t = 3$
    $5x - 3y - 5z + 6t = 8$
    $2x - y + 3z - 2t = 1$
    $-2x + 3y + 2z - 3t = -2$

In Exercises 17 through 22, solve the given problems.

17. Show that the following system of equations has an unlimited number of solutions, and find one of them.

$$x - 2y - 3z = 2$$
$$x - 4y - 13z = 14$$
$$-3x + 5y + 4z = 0$$

18. Show that the following system of equations has no solution.

$$x - 2y - 3z = 2$$
$$x - 4y - 13z = 14$$
$$-3x + 5y + 4z = 2$$

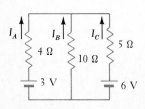

Figure 4–17

19. In applying Kirchhoff's laws (e.g., see Beiser, *Modern Technical Physics*, 4th ed., p. 500) to the electric circuit shown in Fig. 4–17, the following equations are found. Determine the indicated currents. (In Fig. 4–17, I signifies current, in amperes).

$$I_A + I_B + I_C = 0$$
$$4I_A - 10I_B = 3$$
$$-10I_B + 5I_C = 6$$

20. Three oil pumps fill three different tanks. The pumping rates of the pumps in liters per hour are r_1, r_2, and r_3, respectively. Because of malfunctions, they do not operate at capacity each time. Their rates can be found by solving the system of equations

$$r_1 + r_2 + r_3 = 14{,}000$$
$$r_1 + 2r_2 = 13{,}000$$
$$3r_1 + 3r_2 + 2r_3 = 36{,}000$$

Find the pumping rates.

21. The forces acting on a certain girder, as shown in Fig. 4–18, can be found by solving the following system of equations:

$$0.707F_1 - 0.800F_2 = 0$$
$$0.707F_1 + 0.600F_2 - F_3 = 10.0$$
$$3.00F_2 - 3.00F_3 = 20.0$$

Find the forces, in newtons.

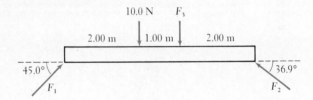

10.0 N F_3

2.00 m   1.00 m   2.00 m

45.0°   36.9°

F_1 F_2

Figure 4–18

22. In a laboratory experiment to measure the acceleration of an object, the distances traveled by the object were recorded for three time intervals. These data led to the following equations:

$$s_0 + 2v_0 + 2a = 20$$
$$s_0 + 4v_0 + 8a = 54$$
$$s_0 + 6v_0 + 18a = 104$$

Here s_0 is the initial displacement (in feet), v_0 is the initial velocity (in feet per second), and a is the acceleration (in feet per second squared). Find s_0, v_0, and a.

In Exercises 23 through 26, set up systems of three linear equations and solve for the indicated quantities.

23. Three machines together produce 64 parts each hour. Three times the production of the first machine equals the production of the other two machines together. Five times the production of the second is 12 parts per hour more than twice the rate of the other two together. Find the production rates of the three machines.

24. Angle A of a triangle equals 20° less than the sum of angles B and C. (The triangle is not a right triangle.) Angle B is one-fifth the sum of angles A and C. Find the angles.

25. By volume, one alloy is 60% copper, 30% zinc, and 10% nickel. A second alloy has percentages 50, 30, and 20, respectively, of the three metals. A third alloy is 30% copper and 70% nickel. How much of each must be mixed so that 100 cm^3 of the resulting alloy has percentages of 40, 15, and 45, respectively?

26. The budget of a certain corporation for three positions in a given department is $70,000. Position *A* pays as much as the other two positions together, and position *A* pays $5000 more than twice position *C*. What do the three positions pay?

4–7 Solving Systems of Three Linear Equations in Three Unknowns by Determinants

Just as systems of two linear equations in two unknowns can be solved by the use of determinants, so can systems of three linear equations in three unknowns. The system as given in Eqs. (4–13) can be solved in general terms by the method of elimination by addition or subtraction. This leads to the following solutions for x, y, and z:

$$x = \frac{d_1b_2c_3 + d_3b_1c_2 + d_2b_3c_1 - d_3b_2c_1 - d_1b_3c_2 - d_2b_1c_3}{a_1b_2c_3 + a_3b_1c_2 + a_2b_3c_1 - a_3b_2c_1 - a_1b_3c_2 - a_2b_1c_3}$$

$$y = \frac{a_1d_2c_3 + a_3d_1c_2 + a_2d_3c_1 - a_3d_2c_1 - a_1d_3c_2 - a_2d_1c_3}{a_1b_2c_3 + a_3b_1c_2 + a_2b_3c_1 - a_3b_2c_1 - a_1b_3c_2 - a_2b_1c_3} \qquad \textbf{(4–14)}$$

$$z = \frac{a_1b_2d_3 + a_3b_1d_2 + a_2b_3d_1 - a_3b_2d_1 - a_1b_3d_2 - a_2b_1d_3}{a_1b_2c_3 + a_3b_1c_2 + a_2b_3c_1 - a_3b_2c_1 - a_1b_3c_2 - a_2b_1c_3}$$

The expression that appears in each of the denominators of Eqs. (4–14) is an example of a **determinant of the third order.** This determinant is denoted by

$$\begin{vmatrix} a_1 & b_1 & c_1 \\ a_2 & b_2 & c_2 \\ a_3 & b_3 & c_3 \end{vmatrix}$$

Therefore, a determinant of the third order is defined by the equation

$$\begin{vmatrix} a_1 & b_1 & c_1 \\ a_2 & b_2 & c_2 \\ a_3 & b_3 & c_3 \end{vmatrix} = a_1b_2c_3 + a_3b_1c_2 + a_2b_3c_1 - a_3b_2c_1 - a_1b_3c_2 - a_2b_1c_3 \qquad \textbf{(4–15)}$$

The elements, rows, columns, and diagonals of a third-order determinant are defined just as are those of a second-order determinant. For example, the principal diagonal is made up of the elements a_1, b_2, and c_3.

Probably the easiest way of remembering the method of determining the value of a third-order determinant is as follows (this method does *not* work for determinants of order higher than three): *Rewrite the first and second columns to the right of the determinant. The products of the elements of the principal diagonal and the two parallel diagonals to the right of it are then added. The products of the elements of the secondary diagonal and the two parallel diagonals to the right of it are subtracted from the first sum. The algebraic sum of these six products* gives the value of the determinant (see Fig. 4–19). Examples A through C illustrate the evaluation of third-order determinants.

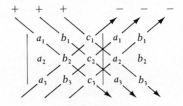

Figure 4–19

Example A

$$\begin{vmatrix} 1 & 5 & 4 \\ -2 & 3 & -1 \\ 2 & -1 & 5 \end{vmatrix} \begin{matrix} 1 & 5 \\ -2 & 3 \\ 2 & -1 \end{matrix} = 15 + (-10) + (+8) - (24) - (1)$$
$$- (-50) = 38$$

Example B

$$\begin{vmatrix} -1 & 4 & -5 \\ 6 & 1 & 0 \\ 9 & -7 & 3 \end{vmatrix} \begin{matrix} -1 & 4 \\ 6 & 1 \\ 9 & -7 \end{matrix} = (-3) + 0 + 210 - (-45) - 0 - 72 = 180$$

Example C

$$\begin{vmatrix} 3 & -2 & 8 \\ -5 & 5 & -2 \\ 4 & 9 & -6 \end{vmatrix} \begin{matrix} 3 & -2 \\ -5 & 5 \\ 4 & 9 \end{matrix} = (-90) + 16 + (-360) - 160 - (-54)$$
$$- (-60) = -480$$

Inspection of Eqs. (4–14) reveals that the numerators of these solutions may also be written in terms of determinants Thus, we may write the general solution to a system of three equations in three unknowns as

$$x = \frac{\begin{vmatrix} d_1 & b_1 & c_1 \\ d_2 & b_2 & c_2 \\ d_3 & b_3 & c_3 \end{vmatrix}}{\begin{vmatrix} a_1 & b_1 & c_1 \\ a_2 & b_2 & c_2 \\ a_3 & b_3 & c_3 \end{vmatrix}} \qquad y = \frac{\begin{vmatrix} a_1 & d_1 & c_1 \\ a_2 & d_2 & c_2 \\ a_3 & d_3 & c_3 \end{vmatrix}}{\begin{vmatrix} a_1 & b_1 & c_1 \\ a_2 & b_2 & c_2 \\ a_3 & b_3 & c_3 \end{vmatrix}} \qquad z = \frac{\begin{vmatrix} a_1 & b_1 & d_1 \\ a_2 & b_2 & d_2 \\ a_3 & b_3 & d_3 \end{vmatrix}}{\begin{vmatrix} a_1 & b_1 & c_1 \\ a_2 & b_2 & c_2 \\ a_3 & b_3 & c_3 \end{vmatrix}} \qquad (4\text{–}16)$$

If the determinant of the denominator is zero and the determinant of the numerator is not zero, the system is **inconsistent.** *If the determinant of the denominator is not equal to zero, then there is a unique solution to the system of equations.*
An analysis of Eqs. (4–16) shows that the situation is precisely the same as it was when we were using determinants to solve systems of two linear equations.

That is, the determinants in the denominators in the expressions for x, y, and z are the same. They consist of elements which are the coefficients of the unknowns. The determinant of the numerator of the solution for x is the same as that of the denominator, except that the column of d's replaces the column of a's. The determinant in the numerator of the solution for y is the same as that of the denominator, except that the column of d's replaces the column of b's. The determinant of the numerator of the solution for z is the same as the determinant of the denominator, except that the column of d's replaces the column of c's. To summarize, we can state that *the determinants in the numerators are the same as those in the denominators, except that the column of d's replaces the column of coefficients of the unknown for which we are solving.* This again is **Cramer's rule.** Remember that the equations must be written in the standard form shown in Eqs. (4–13) before the determinants are formed.

Example D

Solve the following system of equations by determinants:

$$x + 2y + 2z = 1$$
$$2x - y + z = 3$$
$$4x + y + 2z = 0$$

$$x = \frac{\begin{vmatrix} 1 & 2 & 2 \\ 3 & -1 & 1 \\ 0 & 1 & 2 \end{vmatrix}\begin{matrix} 1 & 2 \\ 3 & -1 \\ 0 & 1 \end{matrix}}{\begin{vmatrix} 1 & 2 & 2 \\ 2 & -1 & 1 \\ 4 & 1 & 2 \end{vmatrix}\begin{matrix} 1 & 2 \\ 2 & -1 \\ 4 & 1 \end{matrix}} = \frac{-2 + 0 + 6 - 0 - 1 - 12}{-2 + 8 + 4 - (-8) - 1 - 8}$$

$$= \frac{-9}{+9} = -1$$

Since the value of the denominator is already determined, there is no need to write the denominator in determinant form when solving for y and z.

$$y = \frac{\begin{vmatrix} 1 & 1 & 2 \\ 2 & 3 & 1 \\ 4 & 0 & 2 \end{vmatrix}\begin{matrix} 1 & 1 \\ 2 & 3 \\ 4 & 0 \end{matrix}}{9} = \frac{6 + 4 + 0 - 24 - 0 - 4}{9} = \frac{-18}{9} = -2$$

$$z = \frac{\begin{vmatrix} 1 & 2 & 1 \\ 2 & -1 & 3 \\ 4 & 1 & 0 \end{vmatrix}\begin{matrix} 1 & 2 \\ 2 & -1 \\ 4 & 1 \end{matrix}}{9} = \frac{0 + 24 + 2 - (-4) - 3 - 0}{9} = \frac{27}{9} = 3$$

As a check, we substitute these values into the first equation.

$$-1 + 2(-2) + 2(3) \stackrel{?}{=} 1, \quad 1 = 1. \text{ Thus, it checks.}$$

Example E   Solve the following system by determinants:

$$3x + 2y - 5z = -1$$
$$2x - 3y - z = 11$$
$$5x - 2y + 7z = 9$$

$$x = \frac{\begin{vmatrix} -1 & 2 & -5 \\ 11 & -3 & -1 \\ 9 & -2 & 7 \end{vmatrix} \begin{matrix} -1 & 2 \\ 11 & -3 \\ 9 & -2 \end{matrix}}{\begin{vmatrix} 3 & 2 & -5 \\ 2 & -3 & -1 \\ 5 & -2 & 7 \end{vmatrix} \begin{matrix} 3 & 2 \\ 2 & -3 \\ 5 & -2 \end{matrix}}$$

$$= \frac{21 - 18 + 110 - 135 + 2 - 154}{-63 - 10 + 20 - 75 - 6 - 28} = \frac{-174}{-162} = \frac{29}{27}$$

$$y = \frac{\begin{vmatrix} 3 & -1 & -5 \\ 2 & 11 & -1 \\ 5 & 9 & 7 \end{vmatrix} \begin{matrix} 3 & -1 \\ 2 & 11 \\ 5 & 9 \end{matrix}}{-162} = \frac{231 + 5 - 90 + 275 + 27 + 14}{-162}$$

$$= \frac{462}{-162} = -\frac{77}{27}$$

$$z = \frac{\begin{vmatrix} 3 & 2 & -1 \\ 2 & -3 & 11 \\ 5 & -2 & 9 \end{vmatrix} \begin{matrix} 3 & 2 \\ 2 & -3 \\ 5 & -2 \end{matrix}}{-162} = \frac{-81 + 110 + 4 - 15 + 66 - 36}{-162}$$

$$= \frac{48}{-162} = -\frac{8}{27}$$

Substituting into the second equation, we have

$$2\left(\frac{29}{27}\right) - 3\left(-\frac{77}{27}\right) - \left(-\frac{8}{27}\right) = \frac{58 + 231 + 8}{27} = \frac{297}{27}$$

$$= 11$$

which shows that the solution checks.

Example F

An 8% solution, a 10% solution, and a 20% solution of nitric acid are to be mixed in order to get 100 mL of a 12% solution. If the volume of acid from the 8% solution equals half the volume of acid from the other two solutions, how much of each is needed?

Let x = volume of 8% solution needed, y = volume of 10% solution needed, and z = volume of 20% solution needed.

We first use the fact that the sum of the volumes of the three solutions is 100 mL. This leads to the equation $x + y + z = 100$. Next we note that there are $0.08x$ milliliters of pure nitric acid from the first solution, $0.10y$ milliliters from the second, $0.20z$ milliliters from the third solution, and $0.12(100)$ mL in the final solution. This leads to the equation $0.08x + 0.10y + 0.20z = 12$. Finally, using the last stated condition, we have $0.08x = \frac{1}{2}(0.10y + 0.20z)$. These equations are then rewritten in standard form, simplified, and solved.

$$
\begin{aligned}
x + \quad y + \quad z &= 100 \\
0.08x + 0.10y + 0.20z &= \quad 12 \\
0.08x \qquad\qquad\qquad &= 0.05y + 0.10z
\end{aligned}
$$

$$
\begin{aligned}
x + \quad y + \quad z &= 100 \\
4x + \quad 5y + \quad 10z &= 600 \\
8x - \quad 5y - \quad 10z &= \quad 0
\end{aligned}
$$

$$
x = \frac{\begin{vmatrix} 100 & 1 & 1 \\ 600 & 5 & 10 \\ 0 & -5 & -10 \end{vmatrix} \begin{matrix} 100 & 1 \\ 600 & 5 \\ 0 & -5 \end{matrix}}{\begin{vmatrix} 1 & 1 & 1 \\ 4 & 5 & 10 \\ 8 & -5 & -10 \end{vmatrix} \begin{matrix} 1 & 1 \\ 4 & 5 \\ 8 & -5 \end{matrix}}
$$

$$
= \frac{-5000 + 0 - 3000 - 0 + 5000 + 6000}{-50 + 80 - 20 - 40 + 50 + 40} = \frac{3000}{60} = 50
$$

$$
y = \frac{\begin{vmatrix} 1 & 100 & 1 \\ 4 & 600 & 10 \\ 8 & 0 & -10 \end{vmatrix} \begin{matrix} 1 & 100 \\ 4 & 600 \\ 8 & 0 \end{matrix}}{60}
$$

$$
= \frac{-6000 + 8000 + 0 - 4800 - 0 + 4000}{60} = \frac{1200}{60} = 20
$$

$$z = \frac{\begin{vmatrix} 1 & 1 & 100 \\ 4 & 5 & 600 \\ 8 & -5 & 0 \end{vmatrix} \begin{matrix} 1 & 1 \\ 4 & 5 \\ 8 & -5 \end{matrix}}{60}$$

$$= \frac{0 + 4800 - 2000 - 4000 + 3000 - 0}{60} = \frac{1800}{60} = 30$$

Therefore, 50 mL of the 8% solution, 20 mL of the 10% solution, and 30 mL of the 20% solution are required to make the 12% solution. Substitution into the first equation shows that this answer is correct.

Additional techniques which are useful in solving systems of equations are taken up in Chapter 15. Methods of evaluating determinants which are particularly useful for higher order determinants are also discussed.

Exercises 4–7

In Exercises 1 through 12, evaluate the given third-order determinants.

1. $\begin{vmatrix} 5 & 4 & -1 \\ -2 & -6 & 8 \\ 7 & 1 & 1 \end{vmatrix}$

2. $\begin{vmatrix} -7 & 0 & 0 \\ 2 & 4 & 5 \\ 1 & 4 & 2 \end{vmatrix}$

3. $\begin{vmatrix} 8 & 9 & -6 \\ -3 & 7 & 2 \\ 4 & -2 & 5 \end{vmatrix}$

4. $\begin{vmatrix} -2 & 6 & -2 \\ 5 & -1 & 4 \\ 8 & -3 & -2 \end{vmatrix}$

5. $\begin{vmatrix} -3 & -4 & -8 \\ 5 & -1 & 0 \\ 2 & 10 & -1 \end{vmatrix}$

6. $\begin{vmatrix} 10 & 2 & -7 \\ -2 & -3 & 6 \\ 6 & 5 & -2 \end{vmatrix}$

7. $\begin{vmatrix} 4 & -3 & -11 \\ -9 & 2 & -2 \\ 0 & 1 & -5 \end{vmatrix}$

8. $\begin{vmatrix} 9 & -2 & 0 \\ -1 & 3 & -6 \\ -4 & -6 & -2 \end{vmatrix}$

9. $\begin{vmatrix} 5 & 4 & -5 \\ -3 & 2 & -1 \\ 7 & 1 & 3 \end{vmatrix}$

10. $\begin{vmatrix} 20 & 0 & -15 \\ -4 & 30 & 1 \\ 6 & -1 & 40 \end{vmatrix}$

11. $\begin{vmatrix} 0.1 & -0.2 & 0 \\ -0.5 & 1 & 0.4 \\ -2 & 0.8 & 2 \end{vmatrix}$

12. $\begin{vmatrix} 0.2 & -0.5 & -0.4 \\ 1.2 & 0.3 & 0.2 \\ -0.5 & 0.1 & -0.4 \end{vmatrix}$

In Exercises 13 through 28, solve the given systems of equations by use of determinants. (Exercises 15 through 28 are the same as Exercises 1 through 14 of Section 4–6.)

13. $2x + 3y + z = 4$
    $3x - z = -3$
    $x - 2y + 2z = -5$

14. $4x + y + z = 2$
    $2x - y - z = 4$
    $3y + z = 2$

15. $x + y + z = 2$
    $x - z = 1$
    $x + y = 1$

16. $x + y - z = -3$
    $x + z = 2$
    $2x - y + 2z = 3$

17. $2x + 3y + z = 2$
    $-x + 2y + 3z = -1$
    $-3x - 3y + z = 0$

18. $2x + y - z = 4$
    $4x - 3y - 2z = -2$
    $8x - 2y - 3z = 3$

19. $5x + 6y - 3z = 6$
    $4x - 7y - 2z = -3$
    $3x + y - 7z = 1$

20. $3r + s - t = 2$
    $r - 2s + t = 0$
    $4r - s + t = 3$

21. $2x - 2y + 3z = 5$
    $2x + y - 2z = -1$
    $4x - y - 3z = 0$

22. $2u + 2v + 3w = 0$
    $3u + v + 4w = 21$
    $-u - 3v + 7w = 15$

23. $3x - 7y + 3z = 6$
    $3x + 3y + 6z = 1$
    $5x - 5y + 2z = 5$

24. $8x + y + z = 1$
    $7x - 2y + 9z = -3$
    $4x - 6y + 8z = -5$

25. $x + 2y + 2z = 0$
    $2x + 6y - 3z = -1$
    $4x - 3y + 6z = -8$

26. $3x + 3y + z = 6$
    $2x + 2y - z = 9$
    $4x + 2y - 3z = 16$

27. $2x + 3y - 5z = 7$
    $4x - 3y - 2z = 1$
    $8x - y + 4z = 3$

28. $2x - 4y - 4z = 3$
    $3x + 8y + 2z = -11$
    $4x + 6y - z = -8$

In Exercises 29 through 32, solve the given problems by determinants. In Exercises 31 and 32, set up appropriate equations and then solve them.

29. In analyzing the forces on the bell-crank mechanism shown in Fig. 4–20, the following equations are obtained. Find the indicated forces.

$$A \quad - 0.6F = 80$$
$$B - 0.8F = 0$$
$$6A \quad - 10F = 0$$

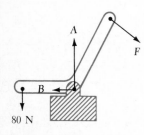

A

F

B

80 N

Figure 4–20

30. In applying Kirchhoff's laws (see Exercise 19 of Section 4–6) to the given electric circuit, these equations result.

$$I_A + I_B + I_C = 0$$
$$- 8I_B + 10I_C = 0$$
$$4I_A - 8I_B \quad = 6$$

Determine the indicated currents, in amperes. (See Fig. 4–21.)

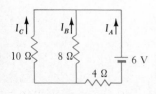

I_C I_B I_A

$10\ \Omega$ $8\ \Omega$

$4\ \Omega$

6 V

Figure 4–21

31. Twenty thousand dollars is invested part at 5.5%, part at 4.5%, and part at 4.0% (all of the amount is invested), yielding an annual interest of $1015. The income from the 5.5% investment yields $305 more annually than the other two investments combined. How much money is invested at each percent?

32. A person traveled a total of 1870 mi from the time he left home until he reached his destination. He averaged 40 mi/h while driving to the airport. The plane averaged 600 mi/h, and he spent twice as long in the plane as in his car. The taxi averaged 30 mi/h between the airport and the destination. Assuming 40 min were used in making connections, what were the times he spent in his car, in the plane, and in the taxi, if the trip took 5.5 h?

4–8 Exercises for Chapter 4

In Exercises 1 through 4, evaluate the given determinants.

1. $\begin{vmatrix} -2 & 5 \\ 3 & 1 \end{vmatrix}$
2. $\begin{vmatrix} 4 & 0 \\ -2 & -6 \end{vmatrix}$
3. $\begin{vmatrix} -8 & -3 \\ -1 & 4 \end{vmatrix}$
4. $\begin{vmatrix} 9 & -1 \\ 7 & -5 \end{vmatrix}$

In Exercises 5 through 8, find the slopes of the lines which pass through the given points.

5. $(2, 0), (4, -8)$
6. $(-1, -5), (-4, 4)$
7. $(4, -2), (-3, -4)$
8. $(-6, 1), (1, -4)$

In Exercises 9 through 12, find the slopes and y-intercepts of the lines with the given equations, and sketch the graphs.

9. $y = -2x + 4$
10. $y = \frac{2}{3}x - 3$
11. $8x - 2y = 5$
12. $3x = 8 + 3y$

In Exercises 13 through 20, solve the given systems of equations graphically.

13. $y = 2x - 4$
    $y = -\frac{3}{2}x + 3$
14. $y = -3x + 3$
    $y = 2x - 6$
15. $4x - y = 6$
    $3x + 2y = 12$

16. $2x - 5y = 10$
    $3x + y = 6$
17. $7x = 2y + 14$
    $y = -4x + 4$
18. $5x = 15 - 3y$
    $y = 6x - 12$

19. $3x + 4y = 6$
    $2x - 3y = 2$
20. $5x + 2y = 5$
    $2x - 4y = 3$

In Exercises 21 through 32, solve the given systems of equations algebraically.

21. $x + 2y = 5$
    $x + 3y = 7$
22. $2x - y = 7$
    $x + y = 2$
23. $4x + 3y = -4$
    $y = 2x - 3$

24. $x = -3y - 2$
    $-2x - 9y = 2$
25. $3x + 4y = 6$
    $9x + 8y = 11$
26. $3x - 6y = 5$
    $7x + 2y = 4$

27. $2x - 5y = 8$
    $5x - 3y = 7$
28. $3x + 4y = 8$
    $2x - 3y = 9$
29. $7x = 2y - 6$
    $7y = 12 - 4x$

30. $3y = 8 - 5x$
    $6x = 8y + 11$
31. $0.9x - 1.1y = 0.4$
    $0.6x - 0.3y = 0.5$
32. $0.42x - 0.56y = 1.26$
    $0.98x - 1.40y = -0.28$

In Exercises 33 through 44, solve the given systems of equations by use of determinants. (These systems are the same as for Exercises 21 through 32.)

33. $x + 2y = 5$
$ x + 3y = 7$

34. $2x - y = 7$
$ x + y = 2$

35. $4x + 3y = -4$
$ y = 2x - 3$

36. $x = -3y - 2$
$ -2x - 9y = 2$

37. $3x + 4y = 6$
$ 9x + 8y = 11$

38. $3x - 6y = 5$
$ 7x + 2y = 4$

39. $2x - 5y = 8$
$ 5x - 3y = 7$

40. $3x + 4y = 8$
$ 2x - 3y = 9$

41. $7x = 2y - 6$
$ 7y = 12 - 4x$

42. $3y = 8 - 5x$
$ 6x = 8y + 11$

43. $0.9x - 1.1y = 0.4$
$ 0.6x - 0.3y = 0.5$

44. $0.42x - 0.56y = 1.26$
$ 0.98x - 1.40y = -0.28$

In Exercises 45 through 48, evaluate the given determinants.

45. $\begin{vmatrix} 4 & -1 & 8 \\ -1 & 6 & -2 \\ 2 & 1 & -1 \end{vmatrix}$

46. $\begin{vmatrix} -5 & 0 & -5 \\ 2 & 3 & -1 \\ -3 & 2 & 2 \end{vmatrix}$

47. $\begin{vmatrix} -2 & -4 & 7 \\ 1 & 6 & -3 \\ -7 & 2 & -1 \end{vmatrix}$

48. $\begin{vmatrix} 3 & 2 & -1 \\ 0 & -3 & 4 \\ 3 & -4 & -2 \end{vmatrix}$

In Exercises 49 through 56, solve the given systems of equations algebraically.

49. $2x + y + z = 4$
$ x - 2y - z = 3$
$ 3x + 3y - 2z = 1$

50. $x + 2y + z = 2$
$ 3x - 6y + 2z = 2$
$ 2x - z = 8$

51. $3x + 2y + z = 1$
$ 9x - 4y + 2z = 8$
$ 12x - 18y = 17$

52. $2x + 2y - z = 2$
$ 3x + 4y + z = -4$
$ 5x - 2y - 3z = 5$

53. $2r + s + 2t = 8$
$ 3r - 2s - 4t = 5$
$ -2r + 3s + 4t = -3$

54. $2u + 2v - w = -2$
$ 4u - 3v + 2w = -2$
$ 8u - 4v - 3w = 13$

55. $4x + 6y - z = -2$
$ 3x - 5y + 4z = 8$
$ 6x + 4y + 2z = 5$

56. $3t + 2u + 6v = 3$
$ 4t - 3u + 2v = 13$
$ 5t + u + v = 0$

In Exercises 57 through 64, solve the given systems of equations by use of determinants. (These systems are the same as for Exercises 49 through 56.)

57. $2x + y + z = 4$
$ x - 2y - z = 3$
$ 3x + 3y - 2z = 1$

58. $x + 2y + z = 2$
$ 3x - 6y + 2z = 2$
$ 2x - z = 8$

59. $3x + 2y + z = 1$
$ 9x - 4y + 2z = 8$
$ 12x - 18y = 17$

60. $2x + 2y - z = 2$
$ 3x + 4y + z = -4$
$ 5x - 2y - 3z = 5$

61. $2r + s + 2t = 8$
$ 3r - 2s - 4t = 5$
$ -2r + 3s + 4t = -3$

62. $2u + 2v - w = -2$
$ 4u - 3v + 2w = -2$
$ 8u - 4v - 3w = 13$

63. $4x + 6y - z = -2$
$ 3x - 5y + 4z = 8$
$ 6x + 4y + 2z = 5$

64. $3t + 2u + 6v = 3$
$ 4t - 3u + 2v = 13$
$ 5t + u + v = 0$

In Exercises 65 through 68, let $1/x = u$ and $1/y = v$. Solve for u and v, and then solve for x and y. In this way we will see how to solve systems of equations involving reciprocals.

65. $\dfrac{1}{x} - \dfrac{1}{y} = \dfrac{1}{2}$ 66. $\dfrac{1}{x} + \dfrac{1}{y} = 3$ 67. $\dfrac{2}{x} + \dfrac{3}{y} = 3$ 68. $\dfrac{3}{x} - \dfrac{2}{y} = 4$

$\ \dfrac{1}{x} + \dfrac{1}{y} = \dfrac{1}{4}$ $\ \dfrac{2}{x} + \dfrac{1}{y} = 1$ $\ \dfrac{5}{x} - \dfrac{6}{y} = 3$ $\ \dfrac{2}{x} + \dfrac{4}{y} = 1$

In Exercises 69 and 70, determine the value of a which makes the system dependent. In Exercises 71 and 72, determine the value of a which makes the system inconsistent.

69. $3x - ay = 6$ 70. $5x + 20y = 15$
    $x + 2y = 2$ $2x + ay = 6$

71. $ax - 2y = 5$ 72. $2x - 5y = 7$
    $4x + 6y = 1$ $ax + 10y = 2$

Solve the systems of equations in Exercises 73 through 76 by any appropriate method.

73. In an experiment to determine the values of two electrical resistors, the following equations were determined:

$$2R_1 + 3R_2 = 16$$
$$3R_1 + 2R_2 = 19$$

Determine the resistances R_1 and R_2 (in ohms).

74. The production of nitric acid makes use of air and nitrogen compounds. In order to determine requirements as to size of equipment, a relationship between the air flow rate m (in moles per hour) and exhaust nitrogen rate n is often used. One particular operation produces the following equations:

$$1.58m + 41.5 = 38.0 + 2.00n$$
$$0.424\,m + 36.4 = 189 + 0.0728n$$

Solve for m and n.

75. A 20-ft crane arm with a supporting cable and with a 100-lb box suspended from its end has forces acting on it as shown in Fig. 4–22. Equations which can be used to find the forces are

$$F_1 + 2F_2 \qquad\ \ = 280$$
$$0.866F_1 \qquad - F_3 = 0$$
$$3F_1 - 4F_2 \qquad = 600$$

Find the forces, in pounds.

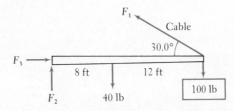

Figure 4–22

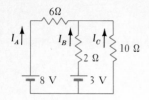

Figure 4–23

76. In applying Kirchhoff's laws (see Exercise 19 of Section 4–6) to the given electric circuit, these equations result:

$$I_A + I_B + I_C = 0$$
$$6I_A \qquad - 10I_C = 8$$
$$6I_A - 2I_B \qquad = 5$$

Determine the indicated currents (in amperes)(see Fig. 4–23).

In Exercises 77 through 84, set up systems of equations and solve by any appropriate method.

77. A plane traveled 2000 mi, with the wind, in 4 h, and it made the return trip in 5 h. Determine the speed of the plane and that of the wind.

78. The relation between Fahrenheit temperature, F, and Celsius temperature, C, can be indicated by $F = aC + b$. If 0°C is equivalent to 32°F and 100°C is equivalent to 212°F, find a and b.

79. In a shipment of two types of parts for a communication satellite, there was a total of 520 parts with a value of $1180. If type A parts cost $2 each and type B parts cost $3 each, how many of each type were in the shipment?

80. The *relative density* of an object may be defined as its weight in air divided by the difference of its weight in air and its weight when submerged in water. If the sum of the weights in water and in air of an object of relative density equal to 10 is 30 lb, what is its weight in air?

81. If a lever is balanced by placing a single support (fulcrum) under a certain point, then the product of a weight on one side of the fulcrum and its distance from the fulcrum equals the product of a weight on the other side of the fulcrum times its distance from the fulcrum. A certain lever is balanced if a weight of 80 lb is put at one end and a weight of 30 lb at the other end. If the 80-lb weight is moved 1 ft closer to the fulcrum, it requires 20 lb at the other end to maintain the balance. How long is the lever? (Neglect the weight of the lever itself.)

82. The weight of a lever may be considered to be at its center. A 20-ft lever of weight w is balanced on a fulcrum 8 ft from one end by a load L at that end. If 4 times the load is placed at that end, it requires a 20-lb weight at the other end to balance the lever. What is the initial load L and the weight w of the lever? (See Exercise 81.)

83. An alloy important in the manufacture of electric transformers contains nickel, iron, and molybdenum. The percentage of nickel is 1% less than five times the percentage of iron. The percentage of iron is 1% more than three times the percentage of molybdenum. Find the percentage of each metal in the alloy.

84. A businesswoman is interested in a building site for a new factory. She determines that at least 30 acres of land are necessary and that over 40 acres would be too expensive. She learns of a 160-acre tract of land, which has been subdivided according to terrain such that one portion is 16 acres less than the sum of the other two and that twice the area of the second portion is 8 acres more than the area of the third portion. Would any of the portions be of interest to her?

CHAPTER 5

Factoring and Fractions

In Chapter 1 we introduced certain fundamental algebraic operations. These have been sufficient for our purposes to this point. However, material which we shall encounter requires algebraic methods beyond those we have developed to this point. Therefore, in this chapter we shall develop certain additional basic algebraic topics, which in turn will allow us to develop other topics having technical and scientific applications.

5–1 Special Products

In working with algebraic expressions, we encounter certain types of products so frequently that we should become extremely familiar with them. These products are obtained by the methods of multiplication of algebraic expressions developed in Chapter 1 and are stated here in general form.

$$a(x + y) = ax + ay \tag{5-1}$$
$$(x + y)(x - y) = x^2 - y^2 \tag{5-2}$$
$$(x + y)^2 = x^2 + 2xy + y^2 \tag{5-3}$$
$$(x - y)^2 = x^2 - 2xy + y^2 \tag{5-4}$$
$$(x + a)(x + b) = x^2 + (a + b)x + ab \tag{5-5}$$
$$(ax + b)(cx + d) = acx^2 + (bc + ad)x + bd \tag{5-6}$$

These **special products** *should be known* **thoroughly** *so that the multi-plications represented are easily and clearly recognized. When this is the case, they allow us to perform many multiplications quickly and easily, often by inspection. We must realize that they are written in their most concise form. Any of the literal numbers appearing in these products may represent an expression which in turn represents a number.

Example A

Using Eq. (5–1) in the following product, we have

$$6(3r + 2s) = 6(3r) + 6(2s) = 18r + 12s$$

Using Eq. (5–2), we have

$$(3r + 2s)(3r - 2s) = (3r)^2 - (2s)^2 = 9r^2 - 4s^2$$

When we use Eq. (5–1) in the first illustration, we see that $a = 6$. In both illustrations $3r = x$ and $2s = y$.

Example B

Using Eqs. (5–3) and (5–4) in the following products, we have

$$(5a + 2)^2 = (5a)^2 + 2(5a)(2) + 2^2 = 25a^2 + 20a + 4$$
$$(5a - 2)^2 = (5a)^2 - 2(5a)(2) + 2^2 = 25a^2 - 20a + 4$$

 In these illustrations, we have let $x = 5a$ and $y = 2$. *It should also be emphasized that $(5a + 2)^2$ is **not** $(5a)^2 + 2^2$, or $25a^2 + 4$.* We must be careful to follow the correct form of Eqs. (5–3) and (5–4) properly and include the middle term, $20a$.

Example C

Using Eqs. (5–5) and (5–6) in the following products, we have

$$(x + 5)(x - 3) = x^2 + [5 + (-3)]x + (5)(-3) = x^2 + 2x - 15$$
$$(4x + 5)(2x - 3) = (4x)(2x) + [5(2) + 4(-3)]x + (5)(-3)$$
$$= 8x^2 - 2x - 15$$

Generally, when we use these special products, we do the middle step as shown in each of the examples above mentally and write down the result directly. This is indicated in the following example.

Example D

$$2(x - 6) = 2x - 12$$
$$(y - 5)(y + 5) = y^2 - 25$$
$$(3x - 2)^2 = 9x^2 - 12x + 4$$
$$(x - 4)(x + 7) = x^2 + 3x - 28$$

At times these special products may appear in combinations. When this happens, it may be necessary to indicate an intermediate step.

Example E

In expanding $7(a + 2)(a - 2)$, we use Eqs. (5–2) and (5–1), preferably in that order. Performing this operation, we have

$$7(a + 2)(a - 2) = 7(a^2 - 4) = 7a^2 - 28$$

Example F

In determining the product $(x + y - 2)^2$, we may group the quantity $(x + y)$ in an intermediate step. This leads to

$$(x + y - 2)^2 = [(x + y) - 2]^2 = (x + y)^2 - 2(x + y)(2) + 2^2$$
$$= x^2 + 2xy + y^2 - 4x - 4y + 4$$

In this example we used Eqs. (5–3) and (5–4).

There are four other special products which occur less frequently. However, they are sufficiently important that they should be readily recognized. They are shown in Eqs. (5–7) through (5–10).

$$(x + y)^3 = x^3 + 3x^2y + 3xy^2 + y^3 \qquad (5-7)$$
$$(x - y)^3 = x^3 - 3x^2y + 3xy^2 - y^3 \qquad (5-8)$$
$$(x + y)(x^2 - xy + y^2) = x^3 + y^3 \qquad (5-9)$$
$$(x - y)(x^2 + xy + y^2) = x^3 - y^3 \qquad (5-10)$$

The following examples illustrate the use of Eqs. (5–7) through (5–10).

Example G

$$(x + 4)^3 = x^3 + 3(x^2)(4) + 3(x)(4^2) + 4^3$$
$$= x^3 + 12x^2 + 48x + 64$$
$$(2x - 5)^3 = (2x)^3 - 3(2x)^2(5) + 3(2x)(5^2) - 5^3$$
$$= 8x^3 - 60x^2 + 150x - 125$$

Example H

$$(x + 3)(x^2 - 3x + 9) = x^3 + 3^3$$
$$= x^3 + 27$$
$$(x - 2)(x^2 + 2x + 4) = x^3 - 2^3$$
$$= x^3 - 8$$

Exercises 5–1

In Exercises 1 through 36, find the indicated products directly *by inspection*. It should not be necessary to write down intermediate steps.

1. $40(x - y)$ 2. $2x(a - 3)$ 3. $2x^2(x - 4)$

4. $3a^2(2a + 7)$ 5. $(y + 6)(y - 6)$ 6. $(s + 2t)(s - 2t)$

7. $(3v - 2)(3v + 2)$ 8. $(ab - c)(ab + c)$ 9. $(4x - 5y)(4x + 5y)$

10. $(7s + 2t)(7s - 2t)$ 11. $(12 + 5ab)(12 - 5ab)$ 12. $(2xy - 11)(2xy + 11)$

13. $(5f + 4)^2$ 14. $(i_1 + 3)^2$ 15. $(2x + 7)^2$

16. $(5a + 2b)^2$ 17. $(x - 1)^2$ 18. $(y - 6)^2$

19. $(4a + 7xy)^2$ 20. $(3x + 10y)^2$ 21. $(x - 2y)^2$

22. $(a - 5p)^2$ 23. $(6s - t)^2$ 24. $(3p - 4q)^2$

25. $(x + 1)(x + 5)$ 26. $(y - 8)(y + 5)$ 27. $(3 + c)(6 + c)$

28. $(1 - t)(7 - t)$ 29. $(3x - 1)(2x + 5)$ 30. $(2x - 7)(2x + 1)$

31. $(4x - 5)(5x + 1)$ 32. $(2y - 1)(3y - 1)$ 33. $(5v - 3)(4v + 5)$

34. $(7s + 6)(2s + 5)$ 35. $(3x + 7y)(2x - 9y)$ 36. $(8x - y)(3x + 4y)$

Use the special products of this section to determine the products of Exercises 37 through 60. You may need to write down one or two intermediate steps.

37. $2(x - 2)(x + 2)$ 38. $5(n - 5)(n + 5)$

39. $2a(2a - 1)(2a + 1)$ 40. $4c(2c - 3)(2c + 3)$

41. $6a(x + 2b)^2$ 42. $4y^2(y + 6)^2$

43. $5n^2(2n + 5)^2$ 44. $7r(5r + 2b)^2$

45. $3s(s - 6)^2$ 46. $8p(p - 7)^2$

47. $4a(2a - 3)^2$ 48. $6t^2(5t - 3s)^2$

49. $(x + y + 1)^2$ 50. $(x + 2 + 3y)^2$

51. $(3 - x - y)^2$ 52. $2(x - y + 1)^2$

53. $(5 - t)^3$ 54. $(2s + 3)^3$

55. $(2x + 5t)^3$ 56. $(x - 5y)^3$

57. $(x + 2)(x^2 - 2x + 4)$ 58. $(a - 3)(a^2 + 3a + 9)$

59. $(4 - 3x)(16 + 12x + 9x^2)$ 60. $(2x + 3a)(4x^2 - 6ax + 9a^2)$

Use the special products of this section to determine the products in Exercises 61 through 68. Each comes from the technical area indicated.

61. $R(i_1 + i_2)^2$ (electricity)

62. $V^2(V - b)^2$ (thermodynamics)

63. $4(p + DA)^2$ (photography)

64. $a^2(kV_1 + V_2)$ (chemistry)

65. $16(4 + t)(3 - t)$ (physics: motion)

66. $Fa(L - a)(L + a)$ (mechanics: force on a beam)

67. $P(1 + r)^3$ (business: compound interest)

68. $p^2(1 - p)^3$ (mathematics: probability)

5-2 Factoring: Common Factor and Difference of Squares

We often find that we want to determine what expressions can be multiplied together to equal a given algebraic expression. We know from Section 1–8 that when an algebraic expression is the product of two or more quantities, each of these quantities is called a **factor** of the expression. *Therefore, determining these factors, which is essentially reversing the process of finding a product, is called* **factoring.**

In our work on factoring we shall consider only the factoring of polynomials (see Section 1–10) which have integers as coefficients for all terms. Also, all factors will have integral coefficients. *A polynomial or a factor is called* **prime** *if it contains no factors other than* $+1$ *or* -1 *and plus or minus itself. We say that an expression is* **factored completely** *if it is expressed as a product of its prime factors.*

Example A

When we factor the expression $12x + 6x^2$ as

$$12x + 6x^2 = 2(6x + 3x^2)$$

we see that it has not been factored completely. The factor $6x + 3x^2$ is not prime, for it may be factored as

$$6x + 3x^2 = 3x(2 + x)$$

Therefore, the expression $12x + 6x^2$ is factored completely as

$$12x + 6x^2 = 6x(2 + x)$$

Here the factors x and $2 + x$ are prime. The numerical coefficient, 6, could be factored into $(2)(3)$, but it is normal practice not to factor numerical coefficients.

 To factor expressions easily, we must be familiar with algebraic multiplication, particularly the special products of the preceding section. The solution of factoring problems is heavily dependent on the recognition of special products. The special products also provide methods of checking answers and deciding whether or not a given factor is prime.

Often an algebraic expression contains a monomial that is common to each term of the expression. Therefore, in accordance with Eq. (5–1), *the first step in factoring any expression should be to factor out any* **common monomial factor** *that may exist.*

In Example A there is a common monomial factor of $6x$. However, the purpose of the example is to show the meaning of factors and complete factoring. The following three examples illustrate how a common monomial factor is identified and factored out of an expression.

Example B

In factoring the expression $6x - 2y$, we note that each term contains the factor 2. Therefore,

$$6x - 2y = 2(3x - y)$$

Here, 2 is the common monomial factor, and $2(3x - y)$ is the required factored form of $6x - 2y$.

In Example B we determined the common factor of 2 by inspection. This is normally the way in which the common factor is found. Once the common factor has been found, the other factor can be determined by dividing the original expression by the common factor.

Example C

Factor: $4ax^2 + 2ax$.

The numerical factor 2 and the literal factors a and x are common to each term. Therefore, the common monomial factor of $4ax^2 + 2ax$ is $2ax$. This means that

$$4ax^2 + 2ax = 2ax(2x + 1)$$

 Note the presence of the 1 in the factored form. When we divide $4ax^2 + 2ax$ by $2ax$ we get $2x + 1$. We may also reason that since each of the factors of the second term is also a factor of the common monomial factor, we must include the factor of 1. Otherwise, if the factored form is multiplied out, we would not obtain the proper expression.

Example D

Factor: $6a^5x^2 - 9a^3x^3 + 3a^3x^2$.

After inspecting each term, we determine that each contains a factor of 3, a^3, and x^2. Thus, the common monomial factor is $3a^3x^2$. This means that

$$6a^5x^2 - 9a^3x^3 + 3a^3x^2 = 3a^3x^2(2a^2 - 3x + 1)$$

Another important form for factoring is based on the special product of Eq. (5–2). In Eq. (5–2) we see that the product of the sum and the difference of two numbers results in the difference between the squares of the numbers. Therefore, *factoring the difference of two squares gives factors which are the sum and the difference of the numbers.*

Example E

In factoring $x^2 - 16$, we note that x^2 is the square of x and that 16 is the square of 4. Therefore,

$$x^2 - 16 = x^2 - 4^2 = (x + 4)(x - 4)$$

Usually in factoring an expression of this type we do not actually write out the middle step as shown.

Example F

Since $4x^2$ is the square of $2x$ and 9 is the square of 3, we may factor $4x^2 - 9$ as

$$4x^2 - 9 = (2x + 3)(2x - 3)$$

In the same way,

$$16x^4 - 25y^2 = (4x^2 + 5y)(4x^2 - 5y)$$

where we note that $16x^4 = (4x^2)^2$ and $25y^2 = (5y)^2$.

 As indicated previously, *if it is possible to factor out a common monomial factor, this factoring should be done first.* We should then inspect the resulting factors to see if further factoring can be done. It is possible, for example, that the resulting factor is a difference of squares. Thus, complete factoring often requires more than one step. Be sure to include all prime factors in writing the result.

Example G

In factoring $20x^2 - 45$, we note that there is a common factor of 5 in each term. Therefore, $20x^2 - 45 = 5(4x^2 - 9)$. However, the factor $4x^2 - 9$ itself is the difference of squares. Therefore, $20x^2 - 45$ is completely factored as

$$20x^2 - 45 = 5(4x^2 - 9) = 5(2x + 3)(2x - 3)$$

In factoring $x^4 - y^4$, we note that we have the difference of squares. Therefore, $x^4 - y^4 = (x^2 + y^2)(x^2 - y^2)$. However, the factor $x^2 - y^2$ is also the difference of squares. This means that

$$x^4 - y^4 = (x^2 + y^2)(x^2 - y^2) = (x^2 + y^2)(x + y)(x - y)$$

 The factor $x^2 + y^2$ is prime.

Exercises 5–2

In Exercises 1 through 36, factor the given expressions completely.

1. $6x + 6y$
2. $3a - 3b$
3. $5a - 5$
4. $2x^2 + 2$
5. $3x^2 - 9x$
6. $4s^2 + 20s$
7. $7b^2y - 28b$
8. $5a^2 - 20ax$
9. $12n^2 + 6n$
10. $18p^3 - 3p^2$
11. $2x + 4y - 8z$
12. $10a - 5b + 15c$
13. $3ab^2 - 6ab + 12ab^3$
14. $4pq - 14q^2 - 16pq^2$
15. $12pq^2 - 8pq - 28pq^3$
16. $27a^2b - 24ab - 9a$
17. $2a^2 - 2b^2 + 4c^2 - 6d^2$
18. $5a + 10ax - 5ay + 20az$
19. $x^2 - 4$
20. $r^2 - 25$
21. $100 - y^2$
22. $49 - z^2$
23. $36a^2 - 1$
24. $81z^2 - 1$

25. $81s^2 - 25t^2$ 26. $36s^2 - 121t^2$ 27. $144n^2 - 169p^4$

28. $36a^2b^2 - 169c^2$ 29. $2x^2 - 8$ 30. $5a^2 - 125$

31. $3x^2 - 27z^2$ 32. $4x^2 - 100y^2$ 33. $x^4 - 16$

34. $y^4 - 81$ 35. $x^8 - 1$ 36. $2x^4 - 8y^4$

In Exercises 37 through 44, the expressions are to be factored by a method known as **factoring by grouping.** An illustration of this method is

$$2x - 2y + ax - ay = (2x - 2y) + (ax - ay)$$
$$= 2(x - y) + a(x - y) = (2 + a)(x - y)$$

The terms are put into groups, a common factor is factored from each group, and then the factoring is continued.

37. $3x - 3y + bx - by$ 38. $am + an + cn + cm$

39. $2r + 2s + ar + as$ 40. $5p + 5q - ap - aq$

41. $a^2 + ax - ab - bx$ 42. $2y - y^2 - 6y^4 + 12y^3$

43. $x^3 + 3x^2 - 4x - 12$ 44. $x^3 - 5x^2 - x + 5$

Factor the expressions given in Exercises 45 through 52. Each comes from the technical area indicated.

45. $iR_1 + iR_2 + ir$ (electricity)   46. $P + Prt$ (business: interest)

47. $mv_1^2 - mv_2^2$ (mechanics: energy)   48. $K - Ka^2$ (chemistry)

49. $PbL^2 - Pb^3$ (architecture)   50. $kD^2 - 4kr^2$ (hydrodynamics)

51. The difference in the expressions for the volume of a cube with edge e and a rectangular solid of edges 2, 2 and e is $e^3 - 4e$. Factor this expression.

52. The difference of the energy radiated by an electric light filament at temperature T_2 and that radiated by a filament at temperature T_1 is given by the formula

$$R = kT_2^4 - kT_1^4$$

Factor the right-hand side of this formula.

5–3 Factoring Trinomials

In the previous section we introduced the concept of factoring and considered factoring based on special products of Eqs. (5–1) and (5–2). We now note that the special products formed from Eqs. (5–3) through (5–6) all result in trinomial (three-term) polynomials. Thus, trinomials of the types formed from these products are important expressions to be factored, and this section is devoted to them.

Factoring expressions based on the special product of Eq. (5–5) will result in factors of the form $x + a$ and $x + b$. The numbers a and b are found by analyzing the constant and the coefficient of x in the expression to be factored. As in Section 5–2, we shall consider only factors in which all terms have integral coefficients.

Example A

In factoring $x^2 + 3x + 2$, the constant term, 2, suggests that the only possibilities for a and b are 2 and 1. The plus sign before the 2 indicates that the sign before the 1 and the 2 in the factors must be the same, either both plus or both minus. Since *the coefficient of the middle term is the sum of a and b*, the plus sign before the 3 tells us that the sign before the 2 and 1 must be plus. Therefore,

$$x^2 + 3x + 2 = (x + 2)(x + 1)$$

In factoring $x^2 - 3x + 2$, the analysis is the same until we note that the middle term is negative. This tells us that both a and b are negative in this case. Therefore,

$$x^2 - 3x + 2 = (x - 2)(x - 1)$$

For a trinomial containing x^2 and 2 to be factorable, the middle term must be $3x$. No other combination of a and b gives the proper middle term. Therefore, the expression

$$x^2 + 4x + 2$$

cannot be factored. The a and b would have to be 2 and 1, but the middle term would not be $4x$.

Example B

In order to factor $x^2 + 7x - 8$, *we must find two integers whose **product** is -8 and whose **sum** is $+7$.* The possible factors of -8 are -8 and $+1$, $+8$ and -1, -4 and $+2$, and $+4$ and -2. Inspecting these we see that only $+8$ and -1 have the sum of $+7$. Therefore,

$$x^2 + 7x - 8 = (x + 8)(x - 1)$$

In the same way, we have

$$x^2 - x - 12 = (x - 4)(x + 3)$$

and

$$x^2 - 5xy + 6y^2 = (x - 3y)(x - 2y)$$

In the last illustration we note that we were to find second terms of each factor such that their product was $6y^2$ and sum was $-5xy$. Thus, each term must contain a factor of y.

In factoring a trinomial in which the second power term is x^2, we may find that the expression fits the form of Eq. (5–3) or Eq. (5–4), as well as Eq. (5–5). The following example illustrates this case.

Example C

In order to factor $x^2 + 10x + 25$, we must find two integers whose sum is $+10$ and whose product is $+25$. Since $5^2 = 25$ we note that this expression may fit the form of Eq. (5–3). This could be the case only if the first and third terms were perfect squares. We see that the sum of $+5$ and $+5$ is $+10$, which means

$$x^2 + 10x + 25 = (x + 5)(x + 5)$$

or

$$x^2 + 10x + 25 = (x + 5)^2$$

Factoring expressions based upon the special product of Eq. (5–6) often requires some trial and error. However, the amount of trial and error can be kept to a minimum with a careful analysis of the coefficients of x^2 and the constant. The coefficient of x^2 gives the possibilities for the coefficients a and c in the factors. The constant gives the possibilities for the numbers b and d in the factors. It is then necessary to try possible combinations to determine which combination provides the middle term of the given expression.

Example D

When factoring the expression $2x^2 + 11x + 5$, the coefficient 2 indicates that the only possibilities for the x-terms in the factors are $2x$ and x. The 5 indicates that only 5 and 1 may be the constants. Therefore, possible combinations are $2x + 1$ and $x + 5$ or $2x + 5$ and $x + 1$. The combination which gives the coefficient of the middle term, 11, is

$$2x^2 + 11x + 5 = (2x + 1)(x + 5)$$

According to this analysis, the expression $2x^2 + 10x + 5$ is not factorable, but the following expression is:

$$2x^2 + 7x + 5 = (2x + 5)(x + 1)$$

Example E

In factoring $4x^2 + 4x - 3$, the $4x^2$ indicates that $4x$ and x or $2x$ and $2x$ are possible. Also, the 3 indicates that only 3 and 1 are possible for constants. The minus sign with the 3 shows that the constants have different signs. The plus sign with the $4x$ tells us that the larger product of ad or bc is positive. This gives us combinations of $4x - 3$ and $x + 1$, $4x - 1$ and $x + 3$, or $2x + 3$ and $2x - 1$. The only one of these which has a middle term of $+4x$ is $2x + 3$ and $2x - 1$. Therefore,

$$4x^2 + 4x - 3 = (2x - 1)(2x + 3)$$

Example F

Other examples of factoring based upon the special product of Eq. (5–6) are as follows:

$$3x^2 - 13x + 12 = (3x - 4)(x - 3)$$
$$6s^2 + 19st - 20t^2 = (6s - 5t)(s + 4t)$$

In the first illustration, other possible factorizations of 12, such as 6×2 and 12×1, can be shown to give improper middle terms. In the second illustration, there are numerous possibilities for the combination of 6 and 20. *We must remember to check carefully that the* **middle term** *of the expression is the proper result of the factors we have chosen.*

Example G

In factoring $9x^2 - 6x + 1$, we note that $9x^2$ is the square of $3x$ and 1 is the square of 1. Therefore, we recognize that this expression might fit the perfect square form of Eq. (5–4). This leads us to factor it tentatively as

$$9x^2 - 6x + 1 = (3x - 1)^2$$

However, before we can be certain that this factorization is correct, we must check to see if the middle term of the expansion of $(3x - 1)^2$ is $-6x$. Since $-6x$ properly fits the form of Eq. (5–4), the factorization is correct.

In the same way, we have

$$36x^2 + 84xy + 49y^2 = (6x + 7y)^2$$

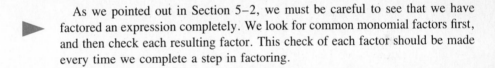

As we pointed out in Section 5–2, we must be careful to see that we have factored an expression completely. We look for common monomial factors first, and then check each resulting factor. This check of each factor should be made every time we complete a step in factoring.

Example H

When factoring $2x^2 + 6x - 8$, we first note the common monomial factor of 2. This leads to

$$2x^2 + 6x - 8 = 2(x^2 + 3x - 4)$$

We now notice that $x^2 + 3x - 4$ is also factorable. Therefore,

$$2x^2 + 6x - 8 = 2(x + 4)(x - 1)$$

Now each factor is prime.

Example I

When factoring $4x^3 + 18x^2 - 10x$, we must first see the factor $2x$ in each term. Factoring out this $2x$, we have

$$4x^3 + 18x^2 - 10x = 2x(2x^2 + 9x - 5)$$

Now we factor $2x^2 + 9x - 5$ as

$$2x^2 + 9x - 5 = (2x - 1)(x + 5)$$

Therefore,

$$4x^3 + 18x^2 - 10x = 2x(2x - 1)(x + 5)$$

Exercises 5–3

In Exercises 1 through 44, factor the given expressions completely.

1. $x^2 + 5x + 4$
2. $x^2 - 5x - 6$
3. $s^2 - s - 42$
4. $a^2 + 14a - 32$
5. $t^2 + 5t - 24$
6. $r^2 - 11r + 18$
7. $x^2 + 2x + 1$
8. $y^2 + 8y + 16$
9. $x^2 - 4x + 4$
10. $b^2 - 12b + 36$
11. $3x^2 - 5x - 2$
12. $2n^2 - 13n - 7$
13. $3y^2 - 8y - 3$
14. $5x^2 + 9x - 2$
15. $2s^2 + 13s + 11$
16. $7y^2 - 12y + 5$
17. $3f^2 - 16f + 5$
18. $5x^2 - 3x - 2$
19. $2t^2 + 7t - 15$
20. $3n^2 - 20n + 20$
21. $3t^2 - 7tu + 4u^2$
22. $3x^2 + xy - 14y^2$
23. $4x^2 - 3x - 7$
24. $2z^2 + 13z - 5$
25. $9x^2 + 7xy - 2y^2$
26. $4r^2 + 11rs - 3s^2$
27. $4m^2 + 20m + 25$
28. $16q^2 + 24q + 9$
29. $4x^2 - 12x + 9$
30. $a^2c^2 - 2ac + 1$
31. $9t^2 - 15t + 4$
32. $6x^2 + x - 12$
33. $8b^2 + 31b - 4$
34. $12n^2 + 8n - 15$
35. $4p^2 - 25pq + 6q^2$
36. $12x^2 + 4xy - 5y^2$
37. $12x^2 + 47xy - 4y^2$
38. $8r^2 - 14rs - 9s^2$
39. $2x^2 - 14x + 12$
40. $6y^2 - 33y - 18$
41. $4x^2 + 14x - 8$
42. $12x^2 + 22x - 4$
43. $x^3 + 4x^2 - 12x$
44. $6x^4 - 13x^3 + 5x^2$

In Exercises 45 through 48, factor the given expression by referring directly to the special products in Eqs. (5–7) through (5–10), respectively. These expressions are not trinomials, but their factorization depends on the proper recognition of their forms, as with Eqs. (5–3) and (5–4).

45. $x^3 + 3x^2 + 3x + 1$ 46. $x^3 - 6x^2 + 12x - 8$

47. $8x^3 + 1$ 48. $x^3 - 27$

Factor the expressions given in Exercises 49 through 54. Each comes from the technical area indicated.

49. $16t^2 - 32t - 128$ (physics: motion)

50. $2p^2 - 108p + 400$ (business)

51. $x^2 - 3Lx + 2L^2$ (mechanics: beams)

52. $V^2 - 2nBV + n^2B^2$ (chemistry)

53. $s^2 + 16s + 48$ (electricity)

54. $bT^2 - 40bT + 400b$ (thermodynamics)

5–4 Equivalent Fractions

When we deal with algebraic expressions, we must be able to work effectively with fractions. Since algebraic expressions are representations of numbers, the basic operations on fractions from arithmetic will form the basis of our algebraic operations. In this section we shall demonstrate a very important property of fractions, and in the following two sections we shall establish the basic algebraic operations with fractions.

This important property of fractions, often referred to as the **fundamental principle of fractions,** is that *the value of a fraction is unchanged if both numerator and denominator are multiplied or divided by the same number, provided this number is not zero.* Two fractions are said to be **equivalent** if one can be obtained from the other by use of the fundamental theorem.

Example A

If we multiply the numerator and denominator of the fraction $\frac{6}{8}$ by 2, we obtain the equivalent fraction $\frac{12}{16}$. If we divide the numerator and denominator of $\frac{6}{8}$ by 2, we obtain the equivalent fraction $\frac{3}{4}$. Therefore, the fractions $\frac{6}{8}$, $\frac{3}{4}$, and $\frac{12}{16}$ are equivalent.

Example B

We may write

$$\frac{ax}{2} = \frac{3a^2x}{6a}$$

since the fraction on the right is obtained from the fraction on the left by multiplying the numerator and the denominator by $3a$. Therefore, the fractions are equivalent.

One of the most important operations to be performed on a fraction is that of reducing it to its **simplest form,** or **lowest terms.** *A fraction is said to be in its simplest form if the numerator and the denominator have no common factors other than +1 or −1.* In reducing a fraction to its simplest form, we use the fundamental theorem by dividing both the numerator and the denominator by all factors which are common to each. (It will be assumed throughout this text that if any of the literal symbols were to be evaluated, numerical values would be restricted so that none of the denominators would be zero. Thereby, we avoid the undefined operation of division by zero.)

Example C

In order to reduce the fraction

$$\frac{16ab^3c^2}{24ab^2c^5}$$

to its lowest terms, we note that both the numerator and the denominator contain the factor $8ab^2c^2$. Therefore, we may write

$$\frac{16ab^3c^2}{24ab^2c^5} = \frac{2b(8ab^2c^2)}{3c^3(8ab^2c^2)} = \frac{2b}{3c^3}$$

Here we divided out the common factor. The resulting fraction is in lowest terms, since there are no common factors in the numerator and the denominator other than +1 or −1.

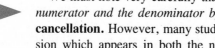 We must note very carefully that in simplifying fractions, *we divide both the numerator and the denominator by the common **factor.** This process is called* **cancellation.** However, many students are tempted to try to remove any expression which appears in both the numerator and the denominator. If a *term* is removed in this way, it is an incorrect application of the cancellation process. The following example illustrates this common error in the simplification of fractions.

Example D

When simplifying the expression

$$\frac{x^2(x-2)}{x^2-4}$$

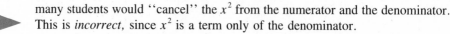 many students would ''cancel'' the x^2 from the numerator and the denominator. This is *incorrect,* since x^2 is a term only of the denominator.

In order to simplify the above fraction properly, we should factor the denominator. We obtain

$$\frac{x^2(x-2)}{(x-2)(x+2)} = \frac{x^2}{x+2}$$

Here, the common *factor* $x-2$ has been divided out.

The following examples illustrate the proper simplification of fractions.

Example E

$$\frac{2a}{2ax} = \frac{1}{x}$$

We divide out the common factor of $2a$.

$$\frac{2a}{2a + x}$$

 This cannot be reduced, since *there are no common **factors** in the numerator and the denominator*.

Example F

$$\frac{2x^2 + 8x}{x + 4} = \frac{2x(x + 4)}{x + 4} = \frac{2x}{1} = 2x$$

In this simplification, the numerator and the denominator were each divided by $x + 4$ after the numerator was factored. Since the only remaining factor in the denominator after the division is 1, it generally is not written in the final answer. Another way of writing the denominator is $1(x + 4)$, which shows the **factor** of 1 more clearly.

Example G

$$\frac{x^2 - 4x + 4}{x^2 - 4} = \frac{(x - 2)(x - 2)}{(x + 2)(x - 2)} = \frac{x - 2}{x + 2}$$

In this simplification, the numerator and the denominator have each been factored first and then the common factor $x - 2$ has been divided out. In the final form, neither the x's nor 2s may be canceled, since they are not common **factors.**

Example H

$$\frac{4x^2 + 14x - 30}{8x - 12} = \frac{2(2x^2 + 7x - 15)}{4(2x - 3)} = \frac{2(2x - 3)(x + 5)}{4(2x - 3)} = \frac{x + 5}{2}$$

Here, the factors common to the numerator and the denominator are 2 and $(2x - 3)$.

In simplifying fractions we must be able to distinguish between factors which differ only in *sign.* Since $-(y - x) = -y + x = x - y$, we have

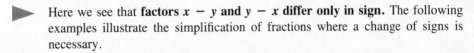

$$x - y = -(y - x) \tag{5–11}$$

Here we see that **factors $x - y$ and $y - x$ differ only in sign.** The following examples illustrate the simplification of fractions where a change of signs is necessary.

Example I

$$\frac{x^2 - 1}{1 - x} = \frac{(x - 1)(x + 1)}{-(x - 1)} = \frac{x + 1}{-1} = -(x + 1)$$

In the second fraction, we replaced $1 - x$ with the equal expression $-(x - 1)$. In the third fraction the common factor $x - 1$ was divided out. Finally, we expressed the result in the more convenient form by dividing $x + 1$ by -1, which makes the quantity $x + 1$ negative.

Example J

$$\frac{2x^3 - 32x}{20 + 7x - 3x^2} = \frac{2x(x^2 - 16)}{(4 - x)(5 + 3x)} = \frac{2x(x - 4)(x + 4)}{-(x - 4)(3x + 5)}$$

$$= -\frac{2x(x + 4)}{3x + 5}$$

Again, the factor $4 - x$ has been replaced by the equal expression $-(x - 4)$. This allows us to recognize the common factor of $x - 4$. Notice also that the order of the terms of the factor $5 + 3x$ has been changed to $3x + 5$. This is merely an application of the commutative law of addition.

Exercises 5–4

In Exercises 1 through 8, multiply the numerator and the denominator of each of the given fractions by the given factor and obtain an equivalent fraction.

1. $\frac{2}{3}$ (by 7)

2. $\frac{7}{5}$ (by 9)

3. $\frac{ax}{y}$ (by $2x$)

4. $\frac{2x^2y}{3n}$ (by $2xn^2$)

5. $\frac{2}{x + 3}$ (by $x - 2$)

6. $\frac{7}{a - 1}$ (by $a + 2$)

7. $\frac{a(x - y)}{x - 2y}$ (by $x + y$)

8. $\frac{x - 1}{x + 1}$ (by $x - 1$)

In Exercises 9 through 16, divide the numerator and the denominator of each of the given fractions by the given factor and obtain an equivalent fraction.

9. $\frac{28}{44}$ (by 4)

10. $\frac{25}{65}$ (by 5)

11. $\frac{4x^2y}{8xy^2}$ (by $2x$)

12. $\frac{6a^3b^2}{9a^5b^4}$ (by $3a^2b^2$)

13. $\frac{2(x - 1)}{(x - 1)(x + 1)}$ (by $x - 1$)

14. $\frac{(x + 5)(x - 3)}{3(x + 5)}$ (by $x + 5$)

15. $\frac{x^2 - 3x - 10}{2x^2 + 3x - 2}$ (by $x + 2$)

16. $\frac{6x^2 + 13x - 5}{6x^3 - 2x^2}$ (by $3x - 1$)

In Exercises 17 through 48, reduce each fraction to its simplest form.

17. $\dfrac{2a}{8a}$

18. $\dfrac{6x}{15x}$

19. $\dfrac{18x^2y}{24xy}$

20. $\dfrac{2a^2xy}{6axyz^2}$

21. $\dfrac{a + b}{5a^2 + 5ab}$

22. $\dfrac{t - a}{t^2 - a^2}$

23. $\dfrac{6a - 4b}{4a - 2b}$

24. $\dfrac{5r - 20s}{10r - 5s}$

25. $\dfrac{4x^2 + 1}{4x^2 - 1}$

26. $\dfrac{x^2 - y^2}{x^2 + y^2}$

27. $\dfrac{3x^2 - 6x}{x - 2}$

28. $\dfrac{5x^2 + 15x}{x + 3}$

29. $\dfrac{2y + 3}{4y^3 + 6y^2}$

30. $\dfrac{3t - 6}{4t^3 - 8t^2}$

31. $\dfrac{x^2 - 8x + 16}{x^2 - 16}$

32. $\dfrac{4a^2 + 12ab + 9b^2}{4a^2 + 6ab}$

33. $\dfrac{2x^2 + 5x - 3}{x^2 + 11x + 24}$

34. $\dfrac{3y^3 + 7y^2 + 4y}{y^2 + 5y + 4}$

35. $\dfrac{5x^2 - 6x - 8}{x^3 + x^2 - 6x}$

36. $\dfrac{4r^2 - 8rs - 5s^2}{6r^2 - 17rs + 5s^2}$

37. $\dfrac{x^4 - 16}{x + 2}$

38. $\dfrac{2x^2 - 8}{4x + 8}$

39. $\dfrac{x^2y^4 - x^4y^2}{y^2 - 2xy + x^2}$

40. $\dfrac{8x^3 + 8x^2 + 2x}{4x + 2}$

41. $\dfrac{(x - 1)(3 + x)}{(3 - x)(1 - x)}$

42. $\dfrac{(2x - 1)(x + 6)}{(x - 3)(1 - 2x)}$

43. $\dfrac{y - x}{2x - 2y}$

44. $\dfrac{x^2 - y^2}{y - x}$

45. $\dfrac{2x^2 - 9x + 4}{4x - x^2}$

46. $\dfrac{3a^2 - 13a - 10}{5 + 4a - a^2}$

47. $\dfrac{(x + 5)(x - 2)(x + 2)(3 - x)}{(2 - x)(5 - x)(3 + x)(2 + x)}$

48. $\dfrac{(2x - 3)(3 - x)(x - 7)(3x + 1)}{(3x + 2)(3 - 2x)(x - 3)(7 + x)}$

In Exercises 49 through 52, reduce each fraction to its simplest form. This will require the use of Eqs. (5–7) through (5–10).

49. $\dfrac{x^3 - y^3}{x^2 - y^2}$

50. $\dfrac{x^3 - 8}{x^2 + 2x + 4}$

51. $\dfrac{x^3 + 3x^2 + 3x + 1}{x^3 + 1}$

52. $\dfrac{a^3 - 6a^2 + 12a - 8}{a^2 - 4a + 4}$

In Exercises 53 through 56, determine which fractions are in simplest form.

53. (a) $\dfrac{x^2(x + 2)}{x^2 + 4}$ (b) $\dfrac{x^4 + 4x^2}{x^4 - 16}$

54. (a) $\dfrac{2x + 3}{2x + 6}$ (b) $\dfrac{2(x + 6)}{2x + 6}$

55. (a) $\dfrac{x^2 - x - 2}{x^2 - x}$ (b) $\dfrac{x^2 - x - 2}{x^2 + x}$

56. (a) $\dfrac{x^3 - x}{1 - x}$ (b) $\dfrac{2x^2 + 4x}{2x^2 + 4}$

5–5 Multiplication and Division of Fractions

From arithmetic we recall that *the product of two fractions is a fraction whose numerator is the product of the numerators and whose denominator is the product of the denominators of the given fractions.* Also, we recall that *we can find the quotient of two fractions by inverting the divisor and proceeding as in multiplication.* Symbolically, multiplication of fractions is indicated by

$$\frac{a}{b} \cdot \frac{c}{d} = \frac{ac}{bd}$$

and division is indicated by

$$\frac{\dfrac{a}{b}}{\dfrac{c}{d}} = \frac{a}{b} \cdot \frac{d}{c} = \frac{ad}{bc}$$

The rule for division may be verified by use of the fundamental principle of fractions. By multiplying the numerator and denominator of the fraction

$$\frac{\dfrac{a}{b}}{\dfrac{c}{d}} \quad \text{by} \quad \frac{d}{c} \quad \text{we obtain} \quad \frac{\dfrac{a}{b} \cdot \dfrac{d}{c}}{\dfrac{c}{d} \cdot \dfrac{d}{c}}$$

In the resulting denominator,

$$\frac{c}{d} \cdot \frac{d}{c}$$

becomes 1, and therefore the fraction is written as *ad / bc*.

Example A

$$\frac{3}{5} \cdot \frac{2}{7} = \frac{(3)(2)}{(5)(7)} = \frac{6}{35}$$

$$\frac{3a}{5b} \cdot \frac{15b^2}{a} = \frac{(3a)(15b^2)}{(5b)(a)} = \frac{45ab^2}{5ab} = \frac{9b}{1} = 9b$$

In the second illustration, we have divided out the common factor of $5ab$ to reduce the resulting fraction to its simplest form.

We shall usually want to express the result in its simplest form, which is generally its most useful form. Since all factors in the numerators and all factors in the denominators are to be multiplied, we should *first only* **indicate** *the multiplication and then factor the numerator and the denominator.* In this way we can easily identify any factors common to both. If we were to multiply out the numerator and the denominator before factoring, it is very possible that we would be unable to factor the result to simplify it. The following example illustrates this point.

Example B

In performing the multiplication

$$\frac{3(x - y)}{(x - y)^2} \cdot \frac{(x^2 - y^2)}{6x + 9y}$$

if we multiply out the numerators and the denominators before performing any factoring, we would have to simplify the fraction

$$\frac{3x^3 - 3x^2y - 3xy^2 + 3y^3}{6x^3 - 3x^2y - 12xy^2 + 9y^3}$$

It is possible to factor the numerator and the denominator, but finding any common factors this way is very difficult. If we first indicate the multiplications and then factor completely, we have

$$\frac{3(x - y)}{(x - y)^2} \cdot \frac{(x^2 - y^2)}{6x + 9y} = \frac{3(x - y)(x^2 - y^2)}{(x - y)^2(6x + 9y)} = \frac{3(x - y)(x + y)(x - y)}{(x - y)^2(3)(2x + 3y)}$$

$$= \frac{3(x - y)^2(x + y)}{3(x - y)^2(2x + 3y)}$$

$$= \frac{x + y}{2x + 3y}$$

The common factor of $3(x - y)^2$ is readily recognized using this procedure.

Example C

$$\frac{2x - 4}{4x + 12} \cdot \frac{2x^2 + x - 15}{3x - 1} = \frac{2(x - 2)(2x - 5)(x + 3)}{4(x + 3)(3x - 1)}$$

$$= \frac{(x - 2)(2x - 5)}{2(3x - 1)}$$

Here the common factor is $2(x + 3)$. It is permissible to multiply out the final form of the numerator and the denominator, but it is often preferable to leave the numerator and denominator in factored form, as indicated.

The following examples illustrate the division of fractions.

Example D

$$\frac{6x}{7} \div \frac{5}{3} = \frac{6x}{7} \cdot \frac{3}{5} = \frac{18x}{35}$$

$$\frac{\dfrac{3a^2}{5c}}{\dfrac{2c^2}{a}} = \frac{3a^2}{5c} \cdot \frac{a}{2c^2} = \frac{3a^3}{10c^3}$$

Example E

$$\frac{x + y}{3} \div \frac{2x + 2y}{6x + 15y} = \frac{x + y}{3} \cdot \frac{6x + 15y}{2x + 2y} = \frac{(x + y)(3)(2x + 5y)}{3(2)(x + y)}$$

$$= \frac{2x + 5y}{2}$$

Example F

$$\frac{\dfrac{4 - x^2}{x^2 - 3x + 2}}{\dfrac{x + 2}{x^2 - 9}} = \frac{4 - x^2}{x^2 - 3x + 2} \cdot \frac{x^2 - 9}{x + 2}$$

$$= \frac{(2 - x)(2 + x)(x - 3)(x + 3)}{(x - 2)(x - 1)(x + 2)}$$

$$= \frac{-(x - 2)(x + 2)(x - 3)(x + 3)}{(x - 2)(x - 1)(x + 2)}$$

$$= -\frac{(x - 3)(x + 3)}{x - 1} \text{ or } \frac{(x - 3)(x + 3)}{1 - x}$$

Note the use of Eq. (5–11) in the simplification and in expressing an alternate form of the result. The factor $2 - x$ was replaced by its equivalent $-(x - 2)$, and then $x - 1$ was replaced by $-(1 - x)$.

Exercises 5–5

In Exercises 1 through 32, perform the indicated operations and simplify.

1. $\dfrac{3}{8} \cdot \dfrac{2}{7}$ 2. $\dfrac{11}{5} \cdot \dfrac{13}{33}$ 3. $\dfrac{4x}{3y} \cdot \dfrac{9y^2}{2}$ 4. $\dfrac{18sy^3}{ax^2} \cdot \dfrac{(ax)^2}{3s}$

5. $\dfrac{2}{9} \div \dfrac{4}{7}$ 6. $\dfrac{5}{16} \div \dfrac{25}{13}$ 7. $\dfrac{xy}{az} \div \dfrac{bz}{ay}$ 8. $\dfrac{sr^2}{2t} \div \dfrac{st}{4}$

9. $\dfrac{4x + 12}{5} \cdot \dfrac{15t}{3x + 9}$ 10. $\dfrac{y^2 + 2y}{6z} \cdot \dfrac{z^3}{y^2 - 4}$

11. $\dfrac{u^2 - v^2}{u + 2v} \cdot \dfrac{3u + 6v}{u - v}$ 12. $(x - y)\dfrac{x + 2y}{x^2 - y^2}$

13. $\dfrac{2a + 8}{15} \div \dfrac{a^2 + 8a + 16}{25}$ 14. $\dfrac{a^2 - a}{3a + 9} \div \dfrac{a^2 - 2a + 1}{a^2 - 9}$

15. $\dfrac{x^2 - 9}{x} \div (x + 3)^2$ 16. $\dfrac{9x^2 - 16}{x + 1} \div (4 - 3x)$

17. $\dfrac{3ax^2 - 9ax}{10x^2 + 5x} \cdot \dfrac{2x^2 + x}{a^2x - 3a^2}$ 18. $\dfrac{2x^2 - 18}{x^3 - 25x} \cdot \dfrac{3x - 15}{2x^2 + 6x}$

19. $\dfrac{x^4 - 1}{8x + 16} \cdot \dfrac{2x^2 - 8x}{x^3 + x}$ 20. $\dfrac{2x^2 - 4x - 6}{x^2 - 3x} \cdot \dfrac{x^3 - 4x^2}{4x^2 - 4x - 8}$

21. $\dfrac{ax + x^2}{2b - cx} \div \dfrac{a^2 + 2ax + x^2}{2bx - cx^2}$ 22. $\dfrac{x^2 - 11x + 28}{x + 3} \div \dfrac{x - 4}{x + 3}$

23. $\dfrac{35a + 25}{12a + 33} \div \dfrac{28a + 20}{36a + 99}$

24. $\dfrac{2a^3 + a^2}{2b^3 + b^2} \div \dfrac{2ab + a}{2ab + b}$

25. $\dfrac{x^2 - 6x + 5}{4x^2 - 17x - 15} \cdot \dfrac{6x + 21}{2x^2 + 5x - 7}$

26. $\dfrac{x^2 + 5x}{3x^2 + 8x - 4} \cdot \dfrac{2x^2 - 8}{x^3 + 2x^2 - 15x}$

27. $\dfrac{7x^2 + 27x - 4}{6x^2 + x - 15} \div \dfrac{4x^2 + 17x + 4}{8x^2 - 10x - 3}$

28. $\dfrac{4x^3 - 9x}{8x^2 + 10x - 3} \div \dfrac{2x^3 - 3x^2}{8x^2 + 18x - 5}$

29. $\dfrac{7x^2}{3a} \div \left(\dfrac{a}{x} \cdot \dfrac{a^2 x}{x^2} \right)$

30. $\left(\dfrac{3u}{8v^2} \div \dfrac{9u^2}{2w^2} \right) \cdot \dfrac{2u^4}{15vw}$

31. $\left(\dfrac{4t^2 - 1}{t - 5} \div \dfrac{2t + 1}{2t} \right) \cdot \dfrac{2t^2 - 50}{4t^2 + 4t + 1}$

32. $\dfrac{2x^2 - 5x - 3}{x - 4} \div \left(\dfrac{x - 3}{x^2 - 16} \cdot \dfrac{1}{3 - x} \right)$

In Exercises 33 through 36, perform the indicated operations and simplify. Exercises 33 and 34 require the use of Eqs. (5–7) through (5–10), and Exercises 35 and 36 require the use of factoring by grouping.

33. $\dfrac{x^3 - y^3}{2x^2 - 2y^2} \cdot \dfrac{x^2 + 2xy + y^2}{x^2 + xy + y^2}$

34. $\dfrac{x^3 + 3x^2 + 3x + 1}{6x - 6} \div \dfrac{5x + 5}{x^2 - 1}$

35. $\dfrac{ax + bx + ay + by}{p - q} \cdot \dfrac{3p^2 + 4pq - 7q^2}{a + b}$

36. $\dfrac{x^4 + x^5 - 1 - x}{x - 1} \div \dfrac{x + 1}{x}$

In Exercises 37 through 40, solve the given problems.

37. In analyzing the stress on a rotating hoop, the expression

$$\dfrac{w\pi rbt}{2btg} \left(\dfrac{2r}{12\pi} \right) \left(\dfrac{144v^2}{r^2} \right)$$

is used. Simplify this expression.

38. In finding the center of mass of a semicircular area, the expression

$$\dfrac{4\pi r^3}{3} \div \left(\dfrac{\pi r^2}{2} \cdot 2\pi \right)$$

is used. Simplify this expression.

39. A rectangular metal plate expands when heated. For small values of Celsius temperature, the length and width of a certain plate, as functions of the temperature, are

$$\dfrac{20{,}000 + 300T + T^2}{1600 - T^2} \quad \text{and} \quad \dfrac{16{,}000 + 360T - T^2}{400 + 2T}$$

respectively. Find the resulting expression for the area of the plate as a function of the temperature.

40. The current in a simple electric circuit is the voltage in the circuit divided by the resistance. Given that the voltage and resistance in a certain circuit are expressed as functions of time

$$V = \dfrac{5t + 10}{2t + 1} \quad \text{and} \quad R = \dfrac{t^2 + 4t + 4}{2t}$$

find the formula for the current as a function of time.

5–6 Addition and Subtraction of Fractions

From arithmetic we recall that *the sum of a set of fractions that all have the same denominator is the sum of the numerators divided by the common denominator.* Since algebraic expressions represent numbers, this fact is also true in algebra. Addition and subtraction of such fractions are illustrated in the following example.

Example A

$$\frac{5}{9} + \frac{2}{9} - \frac{4}{9} = \frac{5 + 2 - 4}{9} = \frac{3}{9} = \frac{1}{3}$$

$$\frac{b}{ax} - \frac{1}{ax} + \frac{2b - 1}{ax} = \frac{b - 1 + (2b - 1)}{ax} = \frac{b - 1 + 2b - 1}{ax}$$

$$= \frac{3b - 2}{ax}$$

If the fractions to be combined do not all have the same denominator, we must first change each to an equivalent fraction so that the resulting fractions do have the same denominator. Normally *the denominator which is most convenient and useful is the* **lowest common denominator.** This is the product of all of the prime factors which appear in the denominators, with each factor raised to the highest power to which it appears in any one of the denominators. Thus, *the lowest common denominator is the simplest algebraic expression into which all the given denominators will divide exactly.* The following two examples illustrate the method used in finding the lowest common denominator of a set of fractions.

Example B

Find the lowest common denominator of the fractions

$$\frac{3}{4a^2b}, \qquad \frac{5}{6ab^3}, \qquad \text{and} \qquad \frac{1}{4ab^2}$$

Expressing each of the denominators in terms of powers of the prime factors, we have

$$4a^2b = 2^2a^2b, \qquad 6ab^3 = 2 \cdot 3 \cdot ab^3, \quad \text{and} \quad 4ab^2 = 2^2ab^2$$

The prime factors to be considered are 2, 3, a, and b. The largest exponent of 2 which appears is 2. This means that 2^2 is a factor of the lowest common denominator. The largest exponent of 3 which appears is 1 (understood in the second denominator). Therefore, 3 is a factor of the lowest common denominator. The largest exponent of a which appears is 2, and the largest exponent of b which appears is 3. Thus, a^2 and b^3 are factors of the lowest common denominator. Therefore, the lowest common denominator of the fractions is $2^2 \cdot 3 \cdot a^2b^3 = 12a^2b^3$. This is the simplest expression into which ***each*** of the denominators above will divide exactly.

Example C

Find the lowest common denominator of the following fractions:

$$\frac{x-4}{x^2-2x+1}, \quad \frac{1}{x^2-1}, \quad \frac{x+3}{x^2-x}$$

Factoring each of the denominators, we find that the fractions are

$$\frac{x-4}{(x-1)^2}, \quad \frac{1}{(x-1)(x+1)}, \quad \text{and} \quad \frac{x+3}{x(x-1)}$$

The factor $(x-1)$ appears in all of the denominators. It is squared in the denominator of the first fraction and appears to the first power only in the other two fractions. Thus, we must have $(x-1)^2$ as a factor of the common denominator. We do not need a higher power of $x-1$ since, as far as this factor is concerned, each denominator will divide into it exactly. Next, the second denominator contains a factor of $(x+1)$. Therefore, the common denominator must also contain a factor of $(x+1)$; otherwise, the second denominator would not divide into it exactly. Finally, the third denominator indicates that a factor of x is also required in the common denominator. The lowest common denominator is, therefore, $x(x+1)(x-1)^2$. All three denominators will divide exactly into this expression, and there is no simpler expression for which this is true.

Once we have found the lowest common denominator for the fractions, we multiply the numerator and denominator of each fraction by the proper quantity to make the resulting denominator in each case the common denominator. After this step, it is necessary only to add the numerators, place this result over the common denominator, and simplify.

Example D

Combine $\dfrac{2}{3r^2} + \dfrac{4}{rs^3} - \dfrac{5}{3s}$.

By looking at the denominators, we see that the factors necessary in the lowest common denominator are 3, r, and s. The 3 appears only to the first power, the largest exponent of r is 2, and the largest exponent of s is 3. Therefore, the lowest common denominator is $3r^2s^3$. We now wish to write each fraction with this quantity as the denominator. Since the denominator of the first fraction already contains factors of 3 and r^2, it is necessary to introduce the factor of s^3. In other words, we must multiply the numerator and denominator of this fraction by s^3. For similar reasons, we must multiply the numerators and the denominators of the second and third fractions by $3r$ and r^2s^2, respectively. This leads to

$$\frac{2}{3r^2} + \frac{4}{rs^3} - \frac{5}{3s} = \frac{2(s^3)}{(3r^2)(s^3)} + \frac{4(3r)}{(rs^3)(3r)} - \frac{5(r^2s^2)}{(3s)(r^2s^2)}$$

$$= \frac{2s^3}{3r^2s^3} + \frac{12r}{3r^2s^3} - \frac{5r^2s^2}{3r^2s^3}$$

$$= \frac{2s^3 + 12r - 5r^2s^2}{3r^2s^3}$$

Example E

$$\frac{a}{x-1} + \frac{a}{x+1} = \frac{a(x+1)}{(x-1)(x+1)} + \frac{a(x-1)}{(x+1)(x-1)}$$

$$= \frac{ax + a + ax - a}{(x+1)(x-1)}$$

$$= \frac{2ax}{(x+1)(x-1)}$$

When we multiply each fraction by the quantity required to obtain the proper denominator, we do not actually have to write the common denominator under each numerator. Placing all the products which appear in the numerators over the common denominator is sufficient. Hence the illustration in this example would appear as

$$\frac{a}{x-1} + \frac{a}{x+1} = \frac{a(x+1) + a(x-1)}{(x-1)(x+1)} = \frac{ax + a + ax - a}{(x-1)(x+1)}$$

$$= \frac{2ax}{(x-1)(x+1)}$$

Example F

$$\frac{x-1}{x^2-25} - \frac{2}{x-5} = \frac{(x-1) - 2(x+5)}{(x-5)(x+5)} = \frac{x - 1 - 2x - 10}{(x-5)(x+5)}$$

$$= \frac{-(x+11)}{(x-5)(x+5)}$$

Example G

$$\frac{3x}{x^2-x-12} - \frac{x-1}{x^2-8x+16} - \frac{6-x}{2x-8}$$

$$= \frac{3x}{(x-4)(x+3)} - \frac{x-1}{(x-4)^2} - \frac{6-x}{2(x-4)}$$

$$= \frac{3x(2)(x-4) - (x-1)(2)(x+3) - (6-x)(x-4)(x+3)}{2(x-4)^2(x+3)}$$

$$= \frac{6x^2 - 24x - 2x^2 - 4x + 6 + x^3 - 7x^2 - 6x + 72}{2(x-4)^2(x+3)}$$

$$= \frac{x^3 - 3x^2 - 34x + 78}{2(x-4)^2(x+3)}$$

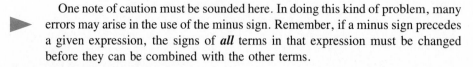

One note of caution must be sounded here. In doing this kind of problem, many errors may arise in the use of the minus sign. Remember, if a minus sign precedes a given expression, the signs of **all** terms in that expression must be changed before they can be combined with the other terms.

Example H

Simplify the fraction

$$\frac{1 + \dfrac{2}{x - 1}}{\dfrac{x^2 + x}{x^2 + x - 2}}$$

Before performing the indicated division, we first perform the indicated addition in the numerator. The numerator becomes

$$\frac{(x - 1) + 2}{x - 1} \quad \text{or} \quad \frac{x + 1}{x - 1}$$

This expression now replaces the numerator of the original fraction. Making this substitution and inverting the divisor, we then proceed with the simplification:

$$\frac{x + 1}{x - 1} \cdot \frac{x^2 + x - 2}{x^2 + x} = \frac{(x + 1)(x + 2)(x - 1)}{(x - 1)(x)(x + 1)} = \frac{x + 2}{x}$$

This is an example of what is known as a **complex fraction**. *In a complex fraction the numerator, the denominator, or both numerator and denominator contain fractions.*

Example I

In simplifying the fraction

$$\frac{\dfrac{1}{x} + \dfrac{1}{x^2 - 4x}}{\dfrac{3}{x^2 - 16} + \dfrac{2}{x^2 + 4x}}$$

we first perform the additions in the numerator and the denominator. We then divide the resulting fraction in the numerator by that in the denominator.

$$\frac{\dfrac{1}{x} + \dfrac{1}{x^2 - 4x}}{\dfrac{3}{x^2 - 16} + \dfrac{2}{x^2 + 4x}} = \frac{\dfrac{1}{x} + \dfrac{1}{x(x - 4)}}{\dfrac{3}{(x - 4)(x + 4)} + \dfrac{2}{x(x + 4)}}$$

$$= \frac{\dfrac{x - 4 + 1}{x(x - 4)}}{\dfrac{3x + 2(x - 4)}{x(x - 4)(x + 4)}} = \frac{\dfrac{x - 3}{x(x - 4)}}{\dfrac{5x - 8}{x(x - 4)(x + 4)}}$$

$$= \frac{x - 3}{x(x - 4)} \cdot \frac{x(x - 4)(x + 4)}{5x - 8}$$

$$= \frac{x(x - 3)(x - 4)(x + 4)}{x(x - 4)(5x - 8)}$$

$$= \frac{(x - 3)(x + 4)}{5x - 8}$$

Exercises 5–6

In Exercises 1 through 40, perform the indicated operations and simplify.

1. $\dfrac{3}{5} + \dfrac{6}{5}$

2. $\dfrac{2}{13} + \dfrac{6}{13}$

3. $\dfrac{1}{x} + \dfrac{7}{x}$

4. $\dfrac{2}{a} + \dfrac{3}{a}$

5. $\dfrac{1}{2} + \dfrac{3}{4}$

6. $\dfrac{5}{9} - \dfrac{1}{3}$

7. $\dfrac{3}{4x} + \dfrac{7a}{4}$

8. $\dfrac{t-3}{a} - \dfrac{t}{2a}$

9. $\dfrac{a}{x} - \dfrac{b}{x^2}$

10. $\dfrac{2}{s^2} + \dfrac{3}{s}$

11. $\dfrac{6}{5x^3} + \dfrac{a}{25x}$

12. $\dfrac{a}{6y} - \dfrac{2b}{3y^4}$

13. $\dfrac{2}{5a} + \dfrac{1}{a} - \dfrac{a}{10}$

14. $\dfrac{2}{a} - \dfrac{6}{b} - \dfrac{9}{c}$

15. $\dfrac{x+1}{x} - \dfrac{x-3}{y} - \dfrac{2-x}{xy}$

16. $5 + \dfrac{1-x}{2} - \dfrac{3+x}{4}$

17. $\dfrac{3}{2x-1} + \dfrac{1}{4x-2}$

18. $\dfrac{5}{6y+3} - \dfrac{a}{8y+4}$

19. $\dfrac{4}{x(x+1)} - \dfrac{3}{2x}$

20. $\dfrac{3}{ax+ay} - \dfrac{1}{a^2}$

21. $\dfrac{s}{2s-6} + \dfrac{1}{4} - \dfrac{3s}{4s-12}$

22. $\dfrac{2}{x+2} - \dfrac{3-x}{x^2+2x} + \dfrac{1}{x}$

23. $\dfrac{3x}{x^2-9} - \dfrac{2}{x+3}$

24. $\dfrac{2}{x^2+4x+4} - \dfrac{3}{x+2}$

25. $\dfrac{3}{x^2-8x+16} - \dfrac{2}{4-x}$

26. $\dfrac{1}{a^2-1} - \dfrac{2}{1-a}$

27. $\dfrac{3}{x^2-11x+30} - \dfrac{2}{x^2-25}$

28. $\dfrac{x-1}{2x^3-4x^2} + \dfrac{5}{x-2}$

29. $\dfrac{x-1}{3x^2-13x+4} - \dfrac{3x+1}{4-x}$

30. $\dfrac{x}{4x^2-12x+5} + \dfrac{2x-1}{4x^2-4x-15}$

31. $\dfrac{t}{t^2-t-6} - \dfrac{2t}{t^2+6t+9} + \dfrac{t}{t^2-9}$

32. $\dfrac{5}{2x^3-3x^2+x} - \dfrac{x}{x^4-x^2} + \dfrac{2-x}{2x^2+x-1}$

33. $\dfrac{1 + \dfrac{1}{x}}{1 - \dfrac{1}{x}}$

34. $\dfrac{x - \dfrac{1}{x}}{1 - \dfrac{1}{x}}$

35. $\dfrac{x - \dfrac{1}{x} - \dfrac{2}{x+1}}{\dfrac{1}{x^2+2x+1} - 1}$

36. $\dfrac{\dfrac{2}{a} - \dfrac{1}{4} - \dfrac{3}{4a-4b}}{\dfrac{1}{4a^2-4b^2} - \dfrac{2}{b}}$

37. $\dfrac{\dfrac{3}{x} + \dfrac{1}{x^2 + x}}{\dfrac{1}{x + 1} - \dfrac{1}{x - 1}}$

38. $\dfrac{\dfrac{1}{2x} - \dfrac{1}{4x^2 - 2x}}{\dfrac{6}{4x^2 - 1} - \dfrac{2}{2x + 1}}$

39. $\dfrac{\dfrac{r + s}{r - s} - \dfrac{r - s}{r + s}}{1 + \dfrac{r - s}{r + s}}$

40. $\dfrac{\dfrac{1}{u - v} + \dfrac{1}{2u + 2v}}{\dfrac{2u}{2u^2 - 3uv + v^2} + \dfrac{2}{2u - v}}$

The expression $f(x + h) - f(x)$ is frequently used in the study of calculus. In Exercises 41 through 44, determine and then simplify this expression for the given functions.

41. $f(x) = \dfrac{x}{x + 1}$

42. $f(x) = \dfrac{3}{2x - 1}$

43. $f(x) = \dfrac{1}{x^2}$

44. $f(x) = \dfrac{2}{x^2 + 4}$

In Exercises 45 through 52, simplify the given expressions.

45. Using the definitions of the trigonometric functions given in Section 3–2, find an expression equivalent to $(\tan \theta)(\cot \theta) + (\sin \theta)^2 - \cos \theta$ in terms of x, y, and r.

46. Using the definitions of the trigonometric functions given in Section 3–2, find an expression equivalent to $\sec \theta - (\cot \theta)^2 + \csc \theta$ in terms of x, y, and r.

47. If $f(x) = 2x - x^2$, find $f(\frac{1}{a})$.

48. If $f(x) = x^2 + x$, find $f(a + \frac{1}{a})$.

49. The analysis of the forces acting on a certain type of concrete slab gives the expression

$$1 - \frac{4c}{\pi l} + \frac{c^3}{3l^3}$$

Combine and simplify.

50. Experimentation to determine the velocity of light may use the expression

$$1 + \frac{v^2}{2c^2} + \frac{3v^4}{4c^4}$$

Combine and simplify.

51. In finding an expression to describe a magnetic field, the following expression is found:

$$\frac{b}{x^2 + y^2} - \frac{2bx^2}{(x^2 + y^2)^2}$$

Perform the indicated subtraction.

52. The expression for the volumetric expansion of liquids in terms of density ρ and temperature T is

$$\frac{\dfrac{1}{\rho_2} - \dfrac{1}{\rho_1}}{(T_2 - T_1)\left(\dfrac{1}{\rho_1} + \dfrac{1}{\rho_2}\right)}{2}$$

Simplify this expression.

5–7 Equations Involving Fractions

Many important equations in science and technology have fractions in them. Although the solution of these equations will still involve the use of the basic operations stated in Section 1–11, an additional procedure can be used to eliminate the fractions and thereby help lead to the solution. The method is to **multiply each term of the equation by the lowest common denominator.** The resulting equation will not involve fractions and can be solved by methods previously discussed. The following examples illustrate how to solve equations involving fractions.

Example A

Solve for x: $\dfrac{x}{12} - \dfrac{1}{8} = \dfrac{x + 2}{6}$.

We first note that the lowest common denominator of the terms of the equation is 24. Therefore, we multiply each term by 24. This gives

$$\frac{24(x)}{12} - \frac{24(1)}{8} = \frac{24(x + 2)}{6}$$

We reduce each term to its lowest terms and solve the resulting equation.

$$2x - 3 = 4(x + 2)$$
$$2x - 3 = 4x + 8$$
$$-2x = 11$$
$$x = -\frac{11}{2}$$

When we check this solution in the original equation, we obtain $-\frac{7}{12}$ on each side of the equal sign. Therefore, the solution is correct.

Example B

Solve for x: $\dfrac{x}{2} - \dfrac{1}{b^2} = \dfrac{x}{2b}$.

We first determine that the lowest common denominator of the terms of the equation is $2b^2$. We then multiply each term by $2b^2$ and continue with the solution.

$$\frac{2b^2(x)}{2} - \frac{2b^2(1)}{b^2} = \frac{2b^2(x)}{2b}$$
$$b^2x - 2 = bx$$
$$b^2x - bx = 2$$

To complete the solution for x, we must factor x from the terms on the left. Therefore, we have

$$x(b^2 - b) = 2$$

$$x = \frac{2}{b^2 - b}$$

Checking shows that each side of the original equation is $1/b^2(b - 1)$.

Example C

When developing the equations which describe the motion of the planets, the equation

$$\frac{1}{2}v^2 - \frac{GM}{r} = -\frac{GM}{2a}$$

is found. Solve for M.

We first determine that the lowest common denominator of the terms of the equation is $2ar$. Multiplying each term by $2ar$ and proceeding, we have

$$\frac{2ar(v^2)}{2} - \frac{2ar(GM)}{r} = -\frac{2ar(GM)}{2a}$$

$$arv^2 - 2aGM = -rGM$$

$$rGM - 2aGM = -arv^2$$

$$M(rG - 2aG) = -arv^2$$

$$M = -\frac{arv^2}{rG - 2aG} \quad \text{or} \quad \frac{arv^2}{2aG - rG}$$

The second form of the result is obtained by using Eq. (5-11). Again, note the use of factoring to arrive at the final result.

Example D

Solve for x: $\dfrac{2}{x + 1} - \dfrac{1}{x} = -\dfrac{2}{x^2 + x}$.

Multiplying each term by the lowest common denominator $x(x + 1)$, we have

$$\frac{2(x)(x + 1)}{x + 1} - \frac{x(x + 1)}{x} = -\frac{2x(x + 1)}{x(x + 1)}$$

Now, simplifying each fraction, we have

$$2x - (x + 1) = -2$$

We now complete the solution.

$$2x - x - 1 = -2$$
$$x = -1$$

(continued on next page)

Checking this solution *in the original equation,* we see that we have zero in the denominators of the first and third terms of the equation. Since division by zero is undefined (see Section 1–3), $x = -1$ cannot be a solution. *Thus there is **no solution** to this equation.* This example points out clearly why it is necessary to check solutions in the original equation. It also shows that *whenever we multiply each term by a common denominator which **contains the unknown**, it is possible to obtain a value which is not a solution of the original equation. Such a value is termed an* **extraneous solution.** Only certain equations will lead to extraneous solutions, but we must be careful to identify them when they occur.

Example E

One pipe can fill a certain oil storage tank in 4 h, while a second pipe can fill it in 6 h. How long will it take to fill the tank if both pipes operate together?

First, we let x = the number of hours required to fill the tank with both pipes operating.

We know that it takes the first pipe 4 h to fill the tank. Therefore, it fills $\frac{1}{4}$ of the tank each hour it operates. This means that it fills $\frac{1}{4}x$ of the tank in x hours. In the same way, the second pipe fills $\frac{1}{6}x$ of the tank in x hours. When x hours have passed, the two pipes will have filled the whole tank (1 represents *one* tank).

$$\frac{x}{4} + \frac{x}{6} = 1$$

Multiplying each term by 12, we have

$$\frac{12x}{4} + \frac{12x}{6} = 12(1)$$

$$3x + 2x = 12$$

$$5x = 12$$

$$x = \frac{12}{5} = 2.4 \text{ h}$$

Therefore, it takes the two pipes 2.4 h to fill the tank when operating together.

Exercises 5–7

In Exercises 1 through 28, solve the given equations and check the results.

1. $\dfrac{x}{2} + 6 = 2x$

2. $\dfrac{x}{5} + 2 = \dfrac{15 + x}{10}$

3. $\dfrac{x}{6} - \dfrac{1}{2} = \dfrac{x}{3}$

4. $\dfrac{3x}{8} - \dfrac{3}{4} = \dfrac{x - 4}{2}$

5. $\dfrac{1}{2} - \dfrac{t - 5}{6} = \dfrac{3}{4}$

6. $\dfrac{2x - 7}{3} + 5 = \dfrac{1}{5}$

7. $\dfrac{3x}{7} - \dfrac{5}{21} = \dfrac{2 - x}{14}$

8. $\dfrac{x - 3}{12} - \dfrac{2}{3} = \dfrac{1 - 3x}{2}$

9. $\dfrac{3}{x} + 2 = \dfrac{5}{3}$

10. $\dfrac{1}{2y} - \dfrac{1}{2} = 4$

11. $3 - \dfrac{x - 2}{x} = \dfrac{1}{3}$

12. $\dfrac{1}{2x} - \dfrac{1}{3} = \dfrac{2}{x}$

13. $\dfrac{2y}{y - 1} = 5$

14. $\dfrac{x}{2x - 3} = 4$

15. $\dfrac{2}{s} = \dfrac{3}{s - 1}$

16. $\dfrac{5}{n + 2} = \dfrac{3}{2n}$

17. $\dfrac{5}{2x + 4} + \dfrac{3}{x + 2} = 2$

18. $\dfrac{3}{4x - 6} + \dfrac{1}{4} = \dfrac{5}{2x - 3}$

19. $\dfrac{4}{4 - x} + 2 - \dfrac{2}{12 - 3x} = \dfrac{1}{3}$

20. $\dfrac{2}{z - 5} - \dfrac{3}{10 - 2z} = 3$

21. $\dfrac{1}{x} + \dfrac{3}{2x} = \dfrac{2}{x + 1}$

22. $\dfrac{3}{t + 3} - \dfrac{1}{t} = \dfrac{5}{2t + 6}$

23. $\dfrac{1}{2x + 3} = \dfrac{5}{2x} - \dfrac{4}{2x^2 + 3x}$

24. $\dfrac{7}{y} = \dfrac{3}{y - 4} + \dfrac{7}{2y^2 - 8y}$

25. $\dfrac{1}{x^2 - x} - \dfrac{1}{x} = \dfrac{1}{x - 1}$

26. $\dfrac{2}{x^2 - 1} - \dfrac{2}{x + 1} = \dfrac{1}{x - 1}$

27. $\dfrac{2}{x^2 - 4} - \dfrac{1}{x - 2} = \dfrac{1}{2x + 4}$

28. $\dfrac{2}{2x^2 + 5x - 3} - \dfrac{1}{4x - 2} + \dfrac{3}{2x + 6} = 0$

In Exercises 29 through 32, solve for the indicated letter.

29. $2 - \dfrac{1}{b} + \dfrac{3}{c} = 0$, for c

30. $\dfrac{2}{3} - \dfrac{h}{x} = \dfrac{1}{2x}$, for x

31. $\dfrac{t - 3}{b} - \dfrac{t}{2b - 1} = \dfrac{1}{2}$, for t

32. $\dfrac{1}{a^2 + 2a} - \dfrac{y}{2a} = \dfrac{2y}{a + 2}$, for y

In Exercises 33 through 44, each of the given formulas arises in the technical or scientific area of study listed. Solve for the indicated letter.

33. $n = n_1 - \dfrac{n_1 v}{V}$, for v (acoustics)

34. $i = \dfrac{1}{a} - \dfrac{1}{s}$, for a (business: annuities)

35. $E = V_0 + \dfrac{(m + M)V^2}{2} + \dfrac{p^2}{2I}$, for M (nuclear physics)

36. $\dfrac{V - 6}{5} + \dfrac{V - 8}{15} + \dfrac{V}{10} = 0$, for V (electricity)

37. $S = \dfrac{P}{A} + \dfrac{Mc}{I}$, for P (machine design)

38. $F = PA + \dfrac{dQ^2}{gA_1} - \dfrac{dQ^2}{gA_2}$, for A (hydrodynamics)

39. $\dfrac{1}{R} = \dfrac{1}{R_1 + r} + \dfrac{1}{R_2}$, for R_1 (electricity)

40. $\dfrac{1}{x} + \dfrac{1}{nx} = \dfrac{1}{f}$, for n (photography)

41. $R = \dfrac{L_1}{kA_1} + \dfrac{L_2}{kA_2}$, for L_1 (air conditioning)

42. $\dfrac{1}{P} = \dfrac{1}{P_1} + \dfrac{X}{P_2} - \dfrac{X}{P_1}$, for P_2 (chemistry)

43. $D = \dfrac{wx^4}{24EI} - \dfrac{wLx^3}{6EI} + \dfrac{wL^2x^2}{4EI}$, for w (mechanics: beams)

44. $\dfrac{1}{f} = (n - 1)\left(\dfrac{1}{R_1} + \dfrac{1}{R_2}\right)$, for R_1 (optics)

In Exercises 45 through 48, set up appropriate equations and solve the given stated problems.

45. One microcomputer can perform a certain number of calculations in 6 s, and a second microcomputer can perform the same number in 9 s. How long would it take the two computers together to perform these calculations?

46. One steamshovel can excavate a certain site in 5 days, while it takes a second steamshovel 8 days. How long would it take the two working together?

47. The width of a particular rectangular land area is $\frac{3}{5}$ that of the length. If the perimeter is 192 m, find the dimensions.

48. The current in a certain stream flows at 3 mi/h. A motorboat can travel downstream 23 mi in the same time it can travel 11 mi upstream. What is the boat's rate in still water?

5–8 Exercises for Chapter 5

In Exercises 1 through 12, find the products *by inspection*. No intermediate steps should be necessary.

1. $3a(4x + 5a)$
2. $-7xy(4x^2 - 7y)$
3. $(2a + 7b)(2a - 7b)$
4. $(x - 4z)(x + 4z)$
5. $(2a + 1)^2$
6. $(4x - 3y)^2$
7. $(b - 4)(b + 7)$
8. $(y - 5)(y - 7)$
9. $(2x + 5)(x - 9)$
10. $(4ax - 3)(5ax + 7)$
11. $(2c + d)(8c - d)$
12. $(3s - 2t)(8s + 3t)$

In Exercises 13 through 44, factor the given expressions completely. Exercises 37 through 40 illustrate Eqs. (5–7) through (5–10), and Exercises 41 through 44 illustrate factoring by grouping.

13. $3s + 9t$
14. $7x - 28y$
15. $a^2x^2 + a^2$
16. $3ax - 6ax^4 - 9a$
17. $x^2 - 144$
18. $900 - n^2$
19. $400r^2 - t^4$
20. $25s^4 - 36t^2$
21. $9t^2 - 6t + 1$
22. $4x^2 - 12x + 9$
23. $25t^2 + 10t + 1$
24. $4x^2 + 36xy + 81y^2$
25. $x^2 + x - 56$
26. $x^2 - 4x - 45$

27. $t^2 - 5t - 36$

28. $n^2 - 11n + 10$

29. $2x^2 - x - 36$

30. $5x^2 + 2x - 3$

31. $4x^2 - 4x - 35$

32. $9x^2 + 7x - 16$

33. $10b^2 + 23b - 5$

34. $12x^2 - 7xy - 12y^2$

35. $4x^2 - 64$

36. $4a^2x^2 + 26a^2x + 36a^2$

37. $x^3 + 9x^2 + 27x + 27$

38. $x^3 - 3x^2 + 3x - 1$

39. $8x^3 + 27$

40. $x^3 - 125$

41. $ab^2 - 3b^2 + a - 3$

42. $axy - ay + ax - a$

43. $nx + 5n - x^2 + 25$

44. $ty - 4t + y^2 - 16$

In Exercises 45 through 68, perform the indicated operations and express results in simplest form.

45. $\dfrac{48ax^3y^6}{9a^3xy^6}$

46. $\dfrac{-39r^2s^4t^8}{52rs^5t}$

47. $\dfrac{6x^2 - 7x - 3}{4x^2 - 8x + 3}$

48. $\dfrac{x^2 - 3x - 4}{x^2 - x - 12}$

49. $\dfrac{4x + 4y}{35x^2} \cdot \dfrac{28x}{x^2 - y^2}$

50. $\dfrac{6x - 3}{x^2} \cdot \dfrac{4x^2 - 12x}{12x - 6}$

51. $\dfrac{18 - 6x}{x^2 - 6x + 9} \div \dfrac{x^2 - 2x - 15}{x^2 - 9}$

52. $\dfrac{6x^2 - xy - y^2}{2x^2 + xy - y^2} \div \dfrac{4x^2 - 16y^2}{x^2 + 3xy + 2y^2}$

53. $\dfrac{\dfrac{3x}{7x^2 + 13x - 2}}{\dfrac{6x^2}{x^2 + 4x + 4}}$

54. $\dfrac{\dfrac{3x - 3y}{2x^2 + 3xy - 2y^2}}{\dfrac{3x^2 - 3y^2}{x^2 + 4xy + 4y^2}}$

55. $\dfrac{x + \dfrac{1}{x} + 1}{x^2 - \dfrac{1}{x}}$

56. $\dfrac{\dfrac{4}{y} - 4y}{2 - \dfrac{2}{y}}$

57. $\dfrac{4}{9x} - \dfrac{5}{12x^2}$

58. $\dfrac{3}{10a^2} + \dfrac{1}{4a^3}$

59. $\dfrac{6}{x} - \dfrac{7}{2x} + \dfrac{3}{xy}$

60. $\dfrac{4}{a^2b} - \dfrac{5}{2ab} + \dfrac{1}{2b}$

61. $\dfrac{a + 1}{a + 2} - \dfrac{a + 3}{a}$

62. $\dfrac{y}{y + 2} - \dfrac{1}{y^2 + 2y}$

63. $\dfrac{2x}{x^2 + 2x - 3} - \dfrac{1}{x^2 + 3x}$

64. $\dfrac{x}{4x^2 + 4x - 3} - \dfrac{3}{4x^2 - 9}$

65. $\dfrac{3x}{2x^2 - 2} - \dfrac{2}{4x^2 - 5x + 1}$

66. $\dfrac{2x - 1}{4 - x} + \dfrac{x + 2}{5x - 20}$

67. $\dfrac{3x}{x^2 + 2x - 3} - \dfrac{2}{x^2 + 3x} + \dfrac{x}{x - 1}$

68. $\dfrac{3}{y^4 - 2y^3 - 8y^2} + \dfrac{y - 1}{y^2 + 2y} - \dfrac{y - 3}{y^2 - 4y}$

In Exercises 69 through 76, solve the given equations.

69. $\dfrac{x}{2} - 3 = \dfrac{x - 10}{4}$

70. $\dfrac{x}{6} - \dfrac{1}{2} = \dfrac{3 - x}{12}$

71. $\dfrac{2x}{c} - \dfrac{1}{2c} = \dfrac{3}{c} - x$, for x

72. $\dfrac{x}{a} - b + \dfrac{x}{c} = \dfrac{a}{b} - c$, for x

73. $\dfrac{2}{t} - \dfrac{1}{at} = 2 + \dfrac{a}{t}$, for t

74. $\dfrac{3}{a^2 y} - \dfrac{1}{ay} = \dfrac{9}{a}$, for y

75. $\dfrac{2x}{x^2 - 3x} - \dfrac{3}{x} = \dfrac{1}{2x - 6}$

76. $\dfrac{3}{x^2 + 3x} - \dfrac{1}{x} = \dfrac{1}{x + 3}$

In Exercises 77 through 96, perform the indicated operations.

77. Show that

$$xy = \frac{1}{4}[(x + y)^2 - (x - y)^2]$$

78. Show that

$$x^2 + y^2 = \frac{1}{2}[(x + y)^2 + (x - y)^2]$$

79. To find the side of a rectangle of a given area, it is necessary to factor the expression $x^2 - 3x - 70$. Factor this expression.

80. Under certain conditions, in order to find the total profit of an article selling for p dollars, one must factor the expression $2p^2 - 126p + 360$. Factor this expression.

81. In finding the center of mass of a certain solid, the expression $4x^3 - 20x^2 + 25x$ arises. Factor this expression.

82. If the edge of one cube is $x + 4$ and the edge of another cube is x, the difference in their volumes is $(x + 4)^3 - x^3$. Expand $(x + 4)^3$, simplify the resulting expression, and then factor.

83. In studying the tension in a certain type of cable, the expression

$$\frac{ws^2}{2d} - \frac{wd}{2}$$

is used. Simplify this expression.

84. An expression which occurs in the study of nuclear physics is

$$pa^2 + (1 - p)b^2 - [pa + (1 - p)b]^2$$

Expand the third term and then factor by grouping.

85. In finding the velocity of an object subject to specified conditions, it is necessary to simplify the expression

$$\frac{(t + 1)^2 - 2t(t + 1)}{(t + 1)^4}$$

Simplify this expression.

86. An expression found in solving a problem related to alternating-current power is

$$\frac{\dfrac{s + 10}{10}}{\left(\dfrac{s + 20}{20}\right)\left(\dfrac{s + 60}{60}\right)}$$

Simplify this expression.

87. An expression found in determining the tension in a certain cable is

$$1 + \frac{w^2 x^2}{6T^2} - \frac{w^4 x^4}{40T^4}$$

Combine and simplify.

88. In determining the amount of rocket fuel which has been used, the expression

$$\frac{Am}{k} - \frac{g}{2}\left(\frac{m}{k}\right)^2 + \frac{AML}{k}$$

is used. Simplify this expression.

89. An expression found in the analysis of the dynamics of missile firing is

$$\frac{1}{s} - \frac{1}{s + 4} + \frac{8}{(s + 4)^2}$$

Perform the indicated operations.

90. A formula for the safety factor, S, of a steel column is

$$S = \frac{5}{3} + \frac{3L}{8Cr} - \frac{L^3}{8C^3 r^3}$$

Simplify the right side of this formula.

91. An expression found in the study of electronic amplifiers is

$$\frac{\left(\dfrac{\mu}{\mu + 1}\right)R}{\dfrac{r}{\mu + 1} + R}$$

Simplify this expression (μ is the Greek letter mu).

92. An expression which arises when finding the path between two points requiring the least time is

$$\frac{\dfrac{u^2}{2g} - x}{\dfrac{1}{2gc^2} - \dfrac{u^2}{2g} + x}$$

Simplify this expression.

93. The focal length f of a lens, in terms of its image distance q and object distance p, is given by

$$\frac{1}{f} = \frac{1}{p} + \frac{1}{q}$$

Solve for q.

94. The combined capacitance, C, of three capacitors connected in series is

$$\frac{1}{C} = \frac{1}{C_1} + \frac{1}{C_2} + \frac{1}{C_3}$$

    Solve for C_1.

95. An equation used in studying the deflection of a beam is

$$\theta = \frac{wL^3}{24EI} - \frac{ML}{6EI}$$

    Solve for M.

96. An equation determined during the study of the characteristics of a certain chemical solution is

$$X = \frac{H}{RT_1} - \frac{H}{RT}$$

    Solve for T.

In Exercises 97 through 102, set up appropriate equations and solve the given stated problems.

97. If one riveter can do a certain job in 12 days, and a second riveter can do it in 16 days, how long will it take them to do it together?

98. Two crews are working on an oil pipeline. Crew A can lay pipe at the rate of 2000 m/day, and crew B can lay pipe at the rate of 2500 m/day. How long will it take the two crews together to lay 10,000 m of pipe?

99. One computer can solve a certain problem in 30 s. With the aid of a second computer, the problem is solved in 10 s. How long would the second computer take to solve the problem alone?

100. An auto mechanic can do a certain motor job in 3.0 h, and with an assistant he can do it in 2.1 h. How long would it take the assistant to do the job alone?

101. One quality control inspector can properly inspect 50 parts per day, and a second inspector can inspect 30 parts per day. They are put on a project together to inspect 200 parts. After 1.5 days the second inspector becomes ill and leaves the project. How long does the project take?

102. A person travels from city A to city B on a train which averages 40 mi/h. She spends 4 h in city B and then returns to city A on a jet which averages 600 mi/h. If the total trip takes 20 h, how far is it from city A to city B?

CHAPTER 6

Quadratic Equations

The solution of simple equations was first introduced in Chapter 1. Then, in Chapter 4, we extended the solution of equations to systems of linear equations. With the development of the algebraic operations in Chapter 5, we are now in a position to solve another important type of equation, the **quadratic equation.**

In the first three sections of this chapter we shall present three algebraic methods of solving quadratic equations. In Section 6–4 we shall discuss the graphical solution.

6–1 Quadratic Equations; Solution by Factoring

Given that a, b, and c are constants, the equation

$$ax^2 + bx + c = 0 \qquad\qquad (6\text{–}1)$$

is called the **general quadratic equation in x.** From Eq. (6–1) we can see that the left side of the equation is a polynomial function of degree 2. *This function,* $ax^2 + bx + c$, *is known as the* **quadratic function.**

Quadratic equations and quadratic functions are found in applied problems of many technical fields of study. For example, in describing projectile motion, the equation $s_0 + v_0 t - 16t^2 = 0$ is found; in analyzing electric power, the function $EI - RI^2$ is found; and in determining the forces on beams, the function $ax^2 + bLx + cL^2$ is used.

Since it is the x^2-term that distinguishes the quadratic equation from other types of equations, the equation is not quadratic if $a = 0$. However, b or c or both may be zero, and the equation is quadratic. We should recognize a quadratic equation even when it does not initially appear in the form of Eq. (6–1). The following examples illustrate the recognition of quadratic equations.

Example A

The following are quadratic equations.

$$x^2 - 4x - 5 = 0 \qquad (a = 1,\ b = -4,\ \text{and } c = -5)$$
$$3x^2 - 6 = 0 \qquad (a = 3,\ b = 0,\ \text{and } c = -6)$$
$$2x^2 + 7x = 0 \qquad (a = 2,\ b = 7,\ \text{and } c = 0)$$
$$(a - 3)x^2 - ax + 7 = 0$$

(The constants in Eq. (6–1) may include literal expressions. In this case, $a - 3$ takes the place of a, $-a$ takes the place of b, and $c = 7$.)

$$4x^2 - 2x = x^2$$

(After all the terms have been collected on the left side, the equation becomes $3x^2 - 2x = 0$.)

$$(x + 1)^2 = 4$$

(Expanding the left side, and collecting all terms on the left, we have $x^2 + 2x - 3 = 0$.)

Example B

The following are not quadratic equations.

$$bx - 6 = 0 \qquad (\text{There is no } x^2\text{-term.})$$
$$x^3 - x^2 - 5 = 0$$

(There should be no term of degree higher than 2. Thus there can be no x^3-term in a quadratic equation.)

$$x^2 + x - 7 = x^2$$

(When terms are collected, there will be no x^2-term.)

From our previous work, we recall that *the solution of an equation consists of all numbers which, when substituted in the equation, produce equality. There are* **two** *such roots for a quadratic equation. Occasionally these roots are equal,* as is shown in Example C, and only one number is actually a solution. Also, due to the presence of the x^2-term, the roots may be imaginary numbers.

Example C

The quadratic equation

$$3x^2 - 7x + 2 = 0$$

has the roots $x = \frac{1}{3}$ and $x = 2$. This can be seen by substituting these values into the equation.

$$3\left(\frac{1}{3}\right)^2 - 7\left(\frac{1}{3}\right) + 2 = 3\left(\frac{1}{9}\right) - \frac{7}{3} + 2 = \frac{1}{3} - \frac{7}{3} + 2 = \frac{0}{3} = 0$$
$$3(2)^2 - 7(2) + 2 = 3(4) - 14 + 2 = 12 - 14 + 2 = 0$$

The quadratic equation

$$4x^2 - 4x + 1 = 0$$

has the **double root** (*both roots are the same*) of $x = \frac{1}{2}$. This can be seen to be a solution by substitution:

$$4\left(\frac{1}{2}\right)^2 - 4\left(\frac{1}{2}\right) + 1 = 4\left(\frac{1}{4}\right) - 2 + 1 = 1 - 2 + 1 = 0$$

The quadratic equation

$$x^2 + 9 = 0$$

has the imaginary roots of $\sqrt{-3}$ and $-\sqrt{-3}$. (At this point all we wish to do is to recognize imaginary roots when they occur.)

In this section we shall deal only with those quadratic equations whose quadratic expression is factorable. Therefore, all roots will be rational. *To solve a quadratic equation by factoring, use the following steps.* (1) *Collect all terms on the left* (the equation will then be in the general form of Eq. (6–1). (2) *Factor the quadratic expression.* (3) *Set each factor equal to zero.* (4) *Solve the resulting linear equations.* Here we are using the fact that a product is zero if any of its factors is zero. The solutions of the resulting linear equations constitute the solution of the quadratic equation.

Example D

$$x^2 - x - 12 = 0$$
$$(x - 4)(x + 3) = 0$$
$$x - 4 = 0, \quad \text{or} \quad x = 4$$
$$x + 3 = 0, \quad \text{or} \quad x = -3$$

The roots are $x = 4$ and $x = -3$. We can check them in the original equation by substitution. For the root $x = 4$, we have

$$(4)^2 - (4) - 12 \overset{?}{=} 0$$
$$0 = 0$$

For the root $x = -3$, we have

$$(-3)^2 - (-3) - 12 \overset{?}{=} 0$$
$$0 = 0$$

Both roots satisfy the original equation.

Example E

$$2x^2 + 7x - 4 = 0$$
$$(2x - 1)(x + 4) = 0$$

$$2x - 1 = 0, \quad \text{or} \quad x = \frac{1}{2}$$

$$x + 4 = 0, \quad \text{or} \quad x = -4$$

Therefore, the roots are $x = \frac{1}{2}$ and $x = -4$. These roots can be checked by the same procedure used in Example D.

Example F

$$x^2 + 4 = 4x$$
$$x^2 - 4x + 4 = 0$$
$$(x - 2)(x - 2) = 0$$
$$x - 2 = 0, \quad \text{or} \quad x = 2$$

Since both factors are the same, there is a double root of $x = 2$.

  It is essential for the expression on the left to be equal to zero, because if a product equals a nonzero number, there is no assurance that either of the factors equals this number. Again, the first step must be to write the equation in the form of Eq. (6–1).

Example G

A car travels to and from a city 180 mi distant in 8.5 h. If the average speed on the return trip is 5 mi/h less than on the trip to the city, what was the average speed of the car when it was going toward the city?

Let x = average speed of car going to the city, and t = time to travel to the city. By our choice of unknowns, we may state that $xt = 180$ (speed times time equals distance). Also, we know that the speed on the return trip was $x - 5$ and that the required time for the return trip was $8.5 - t$. Since the distance traveled returning was also 180 mi, we may state that $(x - 5)(8.5 - t) = 180$. Because we wish to find x, we can eliminate t between the equations by substitution.

$$(x - 5)\left(8.5 - \frac{180}{x}\right) = 180$$

$$(x - 5)(17x - 360) = 360x \qquad \text{(each side multiplied by } 2x)$$
$$17x^2 - 360x - 85x + 1800 = 360x \qquad \text{(remove parentheses)}$$
$$17x^2 - 805x + 1800 = 0 \qquad \text{(collect all terms to one side)}$$
$$(17x - 40)(x - 45) = 0 \qquad \text{(factor the quadratic expression)}$$

$$17x - 40 = 0, \quad \text{or} \quad x = \frac{40}{17}$$

$$x - 45 = 0, \quad \text{or} \quad x = 45$$

The factors lead to two possible solutions, but only one of them has meaning for this problem. The solution $x = \frac{40}{17}$ cannot be the solution, since the return rate of 5 mi/h less would then be negative. Therefore, the solution is $x = 45$ mi/h. By substitution it is found that this solution satisfies the given conditions.

Exercises 6–1

In Exercises 1 through 8, determine whether or not the given equations are quadratic by performing algebraic operations which could put each in the form of Eq. (6–1). If the resulting form is quadratic, identify a, b, and c, with $a > 0$.

1. $x^2 + 5 = 8x$
2. $5x^2 = 9 - x$
3. $x(x - 2) = 4$
4. $(3x - 2)^2 = 2$
5. $x^2 = (x + 2)^2$
6. $x(2x + 5) = 7 + 2x^2$
7. $x(x^2 + x - 1) = x^3$
8. $(x - 7)^2 = (2x + 3)^2$

In Exercises 9 through 44, solve the given quadratic equations by factoring.

9. $x^2 - 4 = 0$
10. $x^2 - 400 = 0$
11. $4y^2 - 9 = 0$
12. $x^2 - 0.16 = 0$
13. $x^2 - 8x - 9 = 0$
14. $s^2 + s - 6 = 0$
15. $x^2 - 7x + 12 = 0$
16. $x^2 - 11x + 30 = 0$
17. $x^2 = -2x$
18. $x^2 = 7x$
19. $27m^2 = 3$
20. $5p^2 = 80$
21. $3x^2 - 13x + 4 = 0$
22. $7x^2 + 3x - 4 = 0$
23. $x^2 + 8x + 16 = 0$
24. $4x^2 - 20x + 25 = 0$
25. $6x^2 = 13x - 6$
26. $6z^2 = 6 + 5z$
27. $4x^2 - 3 = -4x$
28. $10t^2 = 9 - 43t$
29. $x^2 - x - 1 = 1$
30. $2x^2 - 7x + 6 = 3$
31. $x^2 - 4b^2 = 0$
32. $a^2x^2 - 1 = 0$
33. $40x - 16x^2 = 0$
34. $15x = 20x^2$
35. $8s^2 + 16s = 90$
36. $18t^2 - 48t + 32 = 0$
37. $(x + 2)^3 = x^3 + 8$
38. $x(x^2 - 4) = x^2(x - 1)$
39. $(x + a)^2 - b^2 = 0$
40. $x^2(a^2 + 2ab + b^2) - x(a + b) = 0$

41. A projectile is fired vertically into the air. The distance (in feet) above the ground, as a function of the time (in seconds), is given by $s = 160t - 16t^2$. How long will it take the projectile to hit the ground?

42. In a certain electric circuit there is a resistance R of 2 Ω and a voltage E of 60 V. The relationship between current i (in amperes), E, and R is $i^2R + iE = 8000$. What current i ($i > 0$) flows in the circuit?

43. Under certain conditions, the motion of an object suspended by a helical spring requires the solution of the equation $D^2 + 8D + 15 = 0$. Solve for D.

44. In determining the speed, s, in miles per hour of a car under certain conditions, the equation $s^2 - 16s = 3072$ is used. Find s ($s > 0$).

In Exercises 45 through 48, set up the appropriate quadratic equations and solve.

45. In electricity, the equivalent resistance of two resistances connected in parallel is given by

$$\frac{1}{R} = \frac{1}{R_1} + \frac{1}{R_2}$$

Two resistances connected in series have an equivalent resistance given by $R = R_1 + R_2$. If two resistances connected in parallel have an equivalent resistance of 3 Ω and the same two resistances have an equivalent resistance of 16 Ω when connected in series, what are the resistances? (This equation is not quadratic. However, after the proper substitution is made, a fractional equation will exist. After fractions have been cleared, a quadratic equation will exist.)

46. The formula that relates the object distance p, image distance q, and focal length f of a lens is

$$\frac{1}{p} + \frac{1}{q} = \frac{1}{f}$$

Determine the positive value of p if $q = p + 3$ cm and $f = p - 1$ cm. (See Exercise 45.)

47. A certain rectangular machine part has a length which is 4 mm longer than its width. If the area of the part is 96 mm², what are its dimensions?

48. A jet, by increasing its speed by 200 mi/h, could decrease the time needed to cover 4000 mi by 1 h. What is its speed?

6–2 Completing the Square

Many quadratic equations cannot be solved by factoring. This is true of most quadratic equations which arise in applied situations. In this section, therefore, we develop a method which can be used to solve any quadratic equation. *The method is called* **completing the square.** In the following section we shall use completing the square to develop a formula which also may be used to solve any quadratic equation.

In the first example which follows we show the solution of a type of quadratic equation which arises while using the method of completing the square. In the examples which follow it, the method itself is used and described.

Example A

In solving $x^2 = 16$, we may write it as $x^2 - 16 = 0$ and then complete the solution by factoring. This gives us solutions of $x = 4$ and $x = -4$. Therefore, we see that the principal square root of 16 and its negative both satisfy the original equation. Thus, we may solve $x^2 = 16$ by equating x to the principal square root of 16 and to its negative. The roots of $x^2 = 16$ are 4 and -4.

We may solve $(x - 3)^2 = 16$ in a similar way by equating $x - 3$ to 4 and to -4. Thus,

$$x - 3 = 4 \quad \text{or} \quad x - 3 = -4$$

Solving these equations, we obtain the roots 7 and -1.

We may solve $(x - 3)^2 = 17$ in the same way. Thus,

$$x - 3 = \sqrt{17} \quad \text{or} \quad x - 3 = -\sqrt{17}$$

The roots are therefore $3 + \sqrt{17}$ and $3 - \sqrt{17}$. Decimal approximations of these roots are 7.123 and -1.123.

Example B

We wish to find the roots of the quadratic equation

$$x^2 - 6x - 8 = 0$$

First we note that this equation is not factorable. However, we do recognize that $x^2 - 6x$ is part of one of the special products. If 9 were added to this expression, we would have $x^2 - 6x + 9$, which is $(x - 3)^2$. Therefore, we rewrite the original equation as

$$x^2 - 6x = 8$$

 and then add 9 to *both* sides of the equation. The result is

$$x^2 - 6x + 9 = 17$$

The left side of this equation may be rewritten, giving

$$(x - 3)^2 = 17$$

Now, as in the third illustration of Example A, we have

$$x - 3 = \pm\sqrt{17}$$

The $\pm$ sign means that $x - 3 = \sqrt{17}$ and $x - 3 = -\sqrt{17}$.

By adding 3 to each side, we obtain

$$x = 3 \pm \sqrt{17}$$

which means that $x = 3 + \sqrt{17}$ and $x = 3 - \sqrt{17}$ are the two roots of the equation.

Therefore, we see that by creating an expression which is a perfect square and then using the principal square root and its negative, we were finally able to solve the equation as two linear equations.

How do we determine the number which must be added to complete the square? The answer to this question is based on the special products in Eqs. (5–3) and (5–4). We rewrite these in the form

$$(x + a)^2 = x^2 + 2ax + a^2 \qquad (6\text{--}2)$$

and

$$(x - a)^2 = x^2 - 2ax + a^2 \qquad (6\text{--}3)$$

 The coefficient of x in each case is numerically $2a$, and the number added to complete the square is a^2. Thus *if we take half the coefficient of the x-term and square this result, we have the number which completes the square.* In our example, the numerical coefficient of the x-term was 6, and 9 was added to complete the square. We must be certain that the coefficient of the x^2-term is 1 before we start to complete the square. The following example outlines the steps necessary to complete the square.

Example C

Solve the following quadratic equation by the method of completing the square:

$$2x^2 + 16x - 9 = 0$$

First we divide each term by 2 so that the coefficient of the x^2-term becomes 1.

$$x^2 + 8x - \frac{9}{2} = 0$$

Now we put the constant term on the right-hand side by adding $\frac{9}{2}$ to both sides of the equation.

$$x^2 + 8x = \frac{9}{2}$$

Next we divide the coefficient of the x-term, 8, by 2, which gives us 4. We square 4 and obtain 16, which is the number to be added to both sides of the equation.

$$x^2 + 8x + 16 = \frac{9}{2} + 16 = \frac{41}{2}$$

We write the left side as the square of $(x + 4)$.

$$(x + 4)^2 = \frac{41}{2}$$

Equating $x + 4$ to the principal square root of $\frac{41}{2}$ and its negative, we have

$$x + 4 = \pm\sqrt{\frac{41}{2}}$$

Solving for x, we have

$$x = -4 \pm \sqrt{\frac{41}{2}}$$

Since $\sqrt{\frac{41}{2}} = \sqrt{\frac{82}{4}} = \frac{1}{2}\sqrt{82}$, we may write the solution without a radical in the denominator as

$$x = -4 \pm \frac{1}{2}\sqrt{82}$$

$$= \frac{-8 \pm \sqrt{82}}{2}$$

Therefore, the roots are $\frac{1}{2}(-8 + \sqrt{82})$ and $\frac{1}{2}(-8 - \sqrt{82})$. If we approximate $\sqrt{82}$ with 9.055, the approximate values of the roots are 0.528 and -8.528.

Example D

Solve $4x^2 - 12x + 5 = 0$ by completing the square.

$$4x^2 - 12x + 5 = 0$$

$$x^2 - 3x + \frac{5}{4} = 0$$

$$x^2 - 3x \quad\;\; = -\frac{5}{4}$$

$$x^2 - 3x + \frac{9}{4} = -\frac{5}{4} + \frac{9}{4}$$

$$\left(x - \frac{3}{2}\right)^2 = \frac{4}{4} = 1$$

$$x - \frac{3}{2} = \pm 1$$

$$x = \frac{3}{2} \pm 1$$

$$x = \frac{5}{2}, \; x = \frac{1}{2}$$

This equation could have been solved by factoring. However, at this point, we wanted to illustrate the method of completing the square.

Exercises 6-2

In Exercises 1 through 8, solve the given quadratic equations by finding the appropriate square roots as in Example A.

1. $x^2 = 25$
2. $x^2 = 100$
3. $x^2 = 7$
4. $x^2 = 15$
5. $(x - 2)^2 = 25$
6. $(x + 2)^2 = 100$
7. $(x + 3)^2 = 7$
8. $(x - 4)^2 = 15$

In Exercises 9 through 24, solve the given quadratic equations by completing the square. Exercises 9 through 12 and 15 through 18 may be checked by factoring.

9. $x^2 + 2x - 8 = 0$ 10. $x^2 - x - 6 = 0$ 11. $x^2 + 3x + 2 = 0$

12. $t^2 + 5t - 6 = 0$ 13. $x^2 - 4x + 2 = 0$ 14. $x^2 + 10x - 4 = 0$

15. $v^2 + 2v - 15 = 0$ 16. $x^2 - 8x + 12 = 0$ 17. $2s^2 + 5s = 3$

18. $4x^2 + x = 3$ 19. $3y^2 = 3y + 2$ 20. $3x^2 = 3 - 4x$

21. $2y^2 - y - 2 = 0$ 22. $9v^2 - 6v - 2 = 0$ 23. $x^2 + 2bx + c = 0$

24. $px^2 + qx + r = 0$

6–3 The Quadratic Formula

We shall now use the method of completing the square to derive a general formula which may be used for the solution of any quadratic equation.

Consider Eq. (6–1), the general quadratic equation

$$ax^2 + bx + c = 0$$

with $a > 0$. When we divide through by a, we obtain

$$x^2 + \frac{b}{a}x + \frac{c}{a} = 0$$

Subtracting c/a from each side, we have

$$x^2 + \frac{b}{a}x = -\frac{c}{a}$$

Half of b/a is $b/2a$, which squared is $b^2/4a^2$. Adding $b^2/4a^2$ to each side gives us

$$x^2 + \frac{b}{a}x + \frac{b^2}{4a^2} = -\frac{c}{a} + \frac{b^2}{4a^2}$$

Writing the left side as a perfect square, and combining fractions on the right side, we have

$$\left(x + \frac{b}{2a}\right)^2 = \frac{b^2 - 4ac}{4a^2}$$

Now equating $x + \dfrac{b}{2a}$ to the principal square root of the right side and its negative, we have

$$x + \frac{b}{2a} = \frac{\pm\sqrt{b^2 - 4ac}}{2a}$$

When we subtract $b/2a$ from each side and simplify the resulting expression, we obtain the **quadratic formula:**

$$x = \frac{-b \pm \sqrt{b^2 - 4ac}}{2a} \qquad\qquad (6\text{–}4)$$

To solve a quadratic equation by using the quadratic formula, we need only to write the equation in standard form (see Eq. 6–1), identify a, b, and c, and substitute these numbers directly into the formula. We shall use the quadratic formula to solve the quadratic equations in the following examples.

Example A

$$x^2 - 5x + 6 = 0$$

In this equation $a = 1$, $b = -5$, and $c = 6$. Thus, we have

$$x = \frac{-(-5) \pm \sqrt{25 - 4(1)(6)}}{2} = \frac{5 \pm 1}{2} = 3, 2$$

The roots are $x = 3$ and $x = 2$. (This particular equation could have been solved by the method of factoring.)

Example B

$$2x^2 - 7x + 5 = 0$$

In this equation $a = 2$, $b = -7$, and $c = 5$. Hence,

$$x = \frac{7 \pm \sqrt{49 - 4(2)(5)}}{4} = \frac{7 \pm 3}{4} = \frac{5}{2}, 1$$

Thus, the roots are $x = \frac{5}{2}$ and $x = 1$.

Example C

$$9x^2 + 24x + 16 = 0$$

In this example, $a = 9$, $b = 24$, and $c = 16$. Thus,

$$x = \frac{-24 \pm \sqrt{576 - 4(9)(16)}}{18} = \frac{-24 \pm 0}{18} = -\frac{4}{3}$$

Here both roots are $-\frac{4}{3}$, and the answer should be written as $x = -\frac{4}{3}$ and $x = -\frac{4}{3}$.

Example D

$$3x^2 - 5x + 4 = 0$$

In this example, $a = 3$, $b = -5$, and $c = 4$. Therefore,

$$x = \frac{5 \pm \sqrt{25 - 4(3)(4)}}{6} = \frac{5 \pm \sqrt{-23}}{6}$$

We note that the roots contain imaginary numbers. This happens if $b^2 < 4ac$.

Example E

$$2x^2 = 4x + 3$$

First we must put the equation in the proper form. This is

$$2x^2 - 4x - 3 = 0$$

Now we identify $a = 2$, $b = -4$, $c = -3$, which leads to the solution

$$x = \frac{-(-4) \pm \sqrt{(-4)^2 - 4(2)(-3)}}{2(2)} = \frac{4 \pm \sqrt{16 + 24}}{4}$$

$$= \frac{4 \pm \sqrt{40}}{4} = \frac{4 \pm 2\sqrt{10}}{4} = \frac{2(2 \pm \sqrt{10})}{4}$$

$$= \frac{2 \pm \sqrt{10}}{2}$$

Approximate decimal results are $x = 2.581$ and $x = -0.581$. The radical form of the answer is obtained by simplifying the radicals, as shown in Section 1–7.

Example F

$$dx^2 - (3 + d)x + 4 = 0$$

In this example, $a = d$, $b = -(3 + d)$, and $c = 4$. We can use the quadratic formula to solve quadratic equations which have literal coefficients. Thus,

$$x = \frac{3 + d \pm \sqrt{[-(3 + d)]^2 - 4(d)(4)}}{2d} = \frac{3 + d \pm \sqrt{9 - 10d + d^2}}{2d}$$

Example G

A square field has a diagonal which is 10 m longer than one of the sides. What is the length of a side?

Let $x =$ the length of a side of the field, and $y =$ the length of the diagonal. Using the Pythagorean theorem, we know that $y^2 = x^2 + x^2$. From the given information we know that $y = x + 10$. Thus, we have

$$(x + 10)^2 = x^2 + x^2$$

We can now simplify and solve this equation as follows.

$$x^2 + 20x + 100 = 2x^2$$

$$x^2 - 20x - 100 = 0$$

$$x = \frac{20 \pm \sqrt{400 + 400}}{2} = 10 \pm 10\sqrt{2}$$

The negative solution has no meaning in this problem. This means that the solution is $10 + 10\sqrt{2} = 24.1$ m.

 In using the quadratic formula it must be emphasized that the entire expression $-b \pm \sqrt{b^2 - 4ac}$ is divided by $2a$. It is a relatively common error to divide only the radical $\sqrt{b^2 - 4ac}$.

The quadratic formula provides a quick general method for solving quadratic equations. Proper recognition and substitution of the coefficients a, b, and c is all that is required to complete the solution, regardless of the nature of the roots.

Exercises 6–3

See Appendix E for a computer program for solving a quadratic equation by use of the quadratic formula.

In Exercises 1 through 36, solve the given quadratic equations using the quadratic formula. Exercises 1 through 14 are the same as Exercises 9 through 22 of Section 6–2.

1. $x^2 + 2x - 8 = 0$

2. $x^2 - x - 6 = 0$

3. $x^2 + 3x + 2 = 0$

4. $t^2 + 5t - 6 = 0$

5. $x^2 - 4x + 2 = 0$

6. $x^2 + 10x - 4 = 0$

7. $v^2 + 2v - 15 = 0$

8. $x^2 - 8x + 12 = 0$

9. $2s^2 + 5s = 3$

10. $4x^2 + x = 3$

11. $3y^2 = 3y + 2$

12. $3x^2 = 3 - 4x$

13. $2y^2 - y - 2 = 0$

14. $9v^2 - 6v - 2 = 0$

15. $2x^2 - 7x + 4 = 0$

16. $3x^2 - 5x - 4 = 0$

17. $2t^2 + 10t = -15$

18. $2d^2 + 7 = 4d$

19. $3s^2 = s + 9$

20. $6r^2 = 6r + 1$

21. $4x^2 = 9$

22. $6x = x^2$

23. $15 + 4z = 32z^2$

24. $4x^2 - 12x = 7$

25. $x^2 - 0.20x - 0.40 = 0$

26. $3.2x^2 = 2.5x + 7.6$

27. $0.29x^2 - 0.18 = 0.63x$

28. $12.5x^2 + 13.2x = 15.5$

29. $x^2 + 2cx - 1 = 0$

30. $x^2 - 7x + (6 + a) = 0$

31. $b^2x^2 - (b + 1)x + (1 - a) = 0$

32. $c^2x^2 - x - 1 = x^2$

33. Under certain conditions, the partial pressure P of a certain gas (in pascals) is found by solving the equation $P^2 - 3P + 1 = 0$. Solve for P such that $P < 1$ Pa.

34. A missile is fired vertically into the air. The distance (in feet) above the ground as a function of time (in seconds) is given by the formula $s = 300 + 500t - 16t^2$. (a) When will the missile hit the ground? (b) When will the missile be 1000 ft above the ground?

35. The total surface area of a right circular cylinder is found by the formula $A = 2\pi r^2 + 2\pi rh$. If the height of the cylinder is 4 in., how much is the radius if the area is 9π in.^{2}?

36. In calculating the current in an electric circuit with an inductance L (in henrys), a resistance R (in ohms) and capacitance C (in farads), it is necessary to solve the equation $Lx^2 + Rx + \dfrac{1}{C} = 0$. Find x in terms of L, R, and C.

In Exercises 37 through 40, set up appropriate equations and solve the stated problems.

37. The height of a rectangular door is 2.0 ft more than its width. The area of the door is 29 ft^2. What are the dimensions of the door?

38. After a laboratory experiment, a student reported that two particular resistances had a combined resistance of 4 Ω when connected in parallel and a combined resistance of 7 Ω when connected in series. What values would she obtain for the resistances? (See Exercise 45 of Section 6–1.)

39. To cover a given floor with square tiles of a certain size, it is found that 648 tiles are needed. If the tiles were 1 in. larger in both dimensions, only 512 tiles would be required. What is the length of a side of one of the smaller tiles?

40. A jet pilot flies 2400 mi at a given speed. If the speed were increased by 300 mi/h, the trip would take 1 h less. What is the speed of the jet?

6–4 The Graph of the Quadratic Function

We have developed the basic algebraic methods of solving a quadratic equation. In this section we consider the graph of the quadratic function. Then, following the method developed in Section 2–5, we show the graphical solution of a quadratic equation.

In Section 6–1 we noted that $ax^2 + bx + c$ is the quadratic function. As in Chapter 2, by letting $y = ax^2 + bx + c$ we can graph this function. Example A briefly reviews the graph of a quadratic function done in this way.

Example A

Graph the quadratic function $f(x) = x^2 + 2x - 3$.

First we let $y = x^2 + 2x - 3$ and then we set up a table of values. The graph is shown in Fig. 6–1.

x	y
-4	5
-3	0
-2	-3
-1	-4
0	-3
1	0
2	5

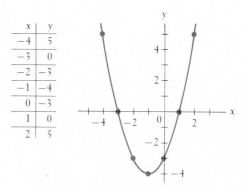

Figure 6–1

*The graph of any quadratic function $y = ax^2 + bx + c$ will have the same general **shape** as that shown in Fig. 6–1, and the graph is called a **parabola**.* (In Section 2–3 we briefly noted that the graphs for Examples B and C were parabolas.) A parabola defined by a quadratic function can open upward (as in Example A) or downward. The exact location of the graph and how it opens depend on the values of a, b, and c.

From Fig. 6–1 we see that the parabola has a minimum point at $(-1, -4)$ and that the curve opens upward. *All parabolas have an **extreme point** of this type. If $a > 0$, the parabola will have a **minimum point** and it will open **upward**. If $a < 0$, the parabola will have a **maximum point** and it will open **downward**.*

Example B

The graph of $y = 2x^2 - 8x + 6$ is shown in Fig. 6–2(a). Here $a = 2$ ($a > 0$) and the graph opens upward. The minimum point is $(2, -2)$.

The graph of $y = -2x^2 + 8x - 6$ is shown in Fig. 6–2(b). Here $a = -2$ ($a < 0$) and the graph opens downward. The maximum point is $(2, 2)$.

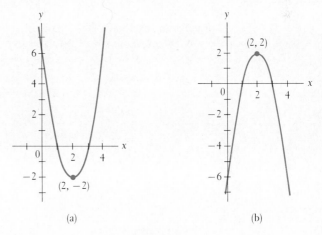

(a)   (b)

Figure 6–2

We can quickly sketch the graph of a parabola by using its basic shape and by knowing the location of two or three points, including the extreme point. The location of the extreme point can be found by using the method of completing the square, in a way similar to that used in deriving the quadratic formula.

In order to find the coordinates of the extreme point we start with the quadratic function

$$y = ax^2 + bx + c$$

Then we factor a from the two terms containing x, obtaining

$$y = a\left(x^2 + \frac{b}{a}x\right) + c$$

Now, completing the square of the terms within parentheses, we have

$$y = a\left(x^2 + \frac{b}{a}x + \frac{b^2}{4a^2}\right) + c - \frac{b^2}{4a}$$

$$= a\left(x + \frac{b}{2a}\right)^2 + c - \frac{b^2}{4a}$$

We now look at the factor $\left(x + \frac{b}{2a}\right)^2$. If $x = -b/2a$, the term is zero. If x is any other value, $\left(x + \frac{b}{2a}\right)^2$ is positive. Thus, if $a > 0$, the value of y increases from that we have for $x = -b/2a$, and if $a < 0$, the value of y decreases from that we have for $x = -b/2a$. *This means that $x = -b/2a$ is the x-coordinate of the extreme point. The y-coordinate can be found by substituting in the function.*

Another easily found point is the y-intercept. As with a linear equation, we find the y-intercept where $x = 0$. For $y = ax^2 + bx + c$, if $x = 0$, then $y = c$. *This means that the point $(0, c)$ is the y-intercept.*

Example C

Figure 6–3

For the function $y = 2x^2 - 8x + 6$, find the extreme point and the y-intercept and sketch the graph. (This function is also used in Example B.)

First, $a = 2$ and $b = -8$. This means that the x-coordinate of the extreme point is

$$\frac{-b}{2a} = \frac{-(-8)}{2(2)} = \frac{8}{4} = 2$$

and the y-coordinate is

$$y = 2(2^2) - 8(2) + 6 = -2$$

Thus, the extreme point is $(2, -2)$. Since $a > 0$, it is a minimum point.

Since $c = 6$, the y-intercept is $(0, 6)$.

We can use the minimum point $(2, -2)$ and the y-intercept $(0, 6)$, along with the fact that the graph is a parabola, to get an approximate sketch of the graph. Noting that a parabola increases (or decreases) away from the extreme point in the same way on each side of it (it is *symmetric* to a vertical line through the extreme point), we sketch the graph in Fig. 6–3. We see that it is the same as that shown in Fig. 6–2(a).

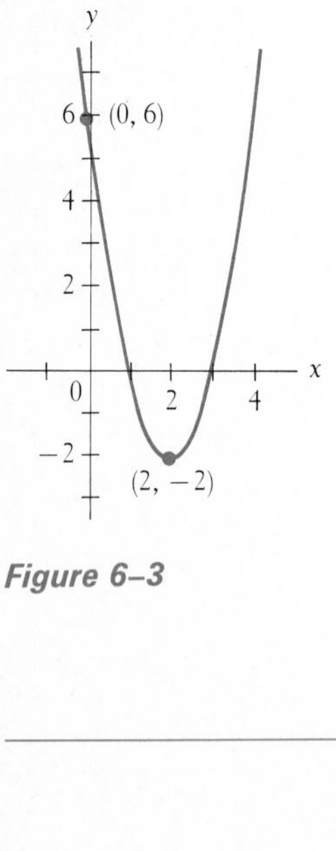

We may need one or two additional points to get a reasonable sketch of a parabola. This would be true if the y-intercept is close to the extreme point. Two points we can find are the x-intercepts, if the parabola crosses the x-axis (one point if the extreme point is on the x-axis). They are found by setting $y = 0$ and solving the quadratic equation $ax^2 + bx + c = 0$. Also, we may simply find one or two points other than the extreme point and the y-intercept.

Example D

Sketch the graph of $y = -x^2 + x + 6$.

We first note that $a = -1$ and $b = 1$. Therefore, the x-coordinate of the maximum point $(a < 0)$ is $-\frac{1}{2(-1)} = \frac{1}{2}$. The y-coordinate is $-(\frac{1}{2})^2 + \frac{1}{2} + 6 = \frac{25}{4}$. This means that the maximum point is $(\frac{1}{2}, \frac{25}{4})$.

The y-intercept is $(0, 6)$.

Using these points in Fig. 6–4, we see that they are close together and do not give a good idea of how wide the parabola opens. Therefore, setting $y = 0$, we solve the equation

$$-x^2 + x + 6 = 0$$

or

$$x^2 - x - 6 = 0$$

This equation is factorable. Thus,

$$(x - 3)(x + 2) = 0$$
$$x = 3, -2$$

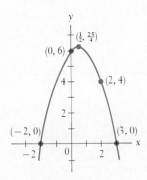

Figure 6–4

This means that the x-intercepts are $(3, 0)$ and $(-2, 0)$, as shown in Fig. 6–4.

Also, rather than finding the x-intercepts, we can let $x = 2$ and then use the point $(2, 4)$.

Example E

Sketch the graph of $y = x^2 + 1$.

Since there is no x-term, $b = 0$. This means that the x-coordinate of the minimum point $(a > 0)$ is 0 and that the minimum point and the y-intercept are both $(0, 1)$. We know that the graph opens upward, since $a > 0$, which in turn means that it does not cross the x-axis. Now letting $x = 2$ and $x = -2$, we find the points $(2, 5)$ and $(-2, 5)$ on the graph, which is shown in Fig. 6–5.

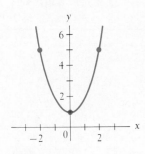

Figure 6–5

The x-intercepts of the quadratic function $y = ax^2 + bx + c$ are the same as the zeros of the function, and we can find them by letting $y = 0$, as we have already noted. We can also use this idea to solve a quadratic equation graphically. This is illustrated in the following example.

Example F

Solve the equation $3x = x(2 - x) + 3$ graphically.

We first algebraically collect all terms on the left side of the equal sign. This leads to the equation $x^2 + x - 3 = 0$. We then let $y = x^2 + x - 3$ and graph the function, as shown in Fig. 6–6. We use the minimum point $(-\frac{1}{2}, -\frac{13}{4})$, the y-intercept $(0, -3)$, and the points $(-3, 3)$ and $(2, 3)$ to sketch the graph. The points $(-2, -1)$ and $(1, -1)$ were then added for extra accuracy when it was determined approximately where the curve crossed the x-axis. Since the x-intercepts are approximately $x = -2.3$ and $x = 1.3$, these are also the solutions to the equation.

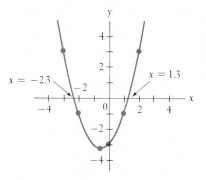

Figure 6–6

If we were to try to solve the equation $x^2 + 1 = 0$ graphically, we would have the graph shown in Fig. 6–5. *Since it does not cross the x-axis, there are no **real** solutions to the equation.*

Example G

A ball is thrown vertically upward from the ground with a velocity of 38 m/s. Its distance above the ground is given by $s = -4.9t^2 + 38t$, where s is the distance in meters and t is the time in seconds. Graph the function, and from the graph determine (a) when the ball will hit the ground, (b) how high it will go, and (c) how long it takes to reach 45 m above the ground.

Since $c = 0$, the s-intercept is at the origin. Using $-b/2a$, we find the maximum point at about $(3.9, 74)$. The t-intercepts are $(0, 0)$ and about $(7.8, 0)$. Using these points, we graph the function, as shown in Fig. 6–7.

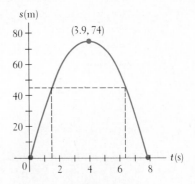

Figure 6–7

(a) The ball will hit the ground when $s = 0$, which is shown by the right t-intercept. Thus, it takes about 7.8 s to hit the ground.

(b) Its maximum height is the s-coordinate of the maximum point. Therefore, it goes to about 74 m above the ground.

(c) We find the time it takes to get 45 m above the ground by drawing a horizontal line through 45 on the s-axis. We see that this line intersects the curve twice—when $t = 1.5$ s and when $t = 6.3$ s. (Two additional points, for $t = 2$ and $t = 6$, are helpful to get better accuracy for this part.)

Exercises 6–4

In Exercises 1 through 8, sketch the graphs of the given parabolas by using only the extreme point and y-intercept.

1. $y = x^2 - 6x + 5$ 2. $y = -x^2 - 4x - 3$
3. $y = -3x^2 + 10x - 4$ 4. $y = 2x^2 + 8x - 5$
5. $y = x^2 - 4x$ 6. $y = -2x^2 - 5x$
7. $y = -2x^2 - 4x - 3$ 8. $y = x^2 - 3x + 4$

In Exercises 9 through 12, sketch the graphs of the given parabolas by using the extreme point, the y-intercept, and the x-intercepts.

9. $y = x^2 - 4$ 10. $y = x^2 + 3x$
11. $y = -2x^2 - 6x + 8$ 12. $y = -3x^2 + 12x - 5$

In Exercises 13 through 16, sketch the graphs of the given parabolas by using the extreme point, the y-intercept, and two other points, not including the x-intercepts.

13. $y = 2x^2 + 3$ 14. $y = x^2 + 2x + 2$
15. $y = -2x^2 - 2x - 6$ 16. $y = -3x^2 - x$

In Exercises 17 through 24, solve the given quadratic equations graphically. If there are no real roots, simply state this as the answer.

17. $2x^2 - 3 = 0$ 18. $5 - x^2 = 0$

19. $-3x^2 + 11x - 5 = 0$ 20. $2x^2 = 7x + 4$

21. $x(2x - 1) = -3$ 22. $2x - 5 = x^2$

23. $6x^2 = 18 - 7x$ 24. $x^2 - 7x + 9 = 0$

In Exercises 25 through 28, solve the given applied problems.

25. The area A of a certain rectangular parcel of land is given by $A = w(7 - 2w)$, where w is the width of the parcel. Sketch the graph of A as a function of w.

26. Under specified conditions, the pressure loss L, in pounds per square inch, in the flow of water through a water line in which the flow is q gallons per minute is given by $L = 0.0002q^2 + 0.005q$. Sketch the graph of L as a function of q, for $q < 100$ gal/min.

27. A missile is fired vertically upward such that its distance s, in feet, above the ground is given by $s = 150 + 250t - 16t^2$, where t is the time in seconds. Sketch the graph, and then determine from the graph (a) when the missile will hit the ground, (b) how high it will go, and (c) how long it takes to reach 800 ft above the ground.

28. In a certain electric circuit, the resistance R, in ohms, that gives resonance is found by solving the equation $25R = 3(R^2 + 4)$. Solve this equation graphically.

6–5 Exercises for Chapter 6

In Exercises 1 through 12, solve the given quadratic equations by factoring.

1. $x^2 + 3x - 4 = 0$ 2. $x^2 + 3x - 10 = 0$

3. $x^2 - 10x + 16 = 0$ 4. $x^2 - 6x - 27 = 0$

5. $3x^2 + 11x = 4$ 6. $6y^2 = 11y - 3$

7. $6t^2 = 13t - 5$ 8. $3x^2 + 5x + 2 = 0$

9. $6s^2 = 25s$ 10. $6n^2 - 23n - 35 = 0$

11. $4x^2 - 8x = 21$ 12. $6x^2 = 8 - 47x$

In Exercises 13 through 24, solve the given quadratic equations by using the quadratic formula.

13. $x^2 - x - 110 = 0$ 14. $x^2 + 3x - 18 = 0$

15. $x^2 + 2x - 5 = 0$ 16. $x^2 - 7x - 1 = 0$

17. $2x^2 - x = 36$ 18. $3x^2 + x = 14$

19. $4x^2 - 3x - 2 = 0$ 20. $5x^2 + 7x - 2 = 0$

21. $2x^2 + 2x + 5 = 0$ 22. $3x^2 - 4x - 1 = 0$

23. $6x^2 = 9 - 4x$ 24. $8x^2 = 5x + 2$

In Exercises 25 through 36, solve the given quadratic equations by any appropriate method.

25. $x^2 + 4x - 4 = 0$ 26. $x^2 + 3x + 1 = 0$

27. $3x^2 + 8x + 2 = 0$ 28. $3p^2 = 28 - 5p$

29. $4v^2 = v + 5$ 30. $6x^2 - x + 2 = 0$

31. $2x^2 + 3x + 7 = 0$ 32. $4y^2 - 5y = 8$

33. $a^2x^2 + 2ax + 2 = 0$ 34. $16r^2 - 8r + 1 = 0$

35. $ax^2 = a^2 - 3x$ 36. $2bx = x^2 - 3b$

In Exercises 37 through 40, solve the given quadratic equations by completing the square.

37. $x^2 - x - 30 = 0$ 38. $x^2 - 2x - 5 = 0$

39. $2x^2 - x - 4 = 0$ 40. $4x^2 - 8x - 3 = 0$

In Exercises 41 through 44, solve the equations involving fractions. After multiplying through by the lowest common denominator, quadratic equations should result.

41. $\dfrac{x-4}{x-1} = \dfrac{2}{x}$ 42. $\dfrac{x-1}{3} = \dfrac{5}{x} + 1$

43. $\dfrac{x^2 - 3x}{x - 3} = \dfrac{x^2}{x + 2}$ 44. $\dfrac{x-2}{x-5} = \dfrac{15}{x^2 - 5x}$

In Exercises 45 through 48, sketch the graphs of the given functions by using the extreme point, the y-intercept, and one or two other points.

45. $y = 2x^2 - x - 1$ 46. $y = -4x^2 - 1$

47. $y = x - 3x^2$ 48. $y = 2x^2 + 8x - 10$

In Exercises 49 through 52, solve the given quadratic equations graphically. If there are no real roots, simply state this as the answer.

49. $2x^2 + x - 4 = 0$ 50. $-4x^2 - x - 1 = 0$

51. $3x^2 = -x - 2$ 52. $x(6x - 5) = 3$

In Exercises 53 through 60, solve the given quadratic equations by any appropriate method.

53. To determine the electric current in a certain alternating-current circuit, it is necessary to solve the equation $m^2 + 10m + 2000 = 0$. Solve for m.

54. Under specified conditions, the deflection of a beam requires the solution of the equation $40x - x^2 - 400 = 0$, where x is the distance (in feet) from one end of the beam. Solve for x.

55. Under specified conditions, the power developed in an element of an electric circuit is $P = EI - RI^2$, where P is the power, E is the specified voltage, and R is a specified resistance. Assuming that P, E, and R are constants, solve for I.

56. For laminar flow of fluids, the coefficient K_e, used to calculate energy loss due to sudden enlargements, is given by

$$K_e = 1.00 - 2.67\frac{S_a}{S_b} + \left(\frac{S_a}{S_b}\right)^2$$

where S_a/S_b is the ratio of cross-sectional areas. If $K_e = 0.500$, what is the value of S_a/S_b?

57. A company determines that the cost C (in dollars) of manufacturing x units of a certain product is given by $C = 0.1x^2 + 0.8x + 7$. How many units can be made for $25?

58. In determining the width w of a parcel of land, the equation $w^2 + 60w = 5000$ is used. What is the width (in meters) of the parcel?

59. In the theory to study the motion of biological cells and viruses, the equation $n = 2.5p - 12.6p^2$ is used. Solve for p in terms of n.

60. A general formula for the distance s traveled by an object, given an initial velocity v and acceleration a in time t, is $s = vt + \frac{1}{2}at^2$. Solve for t.

In Exercises 61 through 68, set up appropriate equations and solve the given stated problems.

61. The sum of two electric voltages is 20 V, and their product is 96 V^2. What are the voltages?

62. The length of one field is 400 m more than the side of a square field. The width is 100 m more than the side of the square field. If the rectangular field has twice the area of the square field, what are the dimensions of each field?

63. A metal cube expands when heated. If the volume changes by 6.00 mm^3 and each edge is 0.20 mm longer after being heated, what was the original length of an edge of the cube?

64. A military jet flies directly over and at right angles to the straight course of a commercial jet. The military jet is flying at 200 mi/h faster than four times the speed of the commercial jet. How fast is each going if they are 2050 mi apart (on a direct line) after 1 h?

65. The manufacturer of a disk-shaped machine part of radius 1.00 in. discovered that he could prevent taking a loss in its production if the amount of material used in each was reduced by 20%. If the thickness remains the same, by how much must the radius be reduced in order to prevent a loss?

66. A roof truss is in the shape of a right triangle with the hypotenuse along the base. If one rafter (neglect overhang) is 4 ft longer than the other, and the base is 36 ft, what are the lengths of the rafters?

67. An electric utility company is placing utility poles along a road. It is determined that five fewer poles per kilometer would be necessary if the distance between poles were increased by 10 m. How many poles are being placed each kilometer?

68. A rectangular duct in a building's ventilating system is made of sheet metal 7 ft wide and has a cross-sectional area of 3 ft^2. What are the cross-sectional dimensions of the duct?

CHAPTER 7

Trigonometric Functions of Any Angle

When we were dealing with trigonometric functions in Chapter 3, we restricted ourselves primarily to the functions of acute angles measured in degrees. Since we did define the functions in general, we can use these same definitions for finding the functions of any possible angle. In this chapter we shall not only find the trigonometric functions of angles measured in degrees, but we shall develop the use of radian measure and its applications as well. As we will see, this will show how the trigonometric functions can be defined for numbers. We start in the first section by determining the signs of the trigonometric functions in each of the four quadrants.

7–1 Signs of the Trigonometric Functions

We recall the definitions of the trigonometric functions which were given in Section 3–2: *Here the point (x, y) is a point on the terminal side of angle θ, and r is the radius vector.*

$$\sin \theta = \frac{y}{r} \qquad \cos \theta = \frac{x}{r} \qquad \tan \theta = \frac{y}{x}$$

$$\cot \theta = \frac{x}{y} \qquad \sec \theta = \frac{r}{x} \qquad \csc \theta = \frac{r}{y} \tag{7–1}$$

We see that we can find the functions if we know the values of the coordinates (x, y) on the terminal side of θ and the radius vector r. Of course, if either x or y is zero in the denominator, the function is undefined, and we will consider this further in the next section. *Remembering that r is always considered to be positive, we can see that the various functions will vary in sign, depending on the signs of x and y.*

203

If the terminal side of the angle is in the first or second quadrant, the value of sin θ will be positive, but if the terminal side is in the third or fourth quadrant, sin θ is negative. This is because y is positive if the point defining the terminal side is above the x-axis, and y is negative if this point is below the x-axis.

Example A

The value of sin 20° is positive, since the terminal side of 20° is in the first quadrant. The value of sin 160° is positive, since the terminal side of 160° is in the second quadrant. The values of sin 200° and sin 340° are negative, since the terminal sides of these angles are in the third and fourth quadrants, respectively.

The sign of tan θ depends upon the ratio of y to x. In the first quadrant both x and y are positive, and therefore the ratio y/x is positive. In the third quadrant both x and y are negative, and therefore the ratio y/x is positive. In the second and fourth quadrants either x or y is positive and the other negative, and so the ratio of y/x is negative.

Example B

The values of tan 20° and tan 200° are positive, since the terminal sides of these angles are in the first and third quadrants, respectively. The values of tan 160° and tan 340° are negative, since the terminal sides of these angles are in the second and fourth quadrants, respectively.

The sign of cos θ depends upon the sign of x. Since x is positive in the first and fourth quadrants, cos θ is positive in these quadrants. In the same way, cos θ is negative in the second and third quadrants.

Example C

The values of cos 20° and cos 340° are positive, since these angles are first- and fourth-quadrant angles, respectively. The values of cos 160° and cos 200° are negative, since these angles are second- and third-quadrant angles, respectively.

Since csc θ is defined in terms of r and y, as is sin θ, the sign of csc θ is the same as that of sin θ. For the same reason, cot θ has the same sign as tan θ, and sec θ has the same sign as cos θ. A method for remembering the signs of the functions in the four quadrants is as follows:

All functions of first-quadrant angles are positive. Sin θ and csc θ are positive for second-quadrant angles. Tan θ and cot θ are positive for third-quadrant angles. Cos θ and sec θ are positive for fourth-quadrant angles. All others are negative.

This discussion does not include the quadrantal angles, those angles with terminal sides on one of the axes. They will be discussed in the following section.

Example D

sin 50°, sin 150°, sin (− 200°), cos 8°, cos 300°, cos (− 40°), tan 220°, tan (− 100°), cot 260°, cot (− 310°), sec 280°, sec (− 37°), csc 140°, and csc (− 190°) are all positive.

Example E

sin 190°, sin 325°, cos 100°, cos (− 95°), tan 172°, tan 295°, cot 105°, cot (− 6°), sec 135°, sec (− 135°), csc 240°, and csc 355° are all negative.

Example F

Determine the trigonometric functions of θ if the terminal side of θ passes through $(-1, \sqrt{3})$.

We know that $x = -1$, $y = +\sqrt{3}$, and from the Pythagorean theorem we find that $r = 2$. Therefore, the trigonometric functions of θ are:

$$\sin \theta = +\frac{\sqrt{3}}{2} \qquad \cos \theta = -\frac{1}{2} \qquad \tan \theta = -\sqrt{3}$$

$$\cot \theta = -\frac{1}{\sqrt{3}} \qquad \sec \theta = -2 \qquad \csc \theta = +\frac{2}{\sqrt{3}}$$

We note that the point $(-1, \sqrt{3})$ is on the terminal side of a second-quadrant angle, and that the signs of the functions of θ are those of a second-quadrant angle.

Exercises 7–1

In Exercises 1 through 8, determine the algebraic sign of the given trigonometric functions.

1. sin 60°, cos 120°, tan 320°
2. tan 185°, sec 115°, sin (− 36°)
3. cos 300°, csc 97°, cot (− 35°)
4. sin 100°, sec (− 15°), cos 188°
5. cot 186°, sec 280°, sin 470°
6. tan (− 91°), csc 87°, cot 103°
7. cos 700°, tan (− 560°), csc 530°
8. sin 256°, tan 321°, cos (− 370°)

In Exercises 9 through 16, find the trigonometric functions of θ, where the terminal side of θ passes through the given point.

9. (2, 1)
10. (− 1, 1)
11. (− 2, − 3)
12. (4, − 3)
13. (− 5, 12)
14. (− 3, − 4)
15. (5, − 2)
16. (3, 5)

In Exercises 17 through 24, determine the quadrant in which the terminal side of θ lies, subject to the given conditions.

17. sin θ positive, cos θ negative
18. tan θ positive, cos θ negative
19. sec θ negative, cot θ negative
20. cos θ positive, csc θ negative
21. csc θ negative, tan θ negative
22. sec θ positive, csc θ positive
23. sin θ negative, tan θ positive
24. cot θ negative, sin θ negative

7–2 Trigonometric Functions of Any Angle

The trigonometric functions of acute angles were discussed in Section 3–3, and in the last section we determined the signs of the trigonometric functions in each of the four quadrants. In this section we shall show how we can find the trigonometric functions of an angle of any magnitude. This information will be very important in Chapter 8, when we discuss vectors and oblique triangles, and in Chapter 9, when we graph the trigonometric functions. Even *a calculator will not always give the required angle for a given value of a function.*

Any angle in standard position is coterminal with some positive angle less than 360°. Since the terminal sides of coterminal angles are the same, the trigonometric functions of coterminal angles are the same. Therefore, we need consider only the problem of finding the values of the trigonometric functions of positive angles less than 360°.

Example A

The following pairs of angles are coterminal.

$$390° \quad \text{and} \quad 30°, \qquad -60° \quad \text{and} \quad 300°$$
$$900° \quad \text{and} \quad 180°, \qquad -150° \quad \text{and} \quad 210°$$

From this we conclude that the trigonometric functions of both angles in these pairs are equal. That is, for example, $\sin 390° = \sin 30°$ and $\tan(-150°) = \tan 210°$.

Considering the definitions of the functions, we see that the values of the functions depend only on the values of x, y, and r. The values of the functions of second-quadrant angles are numerically equal to the functions of corresponding first-quadrant angles. For example, considering the angles shown in Fig. 7–1, for angle θ_2 with terminal side passing through $(-3, 4)$, $\tan \theta_2 = -\frac{4}{3}$, and for angle θ_1 with terminal side passing through $(3, 4)$, $\tan \theta_1 = \frac{4}{3}$. In Fig. 7–1, we see that the triangles containing angles θ_1 and α are congruent, which means that θ_1 and α are equal. We know that the trigonometric functions of θ_1 and θ_2 are numerically equal. This means that

$$|F(\theta_2)| = |F(\theta_1)| = |F(\alpha)| \tag{7–2}$$

where F represents any of the trigonometric functions.

The angle labeled α is called the **reference angle.** *The reference angle of a given angle is the acute angle formed by the terminal side of the angle and the x-axis.*

Using Eq. (7–2) and the fact that $\alpha = 180° - \theta_2$, we may conclude that the value of any trigonometric function of any second-quadrant angle is found from

$$F(\theta_2) = \pm F(180° - \theta_2) = \pm F(\alpha) \qquad (7\text{–}3)$$

The sign to be used depends on whether the *function* is positive or negative in the second quadrant.

Example B

In Fig. 7–1, the trigonometric functions of θ_2 are as follows:

$$\sin \theta_2 = +\sin (180° - \theta_2) = +\sin \alpha = +\sin \theta_1 = \frac{4}{5}$$

$$\cos \theta_2 = -\cos \theta_1 = -\frac{3}{5}$$

$$\tan \theta_2 = -\frac{4}{3}, \qquad \cot \theta_2 = -\frac{3}{4}$$

$$\sec \theta_2 = -\frac{5}{3}, \qquad \csc \theta_2 = +\frac{5}{4}$$

In the same way we may derive the formulas for finding the trigonometric functions of any third- or fourth-quadrant angle. Considering the angles shown in Fig. 7–2, we see that the reference angle α is found by subtracting 180° from θ_3 and that functions of α and θ_1 are numerically equal. Considering the angles shown in Fig. 7–3, we see that the reference angle α is found by subtracting θ_4 from 360°. Therefore, we have

$$F(\theta_3) = \pm F(\theta_3 - 180°) \qquad (7\text{–}4)$$
$$F(\theta_4) = \pm F(360° - \theta_4) \qquad (7\text{–}5)$$

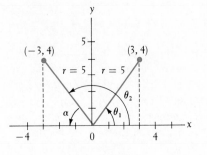

Figure 7–1

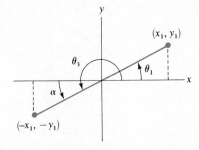

Figure 7–2

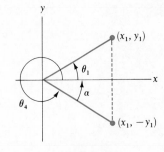

Figure 7–3

Example C

In Fig. 7–2, if $\theta_3 = 210°$, the trigonometric functions of θ_3 are found by using Eq. (7–4) as follows.

$$\sin 210° = -\sin (210° - 180°) = -\sin 30° = -\frac{1}{2} = -0.5000$$
$$\cos 210° = -\cos 30° = -0.8660$$
$$\tan 210° = +0.5774 \qquad \cot 210° = +1.732$$
$$\sec 210° = -1.155 \qquad \csc 210° = -2.000$$

Here we see that the reference angle is $210° - 180° = 30°$, and that we can express the value of a function of $210°$ in terms of the same function of $30°$. We must be careful to attach the correct sign to the result.

Example D

In Fig. 7–3, if $\theta_4 = 315°$, the trigonometric functions of θ_4 are found by using Eq. (7–5) as follows.

$$\sin 315° = -\sin (360° - 315°) = -\sin 45° = -0.7071$$
$$\cos 315° = +\cos 45° = +0.7071$$
$$\tan 315° = -1.000 \qquad \cot 315° = -1.000$$
$$\sec 315° = +1.414 \qquad \csc 315° = -1.414$$

Here we see that the reference angle is $45°$.

Example E

Other illustrations of the use of Eqs. (7–3), (7–4), and (7–5) are as follows.

$$\sin 160° = +\sin (180° - 160°) = \sin 20° = 0.3420$$
$$\tan 110° = -\tan (180° - 110°) = -\tan 70° = -2.747$$
$$\cos 225° = -\cos (225° - 180°) = -\cos 45° = -0.7071$$
$$\cot 260° = +\cot (260° - 180°) = \cot 80° = 0.1763$$
$$\sec 304° = +\sec (360° - 304°) = \sec 56° = 1.788$$
$$\sin 357° = -\sin (360° - 357°) = -\sin 3° = -0.0523$$

In Examples C, D, and E, we have generally rounded off values to four significant digits. We will continue to do this for consistency, unless we are showing calculator display values, which are generally rounded to eight digits.

A calculator can be used directly to find values like those in Examples C, D, and E. We simply enter the angle and then the required function (and use the reciprocal key for $\cot \theta$, $\sec \theta$, and $\csc \theta$, as shown in Section 3–3). The calculator will give us the value of the function of any angle, with the proper sign. However, these examples show how the reference angle is used, and this is important when using a calculator in finding the angle for a given value of a function. As we noted earlier, *if we have the value of the function and want to find the angle, the calculator will not necessarily give us directly the required angle.* It will give us an angle we can use, but whether or not it is the required angle for the problem will depend on the problem being solved.

If a value of a trigonometric function is entered into a calculator, depending on the sign of the value of the function, it is programmed to give the angle as follows. *For positive values of a function, the calculator displays positive acute angles. For negative values of sin θ and tan θ, the calculator displays negative acute angles for* ARCSIN *and* ARCTAN . *For negative values of cos θ, the calculator displays positive angles between 90° and 180° for* ARCCOS . The reason for this type of display is explained in Chapter 19, when these functions are discussed in more detail.

Example F

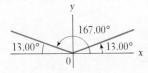

Figure 7–4

For sin θ = 0.2250, the calculator sequence to find θ is (with the calculator in degree mode) .2250 ARCSIN . (Depending on your calculator, SIN⁻¹ or INV SIN may be used for ARCSIN .) The display is 13.002878, which means that θ = 13.00° (rounded off).

This result is correct, but we must remember that

$$\sin(180° - 13.00°) = \sin 167.00° = 0.2250$$

also. If we are dealing only with acute angles, the answer θ = 13.00° is correct. However, if the problem requires a second-quadrant angle for the answer, then θ = 167.00° is the required angle. See Fig. 7–4.

Example G

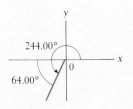

Figure 7–5

Given that tan θ = 2.050 and cos θ < 0, find θ when 0° ≤ θ < 360°.

Since tan θ is positive and cos θ is negative, θ must be a third-quadrant angle. Using the calculator sequence 2.050 ARCTAN , the display is 63.996654, which means that tan 64.00° = 2.050. However, we require a third-quadrant angle, which means that we must add 64.00° to 180°. Thus, the required angle is 244.00°. See Fig. 7–5.

If we are given that tan θ = −2.050 and cos θ < 0, and use the calculator sequence 2.050 +/− ARCTAN , the display is −63.996654. We would have to recognize that the reference angle is 64.00° and subtract it from 180° to get 116.00°, the required second-quadrant angle.

Considering the type of calculator displays described earlier, we see that the calculator gives the reference angle (disregarding any minus signs) in all cases except when cos θ is negative. To avoid confusion from the angle displayed by the calculator, *a good procedure is to find the reference angle first.* Then it can be used to determine the angle required by the problem.

We can find the reference angle by entering the absolute value of the function. The angle displayed will be the reference angle. Then the required angle θ ($0° \leq \theta < 360°$) is found by use of the reference angle α as described previously, and as shown in Eqs. (7–6) for θ in the indicated quadrant.

$$\begin{aligned}
\theta &= \alpha && \text{(first quadrant)} \\
\theta &= 180° - \alpha && \text{(second quadrant)} \\
\theta &= 180° + \alpha && \text{(third quadrant)} \\
\theta &= 360° - \alpha && \text{(fourth quadrant)}
\end{aligned}$$ (7–6)

Example H

Figure 7–6

Given that $\cos \theta = -0.1298$, find θ for $0° \leq \theta < 360°$.

Since $\cos \theta$ is negative, θ is either a second-quadrant angle or a third-quadrant angle. Using the calculator sequence .1298 $\boxed{\text{ARCCOS}}$, the display is 82.541965, which means that the reference angle is 82.54°.

To get the required second-quadrant angle, we subtract 82.54° from 180° and obtain 97.46°. To get the required third-quadrant angle, we add 82.54° to 180° to obtain 262.54°. See Fig. 7–6.

If we were to use the calculator sequence .1298 $\boxed{+/-}$ $\boxed{\text{ARCCOS}}$ to get our answers, the display would be 97.458035. This gives us the required second-quadrant angle of 97.46° directly. However, to get the third-quadrant angle we must then subtract 97.46° from 180° to get the reference angle of 82.54°. The reference angle is then added to 180° to obtain the result of 262.54°.

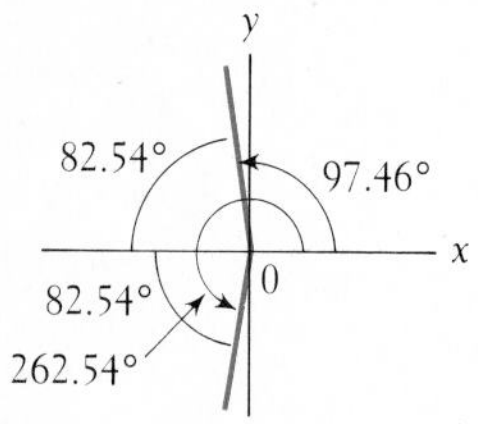

With the use of Eqs. (7–3) through (7–5) we may find the value of any function, as long as the terminal side of the angle lies *in* one of the quadrants. This problem reduces to finding the function of an acute angle. We are left with *the angle for which the terminal side is along one of the axes, a* **quadrantal angle.** Using the definitions of the functions, and remembering that $r > 0$, we arrive at the values in the following table.

θ	$\sin \theta$	$\cos \theta$	$\tan \theta$	$\cot \theta$	$\sec \theta$	$\csc \theta$
0°	0.000	1.000	0.000	undef.	1.000	undef.
90°	1.000	0.000	undef.	0.000	undef.	1.000
180°	0.000	-1.000	0.000	undef.	-1.000	undef.
270°	-1.000	0.000	undef.	0.000	undef.	-1.000
360°	Same as the functions of 0° (same terminal side)					

The values in the table may be verified by referring to the figures in Fig. 7–7.

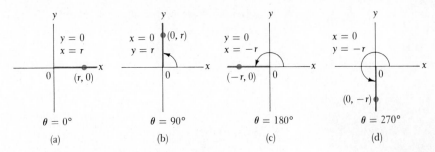

Figure 7–7

Example I

Since $\sin \theta = y/r$, from Fig. 7–7(a) we see that $\sin 0° = 0/r = 0$.

Since $\tan \theta = y/x$, from Fig. 7–7(b) we see that $\tan 90° = r/0$, which is undefined due to the division by zero.

Since $\cos \theta = x/r$, from Fig. 7–7(c) we see that $\cos 180° = -r/r = -1$.

Since $\cot \theta = x/y$, from Fig. 7–7(d) we see that $\cot 270° = 0/-r = 0$.

Exercises 7–2

In Exercises 1 through 8, express the given trigonometric functions in terms of the same function of a positive acute angle.

1. $\sin 160°$, $\cos 220°$

2. $\tan 91°$, $\sec 345°$

3. $\tan 105°$, $\csc 302°$

4. $\cos 190°$, $\cot 290°$

5. $\sin(-123°)$, $\cot 174°$

6. $\sin 98°$, $\sec(-315°)$

7. $\cos 400°$, $\tan(-400°)$

8. $\tan 920°$, $\csc(-550°)$

In Exercises 9 through 16, find the values of the given trigonometric functions by finding the reference angle, using a calculator, and attaching the proper sign.

9. $\sin 195.00°$

10. $\tan 311.00°$

11. $\cos 106.30°$

12. $\sin 103.40°$

13. $\tan 219.15°$

14. $\cos 198.82°$

15. $\sec 328.33°$

16. $\cot 136.53°$

In Exercises 17 through 24, find the values of the given trigonometric functions directly, using a calculator.

17. $\tan 152.40°$

18. $\cos 341.40°$

19. $\sin 310.36°$

20. $\tan 242.68°$

21. $\cos 110.25°$

22. $\sin 163.54°$

23. $\csc 194.82°$

24. $\sec 261.08°$

In Exercises 25 through 32, find θ to the nearest $0.01°$ for $0° \le \theta < 360°$. Use a calculator.

25. $\sin \theta = -0.8480$

26. $\tan \theta = -1.830$

27. $\cos \theta = 0.4003$

28. $\sin \theta = 0.6374$

29. $\tan \theta = 0.2833$

30. $\cos \theta = -0.9287$

31. $\cot \theta = -0.2126$

32. $\csc \theta = -1.096$

In Exercises 33 through 40, find θ to the nearest 0.01° for $0° \le \theta < 360°$. Use a calculator.

33. $\sin \theta = 0.8708$, $\cos \theta < 0$ 34. $\tan \theta = 0.9326$, $\sin \theta < 0$

35. $\cos \theta = -0.1207$, $\tan \theta > 0$ 36. $\sin \theta = -0.1921$, $\tan \theta < 0$

37. $\tan \theta = -1.366$, $\cos \theta > 0$ 38. $\cos \theta = 0.5726$, $\sin \theta < 0$

39. $\sec \theta = 2.047$, $\tan \theta < 0$ 40. $\cot \theta = -0.3256$, $\sin \theta > 0$

In Exercises 41 through 44, determine the function which satisfies the given conditions. Use a calculator.

41. Find $\tan \theta$ when $\sin \theta = -0.5736$ and $\cos \theta > 0$.

42. Find $\sin \theta$ when $\cos \theta = 0.4226$ and $\tan \theta < 0$.

43. Find $\cos \theta$ when $\tan \theta = -0.8098$ and $\csc \theta > 0$.

44. Find $\cot \theta$ when $\sec \theta = 1.122$ and $\sin \theta < 0$.

In Exercises 45 through 48, insert the proper sign, $>$ or $<$ or $=$, between the given expressions.

45. $\sin 90°$ $2 \sin 45°$ 46. $\cos 360°$ $2 \cos 180°$

47. $\tan 180°$ $\tan 0°$ 48. $\sin 270°$ $3 \sin 90°$

In Exercises 49 through 52, evaluate the given expressions. Use a calculator.

49. Under specified conditions, a force F (in pounds) is determined by solving the following equation for F:

$$\frac{F}{\sin 115.0°} = \frac{46.0}{\sin 35.0°}$$

Find the magnitude of the force.

50. A certain AC voltage can be found from the equation $V = 100 \cos 565.0°$. Find the voltage V.

51. A formula for finding the area of a triangle, knowing sides a and b, and angle C is $A = \frac{1}{2}ab \sin C$. Find the area of a triangle for which $a = 37.2$, $b = 57.2$, and $C = 157.0°$.

52. In calculating the area of a triangular tract of land, a surveyor used the formula in Exercise 51. The surveyor used the values $a = 273$ m, $b = 156$ m, and $C = 112.5°$. Find the required area.

In Exercises 53 through 56, the trigonometric functions of negative angles are considered. In Exercises 54, 55, and 56, use the equations derived in Exercise 53.

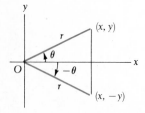

Figure 7–8

53. From Fig. 7–8 we see that $\sin \theta = y/r$ and $\sin(-\theta) = -y/r$. From this we conclude that $\sin(-\theta) = -\sin \theta$. In the same way, verify the remaining equations.

$$\sin(-\theta) = -\sin \theta \qquad \cos(-\theta) = \cos \theta$$
$$\tan(-\theta) = -\tan \theta \qquad \cot(-\theta) = -\cot \theta$$
$$\sec(-\theta) = \sec \theta \qquad \csc(-\theta) = -\csc \theta$$

54. Find (a) $\sin(-60°)$ and   (b) $\cos(-176°)$.

55. Find (a) $\tan(-100°)$ and   (b) $\cot(-215°)$.

56. Find (a) $\sec(-310°)$ and   (b) $\csc(-35°)$.

7–3 Radians

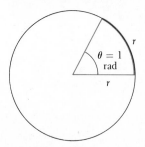

Figure 7–9

For many problems in which trigonometric functions are used, particularly those involving the solution of triangles, degree measurements of angles are quite sufficient. However, in numerous other types of applications and in more theoretical discussions, another way of expressing the measure of angle is more meaningful and convenient. This unit measurement is the **radian,** which we briefly introduced in Chapter 3. *A radian is the measure of an angle with its vertex at the center of a circle and with an intercepted arc on the circle equal in length to the radius of the circle.* See Fig. 7–9.

Since the circumference of any circle in terms of its radius is given by $c = 2\pi r$, the ratio of the circumference to the radius is 2π. This means that the radius may be laid off 2π (about 6.28) times along the circumference, regardless of the length of the radius. Therefore, we see that radian measure is independent of the radius of the circle. In Fig. 7–10 the numbers on each of the radii indicate the number of radians in the angle measured in standard position. The circular arrow shows an angle of 6 radians.

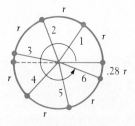

Figure 7–10

Since the radius may be laid off 2π times along the circumference, it follows that there are 2π radians in one complete rotation. Also, there are 360° in one complete rotation. Therefore, 360° is *equivalent* to 2π radians. It then follows that the relation between degrees and radians is 2π rad $= 360°$, or

$$\pi \text{ rad } = 180° \tag{7–7}$$

From this relation we find that

$$1° = \frac{\pi}{180} \text{ rad } = 0.01745 \text{ rad} \tag{7–8}$$

and that

$$1 \text{ rad } = \frac{180°}{\pi} = 57.30° \tag{7–9}$$

We see from Eqs. (7–7) through (7–9) that (1) *to convert an angle measured in degrees to the same angle measured in radians, we multiply the number of degrees by $\pi / 180$, and* (2) *to convert an angle measured in radians to the same angle measured in degrees, we multiply the number of radians by $180/\pi$.*

Example A

$$18.0° = \left(\frac{\pi}{180}\right)(18.0) = \frac{\pi}{10.0} = \frac{3.14}{10.0} = 0.314 \text{ rad}$$

$$120° = \left(\frac{\pi}{180}\right)(120) = \frac{6.28}{3.00} = 2.09 \text{ rad}$$

$$0.400 \text{ rad} = \left(\frac{180°}{\pi}\right)(0.400) = \frac{72.0°}{3.14} = 22.9°$$

$$2.00 \text{ rad} = \left(\frac{180°}{\pi}\right)(2.00) = \frac{360°}{3.14} = 114.6°$$

Due to the nature of the definition of the radian, it is very common to express radians in terms of π. Expressing angles in terms of π is illustrated in the following example.

Example B

$$30° = \left(\frac{\pi}{180}\right)(30) = \frac{\pi}{6} \text{ rad}$$

$$45° = \left(\frac{\pi}{180}\right)(45) = \frac{\pi}{4} \text{ rad}$$

$$\frac{\pi}{2} \text{ rad} = \left(\frac{180°}{\pi}\right)\left(\frac{\pi}{2}\right) = 90°$$

$$\frac{3\pi}{4} \text{ rad} = \left(\frac{180°}{\pi}\right)\left(\frac{3\pi}{4}\right) = 135°$$

We wish now to make a very important point. Since π is a special way of writing the number (slightly greater than 3) that is the ratio of the circumference of a circle to its diameter, it is the ratio of one distance to another. Thus radians really have no units and *radian measure amounts to measuring angles in terms of real numbers*. It is this property of radians that makes them useful in many situations. Therefore, when radians are being used, it is customary that no units are indicated for the angle. *When no units are indicated, the radian is understood to be the unit of angle measurement.*

Example C

$$60.0° = \left(\frac{\pi}{180}\right)(60.0) = \frac{\pi}{3.00} = 1.05$$

$$2.50 = \left(\frac{180°}{\pi}\right)(2.50) = \frac{450°}{3.14} = 143°$$

Since no units are indicated for 1.05 or 2.50 in this example, they are known to be radian measure.

Some calculators have a key which can be used directly to change an angle expressed in degrees to an angle expressed in radians, or from an angle expressed in radians to one expressed in degrees. Generally the key is a second-function key.

We can use a calculator to find the value of a function of an angle in radians. Some calculators use a switch to indicate that either degrees or radians are being used. Other calculators are normally in degree mode, unless the mode is changed by a second-function key. If the calculator is in radian mode, then it uses values in radians directly. Always be careful to use the proper mode.

Example D

To find the value of sin 0.4538, put the calculator in radian mode, and then use the sequence .4538 $\boxed{\text{SIN}}$. The display is 0.4383841, which means that

$$\sin 0.4538 = 0.4384$$

In the same way

$$\tan 0.9977 = 1.550$$

and

$$\cos 1.074 = 0.4766$$

For certain special situations, we may need to know a reference angle in radians. In order to determine the proper quadrant, we should remember that $\frac{1}{2}\pi = 90°$, $\pi = 180°$, $\frac{3}{2}\pi = 270°$, and $2\pi = 360°$. These are shown in Table 7–1 along with the decimal values for angles in radians.

Table 7–1 Quadrantal Angles

Degrees	Radians	Radians (decimal)
90°	$\frac{1}{2}\pi$	1.571
180°	π	3.142
270°	$\frac{3}{2}\pi$	4.712
360°	2π	6.283

Example E

An angle of 3.402 is greater than 3.142, but less than 4.712. This means that it is a third-quadrant angle, and that the reference angle is $3.402 - \pi = 0.260$. The $\boxed{\pi}$ key can be used for this.

An angle of 5.210 is between 4.712 and 6.283. Therefore, it is in the fourth quadrant and the reference angle is $2\pi - 5.210 = 1.073$.

Example F

Express θ in radians, such that $\cos \theta = 0.8829$ and $0 < \theta < 2\pi$.

We are to find θ in radians for the given value of the $\cos \theta$. Also, since θ is restricted to values between 0 and 2π, we must find a first-quadrant angle and a fourth-quadrant angle ($\cos \theta$ is positive in the first and fourth quadrants). With the calculator in radian mode, we use the sequence .8829 $\boxed{\text{ARCCOS}}$ and find that

$$\cos 0.4888 = 0.8829$$

Therefore, for the fourth-quadrant angle,

$$\cos(2\pi - 0.4888) = \cos 5.794$$

This means

$$\theta = 0.4888 \quad \text{or} \quad \theta = 5.794$$

Often when one first encounters radian measure, expressions such as sin 1 and $\sin \theta = 1$ are confused. The first is equivalent to $\sin 57.30°$, since $57.30° = 1$ (radian). The second means that θ is the angle for which the sine is 1. Since $\sin 90° = 1$, we can say that $\theta = 90°$ or that $\theta = \pi/2$. The following examples give additional illustrations of evaluating expressions involving radians.

Example G

$$\sin \frac{\pi}{3} = \frac{\sqrt{3}}{2} \quad \text{since } \frac{\pi}{3} = 60°$$

$$\sin 0.6050 = 0.5688 \quad \text{(We note that } 0.6050 = 34.66°.)$$

$$\tan \theta = 1.709 \quad \text{means that } \theta = 59.67° \text{ (smallest positive } \theta)$$

Since $59.67° = 1.041$, we can state that $\tan 1.041 = 1.709$.

Exercises 7–3

In Exercises 1 through 8, express the given angle measurements in radian measure in terms of π.

1. 15°, 150°   2. 12°, 225°   3. 75°, 330°   4. 36°, 315°

5. 210°, 270°   6. 240°, 300°   7. 160°, 260°   8. 66°, 350°

In Exercises 9 through 16, the given numbers express angle measure. Express the measure of each angle in terms of degrees.

9. $\dfrac{2\pi}{5}, \dfrac{3\pi}{2}$ 10. $\dfrac{3\pi}{10}, \dfrac{5\pi}{6}$ 11. $\dfrac{\pi}{18}, \dfrac{7\pi}{4}$ 12. $\dfrac{7\pi}{15}, \dfrac{4\pi}{3}$

13. $\dfrac{17\pi}{18}, \dfrac{5\pi}{3}$ 14. $\dfrac{11\pi}{36}, \dfrac{5\pi}{4}$ 15. $\dfrac{\pi}{12}, \dfrac{3\pi}{20}$ 16. $\dfrac{7\pi}{30}, \dfrac{4\pi}{15}$

In Exercises 17 through 24, express the given angles in radian measure. Round off results to three significant digits.

17. 23.0°   18. 54.3°   19. 252.0°   20. 104.0°

21. 333.5°   22. 168.7°   23. 178.5°   24. 86.1°

In Exercises 25 through 32, the given numbers express angle measure. Express the measure of each angle in terms of degrees to the nearest 0.1°.

25. 0.750   26. 0.240   27. 3.00   28. 1.70

29. 2.45   30. 34.4   31. 16.4   32. 100

In Exercises 33 through 40, evaluate the given trigonometric functions by first changing the radian measure to degree measure. Round off results to four significant digits.

33. $\sin \dfrac{\pi}{4}$ 34. $\cos \dfrac{\pi}{6}$ 35. $\tan \dfrac{5\pi}{12}$ 36. $\sin \dfrac{7\pi}{18}$

37. $\cos \dfrac{5\pi}{6}$ 38. $\tan \dfrac{4\pi}{3}$ 39. $\sec 4.592$ 40. $\cot 3.273$

In Exercises 41 through 48, evaluate the given trigonometric functions directly, without first changing the radian measure to degree measure. Use a calculator in radian mode.

41. $\tan 0.7359$ 42. $\cos 0.9308$ 43. $\sin 4.24$ 44. $\tan 3.47$

45. $\cos 2.07$ 46. $\sin 2.34$ 47. $\cot 4.86$ 48. $\csc 6.19$

In Exercises 49 through 56, find θ to four significant figures for $0 \le \theta < 2\pi$.

49. $\sin \theta = 0.3090$ 50. $\cos \theta = -0.9135$ 51. $\tan \theta = -0.2126$

52. $\sin \theta = -0.0436$ 53. $\cos \theta = 0.6742$ 54. $\tan \theta = 1.860$

55. $\sec \theta = -1.307$ 56. $\csc \theta = 3.940$

In Exercises 57 through 60, evaluate the given expressions.

57. In optics, when determining the positions of maximum light intensity under specified conditions, the equation $\tan \alpha = \alpha$ is found. Show that a solution to this equation is $\alpha = 1.43\pi$.

58. The instantaneous voltage in a 120-V, 60-Hz power line is given approximately by the equation $V = 170 \sin 377t$, where t is the time (in seconds) after the generator started. Calculate the instantaneous voltage (a) after 0.001 s and (b) after 0.010 s.

59. The velocity v of an object undergoing simple harmonic motion at the end of a spring is given by

$$v = A\sqrt{\frac{k}{m}} \cos \sqrt{\frac{k}{m}}\, t$$

Here m is the mass of the object (in grams), k is a constant depending on the spring, A is the maximum distance the object moves, and t is the time (in seconds). Find the velocity (in centimeters per second) after 0.100 s of a 36.0-g object at the end of a spring for which $k = 400$ g/s^2, if $A = 5.00$ cm.

60. At a point x feet from the base of a building, the angle of elevation of the top of the building is θ. An excellent approximation to the error e in measuring the height of the building due to a small error $(\theta_1 - \theta)$ in measuring θ is given by $e = x(\theta_1 - \theta)\sec^2\theta$. Here it is necessary for $(\theta_1 - \theta)$ to be measured in radians. Calculate the error (in feet), if $x = 180$ ft, $\theta_1 = 30.5°$, and $\theta = 30.0°$.

7–4 Applications of the Use of Radian Measure

Radian measure has numerous applications in mathematics and technology, some of which were indicated in the last four exercises of the previous section. In this section we shall illustrate the usefulness of radian measure in certain specific geometric and physical applications.

From geometry we know that *the length of an arc on a circle is proportional to the central angle* and that the length of the arc of a complete circle is the circumference. Letting s stand for the length of arc, we may state that $s = 2\pi r$ for a complete circle. Since 2π is the central angle (in radians) of the complete circle, *we have as the length of arc*

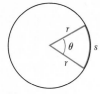

$$s = \theta r \qquad\qquad (7\text{–}10)$$

for any circular arc with central angle θ. If we know the central angle in radians and the radius of a circle, we can find the length of a circular arc directly by using Eq. (7–10). See Fig. 7–11.

Figure 7–11

Example A

If $\theta = \pi/6$ and $r = 3.00$ in.,

$$s = \left(\frac{\pi}{6}\right)(3.00) = \frac{\pi}{2.00} = 1.57 \text{ in.}$$

See Fig. 7–12(a).

If the arc length is 7.20 cm for a central angle of 150° on a certain circle, we may find the radius of the circle by solving $s = \theta r$ for r and then substituting. Thus, $r = s/\theta$. Substituting we have

$$r = \frac{7.20}{150\left(\dfrac{\pi}{180}\right)} = \frac{(7.20)(180)}{150\pi} = 2.75 \text{ cm}$$

See Fig. 7–12(b).

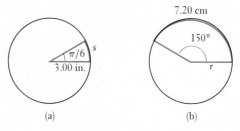

(a)   (b)

Figure 7–12

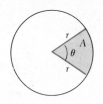

Figure 7–13

Another geometric application of radians is in finding the area of a sector of a circle (see Fig. 7–13). We recall from geometry that areas of sectors of circles are proportional to their central angles. The area of a circle is given by $A = \pi r^2$. This can be written as $A = \frac{1}{2}(2\pi)r^2$. We now note that the angle for a complete circle is 2π, and therefore *the area of any sector of a circle in terms of the radius and the central angle is*

$$A = \frac{1}{2}\theta r^2 \qquad\qquad (7\text{–}11)$$

Example B

The area of a sector of a circle with central angle 218° and a radius of 5.25 in. is

$$A = \frac{1}{2}(218)\left(\frac{\pi}{180}\right)(5.25)^2 = 52.4 \text{ in.}^2$$

See Fig. 7–14(a).

Given that the area of a sector is 75.5 ft^2 and the radius is 12.2 ft, we can find the central angle by solving for θ and then substituting.

$$\theta = \frac{2A}{r^2},$$

$$= \frac{2(75.5)}{(12.2)^2} = 1.01$$

This means that the central angle is 1.01 rad, or 57.9°. See Fig. 7–14(b).

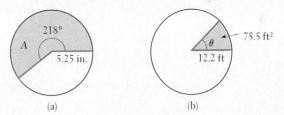

(a)   (b)

Figure 7–14

We should note again that the equations of this section require that the angle θ be expressed in *radians*. A relatively common error is to use θ in degrees.

The next illustration deals with velocity. We know that average velocity is defined by the equation $v = s/t$, where v is the average velocity, s is the distance traveled, and t is the elapsed time. If an object is moving around a circular path with constant speed, the actual distance traveled is the length of arc traversed. Therefore, if we divide both sides of Eq. (7–10) by t, we obtain

$$\frac{s}{t} = \frac{\theta}{t} r$$

or

$$\overline{v = \omega r} \tag{7–12}$$

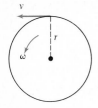

Figure 7–15

Equation (7–12) expresses the relationship between the **linear velocity** v *and the* **angular velocity** ω *of an object moving around a circle of radius* r. See Fig. 7–15. In the figure, v is shown directed tangent to the circle, for that is its direction for the position shown. The direction of v changes constantly.

The most convenient units for ω are radians per unit of time. In this way the formula can be used directly. However, in practice, ω is often given in revolutions per minute, or in some similar unit. In cases like these, it is necessary to convert the units of ω to radians per unit of time before substituting in Eq. (7–12).

Example C

A woman on a hang glider is moving in a horizontal circular arc of radius 90.0 m with an angular velocity of 0.125 rad/s. Her linear velocity is

$$v = (0.125)(90.0) = 11.3 \text{ m/s}$$

(Remember that radians are numbers and are not included in the final set of units.) This means that she is moving along the circumference of the arc at 11.3 m/s (40.7 km/h).

Example D

A flywheel rotates with an angular velocity of 20.0 r/min. If its radius is 18.0 in., find the linear velocity of a point on the rim.

Since there are 2π radians in each revolution,

$$20.0 \text{ r/min} = 40.0\pi \text{ rad/min}$$

Therefore,

$$v = (40.0\ \pi)(18.0) = 2260 \text{ in./min}$$

This means that the linear velocity is 2260 in./min, which is equivalent to 188 ft/min or 3.14 ft/s.

Example E

A pulley belt 10.0 ft long takes 2.00 s to make one complete revolution. The radius of the pulley is 6.00 in. What is the angular velocity (in revolutions per minute) of a point on the rim of the pulley?

Since the linear velocity of a point on the rim of the pulley is the same as the velocity of the belt, $v = 10.0/2.00 = 5.00$ ft/s. The radius of the pulley is $r = 6.00$ in. $= 0.500$ ft, and we can find ω by substituting into Eq. (7–12). This gives us

$$5.00 = \omega(0.500)$$

or

$$\omega = 10.0 \text{ rad/s}$$
$$= 600 \text{ rad/min}$$
$$= 95.5 \text{ r/min}$$

As it is shown in Appendix B, the change of units can be handled algebraically as

$$10.0\frac{\text{rad}}{\text{s}} \times 60\frac{\text{s}}{\text{min}} = 600\frac{\text{rad}}{\text{min}}, \qquad \frac{600 \text{ rad/min}}{2\pi \text{ rad/r}} = 600\frac{\text{rad}}{\text{min}} \times \frac{1}{2\pi}\frac{\text{r}}{\text{rad}}$$

Example F

The current at any time in a certain alternating-current electric circuit is given by $i = I \sin 120\pi t$, where I is the maximum current and t is the time in seconds. Given that $I = 0.0685$ A, find i for $t = 0.00500$ s.

Substituting we get

$$i = 0.0685 \sin(120\pi)(0.00500)$$
$$= 0.0651 \text{ A}$$

With a calculator in radian mode, the sequence of keys is

.0685 $\boxed{\times}$ $\boxed{(}$ 120 $\boxed{\times}$ $\boxed{\pi}$ $\boxed{\times}$ .005 $\boxed{)}$ $\boxed{\text{SIN}}$ $\boxed{=}$.

Exercises 7–4

In Exercises 1 through 34, solve the given problems.

1. In a circle of radius 10.0 in., find the length of arc intercepted on the circumference by a central angle of 60.0°.

2. In a circle of diameter 4.50 ft, find the length of arc intercepted on the circumference by a central angle of 42.0°.

3. Find the area of the circular sector indicated in Exercise 1.

4. Find the area of a sector of a circle, given that the central angle is 120.0° and the diameter is 56.0 cm.

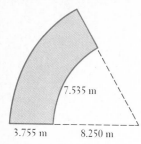

Figure 7–16

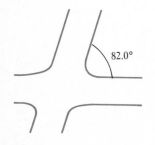

Figure 7–17

5. A section of a natural gas pipeline 3.25 km long is part of a circular arc. If the radius of the circle is 8.50 km, what is the central angle of the arc?

6. A cam is in the shape of a circular sector, as shown in Fig. 7–16. What is the perimeter of (distance around) the cam?

7. A lawn sprinkler can water up to a distance of 65.0 ft. It turns through an angle of 115°. What area can the sprinkler water?

8. A beam of light from a spotlight sweeps through a horizontal angle of 75.0°. If the range of the spotlight is 250 ft, how large an area can it cover?

9. If a car makes a U-turn in 6 s, what is its average angular velocity in the turn?

10. The roller on a computer printer makes 2000 r/min. What is its angular velocity?

11. What is the floor area of the hallway shown in Fig. 7–17? The outside and inside of the hallway are circular arcs.

12. The arm of a car windshield wiper is 12.75 in. long and is attached at the middle of a 15.00 in. blade. (Assume that the arm and blade are in line.) What area of the windshield is cleaned by the wiper if it swings through 110.0° arcs?

13. A pendulum 3.00 ft long oscillates through an angle of 5.0°. Find the distance through which the end of the pendulum swings in going from one extreme position to the other.

14. The radius of the earth is about 3960 mi. What is the length, in miles, of an arc of the earth's equator for a central angle of 1.0°?

15. In turning, an airplane traveling at 540 km/h moves through a circular arc for 2.00 min. What is the radius of the circle, given that the central angle is 8.0°?

16. An ammeter needle is deflected 52.00° by a current of 0.2500 A. The needle is 3.750 in. long and a circular scale is used. How long is the scale for a maximum current of 1.500 A?

17. A flywheel rotates at 300 r/min. If the radius is 6.00 cm, through what total distance does a point on the rim travel in 30.0s?

18. The propeller of the motor on a motor boat is rotating at 130 rad/s. What is the linear velocity of a point on the tip of a blade if it is 22.5 cm long?

19. Two streets meet at an angle of 82.0°. What is the length of the piece of curved curbing at the intersection if it is constructed along the arc of a circle 15.0 ft in radius? See Fig. 7–18.

20. In traveling one-third of the way along a traffic circle, a car travels 0.125 km. What is the radius of the traffic circle?

21. An automobile is traveling at 60.0 mi/h (88.0 ft/s). The tires are 24.5 in. in diameter. What is the angular velocity of the tires in rad/s?

22. Find the velocity, due to the rotation of the earth, of a point on the surface of the earth at the equator (see Exercise 14).

23. An astronaut in a spacecraft circles the moon once each 1.95 h. If his altitude is constant at 70.0 mi, what is his velocity? The radius of the moon is 1080 mi.

24. A man walking at 3.0 ft/s pushes a lawn roller of diameter 2.0 ft. How many revolutions per minute does the roller make?

25. The equation for the displacement y, in centimeters, of an object oscillating up and down at the end of a spring is given by $y = 3.550 \cos 2.580t$, where t is the time in seconds. What is the displacement after 1.575 s?

Figure 7–18

26. The voltage at time t, in seconds, in a certain electric circuit is given by $v = 120 \cos 30\pi t$. Find v, in volts, when $t = 0.0125$ s.

27. The armature of a dynamo is 1.38 ft in diameter and is rotating at 1200 r/min. What is the linear velocity of a point on the rim of the armature?

28. A pulley belt 38.5 cm long takes 2.50 s to make one complete revolution. The diameter of the pulley is 11.0 cm. What is the angular velocity, in r/min, of the pulley?

29. The moon is about 240,000 mi from the earth. It takes the moon about 28 days to make one revolution. What is its angular velocity about the earth, in rad/s?

30. A phonograph record 6.915 in. in diameter rotates 45.00 times per minute. What is the linear velocity of a point on the rim in ft/s?

31. A 1500-kW wind turbine (windmill) rotates at 40.0 r/min. What is the linear velocity of a point on the end of a blade, if the blade is 100 ft. long (from the center of rotation)?

32. The propeller of an airplane is 2.44 m in diameter and rotates at 2200 r/min. What is the linear velocity of a point on the tip of the propeller?

33. A circular sector whose central angle is 210.75° is cut from a circular piece of sheet metal of diameter 12.25 cm. A cone is then formed by bringing the two radii of the sector together. What is the lateral surface area of the cone?

34. A conical tent is made from a circular piece of canvas 15.0 ft in diameter, with a sector of central angle 120° removed. What is the surface area of the tent?

In Exercises 35 through 38, another use of radians is illustrated.

35. It can be shown through advanced mathematics that an excellent approximate method of evaluating $\sin \theta$ or $\tan \theta$ is given by

$$\sin \theta = \tan \theta = \theta \qquad\qquad (7\text{–}13)$$

for small values of θ (the equivalent of a few degrees or less), if θ is expressed in radians. Equation (7–13) can be used for very small values of θ—even some calculators cannot adequately handle very small angles. Using Eq. (7–13), express $1''$ in radians and evaluate $\sin 1''$ and $\tan 1''$.

36. Using Eq. (7–13), evaluate $\tan 0.001°$. Compare with a calculator value.

37. An astronomer observes that a star 12.5 light years away moves through an angle of $0.2''$ in one year. Assuming it moved in a straight line perpendicular to the initial line of observation, how many miles did the star move? (One light year $= 5.88 \times 10^{12}$ mi.) Use Eq. (7–13).

38. In calculating a back line of a lot, a surveyor discovers an error of 0.05° in an angle measurement. If the lot is 136.0 m deep, by how much is the back line calculation in error? See Fig. 7–19. Use Eq. (7–13).

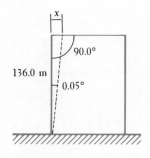

Figure 7–19

7–5 Exercises for Chapter 7

In Exercises 1 through 4, find the trigonometric functions of θ given that the terminal side of θ passes through the given point.

1. (6, 8)   2. (−12, 5)   3. (7, −2)   4. (−2, −3)

In Exercises 5 through 8, express the given trigonometric functions in terms of the same function of a positive acute angle.

5. cos 132°, tan 194°   6. sin 243°, cot 318°

7. sin 289°, sec(−15°)   8. cos 103°, csc(−100°)

In Exercises 9 through 12, express the given angle measurements in terms of π.

9. 40°, 153°   10. 22.5°, 324°   11. 48°, 202.5°   12. 27°, 162°

In Exercises 13 through 20, the given numbers represent angle measure. Express the measure of each angle in terms of degrees.

13. $\dfrac{7\pi}{5}, \dfrac{13\pi}{18}$ 14. $\dfrac{3\pi}{8}, \dfrac{7\pi}{20}$ 15. $\dfrac{\pi}{15}, \dfrac{11\pi}{6}$ 16. $\dfrac{17\pi}{10}, \dfrac{5\pi}{4}$

17. 0.560   18. 1.354   19. 3.607   20. 14.5

In Exercises 21 through 28, express the given angles in radians. (Do not answer in terms of π.)

21. 100°   22. 305°   23. 20.25°   24. 148.38°

25. 262.05°   26. 18.72°   27. 136.2°   28. 385.4°

In Exercises 29 through 48, determine the values of the given trigonometric functions directly on a calculator.

29. cos 245.50°   30. sin 141.33°   31. cot 295.07°   32. tan 184.00°

33. csc 247.82°   34. sec 96.17°   35. sin 205.24°   36. cos 326.72°

37. tan 301.41°   38. sin 103.96°   39. tan 256.42°   40. cos 162.32°

41. $\sin \dfrac{9\pi}{5}$ 42. $\sec \dfrac{5\pi}{8}$ 43. $\cos \dfrac{7\pi}{6}$ 44. $\tan \dfrac{23\pi}{12}$

45. sin 0.5906   46. tan 0.8035   47. csc 2.153   48. cot 5.190

In Exercises 49 through 52, find θ to the nearest 0.01° for $0° \leq \theta < 360°$. Use a calculator.

49. tan θ = 0.1817   50. sin θ = −0.9323

51. cos θ = −0.4730   52. cot θ = 1.196

In Exercises 53 through 56, find θ for $0 \leq \theta < 2\pi$. Use a calculator.

53. cos θ = 0.8387   54. sin θ = 0.1045

55. sin θ = −0.8650   56. tan θ = 2.840

In Exercises 57 through 60, find θ to the nearest 0.01° for $0° \leq \theta < 360°$. Use a calculator.

57. cos θ = −0.7222, sin $\theta < 0$ 58. tan θ = −1.683, cos $\theta < 0$

59. cot θ = 0.4291, cos $\theta < 0$ 60. sin θ = 0.2626, tan $\theta < 0$

In Exercises 61 through 76, solve the given problems.

61. The voltage in a certain alternating-current circuit is given by the equation $v = V \cos 25t$, where V is the maximum possible voltage and t is the time in seconds. Find v for $t = 0.1$ s and $V = 150$ V.

62. The displacement (distance from equilibrium position) of a particle moving with simple harmonic motion is given by $d = A \sin 5t$, where A is the maximum displacement and t is the time. Find d, given that $A = 16.0$ cm and $t = 0.460$ s.

63. A pendulum 5.375 ft long swings through an angle of 4.575°. Through what distance does the bob swing in going from one extreme position to the other?

64. Two pulleys have radii of 10.0 in. and 6.00 in., and their centers are 40.0 in. apart. If the pulley belt is uncrossed, what must be the length of the belt?

65. A piece of circular filter paper 15.0 cm in diameter is folded such that its effective filtering area is the same as that of a sector with central angle of 220°. What is the filtering area?

66. A funnel is made from a circular piece of sheet metal 10.0 in. in radius, from which two pieces were removed. The first piece removed was a circle of radius 1.00 in. at the center, and the second piece removed was a sector of central angle 200°. What is the surface area of the funnel?

67. If the apparatus shown in Fig. 7–20 is rotating at 2.00 r/s, what is the linear velocity of the ball?

68. A lathe is to cut material at the rate of 350 ft/min. Calculate the radius of a cylindrical piece that is turned at the rate of 120 r/min.

69. A thermometer needle passes through 55.25° for a temperature change of 40.00°C. If the needle is 5.250 cm long and the scale is circular, how long must the scale be for a maximum temperature change of 150.00°C?

70. Under certain conditions an electron will travel in a circular path when in a magnetic field. If an electron is moving with a linear velocity of 20,000 m/s in a circular path of radius 0.500 m, how far does it travel in 0.100 s?

71. A cam is constructed so that part of it is a circular arc with a central angle of 72.25° and a radius of 5.315 mm. What is the length of arc along this part of the cam?

72. A horizontal water pipe has a radius of 6.00 ft. If the depth of water in the pipe is 3.00 ft, what percentage of the volume of the pipe is filled?

73. An electric fan blade 15.0 cm in radius rotates at 900 r/min. What is the linear velocity of a point at the end of the blade?

74. A circular saw blade 8.20 in. in diameter rotates at 1500 r/min. What is the linear velocity of a point at the end of one of the teeth?

75. A laser beam is transmitted with a "width" of 0.0008°, and makes a circular spot of radius 2.50 km on a distant object. How far is the object from the source of the laser beam? (See Exercise 35 of Section 7–4.)

76. The planet Venus subtends an angle of 15″ to an observer on Earth. If the distance between Venus and Earth is 1.04×10^8 mi, what is the diameter of Venus? (See Exercise 35 of Section 7–4.)

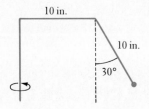

10 in.

10 in.

30°

Figure 7–20

CHAPTER 8

Vectors and Oblique Triangles

We have been dealing with many types of applications in technology. However, to this point we have considered only the magnitude of various quantities. In this chapter we shall consider the very important quantity, a *vector,* for which it is necessary to specify the magnitude and the direction for a complete and meaningful description.

After developing the concept of a vector, we shall then consider the solution of oblique triangles. This will further our study of triangle solution, which we started in Chapter 3.

8–1 Introduction to Vectors

A great many quantities with which we deal may be described by specifying their magnitudes. Generally, one can describe lengths of objects, areas, time intervals, monetary amounts, temperatures, and numerous other quantities by specifying a number: the magnitude of the quantity. *Such quantities are known as* **scalar** *quantities.*

Many other quantities are fully described only when both their magnitude and direction are specified. Such quantities are known as **vectors.** Examples of vectors are velocity, force, and momentum. Vectors are of utmost importance in many fields of science and technology. The following example illustrates a vector quantity and the distinction between scalars and vectors.

Example A

A jet flies over a certain point traveling at 600 mi/h. From this statement alone we know only the *speed* of the jet. *Speed is a scalar quantity,* and it designates only the *magnitude* of the rate.

If we were to add the phrase "in a direction 10° south of west" to the sentence above about the jet, we would be specifying the direction of travel as well as the speed. We then know the *velocity* of the jet; that is, we know the *direction* of travel as well as the *magnitude* of the rate at which the jet is traveling. *Velocity is a vector quantity.*

Let us analyze an example of the action of two vectors: Consider a boat moving in a stream. For purposes of this example, we shall assume that the boat is driven by a motor which can move it at 4 mi/h in still water. We shall assume the current is moving downstream at 3 mi/h. We immediately see that the motion of the boat depends on the direction in which it is headed. If the boat heads downstream, it can travel at 7 mi/h, for the current is going 3 mi/h and the boat moves at 4 mi/h with respect to the water. If the boat heads upstream, however, it progresses at the rate of only 1 mi/h, since the action of the boat and that of the stream are counter to each other. If the boat heads across the stream, the point which it reaches on the other side will not be directly opposite the point from which it started. We can see that this is so because we know that as the boat heads across the stream, the stream is moving the boat downstream *at the same time.*

This last case should be investigated further. Let us assume that the stream is $\frac{1}{2}$ mi wide where the boat is crossing. It will then take the boat $\frac{1}{8}$ h to cross. In $\frac{1}{8}$ h the stream will carry the boat $\frac{3}{8}$ mi downstream. Therefore, when the boat reaches the other side it will be $\frac{3}{8}$ mi downstream. From the Pythagorean theorem, we find that the boat traveled $\frac{5}{8}$ mi from its starting point to its finishing point.

$$d^2 = \left(\frac{4}{8}\right)^2 + \left(\frac{3}{8}\right)^2 = \frac{16 + 9}{64} = \frac{25}{64}; \qquad d = \frac{5}{8} \text{ mi}$$

Since this $\frac{5}{8}$ mi was traveled in $\frac{1}{8}$ h, the magnitude of the velocity of the boat was actually

$$v = \frac{d}{t} = \frac{\frac{5}{8}}{\frac{1}{8}} = \frac{5}{8} \cdot \frac{8}{1} = 5 \text{ mi/h}$$

Also, we see that the direction of this velocity can be represented along a line making an angle θ with the line directed across the stream (see Fig. 8–1).

We have just seen two velocity vectors being *added.* Note that these vectors are not added the way numbers are added. We have to take into account their direction as well as their magnitude. Reasoning along these lines, let us now define the sum of two vectors.

We will represent a vector quantity by a letter printed in boldface type. The same letter in italic (lightface) type represents the magnitude only. Thus, **A** is a vector of magnitude *A*. In handwriting, one usually places an arrow over the letter to represent a vector, such as $\overrightarrow{A}$.

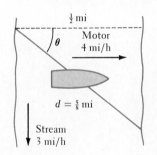

Figure 8–1

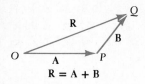

Figure 8–2

Let **A** and **B** represent vectors directed from O to P and P to Q, respectively (see Fig. 8–2). *The vector sum* **A + B** *is the vector* **R**, *from the* **initial point** O *to the* **terminal point** Q. Here, vector **R** is called the **resultant.** *In general, a resultant is a single vector which can replace any number of other vectors and still produce the same physical effect.*

There are two common methods of adding vectors by means of a diagram. The first is illustrated in Fig. 8–3. To add **B** to **A**, we shift **B** parallel to itself until its tail touches the head of **A**. In doing so we are using the meaning of a vector, and we may move it for addition as long as we keep its magnitude and direction unchanged. *The vector sum* **A + B** *is the vector* **R**, *which is drawn from the tail of* **A** *to the head of* **B**. Clearly, when using a diagram to add vectors, the diagram must be drawn with reasonable accuracy.

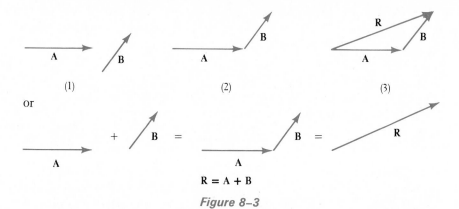

Figure 8–3

Three or more vectors may also be added in the same general manner. We place the initial point of the second vector at the terminal point of the first vector, the initial point of the third vector at the terminal point of the second vector, and so forth. The resultant is the vector from the initial point of the first vector to the terminal point of the last vector. The order in which they are placed together does not matter, as shown in the following example.

Example B

The addition of vectors **A**, **B**, and **C** is shown in Fig. 8–4.

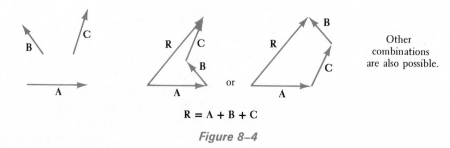

Other combinations are also possible.

$$R = A + B + C$$

Figure 8–4

Another method which is convenient when two vectors are being added is to *let the two vectors being added be the sides of a parallelogram. The resultant is then the diagonal of the parallelogram.* The initial point of the resultant is the common initial point of the vectors being added. In using this method the vectors are first placed tail to tail. See the following example.

Example C

Using the parallelogram method, add vectors **A** and **B** of Fig. 8–3. See Fig. 8–5.

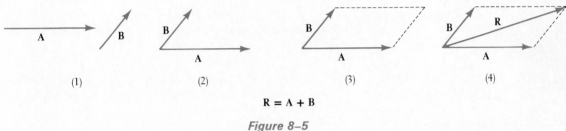

$$R = A + B$$

Figure 8–5

If vector **A** has the same direction as vector **B**, and **A** also has a magnitude n times that of **B**, we may state that $A = nB$. Thus, 2**A** represents a vector twice as long as **A**, but in the same direction.

Example D

For the vectors **A** and **B** in Fig. 8–3, find 3**A** + 2**B**. See Fig. 8–6.

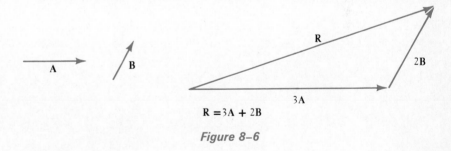

$$R = 3A + 2B$$

Figure 8–6

Vector **B** may be subtracted from vector **A** by reversing the direction of **B** and proceeding as in vector addition. Thus, $A - B = A + (-B)$, where *the minus sign indicates that vector* $-B$ *has the opposite direction of vector* **B**. Vector subtraction is shown in the following example.

Example E

For vectors **A** and **B**, find $2\mathbf{A} - \mathbf{B}$. See Fig. 8–7.

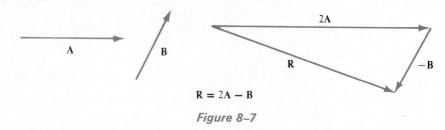

R = 2A − B

Figure 8–7

One very important application of vectors is that of forces acting on an object. The following example illustrates the addition of forces by the parallelogram method.

Example F

Two horizontal ropes are attached to a box. One exerts a force of 50 N directly to the right and the other a force of 35 N at 40° from the first force, as shown in Fig. 8–8(a). Find the resultant force.

We now draw a scale drawing of the forces, as shown in Fig. 8–8(b). We find that the resultant force is about 80 N and acts at an angle of about 16° from the first force.

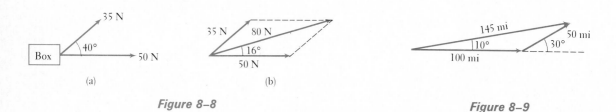

Figure 8–8   *Figure 8–9*

Two other important vector quantities are *velocity* and *displacement*. Velocity as a vector is illustrated in Example A. The displacement of an object is given by the distance from a point of reference and the angle from a reference direction. Displacement is illustrated in the following example.

Example G

A jet travels from city A due east for 100 mi and then turns 30° north of east, and travels another 50 mi. Find its displacement from city A.

We draw a diagram to illustrate the position of the jet, as shown in Fig. 8–9. From the figure we see that the plane is about 145 mi from city A, at an angle of about 10° north of east of the city. By giving both the magnitude and direction, we have given its displacement from the city.

Exercises 8–1

In Exercises 1 through 4, add the given vectors by drawing the appropriate resultant. Use the parallelogram method in Exercises 3 and 4.

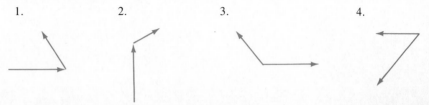

1. 2. 3. 4.

In Exercises 5 through 24, find the indicated vector sums and differences with the given vectors by means of diagrams.

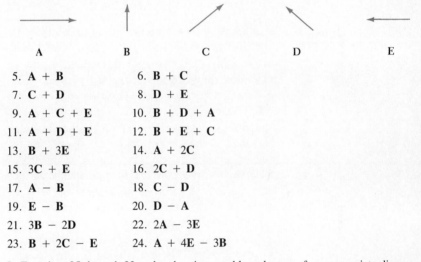

A   B   C   D   E

5. **A + B**   6. **B + C**
7. **C + D**   8. **D + E**
9. **A + C + E**   10. **B + D + A**
11. **A + D + E**   12. **B + E + C**
13. **B + 3E**   14. **A + 2C**
15. **3C + E**   16. **2C + D**
17. **A − B**   18. **C − D**
19. **E − B**   20. **D − A**
21. **3B − 2D**   22. **2A − 3E**
23. **B + 2C − E**   24. **A + 4E − 3B**

In Exercises 25 through 32, solve the given problems by use of an appropriate diagram.

25. Two forces act on a bolt. One is 10 lb and is directed vertically upward, and the other is 12 lb and is directed to the right. Find the resultant force.

26. Two forces act on the end of a beam. One force is 40 N and is directed vertically upward. The second force is 70 N and is directed away from the end of the beam and at an angle of 45° below the horizontal. Find the resultant of these forces.

27. A helium-filled balloon is rising at the rate of 6 ft/s and is in a horizontal wind of 20 ft/s. What is the resultant velocity of the balloon?

28. A small plane travels at 120 mi/h in still air. It is traveling due south in a wind of 30 mi/h from the northeast. What is the resultant velocity of the plane?

29. A person travels 8 mi northwest from her home on a straight road and then turns on another straight road and travels due north for another 10 mi. What is her displacement from home?

30. A ship travels 20 km in a direction of 30° south of east and then turns due south for another 40 km. What is the ship's displacement from its initial position?

31. Three forces act on a small ring in a vertical plane. One force of 20 N acts vertically downward. A second force of 15 N acts horizontally to the right, and the third force of 25 N acts at an angle of 53° above the horizontal to the left. What is the resultant force on the ring?

32. A crate which has a weight of 100 N is suspended by two ropes. The force in one rope is 70 N and is directed to the left at an angle of 60° above the horizontal. What must be the other force in order that the resultant force (including the weight) on the crate is zero?

8–2 Components of Vectors

Adding vectors by means of diagrams is very useful in developing an understanding of vector quantities. However, unless the diagrams are drawn with great care and accuracy, the results we can obtain are quite approximate. Therefore, it is necessary to develop other methods to obtain results of sufficient accuracy.

In this section we show how a given vector can be considered to be the sum of two other vectors, with any required degree of accuracy. In the following section we show how this will allow us to add vectors in order to obtain the sum of vectors with the required accuracy in the result.

Two vectors which, when added together give the original vector, are called **components** *of the original vector.* In the illustration of the boat in Section 8–1, the velocities of 4 mi/h across the stream and 3 mi/h downstream are components of the 5 mi/h vector directed at the angle θ.

In practice, there are certain components of a vector which are of particular importance. If a vector is placed so that its initial point is at the origin of a rectangular coordinate system and its direction is indicated by an angle in standard position, we may find its x- and y-components. These components are vectors directed along the coordinate axes which, when added together, result in the given vector. The initial points of these vectors are at the origin, and the terminal points are located at the points where perpendicular lines from the terminal point of the vector cross the axes. *Finding these component vectors is called* **resolving** *the vector into its components.*

Example A

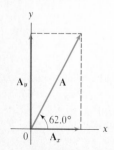

Figure 8–10

Find the x- and y-components of the vector **A** shown in Fig. 8–10. The magnitude of **A** is 7.25.

From the figure we see that A_x, the magnitude of the x-component $\mathbf{A}_x$, is related to **A** by

$$\frac{A_x}{A} = \cos 62.0°$$

or

$$A_x = A \cos 62.0°$$

In the same way, A_y, the magnitude of the y-component $\mathbf{A}_y$, is related to $\mathbf{A}$ ($\mathbf{A}_y$ could be placed along the vertical dashed line) by

$$\frac{A_y}{A} = \sin 62.0°$$

or

$$A_y = A \sin 62.0°$$

From these relations, knowing that $A = 7.25$, we have

$$A_x = 7.25 \cos 62.0° = 3.40$$
$$A_y = 7.25 \sin 62.0° = 6.40$$

This means that the x-component is directed along the x-axis to the right and has a magnitude of 3.40. Also, the y-component is directed along the y-axis upward and its magnitude is 6.40.

Example B

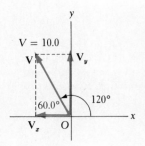

Figure 8–11

Resolve a vector 10.0 units long and directed at an angle of 120.0° into its x- and y-components (see Fig. 8–11).

Placing the initial point of the vector at the origin and putting the angle in standard position, we see that the vector directed along the x-axis, $\mathbf{V}_x$, is related to the vector $\mathbf{V}$ of magnitude V by

$$V_x = V \cos 120.0° = -V \cos 60.0°$$

(The minus sign indicates that the x-component is directed in the negative direction, that is, to the left.) Since the vector directed along the y-axis, $\mathbf{V}_y$, could also be placed along the vertical dashed line, it is related to the vector $\mathbf{V}$ by

$$V_y = V \sin 120.0° = V \sin 60.0°$$

Thus, the vectors $\mathbf{V}_x$ and $\mathbf{V}_y$ have the magnitudes

$$V_x = -10.0(0.5000) = -5.00, \qquad V_y = 10.0(0.8660) = 8.66$$

Therefore, we have resolved the given vector into two components, one directed along the negative x-axis of magnitude 5.00 and the other directed along the positive y-axis of magnitude 8.66.

Example C

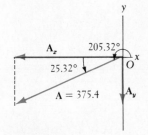

Figure 8–12

Resolve vector $\mathbf{A}$ of magnitude 375.4 and direction $\theta = 205.32°$ into its x- and y-components. See Fig. 8–12.

By placing $\mathbf{A}$ such that θ is in standard position, we see that

$$A_x = A \cos 205.32° = 375.4 \cos 205.32° = -339.3$$

and

$$A_y = A \sin 205.32° = 375.4 \sin 205.32° = -160.5$$

Thus, $\mathbf{A}$ has two components, one directed along the negative x-axis of magnitude 339.3, and the other directed along the negative y-axis of magnitude 160.5.

Example D

The tension in a cable supporting a sign, as shown in Fig. 8–13(a), is 85.0 lb. Find the horizontal and vertical components of the tension.

The tension in the cable is the force that the cable exerts on the sign. Showing the tension in Fig. 8–13(b), we see that

$$T_x = T \cos 53.5° = 85.0 \cos 53.5° = 50.6 \text{ lb}$$
$$T_y = T \sin 53.5° = 85.0 \sin 53.5° = 68.3 \text{ lb}$$

 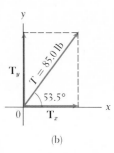

(a)   (b)

Figure 8–13

Exercises 8–2

In Exercises 1 through 4, find the horizontal and vertical components of the vectors shown in the indicated figures. In each the magnitude of the vector is 750.

1.   2.   3.   4.

In Exercises 5 through 16, find the *x*- and *y*-components of the given vectors by use of the trigonometric functions.

5. Magnitude 8.60, $\theta = 68.0°$ 6. Magnitude 9750, $\theta = 243.0°$

7. Magnitude 76.8, $\theta = 145.0°$ 8. Magnitude 0.0998, $\theta = 296.0°$

9. Magnitude 9.04, $\theta = 283.3°$ 10. Magnitude 16,000, $\theta = 156.5°$

11. Magnitude 2.65, $\theta = 197.3°$ 12. Magnitude 67.8, $\theta = 22.5°$

13. Magnitude 0.8734, $\theta = 157.83°$ 14. Magnitude 509.4, $\theta = 221.87°$

15. Magnitude 89,760, $\theta = 7.84°$ 16. Magnitude 1.806, $\theta = 301.83°$

In Exercises 17 through 24, find the horizontal and vertical components of the given vectors.

17. A missile is fired with a velocity of 2250 km/h at an angle 25.0° above the horizontal. What are the components of the velocity of the missile?

18. A football player catches a football, which comes at him with a velocity of 40.0 ft/s at an angle of 35.0° with the horizontal. What are the components of the velocity of the football?

19. The force on an upper hinge of a door is directed downward at an angle of 17.5° with the horizontal. If the force is 45.5 lb, what are the components?

20. A magnetic force of 0.250 N acts on a steel ball. It is directed to the left at an angle 13.6° above the horizontal. What are the components of the force?

21. A jet is 145 km at a position 37.5° north of east of a city. What are the components of the jet's displacement from the city?

22. A helicopter pilot sights a landing position, which has an angle of depression of 15.0° and which is 4500 ft away on a direct line from the helicopter. What are the components of the helicopter's displacement from the landing position?

23. At one point the Pioneer space probe was entering the gravitational field of Jupiter at an angle of 2.55° below the horizontal with a velocity of 18,550 mi/h. What were the components of its velocity?

24. Two upward forces are acting on a bolt. One force of 60.5 lb acts at an angle of 82.4° above the horizontal and the other force of 37.2 lb acts at an angle of 50.5° below the first force. What is the total upward force on the bolt?

8–3 Vector Addition by Components

Now that we have developed the meaning of the components of a vector, we are in a position to add vectors to any degree of required accuracy. To do this we use the vector components, the Pythagorean theorem, and the tangent of the standard position angle of the resultant. In the following example, we show how two vectors at right angles are added.

Example A

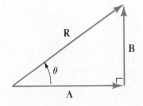

Figure 8–14

Add vectors **A** and **B**, with $A = 14.5$ and $B = 9.10$. The vectors are at right angles, as shown in Fig. 8–14.

We can find the magnitude R of the resultant vector **R** by use of the Pythagorean theorem. This leads to

$$R = \sqrt{A^2 + B^2} = \sqrt{(14.5)^2 + (9.10)^2}$$
$$= 17.1$$

We shall now determine the direction of **R** by specifying its direction as the angle θ, that is, the angle **R** makes with **A**. Therefore,

$$\tan \theta = \frac{B}{A} = \frac{9.10}{14.5} = 0.6276 \quad \text{and} \quad \theta = 32.1°$$

The following calculator sequences can be used to find R and θ.

14.5 $\boxed{x^2}$ $\boxed{+}$ 9.10 $\boxed{x^2}$ $\boxed{=}$ $\boxed{\sqrt{x}}$

and

9.10 $\boxed{\div}$ 14.5 $\boxed{=}$ $\boxed{\text{ARCTAN}}$

Therefore, we see that **R** is a vector of magnitude $R = 17.1$ and in a direction 32.1° from vector **A**.

If vectors are to be added and they are not at right angles, we first place each with tail at the origin. Next, we resolve each vector into its x- and y-components. We then add all the x-components and add the y-components to determine the x- and y-components of the resultant. Then by use of the Pythagorean theorem and the tangent, as in Example A, we find the magnitude and direction of the resultant. **Remember, a vector is not completely specified unless both its magnitude and its direction are given.**

Example B

Find the resultant of two vectors **A** and **B** such that $A = 1200$, $\theta_A = 270.0°$, $B = 1750$, and $\theta_B = 115.0°$.

We first place the vectors on a coordinate system with the tail of each at the origin as shown in Fig. 8–15(a). We then resolve each into its x- and y-components, as shown in Fig. 8–15(b) and as calculated below. (Note that **A** is vertical and has no horizontal component.) Next, the components are combined, as in Fig. 8–15(c) and as calculated. Finally, the magnitude and direction of the resultant, as shown in Fig. 8–15(d), are calculated.

$$A_x = A \cos 270.0° = 0$$
$$A_y = A \sin 270.0° = -1200$$
$$B_x = B \cos 115.0° = 1750 \cos 115.0° = -740$$
$$B_y = B \sin 115.0° = 1750 \sin 115.0° = 1590$$
$$R_x = A_x + B_x = 0 - 740 = -740$$
$$R_y = A_y + B_y = -1200 + 1590 = 390$$
$$R = \sqrt{R_x^2 + R_y^2} = \sqrt{(-740)^2 + (390)^2}$$
$$= 836$$
$$\tan \theta = \frac{R_y}{R_x} = \frac{390}{-740} = -0.5270, \qquad \theta = 152.2°$$

Thus, the resultant has a magnitude of 836 and is directed at an angle of 152.2°. In finding θ from the calculator, the display shows an angle of $-27.8°$. However, we know that θ is a second-quadrant angle, since R_x is negative and R_y is positive. Therefore, we must use 27.8° as a reference angle. Also, in the use of the calculator, several of the above results can be stored in memory and therefore need not actually be recorded.

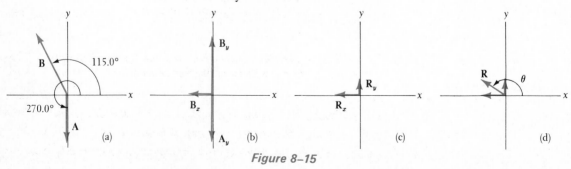

Figure 8–15

Example C

Find the resultant **R** of the two vectors shown in Fig. 8–16(a), **A** of magnitude 8.075 and direction 57.26° and **B** of magnitude 5.437 and direction 322.15°.

$$A_x = 8.075 \cos 57.26° = 4.367$$
$$A_y = 8.075 \sin 57.26° = 6.792$$
$$B_x = 5.437 \cos 322.15° = 4.293$$
$$B_y = 5.437 \sin 322.15° = -3.336$$
$$R_x = A_x + B_x = 4.367 + 4.293 = 8.660$$
$$R_y = A_y + B_y = 6.792 - 3.336 = 3.456$$
$$R = \sqrt{R_x^2 + R_y^2} = \sqrt{(8.660)^2 + (3.456)^2} = 9.324$$
$$\tan \theta = \frac{R_y}{R_x} = \frac{3.456}{8.660} = 0.3991, \qquad \theta = 21.76°$$

The resultant vector is 9.324 units long and is directed at an angle of 21.76°, as shown in Fig. 8–16(c).

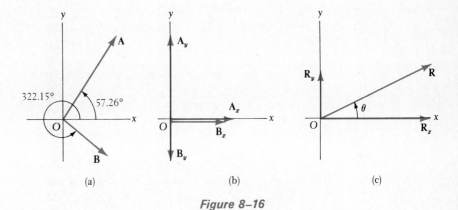

(a)   (b)   (c)

Figure 8–16

Some general formulas can be derived from the previous examples. For a given vector **A**, directed at an angle θ, of magnitude A, and with components A_x and A_y, we have the following relations:

$$A_x = A \cos \theta, \qquad A_y = A \sin \theta \tag{8-1}$$
$$A = \sqrt{A_x^2 + A_y^2} \tag{8-2}$$
$$\tan \theta = \frac{A_y}{A_x} \tag{8-3}$$

From the previous examples we can also see that the following procedure is used for adding vectors.

1. *Resolve the given vectors into their x and y components. Use Eqs. (8–1).*
2. *Add the x-components to obtain* $\mathbf{R}_x$*, and add the y-components to obtain* $\mathbf{R}_y$*.*
3. *Find the magnitude and the direction of the resultant* $\mathbf{R}$ *from the components* $\mathbf{R}_x$ *and* $\mathbf{R}_y$*. Use Eqs. (8–2) and (8–3) in the forms*

$$R = \sqrt{R_x^2 + R_y^2} \quad \text{and} \quad \tan \theta = \frac{R_y}{R_x}$$

Example D

Find the resultant of the three given vectors with $A = 422$, $B = 405$, and $C = 210$, as shown in Fig. 8–17.

We can find the x- and y-components of the vectors by using Eq. (8–1). The following table is helpful for determining the necessary values.

Vector	x-component		y-component	
A	=	$+422$	=	0
B	$-405 \cos 55.0° =$	-232	$-405 \sin 55.0° =$	-332
C	$-210 \cos 70.0° =$	-72	$+210 \sin 70.0° =$	$+197$
R		$+118$		-135

From this table it is possible to compute R and θ:

$$R = \sqrt{(118)^2 + (-135)^2} = 179, \quad \tan \theta = \frac{-135}{118} = -1.144$$

$$\theta = 311.2° \quad \text{(reference angle is } 48.8°)$$

We note that we could have changed the given angles to standard position angles before finding the components. Also, we know that θ is a fourth quadrant angle since R_x is positive and R_y is negative.

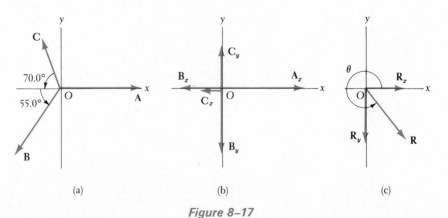

(a)   (b)   (c)

Figure 8–17

Exercises 8–3

In Exercises 1 through 4, vectors **A** and **B** are at right angles. Find the magnitude and direction of the resultant.

1. $A = 14.7$ 2. $A = 592$ 3. $A = 3.086$ 4. $A = 1734$
   $B = 19.2$ $B = 195$ $B = 7.143$ $B = 3297$

In Exercises 5 through 12, with the given sets of components, find R and θ.

5. $R_x = 5.18, R_y = 8.56$

6. $R_x = 89.6, R_y = -52.0$

7. $R_x = -0.982, R_y = 2.56$

8. $R_x = -729, R_y = -209$

9. $R_x = -646, R_y = 2030$

10. $R_x = -31.2, R_y = -41.2$

11. $R_x = 0.6941, R_y = -1.246$

12. $R_x = 7.627, R_y = -6.353$

In Exercises 13 through 24, add the given vectors by using the trigonometric functions and the Pythagorean theorem.

13. $A = 18.0, \theta_A = 0.0°$
    $B = 12.0, \theta_B = 27.0°$

14. $A = 150, \theta_A = 90.0°$
    $B = 128, \theta_B = 43.0°$

15. $A = 56.0, \theta_A = 76.0°$
    $B = 24.0, \theta_B = 200.0°$

16. $A = 6.89, \theta_A = 123.0°$
    $B = 29.0, \theta_B = 260.0°$

17. $A = 9.821, \theta_A = 34.27°$
    $B = 17.45, \theta_B = 752.50°$

18. $A = 1.653, \theta_A = 36.37°$
    $B = 0.9807, \theta_B = 253.06°$

19. $A = 12.653, \theta_A = 98.472°$
    $B = 15.147, \theta_B = 332.092°$

20. $A = 121.36, \theta_A = 292.362°$
    $B = 112.98, \theta_B = 197.892°$

21. $A = 21.9, \theta_A = 236.2°$
    $B = 96.7, \theta_B = 11.5°$
    $C = 62.9, \theta_C = 143.4°$

22. $A = 6300, \theta_A = 189.6°$
    $B = 1760, \theta_B = 320.1°$
    $C = 3240, \theta_C = 75.4°$

23. The vectors shown in Fig. 8–18.   24. The vectors shown in Fig. 8–19.

See Appendix E for a computer program for adding vectors.

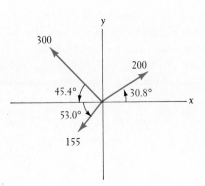

Figure 8–18

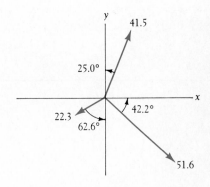

Figure 8–19

8–4 Application of Vectors

In Section 8–1 we introduced the important vector quantities of force, velocity, and displacement, and we found vector sums by use of diagrams. Now we can use the method of Section 8–3 to find sums of these kinds of vectors and others and to use them in various types of applications.

Example A

An object on a horizontal table is acted on by two horizontal forces. The two forces have magnitudes of 6.00 and 8.00 lb, and the angle between their lines of action is 90.0°. What is the resultant of these forces?

By means of an appropriate diagram (Fig. 8–20), we may better visualize the actual situation. We note that a good choice of axes (unless specified, it is often convenient to choose the x- and y-axes to fit the problem) is to have the x-axis in the direction of the 6.00-lb force and the y-axis in the direction of the 8.00-lb force. (This is possible since the angle between them is 90°.) With this choice we note that the two given forces will be the x- and y-components of the resultant. Therefore, we arrive at the following results:

$$F_x = 6.00 \text{ lb}, \qquad F_y = 8.00 \text{ lb},$$
$$F = \sqrt{(6.00)^2 + (8.00)^2} = 10.0 \text{ lb}$$
$$\tan \theta = \frac{F_y}{F_x} = \frac{8.00}{6.00} = 1.333, \qquad \theta = 53.1°$$

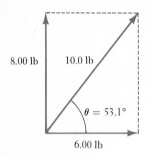

Figure 8–20

We would state that the resultant has a magnitude of 10.0 lb and acts at an angle of 53.1° from the 6.00-lb force.

Example B

A ship sails 32.50 mi due east and then turns 41.25° north of east. After sailing an additional 16.18 mi, where is it with reference to the starting point?

In this problem we are to determine the resultant displacement of the ship from the two given displacements. The problem is diagrammed in Fig. 8–21, where the first displacement has been labeled vector **A** and the second as vector **B**.

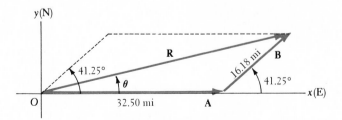

Figure 8–21

Since east corresponds to the positive x-direction, we see that the x-component of the resultant is $\mathbf{A} + \mathbf{B}_x$, and the y-component of the resultant is $\mathbf{B}_y$. Therefore, we have the following results:

$$R_x = A + B_x = 32.50 + 16.18 \cos 41.25°$$
$$= 32.50 + 12.16$$
$$= 44.66 \text{ mi}$$
$$R_y = 16.18 \sin 41.25° = 10.67 \text{ mi}$$
$$R = \sqrt{(44.66)^2 + (10.67)^2} = 45.92 \text{ mi}$$
$$\tan \theta = \frac{10.67}{44.66} = 0.2389$$
$$\theta = 13.44°$$

Therefore, the ship is 45.92 mi from the starting point, in a direction 13.44° north of east.

Example C

An airplane headed due east is in a wind blowing from the southeast. What is the resultant velocity of the plane with respect to the surface of the earth, if the plane's velocity with respect to the air is 600 km/h and that of the wind is 100 km/h (see Fig. 8–22)?

Let $\mathbf{v}_{px}$ be the velocity of the plane in the x-direction (east), $\mathbf{v}_{py}$ the velocity of the plane in the y-direction, $\mathbf{v}_{wx}$ the x-component of the velocity of the wind, $\mathbf{v}_{wy}$ the y-component of the velocity of the wind, and $\mathbf{v}_{pa}$ the velocity of the plane with respect to the air. Therefore,

$$v_{px} = v_{pa} - v_{wx} = 600 - 100(\cos 45.0°) = 529 \text{ km/h}$$
$$v_{py} = v_{wy} = 100(\sin 45.0°) = 70.7 \text{ km/h}$$
$$v = \sqrt{(529)^2 + (70.7)^2} = 534 \text{ km/h}$$
$$\tan \theta = \frac{v_{py}}{v_{px}} = \frac{70.7}{529} = 0.1336, \qquad \theta = 7.6°$$

We have determined that the plane is traveling 534 km/h and is flying in a direction 7.6° north of east. From this we observe that a plane does not necessarily head in the direction of its desired destination.

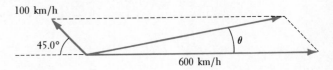

Figure 8–22

Example D

A block is resting on an inclined plane which makes an angle of 30.0° with the horizontal. If the block weighs 100 lb, what is the force of friction between the block and the plane?

The weight of the block is the force exerted on the block due to gravity. Therefore, the weight is directed vertically downward. The frictional force tends to oppose the motion of the block and is directed upward along the plane. The frictional force must be sufficient to counterbalance that component of the weight of the block which is directed down the plane for the block to be at rest. The plane itself "holds up" that component of the weight which is perpendicular to the plane. A convenient set of coordinates (Fig. 8–23) would be one with the x-axis directed up the plane and y-axis perpendicular to the plane. The magnitude of the frictional force $\mathbf{F}_f$ is given by

$$F_f = 100 \sin 30.0° = 100(0.5000) = 50.0 \text{ lb}$$

(Since the component of the weight down the plane is $100 \sin 30.0°$ and is equal to the frictional force, this relation is true.)

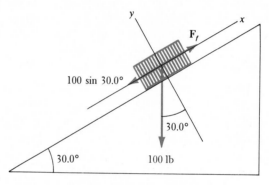

100 sin 30.0°

$\mathbf{F}_f$

30.0°

30.0°

100 lb

All forces are assumed to act
at the center of the block.

Figure 8–23

Example E

A 61.75-lb object hangs from a hook on a wall. If a horizontal force of 42.50 lb pulls the object away from the wall so that the object is in equilibrium (no resultant force in any direction), what is the tension T in the rope attached to the wall?

For the object to be in equilibrium, the tension in the rope must be equal and opposite to the resultant of the 42.50-lb force and the 61.75-lb force which is the weight of the object (see Fig. 8–24). Thus, the magnitude of the x-component of the tension is 42.50 lb and the magnitude of the y-component is 61.75 lb.

$$T = \sqrt{(42.50)^2 + (61.75)^2} = 74.96 \text{ lb}$$

$$\tan \theta = \frac{61.75}{42.50} = 1.453, \qquad \theta = 55.46°$$

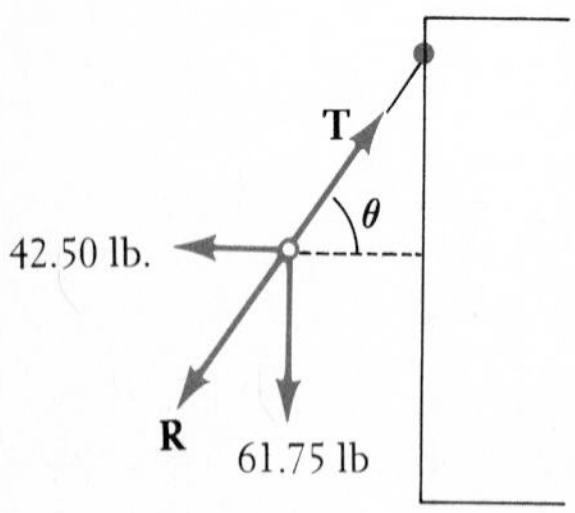

T

θ

42.50 lb.

R   61.75 lb

Figure 8–24

Example F

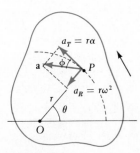

Figure 8–25

If an object rotates about a point O, the tangential component of the acceleration $\mathbf{a}_T$ and the centripetal component of the acceleration $\mathbf{a}_R$ of a point P are given by the expressions shown in Fig. 8–25. The radius of the circle through which P is moving is r, the angular velocity at any instant is ω, and α is the rate at which the angular velocity ω is changing. Given that $r = 2.35$ m, $\omega = 5.60$ rad/s, and $\alpha = 3.75$ rad/s^2, calculate the magnitude of the resultant acceleration and the angle it makes with the tangential component.

$$(\mathbf{a}_R \perp \mathbf{a}_T), \quad a_R = r\omega^2 = (2.35)(5.60)^2 = 73.7 \text{ m/s}^2$$
$$a_T = r\alpha = (2.35)(3.75) = 8.81 \text{ m/s}^2$$
$$a = \sqrt{a_R^2 + a_T^2} = \sqrt{(73.7)^2 + (8.81)^2} = 74.2 \text{ m/s}^2$$

The angle ϕ between $\mathbf{a}$ and $\mathbf{a}_T$ is found from the relation $\tan \phi = a_R / a_T$. Hence,

$$\tan \phi = \frac{73.7}{8.81} = 8.365 \quad \text{or} \quad \phi = 83.2°$$

Exercises 8–4

In Exercises 1 through 24, solve the given problems.

1. Two forces, one of 45.0 lb and the other of 68.0 lb, act on the same object and at right angles to each other. Find the resultant of these forces.

2. Two forces, one of 145.5 lb and the other of 219.3 lb, pull on an object. The angle between these forces is 90.00°. What is the resultant of these forces?

3. Two forces, one of 346.4 N and the other of 521.8 N, pull on an object. The angle between these forces is 25.32°. What is the resultant of these forces?

4. The angle between two forces acting on an object is 63°0′. If the two forces are 86.0 N and 103 N, respectively, what is their resultant?

5. A jet travels 450 mi due west from a city. It then turns and travels 240 mi south. What is its displacement from the city?

6. A ship sails 78.36 km due north after leaving its pier. It then turns and sails 51.21 km east. What is the displacement of the ship from its pier?

7. Town B is 52.10 mi southeast of town A. Town C is 45.25 mi due west of town B. What is the displacement of town C from Town A?

8. A surveyor locates a tree 36.50 m northeast of him. He knows that it is 21.38 m north of a utility pole. What is the surveyor's displacement from the utility pole?

9. A stone is thrown upward into the air with a vertical component of velocity of 72.5 ft/s and a horizontal component of velocity of 57.5 ft/s. What is the resultant velocity of the stone?

10. A rocket is launched with a vertical component of velocity of 2840 km/h and a horizontal component of velocity of 1520 km/h. What is its resultant velocity?

11. A child weighing 60.0 lb sits in a swing and is pulled sideways by a horizontal force of 20.0 lb. What is the tension in each of the supporting ropes? What is the angle between the horizontal and the ropes?

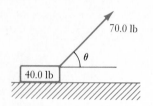

Figure 8–26

12. A rope 12.25 ft long is fastened to supports which are 10.50 ft apart and at the same level. A 175.0 lb weight is then hung from its center. How much is the tension in the supporting rope?

13. A stone is thrown horizontally from a plane traveling at 200 m/s. If the stone is thrown at 50.0 m/s in a direction perpendicular to the direction of the plane, what is the velocity of the stone just after it is released?

14. A plane is headed due north at a velocity of 500 km/h with respect to the air. If the wind is from the southwest at 80.0 km/h, what is the resultant speed of the plane, and in what direction is it traveling?

15. A 70.0-lb force is applied to a 40.0-lb box by a rigid metal rod, as shown in Fig. 8–26. The force is just enough to lift the box from the surface. What angle does the force make with the horizontal, and what is the horizontal component of the force?

16. Assume that the plane in Fig. 8–23 is frictionless and that the angle is not given. If the acceleration due to gravity is 9.80 m/s^2 and the object moves down the plane with an acceleration of 2.75 m/s^2, what is the angle of the plane?

17. In Fig. 8–23, if an additional force of 20.0 lb is exerted on the block upward at an angle of 50.0° with the plane above the block, what is the force exerted on the block by the plane itself?

18. In Fig. 8–23, if a horizontal 10.0-lb force is exerted to the right on the 100-lb object, what would the force of friction have to be so that the object did not move down the plane?

19. An object is dropped from a plane moving at 120 m/s. If the vertical velocity, as a function of time, is given by $v_y = 9.80t$, what is the velocity of the object after 4.50 s? In what direction is it moving?

20. The magnitude of the horizontal and vertical components of displacement of a certain projectile are given by $d_H = 120t$ and $d_V = 160t - 16t^2$, where t is the time in seconds. Find the displacement (in feet) of the object after 3.00 s.

21. In Fig. 8–25, given that $a = 56.41 \text{ ft/s}^2$, $a_R = 37.96 \text{ ft/s}^2$, and $r = 6.255$ ft, find α, the rate of change of angular velocity.

22. A boat travels across a river, reaching the opposite bank at a point directly opposite that from which it left. If the boat travels 6.00 km/h in still water, and the current of the river flows at 3.00 km/h, what was the velocity of the boat in the water?

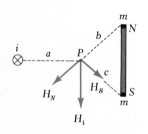

Figure 8–27

23. In Fig. 8–27, a long straight conductor perpendicular to the plane of the paper carries an electric current i. A bar magnet having poles of strength m lies in the plane of the paper. The vectors $\mathbf{H}_i$, $\mathbf{H}_N$, and $\mathbf{H}_S$ represent the components of the magnetic intensity $\mathbf{H}$ due to the current and to the N and S poles of the magnet, respectively. The magnitude of the components of $\mathbf{H}$ are given by

$$H_i = \frac{1}{2\pi}\frac{i}{a}, \qquad H_N = \frac{1}{4\pi}\frac{m}{b^2}, \quad \text{and} \quad H_S = \frac{1}{4\pi}\frac{m}{c^2}$$

Given that $a = 0.300$ m, $b = 0.400$ m, $c = 0.300$ m, the length of the magnet is 0.500 m, $i = 4.00$ A, and $m = 2.00$ A · m, calculate the resultant magnetic intensity $\mathbf{H}$. The component $\mathbf{H}_i$ is parallel to the magnet.

24. Solve the problem of Exercise 23 if $\mathbf{H}_i$ is directed away from the magnet, making an angle of 10.0° with the direction of the magnet.

8–5 Oblique Triangles, the Law of Sines

To this point we have limited our study of triangle solution to right triangles. However, *many triangles which require solution do not contain a right angle. Such triangles are called* **oblique triangles.** Let us now discuss the solutions of oblique triangles.

In Section 3–4 we stated that we need three parts, at least one of them a side, in order to solve any triangle. With this in mind we may determine that there are four possible combinations of parts from which we may solve a triangle. These combinations are as follows:

Case 1.   Two angles and one side.
Case 2.   Two sides and the angle opposite one of them.
Case 3.   Two sides and the included angle.
Case 4.   Three sides.

There are several ways in which oblique triangles may be solved, but we shall restrict our attention to the two most useful methods, the **law of sines** and the **law of cosines.** In this section we shall discuss the law of sines and show that it may be used to solve Case 1 and Case 2.

Let *ABC* be an oblique triangle with sides a, b, and c opposite angles A, B, and C, respectively. By drawing a perpendicular h from B to side b, or its extension, we see from Fig. 8–28(a) that

$$h = c \sin A \quad \text{or} \quad h = a \sin C \tag{8-4}$$

and from Fig. 8–28(b)

$$h = c \sin A \quad \text{or} \quad h = a \sin(180° - C) = a \sin C \tag{8-5}$$

We note that the results are precisely the same in Eqs. (8–4) and (8–5). Setting the results for h equal to each other, we have

$$c \sin A = a \sin C \quad \text{or} \quad \frac{a}{\sin A} = \frac{c}{\sin C} \tag{8-6}$$

By dropping a perpendicular from A to a, we also derive the result

$$c \sin B = b \sin C \quad \text{or} \quad \frac{b}{\sin B} = \frac{c}{\sin C} \tag{8-7}$$

Combining Eqs. (8–6) and (8–7) we have the **law of sines:**

$$\frac{a}{\sin A} = \frac{b}{\sin B} = \frac{c}{\sin C} \tag{8-8}$$

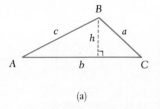

(a)

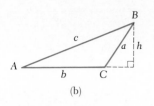

(b)

Figure 8–28

The law of sines is a statement of proportionality between the sides of a triangle and the sines of the angles opposite them.

Now we may see how the law of sines is applied to the solution of a triangle in which two angles and one side are known (Case 1). If two angles are known, the third may be found from the fact that the sum of the angles in a triangle is 180°. At this point we must be able to determine the ratio between the given side and the sine of the angle opposite it. Then, by use of the law of sines, we may find the other sides.

Example A

Given that $c = 6.00$, $A = 60.0°$, and $B = 40.0°$, find a, b, and C.

First we can see that

$$C = 180.0° - (60.0° + 40.0°) = 80.0°$$

Thus,

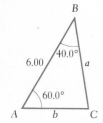

$$\frac{a}{\sin 60.0°} = \frac{6.00}{\sin 80.0°} \quad \text{or} \quad a = \frac{6.00 \sin 60.0°}{\sin 80.0°} = 5.28$$

$$\frac{b}{\sin 40.0°} = \frac{6.00}{\sin 80.0°} \quad \text{or} \quad b = \frac{6.00 \sin 40.0°}{\sin 80.0°} = 3.92$$

Thus, $a = 5.28$, $b = 3.92$, and $C = 80.0°$. See Fig. 8–29.

Since the quotient $6.00/\sin 80.0°$ is used in the solution for both a and b, we can use the memory in a calculator as follows:

Figure 8–29

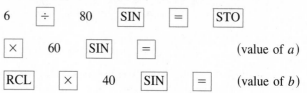

| 6 | ÷ | 80 | SIN | = | STO |

| × | 60 | SIN | = | (value of a) |

| RCL | × | 40 | SIN | = | (value of b) |

Example B

Solve the triangle with the following given parts: $a = 63.7$, $A = 56.0°$, and $B = 97.0°$.

We may immediately determine that $C = 27.0°$. Thus,

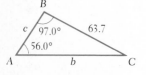

$$\frac{b}{\sin 97.0°} = \frac{63.7}{\sin 56.0°} \quad \text{or} \quad b = \frac{63.7 \sin 97.0°}{\sin 56.0°} = 76.3$$

and

Figure 8–30

$$\frac{c}{\sin 27.0°} = \frac{63.7}{\sin 56.0°} \quad \text{or} \quad c = \frac{63.7 \sin 27.0°}{\sin 56.0°} = 34.9$$

Thus, $b = 76.3$, $c = 34.9$, and $C = 27.0°$. See Fig. 8–30.

Example C

Solve the triangle with the following given parts: $b = 5.063$, $A = 42.25°$, and $C = 28.54°$.

We determine that $B = 109.21°$. Thus,

$$\frac{a}{\sin 42.25°} = \frac{5.063}{\sin 109.21°} \quad \text{or} \quad a = \frac{5.063 \sin 42.25°}{\sin 109.21°} = 3.605$$

and

$$\frac{c}{\sin 28.54°} = \frac{5.063}{\sin 109.21°} \quad \text{or} \quad c = \frac{5.063 \sin 28.54°}{\sin 109.21°} = 2.562$$

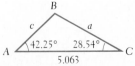

Figure 8–31

Thus, $a = 3.605$, $c = 2.562$, and $B = 109.21°$. See Fig. 8–31.

If the given information is appropriate, the law of sines may be used to solve applied problems. The following example illustrates the use of the law of sines in such a problem.

Example D

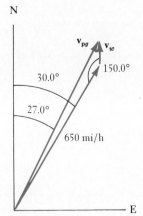

Figure 8–32

A plane traveling at 650 mi/h with respect to the air is headed 30.0° east of north. The wind is blowing from the south, which causes the actual course to be 27.0° east of north. Find the velocity of the wind and the velocity of the plane with respect to the ground.

From the given information the angles are determined, as shown in Fig. 8–32. Then applying the law of sines, we have the relations

$$\frac{v_w}{\sin 3.0°} = \frac{v_{pg}}{\sin 150.0°} = \frac{650}{\sin 27.0°}$$

where v_w is the magnitude of the velocity of the wind and v_{pg} is the magnitude of the velocity of the plane with respect to the ground. Thus,

$$v_w = \frac{650 \sin 3.0°}{\sin 27.0°} = 74.9 \text{ mi/h}$$

and

$$v_{pg} = \frac{650 \sin 150.0°}{\sin 27.0°} = 716 \text{ mi/h}$$

If we have information equivalent to Case 2 (two sides and the angle opposite one of them), we may find that there are *two* triangles which satisfy the given information. The following example illustrates this point.

Example E

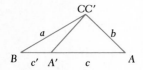

Figure 8–33

Solve the triangle with the following given parts: $a = 60.0$, $b = 40.0$, and $B = 30.0°$.

By making a good scale drawing (Fig. 8–33), we note that the angle opposite a may be either at position A or A'. Both positions of this angle satisfy the given parts. Therefore, there are two triangles which result. Using the law of sines, we solve the case in which A, opposite side a, is an acute angle.

$$\frac{60.0}{\sin A} = \frac{40.0}{\sin 30.0°} \quad \text{or} \quad \sin A = \frac{60.0 \sin 30.0°}{40.0} = 0.7500$$

Therefore, $A = 48.6°$ and $C = 101.4°$. Using the law of sines again to find c, we have

$$\frac{c}{\sin 101.4°} = \frac{40.0}{\sin 30.0°} \quad \text{or} \quad c = \frac{40.0 \sin 101.4°}{\sin 30.0°} = 78.4$$

The other solution is the case in which A', opposite side a, is an obtuse angle. Here we have

$$\frac{60.0}{\sin A'} = \frac{40.0}{\sin 30.0°}$$

which leads to $\sin A' = 0.7500$. Thus, $A' = 131.4°$. For this case we have C' (the angle opposite c when $A' = 131.4°$) as $18.6°$.

Using the law of sines to find c', we have

$$\frac{c'}{\sin 18.6°} = \frac{40.0}{\sin 30.0°} \quad \text{or} \quad c' = \frac{40.0 \sin 18.6°}{\sin 30.0°} = 25.5$$

This means that the second solution is $A' = 131.4°$, $C' = 18.6°$, and $c' = 25.5$.

Example F

In Example E, if $b > 60.0$, only one solution would result. In this case, side b would intercept side c at A. It also intercepts the extension of side c, but this would require that angle B not be included in the triangle (see Fig. 8–34). Thus only one solution may result if $b > a$.

In Example E, there would be no solution if side b were not at least 30.0. For if this were the case, side b would not be sufficiently long to even touch side c. It can be seen that b must at least equal $a \sin B$. If it is just equal to $a \sin B$, there is one solution, a right triangle (Fig. 8–35).

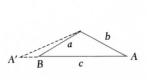

Figure 8–34

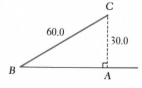

Figure 8–35

Summarizing the results for Case 2 as illustrated in Examples E and F, we make the following conclusions. Given sides a and b, and angle A (assuming here that a and A ($A < 90°$) are the given corresponding parts), we have:

1. *no solution if $a < b \sin A$,*
2. *a right triangle solution if $a = b \sin A$,*
▶ 3. *two solutions if $b \sin A < a < b$,*
4. *one solution if $a > b$.*

See Fig. 8–36, parts (a) through (d), respectively.

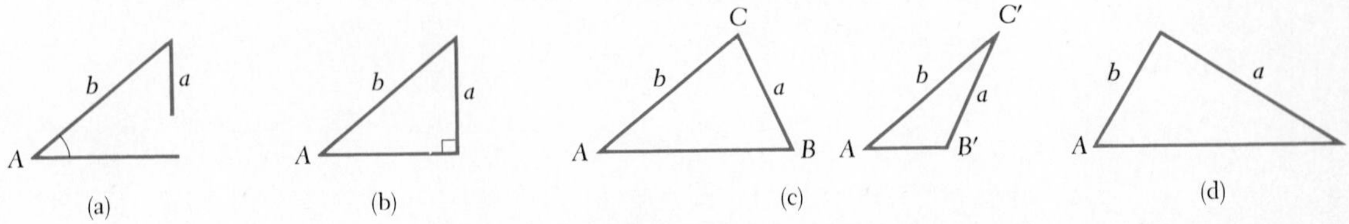

(a)

(b)

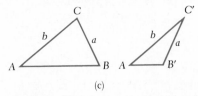

(c)

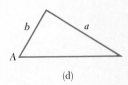

(d)

Figure 8–36

For the reason that two solutions may result from it, Case 2 is often referred to as the **ambiguous case.**

If we attempt to use the law of sines for the solution of Case 3 or Case 4, we find that we do not have sufficient information to complete one of the ratios. These cases can, however, be solved by the law of cosines, which we shall consider in the next section.

Example G

Given the three sides, $a = 5$, $b = 6$, $c = 7$, we would set up the ratios

$$\frac{5}{\sin A} = \frac{6}{\sin B} = \frac{7}{\sin C}$$

However, since there is no way to determine a complete ratio from these equations, we cannot find the solution of the triangle in this manner.

Exercises 8–5

In Exercises 1 through 20, solve the triangles with the given parts.

1. $a = 45.7$, $A = 65.0°$, $B = 49.0°$ 2. $b = 3.07$, $A = 26.0°$, $C = 120.0°$

3. $c = 4380$, $A = 37.4°$, $B = 34.6°$ 4. $a = 93.2$, $B = 17.9°$, $C = 82.6°$

5. $a = 4.601$, $b = 3.107$, $A = 18.23°$ 6. $b = 3.625$, $c = 2.946$, $B = 69.37°$

7. $b = 77.51$, $c = 36.42$, $B = 20.73°$ 8. $a = 150.4$, $c = 250.9$, $C = 76.43°$

9. $b = 0.0742$, $B = 51.0°$, $C = 3.36°$ 10. $c = 729$, $B = 121.0°$, $C = 44.18°$

11. $a = 63.8$, $B = 58.4°$, $C = 22.2°$ 12. $a = 13.0$, $A = 55.2°$, $B = 67.5°$

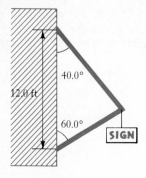

Figure 8–37

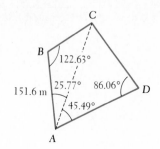

Figure 8–38

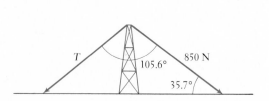

Figure 8–39

13. $b = 4384$, $B = 47.43°$, $C = 64.56°$ 14. $b = 283.2$, $B = 13.79°$, $C = 76.38°$

15. $a = 5.240$, $b = 4.446$, $B = 48.13°$ 16. $a = 89.45$, $c = 37.36$, $C = 15.62°$

17. $b = 2880$, $c = 3650$, $B = 31.4°$ 18. $a = 0.841$, $b = 0.965$, $A = 57.1°$

19. $a = 45.0$, $b = 126$, $A = 64.8°$ 20. $a = 10.0$, $c = 5.00$, $C = 30.0°$

In Exercises 21 through 30, use the law of sines to solve the given problems.

21. Find the lengths of the two steel supports of the sign shown in Fig. 8–37.

22. Find the unknown sides of the four-sided piece of land shown in Fig. 8–38.

23. Find the tension T in the left guy wire attached to the top of the tower shown in Fig. 8–39.

24. Find the distance from city A to city B from the diagram shown in Fig. 8–40.

25. The angles of elevation of an airplane, measured from points A and B, 7540 ft apart on the same side of the airplane (the airplane and points A and B are in the same vertical plane), are $32.0°$ and $44.0°$. How far is point A from the airplane? (See Fig. 8–41.)

26. Resolve vector **A** ($A = 162.4$) into two components in the directions u and v, as shown in Fig. 8–42.

27. A ship leaves a port and travels due west. At a certain point it turns $31.5°$ north of west and travels an additional 42.0 mi to a point 63.0 mi on a direct line from the port. How far from the port is the point where the ship turned?

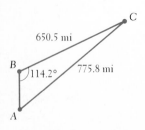

Figure 8–40

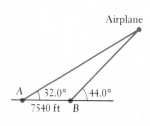

Figure 8–41

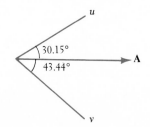

Figure 8–42

28. City B is 43.2° south of east of city A. A pilot wishes to know what direction to head the plane in flying from A to B if the wind is from the west at 40.0 km/h and the plane's velocity with respect to the air is 300 km/h. What should the heading be?

29. A communications satellite is directly above the extension of a line between receiving towers A and B. It is determined from radio signals that the angle of elevation of the satellite from tower A is 89.2°, and the angle of elevation from tower B is 86.5°. If A and B are 1290 km apart, how far is the satellite from A? (Neglect the curvature of the earth.)

30. A person measures a triangular piece of land and reports the following information: "One side is 58.4 ft long and another side is 21.1 ft long. The angle opposite the shorter side is 24.0°." Could this information be correct?

8–6 The Law of Cosines

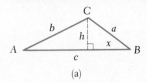

(a)

(b)

Figure 8–43

As we noted at the end of the preceding section, the law of sines cannot be used if the only information given is that of Case 3 or Case 4. Therefore, it is necessary to develop a method of finding at least one more part of the triangle. Here we can use the law of cosines. After obtaining another part by the law of cosines, we can then use the law of sines to complete the solution. We do this because the law of sines generally provides a simpler method of solution than the law of cosines.

Consider any oblique triangle, for example either of the ones in Fig. 8–43. For each we obtain $h = b \sin A$. By using the Pythagorean theorem, we obtain $a^2 = h^2 + x^2$ for each. Thus (with $(\sin A)^2 = \sin^2 A$),

$$a^2 = b^2 \sin^2 A + x^2 \tag{8–9}$$

In Fig. 8–43(a), we have $c - x = b \cos A$, or $x = c - b \cos A$. In Fig. 8–43(b), we have $c + x = b \cos A$, or $x = b \cos A - c$. Substituting these relations into Eq. (8–9), we obtain

$$a^2 = b^2 \sin^2 A + (c - b \cos A)^2$$

and   (8–10)

$$a^2 = b^2 \sin^2 A + (b \cos A - c)^2$$

respectively. When expanded, these give

$$a^2 = b^2 \sin^2 A + b^2 \cos^2 A + c^2 - 2bc \cos A$$

and

$$a^2 = b^2 (\sin^2 A + \cos^2 A) + c^2 - 2bc \cos A \tag{8–11}$$

Recalling the definitions of the trigonometric functions, we know that $\sin \theta = y/r$ and $\cos \theta = x/r$. Thus, $\sin^2 \theta + \cos^2 \theta = (y^2 + x^2)/r^2$. However, $x^2 + y^2 = r^2$, which means that

$$\sin^2 \theta + \cos^2 \theta = 1 \tag{8–12}$$

This equation holds for any angle θ, since we made no assumptions as to the properties of θ. By substituting Eq. (8–12) into Eq. (8–11), we arrive at the **law of cosines:**

$$a^2 = b^2 + c^2 - 2bc \cos A \qquad (8\text{–}13)$$

Using the method above, we may also show that

$$b^2 = a^2 + c^2 - 2ac \cos B$$

and

$$c^2 = a^2 + b^2 - 2ab \cos C$$

Therefore, if we know two sides and the included angle (Case 3) we may directly solve for the side opposite the given angle. Then, by using the law of sines, we may complete the solution. If we are given all three sides (Case 4), we may solve for the angle opposite one of these sides by use of the law of cosines. Again we use the law of sines to complete the solution.

Example A

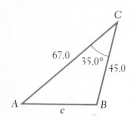

Figure 8–44

Solve the triangle with $a = 45.0$, $b = 67.0$, and $C = 35.0°$ (see Fig. 8–44). Using the law of cosines, we have

$$c^2 = (45.0)^2 + (67.0)^2 - 2(45.0)(67.0)\cos 35.0°$$
$$c = 39.7$$

From the law of sines, we now have

$$\frac{45.0}{\sin A} = \frac{67.0}{\sin B} = \frac{39.7}{\sin 35.0°}$$

which leads to

$$\sin A = 0.6501 \quad \text{or} \quad A = 40.6°$$

We could solve for B from the above relation, or by use of the fact that the sum of all three angles is 180°. Thus,

$$B = 104.4°$$

Example B

If, in Example A, $C = 145.0°$ (see Fig. 8–45), we have

$$c^2 = (45.0)^2 + (67.0)^2 - 2(45.0)(67.0)\cos 145.0°$$
$$c = 107$$

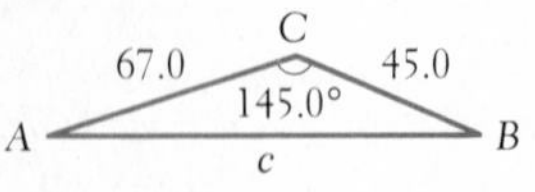

From the law of sines we then find $A = 14.0°$ and $B = 21.0°$.

Figure 8–45

Example C

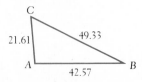

Figure 8-46

Solve the triangle for which $a = 49.33$, $b = 21.61$, and $c = 42.57$ (see Fig. 8-46).

$$\cos A = \frac{b^2 + c^2 - a^2}{2bc} = \frac{(21.61)^2 + (42.57)^2 - (49.33)^2}{2(21.61)(42.57)}$$

$$= -0.08384$$

$$A = 94.81°$$

We note that the calculator shows an obtuse angle, which is consistent with the fact that $\cos A$ is negative.

From the law of sines, we now find that $B = 25.88°$ and $C = 59.31°$.

Example D

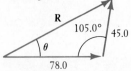

Figure 8-47

Find the resultant of two vectors having magnitudes of 78.0 and 45.0 and directed toward the east and 15.0° east of north, respectively (see Fig. 8-47).

The magnitude of the resultant is given by

$$R = \sqrt{(78.0)^2 + (45.0)^2 - 2(78.0)(45.0)(\cos 105.0°)}$$

$$= 99.6$$

Also, $\theta = 25.9°$ is found from the law of sines.

Example E

A vertical radio antenna is to be built on a hill which makes an angle of 6.0° with the horizontal. If guy wires are to be attached at a point that is 150 ft up on the antenna and at points 100 ft from the base of the antenna, what will be the lengths of guy wires which are positioned directly up and directly down the hill?

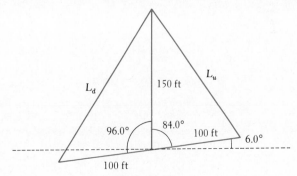

Figure 8-48

Making an appropriate figure, such as Fig. 8-48, we are able to establish the equations necessary for the solution:

$$L_u^2 = (100)^2 + (150)^2 - 2(100)(150)\cos 84.0°$$

$$L_u = 171 \text{ ft}$$

$$L_d^2 = (100)^2 + (150)^2 - 2(100)(150)\cos 96.0°$$

$$L_d = 189 \text{ ft}$$

Exercises 8–6

See Appendix E for a computer program for solving a triangle given three sides.

In Exercises 1 through 20, solve the triangles with the given parts.

1. $a = 6.00$, $b = 7.56$, $C = 54.0°$

2. $b = 87.3$, $c = 34.0$, $A = 130.0°$

3. $a = 4530$, $b = 924$, $C = 98.0°$

4. $a = 0.0845$, $c = 0.116$, $B = 85.0°$

5. $a = 39.53$, $b = 45.22$, $c = 67.15$

6. $a = 23.31$, $b = 27.26$, $c = 29.17$

7. $a = 385.4$, $b = 467.7$, $c = 800.9$

8. $a = 0.2433$, $b = 0.2635$, $c = 0.1538$

9. $a = 320$, $b = 847$, $C = 158.0°$

10. $b = 18.3$, $c = 27.1$, $A = 58.7°$

11. $a = 21.4$, $c = 4.28$, $B = 86.3°$

12. $a = 11.3$, $b = 5.10$, $C = 77.6°$

13. $a = 103.7$, $c = 159.1$, $C = 104.67°$

14. $a = 49.32$, $b = 54.55$, $B = 114.36°$

15. $a = 0.4937$, $b = 0.5956$, $c = 0.6398$

16. $a = 69.72$, $b = 49.30$, $c = 56.29$

17. $a = 723$, $b = 598$, $c = 158$

18. $a = 1.78$, $b = 6.04$, $c = 4.80$

19. $a = 15.6$, $A = 15.1°$, $B = 150.5°$

20. $a = 17.5$, $b = 24.5$, $c = 37.0$

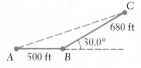

Figure 8–49

In Exercises 21 through 30, use the law of cosines to solve the given problems.

21. To measure the distance AC, a woman walks 500 ft from A to B, turns 30.0° to face C, and walks 680 ft to C. What is the distance AC? See Fig. 8–49.

22. An airplane traveling at 700 km/h leaves the airport at noon going due east. At 2 P.M. the pilot turns 10.0° north of east. How far is the plane from the airport at 3 P.M?

23. In order to get around an obstruction, an oil pipeline is constructed in two straight sections, one 3.756 km long and the other 4.675 km long, with an angle of 168.85° between the sections where they are joined. How much more pipeline was necessary due to the obstruction?

24. A table top in the shape of an isosceles trapezoid is shown in Fig. 8–50. What are the angles between the sides?

25. Two forces, one of 56.14 lb and the other of 67.42 lb, are applied to the same bolt. The resultant force is 82.36 lb. What is the angle between the two forces?

26. A triangular metal frame has sides of 8.00 ft, 12.0 ft, and 16.0 ft. What is the largest angle between parts of the frame?

27. A boat, which can travel 6.00 km/h in still water, heads downstream at an angle of 20.0° with the bank. If the stream is flowing at the rate of 3.00 km/h, how fast is the boat traveling and in what direction?

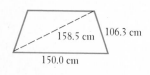

Figure 8–50

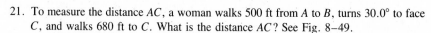

28. An airplane's velocity with respect to the air is 520 mi/h, and it is headed 24.0° north of west. The wind is from due southwest and has a velocity of 55.0 mi/h. What is the true direction of the plane and what is its velocity with respect to the ground?

29. One end of a 13.15-m pole is 15.72 m from an observer's eyes and the other end is 19.33 m from his eyes. Through what angle does the observer see the pole?

30. A triangular piece of land is bounded by 134.13 ft of stone wall, 205.07 ft of road frontage, and 147.25 ft of fencing. What angle does the fence make with the road?

8–7 Exercises for Chapter 8

In Exercises 1 through 4, find the x- and y-components of the given vectors by use of the trigonometric functions.

1. $A = 65.0, \theta_A = 28.0°$ 2. $A = 8.05, \theta_A = 149.0°$

3. $A = 0.9204, \theta_A = 215.59°$ 4. $A = 657.1, \theta_A = 343.74°$

In Exercises 5 through 8, vectors **A** and **B** are at right angles. Find the magnitude and direction of the resultant.

5. $A = 327$ 6. $A = 684$ 7. $A = 4964$ 8. $A = 26.52$
   $B = 505$ $B = 295$ $B = 3298$ $B = 89.86$

In Exercises 9 through 16, add the given vectors by use of the trigonometric functions and the Pythagorean theorem.

9. $A = 780, \theta_A = 28.0°$ 10. $A = 0.0120, \theta_A = 10.5°$
   $B = 346, \theta_B = 320.0°$ $B = 0.0078, \theta_B = 260.0°$

11. $A = 22.51, \theta_A = 130.16°$ 12. $A = 18{,}760, \theta_A = 110.43°$
    $B = 7.604, \theta_B = 200.09°$ $B = 4835, \theta_B = 350.20°$

13. $A = 51.33, \theta_A = 12.25°$ 14. $A = 70.31, \theta_A = 122.54°$
    $B = 42.61, \theta_B = 291.77°$ $B = 30.29, \theta_B = 214.82°$

15. $A = 75.0, \theta_A = 15.0°$ 16. $A = 8100, \theta_A = 141.9°$
    $B = 26.5, \theta_B = 192.4°$ $B = 1540, \theta_B = 165.2°$
    $C = 54.8, \theta_C = 344.7°$ $C = 3470, \theta_C = 296.0°$

In Exercises 17 through 36, solve the triangles with the given parts.

17. $A = 48.0°, B = 68.0°, a = 14.5$

18. $A = 132.0°, b = 7.50, C = 32.0°$

19. $a = 22.8, B = 33.5°, C = 125.3°$

20. $A = 71.0°, B = 48.5°, c = 8.42$

21. $A = 17.85°, B = 154.16°, c = 7863$

22. $a = 1.985, b = 4.189, c = 3.652$

23. $b = 76.07, c = 40.53, B = 110.09°$

24. $A = 77.06°, a = 12.07, c = 5.104$

25. $b = 14.5, c = 13.0, C = 56.6°$

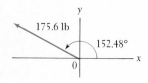

Figure 8–51

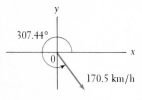

Figure 8–52

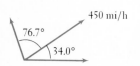

Figure 8–53

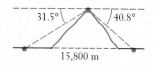

Figure 8–54

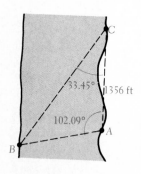

Figure 8–55

26. $B = 40.6°$, $b = 7.00$, $c = 18.0$

27. $a = 186$, $B = 130.0°$, $c = 106$

28. $b = 750$, $c = 1100$, $A = 56.0°$

29. $a = 7.86$, $b = 2.45$, $C = 22.0°$

30. $a = 0.208$, $c = 0.697$, $B = 105.4°$

31. $A = 67.16°$, $B = 96.84°$, $c = 532.9$

32. $A = 43.12°$, $a = 7.893$, $b = 4.113$

33. $a = 17.01$, $b = 12.62$, $c = 25.95$

34. $a = 9064$, $b = 9953$, $c = 1106$

35. $a = 5.30$, $b = 8.75$, $c = 12.5$

36. $a = 47.4$, $b = 40.0$, $c = 45.5$

In Exercises 37 through 56, solve the given problems.

37. Find the horizontal and vertical components of the force shown in Fig. 8–51.

38. Find the horizontal and vertical components of the velocity shown in Fig. 8–52.

39. A jet climbs at an angle of 35.0° while traveling 600 km/h. How long will it take to climb to an altitude of 10,000 m?

40. A bullet is fired into the air at 2000 mi/h at an angle of 25.0° with the horizontal. What is the vertical component of the velocity?

41. A balloon is rising at the rate of 15.0 ft/s and at the same time is being blown horizontally by the wind at the rate of 22.5 ft/s. Find the resultant velocity.

42. A motorboat which travels at 8.00 km/h in still water heads directly across a stream which flows at 3.00 km/h. What is the resultant velocity of the boat?

43. A velocity vector is the resultant of two other vectors. If the given velocity is 450 mi/h and makes angles of 34.0° and 76.7° with the two components, find the magnitudes of the components (see Fig. 8–53).

44. A person on a hill in the middle of a plain relates that the angles of depression of two objects on the plain below (on directly opposite sides of the hill) are 31.5° and 40.8°. She knows that the objects are 15,800 m apart. How far is she from the closer object? (See Fig. 8–54.)

45. In order to find the distance between points A and B on opposite sides of a river, a distance AC is measured off as 1356 ft, where point C is on the same side of the river as A. Angle BAC is measured to be 102.09° and angle ACB is 33.45°. What is the distance between A and B? See Fig. 8–55.

46. A 22.0-ft ladder leans against a wall, making an angle of 29.0° with the wall. If the foot of the ladder is 10.7 ft from the foot of the wall, find the angle of inclination of the wall to the ground.

47. Two points on opposite sides of an obstruction are respectively 117.1 m and 88.16 m from a third point. The lines joining the first two points and the third point intersect at an angle of 115.58° at the third point. How far apart are the original two points?

48. A crate is being held aloft by two ropes which are tied at the same point on the crate. They are 14.5 m and 10.5 m long, respectively, and the angle between them is 104.0°. If the ropes are supported at the same level, how far apart are the supports?

49. Two cars are at the intersection of two straight roads. One travels 5.20 mi on one road, and the other car travels 3.75 mi on the other road. The drivers contact each other on CB radio and find they are at points known to be 4.50 mi apart. What angle do the roads make at the intersection?

50. Determine the angles of the structure indicated in Fig. 8–56.

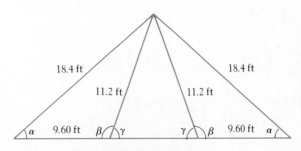

Figure 8–56

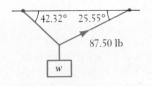

Figure 8–57

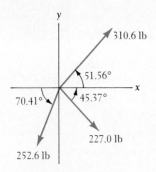

Figure 8–58

51. Atlanta is 290 mi and 51.0° south of east from Nashville. The pilot of an airplane due north of Atlanta radios Nashville and finds the plane is on a line 10.5° south of east from Nashville. How far is the plane from Nashville?

52. The edges of a saw tooth are 2.10 mm and 3.25 mm. The base of the tooth is 2.25 mm. At what angle do the edges of the tooth meet?

53. Determine the weight *w* being supported by the ropes shown in Fig. 8–57. The tension in the right support rope is 87.50 lb.

54. Boston is 650 km and 21.0° south of west from Halifax, Nova Scotia. Radio signals locate a ship 10.5° east of south from Halifax and 5.6° north of east from Boston. How far is the ship from each city?

55. Find the resultant of the forces indicated in Fig. 8–58.

56. A jet plane is traveling horizontally at 1200 ft/s. A missile is fired horizontally from it 30.0° from the direction in which the plane is traveling. If the missile leaves the plane at 2000 ft/s, what is its velocity 10.0 s later if the vertical component is given by $v_V = -32t$ (in feet per second)?

CHAPTER 9

Graphs of the Trigonometric Functions

One of the clearest ways to show the variation of the trigonometric functions is by means of their graphs. The graphs are useful for analyzing properties of the trigonometric functions, and in themselves are valuable in many applications. In this chapter we discuss the graphs of the various trigonometric functions, with an emphasis on the sine and cosine functions. Several of the various applications are indicated in the exercises, particularly in the last two sections of the chapter.

9–1 Graphs of y = a sin x and y = a cos x

The graphs of the trigonometric functions are constructed on the regular rectangular coordinate system. In plotting the trigonometric functions, *it is normal to express the angle in radians. In this way x and the function of x are expressed as real **numbers,** and these numbers may have any desired unit of measurement.* Therefore, in order to determine the graphs, it is necessary to be able to readily use angles expressed in radians. If necessary, Section 7–3 should be reviewed for this purpose.

In this section the graphs of the sine and cosine functions are demonstrated. We begin by constructing a table of values of x and y for the function $y = \sin x$:

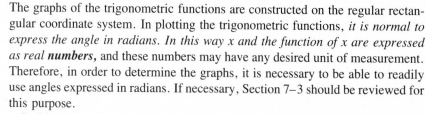

x	0	$\frac{\pi}{6}$	$\frac{\pi}{3}$	$\frac{\pi}{2}$	$\frac{2\pi}{3}$	$\frac{5\pi}{6}$	π	$\frac{7\pi}{6}$	$\frac{4\pi}{3}$	$\frac{3\pi}{2}$	$\frac{5\pi}{3}$	$\frac{11\pi}{6}$	2π
y	0	0.5	0.87	1	0.87	0.5	0	-0.5	-0.87	-1	-0.87	-0.5	0

Plotting these values, we obtain the graph shown in Fig. 9–1.

The graph of $y = \cos x$ may be constructed in the same manner. The following table gives the proper values for the graph of $y = \cos x$. The graph is shown in Fig. 9–2.

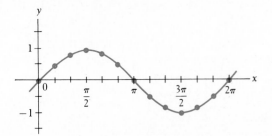

Figure 9-1

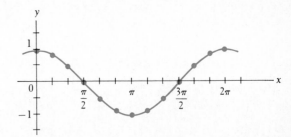

Figure 9-2

x	0	$\frac{\pi}{6}$	$\frac{\pi}{3}$	$\frac{\pi}{2}$	$\frac{2\pi}{3}$	$\frac{5\pi}{6}$	π	$\frac{7\pi}{6}$	$\frac{4\pi}{3}$	$\frac{3\pi}{2}$	$\frac{5\pi}{3}$	$\frac{11\pi}{6}$	2π
y	1	0.87	0.5	0	−0.5	−0.87	−1	−0.87	−0.5	0	0.5	0.87	1

The graphs are continued beyond the values shown in the table to indicate that they continue on indefinitely in each direction. From the values and the graphs, *it can be seen that the two graphs are of exactly the same shape, with the cosine curve displaced* $\pi / 2$ *units to the left of the sine curve.* The shape of these curves should be recognized readily, with special note as to the points at which they cross the axes. This information will be especially valuable in "sketching" similar curves, since the basic shape always remains the same. We shall find it unnecessary to plot numerous points every time we wish to sketch such a curve.

To obtain the graph of $y = a \sin x$, we note that all the y-values obtained for the graph of $y = \sin x$ are to be multiplied by the number a. In this case the greatest value of the sine function is $|a|$ instead of 1. *The number* $|a|$ *is called the* **amplitude** *of the curve and represents the greatest y-value of the curve.* This is also true for $y = a \cos x$.

Example A

Plot the curve of $y = 2 \sin x$.

The table of values to be used is as follows; Fig. 9-3 shows the graph.

x	0	$\frac{\pi}{6}$	$\frac{\pi}{3}$	$\frac{\pi}{2}$	$\frac{2\pi}{3}$	$\frac{5\pi}{6}$	π	$\frac{7\pi}{6}$	$\frac{4\pi}{3}$	$\frac{3\pi}{2}$	$\frac{5\pi}{3}$	$\frac{11\pi}{6}$	2π
y	0	1	1.73	2	1.73	1	0	−1	−1.73	−2	−1.73	−1	0

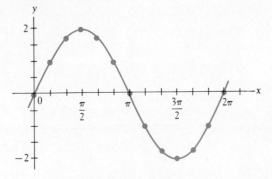

Figure 9-3

Example B

Plot the curve of $y = -3 \cos x$.

The table of values to be used is as follows; Fig. 9–4 shows the graph.

x	0	$\frac{\pi}{6}$	$\frac{\pi}{3}$	$\frac{\pi}{2}$	$\frac{2\pi}{3}$	$\frac{5\pi}{6}$	π	$\frac{7\pi}{6}$	$\frac{4\pi}{3}$	$\frac{3\pi}{2}$	$\frac{5\pi}{3}$	$\frac{11\pi}{6}$	2π
y	-3	-2.6	-1.5	0	1.5	2.6	3	2.6	1.5	0	-1.5	-2.6	-3

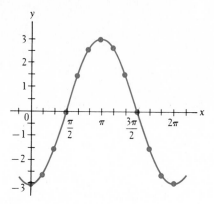

Figure 9–4

Note from Example B that *the effect of the minus sign before the number a is to invert the curve.* The effect of the number a can also be seen readily from these examples.

By knowing the general shape of the sine curve, where it crosses the axes, and the amplitude, *we can rapidly **sketch** curves of form y = a sin x and y = a cos x.* There is generally no need to plot any more points than those corresponding to the values of the amplitude and those where the curve crosses the axes.

Example C

Sketch the graph of $y = 4 \cos x$.

First we set up a table of values for the points where the curve crosses the x-axis and for the highest and lowest points on the curve.

x	0	$\frac{\pi}{2}$	π	$\frac{3\pi}{2}$	2π
y	4	0	-4	0	4

Now we plot the above points and join them, knowing the basic shape of the curve. See Fig. 9–5.

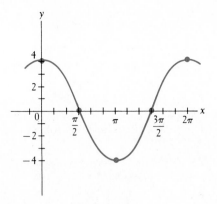

Figure 9–5

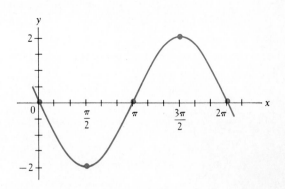

Figure 9–6

Example D

Sketch the curve of $y = -2 \sin x$.

We list here the important values associated with this curve.

x	0	$\frac{\pi}{2}$	π	$\frac{3\pi}{2}$	2π
y	0	-2	0	2	0

Since we know the general shape of the sine curve, we can now sketch the graph, as shown in Fig. 9–6. Note the inversion of the curve due to the minus sign.

Exercises 9–1

In Exercises 1 through 4, complete the following table for the given functions, and then plot the resulting graph.

x	$-\pi$	$-\frac{3\pi}{4}$	$-\frac{\pi}{2}$	$-\frac{\pi}{4}$	0	$\frac{\pi}{4}$	$\frac{\pi}{2}$	$\frac{3\pi}{4}$	π	$\frac{5\pi}{4}$	$\frac{3\pi}{2}$	$\frac{7\pi}{4}$	2π	$\frac{9\pi}{4}$	$\frac{5\pi}{2}$	$\frac{11\pi}{4}$	3π
y																	

1. $y = \sin x$ 2. $y = \cos x$ 3. $y = 3 \cos x$ 4. $y = -4 \sin x$

In Exercises 5 through 20, sketch the curves of the indicated functions.

5. $y = 3 \sin x$ 6. $y = 5 \sin x$ 7. $y = \frac{5}{2} \sin x$

8. $y = 0.5 \sin x$ 9. $y = 2 \cos x$ 10. $y = 3 \cos x$

11. $y = 0.8 \cos x$ 12. $y = \frac{3}{2} \cos x$ 13. $y = -\sin x$

14. $y = -3 \sin x$ 15. $y = -1.5 \sin x$ 16. $y = -0.2 \sin x$

17. $y = -\cos x$ 18. $y = -8 \cos x$ 19. $y = -2.5 \cos x$

20. $y = -0.4 \cos x$

Although units of π are often convenient, we must remember that π is really only a number. Numbers which are not multiples of π may be used as well. In Exercises 21 through 24, plot the indicated graphs by finding the values of y corresponding to the values of 0, 1, 2, 3, 4, 5, 6, and 7 for x by use of a calculator. (Remember, the numbers 0, 1, 2, and so forth represent radian measure.)

21. $y = \sin x$ 22. $y = 3 \sin x$ 23. $y = \cos x$ 24. $y = 2 \cos x$

9–2 Graphs of y = a sin bx and y = a cos bx

In graphing the curve of $y = \sin x$, we note that the values of y start repeating every 2π units of x. This is because $\sin x = \sin(x + 2\pi) = \sin(x + 4\pi)$, and so forth. For any trigonometric function F, we say that it has a **period** P if $F(x) = F(x + P)$. For functions which are periodic, such as the sine and cosine, *the period refers to the x-distance between any point and the next corresponding point for which the values of y start repeating.*

Let us now plot the curve $y = \sin 2x$. This means that we choose a value for x, multiply this value by two, and find the sine of the result. This leads to the following table of values for this function.

x	0	$\frac{\pi}{8}$	$\frac{\pi}{4}$	$\frac{3\pi}{8}$	$\frac{\pi}{2}$	$\frac{5\pi}{8}$	$\frac{3\pi}{4}$	$\frac{7\pi}{8}$	π	$\frac{9\pi}{8}$	$\frac{5\pi}{4}$
$2x$	0	$\frac{\pi}{4}$	$\frac{\pi}{2}$	$\frac{3\pi}{4}$	π	$\frac{5\pi}{4}$	$\frac{3\pi}{2}$	$\frac{7\pi}{4}$	2π	$\frac{9\pi}{4}$	$\frac{5\pi}{2}$
y	0	0.7	1	0.7	0	-0.7	-1	-0.7	0	0.7	1

Plotting these values, we have the curve shown in Fig. 9–7.

From the table and Fig. 9–7, we note that the function $y = \sin 2x$ starts repeating after π units of x. The effect of the 2 before the x has been to make the period of this curve half the period of the curve of $\sin x$. This leads us to the following conclusion: If the period of the trigonometric function $F(x)$ is P, then the period of $F(bx)$ is P/b. Since each of the functions $\sin x$ and $\cos x$ has a period of 2π, *each of the functions $\sin bx$ and $\cos bx$ has a period of $2\pi/b$.*

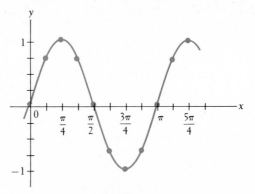

Figure 9–7

Example A

The period of $\sin 3x$ is $\frac{2\pi}{3}$, which means that the curve of the function $y = \sin 3x$ will repeat every $\frac{2\pi}{3}$ (approximately 2.09) units of x.

The period of $\cos 4x$ is $\frac{2\pi}{4} = \frac{\pi}{2}$.

The period of $\sin \frac{1}{2}x$ is $\dfrac{2\pi}{\frac{1}{2}} = 4\pi$. In this case we see that the period is longer than that of the basic sine curve.

Example B

The period of $\sin \pi x$ is $\frac{2\pi}{\pi} = 2$. That is, the curve of the function $\sin \pi x$ will repeat every 2 units. It is then noted that the periods of $\sin 3x$ and $\sin \pi x$ are nearly equal. This is to be expected, since π is only slightly greater than 3.

The period of $\cos 3\pi x$ is $\frac{2\pi}{3\pi} = \frac{2}{3}$.

The period of $\sin \frac{\pi}{4}x$ is $2\pi / \frac{\pi}{4} = 8$.

Combining the result for the period with the results of Section 9–1, we conclude that *each of the functions $y = a \sin bx$ and $y = a \cos bx$ has an amplitude of $|a|$ and a period of $2\pi/b$.* These properties are very useful in sketching these functions, as it is shown in the following examples.

Example C

Sketch the graph of $y = 3 \sin 4x$ for $0 \le x \le \pi$.

We immediately conclude that the amplitude is 3 and the period is $\frac{2\pi}{4} = \frac{\pi}{2}$. Therefore, we know that $y = 0$ when $x = 0$ and $y = 0$ when $x = \frac{\pi}{2}$. Also, we recall that the sine function is zero halfway between these values, which means that $y = 0$ when $x = \frac{\pi}{4}$. The function reaches its amplitude values halfway between the zeros. Therefore, $y = 3$ for $x = \frac{\pi}{8}$ and $y = -3$ for $x = \frac{3\pi}{8}$. A table for these important values of the function $y = 3 \sin 4x$ is as follows.

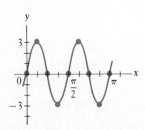

x	0	$\frac{\pi}{8}$	$\frac{\pi}{4}$	$\frac{3\pi}{8}$	$\frac{\pi}{2}$	$\frac{5\pi}{8}$	$\frac{3\pi}{4}$	$\frac{7\pi}{8}$	π
y	0	3	0	-3	0	3	0	-3	0

Using this table and the knowledge of the form of the sine curve, we sketch the function (Fig. 9–8).

Figure 9–8

We see from Example C that an important distance in sketching a sine curve or a cosine curve is one-fourth of the period. For $y = a \sin bx$, it is one-fourth of the period from the origin to the first value of x where y is at the amplitude value. Then, proceeding another one-fourth of the period, there is a zero, another one-fourth period to the next amplitude value, another to the next zero (this is where one period is completed), and so on. Thus, *by finding one-fourth of the period, we can easily find the important values for sketching the curve.* Similarly, one-fourth of the period can be used in sketching $y = a \cos bx$.

The table of important values in sketching $y = a \sin bx$ or $y = a \cos bx$ can be found by:

1. *finding the amplitude, $|a|$,*
2. *finding the period, $2\pi / b$, and*
3. *finding values of the function for each one-fourth period.*

Example D

Sketch the graph of $y = -2 \cos 3x$ for $0 \le x \le 2\pi$.

We note that the amplitude is 2 and that the period is $\frac{2\pi}{3}$. This means that one-fourth of the period is $\frac{1}{4} \cdot \frac{2\pi}{3} = \frac{\pi}{6}$. Since the cosine curve is at its amplitude value for $x = 0$, we have $y = -2$ for $x = 0$ (the negative value is due to the minus sign before the function). The curve then has a zero at $x = \frac{\pi}{6}$, an amplitude value of 2 at $x = 2(\frac{\pi}{6}) = \frac{\pi}{3}$, a zero at $x = 3(\frac{\pi}{6}) = \frac{\pi}{2}$, and its next value of -2 at $x = 4(\frac{\pi}{6}) = \frac{2\pi}{3}$, and so on. Therefore, we have the table of important values.

x	0	$\frac{\pi}{6}$	$\frac{\pi}{3}$	$\frac{\pi}{2}$	$\frac{2\pi}{3}$	$\frac{5\pi}{6}$	π	$\frac{7\pi}{6}$	$\frac{4\pi}{3}$	$\frac{3\pi}{2}$	$\frac{5\pi}{3}$	$\frac{11\pi}{6}$	2π
y	-2	0	2	0	-2	0	2	0	-2	0	2	0	-2

Using this table and the knowledge of the form of the cosine curve, we sketch the function as shown in Fig. 9–9.

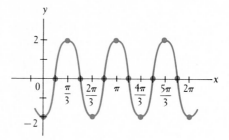

Figure 9–9

Example E

Sketch the function $y = \cos \pi x$ for $0 \le x \le \pi$.

For this function the amplitude is 1; the period is $\frac{2\pi}{\pi} = 2$. Since the value of the period is not in terms of π, it is more convenient to use regular decimal units for x when sketching than to use units in terms of π as in the previous graphs. Therefore, we have the following table.

x	0	0.5	1	1.5	2	2.5	3
y	1	0	-1	0	1	0	-1

The graph of this function is shown in Fig. 9–10.

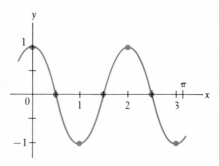

Figure 9–10

Exercises 9–2

See Appendix E for a computer program for sketching the graph of $y = a \sin bx$.

In Exercises 1 through 20, find the period of each of the given functions.

1. $y = 2 \sin 6x$

2. $y = 4 \sin 2x$

3. $y = 3 \cos 8x$

4. $y = \cos 10x$

5. $y = -2 \sin 12x$

6. $y = -\sin 5x$

7. $y = -\cos 16x$

8. $y = -4 \cos 2x$

9. $y = 5 \sin 2\pi x$

10. $y = 2 \sin 3\pi x$

11. $y = 3 \cos 4\pi x$

12. $y = 4 \cos 10\pi x$

13. $y = 3 \sin \frac{1}{3}x$

14. $y = -2 \sin \frac{2}{5}x$

15. $y = -\frac{1}{2} \cos \frac{2}{3}x$

16. $y = \frac{1}{3} \cos \frac{1}{4}x$

17. $y = 0.4 \sin \frac{2\pi x}{3}$

18. $y = 1.5 \cos \frac{\pi x}{10}$

19. $y = 3.3 \cos \pi^2 x$

20. $y = 2.5 \sin \frac{2x}{\pi}$

In Exercises 21 through 40, sketch the graphs of the given functions. For this, use the functions given for Exercises 1 through 20.

In Exercises 41 through 44, sketch the indicated graphs.

41. The electric current in a certain 60-Hz alternating-current circuit is given by $i = 10 \sin 120\pi t$, where i is the current (in amperes) and t is the time (in seconds). Sketch the graph of i vs. t for $0 \le t \le 0.1$ s.

42. A generator produces a voltage given by $V = 200 \cos 50\pi t$, where t is the time (in seconds). Sketch the graph of V vs. t for $0 \le t \le 0.1$ s.

43. The vertical displacement x of a certain object oscillating at the end of a spring is given by $x = 6 \cos 4\pi t$, where x is measured in inches and t in seconds. Sketch the graph of x vs. t for $0 \le t \le 1$ s.

44. The velocity of a piston in an engine is given by $v = 1200 \sin 1200\pi t$, where v is the velocity (in centimeters per second) and t is the time (in seconds). Sketch the graph of v vs. t for $0 \le t \le 0.01$ s.

9–3 Graphs of $y = a \sin(bx + c)$ and $y = a \cos(bx + c)$

There is one more important quantity to be discussed in relation to graphing the sine and cosine functions. *This quantity is the* **phase angle** *of the function. In the function* $y = a \sin(bx + c)$, *c represents this phase angle.* Its meaning is illustrated in the following example.

Example A

Sketch the graph of $y = \sin(2x + \frac{\pi}{4})$.

Note that $c = \frac{\pi}{4}$. This means that in order to obtain the values for the table, we must assume a value of x, multiply it by two, add $\frac{\pi}{4}$ to this value, and then find the sine of this result. In this manner we arrive at the following table.

x	$-\frac{\pi}{8}$	0	$\frac{\pi}{8}$	$\frac{\pi}{4}$	$\frac{3\pi}{8}$	$\frac{\pi}{2}$	$\frac{5\pi}{8}$	$\frac{3\pi}{4}$	$\frac{7\pi}{8}$	π
y	0	0.7	1	0.7	0	-0.7	-1	-0.7	0	0.7

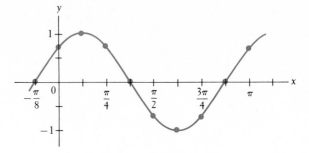

Figure 9–11

We use the value of $x = -\frac{\pi}{8}$ in the table, for we note that it corresponds to finding $\sin 0$. Using the values listed in the table, we plot the graph of $y = \sin(2x + \frac{\pi}{4})$. See Fig. 9–11.

We can see from the table and from the graph in Example A that the curve of

$$y = \sin\left(2x + \frac{\pi}{4}\right)$$

is precisely the same as that of $y = \sin 2x$, except that it is shifted $\frac{\pi}{8}$ units to the left. The effect of c in the equation of $y = a \sin(bx + c)$ is to shift the curve of $y = a \sin bx$ to the left if $c > 0$, and to shift the curve to the right if $c < 0$. Therefore, the amount of this shift is given by $-c/b$. Due to its importance in sketching curves, *the quantity $-c/b$ is called the* **displacement** *(or* **phase shift***)*.

Therefore, the results above combined with the results of Section 9–2 may be used to sketch curves of the functions $y = a \sin(bx + c)$ and $y = a \cos(bx + c)$ where $b > 0$. These are the important quantities to determine:

1. **the amplitude (equal to $|a|$),**
2. **the period $\left(\text{equal to } \dfrac{2\pi}{b}\right)$,**
3. **the displacement $\left(\text{equal to } -\dfrac{c}{b}\right)$.**

By use of these quantities and the one-fourth period distance, the curves of the sine and cosine functions can be readily sketched. A general illustration of the curve of $y = a \sin(bx + c)$ is shown in Fig. 9–12. Note that *the displacement is* **negative** *(to the left) for $c > 0$*, as in Fig. 9–12(a), and that *the displacement is* **positive** *(to the right) for $c < 0$*, as in Fig. 9–12(b).

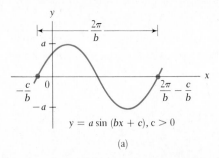

$y = a\sin(bx + c), c > 0$

(a)

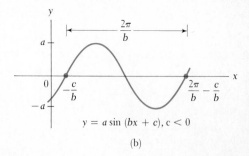

$y = a\sin(bx + c), c < 0$

(b)

$(a > 0, b > 0)$

Figure 9–12

Example B

Sketch the graph of $y = 2\sin(3x - \pi)$ for $0 \le x \le \pi$.

First we note that $a = 2$, $b = 3$, and $c = -\pi$. Therefore, the amplitude is 2, the period is $\frac{2\pi}{3}$, and the displacement is $-(\frac{-\pi}{3}) = \frac{\pi}{3}$.

With this information we can tell that the curve "starts" at $x = \frac{\pi}{3}$ and starts repeating $\frac{2\pi}{3}$ units to the right of this point. (Be sure to grasp this point well. The period tells how many units there are along the x-axis *between* such corresponding points.) One-fourth of the period is $\frac{1}{4}(\frac{2\pi}{3}) = \frac{\pi}{6}$. Therefore, the important values are at $\frac{\pi}{3}, \frac{\pi}{3} + \frac{\pi}{6} = \frac{\pi}{2}, \frac{\pi}{3} + 2(\frac{\pi}{6}) = \frac{2\pi}{3}$, and so on. Therefore, we obtain the table of important values and sketch the graph shown in Fig. 9–13. Extending the curve to the left, we note that—since the period is $\frac{2\pi}{3}$—the curve passes through the origin.

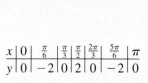

x	0	$\frac{\pi}{6}$	$\frac{\pi}{3}$	$\frac{\pi}{2}$	$\frac{2\pi}{3}$	$\frac{5\pi}{6}$	π
y	0	-2	0	2	0	-2	0

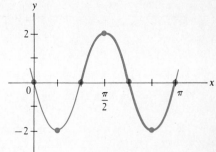

Figure 9–13

Example C

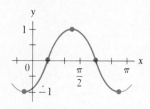

Figure 9–14

Sketch the graph of the function $y = -\cos(2x + \frac{\pi}{6})$.

First we determine that the amplitude is 1, the period is $\frac{2\pi}{2} = \pi$, and that the displacement is $-\frac{\pi}{6} \div 2 = -\frac{\pi}{12}$ (to the left, $c > 0$). From these values we construct the following table, remembering that the curve starts repeating π units to the right of $-\frac{\pi}{12}$.

x	$-\frac{\pi}{12}$	$\frac{\pi}{6}$	$\frac{5\pi}{12}$	$\frac{2\pi}{3}$	$\frac{11\pi}{12}$
y	-1	0	1	0	-1

From this table we sketch the graph, as shown in Fig. 9–14.

Example D

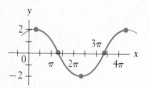

Figure 9–15

Sketch the graph of the function $y = 2\cos(\frac{1}{2}x - \frac{\pi}{6})$.

From the values $a = 2$, $b = \frac{1}{2}$, and $c = -\frac{\pi}{6}$, we determine that the amplitude is 2, the period is $2\pi \div \frac{1}{2} = 4\pi$, and the displacement is $-(-\frac{\pi}{6}) \div \frac{1}{2} = \frac{\pi}{3}$. From these values we construct this table of values.

x	$\frac{\pi}{3}$	$\frac{4\pi}{3}$	$\frac{7\pi}{3}$	$\frac{10\pi}{3}$	$\frac{13\pi}{3}$
y	2	0	-2	0	2

The graph is shown in Fig. 9–15. We note again that when the coefficient of x is less than 1, the period is greater than 2π.

Example E

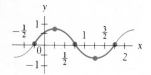

Figure 9–16

Sketch the graph of the function $y = 0.7\sin(\pi x + \frac{\pi}{4})$.

From the values $a = 0.7$, $b = \pi$, and $c = \frac{\pi}{4}$, we can determine that the amplitude is 0.7, the period is $\frac{2\pi}{\pi} = 2$, and the displacement is $-(\frac{\pi}{4}) \div \pi = -\frac{1}{4}$. From these values, we construct the following table of values.

x	$-\frac{1}{4}$	$\frac{1}{4}$	$\frac{3}{4}$	$\frac{5}{4}$	$\frac{7}{4}$
y	0	0.7	0	-0.7	0

Since π is not used in the values of x, it is more convenient to use decimal number units for the graph (Fig. 9–16).

Each of the heavy portions of the graphs in Fig. 9–13, 9–14, 9–15, and 9–16 is called a **cycle** of the curve. *A cycle is the shortest section of the graph which includes one period.*

Exercises 9–3

In Exercises 1 through 24, determine the amplitude, period, and displacement for each of the functions. Then sketch the graphs of the functions.

1. $y = \sin\left(x - \frac{\pi}{6}\right)$

2. $y = 3\sin\left(x + \frac{\pi}{4}\right)$

3. $y = \cos\left(x + \frac{\pi}{6}\right)$

4. $y = 2\cos\left(x - \frac{\pi}{8}\right)$

5. $y = 2\sin\left(2x + \frac{\pi}{2}\right)$

6. $y = -\sin\left(3x - \frac{\pi}{2}\right)$

7. $y = -\cos(2x - \pi)$

8. $y = 4\cos\left(3x + \frac{\pi}{3}\right)$

9. $y = \frac{1}{2}\sin\left(\frac{1}{2}x - \frac{\pi}{4}\right)$

10. $y = 2\sin\left(\frac{1}{4}x + \frac{\pi}{2}\right)$

11. $y = 3\cos\left(\frac{1}{3}x + \frac{\pi}{3}\right)$

12. $y = \frac{1}{3}\cos\left(\frac{1}{2}x - \frac{\pi}{8}\right)$

13. $y = \sin\left(\pi x + \frac{\pi}{8}\right)$

14. $y = -2\sin(2\pi x - \pi)$

15. $y = \dfrac{3}{4} \cos\left(4\pi x - \dfrac{\pi}{5}\right)$

16. $y = 6 \cos\left(3\pi x + \dfrac{\pi}{2}\right)$

17. $y = -0.6 \sin(2\pi x - 1)$

18. $y = 1.8 \sin\left(\pi x + \dfrac{1}{3}\right)$

19. $y = 4 \cos(3\pi x + 2)$

20. $y = 3 \cos(6\pi x - 1)$

21. $y = \sin(\pi^2 x - \pi)$

22. $y = -\dfrac{1}{2} \sin\left(2x - \dfrac{1}{\pi}\right)$

23. $y = -\dfrac{3}{2} \cos\left(\pi x + \dfrac{\pi^2}{6}\right)$

24. $y = \pi \cos\left(\dfrac{1}{\pi}x + \dfrac{1}{3}\right)$

In Exercises 25 through 28, sketch the indicated curves.

25. A wave traveling in a string may be represented by the equation

$$y = A \sin\left(\dfrac{t}{T} - \dfrac{x}{\lambda}\right)$$

Here A is the amplitude, t is the time the wave has traveled, x is the distance from the origin, T is the time required for the wave to travel one *wavelength* λ (the Greek letter lambda). Sketch three cycles of the wave for which $A = 2.00$ cm, $T = 0.100$ s, $\lambda = 20.0$ cm, and $x = 5.00$ cm.

26. The cross-section of a particular water wave is

$$y = 0.5 \sin\left(\dfrac{\pi}{2}x + \dfrac{\pi}{4}\right)$$

where x and y are measured in feet. Sketch two cycles of y vs. x.

27. A certain satellite circles the earth such that its distance y, in miles north or south (altitude is not considered), from the equator is

$$y = 4500 \cos(0.025t - 0.25)$$

where t is the number of minutes after launch. Sketch two cycles of the graph.

28. The voltage in a certain alternating-current circuit is given by

$$y = 120 \cos\left(120\pi t + \dfrac{\pi}{6}\right)$$

where t represents the time (in seconds). Sketch three cycles of the curve.

9–4 Graphs of $y = \tan x$, $y = \cot x$, $y = \sec x$, $y = \csc x$

In this section we shall briefly consider the graphs of the other trigonometric functions. We shall establish the basic form of each curve, and from these we shall be able to sketch other curves for these functions.

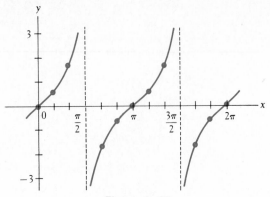

Figure 9–17

Considering the values and signs of the trigonometric functions as established in Chapter 7, we set up the following table for the function $y = \tan x$. The graph is shown in Fig. 9–17.

x	0	$\frac{\pi}{6}$	$\frac{\pi}{3}$	$\frac{\pi}{2}$	$\frac{2\pi}{3}$	$\frac{5\pi}{6}$	π	$\frac{7\pi}{6}$	$\frac{4\pi}{3}$	$\frac{3\pi}{2}$	$\frac{5\pi}{3}$	$\frac{11\pi}{6}$	2π
y	0	0.6	1.7	*	-1.7	-0.6	0	0.6	1.7	*	-1.7	-0.6	0

*Undefined.

Since the curve is not defined for $x = \frac{\pi}{2}$, $x = \frac{3\pi}{2}$, and so forth, we look at the table and note that the value of $\tan x$ becomes very large when x approaches the value $\frac{\pi}{2}$. We must keep in mind, however, that there is no point on the curve corresponding to $x = \frac{\pi}{2}$. We note that *the period of the tangent curve is π*. This differs from the period of the sine and cosine functions.

By following the same procedure, we can set up tables for the graphs of the other functions. In Figures 9–18 through 9–21, we present the graphs of $y = \tan x$, $y = \cot x$, $y = \sec x$, and $y = \csc x$ (the graph of $y = \tan x$ is shown again to illustrate it more completely). *The dashed lines in these figures are* **asymptotes** (see Sections 2–3 and 20–6).

To sketch functions such as $y = a \sec x$, we first sketch $y = \sec x$, and then multiply each y-value by a. *Here a is not an amplitude,* since these functions are not limited in the values they take on, as are the sine and cosine functions.

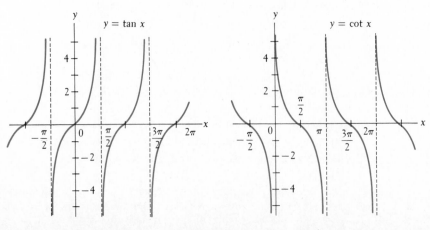

Figure 9–18   **Figure 9–19**

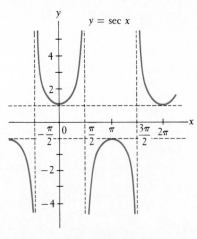

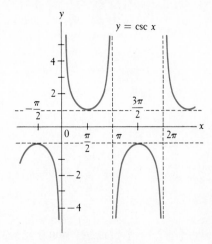

Figure 9–20

Figure 9–21

Example A

Sketch the graph of $y = 2 \sec x$.

First we sketch in $y = \sec x$, shown as the light curve in Fig. 9–22. Now we multiply the y-values of the secant function by 2 (approximately, of course). In this way we obtain the desired curve, shown as the curve in color in Fig. 9–22.

Example B

Sketch the graph of $y = -\frac{1}{2} \cot x$.

We sketch in $y = \cot x$, shown as the light curve in Fig. 9–23. Now we multiply each y-value by $-\frac{1}{2}$. The effect of the negative sign is to invert the curve. The resulting curve is shown as the curve in color in Fig. 9–23.

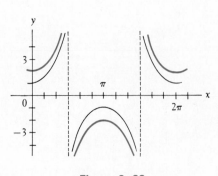

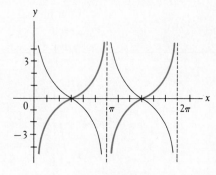

Figure 9–22

Figure 9–23

By knowing the graphs of the sine, cosine, and tangent functions, it is possible to graph the other three functions. This is due to the reciprocal relationships among the functions. In Section 3–3, when we were discussing how to find the values of the cotangent, secant, and cosecant functions on a calculator, we noted that csc x and sin x are reciprocals, sec x and cos x are reciprocals, and cot x and tan x are reciprocals. These relationships come as a result of the definitions of the various functions. We can show these reciprocal relationships by

$$\csc x = \frac{1}{\sin x} \qquad \sec x = \frac{1}{\cos x} \qquad \cot x = \frac{1}{\tan x} \qquad\qquad (9\text{--}1)$$

Thus, to sketch $y = \cot x$, $y = \sec x$, or $y = \csc x$, we sketch the corresponding reciprocal function, and from this graph determine the necessary values.

Example C

Sketch the graph of $y = \csc x$.

We first sketch in the graph of $y = \sin x$ (light curve). Where sin x is 1, csc x will also be 1, since $1/1 = 1$. Where sin x is 0, csc x is undefined, since $1/0$ is undefined. Where sin x is 0.5, csc x is 2, since $1/0.5 = 2$. Thus, as sin x becomes larger, csc x becomes smaller, and as sin x becomes smaller, csc x becomes larger. The two functions always have the same sign. We sketch the graph of $y = \csc x$ from this information, as shown by the curve in color in Fig. 9–24.

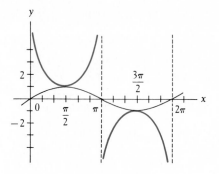

Figure 9–24

Exercises 9–4

In Exercises 1 through 4, fill in the following table for each function and then plot the curve from these points.

x	$-\frac{\pi}{2}$	$-\frac{\pi}{3}$	$-\frac{\pi}{4}$	$-\frac{\pi}{6}$	0	$\frac{\pi}{6}$	$\frac{\pi}{4}$	$\frac{\pi}{3}$	$\frac{\pi}{2}$	$\frac{2\pi}{3}$	$\frac{3\pi}{4}$	$\frac{5\pi}{6}$	π
y													

1. $y = \tan x$ 2. $y = \cot x$ 3. $y = \sec x$ 4. $y = \csc x$

In Exercises 5 through 12, sketch the curves of the given functions by use of the basic curve forms (Figs. 9–18, 9–19, 9–20, 9–21). See Examples A and B.

5. $y = 2 \tan x$ 6. $y = 3 \cot x$ 7. $y = \frac{1}{2} \sec x$

8. $y = \frac{3}{2} \csc x$ 9. $y = -2 \cot x$ 10. $y = -\tan x$

11. $y = -3 \csc x$ 12. $y = -\frac{1}{2} \sec x$

In Exercises 13 through 20, plot the graphs by first making an appropriate table for $0 \leq x \leq \pi$.

13. $y = \tan 2x$ 14. $y = 2 \cot 3x$ 15. $y = \frac{1}{2} \sec 3x$

16. $y = \csc 2x$ 17. $y = 2 \cot\left(2x + \frac{\pi}{6}\right)$ 18. $y = \tan\left(3x - \frac{\pi}{2}\right)$

19. $y = \csc\left(3x - \frac{\pi}{3}\right)$ 20. $y = 3 \sec\left(2x + \frac{\pi}{4}\right)$

In Exercises 21 through 24, sketch the given curves by first sketching the appropriate reciprocal function. See Example C.

21. $y = \sec x$ 22. $y = \cot x$ 23. $y = \csc 2x$ 24. $y = \sec \pi x$

In Exercises 25 through 28, construct the appropriate graphs.

25. For an object sliding down an inclined plane at constant speed, the coefficient of friction μ (the Greek letter mu) between the object and the plane is given by $\mu = \tan \theta$, where θ is the angle between the plane and the horizontal. Sketch a graph of the coefficient of friction vs. the angle of inclination of the plane for $0 \leq \theta \leq 60°$.

26. The tension T at any point in a cable supporting a distributed load is given by $T = T_0 \sec \theta$, where T_0 is the tension where the cable is horizontal, and θ is the angle between the cable and the horizontal at any point. Sketch a graph of T vs. θ for a cable for which $T_0 = 200$ lb.

27. From a point x meters from the base of a building 200 m high, the angle of elevation θ of the top of the building can be found from the equation $x = 200 \cot \theta$. Plot x as a function of θ.

28. An expression relating the initial velocity v_0 of a projectile, the time t of its flight, and the angle θ above the horizontal at which it is fired is $v_0 = gt \csc \theta$, where g is the acceleration due to gravity. Plot v_0 (in feet per second) for $g = 32.0$ ft/s^2 and $t = 10.0$ s.

9–5 Applications of the Trigonometric Graphs

There are a great many applications of the trigonometric functions and their graphs, a few of which have been indicated in the exercises of the previous sections. In this section we shall introduce an important physical concept and indicate some of the technical applications.

In Section 7–4, we discussed the velocity of an object moving in a circular path. *The movement of the* **projection** *on a diameter of a particle revolving about a circle with constant velocity is known as* **simple harmonic motion.** For example, this could be the motion of the shadow of an object which is moving around a circle. Another example is the vertical (or horizontal) position of the end of a spoke of a wheel in motion.

Example A

When we consider Fig. 9–25, let us assume that motion starts with the end of the radius at $(R, 0)$ and that it is moving with constant angular velocity ω. This means that the length of the projection of the radius along the y-axis is given by $d = R \sin \theta$. The length of this projection is shown for a few different positions of the end of the radius. Since $\theta/t = \omega$, or $\theta = \omega t$, we have

$$d = R \sin \omega t \tag{9–2}$$

as the equation for the length of the projection, with time as the independent variable. Normally, the position as a function of time is the important relationship.

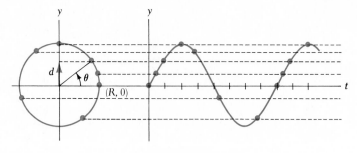

Figure 9–25

For the case where $R = 10.0$ in. and $\omega = 4.00$ rad/s, we have

$$d = 10.0 \sin 4.00t$$

By sketching the graph of this function, we can readily determine the length of the projection d for a given time t. The graph is shown in Fig. 9–26.

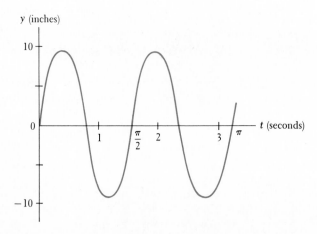

Figure 9–26

Example B

If the end of the radius is at $\left(\frac{R}{\sqrt{2}}, \frac{R}{\sqrt{2}}\right)$, where $\theta = \frac{\pi}{4}$, when $t = 0$, we can express the projection d as a function of the time as

$$d = R \sin\left(\omega t + \frac{\pi}{4}\right)$$

If the end of the radius is at $(0, R)$, where $\theta = \frac{\pi}{2}$, when $t = 0$, we can express the projection d as a function of the time as

$$d = R \sin\left(\omega t + \frac{\pi}{2}\right)$$

or

$$d = R \cos \omega t$$

This can be seen from Fig. 9–25. If the motion started at the first maximum of the indicated curve, the resulting curve would be that of the cosine function.

Other examples of simple harmonic motion are (1) the movement of a pendulum bob through its arc (a very close approximation to simple harmonic motion), (2) the motion of an object on the end of a spring, (3) the motion of an object "bobbing" in the water, and (4) the movement of the end of a vibrating rod (which we hear as sound). Other phenomena which give rise to equations just like those for simple harmonic motion are found in the fields of optics, sound, and electricity. Such phenomena have the same mathematical form because they result from vibratory motion or motion in a circle.

Example C

A very important use of the trigonometric curves arises in the study of alternating current, which is caused by the motion of a wire passing through a magnetic field. If this wire is moving in a circular path, with angular velocity ω, the current i in the wire at time t is given by an equation of the form

$$i = I_m \sin(\omega t + \alpha) \tag{9–3}$$

where I_m is the maximum current attainable and α is the phase angle. The current may be represented by a sine wave, as in the following example.

Example D

In Example C, given that $I_m = 6.00$ A, $\omega = 120\pi$ rad/s, and $\alpha = \pi/6$, we have the equation

$$i = 6.00 \sin\left(120\pi t + \frac{\pi}{6}\right)$$

From this equation we see that the amplitude is 6.00, the period is $\frac{1}{60}$ s, and the displacement is $-\frac{1}{720}$ s. From these values we draw the graph as shown in Fig. 9–27. Since the current takes on both positive and negative values, we conclude that it moves in opposite directions.

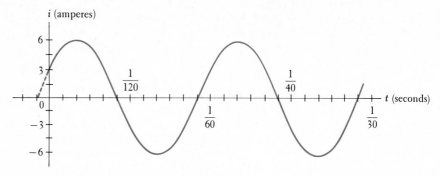

Figure 9–27

It is a common practice to express the rate of rotation in terms of *the* **frequency** *f, the number of cycles per second,* rather than directly in terms of the angular velocity ω, the number of radians per second. *The unit for frequency is the hertz* (*Hz*), *and 1 Hz = 1 cycle/s.* Since there are 2π rad in one cycle, we have

$$\omega = 2\pi f \tag{9-4}$$

Example E

In Example D, we showed the angular velocity ω to be 120π rad/s. The corresponding frequency f is

$$f = \frac{120\pi}{2\pi} = 60 \text{ Hz}$$

This means that 120π rad/s corresponds to 60 cycles/s.

Exercises 9–5

In Exercises 1 and 2, draw two cycles of the curve of the projection of Example A as a function of time for the given values.

  1. $R = 4.00$ in., $\omega = 1.00$ rad/s   2. $R = 8.00$ cm, $f = 0.500$ Hz

In Exercises 3 and 4, for the projection described in Example A, assume that the end of the radius starts at $(0, R)$. Draw two cycles of the projection as a function of time for the given values. (See Example B.)

  3. $R = 2.00$ ft, $f = 1.00$ Hz   4. $R = 2.50$ in., $\omega = 0.300$ rad/s

In Exercises 5 and 6, for the projection described in Example A, assume that the end of the radius starts at the indicated point. Draw two cycles of the projection as a function of time for the given values. (See Example B.)

  5. $R = 6.00$ cm, $f = 2.00$ Hz, starting point $(\frac{\sqrt{3}R}{2}, \frac{R}{2})$

  6. $R = 3.20$ ft, $\omega = 0.200$ rad/s, starting point $(\frac{R}{2}, \frac{\sqrt{3}R}{2})$

In Exercises 7 and 8, for the alternating current discussed in Example C, draw two cycles of the current as a function of time for the given values.

  7. $I_m = 2.00$ A, $f = 60.0$ Hz, $\alpha = 0$

  8. $I_m = 0.600$ A, $\omega = 100$ rad/s, $\alpha = \pi/4$

In Exercises 9 and 10, for an alternating-current circuit in which the voltage is given by

$$e = E \cos(\omega t + \alpha)$$

draw two cycles of the voltage as a function of time for the given values.

  9. $E = 170$ V, $\omega = 50.0$ rad/s, $\alpha = 0$

  10. $E = 110$ V, $f = 60.0$ Hz, $\alpha = -\pi/3$

In Exercises 11 through 16, draw the required curves.

11. Angular displacement θ of a pendulum bob is given in terms of its initial $(t = 0)$ displacement θ_0 by the equation $\theta = \theta_0 \cos \omega t$. If $\omega = 2.00$ rad/s and $\theta_0 = \pi/30$ rad, draw two cycles for the resulting equation.

12. Displacement of the end of a vibrating rod is given by $y = 1.50 \cos 200\pi t$. Sketch two cycles of y (in centimeters) vs. t (in seconds).

13. An object of weight w and cross-sectional area A is depressed a distance x_0 from its equilibrium position when in a liquid of density d and then released; its displacement as a function of time is given by

$$x = x_0 \cos \sqrt{\frac{dgA}{w}} t$$

    where g $(= 32.0$ ft/s$^2)$ is the acceleration due to gravity. If a 4.00-lb object with a cross-sectional area of 2.00 ft^2 is depressed 3.00 ft in water (let $d = 62.4$ lb/ft^3), find the equation which expresses the displacement as a function of time. Draw two cycles of the curve.

14. A wave is traveling in a string. The displacement, as a function of time, from its equilibrium position, is given by $y = A \cos(2\pi/T)t$. T is the period (measured in seconds) of the motion. If $A = 0.200$ in. and $T = 0.100$ s, draw two cycles of the displacement as a function of time.

15. The displacement, as a function of time, from the position of equilibrium, of an object on the end of a spring is given by $x = A \cos(\omega t + \alpha)$. Draw two cycles of the curve for displacement as a function of time for $A = 2.00$ in., $f = 0.500$ Hz, and $\alpha = \pi/6$.

16. The angular displacement θ of a pendulum bob (see Exercise 11) is given by $\theta = \theta_0 \sin(\omega t + \frac{\pi}{6})$. If $\theta_0 = 0.100$ rad and $\omega = \frac{\pi}{2}$ rad/s, sketch two cycles of the graph of θ vs. t.

9–6 Composite Trigonometric Curves

Many applications of trigonometric functions involve the combination of two or more functions. In this section we shall discuss two important methods in which trigonometric curves can be combined.

If we wish to find the curve of a function which itself is the sum of two other functions, *we may find the resulting graph by first sketching the two individual functions and then adding the y-values graphically. This method is called* **addition of ordinates** and is illustrated in the following examples.

Example A

Sketch the graph of $y = 2 \cos x + \sin 2x$.

On the same set of coordinate axes we sketch the curves $y = 2 \cos x$ and $y = \sin 2x$. These are shown as dashed curves in Fig. 9–28. For various values of x, we determine the distance above or below the x-axis of each curve and add these distances, noting that those above the axis are positive and those below the axis are negative. We thereby *graphically **add** the y-values* of these two curves for these values of x to obtain the points on the resulting curve shown as a curve in color in Fig. 9–28. We add the y-values for a sufficient number of x-values to obtain the proper representation. Some points are easily found. Where one curve crosses the x-axis, its y-value is zero, and therefore the resulting curve has its point on the other curve for this value of x. In this example, $\sin 2x$ is zero at $x = 0, \frac{\pi}{2}, \pi$, and so forth. We see that points on the resulting curve lie on the curve of $2 \cos x$. We should also add the values where each curve is at its amplitude values. In this case, $\sin 2x$ equals 1 at $\frac{\pi}{4}$, and the two y-values should be added together here to get a point on the resulting curve. At $x = \frac{5\pi}{4}$, we must take care in adding the values, since $\sin 2x$ is positive and $2 \cos x$ is negative. Reasonable care and accuracy are necessary to obtain a proper resulting curve.

Example B

Sketch the graph of $y = \dfrac{x}{2} - \cos x$.

The method of addition of ordinates is applicable regardless of the kinds of functions being added. Here we note that $y = \frac{x}{2}$ is a straight line, and that it is to be combined with a trigonometric curve.

We could graph the functions $y = \frac{x}{2}$ and $y = \cos x$ and then subtract the ordinates of $y = \cos x$ from the ordinates of $y = \frac{x}{2}$. But it is easier and far less confusing to add values, so we shall sketch $y = \frac{x}{2}$ and $y = -\cos x$ and add the ordinates to obtain points on the resulting curve. These graphs are shown as dashed curves in Fig. 9–29. The important points on the resulting curve are obtained by using the values of x corresponding to the zeros and amplitude values of $y = -\cos x$.

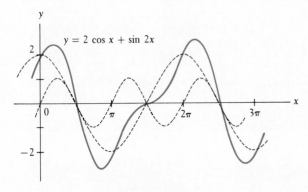

Figure 9–28

Figure 9–29

Example C

Sketch the graph of $y = \cos \pi x - 2 \sin 2x$.

The curves of $y = \cos \pi x$ and $y = -2 \sin 2x$ are shown as dashed curves in Fig. 9–30. Points for the resulting curve are found primarily at the x-values where each of the curves has its zero or amplitude values. Again, special care should be taken where one of the curves is negative and the other is positive.

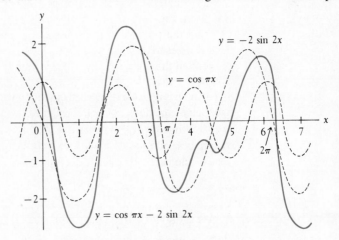

Figure 9–30

Another important application of trigonometric curves is made when they are added at *right angles.* This can be accomplished in practice by applying different voltages to an oscilloscope. Let us consider the following examples.

Example D

Plot the graph for which the values of x and y are given by the equations $y = \sin 2\pi t$ and $x = 2 \cos \pi t$. *(Equations given in this form, x and y in terms of a third variable, are called* **parametric equations.**)

Since both x and y are given in terms of t, by assuming values for t we may find corresponding values of x and y, and use these values to plot the resulting points (see Fig. 9–31).

t	0	$\frac{1}{4}$	$\frac{1}{2}$	$\frac{3}{4}$	1	$\frac{5}{4}$	$\frac{3}{2}$	$\frac{7}{4}$	2	$\frac{9}{4}$
x	2	1.4	0	-1.4	-2	-1.4	0	1.4	2	1.4
y	0	1	0	-1	0	1	0	-1	0	1
Point number	1	2	3	4	5	6	7	8	9	10

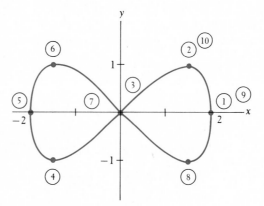

Figure 9–31

Since x and y are trigonometric functions of a third variable t and since the x- and y-axes are at right angles, values of x and y obtained in this manner result in a combination of two trigonometric curves at right angles. *Figures obtained in this manner are called* **Lissajous figures.**

Example E

If we place a circle on the x-axis and another on the y-axis, we may represent the coordinates (x, y) for the curve of Example D by the lengths of the projections (see Example A of Section 9–5) of a point moving around each circle. A careful study of Fig. 9–32 will clarify this. We note that the radius of the circle giving the x-values is 2, whereas the radius of the other is 1. This is due to the manner in which x and y are defined. Also due to the definitions, the point revolves around the y-circle twice as fast as the corresponding point revolves around the x-circle.

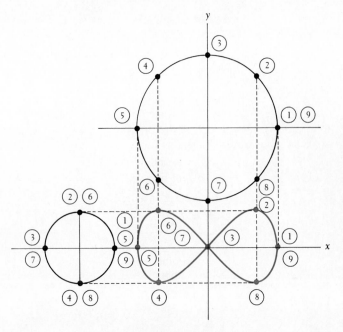

Figure 9-32

Example F

Plot the Lissajous figure for which the x- and y-values are given by the equation $x = 2 \sin 3t$ and $y = 3 \sin(t + \frac{\pi}{3})$.

Since values of t which are multiples of π give convenient values of x and y, the table is constructed with these values of t. Fig. 9-33 shows the graph.

t	x	y	Point number
0	0	2.6	1
$\frac{\pi}{6}$	2	3	2
$\frac{\pi}{3}$	0	2.6	3
$\frac{\pi}{2}$	-2	1.5	4
$\frac{2\pi}{3}$	0	0	5
$\frac{5\pi}{6}$	2	-1.5	6
π	0	-2.6	7
$\frac{7\pi}{6}$	-2	-3	8
$\frac{4\pi}{3}$	0	-2.6	9
$\frac{3\pi}{2}$	2	-1.5	10
$\frac{5\pi}{3}$	0	0	11
$\frac{11\pi}{6}$	-2	1.5	12
2π	0	2.6	13

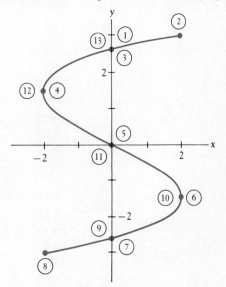

Figure 9-33

Exercises 9–6

In Exercises 1 through 20, use the method of addition of ordinates to sketch the given curves.

1. $y = x + \sin x$

2. $y = x + 2 \cos x$

3. $y = \frac{1}{3}x - \cos x$

4. $y = \frac{1}{4}x - \sin x$

5. $y = \frac{1}{10}x^2 + \sin 2x$

6. $y = \frac{1}{3}x^2 + \cos 3x$

7. $y = \dfrac{1}{x^2 + 1} - \sin \pi x$

8. $y = \frac{1}{5}x^3 - \cos \pi x$

9. $y = \sin x + \cos x$

10. $y = \sin x + \sin 2x$

11. $y = \sin x - \sin 2x$

12. $y = \cos 3x - \sin x$

13. $y = 2 \cos 2x + 3 \sin x$

14. $y = \frac{1}{2} \sin 4x + \cos 2x$

15. $y = 2 \sin x - \cos x$

16. $y = \sin \dfrac{x}{2} - \sin x$

17. $y = 2 \cos 4x - \cos\left(x - \dfrac{\pi}{4}\right)$

18. $y = \sin \pi x - \cos 2x$

19. $y = 2 \sin\left(2x - \dfrac{\pi}{6}\right) + \cos\left(2x + \dfrac{\pi}{3}\right)$

20. $y = 3 \cos 2\pi x + \sin \dfrac{\pi}{2}x$

In Exercises 21 through 28, plot the Lissajous figures.

21. $y = \sin t, \; x = \sin t$

22. $y = 2 \cos t, \; x = \cos t$

23. $y = \sin \pi t, \; x = \cos \pi t$

24. $y = \sin 2t, \; x = \cos\left(t + \dfrac{\pi}{4}\right)$

25. $y = 2 \sin \pi t, \; x = \cos \pi\left(t + \dfrac{1}{6}\right)$

26. $y = \sin^2 \pi t, \; x = \cos 2\pi t$

27. $y = \cos 2t, \; x = 2 \cos 3t$

28. $y = 3 \sin 3\pi t, \; x = 2 \sin \pi t$

In Exercises 29 through 34, sketch the appropriate figures.

29. The current in a certain electric circuit is given by the formula $i = 4 \sin 60\pi t + 2 \cos 120\pi t$. Sketch the curve representing the current (in amperes) as a function of time (in seconds).

30. In optics, two waves are said to interfere destructively if, when they pass through the same medium, the amplitude of the resulting wave is zero. Sketch the curve of $y = \sin x + \cos(x + \frac{\pi}{2})$, and determine whether or not it would represent destructive interference of two waves.

31. An object oscillating on a spring, under specific conditions, has a displacement given by $y = 0.4 \sin 4t + 0.3 \cos 4t$. Plot y (in feet) versus t (in seconds).

32. The resultant voltage in a certain electric circuit is given by the formula $e = 50 \sin 50\pi t + 80 \sin 60\pi t$. Sketch the curve of voltage as a function of time.

33. Two signals are being sent to an oscilloscope, and are seen on the oscilloscope as being at right angles. The equations governing the displacement of these signals are $x = 2 \cos 120\pi t$ and $y = 3 \cos 120\pi t$, respectively. Sketch the figure which would appear on the oscilloscope.

34. In the study of optics, light is said to be elliptically polarized if certain optic vibrations are out of phase. These may be represented by Lissajous figures. Determine the Lissajous figure for two waves of light given by $w_1 = \sin \omega t, \; w_2 = \sin(\omega t + \frac{\pi}{4})$.

9–7 Exercises for Chapter 9

In Exercises 1 through 28, sketch the curves of the given trigonometric functions.

1. $y = \frac{2}{3} \sin x$ 2. $y = -4 \sin x$ 3. $y = -2 \cos x$

4. $y = 2.3 \cos x$ 5. $y = 2 \sin 3x$ 6. $y = 4.5 \sin 12x$

7. $y = 2 \cos 2x$ 8. $y = 4 \cos 6x$ 9. $y = 3 \cos \frac{1}{3}x$

10. $y = 3 \sin \frac{1}{2}x$ 11. $y = \sin \pi x$ 12. $y = 3 \sin 4\pi x$

13. $y = 5 \cos 2\pi x$ 14. $y = -\cos 3\pi x$

15. $y = -0.5 \sin \frac{\pi}{6}x$ 16. $y = 8 \sin \frac{\pi}{4}x$

17. $y = 2 \sin\left(3x - \frac{\pi}{2}\right)$ 18. $y = 3 \sin\left(\frac{x}{2} + \frac{\pi}{2}\right)$

19. $y = -2 \cos(4x + \pi)$ 20. $y = 0.8 \cos\left(\frac{x}{6} - \frac{\pi}{2}\right)$

21. $y = -\sin\left(\pi x + \frac{\pi}{6}\right)$ 22. $y = 2 \sin(3\pi x - \pi)$

23. $y = 8 \cos\left(4\pi x - \frac{\pi}{2}\right)$ 24. $y = 3 \cos(2\pi x + \pi)$

25. $y = 3 \tan x$ 26. $y = \frac{1}{4} \sec x$

27. $y = -\frac{1}{3} \csc x$ 28. $y = -5 \cot x$

In Exercises 29 through 36, sketch the given curves by the method of addition of ordinates.

29. $y = \frac{1}{2} \sin 2x - x$ 30. $y = \frac{1}{2}x - \cos \frac{1}{3}x$

31. $y = \sin 2x + 3 \cos x$ 32. $y = \sin 3x + 2 \cos 2x$

33. $y = 2 \sin x - \cos 2x$ 34. $y = \sin 3x - 2 \cos x$

35. $y = \cos\left(x + \frac{\pi}{4}\right) - 2 \sin 2x$ 36. $y = 2 \cos \pi x + \cos(2\pi x - \pi)$

In Exercises 37 through 40, plot the Lissajous figures.

37. $y = 2 \sin \pi t, \ x = -\cos 2\pi t$ 38. $y = \sin t, \ x = \sin\left(t + \frac{\pi}{6}\right)$

39. $y = \cos \pi t, \ x = \cos\left(2\pi t + \frac{\pi}{4}\right)$ 40. $y = \cos\left(2t + \frac{\pi}{3}\right), \ x = \cos\left(t - \frac{\pi}{6}\right)$

In Exercises 41 through 54, sketch the appropriate figures.

41. A simple pendulum is started by giving it a velocity from its equilibrium position. The angle θ between the vertical and the pendulum is given by

$$\theta = a \sin\left(\sqrt{\frac{g}{l}}t\right)$$

where a is the amplitude (in radians), $g \ (= 32.0 \text{ ft/s}^2)$ is the acceleration due to gravity, l is the length of the pendulum (in feet), and t is the length of time of the motion. Sketch two cycles of θ as a function of t for the pendulum whose length is 2.00 ft and $a = 0.100$ rad.

42. The electric current in a certain circuit is given by $i = i_0 \sin(t/\sqrt{LC})$, where i_0 is the initial current in the circuit, L is an inductance, and C is a capacitance. Sketch two cycles of i as a function of t (in seconds) for the case where $i_0 = 0.500$ A, $L = 1.00$ H, and $C = 100\ \mu$F.

43. A certain object is oscillating at the end of a spring. The displacement as a function of time is given by the relation $y = 0.200 \cos 8t$, where y is measured in meters and t in seconds. Plot the graph of y vs. t.

44. The velocity v, in centimeters per second, of a piston in a certain engine is given by $v = \omega D \cos \omega t$, where ω is the angular velocity of the crankshaft in radians per second and t is the time in seconds. Sketch the graph of v vs. t if the engine is at 3000 r/min and $D = 3.6$ cm.

45. A particular electromagnetic wave is described by the equation

$$y = a \sin\left(8\pi \times 10^{14}t + \frac{\pi}{6}\right)$$

Sketch two cycles of the graph of y (in centimeters) vs. t (in seconds).

46. A circular disk suspended by a thin wire attached to the center at one of its flat faces is twisted through an angle θ. Torsion in the wire tends to turn the disk back in the opposite direction (thus the name *torsion pendulum* is given to this device). The angular displacement as a function of time is given by $\theta = \theta_0 \cos(\omega t + \alpha)$, where θ_0 is the maximum angular displacement, ω is a constant which depends on the properties of the disk and wire, and α is the phase angle. Plot the graph of θ vs. t if $\theta_0 = 0.100$ rad, $\omega = 2.50$ rad/s and $\alpha = \frac{\pi}{4}$.

47. The charge q on a certain capacitor as a function of time is given by $q = 0.001(1 - \cos 100t)$. Sketch two cycles of q as a function of t. Charge is measured in coulombs, and time is measured in seconds.

48. If the upper end of a spring is not fixed and is being moved with a sinusoidal motion, the motion of the bob at the end of the spring is affected. Plot the curve if the motion of the upper end of a spring is being moved by an external force and the bob moves according to the equation $y = 4 \sin 2t - 2 \cos 2t$.

49. Under certain conditions, the path of a certain moving particle is given by $x = 2 \cos 3\pi t$ and $y = \sin 3\pi t$. Plot the graph of the object.

50. Two signals are applied to an oscilloscope. The equations governing the displacement of these signals are $x = 2 \cos 40\pi t$ and $y = \sin 120\pi t$. Sketch the figure which appears on the oscilloscope.

51. The height of a certain rocket ascending vertically is given by the formula $h = 800 \tan \theta$, where θ is the angle of elevation from an observer 800 m from the launch pad. Plot h (in meters) vs. θ.

52. An equation relating a force F and its x-component F_x is $F = F_x \sec \theta$, where θ is the angle between F and F_x. Plot F vs. θ for $F_x = 500$ lb.

53. The area of a rectangle as a function of a diagonal is $A = d^2 \sin \theta \cos \theta$, where θ is the angle between the diagonal and one of the sides. Plot A vs. θ for $d = 10.0$ m.

54. The instantaneous power in an electric circuit is defined as the product of the instantaneous voltage e and the instantaneous current i. If we have $e = 100 \cos 200t$ and $i = 2 \cos(200t + \frac{\pi}{4})$, plot the graph of the voltage and the graph of the current (in amperes), on the same coordinate system, vs. the time (in seconds). Then sketch the power (in watts) vs. time by multiplying appropriate values of e and i.

CHAPTER 10

Exponents and Radicals

In Chapter 1 we introduced exponents and radicals. To this point, only a basic understanding of the meaning and elementary operations with them has been necessary. However, in our future work a more detailed understanding of exponents and radicals, and operations on them, will be required. Therefore, in this chapter we shall develop the necessary operations.

10–1 Integral Exponents

The laws of exponents were given in Section 1–5. We now write them again for reference.

$$a^m \cdot a^n = a^{m+n} \tag{10-1}$$

$$\frac{a^m}{a^n} = a^{m-n} \quad \text{or} \quad \frac{a^m}{a^n} = \frac{1}{a^{n-m}}, \quad a \neq 0 \tag{10-2}$$

$$(a^m)^n = a^{mn} \tag{10-3}$$

$$(ab)^n = a^n b^n, \quad \left(\frac{a}{b}\right)^n = \frac{a^n}{b^n}, \quad b \neq 0 \tag{10-4}$$

$$a^0 = 1, \quad a \neq 0 \tag{10-5}$$

$$a^{-n} = \frac{1}{a^n}, \quad a \neq 0 \tag{10-6}$$

Although Eqs. (10–1) through (10–4) were originally defined for positive integers as exponents, we showed in Section 1–5 that with the definitions given in Eqs. (10–5) and (10–6), they are valid for all integral exponents. Later in this chapter we shall show how fractions may be used as exponents. Since the equations above are very important to the development of the topics in this chapter, they should again be reviewed—and learned thoroughly.

In this section we review the use of exponents in using Eqs. (10–1) through (10–6). Then we show how they are used and handled in somewhat more involved expressions.

Example A

Applying Eq. (10–1), we have

$$a^5 \cdot a^{-3} = a^{5+(-3)} = a^{5-3} = a^2$$

Applying Eq. (10–1) and then Eq. (10–6), we have

$$a^3 \cdot a^{-5} = a^{3-5} = a^{-2} = \frac{1}{a^2}$$

We note that a final result is usually expressed with positive exponents, unless specified otherwise.

Example B

Applying Eq. (10–1), then (10–6), and then (10–4), we have

$$(2^3 \cdot 2^{-4})^2 = (2^{3-4})^2 = (2^{-1})^2 = \left(\frac{1}{2}\right)^2 = \frac{1}{2^2} = \frac{1}{4}$$

Often there are several combinations of the laws that can be used to simplify an expression. For example, this expression can be simplified by using Eq. (10–1), then (10–3), and then (10–6) as follows:

$$(2^3 \cdot 2^{-4})^2 = (2^{3-4})^2 = (2^{-1})^2 = 2^{-2} = \frac{1}{2^2} = \frac{1}{4}$$

The result is in a proper form as either $1/2^2$ or $1/4$. If the exponent is large, then it is common to leave the exponent in the answer.

Example C

Applying Eqs. (10–2) and (10–5), we have

$$\frac{a^2 b^3 c^0}{ab^7} = \frac{a^{2-1}(1)}{b^{7-3}} = \frac{a}{b^4}$$

Applying Eqs. (10–4) and (10–3), we have

$$(x^{-2}y)^3 = (x^{-2})^3(y^3) = x^{-6}y^3 = \frac{y^3}{x^6}$$

Here, the simplification was completed by the use of Eq. (10–6).

Example D

$$\left(\frac{4}{a^2}\right)^{-3} = \frac{1}{\left(\frac{4}{a^2}\right)^3} = \frac{1}{\frac{4^3}{a^6}} = \frac{a^6}{4^3}$$

or

$$\left(\frac{4}{a^2}\right)^{-3} = \frac{4^{-3}}{a^{-6}} = \frac{a^6}{4^3}$$

In the first method we used Eq. (10–6) first, then Eq. (10–4), and finally, we inverted the divisor. In the second method we first used Eq. (10–4) and then Eq. (10–6).

Example E

$$(x^2y)^2\left(\frac{2}{x}\right)^{-2} = \frac{(x^4y^2)}{\left(\frac{2}{x}\right)^2} = \frac{x^4y^2}{\frac{4}{x^2}} = \frac{x^4y^2}{1}\cdot\frac{x^2}{4} = \frac{x^6y^2}{4}$$

or

$$(x^2y)^2\left(\frac{2}{x}\right)^{-2} = (x^4y^2)\left(\frac{2^{-2}}{x^{-2}}\right) = (x^4y^2)\left(\frac{x^2}{2^2}\right) = \frac{x^6y^2}{4}$$

In simplifying expressions, care must be taken to apply the laws of exponents properly. Certain relatively common problems are pointed out in the following two examples.

Example F

The expression $(-5x)^0$ equals 1, whereas the expression $-5x^0$ equals -5. For $(-5x)^0$ the parentheses indicate that the expression $-5x$ is raised to the zero power, whereas for $-5x^0$ only x is raised to the zero power and we have

$$-5x^0 = -5(1) = -5$$

Also, $(-5)^0 = 1$, whereas $-5^0 = -1$. Again note the use of parentheses. For $(-5)^0$ it is -5 which is raised to the zero power, whereas for -5^0 only 5 is raised to the zero power.

Similarly,

$$(-2)^2 = 4 \quad \text{and} \quad -2^2 = -4$$

For the same basic reason,

$$2x^{-1} = \frac{2}{x}$$

whereas

$$(2x)^{-1} = \frac{1}{2x}$$

Example G

$$3a^{-1} - (2a)^{-2} = \frac{3}{a} - \frac{1}{(2a)^2} = \frac{3}{a} - \frac{1}{4a^2}$$

$$= \frac{12a - 1}{4a^2}$$

Example H

Simplify $(2a + b^{-1})^{-2}$.

In simplifying an expression we should give the result with positive exponents. For this expression, we may use various orders of operations. One is as follows:

$$(2a + b^{-1})^{-2} = \frac{1}{(2a + b^{-1})^2} = \frac{1}{\left(2a + \frac{1}{b}\right)^2} = \frac{1}{\left(\frac{2ab + 1}{b}\right)^2}$$

$$= \frac{1}{\frac{(2ab + 1)^2}{b^2}} = \frac{b^2}{(2ab + 1)^2}$$

We may leave the result in this form, or multiply out the denominator and obtain

$$(2a + b^{-1})^{-2} = \frac{b^2}{(2ab + 1)^2} = \frac{b^2}{4a^2b^2 + 4ab + 1}$$

However, a common type of error is sometimes made with this type of expression. An incorrect step which is made is to express

$$(2a + b^{-1})^{-2} \quad \text{as} \quad (2a)^{-2} + (b^{-1})^{-2}, \quad \text{or} \quad \frac{1}{4a^2} + b^2$$

Remember: As noted in Section 5–1, when raising a binomial (or any multinomial) to a power, we cannot simply raise each term to the power to obtain the result.

From the above examples, we see that *when a factor is moved from the denominator to the numerator of a fraction, or conversely, the sign of the exponent is changed.* We should heed carefully the word *factor;* this rule does not apply to moving terms in the numerator or denominator.

Example I

$$3^{-1}\left(\frac{4^{-2}}{3 - 3^{-1}}\right) = \frac{1}{3}\left(\frac{1}{4^2}\right)\left(\frac{1}{3 - \frac{1}{3}}\right) = \frac{1}{3 \cdot 4^2}\left(\frac{1}{\frac{9 - 1}{3}}\right)$$

$$= \frac{1}{3 \cdot 4^2}\left(\frac{3}{8}\right) = \frac{1}{128}$$

Example J

$$\frac{1}{x^{-1}}\left(\frac{x^{-1}-y^{-1}}{x^2-y^2}\right) = \frac{x}{1}\left(\frac{\frac{1}{x}-\frac{1}{y}}{x^2-y^2}\right) = x\left(\frac{\frac{y-x}{xy}}{x^2-y^2}\right)$$

$$= \frac{\frac{x(y-x)}{xy}}{(x-y)(x+y)} = \frac{x(y-x)}{xy}\cdot\frac{1}{(x-y)(x+y)}$$

$$= \frac{x(y-x)}{xy(x-y)(x+y)} = \frac{-(x-y)}{y(x-y)(x+y)}$$

$$= -\frac{1}{y(x+y)}$$

▶ Note that in this example the x^{-1} and y^{-1} in the numerator could not be moved directly to the denominator with positive exponents because they are only terms of the original numerator.

Example K

$$3(x+4)^2(x-3)^{-2} - 2(x-3)^{-3}(x+4)^3$$

$$= \frac{3(x+4)^2}{(x-3)^2} - \frac{2(x+4)^3}{(x-3)^3} = \frac{3(x-3)(x+4)^2 - 2(x+4)^3}{(x-3)^3}$$

$$= \frac{(x+4)^2[3(x-3) - 2(x+4)]}{(x-3)^3} = \frac{(x+4)^2(x-17)}{(x-3)^3}$$

Expressions such as the one in this example are commonly found in problems in calculus.

Exercises 10–1

In Exercises 1 through 56, express each of the given expressions in the simplest form which contains only positive exponents.

1. $x^7 \cdot x^{-4}$
2. $y^9 \cdot y^{-2}$
3. $a^2 \cdot a^{-6}$
4. $s \cdot s^{-5}$

5. $5 \cdot 5^{-3}$
6. $2 \cdot 7^4 \cdot 7^{-2}$
7. $(2^{-1} \cdot 5)^2$
8. $(3^2 \cdot 4^{-3})^3$

9. $(2ax^{-1})^2$
10. $(3xy^{-2})^3$
11. $(5an^{-2})^{-1}$
12. $(6s^2t^{-1})^{-2}$

13. $(-4)^0$
14. -4^0
15. $-7x^0$
16. $(-7x)^0$

17. $3x^{-2}$
18. $(3x)^{-2}$
19. $(7ax)^{-3}$
20. $7ax^{-3}$

21. $\left(\frac{2}{n^3}\right)^{-1}$
22. $\left(\frac{3}{x^3}\right)^{-2}$
23. $\left(\frac{a}{b^{-2}}\right)^{-3}$
24. $\left(\frac{2n^{-2}}{m^{-1}}\right)^{-2}$

25. $(a+b)^{-1}$
26. $a^{-1} + b^{-1}$
27. $3x^{-2} + 2y^{-2}$
28. $(3x+2y)^{-2}$

29. $(2 \cdot 3^{-2})^2\left(\frac{3}{2}\right)^{-1}$
30. $(3^{-1} \cdot 7^2)^{-1}\left(\frac{3}{7}\right)^2$

31. $(ab)^2\left(\frac{3}{a}\right)^{-1}$
32. $(3st^{-1})^3\left(\frac{s}{t}\right)^{-2}$

33. $\left(\frac{3a^2}{4b}\right)^{-3}\left(\frac{4}{a}\right)^{-5}$
34. $\left(\frac{2n}{p^2}\right)^{-2}\left(\frac{p}{4}\right)^{-1}$

35. $\left(\dfrac{v^{-1}}{2t}\right)^{-2}\left(\dfrac{t^2}{v^{-2}}\right)^{-3}$

36. $\left(\dfrac{a^{-2}}{b^2}\right)^{-3}\left(\dfrac{a^{-3}}{b^5}\right)^2$

37. $(x^2y^{-1})^2 - x^{-4}$

38. $(st^{-2})^{-1} - s^{-2}$

39. $4a^{-2} + (3a^2)^{-2}$

40. $3(a^{-1}z^2)^{-3} + c^{-2}z^{-1}$

41. $2\cdot 3^{-1} + 4\cdot 3^{-2}$

42. $5\cdot 2^{-2} - 3^{-1}\cdot 2^3$

43. $(a^{-1} + b^{-1})^{-1}$

44. $(2a - b^{-2})^{-1}$

45. $(n^{-2} - 2n^{-1})^2$

46. $(2^{-3} - 4^{-1})^2$

47. $\dfrac{3 - 2^{-1}}{3^{-2}}$

48. $\dfrac{6^{-1}}{4^{-2} + 2}$

49. $\dfrac{x - y^{-1}}{x^{-1} - y}$

50. $\dfrac{x^{-2} - y^{-2}}{x^{-1} - y^{-1}}$

51. $\dfrac{ax^{-2} + a^{-2}x}{a^{-1} + x^{-1}}$

52. $\dfrac{2x^{-2} - 2y^{-2}}{(xy)^{-3}}$

53. $2t^{-2} + t^{-1}(t + 1)$

54. $3x^{-1} + x^{-3}(y + 2)$

55. $(x - 1)^{-1} + (x + 1)^{-1}$

56. $4(2x - 1)(x + 2)^{-1} - (2x - 1)^2(x + 2)^{-2}$

In Exercises 57 through 62, perform the indicated operations.

57. When discussing electronic amplifiers, the expression $\left(\dfrac{1}{r} + \dfrac{1}{R}\right)^{-1}$ is found. Simplify this expression.

58. Physical units associated with numbers are often expressed in terms of negative exponents (see Appendix B). If the units of a certain quantity are $(m\cdot s^{-1})^2$, express these units without the use of negative exponents.

59. An expression which is used for the focal length of a certain lens is $[(\mu - 1)(r_1^{-1} - r_2^{-1})]^{-1}$. Rewrite this expression without the use of negative exponents.

60. An expression encountered in the mathematics of finance is

$$\dfrac{p(1 + i)^{-1}[(1 + i)^{-n} - 1]}{(1 + i)^{-1} - 1}$$

where n is an integer. Simplify this expression.

61. By use of Eqs. (10–4) and (10–6) show that

$$\left(\dfrac{a}{b}\right)^{-n} = \left(\dfrac{b}{a}\right)^n$$

62. Using a calculator, verify the equation in Exercise 61 by evaluating the expression on each side with $a = 3.576$, $b = 8.091$, and $n = 7$.

10–2 Fractional Exponents

In Section 10–1 we reviewed the use of integral exponents, including exponents which are negative integers and zero. We now show how rational numbers may be used as exponents. With the appropriate definitions, all the laws of exponents are valid for all rational numbers as exponents.

Equation (10–3) states that $(a^m)^n = a^{mn}$. If we were to let $m = \frac{1}{2}$ and $n = 2$, we would have $(a^{1/2})^2 = a^1$. However, we already have a way of writing a quantity which when squared equals a. This is written as $\sqrt{a}$. To be consistent with previous definitions and to allow the laws of exponents to hold, we define

$$a^{1/n} = \sqrt[n]{a} \qquad\qquad (10\text{–}7)$$

In order that Eqs. (10–3) and (10–7) may hold at the same time, we define

$$a^{m/n} = \sqrt[n]{a^m} = (\sqrt[n]{a})^m \qquad\qquad (10\text{–}8)$$

It can be shown that these definitions are valid for all the laws of exponents.

Example A

We shall verify here that Eq. (10–1) holds for the above definitions:

$$a^{1/4}a^{1/4}a^{1/4}a^{1/4} = a^{(1/4)+(1/4)+(1/4)+(1/4)} = a^1$$

Now $a^{1/4} = \sqrt[4]{a}$, by definition. Also, by definition $\sqrt[4]{a}\ \sqrt[4]{a}\ \sqrt[4]{a}\ \sqrt[4]{a} = a$. Equation (10–1) is thereby verified for $n = 4$. Equation (10–3) is verified by the following:

$$a^{1/4}\ a^{1/4}\ a^{1/4}\ a^{1/4} = a^{4(1/4)} = a = \sqrt[4]{a^4}$$

We note that we may interpret $a^{m/n}$ in Eq. (10–8) as the mth power of the nth root of a, as well as the nth root of the mth power of a. This is illustrated in the following example.

Example B

$$(\sqrt[3]{a})^2 = \sqrt[3]{a^2} = a^{2/3}$$
$$8^{2/3} = (\sqrt[3]{8})^2 = (2)^2 = 4 \quad \text{or} \quad 8^{2/3} = \sqrt[3]{8^2} = \sqrt[3]{64} = 4$$

We must note that Eq. (10–8) is valid as long as $\sqrt[n]{a}$ does not involve the even root of a negative number. Such numbers are imaginary and are considered in Chapter 11.

Although both interpretations of Eq. (10–8) are possible, as indicated in Example B, in evaluating numerical expressions involving fractional exponents without a calculator, it is almost always best to find the root first, as indicated by the denominator of the fractional exponent. This will allow us to find the root of the smaller number, which is normally easier to find.

Example C

To evaluate $(64)^{5/2}$, we should proceed as follows:

$$(64)^{5/2} = [(64)^{1/2}]^5 = 8^5 = 32,768$$

If we raised 64 to the fifth power first, we would have

$$(64)^{5/2} = (64^5)^{1/2} = (1,073,741,824)^{1/2}$$

We would now have to evaluate the indicated square root. This demonstrates why it is preferable to find the indicated root first.

Example D

$$(16)^{3/4} = (16^{1/4})^3 = 2^3 = 8$$

$$4^{-1/2} = \frac{1}{4^{1/2}} = \frac{1}{2}, \qquad 9^{3/2} = (9^{1/2})^3 = 3^3 = 27$$

We note in the illustration of $4^{-1/2}$ that Eq. (10–6) must also hold for negative rational exponents. The only change in the exponent which is made in writing it as $1/4^{1/2}$ is that the sign of the exponent is changed.

Fractional exponents allow us to find roots of numbers on a calculator. By use of the $\boxed{x^y}$ (or $\boxed{y^x}$) key, we may raise any positive number to any power. For fractional exponents, the decimal form is used.

Example E

Evaluate $(235.6)^{3/7}$ on a calculator.

One way to find this value is to first find the decimal form of 3/7 and put its value in memory. Doing this we have the following sequence of keys.

$$3 \quad \boxed{\div} \quad 7 \quad \boxed{=} \quad \boxed{\text{STO}} \quad 235.6 \quad \boxed{x^y} \quad \boxed{\text{RCL}} \quad \boxed{=}$$

The display is 10.390742, which means that

$$(235.6)^{3/7} = 10.39$$

Another way to evaluate $(235.6)^{3/7}$ is to use parentheses. This sequence is

$$235.6 \quad \boxed{x^y} \quad \boxed{(} \quad 3 \quad \boxed{\div} \quad 7 \quad \boxed{)} \quad \boxed{=}$$

Another reason for developing fractional exponents is that they are often easier to use in more-complex expressions involving roots. Therefore, any expression involving radicals can also be expressed with fractional exponents and then simplified. We now show some additional examples with fractional exponents.

Example F

$$(8a^2b^4)^{1/3} = [(8^{1/3})(a^2)^{1/3}(b^4)^{1/3}] = 2a^{2/3}b^{4/3}$$

$$a^{3/4}a^{4/5} = a^{3/4+4/5} = a^{31/20}$$

$$(25a^{-2}c^4)^{3/2} = [(25a^{-2}c^4)^{1/2}]^3 = \left(\frac{(25)^{1/2}(c^4)^{1/2}}{(a^2)^{1/2}}\right)^3 = \left(\frac{5c^2}{a}\right)^3$$

$$= \frac{125c^6}{a^3}$$

Example G

$$\left(\frac{4^{-3/2}x^{2/3}y^{-7/4}}{2^{3/2}x^{-1/3}y^{3/4}}\right)^{2/3} = \left(\frac{x^{2/3+1/3}}{2^{3/2}4^{3/2}y^{3/4+7/4}}\right)^{2/3}$$

$$= \frac{x^{(1)(2/3)}}{2^{(3/2)(2/3)}4^{(3/2)(2/3)}y^{(10/4)(2/3)}} = \frac{x^{2/3}}{8y^{5/3}}$$

Example H

$$(4x^4)^{-1/2} - 3x^{-3} = \frac{1}{(4x^4)^{1/2}} - \frac{3}{x^3}$$

$$= \frac{1}{2x^2} - \frac{3}{x^3}$$

$$= \frac{x - 6}{2x^3}$$

Example I

$$(2x + 1)^{1/2} + (x + 3)(2x + 1)^{-1/2} = (2x + 1)^{1/2} + \frac{x + 3}{(2x + 1)^{1/2}}$$

$$= \frac{(2x + 1)^{1/2}(2x + 1)^{1/2} + (x + 3)}{(2x + 1)^{1/2}}$$

$$= \frac{(2x + 1) + (x + 3)}{(2x + 1)^{1/2}} = \frac{3x + 4}{(2x + 1)^{1/2}}$$

Exercises 10–2

In Exercises 1 through 28, evaluate the given expressions.

1. $(25)^{1/2}$
2. $(49)^{1/2}$
3. $(27)^{1/3}$

4. $(81)^{1/4}$
5. $8^{4/3}$
6. $(125)^{2/3}$

7. $(100)^{25/2}$
8. $(16)^{5/4}$
9. $8^{-1/3}$

10. $16^{-1/4}$
11. $(64)^{-2/3}$
12. $(32)^{-4/5}$

13. $5^{1/2}5^{3/2}$
14. $8^{1/3}4^{1/2}$
15. $(4^4)^{3/2}$

16. $(3^6)^{2/3}$
17. $\dfrac{121^{-1/2}}{100^{1/2}}$
18. $\dfrac{1000^{1/3}}{400^{-1/2}}$

19. $\dfrac{7^{-1/2}}{6^{-1}7^{1/2}}$
20. $\dfrac{15^{2/3}}{5^2 15^{-1/3}}$
21. $\dfrac{(-27)^{1/3}}{6}$

22. $\dfrac{(-8)^{2/3}}{-2}$ 23. $\dfrac{-8}{(-27)^{-1/3}}$ 24. $\dfrac{-4}{(-64)^{-2/3}}$

25. $(125)^{-2/3} - (100)^{-3/2}$ 26. $36^{-1/2} + 27^{-2/3}$

27. $\dfrac{25^{-1/2}}{5} + \dfrac{20^{-1/2}}{20^{1/2}}$ 28. $\dfrac{4^{-1}}{(36)^{-1/2}} - \dfrac{5^{-1/2}}{5^{1/2}}$

In Exercises 29 through 32, evaluate the given expressions by use of a calculator.

29. $(17.98)^{1/4}$ 30. $(750.8)^{2/3}$

31. $(4.0187)^{-4/9}$ 32. $(0.18632)^{-1/6}$

In Exercises 33 through 60, use the laws of exponents to simplify the given expressions. Express all answers with positive exponents.

33. $a^{2/3}a^{1/2}$ 34. $x^{5/6}x^{-1/3}$ 35. $\dfrac{y^{-1/2}}{y^{2/5}}$

36. $\dfrac{2r^{4/5}}{r^{-1}}$ 37. $\dfrac{s^{1/4}s^{2/3}}{s^{-1}}$ 38. $\dfrac{x^{3/10}}{x^{-1/5}x^2}$

39. $\dfrac{y^{-1}}{y^{1/3}y^{-1/4}}$ 40. $\dfrac{a^{-2/5}a^2}{a^{-3/10}}$ 41. $(8a^3b^6)^{1/3}$

42. $(8b^{-4}c^2)^{2/3}$ 43. $(16a^4b^3)^{-3/4}$ 44. $(32x^5y^4)^{-2/5}$

45. $\frac{1}{2}(4x^2 + 1)^{-1/2}(8x)$ 46. $\frac{2}{3}(x^3 + 1)^{-1/3}(3x^2)$

47. $\left(\dfrac{9t^{-2}}{16}\right)^{3/2}$ 48. $\left(\dfrac{a^{5/7}}{a^{2/3}}\right)^{7/4}$ 49. $\left(\dfrac{4a^{5/6}b^{-1/5}}{a^{2/3}b^2}\right)^{-1/2}$

50. $\left(\dfrac{a^0b^8c^{-1/8}}{ab^{63/64}}\right)^{32/3}$ 51. $\dfrac{6x^{-1/2}y^{2/3}}{18x^{-1}}\cdot\dfrac{2y^{1/4}}{x^{1/3}}$ 52. $\dfrac{3^{-1}a^{1/2}}{4^{-1/2}b} \div \dfrac{9^{1/2}a^{-1/3}}{2b^{-1/4}}$

53. $(x^{-1} + 2x^{-2})^{-1/2}$ 54. $(a^{-2} - a^{-4})^{-1/4}$ 55. $(a^3)^{-4/3} + a^{-2}$

56. $(4x^6)^{-1/2} - 2x^{-1}$ 57. $[(a^{1/2} - a^{-1/2})^2 + 4]^{1/2}$ 58. $4x^{1/2} + \frac{1}{2}x^{-1/2}(4x + 1)$

59. $x^2(2x - 1)^{-1/2} + 2x(2x - 1)^{1/2}$

60. $(3x - 1)^{-2/3}(1 - x) - (3x - 1)^{1/3}$

In Exercises 61 through 64, perform the indicated operations.

61. In determining the number of electrons involved in a certain calculation with semi-conductors, the expression $9.60 \times 10^{18}T^{3/2}$ is used. Evaluate this expression for $T = 289$ K.

62. In studying the properties of biological fluids, the expression $0.036M^{3/4}$ is used. Evaluate this expression for $M = 1.6 \times 10^5$.

63. An approximate expression for the efficiency of an engine is given by $E = 100(1 - R^{-2/5})$, where R is the compression ratio. What is the efficiency (in percent) of an engine for which $R = 7.35$?

64. An estimate of gas diffusivity may be made by the equation

$$D_m = 0.01\dfrac{T^{1/2}}{(v_a^{1/3} + v_b^{1/3})^2}\left(\dfrac{1}{m_a} + \dfrac{1}{m_b}\right)^{1/2}$$

where T is the temperature in degrees Fahrenheit and the other symbols are constants which depend on the gases under consideration. Calculate the diffusivity of a gas in air at 475°F if $v_a = 26.5$, $v_b = 122$, $m_a = 32.1$, and $m_b = 136$. (The units of diffusivity are in lb·mol/ft·h.)

10–3 Simplest Radical Form

Radicals were first introduced in Section 1–7, and we used them again in developing the concept of a fractional exponent. As we mentioned in the preceding section, it is possible to use fractional exponents for any operation required with radicals. For operations involving multiplication and division, this method has certain advantages. But for adding and subtracting radicals, there is normally little advantage in changing form.

We shall now define the operations with radicals so that these definitions are consistent with the laws of exponents. This will enable us from now on to use either fractional exponents or radicals, whichever is more convenient.

$$\sqrt[n]{a^n} = (\sqrt[n]{a})^n = a \tag{10–9}$$

$$\sqrt[n]{a}\sqrt[n]{b} = \sqrt[n]{ab} \tag{10–10}$$

$$\sqrt[m]{\sqrt[n]{a}} = \sqrt[mn]{a} \tag{10–11}$$

$$\frac{\sqrt[n]{a}}{\sqrt[n]{b}} = \sqrt[n]{\frac{a}{b}}, \qquad b \neq 0 \tag{10–12}$$

The number under the radical is called the **radicand,** *and the number indicating the root being taken is called the* **order** *of the radical.* To avoid difficulties with imaginary numbers (which are considered in the next chapter), we shall assume that all letters represent positive numbers.

Example A

Following are illustrations of each of Eqs. (10–9) through (10–12).

$$\sqrt[5]{4^5} = (\sqrt[5]{4})^5 = 4 \qquad \text{(Eq. (10–9))}$$

$$\sqrt[3]{2}\sqrt[3]{3} = \sqrt[3]{2 \cdot 3} = \sqrt[3]{6} \qquad \text{(Eq. (10–10))}$$

$$\sqrt[3]{\sqrt{5}} = \sqrt[3 \cdot 2]{5} = \sqrt[6]{5} \qquad \text{(Eq. (10–11))}$$

$$\frac{\sqrt{7}}{\sqrt{3}} = \sqrt{\frac{7}{3}} \qquad \text{(Eq. (10–12))}$$

There are certain operations which are performed on radicals in order to put them in their simplest form. The following two examples illustrate one of these operations.

Example B

To simplify $\sqrt{75}$, we recall that $75 = (25)(3)$ and that $\sqrt{25} = 5$. Using Eq. (10–10), we write

$$\sqrt{75} = \sqrt{25}\sqrt{3} = 5\sqrt{3}$$

This illustrates one step which should always be carried out in simplifying radicals. *Always remove all perfect nth-power factors from the radicand of a radical of order n.*

Example C

$$\sqrt{72} = \sqrt{(36)(2)} = \sqrt{36}\sqrt{2} = 6\sqrt{2}$$
$$\sqrt{a^3b^2} = \sqrt{a^2}\sqrt{a}\sqrt{b^2} = ab\sqrt{a}$$
$$\sqrt[3]{40} = \sqrt[3]{8}\sqrt[3]{5} = 2\sqrt[3]{5}$$
$$\sqrt[5]{64x^8y^{12}} = \sqrt[5]{(32)(2)(x^5)(x^3)(y^{10})(y^2)} = \sqrt[5]{(32)(x^5)(y^{10})}\sqrt[5]{2x^3y^2}$$
$$= 2xy^2\,\sqrt[5]{2x^3y^2}$$

The following examples illustrate another procedure which can be used to simplify certain radicals. The procedure is to *reduce the order of the radical,* when possible.

Example D

$$\sqrt[6]{8} = \sqrt[6]{2^3} = 2^{3/6} = 2^{1/2} = \sqrt{2}$$

In this example we started with a sixth root and ended with a square root. Thus the order of the radical was reduced. Fractional exponents are often helpful when we perform this operation.

Example E

$$\sqrt[8]{16} = \sqrt[8]{2^4} = 2^{4/8} = 2^{1/2} = \sqrt{2}$$
$$\frac{\sqrt[4]{9}}{\sqrt{3}} = \frac{\sqrt[4]{3^2}}{\sqrt{3}} = \frac{3^{2/4}}{3^{1/2}} = 1$$
$$\frac{\sqrt[6]{8}}{\sqrt{7}} = \frac{\sqrt[6]{2^3}}{\sqrt{7}} = \frac{2^{1/2}}{7^{1/2}} = \sqrt{\frac{2}{7}}$$
$$\sqrt[9]{27x^6y^{12}} = \sqrt[9]{3^3x^6y^9y^3} = 3^{3/9}x^{6/9}y^{9/9}y^{3/9} = 3^{1/3}x^{2/3}yy^{1/3}$$
$$= y\sqrt[3]{3x^2y}$$

If a radical is to be written in *simplest form,* the two operations illustrated in the last four examples must be performed. They are:

1. *all perfect nth-power factors are removed from a radical of order n,*
2. *if possible, the order of the radical is reduced.*

When working with fractions, it has traditionally been the practice to write a fraction with radicals in a form in which the denominator contains no radicals.

Such a fraction was not considered to be in simplest form unless this was done. This step of simplification was performed primarily for ease of calculation, but with a calculator it does not matter to any extent that there is a radical in the denominator. However, the procedure of writing a radical in this form, called **rationalizing the denominator,** is at times useful for other purposes. Therefore, the following examples show how the process of rationalizing the denominator is carried out.

Example F

To write $\sqrt{\frac{2}{5}}$ in an equivalent form in which the denominator is not included under the radical sign, we create a perfect square in the denominator by multiplying the numerator and the denominator under the radical by 5. This give us $\sqrt{\frac{10}{25}}$, which may be written as $\frac{1}{5}\sqrt{10}$ or $\frac{\sqrt{10}}{5}$. These steps are written as follows:

$$\sqrt{\frac{2}{5}} = \sqrt{\frac{2 \cdot 5}{5 \cdot 5}} = \sqrt{\frac{10}{25}} = \frac{\sqrt{10}}{\sqrt{25}} = \frac{\sqrt{10}}{5}$$

Example G

$$\sqrt{\frac{5}{7}} = \sqrt{\frac{5 \cdot 7}{7 \cdot 7}} = \frac{\sqrt{35}}{\sqrt{49}} = \frac{\sqrt{35}}{7}, \qquad \frac{3}{\sqrt{8}} = \frac{3\sqrt{2}}{\sqrt{8 \cdot 2}} = \frac{3\sqrt{2}}{\sqrt{16}} = \frac{3\sqrt{2}}{4}$$

$$\sqrt[3]{\frac{2}{3}} = \sqrt[3]{\frac{2 \cdot 9}{3 \cdot 9}} = \sqrt[3]{\frac{18}{27}} = \frac{\sqrt[3]{18}}{\sqrt[3]{27}} = \frac{\sqrt[3]{18}}{3}$$

In the second illustration, a perfect square was made by multiplying by $\sqrt{2}$. We should try to choose the smallest possible number which can be used. In the third illustration we wanted a perfect cube in the denominator, since a cube root is being found.

Example H

Simplify the radical $\sqrt{\dfrac{3a}{4b} - 2 + \dfrac{4b}{3a}}$, for $3a \geq 4b$, and rationalize the denominator.

$$\sqrt{\frac{3a}{4b} - 2 + \frac{4b}{3a}} = \sqrt{\frac{(3a)(3a) - 2(3a)(4b) + (4b)(4b)}{(3a)(4b)}}$$

$$= \sqrt{\frac{(3a - 4b)^2}{4(3ab)}} = \frac{3a - 4b}{2}\sqrt{\frac{1}{3ab}}$$

Rationalizing the denominator, we have

$$\frac{3a - 4b}{2}\sqrt{\frac{1}{3ab}} = \frac{3a - 4b}{2}\sqrt{\frac{1(3ab)}{3ab(3ab)}} = \frac{(3a - 4b)\sqrt{3ab}}{6ab}$$

Exercises 10–3

In Exercises 1 through 56, write each expression in simplest radical form. Where a radical appears in a denominator, rationalize the denominator.

1. $\sqrt{24}$ 2. $\sqrt{150}$ 3. $\sqrt{45}$

4. $\sqrt{98}$ 5. $\sqrt{x^2y^5}$ 6. $\sqrt{s^3t^6}$

7. $\sqrt{pq^2r^7}$ 8. $\sqrt{x^2y^4z^3}$ 9. $\sqrt{5x^2}$

10. $\sqrt{12ab^2}$ 11. $\sqrt{18a^3bc^4}$ 12. $\sqrt{54m^5n^3}$

13. $\sqrt[3]{16}$ 14. $\sqrt[4]{48}$ 15. $\sqrt[5]{96}$

16. $\sqrt[3]{-16}$ 17. $\sqrt[3]{8a^2}$ 18. $\sqrt[5]{5a^4b^2}$

19. $\sqrt[4]{64r^3s^4t^5}$ 20. $\sqrt[5]{16x^5y^3z^{11}}$ 21. $\sqrt[5]{8}\sqrt{4}$

22. $\sqrt[7]{4}\sqrt[7]{64}$ 23. $\sqrt[3]{ab^4}\sqrt[3]{a^2b}$ 24. $\sqrt[6]{3m^4n^5}\sqrt[6]{9m^2n^8}$

25. $\sqrt{\dfrac{3}{2}}$ 26. $\sqrt{\dfrac{6}{5}}$ 27. $\sqrt{\dfrac{a}{b}}$

28. $\sqrt{\dfrac{a}{b^3}}$ 29. $\sqrt[3]{\dfrac{3}{4}}$ 30. $\sqrt[4]{\dfrac{2}{5}}$

31. $\sqrt[5]{\dfrac{1}{9}}$ 32. $\sqrt[6]{\dfrac{5}{4}}$ 33. $\sqrt[4]{400}$

34. $\sqrt[8]{81}$ 35. $\sqrt[6]{64}$ 36. $\sqrt[9]{27}$

37. $\sqrt{4 \times 10^4}$ 38. $\sqrt{4 \times 10^5}$ 39. $\sqrt{4 \times 10^6}$

40. $\sqrt{16 \times 10^5}$ 41. $\sqrt[4]{4a^2}$ 42. $\sqrt[6]{b^2c^4}$

43. $\sqrt[4]{\dfrac{1}{4}}$ 44. $\dfrac{\sqrt[4]{80}}{\sqrt[4]{5}}$ 45. $\sqrt[4]{\sqrt[3]{16}}$

46. $\sqrt[5]{\sqrt[4]{9}}$ 47. $\sqrt{\sqrt{\sqrt{2}}}$ 48. $\sqrt{b^4\sqrt{a}}$

49. $\sqrt{\dfrac{1}{2} - \dfrac{1}{3}}$ 50. $\sqrt{\dfrac{5}{4} - \dfrac{1}{8}}$ 51. $\sqrt{\dfrac{1}{a^2} + \dfrac{1}{b}}$

52. $\sqrt{\dfrac{x}{y} + \dfrac{y}{x}}$ 53. $\sqrt{a^2 + 2ab + b^2}$ 54. $\sqrt{a^2 + b^2}$

55. $\sqrt{x^2 + \dfrac{1}{4}}$ 56. $\sqrt{\dfrac{1}{2} + 2r + 2r^2}$

In Exercises 57 through 60, perform the required operation.

57. The period (in seconds) for one cycle of a simple pendulum is given by $T = 2\pi\sqrt{L/g}$, where L is the length of the pendulum (in feet) and g is the acceleration due to gravity ($g = 32$ ft/s^2). If L is 3 ft, express the period of the pendulum in rationalized form, and then find the decimal form on a calculator.

58. Under certain circumstances, the frequency in an electric circuit containing an inductance L and capacitance C is given by $f = \dfrac{1}{2\pi\sqrt{LC}}$. If $L = 0.1$ H and $C = 250 \times 10^{-6}$ F, express f in rationalized form, and then find the decimal form on a calculator.

59. When analyzing the velocity of an object which falls through a very great distance, the expression $a\sqrt{2g/a}$ is derived. Show by rationalizing the denominator that this expression takes on a simpler form.

60. When dealing with the voltage of an electronic device, the expression $V(x/d)^{4/3}$ arises. Write this expression in rationalized radical form.

10–4 Addition and Subtraction of Radicals

When we first introduced the concept of adding algebraic expressions, we found that it was possible to combine similar terms, that is, those which differed only in numerical coefficients. The same is true of adding radicals. *We must have similar radicals in order to perform the addition,* rather than simply to be able to indicate addition. *By similar radicals we mean radicals which differ only in their numerical coefficients,* and which must therefore be of the same order and have the same radicand.

In order to add radicals, we first express each radical in its simplest form, rationalize any denominators, and then add those which are similar. For those which are not similar, we can only indicate the addition.

Example A

$$2\sqrt{7} - 5\sqrt{7} + \sqrt{7} = -2\sqrt{7}$$
$$\sqrt[5]{6} + 4\sqrt[5]{6} - 2\sqrt[5]{6} = 3\sqrt[5]{6}$$
$$\sqrt{5} + 2\sqrt{3} - 5\sqrt{5} = 2\sqrt{3} - 4\sqrt{5}$$

We note that in the last illustration we are able only to indicate the final subtraction.

Example B

$$\sqrt{2} + \sqrt{8} = \sqrt{2} + \sqrt{4\cdot2} = \sqrt{2} + \sqrt{4}\sqrt{2} = \sqrt{2} + 2\sqrt{2} = 3\sqrt{2}$$
$$\sqrt[3]{24} + \sqrt[3]{81} = \sqrt[3]{8\cdot3} + \sqrt[3]{27\cdot3} = \sqrt[3]{8}\sqrt[3]{3} + \sqrt[3]{27}\sqrt[3]{3}$$
$$= 2\sqrt[3]{3} + 3\sqrt[3]{3} = 5\sqrt[3]{3}$$

 Notice that $\sqrt{8}$, $\sqrt[3]{24}$, and $\sqrt[3]{81}$ were simplified before performing the addition. We also note that $\sqrt{2} + \sqrt{8}$ is ***not*** equal to $\sqrt{2 + 8}$.

We note in the illustrations of Example B that the radicals do not initially appear to be similar. However, after each is simplified we are able to recognize the similar radicals.

Example C

$$6\sqrt{7} - \sqrt{28} + 3\sqrt{63} = 6\sqrt{7} - \sqrt{4\cdot7} + 3\sqrt{9\cdot7}$$
$$= 6\sqrt{7} - 2\sqrt{7} + 3(3\sqrt{7})$$
$$= 6\sqrt{7} - 2\sqrt{7} + 9\sqrt{7} = 13\sqrt{7}$$
$$3\sqrt{125} - \sqrt{20} + \sqrt{27} = 3\sqrt{25\cdot5} - \sqrt{4\cdot5} + \sqrt{9\cdot3}$$
$$= 3(5\sqrt{5}) - 2\sqrt{5} + 3\sqrt{3}$$
$$= 13\sqrt{5} + 3\sqrt{3}$$

Example D

$$\sqrt{24} + \sqrt{\frac{3}{2}} = 2\sqrt{6} + \frac{\sqrt{6}}{2} = \frac{4\sqrt{6} + \sqrt{6}}{2} = \frac{5}{2}\sqrt{6}$$

One radical was simplified by removing the perfect square factor and in the other we rationalized the denominator. Note that we would not be able to combine the radicals if we did not rationalize the denominator of the second radical.

Our main purpose in this section is to add radicals in radical form. However, a decimal value can be obtained by use of a calculator, and in the following example we use decimal values to verify the result of the addition.

Example E

Perform the addition $\sqrt{32} + 7\sqrt{18} - 2\sqrt{200}$ and use a calculator to verify the result.

$$\begin{aligned}
\sqrt{32} + 7\sqrt{18} - 2\sqrt{200} &= \sqrt{16 \cdot 2} + 7\sqrt{9 \cdot 2} - 2\sqrt{100 \cdot 2} \\
&= 4\sqrt{2} + 7(3\sqrt{2}) - 2(10\sqrt{2}) \\
&= 4\sqrt{2} + 21\sqrt{2} - 20\sqrt{2} \\
&= 5\sqrt{2}
\end{aligned}$$

Using the calculator sequence

$$32 \quad \boxed{\sqrt{x}} \quad \boxed{+} \quad 7 \quad \boxed{\times} \quad 18 \quad \boxed{\sqrt{x}}$$

$$\boxed{-} \quad 2 \quad \boxed{\times} \quad 200 \quad \boxed{\sqrt{x}} \quad \boxed{=}$$

we find that the decimal value of the original expression is 7.0710678, which is the same as the value of $5\sqrt{2}$.

We now show two examples of adding radical expressions which contain literal symbols.

Example F

$$\begin{aligned}
\sqrt{\frac{2}{3a}} - 2\sqrt{\frac{3}{2a}} &= \frac{1}{3a}\sqrt{6a} - \frac{2}{2a}\sqrt{6a} = \frac{1}{3a}\sqrt{6a} - \frac{1}{a}\sqrt{6a} \\
&= \frac{\sqrt{6a} - 3\sqrt{6a}}{3a} = \frac{-2\sqrt{6a}}{3a} = -\frac{2}{3a}\sqrt{6a}
\end{aligned}$$

Example G

$$\sqrt{\frac{4}{a} - 4 + a} + \sqrt{\frac{1}{a}} - \sqrt{16a^3} = \sqrt{\frac{4 - 4a + a^2}{a}} + \sqrt{\frac{1}{a}} - 4a\sqrt{a}$$

$$= \sqrt{\frac{(2 - a)^2 \cdot a}{a \cdot a}} + \sqrt{\frac{1 \cdot a}{a \cdot a}} - 4a\sqrt{a}$$

$$= \frac{2 - a}{a}\sqrt{a} + \frac{1}{a}\sqrt{a} - 4a\sqrt{a}$$

$$= \sqrt{a}\left(\frac{2 - a}{a} + \frac{1}{a} - 4a\right) = \sqrt{a}\left(\frac{2 - a + 1 - 4a^2}{a}\right)$$

$$= \frac{(3 - a - 4a^2)\sqrt{a}}{a}$$

This simplification is valid for $a \leq 2$, since we let $\sqrt{(2 - a)^2} = 2 - a$.

Exercises 10–4

In Exercises 1 through 36, express each radical in simplest form, rationalize denominators, and perform the indicated operations.

1. $2\sqrt{3} + 5\sqrt{3}$
2. $8\sqrt{11} - 3\sqrt{11}$
3. $2\sqrt{7} + \sqrt{5} - 3\sqrt{7}$
4. $8\sqrt{6} - 2\sqrt{3} - 5\sqrt{6}$
5. $\sqrt{5} + \sqrt{20}$
6. $\sqrt{7} + \sqrt{63}$
7. $2\sqrt{3} - 3\sqrt{12}$
8. $4\sqrt{2} - \sqrt{50}$
9. $\sqrt{8} - \sqrt{32}$
10. $\sqrt{27} + 2\sqrt{18}$
11. $2\sqrt{28} + 3\sqrt{175}$
12. $5\sqrt{300} - 7\sqrt{48}$
13. $2\sqrt{20} - \sqrt{125} - \sqrt{45}$
14. $2\sqrt{44} - \sqrt{99} + \sqrt{176}$
15. $3\sqrt{75} + 2\sqrt{48} - 2\sqrt{18}$
16. $2\sqrt{28} - \sqrt{108} - 2\sqrt{175}$
17. $\sqrt{60} + \sqrt{\frac{5}{3}}$
18. $\sqrt{84} - \sqrt{\frac{3}{7}}$
19. $\sqrt{\frac{1}{2}} + \sqrt{\frac{25}{2}} - \sqrt{18}$
20. $\sqrt{6} - \sqrt{\frac{2}{3}} - \sqrt{18}$
21. $\sqrt[3]{81} + \sqrt[3]{3000}$
22. $\sqrt[3]{-16} + \sqrt[3]{54}$
23. $\sqrt[4]{32} - \sqrt[8]{4}$
24. $\sqrt[6]{\sqrt{2}} - \sqrt[12]{2^{13}}$
25. $\sqrt{a^3b} - \sqrt{4ab^5}$
26. $\sqrt{2x^2y} + \sqrt{8y^3}$
27. $\sqrt{6}\sqrt{5}\sqrt{3} - \sqrt{40a^2}$
28. $\sqrt{60n} + 2\sqrt{15b^2n} - b\sqrt{135n}$
29. $\sqrt[3]{24a^2b^4} - \sqrt[3]{3a^5b}$
30. $\sqrt[5]{32a^6b^4} + 3a\sqrt[5]{243ab^9}$
31. $\sqrt{\frac{a}{c^5}} - \sqrt{\frac{c}{a^3}}$
32. $\sqrt{\frac{2x}{3y}} + \sqrt{\frac{27y}{8x}}$
33. $\sqrt[3]{\frac{a}{b}} - \sqrt[3]{\frac{8b^2}{a^2}}$
34. $\sqrt[4]{\frac{c}{b}} - \sqrt[4]{bc}$
35. $\sqrt{\frac{a - b}{a + b}} - \sqrt{\frac{a + b}{a - b}}$
36. $\sqrt{\frac{16}{x} + 8 + x} - \sqrt{1 - \frac{1}{x}}$

In Exercises 37 through 40, express each radical in simplified form, rationalize denominators, and perform the indicated operations. Then use a calculator to verify the result.

37. $3\sqrt{45} + 3\sqrt{75} - 2\sqrt{500}$ 38. $2\sqrt{40} + 3\sqrt{90} - 5\sqrt{250}$

39. $2\sqrt{\frac{2}{3}} + \sqrt{24} - 5\sqrt{\frac{3}{2}}$ 40. $\sqrt{\frac{2}{7}} - 2\sqrt{\frac{7}{2}} + 5\sqrt{56}$

In Exercises 41 and 42, solve the given problems.

41. Find the sum of the two roots of the quadratic equation $ax^2 + bx + c = 0$.

42. In the study of the kinetic theory of gases, the expression

$$a\sqrt{\frac{h^3 m^5}{\pi}} + b\sqrt{\frac{h^3 m^3}{\pi}}$$

is found. Perform the indicated addition.

10–5 Multiplication of Radicals

When multiplying expressions containing radicals, we use Eq. (10–10), along with the normal procedures of algebraic multiplication. Note that the orders of the radicals being multiplied in Eq. (10–10) are the same. The following examples illustrate the method.

Example A

$$\sqrt{5}\sqrt{2} = \sqrt{10},$$
$$\sqrt{33}\sqrt{3} = \sqrt{99} = \sqrt{9(11)} = \sqrt{9}\sqrt{11} = 3\sqrt{11}$$

We note that we must be careful to express the resulting radical in simplest form.

Example B

$$\sqrt[3]{6}\sqrt[3]{4} = \sqrt[3]{24} = \sqrt[3]{8}\sqrt[3]{3} = 2\sqrt[3]{3}$$
$$\sqrt[5]{8a^3b^4}\sqrt[5]{8a^2b^3} = \sqrt[5]{64a^5b^7} = \sqrt[5]{32a^5b^5}\sqrt[5]{2b^2} = 2ab\sqrt[5]{2b^2}$$

Example C

$$\sqrt{2}(3\sqrt{5} - 4\sqrt{2}) = 3\sqrt{2}\sqrt{5} - 4\sqrt{2}\sqrt{2} = 3\sqrt{10} - 4\sqrt{4}$$
$$= 3\sqrt{10} - 4(2) = 3\sqrt{10} - 8$$

When raising a single-term radical expression to a power, we use the basic meaning of the power. When raising a binomial to a power, we proceed as with any binomial and use Eq. (10–10).

Example D

$$(2\sqrt{7})^2 = 2^2(\sqrt{7})^2 = 4(7) = 28$$
$$(\sqrt{a} - \sqrt{b})^2 = (\sqrt{a})^2 - 2\sqrt{a}\sqrt{b} + (\sqrt{b})^2 = a + b - 2\sqrt{ab}$$

Example E
$$(5\sqrt{7} - 2\sqrt{3})(4\sqrt{7} + 3\sqrt{3}) = 20\sqrt{7}\sqrt{7} + 15\sqrt{7}\sqrt{3}$$
$$- 8\sqrt{3}\sqrt{7} - 6\sqrt{3}\sqrt{3}$$
$$= 20(7) + 15\sqrt{21} - 8\sqrt{21} - 6(3)$$
$$= 140 + 7\sqrt{21} - 18$$
$$= 122 + 7\sqrt{21}$$

Using a calculator we find that the decimal value of the original expression is 154.07803, which is the same as that for $122 + 7\sqrt{21}$.

Example F
$$(\sqrt{6} - \sqrt{2} - \sqrt{3})(\sqrt{6} + \sqrt{2}) = (\sqrt{6} - \sqrt{2})(\sqrt{6} + \sqrt{2})$$
$$- \sqrt{3}(\sqrt{6} + \sqrt{2})$$
$$= (6 - 2) - \sqrt{18} - \sqrt{6}$$
$$= 4 - 3\sqrt{2} - \sqrt{6}$$

Example G
$$\left(3\sqrt{\frac{a}{b}} - \sqrt{ab}\right)\left(2\sqrt{\frac{a}{b}} - \sqrt{ab}\right) = 6\frac{a}{b} - 5\sqrt{\frac{a^2b}{b}} + ab$$
$$= \frac{6a}{b} - 5a + ab$$
$$= \frac{6a - 5ab + ab^2}{b}$$
$$= \frac{a(6 - 5b + b^2)}{b}$$
$$= \frac{a(3 - b)(2 - b)}{b}$$

Again, we note that *to multiply radicals and combine them under one radical sign, it is necessary that the order of the radicals be the same.* If necessary we can make the order of each radical the same by appropriate operations on each radical separately. Fractional exponents are frequently useful for this purpose.

Example H
$$\sqrt[3]{2}\sqrt{5} = 2^{1/3}5^{1/2} = 2^{2/6}5^{3/6} = (2^2 5^3)^{1/6} = \sqrt[6]{500}$$
$$\sqrt[3]{4a^2b}\,\sqrt[4]{8a^3b^2} = (2^2a^2b)^{1/3}(2^3a^3b^2)^{1/4} = (2^2a^2b)^{4/12}(2^3a^3b^2)^{3/12}$$
$$= (2^8a^8b^4)^{1/12}(2^9a^9b^6)^{1/12} = (2^{17}a^{17}b^{10})^{1/12}$$
$$= 2a(2^5a^5b^{10})^{1/12}$$
$$= 2a\sqrt[12]{32a^5b^{10}}$$

Exercises 10–5

In Exercises 1 through 48, perform the indicated multiplications, expressing answers in simplest form with rationalized denominators.

1. $\sqrt{3}\sqrt{10}$ 2. $\sqrt{2}\sqrt{51}$ 3. $\sqrt{6}\sqrt{2}$ 4. $\sqrt{7}\sqrt{14}$

5. $\sqrt[3]{4}\sqrt[3]{2}$ 6. $\sqrt[4]{3}\sqrt[4]{27}$ 7. $\sqrt[5]{4}\sqrt[5]{16}$ 8. $\sqrt[3]{25}\sqrt[3]{50}$

9. $(5\sqrt{2})^2$ 10. $(3\sqrt{3})^2$ 11. $(2\sqrt[3]{2})^3$ 12. $(3\sqrt[3]{5})^3$

13. $\sqrt{\frac{2}{3}}\sqrt{5}$ 14. $\sqrt{8}\sqrt{\frac{5}{2}}$ 15. $\sqrt{\frac{5}{6}}\sqrt{\frac{2}{11}}$ 16. $\sqrt{\frac{6}{7}}\sqrt{\frac{2}{3}}$

17. $\sqrt{3}(\sqrt{2} - \sqrt{5})$ 18. $\sqrt{5}(\sqrt{7} + \sqrt{2})$

19. $2\sqrt{2}(\sqrt{8} - 3\sqrt{6})$ 20. $3\sqrt{5}(\sqrt{15} - 2\sqrt{5})$

21. $(2 - \sqrt{5})(2 + \sqrt{5})$ 22. $(6 - \sqrt{3})(6 + \sqrt{3})$

23. $(6 - \sqrt{3})^2$ 24. $(2 - \sqrt{5})^2$

25. $(3\sqrt{5} - 2\sqrt{3})(6\sqrt{5} + 7\sqrt{3})$ 26. $(3\sqrt{7} - \sqrt{8})(\sqrt{7} + \sqrt{2})$

27. $(3\sqrt{11} - \sqrt{6})(2\sqrt{11} + 5\sqrt{6})$ 28. $(2\sqrt{10} + 3\sqrt{15})(\sqrt{10} - 7\sqrt{15})$

29. $\sqrt{a}(\sqrt{ab} + \sqrt{c})$ 30. $\sqrt{3x}(\sqrt{3x} - \sqrt{xy})$

31. $\sqrt{5n}(\sqrt{15n} + \sqrt{20m})$ 32. $\sqrt{2x}(\sqrt{8xy} - 3\sqrt{y})$

33. $(\sqrt{2a} - \sqrt{b})(\sqrt{2a} + 3\sqrt{b})$ 34. $(2\sqrt{mn} - 3\sqrt{n})(3\sqrt{mn} + 2\sqrt{n})$

35. $(\sqrt{2} + \sqrt{3} + \sqrt{5})(\sqrt{3} - \sqrt{5})$ 36. $(2\sqrt{7} - \sqrt{5})(\sqrt{14} - 2\sqrt{5} + \sqrt{7})$

37. $\sqrt{2}\sqrt[3]{3}$ 38. $\sqrt[5]{16}\sqrt[3]{8}$

39. $\sqrt[4]{ab}\sqrt[3]{bc}$ 40. $\sqrt{2x}\sqrt[5]{16x}$

41. $(\sqrt[5]{\sqrt{6}} - \sqrt{5})(\sqrt[5]{\sqrt{6}} + \sqrt{5})$ 42. $\sqrt[3]{5} - \sqrt{17}\sqrt[3]{5} + \sqrt{17}$

43. $(\sqrt{2x^2} - \sqrt[3]{y})(\sqrt{4x} + \sqrt[3]{y^2})$ 44. $(\sqrt{a} - \sqrt[3]{b})(2\sqrt{a} - \sqrt[3]{b})$

45. $\left(\sqrt{\frac{2}{a}} + \sqrt{\frac{a}{2}}\right)\left(\sqrt{\frac{2}{a}} - 2\sqrt{\frac{a}{2}}\right)$ 46. $\left(\sqrt{\frac{x}{y}} - \sqrt{xy}\right)\left(\sqrt{\frac{y}{x}} + \sqrt{xy} - 1\right)$

47. $(2x - \sqrt{x - 2y})^2$ 48. $(3 + \sqrt{6 - 2a})(2 - \sqrt{6 - 2a})$

In Exercises 49 through 52, perform the indicated multiplications, expressing answers in simplest form. Then use a calculator to verify the result.

49. $(\sqrt{11} + \sqrt{6})(\sqrt{11} - 2\sqrt{6})$ 50. $(2\sqrt{5} - \sqrt{7})(3\sqrt{5} + \sqrt{7})$

51. $(5\sqrt{13} + 2\sqrt{14})(2\sqrt{13} + \sqrt{14})$ 52. $(3\sqrt{3} - 5\sqrt{6})(8\sqrt{3} - 7\sqrt{6})$

In Exercises 53 and 54, combine the terms into a single fraction, but do not rationalize the denominator.

53. $\dfrac{x^2}{\sqrt{2x + 1}} + 2x\sqrt{2x + 1}$ 54. $4\sqrt{x^2 + 1} - \dfrac{4x}{\sqrt{x^2 + 1}}$

In Exercises 55 and 56, perform the indicated operations.

55. Determine the product of the two roots of the quadratic equation

$$ax^2 + bx + c = 0.$$

56. Relationships involving mass transfer of liquid involve the expression

$$\sqrt{\frac{dG}{u}}\sqrt[3]{\frac{MD}{u}}$$

Express this in simplest radical form.

10–6 Division of Radicals

We have already dealt with some cases of division of radicals in the previous sections. When we have had the indicated division of one radical by another, we have generally expressed the answer with no radicals in the denominator by rationalizing the denominator. Therefore, if a change is to be made in an expression involving division by a radical, rationalization of the denominator is the principal step to be carried out. When we rationalize the denominator, we change the fraction to an equivalent form in which the denominator is free of radicals. In doing so, multiplication of the numerator and the denominator by the appropriate quantity is a primary step.

Example A

$$\frac{\sqrt{3}}{\sqrt{5}} = \frac{\sqrt{3}\sqrt{5}}{\sqrt{5}\sqrt{5}} = \frac{\sqrt{15}}{5}$$

$$\frac{\sqrt{a}}{\sqrt[3]{b}} = \frac{\sqrt{a}}{\sqrt[3]{b}} \cdot \frac{\sqrt[3]{b^2}}{\sqrt[3]{b^2}} = \frac{\sqrt{a}\sqrt[3]{b^2}}{b} = \frac{a^{3/6}b^{4/6}}{b} = \frac{\sqrt[6]{a^3b^4}}{b}$$

Notice in the second illustration that the denominator was rationalized and that the factors of the numerator were written in terms of fractional exponents so they could be combined under one radical.

If the denominator is the sum (or difference) of two terms, at least one of which is a radical, the fraction can be rationalized by multiplying both the numerator and the denominator by the difference (or sum) of the same two terms, if the radicals are square roots.

Example B

The fraction

$$\frac{1}{\sqrt{3} - \sqrt{2}}$$

can be rationalized by multiplying the numerator and the denominator by $\sqrt{3} + \sqrt{2}$. In this way the radicals will be removed from the denominator.

$$\frac{1}{\sqrt{3} - \sqrt{2}} \cdot \frac{\sqrt{3} + \sqrt{2}}{\sqrt{3} + \sqrt{2}} = \frac{\sqrt{3} + \sqrt{2}}{(\sqrt{3})^2 - (\sqrt{2})^2} = \frac{\sqrt{3} + \sqrt{2}}{3 - 2} = \sqrt{3} + \sqrt{2}$$

The reason this technique works is that an expression of the form $a^2 - b^2$ is created in the denominator, where a or b (or both) is a radical. We see that the result is a denominator free of radicals.

Example C

$$\frac{\sqrt{2}}{2\sqrt{5}+\sqrt{3}}=\frac{\sqrt{2}}{2\sqrt{5}+\sqrt{3}}\cdot\frac{2\sqrt{5}-\sqrt{3}}{2\sqrt{5}-\sqrt{3}}=\frac{2\sqrt{2}\sqrt{5}-\sqrt{2}\sqrt{3}}{(2\sqrt{5})^2-(\sqrt{3})^2}$$

$$=\frac{2\sqrt{10}-\sqrt{6}}{2^2(\sqrt{5})^2-(\sqrt{3})^2}=\frac{2\sqrt{10}-\sqrt{6}}{20-3}=\frac{2\sqrt{10}-\sqrt{6}}{17}$$

Calculating the decimal value of the original expression and that of the answer, we find that they are both 0.2279450.

Example D

$$\frac{\sqrt{x-y}}{1-\sqrt{x-y}}=\frac{\sqrt{x-y}(1+\sqrt{x-y})}{(1-\sqrt{x-y})(1+\sqrt{x-y})}$$

$$=\frac{\sqrt{x-y}+x-y}{1-x+y}\qquad(x>y)$$

Example E

$$\frac{1+\dfrac{\sqrt{3}}{2}}{1-\dfrac{\sqrt{3}}{2}}=\frac{\dfrac{2+\sqrt{3}}{2}}{\dfrac{2-\sqrt{3}}{2}}=\frac{2+\sqrt{3}}{2}\cdot\frac{2}{2-\sqrt{3}}=\frac{2+\sqrt{3}}{2-\sqrt{3}}$$

$$=\frac{(2+\sqrt{3})(2+\sqrt{3})}{(2-\sqrt{3})(2+\sqrt{3})}=\frac{4+4\sqrt{3}+3}{4-3}=7+4\sqrt{3}$$

We note that in this particular case, the result has a much simpler form after rationalizing the denominator than the original expression.

Exercises 10–6

In Exercises 1 through 40, perform the indicated operations and express the answers in simplest form with rationalized denominators.

1. $\dfrac{\sqrt{21}}{\sqrt{3}}$

2. $\dfrac{\sqrt{105}}{\sqrt{5}}$

3. $\sqrt{7}\div\sqrt{2}$

4. $3\sqrt{2}\div2\sqrt{3}$

5. $\dfrac{\sqrt[3]{x^2}}{\sqrt[3]{24}}$

6. $\dfrac{\sqrt[4]{3}}{\sqrt[4]{2}}$

7. $\dfrac{\sqrt[3]{5}}{\sqrt{2}}$

8. $\dfrac{\sqrt{6}}{\sqrt[3]{2}}$

9. $\dfrac{\sqrt{a}}{\sqrt[3]{4}}$

10. $\dfrac{\sqrt[4]{32}}{\sqrt[5]{b^3}}$

11. $\dfrac{\sqrt{6}-3}{\sqrt{6}}$

12. $\dfrac{5-\sqrt{10}}{\sqrt{10}}$

13. $\dfrac{\sqrt{2a}-b}{\sqrt{a}}$

14. $\dfrac{\sqrt{8x}+\sqrt{2}}{\sqrt{2}}$

15. $\dfrac{\sqrt{3a}-\sqrt{b}}{\sqrt{3}}$

16. $\dfrac{\sqrt{7x}-\sqrt{14}}{\sqrt{7}}$

17. $\dfrac{1}{\sqrt{7}+\sqrt{3}}$

18. $\dfrac{4}{\sqrt{6}+\sqrt{2}}$

19. $\dfrac{\sqrt{7}}{\sqrt{5}-\sqrt{2}}$

20. $\dfrac{\sqrt{8}}{2\sqrt{3}-\sqrt{5}}$

21. $\dfrac{3}{2\sqrt{5}-6}$

22. $\dfrac{\sqrt{7}}{4 - 2\sqrt{7}}$

23. $\dfrac{2\sqrt{3}}{3\sqrt{3} - 1}$

24. $\dfrac{6\sqrt{5}}{5 - 2\sqrt{5}}$

25. $\dfrac{\sqrt{2} - 1}{\sqrt{7} - 3\sqrt{2}}$

26. $\dfrac{3 - \sqrt{5}}{2\sqrt{2} + \sqrt{5}}$

27. $\dfrac{2 - \sqrt{3}}{5 - 2\sqrt{3}}$

28. $\dfrac{2\sqrt{15} - 3}{\sqrt{15} + 4}$

29. $\dfrac{2\sqrt{3} - 5\sqrt{5}}{\sqrt{3} + 2\sqrt{5}}$

30. $\dfrac{2\sqrt{6} + \sqrt{11}}{\sqrt{6} - 3\sqrt{11}}$

31. $\dfrac{\sqrt{7} - \sqrt{14}}{2\sqrt{7} - 3\sqrt{14}}$

32. $\dfrac{\sqrt{15} - 3\sqrt{5}}{2\sqrt{15} - \sqrt{5}}$

33. $\dfrac{2\sqrt{x}}{\sqrt{x} - \sqrt{y}}$

34. $\dfrac{6\sqrt{a}}{2\sqrt{a} - b}$

35. $\dfrac{8}{3\sqrt{a} - 2\sqrt{b}}$

36. $\dfrac{6}{1 + 2\sqrt{x}}$

37. $\dfrac{\sqrt{2c} + 3\sqrt{d}}{\sqrt{2c} - \sqrt{d}}$

38. $\dfrac{3\sqrt{2x} + 2\sqrt{x}}{\sqrt{2x} - \sqrt{x}}$

39. $\dfrac{\sqrt{x} + y}{\sqrt{x} - y - \sqrt{x}}$

40. $\dfrac{\sqrt{1 + a}}{a - \sqrt{1 - a}}$

In Exercises 41 through 44, perform the indicated operations and express the answers in simplest form with rationalized denominators. Then use a calculator to verify the result.

41. $\dfrac{2\sqrt{7}}{\sqrt{7} + \sqrt{6}}$

42. $\dfrac{3\sqrt{13}}{\sqrt{13} - \sqrt{11}}$

43. $\dfrac{2\sqrt{6} - \sqrt{5}}{3\sqrt{6} - 4\sqrt{5}}$

44. $\dfrac{\sqrt{7} - 4\sqrt{2}}{5\sqrt{7} - 4\sqrt{2}}$

In Exercises 45 and 46, perform the indicated operations.

45. An expression used in determining the characteristics of a spur gear is $\dfrac{50}{50 + \sqrt{V}}$. Write this expression in rationalized form.

46. In the theory of waves in wires, the following expression is found:

$$\dfrac{\sqrt{d_1} - \sqrt{d_2}}{\sqrt{d_1} + \sqrt{d_2}}$$

Evaluate this expression if $d_1 = 10$ and $d_2 = 3$.

10–7 Exercises for Chapter 10

In Exercises 1 through 28, express each of the given expressions in the simplest form which contains only positive exponents.

1. $2a^{-2}b^0$

2. $(2c)^{-1}z^{-2}$

3. $\dfrac{2c^{-1}}{d^{-3}}$

4. $\dfrac{-5x^0}{3y^{-1}}$

5. $3(25)^{3/2}$

6. $32^{2/5}$

7. $400^{-3/2}$

8. $1000^{-2/3}$

9. $\left(\dfrac{3}{t^2}\right)^{-2}$

10. $\left(\dfrac{2x^3}{3}\right)^{-3}$

11. $\dfrac{-8^{2/3}}{49^{-1/2}}$

12. $\dfrac{81^{-3/4}}{7^{-1}}$

13. $(2a^{1/3}b^{5/6})^6$

14. $(ax^{-1/2}y^{1/4})^8$

15. $(-32m^{15}n^{10})^{3/5}$

16. $(27x^{-6}y^9)^{2/3}$

17. $2x^{-2} - y^{-1}$

18. $a^{-1} + b^{-2}$

19. $\dfrac{2x^{-1}}{x^{-1} + y^{-1}}$

20. $\dfrac{3a}{(2a)^{-1} - a}$

21. $(a - 3b^{-1})^{-1}$

22. $(2s^{-2} + t)^{-2}$

23. $(x^3 - y^{-3})^{1/3}$

24. $(x^2 + 2xy + y^2)^{-1/2}$

25. $(8a^3)^{2/3}(4a^{-2} + 1)^{1/2}$

26. $\left[\dfrac{(9a)^0(4x^2)^{1/3}(3b^{1/2})}{(2b^0)^2}\right]^{-6}$

27. $2x(x - 1)^{-2} - 2(x^2 + 1)(x - 1)^{-3}$

28. $4(1 - x^2)^{1/2} - (1 - x^2)^{-1/2}$

In Exercises 29 through 72, perform the indicated operations and express the answer in simplest radical form with rationalized denominators.

29. $\sqrt{68}$

30. $\sqrt{96}$

31. $\sqrt{ab^5c^2}$

32. $\sqrt{x^3y^4z^6}$

33. $\sqrt{9a^3b^4}$

34. $\sqrt{8x^5y^2}$

35. $\sqrt{84st^3u^2}$

36. $\sqrt{52x^2y^5}$

37. $\dfrac{5}{\sqrt{2s}}$

38. $\dfrac{3a}{\sqrt{5x}}$

39. $\sqrt{\dfrac{11}{27}}$

40. $\sqrt{\dfrac{7}{8}}$

41. $\sqrt[4]{8m^6n^9}$

42. $\sqrt[3]{9a^7b^{-3}}$

43. $\sqrt[4]{\sqrt[3]{64}}$

44. $\sqrt{a^{-3}\sqrt[5]{b^{12}}}$

45. $\sqrt{200} + \sqrt{32}$

46. $2\sqrt{68} - \sqrt{153}$

47. $\sqrt{63} - 2\sqrt{112} - \sqrt{28}$

48. $2\sqrt{20} - \sqrt{80} - 2\sqrt{125}$

49. $a\sqrt{2x^3} + \sqrt{8a^2x^3}$

50. $2\sqrt{m^2n^3} - \sqrt{n^5}$

51. $\sqrt[3]{8a^4} + b\sqrt[3]{a}$

52. $\sqrt[4]{2xy^5} - \sqrt[4]{32xy}$

53. $\sqrt{5}(2\sqrt{5} - \sqrt{11})$

54. $2\sqrt{8}(5\sqrt{2} - \sqrt{6})$

55. $2\sqrt{2}(\sqrt{6} - \sqrt{10})$

56. $3\sqrt{5}(\sqrt{15} + 2\sqrt{35})$

57. $(2 - 3\sqrt{17})(3 + \sqrt{17})$

58. $(5\sqrt{6} - 4)(3\sqrt{6} + 5)$

59. $(2\sqrt{7} - 3\sqrt{3})(3\sqrt{7} + \sqrt{3})$

60. $(3\sqrt{2} - \sqrt{13})(5\sqrt{2} + 3\sqrt{13})$

61. $\dfrac{\sqrt{3x}}{2\sqrt{3x} - \sqrt{y}}$

62. $\dfrac{5\sqrt{a}}{2\sqrt{a} - c}$

63. $\dfrac{\sqrt{2}}{\sqrt{3} - 4\sqrt{2}}$

64. $\dfrac{4}{3 - 2\sqrt{7}}$

65. $\dfrac{\sqrt{7} - \sqrt{5}}{\sqrt{5} + 3\sqrt{7}}$

66. $\dfrac{4 - 2\sqrt{6}}{3 + 2\sqrt{6}}$

67. $\dfrac{2\sqrt{x} - a}{3\sqrt{x} + 5a}$

68. $\dfrac{3\sqrt{y} - \sqrt{z}}{2\sqrt{y} + 5\sqrt{z}}$

69. $\sqrt{a^{-1} + b^2}$

70. $\sqrt{a^{-2} + \dfrac{1}{b^2}}$

71. $\left(\dfrac{2 - \sqrt{15}}{2}\right)^2 - \left(\dfrac{2 - \sqrt{15}}{2}\right)$

72. $\sqrt{2 + \dfrac{b}{a} + \dfrac{a}{b}} + \sqrt{a^4b^2 + 2a^3b^2 + a^2b^2}$

In Exercises 73 through 76, perform the indicated operations and express the answer in simplest radical form with rationalized denominators. Then verify the results with a calculator.

73. $\sqrt{52} + 4\sqrt{24} - \sqrt{54}$

74. $2\sqrt{5}(6\sqrt{5} - 5\sqrt{6})$

75. $(\sqrt{7} - 2\sqrt{15})(3\sqrt{7} - \sqrt{15})$

76. $\dfrac{2\sqrt{3} - 7\sqrt{14}}{3\sqrt{3} + 2\sqrt{14}}$

In Exercises 77 through 86, perform the indicated operations.

77. In the study of electricity, the expression $e^{-i(\omega t - \alpha t)}$ is found. Rewrite this expression so that it contains no minus signs in the exponent.

78. An expression found when convection of heat is discussed is $k^{-1}x + h^{-1}$. Write this expression without the use of negative exponents.

79. In the study of fluid flow in pipes, the expression $0.220N^{-1/6}$ is found. Evaluate this expression for $N = 64 \times 10^6$.

80. In the theory of semiconductors, the expression $km^{3/2}(E - E_1)^{1/2}$ is found. Write this expression in simplified radical form.

81. The distance between ion layers in a crystalline solid such as table salt is given by the expression $\sqrt[3]{M/2N\rho}$, where M is the molecular weight, N is called Avogadro's number, and ρ is the density. Express this in rationalized form.

82. In the study of biological effects of sound, the expression $(2n/\omega r)^{-1/2}$ is found. Express this in simplest rationalized radical form.

83. The root-mean-square velocity of a gas molecule is given by $v = \sqrt{3RT/M}$, where R is the gas constant, T is the thermodynamic temperature, and M is the molecular weight. Express the velocity (in meters per second) of an oxygen molecule in simplest rationalized radical form. Then calculate the value of the expression if $T = 300$ K, $R = 8.31$ J/mol·K, and $M = 0.032$ kg/mol.

84. A surveyor measuring distances with a steel tape must be careful to correct for the tension which is applied to the tape and for the sag in the tape. If she applies what is known as "normal tension," these two effects will cancel each other. An expression involving the normal tension T_n which is found for a certain tape is

$$\frac{0.2W\sqrt{2.7 \times 10^5}}{\sqrt{T_n - 20}}$$

Express this in simplest rationalized radical form.

85. The frequency of a certain electric circuit is given by

$$\frac{1}{2\pi\sqrt{\dfrac{LC_1C_2}{C_1 + C_2}}}$$

Express this in simplest rationalized radical form.

86. In determining the deflection of a certain type of beam, the expression

$$l\sqrt{\frac{a}{2l + a}}$$

is used. Express this in simplest rationalized radical form.

CHAPTER 11

The j-Operator

In Chapter 1, when we were introducing the topic of numbers, imaginary numbers were mentioned. Again, when we considered quadratic equations and their solutions in Chapter 6, we briefly came across this type of number. However, until now we have purposely avoided any extended discussion of imaginary numbers. In this chapter we shall discuss the properties of these numbers and show some of the ways in which they may be applied.

11–1 Complex Numbers

When we defined radicals we were able to define square roots of positive numbers easily, since any positive or negative number squared equals a positive number. For this reason we can see that it is impossible to square any real number and have the product equal a negative number. We must define a new number system if we wish to include square roots of negative numbers. With the proper definitions, we shall find that these numbers can be used to great advantage in certain applications.

If the radicand in a square root is negative, we can express the indicated root as the product of $\sqrt{-1}$ and the square root of a positive number. *The symbol $\sqrt{-1}$ is defined as the* **imaginary** **unit** *and is denoted by the symbol j.* (The symbol i is also often used for this purpose, but in electrical work i usually represents current. Therefore, we shall use j for the imaginary unit to avoid confusion.) In keeping with the definition of j, we have

$$j^2 = -1 \tag{11–1}$$

Example A

$$\sqrt{-9} = \sqrt{(9)(-1)} = \sqrt{9}\sqrt{-1} = 3j$$
$$\sqrt{-16} = \sqrt{16}\sqrt{-1} = 4j$$
$$\sqrt{-0.25} = \sqrt{0.25}\sqrt{-1} = 0.5j$$
$$\sqrt{-5} = \sqrt{5}\sqrt{-1} = \sqrt{5}j = j\sqrt{5}$$

When a radical appears in the final result, we will write this result in a form with the j before the radical as in the last illustration. This clearly shows that the j is not *under* the radical.

Example B

$$(\sqrt{-4})^2 = (j\sqrt{4})^2 = 4j^2 = -4$$

We note that the simplification of this expression does not follow Eq. (10–10), which states that $\sqrt{ab} = \sqrt{a}\sqrt{b}$ for square roots. This is the reason it was noted as being valid only if a and b are positive. In fact, *Eq. (10–10) does not necessarily hold in general for negative values for a or b.* If $(\sqrt{-4})^2$ did follow Eq. (10–10) we would have

$$(\sqrt{-4})^2 = \sqrt{-4}\sqrt{-4} = \sqrt{16} = 4$$

We note that we obtain 4, and do not obtain the correct result of -4.

Example C

To further illustrate the method of handling square roots of negative numbers, consider the difference between $\sqrt{-3}\sqrt{-12}$ and $\sqrt{(-3)(-12)}$. For these expressions we have

$$\sqrt{-3}\sqrt{-12} = (j\sqrt{3})(j\sqrt{12}) = (\sqrt{3}\sqrt{12})j^2 = (\sqrt{36})j^2$$
$$= 6(-1) = -6$$

and

$$\sqrt{(-3)(-12)} = \sqrt{36} = 6$$

For $\sqrt{-3}\sqrt{-12}$ we have the product of square roots of negative numbers, whereas for $\sqrt{(-3)(-12)}$ we have the product of negative numbers under the radical. We must be careful to note the difference.

From Examples B and C we see that *when we are dealing with the square roots of negative numbers,* **each should be expressed in terms of** j **before proceeding.** To do this, for any positive real number a we write

$$\sqrt{-a} = j\sqrt{a}, \quad (a > 0) \tag{11–2}$$

Example D

$$\sqrt{-6} = \sqrt{(6)(-1)} = \sqrt{6}\sqrt{-1} = j\sqrt{6}$$
$$\sqrt{-18} = \sqrt{(18)(-1)} = \sqrt{(9)(2)}\sqrt{-1} = 3j\sqrt{2}$$
$$-\sqrt{-75} = -\sqrt{(25)(3)(-1)} = -\sqrt{(25)(3)}\sqrt{-1} = -5j\sqrt{3}$$

In working with imaginary numbers, we often need to be able to raise these numbers to some power. Therefore, using the definitions of exponents and of j, we have the following results:

$$j = j, \qquad\qquad j^4 = j^2 j^2 = (-1)(-1) = 1$$
$$j^2 = -1, \qquad\qquad j^5 = j^4 j = j$$
$$j^3 = j^2 j = -j, \qquad j^6 = j^4 j^2 = (1)(-1) = -1$$

The powers of j go through the cycle of $j, -1, -j, 1, j, -1, -j, 1$, and so forth. Noting this and the fact that j raised to a power which is a multiple of 4 equals 1 allows us to raise j to any integral power almost on sight.

Example E

$$j^{10} = j^8 j^2 = (1)(-1) = -1$$
$$j^{45} = j^{44} j = (1)(j) = j$$
$$j^{531} = j^{528} j^3 = (1)(-j) = -j$$

Using real numbers and the imaginary unit j, we define a new kind of number. *A* **complex number** *is any number which can be written in the form $a + bj$, where a and b are real numbers. If $a = 0$, we have a number of the form bj, which is a* **pure imaginary number.** *If $b = 0$, then $a + bj$ is a real number. The form $a + bj$ is known as the* **rectangular form** *of a complex number, where a is known as the* **real part** *and b is known as the* **imaginary part.** We can see that complex numbers include all the real numbers and all of the pure imaginary numbers.

A comment here about the words *imaginary* and *complex* is in order. The choice of the names of these numbers is historical in nature, and unfortunately it leads to some misconceptions about the numbers. The use of imaginary does not imply that the numbers do not exist. Imaginary numbers do in fact exist, as they are defined above. In the same way, the use of complex does not imply that the numbers are complicated and therefore difficult to understand. With the appropriate definitions and operations, we can work with complex numbers, just as with any type of number.

For complex numbers written in terms of j to follow all the operations defined in algebra, we define equality of two complex numbers in a special way. Complex numbers are not positive or negative in the ordinary sense of the terms, but the real and imaginary parts of complex numbers *are* positive or negative. *We define two complex numbers to be equal if the real parts are equal and the imaginary parts are equal.* That is, two complex numbers, $a + bj$ and $x + yj$, are equal if $a = x$ and $b = y$.

Example F

$$a + bj = 3 + 4j \qquad \text{if } a = 3 \quad \text{and} \quad b = 4$$
$$x + yj = 5 - 3j \qquad \text{if } x = 5 \quad \text{and} \quad y = -3$$

Example G

What values of x and y satisfy the equation $4 - 6j - x = j + jy$?

One way to solve this is to rearrange the terms so that all the known terms are on the right and all the terms containing the unknowns x and y are on the left. This leads to $-x - jy = -4 + 7j$. From the definition of equality of complex numbers, $-x = -4$ and $-y = 7$, or $x = 4$ and $y = -7$.

Example H

What values of x and y satisfy the equation

$$x + 3(xj + y) = 5 - j - jy$$

Rearranging the terms so that the known terms are on the right and the terms containing x and y are on the left, we have

$$x + 3y + 3jx + jy = 5 - j$$

Next, factoring j from the two terms on the left will put the expression on the left into proper form. This leads to

$$(x + 3y) + (3x + y)j = 5 - j$$

Using the definition of equality, we have

$$x + 3y = 5 \quad \text{and} \quad 3x + y = -1$$

We now solve this system of equations. The solution is $x = -1$ and $y = 2$. Actually, the solution can be obtained at any point by writing each side of the equation in the form $a + bj$ and then equating first the real parts and then the imaginary parts.

The **conjugate** *of the complex number* $a + bj$ *is the complex number* $a - bj$. We see that the sign of the imaginary part of a complex number is changed to obtain its conjugate.

Example I

$3 - 2j$ is the conjugate of $3 + 2j$. We may also say that $3 + 2j$ is the conjugate of $3 - 2j$. Thus each is the conjugate of the other.

Exercises 11–1

In Exercises 1 through 12, express each number in terms of j.

1. $\sqrt{-81}$ 2. $\sqrt{-121}$ 3. $-\sqrt{-4}$ 4. $-\sqrt{-49}$

5. $\sqrt{-0.36}$ 6. $-\sqrt{-0.01}$ 7. $\sqrt{-8}$ 8. $\sqrt{-48}$

9. $\sqrt{-\frac{7}{4}}$ 10. $-\sqrt{-\frac{5}{9}}$ 11. $-\sqrt{-\frac{2}{5}}$ 12. $\sqrt{-\frac{5}{3}}$

In Exercises 13 through 16, simplify each of the given expressions.

13. $(\sqrt{-7})^2$; $\sqrt{(-7)^2}$

14. $\sqrt{(-15)^2}$; $(\sqrt{-15})^2$

15. $\sqrt{(-2)(-8)}$; $\sqrt{-2}\sqrt{-8}$

16. $\sqrt{-9}\sqrt{-16}$; $\sqrt{(-9)(-16)}$

In Exercises 17 through 24, simplify the given expressions.

17. j^7

18. j^{49}

19. $-j^{22}$

20. j^{408}

21. $j^2 - j^6$

22. $2j^5 - j^7$

23. $j^{15} - j^{13}$

24. $3j^{48} + j^{200}$

In Exercises 25 through 32, perform the indicated operations and simplify each complex number to its rectangular form $a + bj$.

25. $2 + \sqrt{-9}$

26. $-6 + \sqrt{-64}$

27. $2j^2 + 3j$

28. $j^3 - 6$

29. $\sqrt{18} - \sqrt{-8}$

30. $\sqrt{-27} + \sqrt{12}$

31. $(\sqrt{-2})^2 + j^4$

32. $(2\sqrt{2})^2 - (\sqrt{-1})^2$

In Exercises 33 through 36, find the conjugate of each complex number.

33. $6 - 7j$ 34. $-3 + 2j$ 35. $2j$ 36. 6

In Exercises 37 through 44, find the values of x and y which satisfy the given equations.

37. $7x - 2yj = 14 + 4j$

38. $2x + 3jy = -6 + 12j$

39. $6j - 7 = 3 - x - yj$

40. $9 - j = xj + 1 - y$

41. $x - y = 1 - xj - yj - j$

42. $2x - 2j = 4 - 2xj - yj$

43. $x + 2 + 7j = yj - 2xj$

44. $2x + 6xj + 3 = yj - y + 7j$

In Exercises 45 through 48, answer the given questions.

45. Are $8j$ and $-8j$ the solutions to the equation $x^2 + 64 = 0$?

46. Are $2j\sqrt{5}$ and $-2j\sqrt{5}$ the solutions to the equation $x^2 + 20 = 0$?

47. What condition must be satisfied if a complex number and its conjugate are to be equal?

48. What type of number is a complex number if it is equal to the negative of its conjugate?

11–2 Basic Operations with Complex Numbers

The basic operations of addition, subtraction, multiplication, and division are defined in the same way for complex numbers in rectangular form as they are for real numbers. These operations are performed without regard for the fact that j has a special meaning. However, we must **be careful to express all complex numbers in terms of j before performing these operations.** Once this is done, we may proceed as with real numbers. We have the following definitions for these operations on complex numbers.

Addition (and subtraction):

$$(a + bj) + (c + dj) = (a + c) + (b + d)j \tag{11-3}$$

Multiplication:

$$(a + bj)(c + dj) = (ac - bd) + (ad + bc)j \tag{11-4}$$

Division:

$$\frac{a + bj}{c + dj} = \frac{(a + bj)(c - dj)}{(c + dj)(c - dj)} = \frac{(ac + bd) + (bc - ad)j}{c^2 + d^2} \tag{11-5}$$

Recalling Examples B and C of Section 11–1, we see the reason for expressing all complex numbers in terms of j before proceeding with any indicated operations.

We note from Eq. (11–3) that the addition or subtraction of complex numbers is accomplished by combining the real parts and combining the imaginary parts. Consider the following examples.

Example A

$$(3 - 2j) + (-5 + 7j) = (3 - 5) + (-2 + 7)j = -2 + 5j$$

Example B

$$(7 + 9j) - (6 - 4j) = 7 + 9j - 6 + 4j = 1 + 13j$$

Example C

$$(3\sqrt{-4} - 4) - (6 - 2\sqrt{-25}) - \sqrt{-81}$$
$$= [3(2j) - 4] - [6 - 2(5j)] - 9j$$
$$= [6j - 4] - [6 - 10j] - 9j$$
$$= 6j - 4 - 6 + 10j - 9j$$
$$= -10 + 7j$$

Here we note that our first step was to express the numbers in terms of j.

When complex numbers are multiplied, Eq. (11–4) indicates that we proceed as in any algebraic multiplication, properly expressing numbers in terms of j and evaluating powers of j. This is illustrated in Examples D and E.

Example D

$$(6 - \sqrt{-4})(\sqrt{-9}) = (6 - 2j)(3j) = 18j - 6j^2$$
$$= 18j - 6(-1) = 6 + 18j$$

Example E

$$(-9.4 - 6.2j)(2.5 + 1.5j) = -23.5 - 14.1j - 15.5j - 9.3j^2$$
$$= -23.5 - 29.6j - 9.3(-1)$$
$$= -14.2 - 29.6j$$

We note that our procedure in dividing two complex numbers is the same procedure that we used for rationalizing the denominator of a fraction with a radical in the denominator. We use this procedure so that we can express any answer in the form of a complex number. We need merely to multiply numerator and denominator by the conjugate of the denominator in order to perform this operation.

Example F

$$\frac{7 - 2j}{3 + 4j} = \frac{7 - 2j}{3 + 4j} \cdot \frac{3 - 4j}{3 - 4j} = \frac{21 - 28j - 6j + 8j^2}{9 - 16j^2}$$

$$= \frac{21 - 34j + 8(-1)}{9 - 16(-1)} = \frac{13 - 34j}{25}$$

This could be written in the form $a + bj$ as $\frac{13}{25} - \frac{34}{25}j$, but is generally left as a single fraction. In decimal form, the result could be expressed as $0.52 - 1.36j$.

Example G

$$\frac{6 + j}{2j} = \frac{6 + j}{2j} \cdot \frac{-2j}{-2j} = \frac{-12j - 2j^2}{4} = \frac{2 - 12j}{4} = \frac{1 - 6j}{2}$$

Example H

$$\frac{j^3 + 2j}{1 - j^5} = \frac{-j + 2j}{1 - j} = \frac{j}{1 - j} \cdot \frac{1 + j}{1 + j} = \frac{-1 + j}{2}$$

Exercises 11–2

In Exercises 1 through 48, perform the indicated operations, expressing all answers in the form $a + bj$.

1. $(3 - 7j) + (2 - j)$
2. $(-4 - j) + (-7 - 4j)$
3. $(7j - 6) - (3 + j)$
4. $(0.23 + 0.67j) - (0.46 - 0.19j)$
5. $(4 + \sqrt{-16}) + (3 - \sqrt{-81})$
6. $(-1 + 3\sqrt{-4}) + (8 - 4\sqrt{-49})$
7. $(5 - \sqrt{-9}) - (\sqrt{-4} + 5)$
8. $(\sqrt{-25} - 1) - \sqrt{-9}$
9. $j - (j - 7) - 8$
10. $(7 - j) - (4 - 4j) + (6 - j)$
11. $(2\sqrt{-25} - 3) - (5 - 3\sqrt{-36}) - (\sqrt{-49})$
12. $(6 - 2\sqrt{-64}) - \sqrt{-100} - (\sqrt{-81} - 5)$
13. $(7 - j)(7j)$
14. $(-2.2j)(1.5j - 4.0)$
15. $\sqrt{-16}(2.8\sqrt{-4.0} + 1.6)$
16. $(\sqrt{-4} - 1)(\sqrt{-9})$
17. $(4 - j)(5 + 2j)$
18. $(3 - 5j)(6 + 7j)$
19. $(2\sqrt{-9} - 3)(3\sqrt{-4} + 2)$
20. $(5\sqrt{-64} - 5)(7 + \sqrt{-16})$
21. $\sqrt{-18}\sqrt{-4}\sqrt{-9}$
22. $\sqrt{-6}\sqrt{-12}\sqrt{3}$
23. $(\sqrt{-5})^5$
24. $(\sqrt{-36})^4$
25. $\sqrt{-108} - \sqrt{-27}$
26. $2\sqrt{-54} + \sqrt{-24}$
27. $3\sqrt{-28} - 2\sqrt{12}$
28. $5\sqrt{24} - 3\sqrt{-45}$

29. $7j^3 - 7\sqrt{-9}$

30. $6j - 5j^2\sqrt{-63}$

31. $j\sqrt{-7} - j^6\sqrt{112} + 3j$

32. $j^2\sqrt{-7} - \sqrt{-28} + 8$

33. $(3 - 7j)^2$ 34. $(4j + 5)^2$ 35. $(1 - j)^3$ 36. $(1 + j)(1 - j)^2$

37. $\dfrac{6j}{2 - 5j}$

38. $\dfrac{4}{3 + 7j}$

39. $\dfrac{0.25}{3.0 - \sqrt{-1.0}}$

40. $\dfrac{\sqrt{-4}}{2 + \sqrt{-9}}$

41. $\dfrac{1 - j}{1 + j}$ 42. $\dfrac{9 - 8j}{j - 1}$ 43. $\dfrac{\sqrt{-2} - 5}{\sqrt{-2} + 3}$ 44. $\dfrac{2 + 3\sqrt{-3}}{5 - \sqrt{-3}}$

45. $\dfrac{\sqrt{-16} - \sqrt{2}}{\sqrt{2} + j}$ 46. $\dfrac{1 - \sqrt{-4}}{2 + 9j}$ 47. $\dfrac{j^2 - j}{2j - j^8}$ 48. $\dfrac{j^5 - j^3}{3 + j}$

In Exercises 49 through 52, perform the indicated operations.

49. Under certain conditions, the voltage E in an electric circuit is given by $E = IZ$, where I is the current and Z is the impedance. If $I = 8 - 3j$ amperes and $Z = 4 + 2j$ ohms, find the expression for E.

50. In Exercise 49, if $E = 15 + 2j$ volts and $Z = 1 - 3j$ ohms, find the expression for I.

51. Show that $-1 + j$ is a solution to the equation $x^2 + 2x + 2 = 0$.

52. Show that $-1 - j$ is a solution to the equation $x^2 + 2x + 2 = 0$.

In Exercises 53 through 56, demonstrate the indicated properties.

53. Show that the sum of a complex number and its conjugate is a real number.

54. Show that the product of a complex number and its conjugate is a real number.

55. Show that the difference between a complex number and its conjugate is an imaginary number.

56. Show that the reciprocal of the imaginary unit is the negative of the imaginary unit.

11–3 Graphical Representation of Complex Numbers

We showed in Section 1–1 how we could represent real numbers as points on a line. Because complex numbers include all real numbers as well as imaginary numbers, it is necessary to represent them graphically in a different way. Since there are two numbers associated with each complex number (the real part and the imaginary part), we find that we can represent complex numbers by representing the real parts by the x-values of the rectangular coordinate system, and the imaginary parts by the y-values. In this way *each complex number is represented as a point in the plane*, the point being designated as $a + bj$. When the rectangular coordinate system is used in this manner, it is called the **complex plane.** *The horizontal axis is called the* **real axis,** *and the vertical axis is called the* **imaginary axis.**

Example A

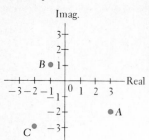

Figure 11–1

In Fig. 11–1, the point *A* represents the complex number $3 - 2j$; point *B* represents $-1 + j$; point *C* represents $-2 - 3j$. We note that these complex numbers are represented by the points $(3, -2)$, $(-1, 1)$, and $(-2, -3)$, but we must keep in mind that the meaning is different from that used in the standard rectangular coordinate system. A point in the complex plane represents a single complex number.

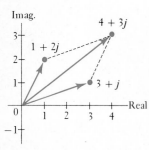

Figure 11–2

Let us represent two complex numbers and their sum in the complex plane. Consider, for example, the two complex numbers $1 + 2j$ and $3 + j$. By algebraic addition the sum is $4 + 3j$. When we draw lines from the origin to these points (see Fig. 11–2), we note that if we think of the complex numbers as being vectors, their sum is the vector sum. Because complex numbers can be used to represent vectors, these numbers are particularly important. *Any complex number can be thought of as representing a vector from the origin to its point in the complex plane.* To add two complex numbers graphically, we find the point corresponding to one of them and draw a line from the origin to this point. We repeat this process for the second point. Next we complete a parallelogram with the lines drawn as adjacent sides. The resulting fourth vertex is the point representing the sum of the two complex numbers. Note that this is equivalent to adding vectors by graphical means.

Example B

Add the complex numbers $5 - 2j$ and $-2 - j$ graphically.

The solution is indicated in Fig. 11–3. We can see that the fourth vertex of the parallelogram is very near $3 - 3j$, which is, of course, the algebraic sum.

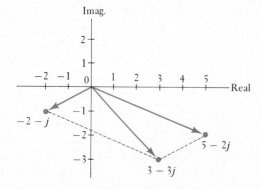

Figure 11–3

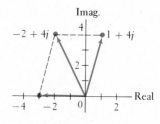

Figure 11–4

Example C

Add the complex numbers -3 and $1 + 4j$ graphically.

First we note that $-3 = -3 + 0j$, which means that the point representing -3 is on the negative real axis. In Fig. 11–4, we show the numbers -3 and $1 + 4j$ on the graph and complete the parallelogram. From the graph we see that the sum is $-2 + 4j$.

Example D

Subtract $4 - 2j$ from $2 - 3j$ graphically.

Subtracting $4 - 2j$ is equivalent to adding $-4 + 2j$. Thus, we complete the solution by adding $-4 + 2j$ and $2 - 3j$ (see Fig. 11–5). The result is $-2 - j$.

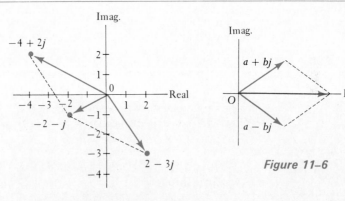

Figure 11–5   *Figure 11–6*

Example E

Show graphically that the sum of a complex number and its conjugate is a real number.

If we choose the complex number $a + bj$, we know that its conjugate is $a - bj$. The imaginary coordinate for the conjugate is as far below the real axis as the imaginary coordinate of $a + bj$ is above it. Therefore, the sum of the imaginary parts must be zero and the sum of the two numbers must therefore lie on the real axis, as shown in Fig. 11–6. Therefore we have shown that the sum of $a + bj$ and $a - bj$ is real.

Exercises 11–3

In Exercises 1 through 4, locate the given complex numbers in the complex plane.

1. $2 + 6j$ 2. $-5 + j$ 3. $-4 - 3j$ 4. $3 - 4j$

In Exercises 5 through 24, perform the indicated operations graphically; check them algebraically.

5. $2 + (3 + 4j)$ 6. $2j + (-2 + 3j)$

7. $(5 - j) + (3 + 2j)$ 8. $(3 - 2j) + (-1 - j)$

9. $5 - (1 - 4j)$ 10. $(2 - j) - j$

11. $(2 - 4j) + (-2 + j)$ 12. $(-1 - 2j) + (6 - j)$

13. $(3 - 2j) - (4 - 6j)$ 14. $(-2 - 4j) - (2 - 5j)$

15. $(1 + 4j) - (3 + j)$ 16. $(-j - 2) - (-1 - 3j)$

17. $(4 - j) + (3 + 2j)$ 18. $(5 + 2j) - (-4 - 2j)$

19. $(3 - 6j) - (-1 + 5j)$ 20. $(-6 - 3j) + (2 - 7j)$

21. $(2j + 1) - 3j - (j + 1)$ 22. $(6 - j) - 9 - (2j - 3)$

23. $(j - 6) - j + (j - 7)$ 24. $j - (1 - j) + (3 + 2j)$

In Exercises 25 through 28, show the given number, its negative, and its conjugate on the same coordinate system.

25. $3 + 2j$ 26. $-2 + 4j$ 27. $-3 - 5j$ 28. $5 - j$

In Exercises 29 through 32, show the given number $a + bj$, $3(a + bj)$, and $-3(a + bj)$ on the same coordinate system. The multiplication of a complex number by a real number is called the **scalar multiplication** of the complex number.

29. $-2 + j$ 30. $-1 - 3j$

31. $3 - j$ 32. $2 + j$

11–4 Polar Form of a Complex Number

We have just seen the relationship between complex numbers and vectors. Since one can be used to represent the other, we shall use this fact to write complex numbers in another way. The new form has certain advantages when basic operations are performed on complex numbers.

By drawing a vector from the origin to the point in the complex plane which represents the number $x + yj$, we see the relation between vectors and complex numbers. Further observation indicates an angle in standard position has been formed. Also, the point $x + yj$ is r units from the origin. In fact, *we can find any point in the complex plane by knowing this angle θ and the value of r.* We have already developed the relations between x, y, r, and θ, in Eqs. (8–1) to (8–3). Let us rewrite these equations in a slightly different form. By referring to Eqs. (8–1) through (8–3) and to Fig. 11–7, we see that

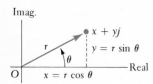

Figure 11–7

$$x = r \cos \theta \qquad y = r \sin \theta \qquad (11\text{–}6)$$

$$r^2 = x^2 + y^2 \qquad \tan \theta = \frac{y}{x} \qquad (11\text{–}7)$$

Substituting Eqs. (11–6) into the rectangular form $x + yj$ of a complex number, we have

$$x + yj = r \cos \theta + j(r \sin \theta)$$

or

$$x + yj = r(\cos \theta + j \sin \theta) \qquad (11\text{–}8)$$

The right side of Eq. (11–8) is called the **polar form** *of a complex number. Sometimes it is referred to as the* **trigonometric form.** An abbreviated form of writing the polar form which is sometimes used is r cis θ. *The length r is called the* **absolute value,** *or the* **modulus,** *and the angle θ is called the* **argument** *of the complex number.* Therefore, Eq. (11–8), along with Eqs. (11–7), defines the polar form of a complex number.

Example A

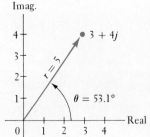

Figure 11–8

Represent the complex number $3 + 4j$ graphically, and give its polar form.

From the rectangular form $3 + 4j$, we see that $x = 3$ and $y = 4$. Using Eqs. (11–7), we have

$$r = \sqrt{3^2 + 4^2} = 5, \qquad \tan \theta = \frac{4}{3} = 1.333, \qquad \theta = 53.1°$$

Thus, the polar form is

$$5(\cos 53.1° + j \sin 53.1°)$$

The graphical representation is shown in Fig. 11–8.

A note on significant digits is in order here. In writing a complex number as $3 + 4j$ in Example A, no approximate values are intended. However, in expressing the polar form as $5(\cos 53.1° + j \sin 53.1°)$, we rounded off the angle to the nearest $0.1°$, as it is not possible to express the result exactly in degrees. Thus, in dealing with nonexact numbers, we shall express angles to the nearest $0.1°$. Other results, when approximate, will be expressed to three significant digits, unless a different accuracy is given in the problem. Of course, in applied situations most numbers used are derived through measurement and are therefore approximate.

Another convenient and widely used notation for the polar form is $r \angle \theta$. We must remember in using this form that it represents a complex number and is simply a shorthand way of writing $r(\cos \theta + j \sin \theta)$. Therefore,

$$r \angle \theta = r(\cos \theta + j \sin \theta) \qquad (11\text{–}9)$$

Example B

$$3(\cos 40° + j \sin 40°) = 3\,\underline{/40°}$$

$$6.26(\cos 217.3° + j \sin 217.3°) = 6.26\,\underline{/217.3°}$$

$$5\,\underline{/120°} = 5(\cos 120° + j \sin 120°)$$

$$14.5\,\underline{/306.2°} = 14.5(\cos 306.2° + j \sin 306.2°)$$

Example C

Figure 11–9

Represent the complex number $2.08 - 3.12j$ graphically, and give its polar forms.

From Eqs. (11–7), we have

$$r = \sqrt{(2.08)^2 + (3.12)^2} = 3.75$$

$$\tan \theta = \frac{-3.12}{2.08} = -1.500$$

Since we know that θ is a fourth-quadrant angle, from the calculator value of $-56.3°$, we have $\theta = 303.7°$. In this case $-56.3°$ is actually correct, but for consistency we will give angles from $0°$ to $360°$. Therefore, the polar forms are

$$3.75(\cos 303.7° + j \sin 303.7°) = 3.75\,\underline{/303.7°}$$

See Fig. 11–9.

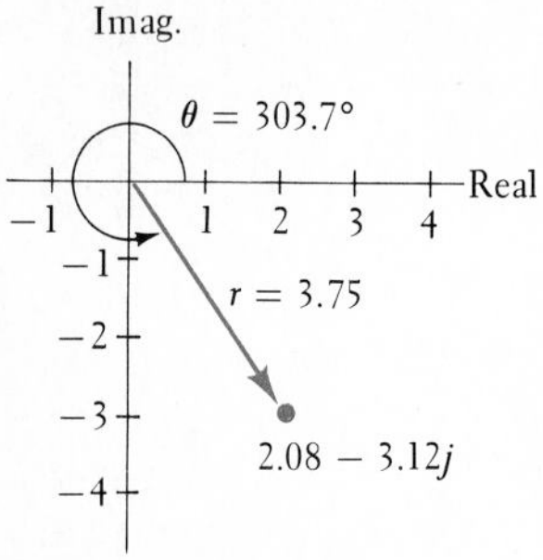

Example D

Figure 11–10

Express the complex number $3.00(\cos 120.0° + j \sin 120.0°)$ in rectangular form.

From the given polar form, we know that $r = 3.00$ and $\theta = 120.0°$. Using Eqs. (11–6), we have

$$x = 3.00 \cos 120.0° = -1.50$$

$$y = 3.00 \sin 120.0° = 2.60$$

Therefore, the rectangular form is

$$-1.50 + 2.60j$$

See Fig. 11–10.

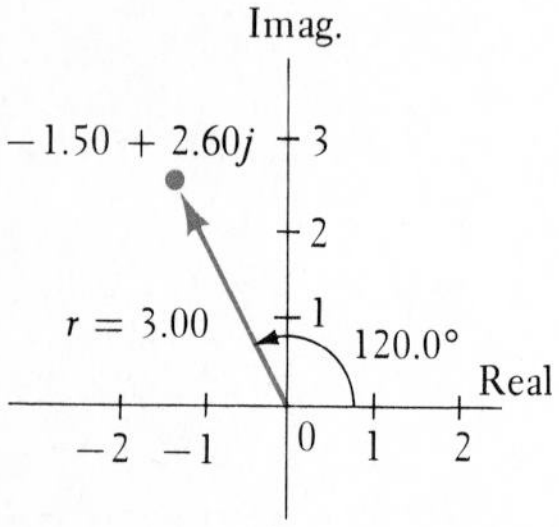

Example E

Figure 11–11

Express the complex number $65.17\,\underline{/214.48°}$ in rectangular form.

From the polar form, we have $r = 65.17$ and $\theta = 214.48°$. This means that

$$x = 65.17 \cos 214.48° = -53.72$$

$$y = 65.17 \sin 214.48° = -36.89$$

Therefore, the rectangular form is

$$-53.72 - 36.89j$$

See Fig. 11–11.

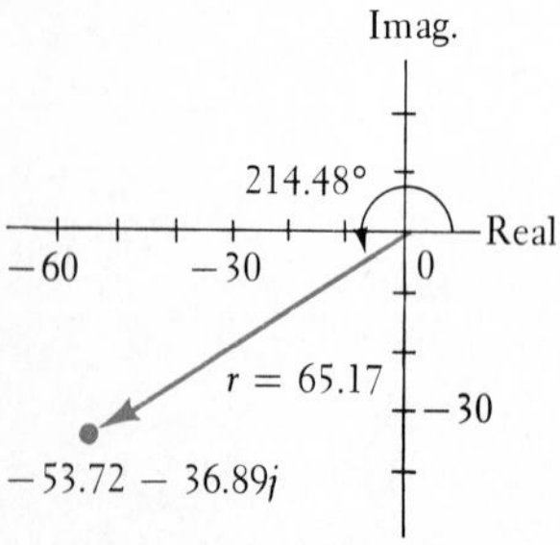

Example F

Represent the numbers $5, -5, 7j$, and $-7j$ in polar form.

Since any positive real number lies on the positive real axis in the complex plane, real numbers are expressed in polar form by

$$a = a(\cos 0° + j \sin 0°) = a \angle 0°$$

Negative real numbers, being on the negative real axis, are written as

$$a = |a|(\cos 180° + j \sin 180°) = |a| \angle 180°$$

Thus,

$$5 = 5(\cos 0° + j \sin 0°) = 5 \angle 0°$$

and

$$-5 = 5(\cos 180° + j \sin 180°) = 5 \angle 180°$$

Positive pure imaginary numbers lie on the positive imaginary axis and are expressed in polar form by

$$bj = b(\cos 90° + j \sin 90°) = b \angle 90°$$

Similarly, negative pure imaginary numbers, being on the negative imaginary axis, are written as

$$bj = |b|(\cos 270° + j \sin 270°) = |b| \angle 270°$$

Thus,

$$7j = 7(\cos 90° + j \sin 90°) = 7 \angle 90°$$

and

$$-7j = 7(\cos 270° + j \sin 270°) = 7 \angle 270°$$

The graphical representations of the *complex numbers* $5, -5, 7j$, and $-7j$ are shown in Fig. 11–12.

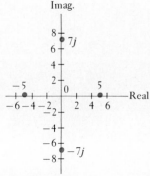

Figure 11–12

Exercises 11–4

See Appendix E for a computer program for changing a complex number to polar form.

In Exercises 1 through 16, represent each of the complex numbers graphically, and give the polar form of each number.

1. $8 + 6j$
2. $-8 - 15j$
3. $3 - 4j$
4. $-5 + 12j$
5. $-2.00 + 3.00j$
6. $7.00 - 5.00j$
7. $-5.50 - 2.40j$
8. $4.60 - 4.60j$
9. $1 + j\sqrt{3}$
10. $\sqrt{2} - j\sqrt{2}$
11. $3.514 - 7.256j$
12. $6.231 + 9.527j$
13. -3
14. 6
15. $9j$
16. $-2j$

In Exercises 17 through 36, represent each of the complex numbers graphically, and give the rectangular form of each number.

17. $5.00(\cos 54.0° + j \sin 54.0°)$

18. $3.00(\cos 232.0° + j \sin 232.0°)$

19. $1.60(\cos 150.0° + j \sin 150.0°)$

20. $2.50(\cos 315.0° + j \sin 315.0°)$

21. $6(\cos 180° + j \sin 180°)$

22. $12(\cos 270° + j \sin 270°)$

23. $8(\cos 360° + j \sin 360°)$

24. $15(\cos 0° + j \sin 0°)$

25. $12.36(\cos 345.56° + j \sin 345.56°)$

26. $220.8(\cos 155.13° + j \sin 155.13°)$

27. $\cos 240.0° + j \sin 240.0°$

28. $\cos 99.0° + j \sin 99.0°$

29. $4.75 \angle 172.8°$

30. $1.50 \angle 62.3°$

31. $0.9326 \angle 229.54°$

32. $277.8 \angle 342.63°$

33. $7.32 \angle 90.0°$

34. $18.3 \angle 180.0°$

35. $86.42 \angle 94.62°$

36. $4629 \angle 182.44°$

In Exercises 37 through 40, solve the given problems.

37. Considering the relationship between complex numbers and vectors, what are the magnitude and direction of a displacement vector which is represented by $0.120 + 0.160j$ millimeters?

38. What are the magnitude and direction of a force which is represented by $3.72 - 5.12j$ pounds? (See Exercise 37.)

39. Under certain conditions, the voltage, in kilovolts, of a generator is given by $E = 28.2 \angle 6.1°$. Write this in rectangular form.

40. The current, in microamperes, in a certain microprocessor circuit is given by $3.75 \angle 15.0°$. Write this in rectangular form.

11–5 Exponential Form of a Complex Number

Another important form of a complex number is known as the **exponential form,** which is written $re^{j\theta}$. In this expression, r and θ have the same meanings as given in the last section, although in $re^{j\theta}$, θ is expressed in radians. *The number e represents a very special irrational number, where an approximate value of e is*

$$e = 2.7182818$$

This number e is very important in mathematics, and we shall see it again in the next chapter. At this point it is necessary to accept the value for e, although in calculus its basic meaning is shown along with the reason it has the above value.
    We now define

$$re^{j\theta} = r(\cos \theta + j \sin \theta) \qquad (11\text{–}10)$$

By expressing θ in radians, the expression $j\theta$ is an exponent, and it can be shown to obey all of the laws of exponents which we discussed in Chapter 10. Therefore, *we shall always express θ in radians when using the exponential form.* The following examples show how complex numbers can be changed to and from exponential form.

Example A

Express the number $3 + 4j$ in exponential form.
   From Example A of Section 11–4, we know that this complex number may be written in polar form as $5(\cos 53.1° + j \sin 53.1°)$. Therefore, we know that $r = 5$. We now express $53.1°$ in terms of radians as

$$\frac{53.1\pi}{180} = 0.927 \text{ rad}$$

Thus, the exponential form is $5e^{0.927j}$. This means that

$$3 + 4j = 5(\cos 53.1° + j \sin 53.1°) = 5e^{0.927j}$$

Example B

Express the number $8.50\angle 136.3°$ in exponential form.
   Since this complex number is in polar form we note that $r = 8.50$ and that we must express $136.3°$ in radians. Changing $136.3°$ to radians, we have

$$\frac{136.3\pi}{180} = 2.38 \text{ rad}$$

Therefore, the required exponential form is $8.50e^{2.38j}$. This means that

$$8.50\angle 136.3° = 8.50e^{2.38j}$$

Example C

Express the number $3.07 - 7.43j$ in exponential form.
   From the rectangular form of the number, we have $x = 3.07$ and $y = -7.43$. Therefore,

$$r = \sqrt{(3.07)^2 + (-7.43)^2} = 8.04$$

$$\tan \theta = \frac{-7.43}{3.07} = -2.420, \qquad \theta = 292.5°$$

Changing $292.5°$ to radians, we have $292.5° = 5.10$ rad. Therefore, the exponential form is $8.04e^{5.10j}$. This means that

$$3.07 - 7.43j = 8.04e^{5.10j}$$

Example D

Express the complex number $2.00e^{4.80j}$ in polar and rectangular forms.

We first express 4.80 rad as 275.0°. From the exponential form we know that $r = 2.00$. Thus, the polar form is $2.00 (\cos 275.0° + j \sin 275.0°)$. Next we find that $\cos 275.0° = 0.0872$ and $\sin 275.0° = -0.9962$. The rectangular form is $2.00 (0.0872 - 0.9962j) = 0.174 - 1.99j$. This means that

$$2.00e^{4.80j} = 2.00(\cos 275.0° + j \sin 275.0°) = 0.174 - 1.99j$$

Example E

Express the complex number $3.408e^{2.457j}$ in polar and rectangular forms.

We first express 2.457 rad as 140.78°. From the exponential form, we know that $r = 3.408$. Thus, the polar form is $3.408 \angle 140.78°$. Next we find that $\cos 140.78° = -0.77472$ and $\sin 140.78° = 0.63230$. The rectangular form is $3.408(-0.77472 + 0.63230j) = -2.640 + 2.155j$. This means that

$$3.408e^{2.457j} = 3.408 \angle 140.78° = -2.640 + 2.155j.$$

An important application of the use of complex numbers is in alternating current. When an alternating current flows through a given circuit, usually the current and voltage have different phases. That is, they do not reach their peak values at the same time. Therefore, one way of accounting for the magnitude as well as the phase of an electric current or voltage is to write it as a complex number. Here the modulus is the actual value of the current or voltage, and the argument is a measure of the phase.

Example F

A current of $2.00 - 4.00j$ amperes flows through a given circuit. Write this current in exponential form and determine the magnitude of the current in the circuit.

From the rectangular form, we have $x = 2.00$ and $y = -4.00$. Therefore,

$$r = \sqrt{(2.00)^2 + (-4.00)^2} = 4.47.$$

Also,

$$\tan \theta = -\frac{4.00}{2.00} = -2.000$$

This means that $\theta = -63.4°$ (it is normal to express the phase in terms of negative angles). Changing 63.4° to radians, we have $63.4° = 1.11$ rad. Therefore, the exponential form of the current is $4.47e^{-1.11j}$. The modulus is 4.47, meaning the magnitude of the current is 4.47 A.

At this point we shall summarize the three important forms of a complex number.

Rectangular: $x + yj$
Polar: $r(\cos\theta + j\sin\theta) = r \underline{/\theta}$
Exponential: $re^{j\theta}$

It follows that

$$x + yj = r(\cos\theta + j\sin\theta) = r\underline{/\theta} = re^{j\theta} \tag{11–11}$$

where

$$r^2 = x^2 + y^2 \qquad \tan\theta = \frac{y}{x} \tag{11–7}$$

Exercises 11–5

In Exercises 1 through 20, express the given complex numbers in exponential form.

1. $3.00(\cos 60.0° + j\sin 60.0°)$ 2. $5.00(\cos 135.0° + j\sin 135.0°)$

3. $4.50(\cos 282.3° + j\sin 282.3°)$ 4. $2.10(\cos 228.7° + j\sin 228.7°)$

5. $375.5(\cos 95.46° + j\sin 95.46°)$ 6. $16.72(\cos 7.14° + j\sin 7.14°)$

7. $0.515\underline{/198.3°}$ 8. $4650\underline{/326.5°}$

9. $4.06\underline{/101.6°}$ 10. $0.0192\underline{/76.7°}$

11. $9245\underline{/296.32°}$ 12. $827.6\underline{/110.09°}$

13. $3 - 4j$ 14. $-1 - 5j$

15. $-3 + 2j$ 16. $6 + j$

17. $5.90 + 2.40j$ 18. $47.3 - 10.9j$

19. $-634.6 - 528.2j$ 20. $-8573 + 5477j$

In Exercises 21 through 28, express the given complex numbers in polar and rectangular forms.

21. $3.00e^{0.500j}$ 22. $2.00e^{1.00j}$ 23. $4.64e^{1.85j}$ 24. $2.50e^{3.84j}$

25. $3.20e^{5.41j}$ 26. $0.800e^{3.00j}$ 27. $0.1724e^{2.391j}$ 28. $820.7e^{3.492j}$

In Exercises 29 and 30, perform the indicated operations.

29. The electric current in a certain alternating-current circuit is given by $0.500 + 0.220j$ amperes. Write this current in exponential form and determine the magnitude of the current in the circuit.

30. The voltage in a certain alternating-current circuit is $125e^{1.31j}$. Determine the magnitude of the voltage in the circuit and the in-phase component (the real part) of the voltage.

11–6 Products, Quotients, Powers, and Roots of Complex Numbers

We have previously performed products and quotients using the rectangular forms of the given numbers. However, these operations can also be performed with complex numbers in polar and exponential forms. We find that these operations are convenient, and also useful for purposes of finding powers and roots of complex numbers.

We may find the product of two complex numbers by using the exponential form and the laws of exponents. Multiplying $r_1 e^{j\theta_1}$ by $r_2 e^{j\theta_2}$, we have

$$r_1 e^{j\theta_1} \cdot r_2 e^{j\theta_2} = r_1 r_2 e^{j\theta_1 + j\theta_2} = r_1 r_2 e^{j(\theta_1 + \theta_2)}$$

We use this equation to express the product of two complex numbers in polar form:

$$r_1 e^{j\theta_1} \cdot r_2 e^{j\theta_2} = r_1(\cos \theta_1 + j \sin \theta_1) \cdot r_2(\cos \theta_2 + j \sin \theta_2)$$

and

$$r_1 r_2 e^{j(\theta_1 + \theta_2)} = r_1 r_2 [\cos(\theta_1 + \theta_2) + j \sin(\theta_1 + \theta_2)]$$

Therefore, the polar expressions are equal, which means that *the product of two complex numbers is*

$$r_1(\cos \theta_1 + j \sin \theta_1) r_2(\cos \theta_2 + j \sin \theta_2)$$
$$= r_1 r_2 [\cos(\theta_1 + \theta_2) + j \sin(\theta_1 + \theta_2)] \qquad (11\text{--}12)$$
$$(r_1 \angle \theta_1)(r_2 \angle \theta_2) = r_1 r_2 \angle \theta_1 + \theta_2$$

Example A

Multiply the complex numbers $2 + 3j$ and $1 - j$ by using the polar form of each.

$$r_1 = \sqrt{4 + 9} = 3.61 \qquad \tan \theta_1 = 1.500 \qquad \theta_1 = 56.3°$$
$$r_2 = \sqrt{1 + 1} = 1.41 \qquad \tan \theta_2 = -1.000 \qquad \theta_2 = 315.0°$$

$$(3.61)(\cos 56.3° + j \sin 56.3°)(1.41)(\cos 315.0° + j \sin 315.0°)$$
$$= (3.61)(1.41)[\cos(56.3° + 315.0°) + j \sin(56.3° + 315.0°)]$$
$$= 5.09(\cos 371.3° + j \sin 371.3°)$$
$$= 5.09(\cos 11.3° + j \sin 11.3°)$$

Example B

When we use the $r \, \underline{/\theta}$ polar form to multiply the two complex numbers in Example A, we have

$$r_1 = 3.61 \qquad \theta_1 = 56.3°$$
$$r_2 = 1.41 \qquad \theta_2 = 315.0°$$
$$(3.61 \, \underline{/56.3°})(1.41 \, \underline{/315.0°}) = (3.61)(1.41) \, \underline{/56.3° + 315.0°}$$
$$= 5.09 \, \underline{/371.3°}$$
$$= 5.09 \, \underline{/11.3°}$$

If we wish to *divide* one complex number in exponential form by another, we arrive at the following result:

$$r_1 e^{j\theta_1} \div r_2 e^{j\theta_2} = \frac{r_1}{r_2} e^{j(\theta_1 - \theta_2)} \tag{11–13}$$

Therefore, *the result of dividing one number in polar form by another is given by:*

$$\frac{r_1(\cos \theta_1 + j \sin \theta_1)}{r_2(\cos \theta_2 + j \sin \theta_2)} = \frac{r_1}{r_2}[\cos(\theta_1 - \theta_2) + j \sin(\theta_1 - \theta_2)]$$
$$\tag{11–14}$$
$$\frac{r_1 \, \underline{/\theta_1}}{r_2 \, \underline{/\theta_2}} = \frac{r_1}{r_2} \, \underline{/\theta_1 - \theta_2}$$

Example C

Divide the first complex number of Example A by the second. Using polar form, we have the following:

$$\frac{3.61(\cos 56.3° + j \sin 56.3°)}{1.41(\cos 315.0° + j \sin 315.0°)} = \frac{3.61}{1.41}[\cos(56.3° - 315.0°)$$
$$+ j \sin(56.3° - 315.0°)]$$
$$= 2.56[\cos(-258.7°) + j \sin(-258.7°)]$$
$$= 2.56(\cos 101.3° + j \sin 101.3°)$$

Example D

Repeating Example C using the $r \, \underline{/\theta}$ polar form, we have

$$\frac{3.61 \, \underline{/56.3°}}{1.41 \, \underline{/315.0°}} = \frac{3.61}{1.41} \, \underline{/56.3° - 315.0°} = 2.56 \, \underline{/-258.7°}$$
$$= 2.56 \, \underline{/101.3°}$$

We have just seen that multiplying and dividing numbers in polar form can be very readily performed directly. However, if we are to add or subtract numbers in polar form, we must do the addition or subtraction by using rectangular form. This is illustrated in the following example.

Example E

Perform the addition $1.563\,\underline{/37.56°} + 3.827\,\underline{/146.23°}$.
In order to do this addition, we must change each number to rectangular form.

$$1.563\,\underline{/37.56°} + 3.827\,\underline{/146.23°}$$
$$= 1.563(\cos 37.56° + j \sin 37.56°)$$
$$+ 3.827(\cos 146.23° + j \sin 146.23°)$$
$$= 1.2390 + 0.9528j - 3.1813 + 2.1273j$$
$$= -1.9423 + 3.0801j$$

Now we change this to polar form.

$$r = \sqrt{(-1.9423)^2 + (3.0801)^2} = 3.641$$

$$\tan \theta = \frac{3.0801}{-1.9423} = -1.5858, \qquad \theta = 122.24°$$

Therefore,

$$1.563\,\underline{/37.56°} + 3.827\,\underline{/146.23°} = 3.641\,\underline{/122.24°}$$

To raise a complex number to a power, we simply multiply one complex number by itself the required number of times. For example, squaring a number in exponential form, we have

$$(re^{j\theta})^2 = r^2 e^{j2\theta} \tag{11--15}$$

Multiplying the expression in Eq. (11–15) by $re^{j\theta}$ gives $r^3 e^{j3\theta}$. This leads to the general expression for raising a complex number to the nth power,

$$(re^{j\theta})^n = r^n e^{jn\theta} \tag{11--16}$$

Extending this to polar form, we have

$$[r\,(\cos \theta + j \sin \theta)]^n = r^n(\cos n\theta + j \sin n\theta)$$
$$(r\,\underline{/\theta})^n = r^n\,\underline{/n\theta} \tag{11--17}$$

Equation (11–17) is known as **DeMoivre's theorem.** *It is valid for all real values of n and may also be used for finding the roots of complex numbers if n is a fractional exponent.*

Example F

Using DeMoivre's theorem, find $(2 + 3j)^3$.
From Example A of this section, we know $r = 3.61$ and $\theta = 56.3°$. Thus, we have

$$[3.61(\cos 56.3° + j \sin 56.3°)]^3$$
$$= (3.61)^3[\cos (3 \times 56.3°) + j \sin(3 \times 56.3°)]$$
$$= 47.0(\cos 168.9° + j \sin 168.9°)$$
$$= 47.0\,\underline{/168.9°}$$

Expressing θ in radians, we have $\theta = 0.983$ rad. Thus, in exponential form,

$$(3.61e^{0.983j})^3 = (3.61)^3 e^{3 \times 0.983j} = 47.0e^{2.95j}$$

Therefore,

$$(2 + 3j)^3 = 47.0(\cos 168.9° + j \sin 168.9°)$$
$$= 47.0 \angle 168.9° = 47.0e^{2.95j}$$

Example G

Find the cube root of -1.

Since we know that -1 is a real number, we can find its cube root by means of the definition. That is, $(-1)^3 = -1$. We shall check this by DeMoivre's theorem. Writing -1 in polar form, we have

$$-1 = 1(\cos 180° + j \sin 180°)$$

Applying DeMoivre's theorem, with $n = \frac{1}{3}$, we obtain

$$(-1)^{1/3} = 1^{1/3}\left(\cos \frac{1}{3}180° + j \sin \frac{1}{3}180°\right) = \cos 60° + j \sin 60°$$
$$= 0.5000 + 0.8660j$$

Observe that we did not obtain -1 as an answer. If we check the answer which was obtained, in the form $\frac{1}{2}(1 + \sqrt{3}j)$, by actually cubing it, we obtain -1! Thus it is a correct answer.

We should note that it is possible to take $\frac{1}{3}$ of any angle up to 1080° and still have an angle less than 360°. Since 180° and 540° have the same terminal side, let us try writing -1 as $1(\cos 540° + j \sin 540°)$. Using DeMoivre's theorem, we have

$$(-1)^{1/3} = 1^{1/3}\left(\cos \frac{1}{3}540° + j \sin \frac{1}{3}540°\right)$$
$$= \cos 180° + j \sin 180° = -1$$

We have found the answer we originally anticipated.

Angles of 180° and 900° also have the same terminal side, so we try

$$(-1)^{1/3} = 1^{1/3}\left(\cos \frac{1}{3}900° + j \sin \frac{1}{3}\sin 900°\right) = \cos 300° + j \sin 300°$$
$$= 0.5000 - 0.8660j$$

Checking this, we find that it is also a correct root. We may try 1260°, but $\frac{1}{3}(1260°) = 420°$, which has the same functional values as 60°, and would give us the answer $0.5000 + 0.8660j$ again.

We have found, therefore, *three cube roots* of -1. They are

$$-1, \quad 0.5000 + 0.8660j, \quad \text{and} \quad 0.5000 - 0.8660j$$

When we generalize on the results of Example G, it can be proved that

there are *n* *n*th roots of a complex number

▶ The method for finding the *n* roots is to *use θ to find one root and then add 360° to θ n − 1 times in order to find the other roots.*

Example H

Find the two square roots of *j*.

We must first properly write *j* in polar form so that we may use DeMoivre's theorem to find the roots. In polar form, *j* is

$$j = 1(\cos 90° + j \sin 90°)$$

To find the square roots, we apply DeMoivre's theorem with $n = \frac{1}{2}$.

$$j^{1/2} = 1^{1/2}\left(\cos\frac{90°}{2} + j \sin\frac{90°}{2}\right) = \cos 45° + j \sin 45° = 0.7071 + 0.7071j$$

To find the other square root using DeMoivre's theorem, we must write *j* in polar form as

$$j = 1[\cos(90° + 360°) + j \sin(90° + 360°)] = 1(\cos 450° + j \sin 450°)$$

Applying DeMoivre's theorem to *j* in this form, we have

$$j^{1/2} = 1^{1/2}\left(\cos\frac{450°}{2} + j \sin\frac{450°}{2}\right) = \cos 225° + j \sin 225°$$

$$= -0.7071 - 0.7071j$$

Thus, the two square roots of *j* are $0.7071 + 0.7071j$ and $-0.7071 - 0.7071j$.

Example I

Find the six sixth roots of 1.

Here we shall directly use the method for finding the roots of a number, as outlined at the end of Example G.

$$1 = 1(\cos 0° + j \sin 0°)$$

First root: $\quad 1^{1/6} = 1^{1/6}\left(\cos\frac{0°}{6} + j \sin\frac{0°}{6}\right) = \cos 0° + j \sin 0° = 1$

Second root: $\quad 1^{1/6} = 1^{1/6}\left(\cos\frac{0° + 360°}{6} + j \sin\frac{0° + 360°}{6}\right)$

$$= \cos 60° + j \sin 60° = \frac{1}{2} + j\frac{\sqrt{3}}{2}$$

Third root: $\quad 1^{1/6} = 1^{1/6}\left(\cos\frac{0° + 720°}{6} + j \sin\frac{0° + 720°}{6}\right)$

$$= \cos 120° + j \sin 120° = -\frac{1}{2} + j\frac{\sqrt{3}}{2}$$

Fourth root: $\quad 1^{1/6} = 1^{1/6}\left(\cos\frac{0° + 1080°}{6} + j \sin\frac{0° + 1080°}{6}\right)$

$$= \cos 180° + j \sin 180° = -1$$

Fifth root: $1^{1/6} = 1^{1/6}\left(\cos\dfrac{0° + 1440°}{6} + j\sin\dfrac{0° + 1440°}{6}\right)$

$$= \cos 240° + j \sin 240° = -\frac{1}{2} - j\frac{\sqrt{3}}{2}$$

Sixth root: $1^{1/6} = 1^{1/6}\left(\cos\dfrac{0° + 1800°}{6} + j\sin\dfrac{0° + 1800°}{6}\right)$

$$= \cos 300° + j \sin 300° = \frac{1}{2} - j\frac{\sqrt{3}}{2}$$

At this point we can see advantages for the various forms of writing complex numbers. Rectangular form lends itself best to addition and subtraction. Polar form is generally used for multiplying, dividing, raising to powers, and finding roots. Exponential form is used for theoretical purposes (e.g., deriving DeMoivre's theorem).

Exercises 11–6

In Exercises 1 through 16, perform the indicated operations. Leave the result in polar form.

1. $[4(\cos 60° + j \sin 60°)][2(\cos 20° + j \sin 20°)]$

2. $[3(\cos 120° + j \sin 120°)][5(\cos 45° + j \sin 45°)]$

3. $(0.5 \,\angle 140°)(6 \,\angle 110°)$ 4. $(0.4 \,\angle 320°)(5.5 \,\angle 150°)$

5. $\dfrac{8(\cos 100° + j \sin 100°)}{4(\cos 65° + j \sin 65°)}$ 6. $\dfrac{9(\cos 230° + j \sin 230°)}{3(\cos 80° + j \sin 80°)}$

7. $\dfrac{12 \,\angle 320°}{5 \,\angle 210°}$ 8. $\dfrac{2 \,\angle 90°}{4 \,\angle 75°}$

9. $[2(\cos 35° + j \sin 35°)]^3$ 10. $[3(\cos 120° + j \sin 120°)]^4$

11. $(2 \,\angle 135°)^8$ 12. $(1 \,\angle 142°)^{10}$

13. $2.78 \,\angle 56.8° + 1.37 \,\angle 207.3°$ 14. $15.9 \,\angle 142.6° - 18.5 \,\angle 71.4°$

15. $7085(\cos 115.62° + j \sin 115.62°) - 4667(\cos 296.34° + j \sin 296.34°)$

16. $307.5(\cos 326.54° + j \sin 326.54°) + 726.3(\cos 96.41° + j \sin 96.41°)$

In Exercises 17 through 28, change each number to polar form and then perform the indicated operations. Express the final result in rectangular and polar forms. Check by performing the same operation in rectangular form.

17. $(3 + 4j)(5 - 12j)$ 18. $(-2 + 5j)(-1 - j)$ 19. $(7 - 3j)(8 + j)$

20. $(1 + 5j)(4 + 2j)$ 21. $\dfrac{7}{1 - 3j}$ 22. $\dfrac{8j}{7 + 2j}$

23. $\dfrac{3 + 4j}{5 - 12j}$ 24. $\dfrac{-2 + 5j}{-1 - j}$ 25. $(3 + 4j)^4$

26. $(-1 - j)^8$ 27. $(2 + 3j)^5$ 28. $(1 - 2j)^6$

In Exercises 29 through 36, use DeMoivre's theorem to find the indicated roots. Be sure to find all roots.

29. The two square roots of $4(\cos 60° + j \sin 60°)$.

30. The three cube roots of $27(\cos 120° + j \sin 120°)$.

31. The three cube roots of $3 - 4j$.

32. The two square roots of $-5 + 12j$.

33. The fourth roots of 1.   34. The cube roots of 8.

35. The cube roots of $-j$.   36. The fourth roots of j.

In Exercises 37 and 38, perform the indicated multiplications.

37. In Example G we showed that one cube root of -1 is $0.500 - 0.866j$. The exact form of this root is $\frac{1}{2}(1 - \sqrt{3}j)$. Cube this expression in rectangular form and show that the result is -1.

38. In Example H we showed that one of the square roots of j is equal to $0.7071 + 0.7071j$. The exact form of this root is $\frac{1}{2}\sqrt{2}(1 + j)$. Square this expression and show that the result is j.

11–7 An Application to Alternating-Current (AC) Circuits

We shall complete our study of the j-operator by showing its use in one aspect of alternating-current circuit theory. This application will be made to measuring voltage between any two points in a simple AC circuit, similar to the application mentioned in Section 11–5. We shall consider a circuit containing a resistance, a capacitance, and an inductance.

Briefly, a *resistance* is any part of a circuit which tends to obstruct the flow of electric current through the circuit. It is denoted by R (units in ohms, Ω) and in diagrams by -\/\/\-. In essence, a *capacitance* is two nonconnected plates in a circuit; no current actually flows across the gap between them. In an AC circuit, an electric charge is continually going to and from each plate and, therefore, the current in the circuit is not effectively stopped. It is denoted by C (units in farads, F) and in diagrams by -||- (Fig. 11–13). An *inductance,* basically, is a coil of wire in which current is induced because the current is continuously changing in the circuit. It is denoted by L (units in henrys, H) and in diagrams by -∞∞-. All these elements affect the voltage in an alternating-current circuit. We shall state here the relation each has to the voltage and current in the circuit.

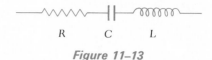

$\qquad R \qquad\qquad C \qquad\qquad L$

Figure 11–13

In Chapter 9, when we were discussing the graphs of the trigonometric functions, we noted that the current and voltage in an AC circuit could be represented by a sine or cosine curve. Therefore, each reaches peak values periodically. *If* *they reach their respective peak values at the same time, we say they are **in phase.** If the voltage reaches its peak before the current, we say that the voltage **leads** the current. If the voltage reaches its peak after the current, we say that the voltage **lags** the current.* In the study of electricity, it is shown that the voltage across a resistance is in phase with the current. The voltage across a capacitor lags the current by 90°, and the voltage across an inductance leads the current by 90°.

Each element in an AC circuit tends to offer a type of resistance to the flow of current. *The effective resistance of any part of the circuit is called the* **reactance,** and it is denoted by X. The voltage across any part of the circuit whose reactance is X is given by $V = IX$, where I is the current (in amperes) and V is the voltage (in volts). Therefore, the voltage across a resistor, capacitor, and inductor, is, respectively

$$V_R = IX_R \qquad V_C = IX_C \qquad V_L = IX_L \tag{11–18}$$

To determine the voltage across a combination of these elements of a circuit, we must account for the reactance, as well as the phase of the voltage across the individual elements. Since the voltage across a resistor is in phase with the current, we shall represent X_R along the positive real axis as a real number R (the actual value of the resistance). Since the voltage across an inductance *leads* the current by 90°, we shall represent this voltage as a positive, pure imaginary number. In the same way, by representing the voltage across a capacitor as a negative, pure imaginary number, we show that the voltage *lags* the current by 90°. These representations are meaningful since the positive imaginary axis is $+90°$ from the positive real axis, and the negative imaginary axis is $-90°$ from the positive real axis.

The total voltage across a combination of all three elements is given by $V_R + V_L + V_C$, which we shall represent by V_{RLC}. Therefore,

$$V_{RLC} = IR + IX_L j - IX_C j = I[R + j(X_L - X_C)]$$

This expression is also written as

$$V_{RLC} = IZ \tag{11–19}$$

where the symbol Z is called the **impedance** *of the circuit. It is the total effective resistance to the flow of current by a combination of the elements in the circuit,* taking into account the phase of the voltage in each element. From its definition, we see that Z is a complex number.

$$Z = R + j(X_L - X_C) \tag{11–20}$$

with a magnitude

$$|Z| = \sqrt{R^2 + (X_L - X_C)^2} \tag{11–21}$$

Also, as a complex number, it makes an angle θ with the x-axis, given by

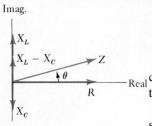

Figure 11–14

$$\tan \theta = \frac{X_L - X_C}{R} \qquad (11\text{–}22)$$

All these equations are based on phase relations of voltages with respect to the current. Therefore, the angle θ represents the phase angle between the current and the voltage (see Fig. 11–14).

In the examples and exercises of this section, the commonly used units and symbols for them are used. For a summary of these units and symbols, including prefixes, see Appendix B.

Example A

In the series circuit shown in Fig. 11–15(a), $R = 12.0\ \Omega$ and $X_L = 5.00\ \Omega$. A current of 2.00 A is in the circuit. Find the voltage across each element, the impedance, the voltage across the combination, and the phase angle between the current and voltage.

Since the voltage across any element is the product of the current and reactance, we have the voltage across the resistor (between points a and b) as $V_R = (2.00)(12.0) = 24.0$ V. The voltage across the inductor (between points b and c) is $V_L = (2.00)(5.00) = 10.0$ V. To find the voltage across the combination, between points a and c, we must first find the magnitude of the impedance. The voltage is *not* the arithmetic sum of V_R and V_L; we must account for the phase. By Eq. (11–20), the impedance is $Z = 12.0 + 5.00j$ with magnitude

$$|Z| = \sqrt{R^2 + X_L^2} = \sqrt{(12.0)^2 + (5.00)^2} = 13.0\ \Omega$$

Therefore, the voltage across the combination is

$$V_{RL} = (2.00)(13.0) = 26.0\text{ V}$$

The phase angle between the voltage and current is found by Eq. (11–22). This gives $\tan \theta = \frac{5.00}{12.0} = 0.4167$ which means that $\theta = 22.6°$. The voltage leads the current by 22.6°, as shown in Fig. 11–15(b).

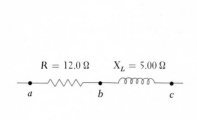

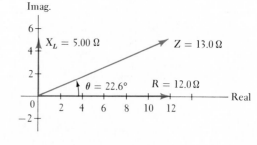

(a)   (b)

Figure 11–15

Example B

For a circuit in which $R = 8.00 \, \Omega$, $X_L = 7.00 \, \Omega$, and $X_C = 13.0 \, \Omega$, find the impedance and the phase angle between the current and the voltage.

By the definition of impedance, Eq. (11–20), we have

$$Z = 8.00 + (7.00 - 13.0)j = 8.00 - 6.00j$$

where the magnitude of the impedance is

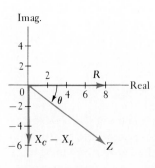

$$|Z| = \sqrt{(8.00)^2 + (-6.00)^2} = 10.0 \, \Omega$$

The phase angle is found by

$$\tan \theta = \frac{-6.00}{8.00} = -0.7500$$

Therefore, $\theta = -36.9°$ (negative angles are used, having a useful purpose in this type of problem). This means that the voltage lags the current by 36.9° (see Fig. 11–16).

Figure 11–16

Example C

Let $R = 6.145 \, \Omega$, $X_L = 8.304 \, \Omega$, and $X_C = 4.018 \, \Omega$. Find the impedance and the phase angle between the current and voltage.

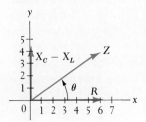

$$Z = 6.145 + (8.304 - 4.018)j = 6.145 + 4.286j$$

$$|Z| = \sqrt{(6.145)^2 + (4.286)^2} = 7.492 \, \Omega$$

$$\tan \theta = \frac{4.286}{6.145} = 0.69748$$

$$\theta = 34.89°$$

The voltage leads the current by 34.89° (see Fig. 11–17).

Figure 11–17

Note that the resistance is represented in the same way as a vector along the positive x-axis. Actually resistance is not a vector quantity, but is represented in this manner in order to assign an angle as the phase of the current. The important concept in this analysis is that *the phase **difference** between the current and voltage is constant,* and therefore any direction may be chosen arbitrarily for one of them. Once this choice is made, other phase angles are measured with respect to this direction. A common choice, as above, is to make the phase angle of the current zero. If an arbitrary angle is chosen, it is necessary to treat the current, voltage, and impedance as complex numbers.

Example D

In a particular circuit, the current is $2.00 - 3.00j$ amperes and the impedance is $6.00 + 2.00j$ ohms. The voltage across this part of the circuit is

$$V = (2.00 - 3.00j)(6.00 + 2.00j) = 12.0 - 14.0j - 6.00j^2$$
$$= 12.0 - 14.0j + 6.00$$
$$= 18.0 - 14.0 \text{ volts}$$

The magnitude of the voltage is

$$|V| = \sqrt{(18.0)^2 + (-14.0)^2} = 22.8 \text{ V}$$

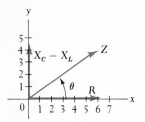

Figure 11–17

Since the voltage across a resistor is in phase with the current, this voltage can be represented as having a phase difference of zero with respect to the current. Therefore, the resistance is indicated as an arrow in the positive real direction, denoting the fact that the current and voltage are in phase. *Such a representation is called a* **phasor.** The arrow denoted by R, as in Fig. 11–17, is actually the phasor representing the voltage across the resistor. Remember, the positive real axis is arbitrarily chosen as the direction of the phase of the current.

To show properly that the voltage across an inductance leads the current by 90°, its reactance (effective resistance) is multiplied by j. We know that there is a positive 90° angle between a real number and a positive imaginary number. In the same way, by multiplying the capacitive reactance by $-j$, we show the 90° difference in phase between the voltage and current in a capacitor, with the current leading. Therefore, jX_L represents the phasor for the voltage across an inductor and $-jX_C$ is the phasor for the voltage across the capacitor. The phasor for the voltage across the combination of the resistance, inductance, and capacitance is Z, where the phase difference between the voltage and current for the combination is the angle θ.

This also points out well the significance of the word *operator* in the term *j-operator. Multiplying a phasor by j means to perform the operation of rotating it through 90°.* We have seen that this is the same result obtained when a real number is multiplied by j.

An alternating current is produced by a coil of wire rotating through a magnetic field. If the angular velocity of this wire is ω, the capacitive and inductive reactances are given by the relations

$$X_C = \frac{1}{\omega C} \quad \text{and} \quad X_L = \omega L \tag{11–23}$$

Therefore, if ω, C, and L are known, the reactance of the circuit may be determined.

Example E

Given that $R = 12.0\ \Omega$, $L = 0.300\ \text{H}$, $C = 250\ \mu\text{F}$, $\omega = 80.0\ \text{rad/s}$, determine the impedance and the phase difference between the current and voltage.

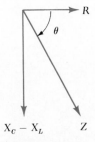

$$X_C = \frac{1}{(80.0)(250 \times 10^{-6})} = 50.0\ \Omega$$

$$X_L = (0.300)(80.0) = 24.0\ \Omega$$

$$Z = 12.0 + (24.0 - 50.0)j = 12.0 - 26.0j$$

$$|Z| = \sqrt{(12.0)^2 + (-26.0)^2} = 28.6\ \Omega$$

$$\tan \theta = \frac{-26.0}{12.0} = -2.167,$$

$$\theta = -65.2°$$

$X_C - X_L$ Z

Figure 11–18

The voltage lags the current (see Fig. 11–18).

We recall from Section 9–5 that the angular velocity ω is related to the frequency f by the relation $\omega = 2\pi f$. It is very common to use frequency when discussing alternating current.

An important concept in the application of this theory is that of **resonance.** *For resonance, the impedance of any circuit is a minimum, or the total impedance is* R. *Thus,* $X_L - X_C = 0$. Also, it can be seen that the current and voltage are in phase under these conditions. Resonance is required for the tuning of radio and television receivers.

Example F

In the antenna circuit of a radio receiver, the inductance is 4.20 mH and the capacitance can be varied. What should be the capacitance in order to receive a 1150-kHz radio signal?

For proper tuning, the circuit should be in resonance, or $X_L = X_C$. This means that

$$2\pi f L = \frac{1}{2\pi f C}$$

In this case $f = 1.15 \times 10^6$ Hz and $L = 4.20 \times 10^{-3}$ H. Therefore,

$$C = \frac{1}{(2\pi f)^2 L} = \frac{1}{(2\pi)^2 (1.15 \times 10^6)^2 (4.20 \times 10^{-3})}$$

$$= 4.56 \times 10^{-12}\ \text{F}$$

$$= 4.56\ \text{pF}$$

Exercises 11–7

For Exercises 1 through 4, use the circuit shown in Fig. 11–19. A current of 3.00 A flows through the circuit. Determine the indicated quantities.

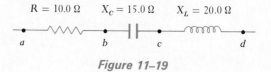

$$R = 10.0 \ \Omega \qquad X_C = 15.0 \ \Omega \qquad X_L = 20.0 \ \Omega$$

Figure 11–19

1. Find the voltage across the resistor (between points a and b).
2. Find the voltage across the capacitor (between points b and c).
3. (a) Find the magnitude of the impedance across the resistor and the capacitor (between points a and c).
   (b) Find the phase angle between the current and voltage for this combination.
   (c) Find the voltage across this combination.
4. (a) Find the magnitude of the impedance across the resistor, capacitor, and inductor (between points a and d).
   (b) Find the phase angle between the current and voltage for this combination.
   (c) Find the voltage across this combination.

In Exercises 5 through 8, use the following information to find the required quantities. In a given circuit $R = 6.00 \ \Omega$, $X_C = 10.0 \ \Omega$, and $X_L = 7.00 \ \Omega$. Find (a) the magnitude of the impedance and (b) the phase angle between the current and voltage, under the specified conditions.

5. With the resistor removed (an LC circuit).
6. With the capacitor removed (an RL circuit).
7. With the inductor removed (an RC circuit).
8. With all elements present (an RLC circuit).

In Exercises 9 through 20, find the required quantities.

9. Given that the current in a given circuit is $8.00 - 2.00j$ A and the impedance is $2.00 + 5.00j \ \Omega$, find the magnitude of the voltage.
10. Given that the voltage in a given circuit is $8.375 - 3.140j$ V and the impedance is $2.146 - 1.114j \ \Omega$, find the magnitude of the current.
11. Given that $\omega = 1875 \ \text{rad/s}$, $C = 0.5130 \ \mu\text{F}$, and $L = 3.263$ H, find the capacitive and inductive reactances.
12. A coil of wire is rotating through a magnetic field at 60.0 Hz. Determine the capacitive reactance for a capacitor of 2.00 μF which is in the circuit.
13. If the capacitor and inductor in Exercise 11 are put in a series circuit with a resistance of 4275 Ω, what is the impedance of this combination? Find the phase angle.
14. Rework Exercise 13, assuming that the capacitor is removed.
15. If $\omega = 100 \ \text{rad/s}$ and $L = 0.500$ H, what must be the value of C to produce resonance?
16. What is the frequency f (in hertz) for resonance in a circuit for which $L = 2.00$ H and $C = 25.0 \ \mu\text{F}$?

17. In Example F, what should be the capacitance in order to receive a 680-kHz radio signal?

18. What are the maximum and minimum values of the capacitance of the tuning capacitor of a radio which has an inductance of 8.75 mH, if it is to receive all radio stations on the band from 540 kHz to 1600 kHz?

19. The average power supplied to any combination of components in an AC circuit is given by the relation $P = VI \cos \theta$, where P is the power (in watts), V is the effective voltage, I is the effective current, and θ is the phase angle between the current and voltage. Assuming that the effective voltage across the resistor, capacitor, and inductor combination in Exercise 13 is 200 V, determine the power supplied to these elements.

20. Find the power supplied to a resistor of 120 Ω and a capacitor of 0.700 μF if $\omega = 1000$ rad/s and the effective voltage is 110 V.

11–8 Exercises for Chapter 11

In Exercises 1 through 16, perform the indicated operations, expressing all answers in the simplest rectangular form.

1. $(6 - 2j) + (4 + j)$

2. $(12 + 7j) + (-8 + 6j)$

3. $(18 - 3j) - (12 - 5j)$

4. $(-4 - 2j) - (-6 - 7j)$

5. $(2 + j)(4 - j)$

6. $(-5 + 3j)(8 - 4j)$

7. $(2j)(6 - 3j)(4 + 3j)$

8. $j(3 - 2j) - (j^3)(5 + j)$

9. $\dfrac{3}{7 - 6j}$

10. $\dfrac{4j}{2 + 9j}$

11. $\dfrac{6 - 4j}{7 - 2j}$

12. $\dfrac{3 - 2j}{4 + j}$

13. $\dfrac{5j - (3 - j)}{4 - 2j}$

14. $\dfrac{2 + (j - 6)}{5 - 2j}$

15. $\dfrac{j(7 - 3j)}{2 + j}$

16. $\dfrac{(2 - j)(3 + 2j)}{4 - 3j}$

In Exercises 17 through 20, find the values of x and y for which the equations are valid.

17. $3x - 2j = yj - 2$

18. $2xj - 2y = (y + 3)j - 3$

19. $2x - j + 4 = 6y + 2xj$

20. $3yj + xj = 6 + 3x + y$

In Exercises 21 through 24, perform the indicated operations graphically; check them algebraically.

21. $(-1 + 5j) + (4 + 6j)$

22. $(7 - 2j) + (5 + 4j)$

23. $(9 + 2j) - (5 - 6j)$

24. $(1 + 4j) - (-3 - 3j)$

In Exercises 25 through 32, give the polar and exponential forms of each of the complex numbers.

25. $1 - j$

26. $4 + 3j$

27. $-2 - 7j$

28. $6 - 2j$

29. $1.07 + 4.55j$

30. $-327 + 158j$

31. 10

32. $-4j$

In Exercises 33 through 44, give the rectangular form of each of the complex numbers.

33. $2(\cos 225° + j \sin 225°)$

34. $4(\cos 60° + j \sin 60°)$

35. $5.011(\cos 123.82° + j \sin 123.82°)$

36. $2.417(\cos 296.26° + j \sin 296.26°)$

37. $0.6 \angle 72°$

38. $20 \angle 160°$

39. $27.08 \angle 346.27°$

40. $1.689 \angle 194.36°$

41. $2.00e^{0.25j}$

42. $e^{3.62j}$

43. $25.37e^{1.906j}$

44. $44.47e^{6.046j}$

In Exercises 45 through 56, perform the indicated operations. Leave the result in polar form.

45. $[3(\cos 32° + j \sin 32°)][5(\cos 52° + j \sin 52°)]$

46. $[2.5(\cos 162° + j \sin 162°)][8(\cos 115° + j \sin 115°)]$

47. $(40 \angle 18°)(0.5 \angle 245°)$

48. $(0.1254 \angle 172.38°)(27.17 \angle 204.34°)$

49. $\dfrac{24(\cos 165° + j \sin 165°)}{3(\cos 106° + j \sin 106°)}$

50. $\dfrac{18(\cos 403° + j \sin 403°)}{4(\cos 192° + j \sin 192°)}$

51. $\dfrac{245.6 \angle 326.44°}{17.19 \angle 192.83°}$

52. $\dfrac{100 \angle 206°}{4 \angle 320°}$

53. $0.983 \angle 47.2° + 0.366 \angle 95.1°$

54. $17.8 \angle 110.4° - 14.9 \angle 226.3°$

55. $7644 \angle 294.36° - 6871 \angle 17.86°$

56. $4.944 \angle 327.49° + 8.009 \angle 7.37°$

57. $[2(\cos 16° + j \sin 16°)]^{10}$

58. $[3(\cos 36° + j \sin 36°)]^{6}$

59. $(3 \angle 110.5°)^{3}$

60. $(5.36 \angle 220.3°)^{4}$

In Exercises 61 through 64, change each number to polar form and then perform the indicated operations. Express the final result in rectangular and polar forms. Check by performing the same operation in rectangular form.

61. $(1 - j)^{10}$

62. $(\sqrt{3} + j)^{8}(1 + j)^{5}$

63. $\dfrac{(5 + 5j)^{4}}{(-1 - j)^{6}}$

64. $(\sqrt{3} - j)^{-8}$

In Exercises 65 through 68, use DeMoivre's theorem to find the indicated roots. Be sure to find all roots.

65. The cube roots of -8.

66. The cube roots of 1.

67. The fourth roots of $-j$.

68. The fifth roots of -32.

In Exercises 68 through 80, find the required quantities.

69. In a given circuit $R = 7.50 \ \Omega$ and $X_C = 10.0 \ \Omega$. Find the magnitude of the impedance and the phase angle between the current and voltage.

70. In a given circuit $R = 15.0 \ \Omega$, $X_C = 27.0 \ \Omega$, and $X_L = 35.0 \ \Omega$. Find the magnitude of the impedance and the phase angle between the current and voltage.

71. A coil of wire is going around a circle at $60.0 \ r/s$. If this coil generates a current in a circuit containing a resistance of $10.0 \ \Omega$, an inductance of 0.010 H, and a capacitance of $500 \ \mu F$, what is the magnitude of the impedance of the circuit? What is the angle between the current and voltage?

72. A coil of wire rotates at $120.0 \ r/s$. If the coil generates a current in a circuit containing a resistance of $12.07 \ \Omega$, an inductance of 0.1405 H and an impedance of $22.35 \ \Omega$, what must be the value of a capacitor (in farads) in the circuit?

73. In a given circuit the current is $5.136 - 2.172j$ amperes and the impedance is $6.454 + 3.166j$ ohms. Find the magnitude of the voltage.

74. In a given circuit $I = 6.075$ amperes and $V = 3.144 - 2.532j$ volts. Find the magnitude of the impedance.

75. What are the magnitude and direction of a force which is represented by $600 - 550j$ pounds?

76. What are the magnitude and direction of a velocity which is represented by $2500 + 1500j$ kilometers per hour?

77. In the study of shearing effects in the spinal column, the expression

$$\frac{1}{u + j\omega n}$$

is found. Express this in rectangular form.

78. In the theory of light reflection on metals, the expression

$$\frac{\mu(1 - kj) - 1}{\mu(1 - kj) + 1}$$

is encountered. Simplify this expression.

79. Show that $e^{j\pi} = -1$.

80. Show that $(e^{j\pi})^{1/2} = j$.

CHAPTER 12

Exponential and Logarithmic Functions

In this chapter we introduce two more important types of functions, the *exponential function* and the *logarithmic function*. Historically, logarithms were first used for computational work, but with the extensive use of calculators and computers, they are not directly used for this purpose to any extent today. However, these functions are very important in many scientific and technical applications, as well as in many areas of advanced mathematics.

To illustrate the importance of logarithms, many applications may be cited. The basic units used to measure the intensity of sound and those used to measure the intensity of earthquakes are defined in terms of logarithms. In chemistry, the distinction between a base and an acid is defined in terms of logarithms. In electrical transmission lines, power gains and losses are measured in terms of logarithmic units. In electronics and in mechanical systems, the use of exponential functions, which are closely related to logarithms, is extensive. Many of these applications are illustrated throughout the chapter.

12–1 The Exponential Function; Logarithms

Chapter 10 dealt with exponents in expressions of the form x^n, where we showed that n could be any rational number. Here we shall deal with expressions of the form b^x, where x is any real number. When we look at these expressions, we note the primary difference is that in the second expression *the exponent is variable*. We have not previously dealt with variable exponents. *Thus let us define the* **exponential function** *to be*

$$y = b^x \qquad (12\text{--}1)$$

In Eq. (12–1), x is called the **logarithm** of the number y to the base b. In our work with logarithms we shall restrict all numbers to the real number system. *This leads us to choose the base as a positive number other than* 1. We know that 1 raised to any power will result in 1, which would make y a constant regardless of the value of x. Negative numbers for b would result in imaginary values for y if x were any fractional exponent with an even integer for its denominator.

Example A

$y = 2^x$ is an exponential function, where x is the logarithm of y to the base 2. This means that 2 raised to a given power gives us the corresponding value of y.

If $x = 2$, $y = 2^2 = 4$; this means that 2 is the logarithm of 4 to the base 2.
If $x = 4$, $y = 2^4 = 16$; this means that 4 is the logarithm of 16 to the base 2.
If $x = \frac{1}{2}$, $y = 2^{1/2} = 1.41$; this means that $\frac{1}{2}$ is the logarithm of 1.41 to the base 2.

Using the definition of a logarithm, Eq. (12–1) may be solved for x, and is written in the form

$$x = \log_b y \qquad\qquad (12\text{–}2)$$

This equation is read in accordance with the definition of x in Eq. (12–1): *x equals the logarithm of y to the base b.* This means that x is the power to which the base b must be raised in order to equal the number y; that is, x is a logarithm, and *a logarithm is an exponent.* Note that Eqs. (12–1) and (12–2) state the same relationship, but in a different manner. Equation (12–1) is the **exponential form,** and Eq. (12–2) is the **logarithmic form.**

Example B

The equation $y = 2^x$ would be written as $x = \log_2 y$ if we put it in logarithmic form. When we choose values of y to find the corresponding values of x from this equation, we ask ourselves, "2 raised to what power gives y?" Hence if $y = 4$, we know that 2^2 is 4, and x would be 2. If $y = 8$, $2^3 = 8$, or $x = 3$.

Example C

$3^2 = 9$ in logarithmic form is $2 = \log_3 9$; $4^{-1} = \frac{1}{4}$ in logarithmic form is $-1 = \log_4(\frac{1}{4})$. Remember, the exponent may be negative. The base must be positive.

Example D

$(64)^{1/3} = 4$ in logarithmic form is $\frac{1}{3} = \log_{64} 4$,
$(32)^{3/5} = 8$ in logarithmic form is $\frac{3}{5} = \log_{32} 8$

Example E

$\log_2 32 = 5$ in exponential form is $32 = 2^5$,
$\log_6(\frac{1}{36}) = -2$ in exponential form is $\frac{1}{36} = 6^{-2}$

Example F

Find b, given that $-4 = \log_b(\frac{1}{81})$.
   Writing this in exponential form, we have $\frac{1}{81} = b^{-4}$. Thus, $\frac{1}{81} = \frac{1}{b^4}$ or $\frac{1}{3^4} = \frac{1}{b^4}$. Therefore, $b = 3$.

Example G

Find y, given that $\log_4 y = \frac{1}{2}$.
   In exponential form we have $y = 4^{1/2}$, or $y = 2$.

We see that exponential form is very useful for determining values written in logarithmic form. For this reason it is important that you learn to transform readily from one form to the other.

Example H

The power supply P, in watts, of a certain satellite is given by $P = 75e^{-0.005t}$, where t is the time in days after launch. By writing this equation in logarithmic form, solve for t.
   In order to have the equation in the exponential form of Eq. (12–1), we must have only $e^{-0.005t}$ on the right. Therefore, by dividing by 75, we have

$$\frac{P}{75} = e^{-0.005t}$$

Writing this in logarithmic form, we have

$$\log_e\left(\frac{P}{75}\right) = -0.005t$$

or

$$t = \frac{\log_e\left(\frac{P}{75}\right)}{-0.005} = -200 \log_e\left(\frac{P}{75}\right)$$

Exercises 12–1

In Exercises 1 through 4, evaluate the exponential function $y = 9^x$ for the given values of x.

1. $x = 0.5$ 2. $x = 4$ 3. $x = -2$ 4. $x = -0.5$

In Exercises 5 through 16, express the given equations in logarithmic form.

5. $3^3 = 27$ 6. $5^2 = 25$ 7. $4^4 = 256$ 8. $8^2 = 64$

9. $4^{-2} = \frac{1}{16}$ 10. $3^{-2} = \frac{1}{9}$ 11. $2^{-6} = \frac{1}{64}$ 12. $(12)^0 = 1$

13. $8^{1/3} = 2$ 14. $(81)^{3/4} = 27$ 15. $(\frac{1}{4})^2 = \frac{1}{16}$ 16. $(\frac{1}{2})^{-2} = 4$

In Exercises 17 through 28, express the given equations in exponential form.

17. $\log_3 81 = 4$ 18. $\log_{11} 121 = 2$ 19. $\log_9 9 = 1$

20. $\log_{15} 1 = 0$ 21. $\log_{25} 5 = \frac{1}{2}$ 22. $\log_8 16 = \frac{4}{3}$

23. $\log_{243} 3 = \frac{1}{5}$ 24. $\log_{1/32}(\frac{1}{8}) = \frac{3}{5}$ 25. $\log_{10} 0.1 = -1$

26. $\log_7(\frac{1}{49}) = -2$ 27. $\log_{0.5} 16 = -4$ 28. $\log_{1/3} 3 = -1$

In Exercises 29 through 44 determine the value of the unknown.

29. $\log_4 16 = x$ 30. $\log_5 125 = x$ 31. $\log_{10} 0.01 = x$

32. $\log_{16}(\frac{1}{4}) = x$ 33. $\log_7 y = 3$ 34. $\log_8 N = 3$

35. $\log_8 y = -\frac{2}{3}$ 36. $\log_7 y = -2$ 37. $\log_b 81 = 2$

38. $\log_b 625 = 4$ 39. $\log_b 4 = -\frac{1}{3}$ 40. $\log_b 4 = \frac{2}{3}$

41. $\log_{10} 10^{0.2} = x$ 42. $\log_5 5^{1.3} = x$ 43. $\log_3 27^{-1} = x$

44. $\log_b(\frac{1}{4}) = -\frac{1}{2}$

In Exercises 45 through 48, perform the indicated operations.

45. If there are initially 1000 bacteria in a culture, and the number of bacteria then doubles each hour, the number of bacteria as a function of time is $N = 1000(2^t)$. By writing this equation in logarithmic form, solve for t.

46. Under specified conditions, the instantaneous voltage E in a given circuit can be expressed as

$$E = E_m e^{-Rt/L}$$

By rewriting this equation in logarithmic form, solve for t.

47. An equation relating the number N of atoms of radium at any time t in terms of the number of atoms at $t = 0$, N_0, is $\log_e(N/N_0) = -kt$, where k is a constant. By expressing this equation in exponential form, solve for N.

48. In the theory dealing with the optical brightness of objects, the equation $D = \log_{10}(I_0/I)$ is found. By writing this equation in exponential form, solve for I.

12–2 Graphs of $y = b^x$ and $y = \log_b x$

When we are working with functions, we must keep in mind that a function is defined by the operation being performed on the independent variable, and not by the letter chosen to represent it. However, for consistency, it is standard practice to let y represent the dependent variable and x represent the independent variable. Therefore, *the* **logarithmic function** *is*

$$y = \log_b x \qquad\qquad (12\text{--}3)$$

Equations (12–2) and (12–3) express the same *function*, the logarithmic function. They do not represent different functions, due to the difference in location of the variables, since they represent the same operation on the independent variable which appears in each. However, Eq. (12–3) expresses the function with the standard dependent and independent variables.

We note that for the exponential function $y = b^x$ and the logarithmic function $y = \log_b x$, if we change the form of one and then interchange the variables x and y, we then obtain the other function. *Such functions are called* **inverse functions.** (This was essentially the procedure we used in defining the logarithmic function.)

Graphical representation of functions is often valuable when we wish to demonstrate their properties. We shall now show the graphs of the exponential function [Eq. (12–1)] and the logarithmic function [Eq. (12–3)].

Example A

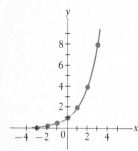

Figure 12–1

Plot the graph of $y = 2^x$.

Assuming values for x and then finding the corresponding values for y, we obtain the following table.

x	-3	-2	-1	0	1	2	3	4
y	$\frac{1}{8}$	$\frac{1}{4}$	$\frac{1}{2}$	1	2	4	8	16

From these values we plot the curve, as shown in Fig. 12–1. We note that the x-axis is an asymptote of the curve.

Example B

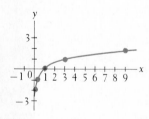

Figure 12–2

Plot the graph of $y = \log_3 x$.

We can find the points for this graph more easily if we first put the equation in exponential form: $x = 3^y$. By assuming values for y, we can find the corresponding values for x.

x	$\frac{1}{9}$	$\frac{1}{3}$	1	3	9	27
y	-2	-1	0	1	2	3

Using these values, we construct the graph seen in Fig. 12–2.

Any exponential or logarithmic curve, where $b > 1$, will be similar in shape to those of Examples A and B. From these curves we can draw certain conclusions:

1. If $0 < x < 1$, $\log_b x < 0$; if $x = 1$, $\log_b 1 = 0$; if $x > 1$, $\log_b x > 0$.
2. If $x > 1$, x increases more rapidly than $\log_b x$.
3. For all values of x, $b^x > 0$.
4. If $x > 1$, b^x increases more rapidly than x.

Although the bases important to applications are greater than 1, to understand how the curve of the exponential function differs somewhat if $b < 1$, let us consider the following example.

Example C

Plot the graph of $y = \left(\frac{1}{2}\right)^x$.

The values are found for the following table; the graph is plotted in Fig. 12–3.

x	-3	-2	-1	0	1	2	3	4
y	8	4	2	1	$\frac{1}{2}$	$\frac{1}{4}$	$\frac{1}{8}$	$\frac{1}{16}$

We note that as the values of x increase, the values of $\left(\frac{1}{2}\right)^x$ decrease. This is different from the behavior of $y = b^x$, where $b > 1$.

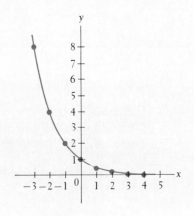

Figure 12–3

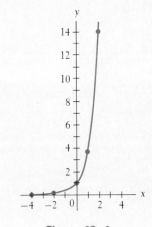

Figure 12–4

In the following example, we use a calculator to find the values of an exponential function in order to sketch its graph.

Example D

Sketch the graph of the function $y = 3.75^x$.

By use of the $\boxed{x^y}$ key on a calculator, we can easily raise 3.75 to any power we choose. Also, since the base 3.75 is greater than 1, we know that the graph will have the same basic shape as that in Fig. 12–1. Therefore, only a few points are needed for the graph shown in Fig. 12–4.

x	-4	-2	0	1	2
y	0.005	0.07	1	3.75	14.1

The calculator sequence for $x = -4$ is

$$3.75 \quad \boxed{x^y} \quad 4 \quad \boxed{+/-} \quad \boxed{=}$$

Example E

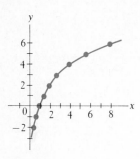

Figure 12–5

Plot the graph of $y = 2 \log_2 x$.

As in Example B, it is generally more convenient to work with the exponential form. In order to change this equation to exponential form, we first divide each side by 2. Thus, we have

$$\frac{y}{2} = \log_2 x, \qquad x = 2^{y/2}$$

Choosing values of y, we then calculate corresponding values of x and obtain the following table.

x	0.5	0.7	1	1.4	2	2.8	4	5.7	8
y	-2	-1	0	1	2	3	4	5	6

See Fig. 12–5.

The following example illustrates an application in which there is an exponential function and for which we show its graph.

Example F

In an electric circuit in which there is a battery, an inductor, and a resistor, the current i, in amperes—as a function of the time t, in seconds—is given by $i = 0.8(1 - e^{-4t})$. Here e is the same irrational number which was used in Section 11–5. Its value is approximately $e = 2.718$.

Here we see that values of the exponential function e^{-4t} are to be subtracted from 1 and the result is then multiplied by 0.8. Also, e^{-4t} becomes very small after a short period of time, and this means that i quickly becomes nearly 0.8 A, although theoretically it never quite reaches this value. We shall calculate values for i using the $\boxed{e^x}$ key on the calculator and values of t from 0 to 1 s.

t	0	0.1	0.2	0.4	0.6	0.8	1.0
i	0	0.26	0.44	0.64	0.73	0.77	0.79

Figure 12–6

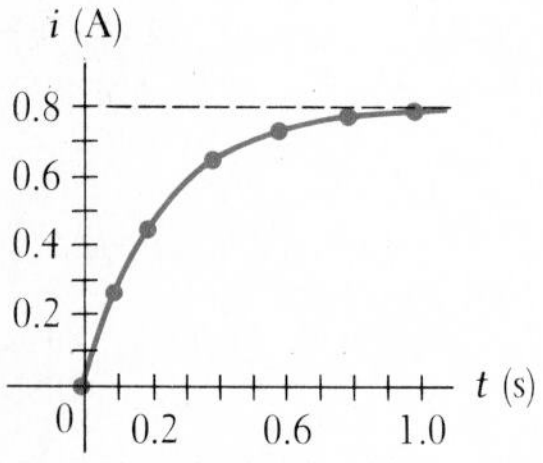

See Fig. 12–6.

Exercises 12–2

In Exercises 1 through 12, plot graphs of the given functions. Values of x from -3 or -2 through 2 or 3 are appropriate.

1. $y = 3^x$ 2. $y = 4^x$ 3. $y = 6^x$

4. $y = 10^x$ 5. $y = (1.65)^x$ 6. $y = (2.72)^x$

7. $y = (\frac{1}{3})^x$ 8. $y = (\frac{1}{4})^x$ 9. $y = 2(2^x)$

10. $y = 1.5(4.15)^x$ 11. $y = 0.5(3.06)^x$ 12. $y = 0.1(10^x)$

In Exercises 13 through 24, plot graphs of the given functions. Values of y from -3 or -2 through 2 or 3 are appropriate for Exercises 13 through 20. Care should be taken in selecting appropriate values used in Exercises 21 through 24.

13. $y = \log_2 x$ 14. $y = \log_4 x$ 15. $y = \log_6 x$

16. $y = \log_{10} x$ 17. $y = \log_{1.65} x$ 18. $y = \log_{2.72} x$

19. $y = \log_{32} x$ 20. $y = \log_{0.5} x$ 21. $y = 2 \log_3 x$

22. $y = 3 \log_2 x$ 23. $y = 0.2 \log_4 x$ 24. $y = 5 \log_{10} x$

In Exercises 25 through 30, plot the indicated graphs.

25. The electric current in a certain type of circuit is given by $i = I_0 e^{-Rt/L}$, where I_0 is the initial current, R is a resistance, and L is an inductance (see Section 11–7). Sketch the graph for i vs. t for a circuit in which $I_0 = 5.0$ A, $R = 10\ \Omega$, and $L = 5.0$ H for $0 \leq t < 4$ s.

26. If \$1000 is placed in a bank account in which interest is compounded daily with an effective 8% annual interest, the amount in the account after t years is $A = 1000(1.08)^t$. Sketch A as a function of t for $0 \leq t \leq 5$ years.

27. Under certain conditions the temperature T (in degrees Celsius) of a cooling object as a function of time (in minutes) is $T = 50.0(10^{-0.1t})$. Sketch T as a function of t.

28. Considering air resistance and other conditions, the velocity v, in meters per second, of a certain falling object is given by $v = 95(1 - e^{-0.1t})$, where t is the time of fall in seconds. Sketch the graph of this function.

29. The current i, in amperes, in a certain electric circuit is given by $i = 16(1 - e^{-250t})$, where t is the time in seconds. Using appropriate values of t, sketch the graph of this function.

30. In Exercise 20, the graph of $y = \log_{0.5} x$ is plotted. By inspecting the graph and noting the properties of $\log_{0.5} x$, determine some of the differences of logarithms to a base less than 1 from those to a base greater than 1.

12–3 Properties of Logarithms

Since a logarithm is an exponent, it must follow the laws of exponents. Those laws which are of the greatest importance at this time are listed here for reference.

$$b^u \cdot b^v = b^{u+v} \tag{12–4}$$

$$b^u / b^v = b^{u-v} \tag{12–5}$$

$$(b^u)^n = b^{nu} \tag{12–6}$$

We shall now show how these laws for exponents give certain useful properties to logarithms.

If we let $u = \log_b x$ and $v = \log_b y$ and write these equations in exponential form, we have $x = b^u$ and $y = b^v$. Therefore, forming the product of x and y, we obtain

$$xy = b^u b^v = b^{u+v} \quad \text{or} \quad xy = b^{u+v}$$

Writing this last equation in logarithmic form yields

$$u + v = \log_b xy$$

or

$$\log_b x + \log_b y = \log_b xy \tag{12--7}$$

Equation (12–7) states the property that *the logarithm of the product of two numbers is equal to the sum of the logarithms of the numbers.*

Using the same definitions of u and v to form the quotient of x and y, we then have

$$\frac{x}{y} = \frac{b^u}{b^v} = b^{u-v} \quad \text{or} \quad \frac{x}{y} = b^{u-v}$$

Writing this last equation in logarithmic form, we have

$$u - v = \log_b\left(\frac{x}{y}\right)$$

or

$$\log_b x - \log_b y = \log_b\left(\frac{x}{y}\right) \tag{12--8}$$

Equation (12–8) states the property that *the logarithm of the quotient of two numbers is equal to the logarithm of the numerator minus the logarithm of the denominator.*

If we again let $u = \log_b x$ and write this in exponential form, we have $x = b^u$. To find the nth power of x, we write

$$x^n = (b^u)^n = b^{nu}$$

Expressing this equation in logarithmic form yields

$$nu = \log_b(x^n)$$

or

$$n \log_b x = \log_b(x^n) \tag{12--9}$$

This last equation states that *the logarithm of the nth power of a number is equal to n times the logarithm of the number.* The exponent n may be integral or fractional and, therefore, we may use Eq. (12–9) for finding powers and roots of numbers.

We now recall from Section 12–1 that we showed that the base b of logarithms must be a positive number. Since $x = b^u$ and $y = b^v$, this means that x and y are also positive numbers. Therefore, *the properties of logarithms which have just been derived are valid only for positive values of x and y.*

Example A

Using Eq. (12–7) we may express $\log_4 15$ as a sum of logarithms as follows:

$$\log_4 15 = \log_4(3 \cdot 5) = \log_4 3 + \log_4 5$$

Using Eq. (12–8) we may express $\log_4(\frac{5}{3})$ as the difference of logarithms as follows:

$$\log_4\left(\frac{5}{3}\right) = \log_4 5 - \log_4 3$$

Using Eq. (12–9) we may express $\log_4 9$ as twice $\log_4 3$ as follows:

$$\log_4 9 = \log_4(3^2) = 2 \log_4 3$$

Example B

Using Eqs. (12–7) through (12–9), we may express a sum or difference of logarithms as the logarithm of a single quantity.

$$\log_4 3 + \log_4 x = \log_4(3 \cdot x) = \log_4 3x$$

$$\log_4 3 - \log_4 x = \log_4\left(\frac{3}{x}\right)$$

$$\log_4 3 + 2 \log_4 x = \log_4 3 + \log_4(x^2) = \log_4 3x^2$$

$$\log_4 3 + 2 \log_4 x - \log_4 y = \log_4\left(\frac{3x^2}{y}\right)$$

We may use Eq. (12–9) to find another important property of logarithms. Since $b = b^1$ in logarithmic form is $\log_b b = 1$, we have

$$\log_b(b^n) = n \log_b b = n(1) = n$$

or

$$\underline{\log_b(b^n) = n} \qquad\qquad (12\text{–}10)$$

Equation (12–10) may be used to evaluate certain logarithms when the exact values may be determined.

Example C

We may evaluate $\log_3 9$ in the following manner: Using Eq. (12–10), we have

$$\log_3 9 = \log_3(3^2) = 2$$

We can establish the exact value since the base of logarithms and the number being raised to the power are the same. Of course, this could have been evaluated directly from the definition of a logarithm.

However, we cannot establish an exact value of $\log_4 9$ in this way. We have

$$\log_4 9 = \log_4(3^2) = 2\log_4 3$$

We must leave it in this form until we are able to evaluate $\log_4 3$. This can be done, and we shall consider this type of expression later in the chapter.

Example D

Calculate $\log_3(3^{0.4})$.

Using Eq. (12–10) we may calculate this logarithm. This gives us

$$\log_3(3^{0.4}) = 0.4$$

Although we did not calculate $3^{0.4}$ at this point, we were able to calculate $\log_3 3^{0.4}$.

Example E

We may also use Eq. (12–10) in combination with other properties of logarithms. The following illustration shows how it can be used alone or in combination with Eq. (12–8) to evaluate the indicated logarithm.

$$\log_5(\tfrac{1}{25}) = \log_5 1 - \log_5 25 = 0 - \log_5(5^2) = -2$$
$$\log_5(\tfrac{1}{25}) = \log_5(5^{-2}) = -2$$

Either method is appropriate.

In the following examples, the properties of logarithms of Eqs. (12–7) through (12–10) are further illustrated.

Example F

$$\log_2 6 = \log_2(2 \cdot 3) = \log_2 2 + \log_2 3 = 1 + \log_2 3$$
$$\log_3(\tfrac{2}{9}) = \log_3 2 - \log_3 9 = \log_3 2 - \log_3(3^2) = \log_3 2 - 2$$
$$= -2 + \log_3 2$$

Example G

$$\log_{10}\sqrt{7} = \log_{10}(7^{1/2}) = \tfrac{1}{2}\log_{10} 7$$

This demonstrates the property which is especially useful for finding roots of numbers. We see that we need merely to multiply the logarithm of the number by the fractional exponent representing the root to obtain the logarithm of the root.

Similarly,

$$\log_{10}\sqrt[3]{10x} = \log_{10}(10x)^{1/3} = \tfrac{1}{3}\log_{10}(10x)$$
$$= \tfrac{1}{3}(\log_{10}10 + \log_{10}x)$$
$$= \tfrac{1}{3}(1 + \log_{10}x)$$

Example H

Use the basic properties of logarithms to solve for y in terms of x: $\log_b y = 2\log_b x + \log_b a$.

Using Eq. (12–9) and then Eq. (12–7), we have

$$\log_b y = \log_b(x^2) + \log_b a = \log_b(ax^2)$$

Now, since we have the logarithm to the base b of different expressions on each side of the resulting equation, the expressions must be equal. Therefore, $y = ax^2$.

Example I

Solve for y in terms of x: $2\log_3 x - \log_3 y = 1$.

Using the properties of logarithms, we can rewrite the left side of this equation as

$$\log_3 x^2 - \log_3 y = 1$$

$$\log_3\frac{x^2}{y} = 1$$

Rewriting this in exponential form, we have

$$\frac{x^2}{y} = 3^1$$
$$x^2 = 3y$$

or

$$y = \frac{1}{3}x^2$$

Exercises 12–3

In Exercises 1 through 12, express each as a sum, difference, or a multiple of logarithms. See Example A.

See Appendix E for a computer program for the properties of logarithms.

1. $\log_5 xy$

2. $\log_3 7y$

3. $\log_7\left(\dfrac{5}{a}\right)$

4. $\log_3\left(\dfrac{r}{s}\right)$

5. $\log_2(a^3)$

6. $\log_8(n^5)$

7. $\log_6 abc$

8. $\log_2\left(\dfrac{xy}{z}\right)$

9. $\log_5\sqrt[4]{y}$

10. $\log_4\sqrt[7]{x}$

11. $\log_2\left(\dfrac{\sqrt{x}}{a^2}\right)$

12. $\log_3\left(\dfrac{\sqrt{y}}{8}\right)$

In Exercises 13 through 20, express each as the logarithm of a single quantity. See Example B.

13. $\log_b a + \log_b c$

14. $\log_2 3 + \log_2 x$

15. $\log_5 9 - \log_5 3$

16. $\log_8 6 - \log_8 a$

17. $\log_b x^2 - \log_b \sqrt{x}$

18. $\log_4 3^3 + \log_4 9$

19. $2\log_e 2 + 3\log_e n$

20. $\frac{1}{2}\log_b a - 2\log_b 5$

In Exercises 21 through 28, determine the exact value of each of the given logarithms.

21. $\log_2\left(\frac{1}{32}\right)$

22. $\log_3\left(\frac{1}{81}\right)$

23. $\log_2(2^{2.5})$

24. $\log_5(5^{0.1})$

25. $\log_7\sqrt{7}$

26. $\log_6\sqrt[3]{6}$

27. $\log_3\sqrt[4]{27}$

28. $\log_5\sqrt[3]{25}$

In Exercises 29 through 40, express each as a sum, difference, or multiple of logarithms. In each case part of the logarithm may be determined exactly.

29. $\log_3 18$

30. $\log_5 75$

31. $\log_2\left(\frac{1}{6}\right)$

32. $\log_{10}(0.05)$

33. $\log_3\sqrt{6}$

34. $\log_2\sqrt[3]{24}$

35. $\log_2(4^2 \cdot 3^3)$

36. $\log_7(7^4 \cdot 3^5)$

37. $\log_{10}3000$

38. $\log_{10}(40^2)$

39. $\log_{10}\left(\frac{27}{100}\right)$

40. $\log_5\left(\frac{4}{125}\right)$

In Exercises 41 through 52, solve for y in terms of x.

41. $\log_b y = \log_b 2 + \log_b x$

42. $\log_b y = \log_b 6 + \log_b x$

43. $\log_4 y = \log_4 x - \log_4 5$

44. $\log_3 y = \log_3 7 - \log_3 x$

45. $\log_{10} y = 2\log_{10}7 - 3\log_{10}x$

46. $\log_b y = 3\log_b\sqrt{x} + 2\log_b 10$

47. $5\log_2 y - \log_2 x = 3\log_2 4 + \log_2 a$

48. $4\log_2 x - 3\log_2 y = \log_2 27$

49. $\log_2 x + \log_2 y = 1$

50. $3\log_4 x + \log_4 y = 1$

51. $2\log_5 x - \log_5 y = 2$

52. $\log_8 x - 2\log_8 y = 4$

In Exercises 53 through 56, evaluate each of the given expressions using $\log_{10}2 = 0.301$. (Hint: Each number can be expressed in terms of 2 or 10 or both.)

53. $\log_{10}4$

54. $\log_{10}20$

55. $\log_{10}(0.5)$

56. $\log_{10}8000$

In Exercises 57 through 60, plot the indicated graphs and perform the indicated operations.

57. Plot the graphs of $y = 2\log_2 x$ and $y = \log_2 x^2$, and show that they are the same.

58. Plot the graphs of $y = \log_2 4x$ and $y = 2 + \log_2 x$, and show that they are the same.

59. An equation used in thermodynamics is $S = C\log_e T - nR\log_e P$. Express this equation with a single logarithm on the right side.

60. An equation used for a certain electric circuit is $\log_e i - \log_e I = -t/RC$. Solve for i.

12–4 Logarithms to the Base 10

In Section 12–1 we stated that a base of logarithms must be a positive number, not equal to one. In the examples and exercises of the previous sections we used a number of different bases. There are, however, only two bases which are generally used. They are 10 and e, where e is the irrational number approximately equal to 2.718 that we introduced in Section 11–5 and used again in Section 12–2.

Base 10 logarithms were developed for calculational purposes and were used a great deal for making calculations until the 1970s, when the modern electronic calculator became widely available. However, there are a number of scientific measurements in which base 10 logarithms are still used, and a need therefore still exists for them. Base e logarithms are used extensively in technical and scientific work, and they are considered in Section 12–6. In this section we discuss how the base 10 logarithm of a number is determined.

Logarithms to the base 10 are called **common logarithms.** They may be found directly by use of a calculator. The $\boxed{\text{LOG}}$ key is used for this purpose. This calculator key indicates the common notation: *When no base is shown, it is assumed to be base 10.*

Example A

Find log 426 by use of a calculator.

By using the sequence 426 $\boxed{\text{LOG}}$, the display shows 2.6294096, which means

$$\log 426 = 2.629$$

when the result is rounded off. The decimal part of a logarithm is normally expressed to the same accuracy as that of the number of which it is the logarithm, although showing one additional digit in the logarithm is generally acceptable.

Example B

Find log 0.03654 by use of a calculator.

In this case the calculator display is -1.4372315, which means that

$$\log 0.03654 = -1.4372$$

We note that the logarithm here is negative. This should be the case when we recall the meaning of a logarithm. Raising 10 to a negative power gives us a number between 0 and 1, and here we have

$$10^{-1.4372} = 0.03654$$

We may also use the calculator to find a number N if we know log N. *In this case we refer to N as the* **antilogarithm** *of log N.* On the calculator we use either $\boxed{\text{INV}}$ $\boxed{\text{LOG}}$ key sequence or the $\boxed{10^x}$ key, depending on the calculator. Note that the $\boxed{\text{INV}}$ $\boxed{\text{LOG}}$ sequence uses the fact that the exponential and logarithmic functions are inverse functions, as we stated in Section 12–2. The $\boxed{10^x}$ key uses the basic definition of a logarithm.

Example C

Given $\log N = 2.1854$, find N.

Using the sequence 2.1854 $\boxed{\text{INV}}$ $\boxed{\text{LOG}}$ or 2.1854 $\boxed{10^x}$, the display shows 153.24983, which means that

$$N = 153.2$$

where the result has been rounded off.

We can also use a table of logarithms to find the common logarithm of a number or to find an antilogarithm. Table 2 in Appendix F is such a table. We now develop the method of using this table.

In Section 1–6, we showed that any number may be expressed in scientific notation as the product of a number greater than or equal to 1 and less than 10 and a power of 10. Writing this as $N = P \times 10^k$ and taking logarithms of both sides of this equation, we have

$$\log_b N = \log_b (P \times 10^k) = \log_b P + \log_b 10^k = \log_b P + k \log_b 10$$

If we let $b = 10$, then $k \log_{10} 10 = k$, and this equation becomes

$$\underline{\log N = k + \log P} \qquad \qquad (12\text{–}11)$$

In Eq. (12–11), k *is called the* **characteristic,** *and log P is known as the* **mantissa.** Remember, k is the power of 10 of the number, when it is written in scientific notation, and the term $\log P$ is the logarithm of the number between 1 and 10. Equation (12–11) shows us that if we have a method for finding logarithms to the base 10 of numbers between 1 and 10, then we can find the logarithm of *any* number to base 10.

Example D

For $N = 3600 = 3.6 \times 10^3$, we see that the characteristic $k = 3$, and the mantissa $\log P = \log 3.6$. Therefore, $\log 3600 = 3 + \log 3.6$.

For $N = 80.9 = 8.09 \times 10^1$, we see that $k = 1$ and $\log P = \log 8.09$. Therefore, $\log 80.9 = 1 + \log 8.09$.

Example E

For $N = 0.00543 = 5.43 \times 10^{-3}$, we see that $k = -3$ and $\log P = \log 5.43$. Therefore, $\log 0.00543 = -3 + \log 5.43$.

For $N = 0.741 = 7.41 \times 10^{-1}$, we see that $k = -1$ and $\log P = \log 7.41$. Therefore, $\log 0.741 = -1 + \log 7.41$.

To find $\log P$ we use Table 2 in Appendix F. The following examples illustrate how to use this table.

Example F

Find log 572.

We first write the number in scientific notation as 5.72×10^2. The characteristic is 2, and we must now find log 5.72. We look in the column headed N and find 57 (the first two significant digits). Then, to the right of this, we look under the column headed 2 (the third significant digit) and we find 7574. All numbers between 1 and 10 will have common logarithms between 0 and 1 (log 1 = 0 and log 10 = 1). Therefore, log 5.72 = 0.7574, and the logarithm of 572 = 2 + 0.7574. We then write this in the usual form of 2.7574 and write this result as

$$\log 572 = 2.7574$$

It should be emphasized that both the mantissa and characteristic of a logarithm must be found before the logarithm of a number is complete. *The mantissa is found from Table 2, and the characteristic is found from the location of the decimal point.* A rather common error is to assume the logarithm has been found when the mantissa is determined. However, the characteristic is just as much a proper part of the logarithm as the mantissa.

Example G

Find log 0.00485.

Writing this number in scientific notation gives us 4.85×10^{-3}, and we see that $k = -3$. From the tables we find that log 4.85 = 0.6857. Thus, log 0.00485 = -3 + 0.6857. We do *not* write this as -3.6857, for this would say that the mantissa was also negative, which it is not. To avoid this possible confusion, we shall write it in the form 0.6857 - 3. We shall follow this policy whenever the characteristic is negative. That is, we shall write a negative characteristic after the mantissa. Thus,

$$\log 0.00485 = 0.6857 - 3$$

If we actually subtract 3 from 0.6857, we would have the value of the logarithm, which in this case is negative, as it would be found on a calculator. That is, -2.3143 is the rounded-off value of log 0.00485 which would be displayed by a calculator.

Example H

Other examples of logarithms are as follows:

$$89,000 = 8.9 \times 10^4: k = 4, \log 8.9 = 0.9494; \quad \log 89,000 = 4.9494$$
$$0.307 = 3.07 \times 10^{-1}: \quad k = -1; \quad \log 3.07 = 0.4871$$
$$\log 0.307 = 0.4871 - 1$$
$$0.00629 = 6.29 \times 10^{-3}: \quad k = -3; \quad \log 6.29 = 0.7987$$
$$\log 0.00629 = 0.7987 - 3$$

Example I

Given log $N = 1.5265$, find N.

Direct observation of the given logarithm tells us that the characteristic is 1 and that $N = P \times 10^1$. In Table 2 we find 5263 (the number nearest 5265) opposite 33 and under 6. Thus, $P = 3.36$. This means that $N = 3.36 \times 10^1$, or in ordinary notation, $N = 33.6$.

In the previous examples we have used the value of a logarithm which is closest to that shown in the table. It is possible to use interpolation (see Section 2–5) to increase the accuracy of the values used from the table. This was common practice before the extensive use of calculators.

The following example illustrates an application in which a measurement requires the direct use of the value of a logarithm.

Example J

The power gain G, in decibels, of an electronic device is given by $G = 10 \log(P_0/P_i)$, where P_0 is the output power, in watts, and P_i is the input power. Determine the power gain for an amplifier for which $P_0 = 15.8$ W and $P_i = 0.625$ W.

Substituting the given values, we have

$$G = 10 \log \frac{15.8}{0.625}$$

$$= 10 \log 25.28$$

$$= 14.0 \text{ dB}$$

where the result has been rounded off to three significant digits, the accuracy of the given data.

Exercises 12–4

In Exercises 1 through 12, find the common logarithm of each of the given numbers by use of a calculator.

1. 567	2. 60.5	3. 0.0640	4. 0.000566
5. 9.24×10^6	6. 3.19×10^{15}	7. 1.172×10^{-4}	8. 8.043×10^{-8}
9. 73.27	10. 0.008726	11. 0.18643	12. 164,850

In Exercises 13 through 24, find the antilogarithm from the given logarithms by use of a calculator.

13. 4.437	14. 0.929	15. -1.3045	16. -6.9788
17. 3.30112	18. 8.82436	19. $0.8594 - 1$	20. $0.4412 - 3$
21. 0.15485	22. 10.27562	23. -2.23746	24. -10.33577

In Exercises 25 through 32, find the common logarithm of each of the given numbers by use of Table 2.

25. 328   26. 15.7   27. 0.0529   28. 0.184

29. 8.78×10^5 30. 2.07×10^7 31. 9.65×10^{-2} 32. 6.38×10^{-8}

In Exercises 33 through 40, find the antilogarithm from each of the given logarithms by use of Table 2. Use the nearest value listed.

33. 1.7574   34. 5.4713   35. $0.3767 - 2$ 36. $0.9719 - 1$

37. 8.7262   38. 10.2260   39. $0.9182 - 10$ 40. $0.4333 - 6$

In Exercises 41 through 48, find the logarithms of the given numbers.

41. A certain radar signal has a frequency of 1.15×10^9 Hz.

42. The earth travels about 595,000,000 mi in 1 year.

43. A certain bank charges 13.5% interest on loans that it makes.

44. The coefficient of thermal expansion of steel is about 1.2×10^{-5} per degree Celsius.

45. A typical X-ray tube operates with a voltage of 150,000 V.

46. The bending moment of a particular concrete column is 4.60×10^6 lb-in.

47. Electronic calculators which display numbers in scientific notation can calculate results which are as numerically small as 10^{-99}.

48. In an air sample taken in an urban area, $5/10^6$ of the air was carbon monoxide.

In Exercises 49 and 50, solve the given problems by finding the appropriate logarithms.

49. A stereo amplifier has an input power of 0.750 W and an output power of 25.0 W. What is the power gain? (See Example J.)

50. Measured on the Richter scale, the magnitude of an earthquake of intensity I is defined as $R = \log(I/I_0)$, where I_0 is a minimum level for comparison. What is the Richter scale reading for an earthquake for which $I = 75,000 I_0$?

12–5 Computations Using Logarithms

Logarithms were developed in the seventeenth century for the purpose of making tedious and complicated calculations which arose in astronomy and navigation. Until the calculator came into extensive use, logarithms were commonly used for calculational purposes. Logarithms can be used to make more complicated calculations by means of basic additions, subtractions, multiplications, and divisions. In this section we show how logarithms are used to make such calculations. Performing these calculations provides an opportunity to better understand the meaning and properties of logarithms, even if logarithms are no longer generally used for calculations. The following examples illustrate the use of logarithms from Table 2 in making basic computations.

Example A

By the use of logarithms, calculate the value of (42.8)(215).

Equation (12–7) tells us that $\log xy = \log x + \log y$. If we find log 42.8 and log 215 and add them, we shall have the logarithm of the product. Using this result, we look up the antilogarithm, which is the desired product.

$$
\begin{aligned}
\log 42.8 &= 1.6314 \\
\log 215 &= \underline{2.3324} \\
\log[(42.8)(215)] &= 3.9638 \\
\log 9200 &= 3.9638
\end{aligned}
$$

Thus, (42.8)(215) = 9200.

Example B

By the use of logarithms, calculate the value of 8.64 ÷ 45.6.

From Eq. (12–8), we know that $\log (x/y) = \log x - \log y$. Therefore, by subtracting log 45.6 from log 8.64, we shall have the logarithm of the quotient. The antilogarithm gives the desired result.

$$
\begin{aligned}
\log 8.64 &= 1.9365 - 1 \\
\log 45.6 &= \underline{1.6590} \\
\log(8.64/45.6) &= 0.2775 - 1
\end{aligned}
$$

We wrote the characteristic of log 8.64 as $1 - 1$, so that when we subtracted, the part of the result containing the mantissa would be positive, although the characteristic was negative. The antilogarithm of $0.2775 - 1$ is 0.189, using the nearest value which appears in the table. Therefore,

$$
\frac{8.64}{45.6} = 0.189
$$

Example C

By the use of logarithms, calculate the value of $\sqrt[5]{0.0376}$.

From Eq. (12–9), we know that $\log x^n = n \log x$. Therefore, by writing $\sqrt[5]{0.0376} = (0.0376)^{1/5}$, we know we can find the logarithm of the result by multiplying log 0.0376 by $\frac{1}{5}$. Now, we determine that $\log 0.0376 = 0.5752 - 2$. Since we wish to multiply this by $\frac{1}{5}$, we shall write this logarithm as

$$
\log 0.0376 = 3.5752 - 5
$$

by adding and subtracting 3. Multiplying by $\frac{1}{5}$, we have

$$
\frac{1}{5} \log 0.0376 = \frac{1}{5}(3.5752 - 5) = 0.7150 - 1
$$

The antilogarithm of $0.7150 - 1$ is 0.519. Therefore,

$$
\sqrt[5]{0.0376} = 0.519
$$

The following example illustrates an application in which the calculation requires a combination of the basic properties of logarithms.

Example D

The time t, in seconds, during which an object falls under gravity is related to the distance h, in meters, through which it falls by the function $t = \sqrt{\dfrac{2h}{g}}$.

Here g is the acceleration due to gravity and has the value 9.81 m/s^2. Calculate the time it takes for an object to fall 375 m.

Substituting the given values into the equation, we have

$$t = \sqrt{\frac{2(375)}{9.81}}$$

Using logarithms for the calculation, we have

$$\log t = \tfrac{1}{2}(\log 2 + \log 375 - \log 9.81)$$

$$\begin{aligned}
\log 2 &= 0.3010 \\
\log 375 &= \underline{2.5740} \\
&\ 2.8750
\end{aligned}$$

$$\log 9.81 = \underline{0.9917}$$

$$\log \frac{2(375)}{9.81} = 1.8833$$

$$\tfrac{1}{2} \log \frac{2(375)}{9.81} = 0.9417$$

$$t = 8.74 \text{ s}$$

Exercises 12–5

In Exercises 1 through 32, use logarithms to perform the indicated calculations.

1. $(5.98)(14.3)$
2. $(0.764)(551)$
3. $(0.825)(0.0453)$
4. $(0.000808)(2620)$
5. $\dfrac{790}{8.02}$
6. $\dfrac{31.6}{0.454}$
7. $\dfrac{76.9}{43.8}$
8. $\dfrac{0.00867}{0.652}$
9. $(6.75)^6$
10. $(0.904)^5$
11. $(89.0)^{0.3}$
12. $(0.0403)^{0.6}$
13. $\sqrt[5]{7.60}$
14. $\sqrt[3]{95.4}$
15. $\sqrt{641}$
16. $\sqrt[8]{308}$
17. $\dfrac{(4510)(0.612)}{738}$
18. $\dfrac{87.4}{(11.5)(0.931)}$
19. $\dfrac{\sqrt{0.0753}}{86.0}$
20. $(\sqrt{5.27})(\sqrt[3]{42.1})$
21. $(47.3)(22.8)^{250}$
22. $\dfrac{895}{73.4^{86}}$
23. $(\sqrt[10]{7.32})(2470)^{30}$
24. $\dfrac{126{,}000^{20}}{2.63^{2.5}}$

25. What is the area of a rectangular field 325 m by 246 m?

26. In testing a new engine in order to determine its fuel economy, a car traveled 426 mi on 11.4 gal of gasoline. What is the miles-per-gallon rating of the car's engine?

27. Plutonium is radioactive and disintegrates such that of 1000 mg originally present, $1000(0.5)^{0.0000410t}$ mg will remain after t years. Calculate the amount present after 10,000 years.

28. The molecular mass M of a gas may be calculated from the formula $PV = mRT/M$, where P is the pressure, V is the volume, m is the mass, R is a constant for all gases, and T is the thermodynamic temperature. Determine M (in kilograms) if you are given that $P = 1.08 \times 10^5$ Pa, $V = 2.48 \times 10^{-4}$ m^3, $R = 8.31$ J/mol·K, $T = 373$ K, and $m = 1.26 \times 10^{-3}$ kg.

29. The velocity of sound in air is given by $v = \sqrt{1.41\,p/d}$, where p is the pressure and d is the density. Given that $p = 1.01 \times 10^5$ Pa and $d = 1.29$ kg/m^3, find v (in meters per second).

30. In undergoing an adiabatic (no *heat* gained or lost) expansion, the relation between the initial and final temperatures and volumes is given by $T_f = T_i(V_i/V_f)^{0.4}$, where the temperatures are expressed in kelvins. Given that $V_i = 1.50$ cm^3, $V_f = 0.129$ cm^3 and $T_i = 373$ K, find T_f.

31. Given the density of iron as 491 lb/ft^3, find the radius of a spherical iron ball which weighs 25.6 lb.

32. When a light ray is incident on glass, the percentage of light reflected is given by

$$I_r = 100\left(1 - \frac{4n_a n_g}{(n_a + n_g)^2}\right)$$

where n_a and n_g are the indices of refraction of air and glass, respectively. What percentage of light is reflected if $n_g = 1.53$ and $n_a = 1.00$?

12–6 Logarithms to Bases Other Than 10; Natural Logarithms

As we mentioned earlier, another number which is important as a base of logarithms is the number e. *Logarithms to the base e are called* **natural logarithms.** Since e is an irrational number equal to approximately 2.718, it may appear to be a very unnatural choice as a base of logarithms. However, in the development of the calculus, the reason for its choice and the fact that it is a very natural number for a base of logarithms are shown.

Just as the notation $\log x$ refers to logarithms to the base 10, the notation $\ln x$ is used to denote logarithms to the base e. Due to the extensive use of natural logarithms, the notation $\ln x$ is more convenient than $\log_e x$, although they mean the same thing.

Since more than one base is important, there are times when it is useful to be able to change a logarithm in one base to another base. If $u = \log_b x$, then $b^u = x$. Taking logarithms of both sides of this last expression to the base a, we have

$$\log_a b^u = \log_a x$$
$$u \log_a b = \log_a x$$

Solving this last equation for u, we have

$$u = \frac{\log_a x}{\log_a b}$$

However, $u = \log_b x$, which means that

$$\log_b x = \frac{\log_a x}{\log_a b} \qquad\qquad (12\text{–}12)$$

Equation (12–12) allows us to change a logarithm in one base to a logarithm in another base. The following examples illustrate the method of performing this operation.

Example A

Change $\log 20 = 1.3010$ to a logarithm with base e; that is, find $\ln 20$.
 In Eq. (12–12), if we let $b = e$ and $a = 10$, we have

$$\log_e x = \frac{\log_{10} x}{\log_{10} e}$$

or

$$\ln x = \frac{\log x}{\log e}$$

In this example, $x = 20$. Therefore,

$$\ln 20 = \frac{\log 20}{\log e} = \frac{\log 20}{\log 2.718} = \frac{1.3010}{0.4343} = 2.996$$

Example B

Find $\log_5 560$.
 In Eq. (12–12), if we let $b = 5$ and $a = 10$, we have

$$\log_5 x = \frac{\log x}{\log 5}$$

In this example, $x = 560$. Therefore, we have

$$\log_5 560 = \frac{\log 560}{\log 5} = \frac{2.7482}{0.6990} = 3.932$$

Therefore, we have found that $\log_5 560 = 3.932$. From the definition of a logarithm this means that

$$5^{3.932} = 560$$

Since natural logarithms are used extensively, it is often convenient to have Eq. (12–12) written specifically for use with logarithms to the base 10 and natural logarithms. Since log e = 0.4343 and ln 10 = 2.3026, we can write

$$\text{ln } x = 2.3026 \text{ log } x \qquad\qquad (12\text{–}13)$$

and

$$\text{log } x = 0.4343 \text{ ln } x \qquad\qquad (12\text{–}14)$$

Example C

Using Eq. (12–13), find ln 0.811.
  From Eq. (12–13), we have

$$\text{ln } 0.811 = 2.3026 \text{ log } 0.811$$

In applications of natural logarithms, it is generally preferable to write the logarithms in their explicit form, including when they are negative. Therefore, we express log 0.811 as

$$\text{log } 0.811 = 0.9090 - 1 = -0.0910$$

Thus,

$$\text{ln } 0.811 = 2.3026(-0.0910)$$
$$= -0.2095$$

Values of natural logarithms can be found directly on a scientific calculator. The $\boxed{\text{LN}}$ key is used for this purpose. In order to find the antilogarithm of a natural logarithm we use the $\boxed{e^x}$ key or the key sequence $\boxed{\text{INV}}$ $\boxed{\text{LN}}$, depending on the calculator.

Example D

Find ln 236.5 by use of a calculator.
  The calculator key sequence of 236.5 $\boxed{\text{LN}}$ gives a display of 5.4659482, which means

$$\text{ln } 236.5 = 5.466$$

Example E

Find N if ln $N = -0.8729$ by use of a calculator.
  The calculator key sequence .8729 $\boxed{+/-}$ $\boxed{e^x}$ or the sequence .8729 $\boxed{+/-}$ $\boxed{\text{INV}}$ $\boxed{\text{LN}}$ gives a display of 0.4177384. This means that the natural antilogarithm of -0.8729 is 0.4177. Thus,

$$N = 0.4177$$

Natural logarithms can be found by use of tables such as Table 4 in Appendix F. However, this table cannot be used in the same way as we did the table of common logarithms. This can be seen by taking natural logarithms of a number written in scientific notation. For $N = P \times 10^k$, we have

$$\ln N = \ln P + \ln 10^k$$
$$= \ln P + k \ln 10$$

or

$$\ln N = k \ln 10 + \ln P \tag{12–15}$$

Thus, a table of natural logarithms can be used directly only for the numbers which are tabulated. Its use can be extended by use of Eq. (12–15). The following example illustrates the use of Table 4 and Eq. (12–15).

Example F

Using Table 4 and Eq. (12–15), determine the value of ln 820.

Since 820 does not appear in the column labeled n in Table 4, we cannot find its value directly from the table. However, we can write $820 = 8.2 \times 10^2$ and use Eq. (12–15). Here we have

$$\ln 820 = 2 \ln 10 + \ln 8.2$$

We determine ln 8.2 from the table, and note that the value of $2 \ln 10$ is also given. Thus,

$$\ln 820 = 4.6052 + 2.1041$$
$$= 6.7093$$

It is possible to use natural logarithms for calculations in the same way that we use common logarithms. Other applications of natural logarithms are found in many fields of technology. One such application is shown in the following example, and others are found in the exercises.

Example G

Under certain conditions, the electric current i in a circuit containing a resistance and an inductance (see Section 11–7) is given by

$$\ln \frac{i}{I} = -\frac{Rt}{L}$$

where I is the current at $t = 0$, R is the resistance, t is the time, and L is the inductance. Calculate how long (in seconds) it takes i to reach 0.430 A, if $I = 0.750$ A, $R = 7.50$ Ω, and $L = 1.25$ H.

(continued on next page)

Solving for t, we have

$$t = -\frac{L \ln(i/I)}{R} = -\frac{L(\ln i - \ln I)}{R}$$

Thus, for the given values, we have

$$t = -\frac{1.25(\ln 0.430 - \ln 0.750)}{7.50}$$

$$= -\frac{1.25(-0.8440 + 0.2877)}{7.50} = 0.0927 \text{ s}$$

Therefore, the current changes from 0.750 A to 0.430 A in 0.0927 s.

Exercises 12–6

In Exercises 1 through 8, use logarithms to the base 10 to find the natural logarithms of the given numbers.

1. 26.0   2. 631   3. 1.562   4. 45.73

5. 0.5017   6. 0.05294   7. 0.0073267   8. 0.00044348

In Exercises 9 through 16, use logarithms to the base 10 to find the indicated logarithms.

9. $\log_7 42$ 10. $\log_2 86$ 11. $\log_5 245$ 12. $\log_3 706$

13. $\log_{12} 122$ 14. $\log_{20} 86$ 15. $\log_{40} 750$ 16. $\log_{100} 3720$

In Exercises 17 through 24, use Eq. (12–13) to find the natural logarithms of the indicated numbers.

17. 51.4   18. 293   19. 1.394   20. 65.62

21. 0.9917   22. 0.002086   23. 0.012937   24. 0.000060808

In Exercises 25 through 28, use a calculator to find the natural logarithms of the given numbers.

25. 45.17   26. 8765   27. 0.68528   28. 0.0014298

In Exercises 29 through 32, use a calculator to find the natural antilogarithms of the given logarithms.

29. 2.19   30. 0.632   31. -0.7429 32. -2.94218

In Exercises 33 through 36, perform the indicated calculations by use of natural logarithms from Table 4.

33. 2.50×4700 34. $\dfrac{380}{0.900}$ 35. $\sqrt{75.0}$ 36. $(2.9)^{10}$

In Exercises 37 through 44, solve the given problems.

37. Solve for y in terms of x: $\ln y - \ln x = 1.0986$.

38. Solve for y in terms of x: $\ln y + 2 \ln x = 1 + \ln 5$.

39. If interest is compounded continuously (daily compounded interest closely approximates this), a bank account can double in t years according to the equation $i = \dfrac{\ln 2}{t}$, where i is the interest rate. What interest rate is required for an account to double in 8.5 years?

40. One approximate formula for world population growth is $T = 50.0 \ln 2$, where T is the number of years for the population to double. According to this formula, how long does it take for the population to double?

41. For the electric circuit of Example G, find how long it takes the current to reach 0.1 of the initial value of 0.750 A.

42. Under specific conditions, an equation relating the pressure P and volume V of a gas is $\ln P = C - \gamma \ln V$, where C and γ (the Greek letter gamma) are constants. Find P (in kilopascals) if $C = 3.000$, $\gamma = 1.50$, and $V = 0.220$ m^3.

43. If 100 mg of radium radioactively decays, an equation relating the amount Q which remains, and the time t is

$$\ln Q - \ln 100 = kt$$

where k is a constant. If $Q = 90.0$ mg, and $k = -0.000410$ per year, find t.

44. For a certain electric circuit, the voltage v is given by $v = e^{-0.1t}$. What is $\ln v$ after 2.00 s?

12-7 Exponential and Logarithmic Equations

In solving equations in which there is an unknown exponent, it is often advantageous to take logarithms of both sides of the equation and then proceed. In solving logarithmic equations, one should keep in mind the basic properties of logarithms, since these can often help transform the equation into a solvable form. There is, however, no general algebraic method for solving such equations, and here we shall solve only some special cases.

Example A

Solve the equation $2^x = 8$.

By writing this in logarithmic form, we have

$$x = \log_2 8 = 3$$

We could also solve this equation by taking logarithms of both sides. This would yield

$$\log 2^x = \log 8$$
$$x \log 2 = \log 8$$
$$x = \frac{\log 8}{\log 2} = \frac{0.9031}{0.3010} = 3.00$$

This last method is more generally applicable, since the first method is good only if we can directly evaluate the logarithm which results.

Example B

Solve the equation $3^{x-2} = 5$.

Taking logarithms of both sides, we have

$$\log 3^{x-2} = \log 5 \quad \text{or} \quad (x-2)\log 3 = \log 5$$

Solving this last equation for x, we have

$$x = 2 + \frac{\log 5}{\log 3} = 2 + \frac{0.6990}{0.4771} = 3.465$$

Thus, the solution to this equation is $x = 3.465$.

Example C

Solve the equation $2(4^{x-1}) = 17^x$.

By taking logarithms of both sides, we have the following:

$$\log 2 + (x-1) \log 4 = x \log 17$$

$$x \log 4 - x \log 17 = \log 4 - \log 2$$

$$x(\log 4 - \log 17) = \log 4 - \log 2$$

$$x = \frac{\log 4 - \log 2}{\log 4 - \log 17} = \frac{\log(4/2)}{\log 4 - \log 17}$$

$$= \frac{\log 2}{\log 4 - \log 17} = \frac{0.3010}{0.6021 - 1.2304}$$

$$= -0.479$$

The equations in these examples could also have been solved by using natural logarithms. The following example illustrates an applied problem which is most easily solved by using natural logarithms.

Example D

Under the condition of constant temperature, the atmospheric pressure p, in pascals, at an altitude h, in meters, is given by $p = p_0 e^{kh}$, where p_0 is the pressure where $h = 0$ (usually taken as sea level). Given that $p_0 = 101.3$ kPa (atmospheric pressure at sea level) and $p = 68.9$ kPa for $h = 3050$ m, find the value of k.

Since the equation is defined in terms of e, we can solve it most easily by taking natural logarithms of both sides. By doing this we have the following solution.

$$\ln p = \ln(p_0 e^{kh}) = \ln p_0 + \ln e^{kh}$$

$$= \ln p_0 + kh \ln e = \ln p_0 + kh$$

$$k = \frac{\ln p - \ln p_0}{h}$$

Substituting the given values, we have

$$k = \frac{\ln(68.9 \times 10^3) - \ln(101.3 \times 10^3)}{3050}$$

$$= -0.000126/\text{m}$$

Some of the important measurements in scientific and technical work are defined in terms of logarithms. We shall consider here some of the important applications which are basic logarithmic formulas. In using them we shall solve some equations in which logarithms are involved. The following example illustrates one such area of application, and others are found in the exercises.

Example E

It has been found that the human ear responds to sound on a scale which is approximately proportional to the logarithm of the intensity of the sound. Thus, the loudness of sound, measured in decibels, is defined by the equation $b = 10 \log (I/I_0)$, where I is the intensity of the sound and I_0 is the minimum intensity detectable.

A busy city street has a loudness of 70 dB, and riveting has a loudness of 100 dB. How many times greater is the intensity of the sound of riveting I_r than the sound of the city street I_c?

First, we substitute the decibel readings into the above definition. This gives us

$$70 = 10 \log\left(\frac{I_c}{I_0}\right) \quad \text{and} \quad 100 = 10 \log\left(\frac{I_r}{I_0}\right)$$

To solve these equations for I_c and I_r, we divide each side by 10 and then use the exponential form. Thus, we have

$$7.0 = \log\left(\frac{I_c}{I_0}\right) \quad \text{and} \quad 10 = \log\left(\frac{I_r}{I_0}\right)$$

$$\frac{I_c}{I_0} = 10^{7.0} \qquad \frac{I_r}{I_0} = 10^{10}$$

$$I_c = I_0(10^{7.0}) \qquad I_r = I_0(10^{10})$$

Since we want the number of times I_r is greater than I_c, we divide I_r by I_c. This gives us

$$\frac{I_r}{I_c} = \frac{I_0(10^{10})}{I_0(10^{7.0})} = \frac{10^{10}}{10^{7.0}} = 10^{3.0}$$

or

$$I_r = 10^{3.0}I_c = 1000I_c$$

Thus, the sound of riveting is 1000 times as intense as the sound of the city street. This demonstrates that sound intensity levels are considerably greater than loudness levels. (See Exercise 42.)

The following examples illustrate the solution of other logarithmic equations.

Example F

Solve the equation $\log_2 7 - \log_2 14 = x$.
    Using the basic properties of logarithms, we arrive at the following result:

$$\log_2\left(\tfrac{7}{14}\right) = x$$

$$\log_2\left(\tfrac{1}{2}\right) = x \quad \text{or} \quad \tfrac{1}{2} = 2^x$$

Thus, $x = -1$.

Example G

Solve the equation $2 \ln 2 + \ln x = \ln 3$.
    Using the properties of logarithms we have the following solution.

$$2 \ln 2 + \ln x = \ln 3$$

$$\ln 2^2 + \ln x - \ln 3 = 0$$

$$\ln \frac{4x}{3} = 0$$

$$\frac{4x}{3} = e^0 = 1$$

$$4x = 3$$

$$x = \frac{3}{4}$$

Example H

Solve the equation $2 \log x - 1 = \log(1 - 2x)$.

$$\log x^2 - \log(1 - 2x) = 1$$

$$\log \frac{x^2}{1 - 2x} = 1$$

$$\frac{x^2}{1 - 2x} = 10^1$$

$$x^2 = 10 - 20x$$

$$x^2 + 20x - 10 = 0$$

$$x = \frac{-20 \pm \sqrt{400 + 40}}{2} = -10 \pm \sqrt{110}$$

Since logarithms of negative numbers are not defined, we have the result that $x = \sqrt{110} - 10 = 0.488$.

Exercises 12-7

In Exercises 1 through 36, solve the given equations.

1. $2^x = 16$

2. $3^x = \frac{1}{81}$

3. $5^x = 4$

4. $6^x = 15$

5. $3^{-x} = 0.525$

6. $15^{-x} = 1.326$

7. $e^{2x} = 3.625$

8. $e^{-x} = 17.54$

9. $6^{x+1} = 10$

10. $5^{x-1} = 2$

11. $4(3^x) = 5$

12. $14^x = 40$

13. $0.8^x = 0.4$

14. $0.6^x = 100$

15. $(15.6)^{x+2} = 23^x$

16. $5^{x+2} = e^{2x}$

17. $2 \log_2 x = 4$

18. $3 \log_8 x = 1$

19. $3 \log_8 x = -2$

20. $5 \log_{32} x = -3$

21. $2 \ln x = 1$

22. $3 \ln 2x = 2$

23. $\log_2 x + \log_2 7 = \log_2 21$

24. $2 \log_2 3 - \log_2 x = \log_2 45$

25. $2 \log(3 - x) = 1$

26. $3 \log(2x - 1) = 1$

27. $\ln x + \ln 3 = 1$

28. $\ln 5 - \ln x = -1$

29. $3 \ln 2 + \ln(x - 1) = \ln 24$

30. $\ln(2x - 1) - 2 \ln 4 = 3 \ln 2$

31. $\frac{1}{2} \log(x + 2) + \log 5 = 1$

32. $\frac{1}{2} \log(x - 1) - \log x = 0$

33. $\log_5(x - 3) + \log_5 x = \log_5 4$

34. $\log_7 x + \log_7(2x - 5) = \log_7 3$

35. $\log(2x - 1) + \log(x + 4) = 1$

36. $\log_2 x + \log_2(x + 2) = 3$

In Exercises 37 through 48, determine the required quantities.

37. For a certain electric circuit, the current i is given by $i = 1.50e^{-200t}$. For what value of t (in seconds) is $i = 1.00$ A?

38. If A_0 dollars are invested at 8% compounded continuously for t years, the value A of the investment is given by $A = A_0 e^{0.08t}$. Determine how long it takes for the investment to double in value.

39. The temperature T, in degrees Celsius, of a certain cooling object is given by $T = T_0 + 70(0.40)^{0.20t}$, where T_0 is the temperature of the surroundings of the object and t is the time in minutes. Determine how long it takes the object to cool to 50°C if $T_0 = 30$°C.

40. The amount q of a certain radioactive substance remaining after t years is given by $q = 100(0.900)^t$. After how many years are there 50.0 mg of the substance remaining?

41. Referring to Example E, how many times I_0 is the intensity of sound of a jet plane which has a loudness of 110 dB?

42. Referring to Example E, show that if the difference in loudness of two sounds is d decibels, the louder sound is $10^{d/10}$ more intense than the quieter sound.

43. Measured on the Richter scale, the magnitude of an earthquake of intensity I is defined as $R = \log(I / I_0)$, where I_0 is a minimum level for comparison. How many times I_0 was the 1906 San Francisco earthquake whose magnitude was 8.25 on the Richter scale?

44. How many more times intense was the 1964 Alaska earthquake, $R = 7.5$, than the 1971 Los Angeles earthquake, $R = 6.7$? (See Exercise 43.)

45. Pure water is running into a certain brine solution, and the same amount of solution is running out. The number n of kilograms of salt in the solution after t minutes is found by solving the equation $\ln n = -0.04t + \ln 20$. Solve for n as a function of t.

46. In an electric circuit containing a resistor and a capacitor with an initial charge q_0, the charge q on the capacitor at any time t after closing the switch can be found by solving the equation $\ln q = -\dfrac{t}{RC} + \ln q_0$. Here R is the resistance and C is the capacitance. Solve for q as a function of t.

47. In chemistry, the pH value of a solution is a measure of its acidity. The pH value is defined by the relation $\text{pH} = -\log(\text{H}^+)$, where H^+ is the hydrogen ion concentration. If the pH of a certain wine is 3.4065, find the hydrogen ion concentration. (If the pH value is less than 7, the solution is acid. If the pH value is above 7, the solution is basic.)

48. Referring to Exercise 47, find the hydrogen ion concentration for ammonia for which the pH is 10.8.

To solve more complicated problems, we may use graphical methods. For example, if we wish to solve the equation $2^x + 3^x = 50$, we can set up the function $y = 2^x + 3^x - 50$ and then determine its zeros graphically. Note that the given equation can be written as $2^x + 3^x - 50 = 0$, and therefore the zeros of the function which has been set up will give the desired solution. In Exercises 49 through 52, solve the given equations in this way.

49. $2^x + 3^x = 50$ 50. $3^{x+1} - 4^x = 1$

51. $3^x - 2x = 40$ 52. $4^x + x^2 = 25$

12–8 Graphs on Logarithmic and Semilogarithmic Paper

If, when we are graphing, the range of values of one variable is much greater than the corresponding range of values of the other variable, it is often convenient to use what is known as **semilogarithmic** paper. On this paper the y-axis (usually) is marked off in distances proportional to the logarithms of numbers. This means that the distances between numbers on this axis are not even, but this system does allow for a much greater range of values, and with much greater accuracy for many of the numbers. There is another advantage to this paper: Many equations which would exhibit more complex curves on ordinary graph paper will work out as straight lines on semilogarithmic paper. In many instances this makes the analysis of the curve easier.

If we wish to indicate a large range of values for each of the variables, we use what is known as **logarithmic** paper, or as **log-log** paper. Both axes are marked off with logarithmic scales. Again, the more complicated equations give simple curves or straight lines on this paper.

The following examples will illustrate the use of semilogarithmic and logarithmic paper.

Example A

Construct the graph of $y = 4(3)^x$ on semilogarithmic paper.

First we construct a table of values.

x	-1	0	1	2	3	4	5
y	1.3	4	12	36	108	324	972

From the table we see that the range of y-values is large. If we plotted this curve on a regular coordinate system we would have large units for each interval along the y-axis. This would make the values of 1.3, 4, 12, and 36 appear at practically the same level. However, if we use semilogarithmic graph paper, we can label each axis such that all y-values are accurately plotted as well as the x-values.

The logarithmic scale is shown in cycles, and we must label the base line of the first cycle as 1 times a power of ten (0.01, 0.1, 1, 10, 100, and so on) with the following cycle labeled with the next power of ten. The lines between are labeled with 2, 3, 4, and so on, times the proper power of ten. See the vertical scale in Fig. 12–7. We now plot the points in the table on the graph. The resulting graph is a straight line, as we see in Fig. 12–7. Taking logarithms of both sides of the equation, we have

$$\log y = \log 4 + x \log 3$$

However, since $\log y$ was plotted automatically (because we used semilogarithmic paper), the graph really represents

$$u = \log 4 + x \log 3$$

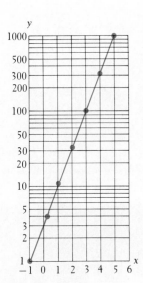

Figure 12–7

where $u = \log y$; $\log 3$ and $\log 4$ are constants, and therefore this equation is of the form $u = mx + b$, which is a straight line (see Section 4–2). If we had sketched this graph on regular coordinate paper, the scale would be so reduced that the values of 1.3, 4, 12, and 36 would appear at practically the same level.

Example B

Construct the graph of $x^4y^2 = 1$ on logarithmic paper.

First we solve for y and then construct a table of values. Considering positive values of x and y, we have

$$y = \sqrt{\frac{1}{x^4}} = \frac{1}{x^2}$$

x	0.5	1	2	8	20
y	4	1	0.25	0.0156	0.0025

We now plot these values on log-log paper on which both scales are logarithmic, as shown in Fig. 12–8. We again note that we have a straight line. Taking logarithms of both sides of the equation, we have

$$4 \log x + 2 \log y = 0$$

(continued on next page)

If we let $u = \log y$ and $v = \log x$, we then have

$$4v + 2u = 0 \quad \text{or} \quad u = -2v$$

which is the equation of a straight line as shown in Fig. 12–8. It should be pointed out that not all graphs on logarithmic paper are straight lines.

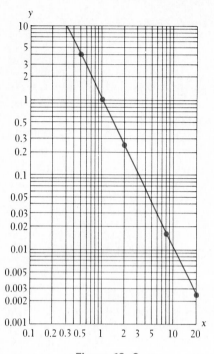

Figure 12–8

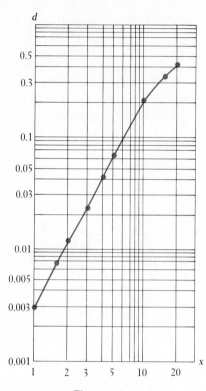

Figure 12–9

Example C

The deflection (in feet) of a certain cantilever beam as a function of the distance x from one end is

$$d = 0.0001(30x^2 - x^3)$$

If the beam is 20.0 ft long, plot a graph of d vs. x on log-log paper. Constructing a table of values we have

x (ft)	1.00	1.50	2.00	3.00	4.00	5.00	10.0	15.0	20.0
d (ft)	0.00290	0.00641	0.0112	0.0243	0.0416	0.0625	0.200	0.338	0.400

The graph is shown in Fig. 12–9.

Logarithmic and semilogarithmic paper are often useful for plotting data derived from experimentation. Often the data cover too large a range of values to be plotted on ordinary graph paper. The following example illustrates how we use semilogarithmic paper to plot data.

Example D

The vapor pressure of water depends on the temperature. The following table gives the vapor pressure (in kilopascals) for corresponding values of temperature (in degrees Celsius).

Pressure	1.19	2.33	7.34	19.9	47.3	101	199	361	617
Temp.	10	20	40	60	80	100	120	140	160

These data are then plotted on semilogarithmic paper, as shown in Fig. 12–10. Intermediate values of temperature and pressure can then be read directly from the graph.

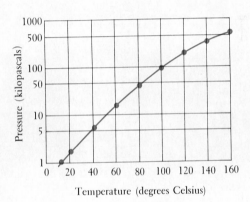

Figure 12–10

Exercises 12–8

In Exercises 1 through 12, plot the graphs of the given functions on semilogarithmic paper.

1. $y = 2^x$
2. $y = 5^x$
3. $y = 2(4^x)$
4. $y = 5(10^x)$
5. $y = 3^{-x}$
6. $y = 2^{-x}$
7. $y = x^3$
8. $y = x^5$
9. $y = 3x^2$
10. $y = 2x^4$
11. $y = 2x^3 + 4x$
12. $y = 4x^3 + 2x^2$

In Exercises 13 through 24, plot the graphs of the given functions on log-log paper.

13. $y = 0.01x^4$
14. $y = 0.02x^3$
15. $y = \sqrt{x}$
16. $y = x^{2/3}$
17. $y = x^2 + 2x$
18. $y = x + \sqrt{x}$
19. $xy = 4$
20. $xy^2 = 10$
21. $y^2x = 1$
22. $x^2y^3 = 1$
23. $x^2y^2 = 25$
24. $x^3y = 8$

In Exercises 25 through 32, plot the indicated graphs.

25. The atmospheric pressure p at a given height h is given by $p = p_0 e^{-kh}$, where p_0 and k are constants. On semilogarithmic paper plot p (in atmospheres) vs. h (in feet) for $0 \le h \le 10^5$. Use $e = 2.7$, $p_0 = 1$ atm, and $k = 10^{-5}$ per foot.

26. Strontium 90 decays according to the equation $N = N_0 e^{-0.028t}$, where N is the amount present after t years, and N_0 is the original amount. Plot N vs. t on semilogarithmic paper if $N_0 = 1000$ g.

27. A company estimates that the value of a piece of machinery is $V = 75{,}000(2^{-0.15t})$, where V is the value (in dollars) t years after the purchase. Plot V vs. t on semilogarithmic paper.

28. At constant temperature, the relation between the volume V and pressure P of a gas is $PV = c$, where c is a constant. On logarithmic paper, plot the graph of P (in atmospheres) vs. V (in cubic feet) for $c = 4$ atm-ft^3. Use values of $0.1 \le P \le 10$.

29. One end of a very hot steel bar is sprayed with a stream of cool water. The rate of cooling (in degrees Fahrenheit per second) as a function of the distance (in inches) from the end of the bar is then measured. The following results are obtained.

Cooling rate	600	190	100	72	46	29	17	10	6
Distance	0.063	0.13	0.19	0.25	0.38	0.50	0.75	1.0	1.5

Plot the data on logarithmic paper. Such experiments are made to determine the hardenability of steel.

30. The magnetic intensity H (in amperes per meter) and flux density B (in teslas) of annealed iron are given in the following table.

H	10	50	100	150	200	500	1000	10,000	100,000
B	0.0042	0.043	0.67	1.01	1.18	1.44	1.58	1.72	2.26

Plot H vs. B on logarithmic paper.

In Exercises 31 and 32, plot the indicated semilogarithmic graphs for the following application.

In a particular electric circuit, called a low-pass filter, the input voltage V_i is across a resistor and a capacitor, and the output voltage V_0 is across the capacitor (see Fig. 12–11). The voltage gain G, in decibels, in such a circuit is given by

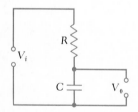

Figure 12–11

$$G = 20 \log \frac{1}{\sqrt{1 + (\omega T)^2}} \quad \text{where } \tan \phi = -\omega T$$

Here ϕ is the phase angle of V_0 / V_i. For values of ωT of 0.01, 0.1, 0.3, 1.0, 3.0, 10.0, 30.0, and 100, plot the indicated graphs. These graphs are called a *Bode diagram* for the circuit.

31. Calculate values of G for the given values of ωT and plot a semilogarithmic graph of G vs. ωT.

32. Calculate values of ϕ (as negative angles) for the given values of ωT, and plot a semilogarithmic graph of ϕ vs. ωT.

12–9 Exercises for Chapter 12

In Exercises 1 through 12, determine the value of x.

1. $\log_{10} x = 4$ 2. $\log_9 x = 3$ 3. $\log_5 x = -1$

4. $\log_4 x = -\frac{1}{2}$ 5. $\log_2 64 = x$ 6. $\log_{12} 144 = x$

7. $\log_8 32 = x$ 8. $\log_9 27 = x$ 9. $\log_x 36 = 2$

10. $\log_x 243 = 5$ 11. $\log_x 10 = \frac{1}{2}$ 12. $\log_x 8 = \frac{3}{4}$

In Exercises 13 through 24, express each as a sum, difference, or multiple of logarithms. Wherever possible, evaluate logarithms of the result.

13. $\log_3 2x$ 14. $\log_5 \left(\dfrac{7}{a} \right)$ 15. $\log_3 (t^2)$

16. $\log_6 \sqrt{5}$ 17. $\log_2 28$ 18. $\log_7 98$

19. $\log_3 \left(\dfrac{9}{x} \right)$ 20. $\log_6 \left(\dfrac{5}{36} \right)$ 21. $\log_4 \sqrt{48}$

22. $\log_6 \sqrt{72y}$ 23. $\log_{10}(1000x^4)$ 24. $\log_3 (9^2 \cdot 6^3)$

In Exercises 25 through 32, solve for y in terms of x.

25. $\log_6 y = \log_6 4 - \log_6 x$ 26. $\log_3 y = \frac{1}{2} \log_3 7 + \frac{1}{2} \log_3 x$

27. $\log_2 y + \log_2 x = 3$ 28. $6 \log_4 y = 8 \log_4 4 - 3 \log_4 x$

29. $\log_5 x + \log_5 y = \log_5 3 + 1$ 30. $\log_7 y = 2 \log_7 5 + \log_7 x + 2$

31. $3(\log_8 y - \log_8 x) = 1$ 32. $2(\log_9 y + 2 \log_9 x) = 1$

In Exercises 33 through 40, graph the given functions.

33. $y = 0.5(5^x)$ 34. $y = 3(2^x)$

35. $y = 0.5 \log_4 x$ 36. $y = 10 \log_{16} x$

37. $y = \log_{3.15} x$ 38. $y = 0.1 \log_{4.65} x$

39. $y = 1 - e^{-x}$ 40. $y = 2(1 - e^{-0.2x})$

In Exercises 41 through 44, use logarithms to perform the indicated calculations.

41. $(13.6)(0.693)$ 42. $(0.00624)^{0.2}$

43. $\dfrac{\sqrt{8640}}{19.5}$ 44. $45.1(60.7)^{64}$

In Exercises 45 through 48, by using logarithms to the base 10, find the natural logarithms of the given numbers.

45. 8.86 46. 33.0 47. 2.07 48. 0.542

In Exercises 49 through 56, solve the given equations.

49. $e^{2x} = 5$ 50. $5^x = 10$

51. $3^{x+2} = 5^x$ 52. $6^{x+2} = 12^{x-1}$

53. $\log_4 x + \log_4 6 = \log_4 12$ 54. $2 \log_3 2 - \log_3 (x + 1) = \log_3 5$

55. $\log_8 (x + 2) + \log_8 2 = 2$ 56. $\log(x + 2) + \log x = 0.4771$

In Exercises 57 and 58, plot the graphs of the given functions on semilogarithmic paper. In Exercises 59 and 60, plot the graphs of the given functions on log-log paper.

57. $y = 6^x$ 58. $y = 5x^3$ 59. $y = \sqrt[3]{x}$ 60. $xy^4 = 16$

In Exercises 61 through 76, solve the given problems.

61. The vapor pressure P over a liquid may be related to temperature by the formula $\log P = a/T + b$, where a and b are constants. Solve for P.

62. The Beer-Lambert law of light absorption may be expressed as

$$\log\left(\frac{I}{I_0}\right) = -\alpha x$$

where I/I_0 is that fraction of the incident light beam which is transmitted, α is a constant, and x is the distance the light travels through the medium. Solve for I.

63. In a certain electric circuit, the current i is given by $i = i_0 e^{-5t}$, where i_0 is the current for $t = 0$ and t is the time. Solve for t.

64. The number of pounds of salt in a tank filled with brine is given by $x = 50(2 + e^{-0.02t})$, where t is the time. Brine of differing concentrations flows into the tank and from the tank. Solve for t.

65. Under certain circumstances the efficiency of an internal combustion engine is given by

$$\text{eff (in percent)} = 100\left(1 - \frac{1}{(V_1/V_2)^{0.4}}\right)$$

where V_1 and V_2 are, respectively, the maximum and minimum volumes of air in a cylinder. The ratio V_1/V_2 is called the *compression ratio*. Use logarithms to compute the efficiency of an engine with a compression ratio of 6.55.

66. If P dollars are invested at an interest rate r which is compounded n times a year, the value A of the investment t years later is given by the formula $A = P\left(1 + \frac{r}{n}\right)^{nt}$. What is the value after 5 years of \$5630 invested at 7.5% compounded quarterly? Use logarithms for the calculation.

67. In thermodynamics, when studying the pressure p and volume V of a gas under certain conditions, the equation $\ln p + \gamma \ln V = \ln k$ is used. Here, γ and k are constants. Solve for p as a function of V.

68. Taking into account the weight loss of fuel, the maximum velocity v_m of a rocket is given by the equation

$$v_m = u(\ln m_0 - \ln m_s) - gt_f$$

where m_0 is the initial mass of the rocket and fuel, m_s is the mass of the rocket shell, t_f is the time during which fuel is expended, u is the velocity of the expelled fuel, and g is the acceleration due to gravity. Solve for m_0.

69. Under certain conditions, the potential (in volts) due to a magnet is given by $V = -k \ln\left(1 + \frac{l}{b}\right)$, with l the length of the magnet and b the distance from the point where the potential is measured. Find V, if $k = 2$ units, $l = 5.00$ cm, and $b = 2.00$ cm.

70. The Nernst equation,

$$E = E_0 - \frac{0.05910}{n} \log Q$$

is used for oxidation-reduction reactions. In the equation, E and E_0 are voltages, n is the number of electrons involved in the reaction, and Q is a measure of the activity of reaction. Given that $E_0 = 1.1000$ V, $E = 1.1300$ V, and $n = 2$, what is the value of Q?

71. For first-order chemical reactions, concentration of a reacting chemical species is related to time by the expression

$$\log\left(\frac{x_0}{x}\right) = kt$$

where x_0 is the initial concentration and x is the concentration after time t. Determine the quantity of sucrose remaining after 3 h, if the initial concentration is 9.00 g·mol/L and $k = 0.00158$ per minute.

72. The power gain of an electronic device such as an amplifier is defined as $n = 10 \log (P_0/P_i)$, where n is measured in decibels, P_0 is the power output, and P_i is the power input. If $P_0 = 10.0$ W and $P_i = 0.125$ W, calculate the power gain. (See Example E of Section 12–7.)

73. In 1980 it was estimated that the total annual world demand for copper was $C = 12\, e^{0.08t}$, where C is in millions of tons of copper and t is the number of years after 1980. When will copper demand be 20 million tons?

74. The bacteria population in a certain culture is given by $N = 1000(1.5)^t$. How long does it take for the population to reach 10,000 if t is measured in hours?

75. Plot a semilogarithmic graph of N vs. t for the bacteria culture of Exercise 74.

76. The luminous efficiency (measured in lumens per watt) of a tungsten lamp as a function of its input power (in watts) is given by the following table. On semilogarithmic paper, plot efficiency vs. power.

Efficiency	7.8	10.4	11.7	13.9	16.3	18.3	19.9	21.5
Power	10	25	40	60	100	200	500	1000

CHAPTER 13

Additional Types of Equations and Systems of Equations

In Chapter 2 we determined how to graph a function, as well as how to solve equations graphically. Since then we have discussed the graphs of linear and quadratic functions and we have dealt with methods for solving quadratic equations and systems of linear equations. Also, we have graphed the trigonometric, exponential, and logarithmic functions. In the first section of this chapter, we shall introduce one more general type of equation and then discuss graphical solutions of systems of equations involving quadratic equations, as well as other types of equations.

In the second section we shall discuss the algebraic solution of certain types of equations involving quadratic equations. In the final sections we shall consider equations which are not quadratic, but which can be solved by methods developed for quadratic equations, and equations which involve radicals.

13–1 Graphical Solution of Systems of Equations

As we noted, in this section we introduce another general type of equation: the general quadratic equation. Then we consider graphical solutions of systems of equations.

An equation of the form

$$ax^2 + bxy + cy^2 + dx + ey + f = 0 \tag{13–1}$$

is called a **general quadratic equation in *x* and *y*.** We shall be interested primarily in some special cases of this equation. *The graphs of the various possible forms of this equation result in curves known as* **conic sections.** These curves are the circle, parabola, ellipse, and hyperbola. We were previously introduced to the parabola when we discussed the graph of the quadratic function in Chapter 6. The following examples illustrate these curves. A more complete discussion is found in Chapter 20.

Example A

Graph the equation $y = 3x^2 - 6x$.

We graphed equations of this form in Section 2–3 and in Section 6–4. Using the method of Section 6–4 (and the constants a, b, and c as defined there) we see that $a = 3$, $b = -6$, and $c = 0$. This means that $-b/2a = -(-6)/2(3) = 1$, which tells us that the x-coordinate of the extreme point is 1. For $x = 1$, $y = -3$, which means that the extreme point is $(1, -3)$. Also, we know that it is a minimum point, since $a > 0$.

Since $c = 0$, the y-intercept is $(0, 0)$. Setting $y = 0$, we have $3x^2 - 6x = 0$, or $3x(x - 2) = 0$. This means that $y = 0$ for $x = 0$ and $x = 2$ and the points $(0, 0)$ and $(2, 0)$ are on the curve. (We have found the point $(0, 0)$ in two ways.) To define the curve a little better, we find the points $(-1, 9)$ and $(3, 9)$. These values are summarized in the following table, and the graph is shown in Fig. 13–1.

x	-1	0	1	2	3
y	9	0	-3	0	9

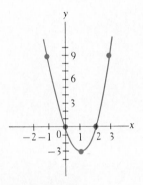

Figure 13–1

As we showed in Section 6–4, the curve is a parabola. Using the constants of the general quadratic equation in Eq. (13–1), a parabola always results if the equation is of the form $y = ax^2 + dx + f$.

Example B

Plot the graph of the equation $x^2 + y^2 = 25$.

We first solve this equation for y, and we obtain $y = \sqrt{25 - x^2}$, or $y = -\sqrt{25 - x^2}$, which we write as $y = \pm\sqrt{25 - x^2}$. We now assume values for x and find the corresponding values for y.

x	0	± 1	± 2	± 3	± 4	± 5
y	± 5	± 4.9	± 4.6	± 4	± 3	0

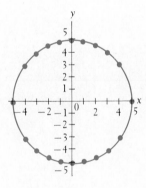

Figure 13–2

If we try values greater than 5, we have imaginary numbers. These cannot be plotted, for we assume that both x and y are real. (The complex plane is only for *numbers* of the form $a + bj$ and does not represent pairs of numbers representing two variables.) When we give the value $x = \pm 4$ when $y = \pm 3$, this is simply a short way of representing 4 points. These points are $(4, 3)$, $(4, -3)$, $(-4, 3)$, $(-4, -3)$. We note in Fig. 13–2 that *the resulting curve is a* **circle.** A circle always results from an equation of the form $x^2 + y^2 = r^2$, and r is the radius of the circle with its center at the origin.

Example C

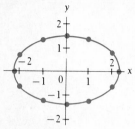

Figure 13–3

Plot the graph of the equation $2x^2 + 5y^2 = 10$.

We first solve for y, then we construct the table of values.

$$y = \pm\sqrt{\frac{10 - 2x^2}{5}}$$

x	0	± 1	± 2	$\pm\sqrt{5}(= \pm 2.2)$
y	± 1.4	± 1.3	± 0.6	0

Values of x greater than $\sqrt{5}$ result in imaginary values of y. *The curve* (Fig. 13–3) *is an* **ellipse.** An ellipse results from an equation that can be written in the form $ax^2 + cy^2 = k$. (Constants a, c, and k must all have the same sign.)

Example D

Plot the graph of the equation $xy = 4$.

Solving for y, we obtain $y = 4/x$. Now, constructing the table of values, we have the following points.

x	-8	-4	-2	-1	$-\frac{1}{2}$	$\frac{1}{2}$	1	2	4	8
y	$-\frac{1}{2}$	-1	-2	-4	-8	8	4	2	1	$\frac{1}{2}$

Plotting these points, we obtain the curve in Fig. 13–4. *This curve is called a* **hyperbola.** A hyperbola always results if the equation is of the form $xy = k$. We also obtain a hyperbola if the equation can be written in the form $ax^2 + cy^2 = k$. (Constants a and c must have *different* signs.)

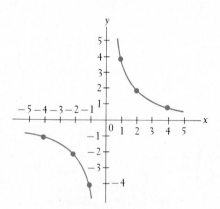

Figure 13–4

As in solving systems of linear equations, we obtain the desired solutions of any system by finding the values of x and y which satisfy both equations at the same time. To solve a system of equations graphically, we find all points which the graphs have in common. This means we need only to graph the equations and then locate the points of intersection. If the curves do not intersect, the system has no real solutions.

Example E

Graphically solve the system of equations

$$2x^2 - y^2 = 4$$
$$x - 3y = 6$$

We should recognize the second equation as that of a straight line. Now constructing the tables, we solve $2x^2 - y^2 = 4$ for y and get $y = \pm\sqrt{2x^2 - 4}$. Therefore, we obtain the following table.

x	± 1.4	± 2	± 4	± 6
y	0	± 2	± 5.3	± 8.2

The straight line has intercepts of $(0, -2)$ and $(6, 0)$ and a slope of $\frac{1}{3}$. The solutions, as indicated on the graph in Fig. 13–5, are approximately $x = 1.8$, $y = -1.4$ and $x = -2.4$, $y = -2.8$.

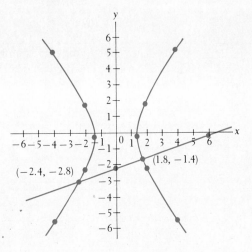

Figure 13–5

The following example illustrates the graphical solution of a system of equations in which one of the equations is not a general quadratic type.

Example F

Graphically solve the system of equations

$$9x^2 + 4y^2 = 36$$
$$y = 3^x$$

The first equation is of the form represented by an ellipse, as indicated in Example C. The second equation is an exponential function, as discussed in Chapter 12. Solving the first equation for y, we have $y = \pm\frac{1}{2}\sqrt{36 - 9x^2}$. Substituting values for x, we obtain the following table.

x	0	± 1	± 2
y	± 3	± 2.6	0

For the exponential function, we obtain the following table.

x	-3	-2	-1	0	1	2
y	$\frac{1}{27}$	$\frac{1}{9}$	$\frac{1}{3}$	1	3	9

We plot these curves as shown in Fig. 13–6. The points of intersection are approximately $x = -1.9$, $y = 0.1$ and $x = 0.9$, $y = 2.7$.

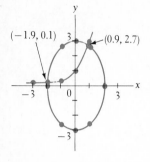

Figure 13–6

Example G

Graphically solve the system of equations

$$x^2 = 2y$$
$$3x - y = 5$$

We note that the two curves in this system are a parabola and a straight line. Solving the equation of the parabola for y, we obtain $y = \frac{1}{2}x^2$. We construct the following table.

x	0	± 1	± 2	± 3	± 4
y	0	$\frac{1}{2}$	2	$\frac{9}{2}$	8

For the straight line we have the following points.

x	0	$\frac{5}{3}$	3
y	-5	0	4

We plot these curves and see in Fig. 13–7 that they do not intersect, so we conclude that there are no real solutions to the system.

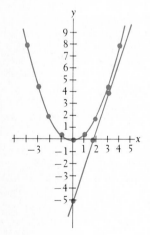

Figure 13–7

Exercises 13–1

In Exercises 1 through 24, solve the given systems of equations graphically.

1. $y = 2x$
$\quad x^2 + y^2 = 16$

2. $3x - y = 4$
$\quad y = 6 - 2x^2$

3. $x^2 + 2y^2 = 8$
$\quad x - 2y = 4$

4. $y = 3x - 6$
   $xy = 6$

5. $y = x^2 - 2$
   $4y = 12x - 17$

6. $x^2 + 4y^2 = 4$
   $2y = 12 - x$

7. $y = x^2$
   $xy = 4$

8. $y = -2x^2$
   $y = x^2 - 6$

9. $y = -x^2 + 4$
   $x^2 + y^2 = 9$

10. $y = 2x^2 - 1$
    $x^2 + 2y^2 = 16$

11. $x^2 - 4y^2 = 16$
    $x^2 + y^2 = 1$

12. $y = 2x^2 - 4x$
    $xy = -4$

13. $2x^2 + 3y^2 = 19$
    $x^2 + y^2 = 9$

14. $x^2 - y^2 = 4$
    $2x^2 + y^2 = 16$

15. $x^2 + y^2 = 1$
    $xy = \frac{1}{2}$

16. $x^2 + y^2 = 25$
    $x^2 - y^2 = 7$

17. $y = x^2$
    $y = \sin x$

18. $y = 4x - x^2$
    $y = 2 \cos x$

19. $y = e^{-x}$
    $x + y = 2$

20. $y = 2^x$
    $x^2 + y^2 = 4$

21. $x^2 - y^2 = 1$
    $y = \log_2 x$

22. $x^2 + 4y^2 = 16$
    $y = 2 \ln x$

23. $y = \ln(x - 1)$
    $y = \sin \frac{1}{2}x$

24. $y = \cos x$
    $y = \log_3 x$

In Exercises 25 through 28, solve the indicated systems of equations graphically. In Exercises 27 and 28, the necessary systems of equations must be properly set up.

25. A rectangular field has a perimeter of 1140 m and an area of 75,600 m². Show that the two equations necessary to find the dimensions l and w are $l + w = 570$, $lw = 75,600$. Graphically solve for l and w.

26. The diagonal of a rectangular metal plate is 10 in. and the area is 50 in.². Show that the two equations necessary to find the dimensions l and w are $l^2 + w^2 = 100$, $lw = 50$. Graphically solve for l and w.

27. The power developed in an electric resistor R is i^2R, where i is the current. If one current passes through a 2-Ω resistor, and the second current passes through a 3-Ω resistor, the total power produced is 12 W. The sum of the currents is 3 A. Find the currents.

28. A circular hole y cm in radius is cut from a square piece of wood x cm on a side, leaving an area of 10 cm². If $x - y = 4$ cm, find x and y graphically.

13–2 Algebraic Solution of Systems of Equations

Often the graphical method is the easiest way to solve a system of equations. However, this method does not usually give an exact answer. Using algebraic methods to find exact solutions for some systems of equations is either not possible or quite involved. There are some systems, however, which do lend themselves to relatively simple solutions by algebraic means. In this section we shall consider two useful methods, both of which we discussed before when we were studying systems of linear equations.

The first method is substitution. If we can solve one equation for one of its variables, we can substitute this solution into the other equation. We then have only one unknown in the resulting equation, and we can solve this equation by methods discussed in earlier chapters.

Example A

Solve, by substitution, the system of equations

$$2x - y = 4$$
$$x^2 - y^2 = 4$$

We solve the first equation for y, obtaining $y = 2x - 4$. We now substitute this into the second equation, getting $x^2 - (2x - 4)^2 = 4$. When simplified, this gives a quadratic equation.

$$x^2 - (4x^2 - 16x + 16) = 4$$
$$-3x^2 + 16x - 20 = 0$$
$$x = \frac{-16 \pm \sqrt{256 - 4(-3)(-20)}}{-6} = \frac{-16 \pm \sqrt{16}}{-6} = \frac{-16 \pm 4}{-6} = \frac{10}{3}, 2$$

We now find the corresponding values of y by substituting into $y = 2x - 4$. Thus, we have the solutions $x = \frac{10}{3}$, $y = \frac{8}{3}$, and $x = 2$, $y = 0$. By substitution, these values also satisfy the equation $x^2 - y^2 = 4$. (We do this as a check.)

Example B

Solve, by substitution, the system of equations

$$xy = -2$$
$$2x + y = 2$$

From the first equation we have $y = -2/x$. Substituting this into the second equation, we have

$$2x - \left(\frac{2}{x}\right) = 2$$
$$2x^2 - 2 = 2x$$
$$x^2 - x - 1 = 0$$
$$x = \frac{1 \pm \sqrt{1 + 4}}{2} = \frac{1 \pm \sqrt{5}}{2}$$

We find the corresponding values of y, and we have the solutions

$$x = \frac{1 + \sqrt{5}}{2}, \quad y = 1 - \sqrt{5} \quad \text{and} \quad x = \frac{1 - \sqrt{5}}{2}, \quad y = 1 + \sqrt{5}$$

Now let us use the other algebraic method of solution, that of addition or subtraction. This method can be used to great advantage if both equations have only squared terms and constants.

Example C

Solve, by addition or subtraction, the system of equations

$$2x^2 + y^2 = 9$$
$$x^2 - y^2 = 3$$

We note that if we add the two equations we get $3x^2 = 12$. Thus, $x = \pm 2$. For $x = 2$, we have two corresponding y-values, $y = \pm 1$. Also for $x = -2$, we have two corresponding y-values, $y = \pm 1$. Thus, we have four solutions: $x = 2$, $y = 1$; $x = 2$, $y = -1$; $x = -2$, $y = 1$; and $x = -2$, $y = -1$.

Example D

Solve, by addition or subtraction, the system of equations

$$3x^2 - 2y^2 = 5$$
$$x^2 + y^2 = 5$$

If we multiply the second equation by 2 and then add the two resulting equations, we get $5x^2 = 15$. Thus, $x = \pm\sqrt{3}$. The corresponding values of y for each value of x are $\pm\sqrt{2}$. Again we have four solutions: $x = \sqrt{3}$, $y = \sqrt{2}$; $x = \sqrt{3}$, $y = -\sqrt{2}$; $x = -\sqrt{3}$, $y = \sqrt{2}$; and $x = -\sqrt{3}$, $y = -\sqrt{2}$.

Example E

A certain number of machine parts cost \$1000. If they cost \$5 less per part, ten additional parts could be purchased for the same amount of money. What is the cost of each?

Since the cost of each part is required, we let $c =$ the cost per part. Also, we let $n =$ the number of parts. From the first statement of the problem, we see that $cn = 1000$. Also, from the second statement, we have $(c - 5)(n + 10) = 1000$. Therefore, we are to solve the system of equations

$$cn = 1000$$
$$(c - 5)(n + 10) = 1000$$

Solving the first equation for n, and multiplying out the second equation, we have $n = \dfrac{1000}{c}$ and $cn + 10c - 5n - 50 = 1000$. Now, substituting the expression for n into the second equation, we solve for c.

$$c\left(\frac{1000}{c}\right) + 10c - 5\left(\frac{1000}{c}\right) - 50 = 1000$$

$$1000 + 10c - \frac{5000}{c} - 50 = 1000$$

$$10c - \frac{5000}{c} - 50 = 0$$

$$c^2 - 5c - 500 = 0$$

$$(c + 20)(c - 25) = 0$$

$$c = -20, 25$$

Since a negative answer has no significance in this particular situation, we see that the solution is $c = \$25$ per part. Checking with the original statement of the problem, we see that this is correct.

Exercises 13-2

In Exercises 1 through 24, solve the given systems of equations algebraically.

1. $y = x + 1$
   $y = x^2 + 1$

2. $y = 2x - 1$
   $y = 2x^2 + 2x - 3$

3. $x + 2y = 3$
   $x^2 + y^2 = 26$

4. $y = x + 1$
   $x^2 + y^2 = 25$

5. $x + y = 1$
   $x^2 - y^2 = 1$

6. $x + y = 2$
   $2x^2 - y^2 = 1$

7. $2x - y = 2$
   $2x^2 + 3y^2 = 4$

8. $6y - x = 6$
   $x^2 + 3y^2 = 36$

9. $xy = 1$
   $x + y = 2$

10. $xy = 2$
    $x + y = 3$

11. $xy = 3$
    $3x - 2y = -7$

12. $xy = -4$
    $2x + y = -2$

13. $y = x^2$
    $y = 3x^2 - 8$

14. $y = x^2 - 1$
    $2x^2 - y^2 = 2$

15. $x^2 - y = -1$
    $x^2 + y^2 = 5$

16. $x^2 + y = 5$
    $x^2 + y^2 = 25$

17. $x^2 - 1 = y$
    $x^2 - 2y^2 = 1$

18. $2y^2 - 4x = 7$
    $y^2 + 2x^2 = 3$

19. $x^2 + y^2 = 25$
    $x^2 - 2y^2 = 7$

20. $3x^2 - y^2 = 4$
    $x^2 + 4y^2 = 10$

21. $y^2 - 2x^2 = 6$
    $5x^2 + 3y^2 = 20$

22. $y^2 - 2x^2 = 17$
    $2y^2 + x^2 = 54$

23. $x^2 + 3y^2 = 37$
    $2x^2 - 9y^2 = 14$

24. $5x^2 - 4y^2 = 15$
    $3y^2 + 4x^2 = 12$

In Exercises 25 through 32, solve the indicated systems of equations algebraically. In Exercises 27 through 32, it is necessary to properly set up the systems of equations.

25. The vertical distance which a certain projectile travels from its starting point is given by $y = 60t - 16t^2$. When the horizontal distance it has traveled equals twice the vertical distance, $y = 40t$. Find the values of y (in feet) and t (in seconds) which satisfy these equations.

26. A 300-g block and a 200-g block collide. Using the physical laws of conservation of energy and conservation of momentum, along with certain given conditions, we can establish the following equations involving velocities of each block after collision:

$$150v_1^2 + 100v_2^2 = 1,375,000$$
$$300v_1 + 200v_2 = -5000$$

Find these velocities (in centimeters per second).

27. Find two positive numbers such that the sum of their squares is 233 and the difference between their squares is 105.

28. The length of a table is three times the width, and the area is 48 ft^2. Find the dimensions of the table.

29. Two ships leave a port, one traveling due south and the other due east. The ship going east travels twice as far as the other ship, at which time they are 10 km apart. How far does each travel?

30. To enclose a rectangular field of 11,200 ft^2 in area, 440 ft of fence are required. What are the dimensions of the field?

31. The radii of two spheres differ by 4 in., and the difference between the spherical surfaces is 320π in.^{2}. Find the radii. (The surface area of a sphere is $4\pi r^2$.)

32. Two cities are 2000 mi apart. If an airplane increases its usual speed between these two cities by 100 mi/h, the trip would take 1 h less. Find the normal speed of the plane and the normal time of the flight.

13–3 Equations in Quadratic Form

 Often we encounter equations which can be solved by methods applicable to quadratic equations, even though these equations are not actually quadratic. They do have the property, however, that *with a proper substitution they may be written in the form of a quadratic equation.* All that is necessary is that the equation have terms including some quantity, its square, and perhaps a constant term. The following example illustrates these types of equations.

Example A

The equation $x - 2\sqrt{x} - 5 = 0$ is an equation in quadratic form, because if we let $y = \sqrt{x}$, we have the resulting equivalent equation $y^2 - 2y - 5 = 0$.

Other examples of equations in quadratic form are as follows:

$$t^{-4} - 5t^{-2} + 3 = 0$$
By letting $y = t^{-2}$, we have $\quad y^2 - 5y + 3 = 0.$
$$t^3 - 3t^{3/2} - 7 = 0$$
By letting $y = t^{3/2}$, we have $\quad y^2 - 3y - 7 = 0.$
$$(x + 1)^4 - (x + 1)^2 - 1 = 0$$
By letting $y = (x + 1)^2$, we have $\quad y^2 - y - 1 = 0.$
$$x^{10} - 2x^5 + 1 = 0$$
By letting $y = x^5$, we have $\quad y^2 - 2y + 1 = 0.$
$$(x - 3) + \sqrt{x - 3} - 6 = 0$$
By letting $y = \sqrt{x - 3}$, we have $\quad y^2 + y - 6 = 0.$

The following examples illustrate the method of solving equations in quadratic form.

Example B

Solve the equation $x^4 - 5x^2 + 4 = 0$.

We first let $y = x^2$, and obtain the resulting equivalent equation $y^2 - 5y + 4 = 0$. This may be factored as $(y - 4)(y - 1) = 0$. Thus, we have the solutions $y = 4$ and $y = 1$. Therefore, $x^2 = 4$ and $x^2 = 1$, which means that we have $x = \pm 2$ and $x = \pm 1$. Substitution into the original equation verifies that each of these is a solution.

Example C

Solve the equation $2x^4 + 7x^2 = 4$.

As in Example B, we let $y = x^2$, and then write the resulting equation in quadratic form, $2y^2 + 7y - 4 = 0$. This factors into $(2y - 1)(y + 4) = 0$, which leads to $y = \frac{1}{2}$ and $y = -4$. This means that $x^2 = \frac{1}{2}$ and $x^2 = -4$, which in turn means that $x = \frac{1}{\sqrt{2}}, -\frac{1}{\sqrt{2}}, 2j, -2j$. Substitution into the original equation shows that each is a solution.

Two of the solutions in Example C are complex numbers. We were able to find these solutions directly from the definition of the square root of a negative number. In some cases (see Exercises 23 and 24 of this section) it is necessary to use the method of Section 11–6 to find such complex number solutions.

Example D

Solve the equation $x - \sqrt{x} - 2 = 0$.

By letting $y = \sqrt{x}$, we have the equivalent equation $y^2 - y - 2 = 0$. This is factorable into $(y - 2)(y + 1) = 0$. Therefore, we have $y = 2$ and $y = -1$. Since $y = \sqrt{x}$, we note that y cannot be negative, and this in turn tells us that $y = -1$ cannot lead to a solution. For $y = 2$ we have $x = 4$. Checking, we find that $x = 4$ satisfies the original equation. Thus, the only solution is $x = 4$.

Example D illustrates a very important point: *Whenever any operation involving the unknown is performed on an equation, this operation may introduce roots into a subsequent equation which are not roots of the original equation. Therefore, we must check all answers in the original equation.* Only operations involving constants—that is, adding, subtracting, multiplying by, or dividing by constants—are certain not to introduce these **extraneous roots.** Squaring both sides of an equation is a common way of introducing extraneous roots. We first encountered the concept of an extraneous root in Section 5–7, when we were discussing equations involving fractions.

Example E

Solve the equation $x^{-2} + 3x^{-1} + 1 = 0$.

By substituting $y = x^{-1}$, we have $y^2 + 3y + 1 = 0$. To solve this equation we may use the quadratic formula:

$$y = \frac{-3 \pm \sqrt{9 - 4}}{2} = \frac{-3 \pm \sqrt{5}}{2}$$

Thus, since $x = 1/y$,

$$x = \frac{2}{-3 + \sqrt{5}}, \qquad \frac{2}{-3 - \sqrt{5}}$$

Rationalizing the denominators of the values of x, we have

$$x = \frac{-6 - 2\sqrt{5}}{4} = \frac{-3 - \sqrt{5}}{2} \quad \text{and} \quad x = \frac{-3 + \sqrt{5}}{2}$$

Checking these solutions, we have

$$\left(\frac{-3-\sqrt{5}}{2}\right)^{-2} + 3\left(\frac{-3-\sqrt{5}}{2}\right)^{-1} + 1 \stackrel{?}{=} 0 \quad \text{or} \quad 0 = 0$$

and

$$\left(\frac{-3+\sqrt{5}}{2}\right)^{-2} + 3\left(\frac{-3+\sqrt{5}}{2}\right)^{-1} + 1 \stackrel{?}{=} 0 \quad \text{or} \quad 0 = 0$$

Thus, these solutions check.

Example F

Solve the equation $(x^2 - x)^2 - 8(x^2 - x) + 12 = 0$.

By substituting $y = x^2 - x$, we have $y^2 - 8y + 12 = 0$. This is solved by factoring, which gives us the solutions $y = 2$ and $y = 6$. Thus, $x^2 - x = 2$ and $x^2 - x = 6$. Solving these, we find that $x = 2, -1, 3, -2$. Substituting these in the original equation, we find that all are solutions.

Example G illustrates a problem that leads to an equation in quadratic form.

Example G

A rectangular plate has an area of 60 cm^2. The diagonal of the plate is 13 cm. Find the length and width of the plate.

Since the required quantities are the length and width, let $l =$ the length of the plate and $w =$ the width of the plate. Since the area is 60 cm^2, $lw = 60$. Also, using the Pythagorean theorem and the fact that the diagonal is 13 cm, we have $l^2 + w^2 = 169$. Therefore, we are to solve the system of equations

$$lw = 60, \qquad l^2 + w^2 = 169$$

Solving the first equation for l, we have $l = 60/w$. Substituting this expression into the second equation we have

$$\left(\frac{60}{w}\right)^2 + w^2 = 169$$

We now solve for w as follows:

$$\frac{3600}{w^2} + w^2 = 169$$

$$3600 + w^4 = 169w^2$$

$$w^4 - 169w^2 + 3600 = 0$$

Let $x = w^2$.

$$x^2 - 169x + 3600 = 0$$

$$(x - 144)(x - 25) = 0$$

$$x = 25, 144$$

(continued on next page)

Therefore,

$$w^2 = 25, 144$$

Solving for w, we obtain $w = \pm 5$ or $w = \pm 12$. Only the positive values of w are meaningful in this problem. Therefore, if $w = 5$ cm, then $l = 12$ cm. Normally, we designate the length as the longer dimension. Checking in the original equation, we find that this solution is correct.

Exercises 13–3

In Exercises 1 through 24, solve the given equations.

1. $x^4 - 13x^2 + 36 = 0$ 2. $x^4 - 20x^2 + 64 = 0$

3. $x^4 - 3x^2 - 4 = 0$ 4. $4x^4 + 15x^2 = 4$

5. $x^{-2} - 2x^{-1} - 8 = 0$ 6. $10x^{-2} + 3x^{-1} - 1 = 0$

7. $x^{-4} + 2x^{-2} = 24$ 8. $x^{-4} + 1 = 2x^{-2}$

9. $x - 4\sqrt{x} + 3 = 0$ 10. $2x + \sqrt{x} - 1 = 0$

11. $3\sqrt[3]{x} - 5\sqrt[6]{x} + 2 = 0$ 12. $\sqrt{x} + 3\sqrt[4]{x} = 28$

13. $x^{2/3} - 2x^{1/3} - 15 = 0$ 14. $x^3 + 2x^{3/2} - 80 = 0$

15. $x^{1/2} + x^{1/4} = 20$ 16. $4x^{4/3} + 9 = 13x^{2/3}$

17. $(x - 1) - \sqrt{x - 1} - 2 = 0$ 18. $(x + 1)^{-2/3} + 5(x + 1)^{-1/3} - 6 = 0$

19. $(x^2 - 2x)^2 - 11(x^2 - 2x) + 24 = 0$

20. $3(x^2 + 3x)^2 - 2(x^2 + 3x) - 5 = 0$

21. $x - 3\sqrt{x - 2} = 6$ (Let $y = \sqrt{x - 2}$.)

22. $(x^2 - 1)^2 + (x^2 - 1)^{-2} = 2$

23. $x^6 + 7x^3 - 8 = 0$ 24. $x^6 - 19x^3 - 216 = 0$

In Exercises 25 through 28, solve the indicated equations. In Exercises 27 and 28, it is necessary to set up the required equation.

25. A manufacturer determines that the total profit P when producing x thousand television sets monthly is given by $P = 10,000(-x^4 + 8x^2 - 2)$, where P is measured in dollars. Determine the number of sets which can be produced for a profit of $140,000.

26. In optics, in the theory which deals with interferometers, the equation $\sqrt{F} = \dfrac{2\sqrt{p}}{1 - p}$ is found. Solve for p if $F = 16$.

27. Find the dimensions of a rectangular area having a diagonal of 40 ft and an area of 768 ft^2.

28. A metal plate is in the shape of an isosceles triangle. The length of the base equals the square root of one of the equal sides. Determine the lengths of the sides if the perimeter of the plate is 55 cm.

13–4 Equations with Radicals

Equations with radicals in them are normally solved by squaring both sides of the equation, or by a similar operation. However, when we do this, we often introduce extraneous roots. Thus, it is very important that all solutions be checked in the original equation.

Example A

Solve the equation $\sqrt{x - 4} = 2$.

By squaring both sides of the equation, we have

$$(\sqrt{x - 4})^2 = 2^2 \quad \text{or} \quad x - 4 = 4 \quad \text{or} \quad x = 8$$

This solution checks when put into the original equation.

Example B

Solve the equation $\sqrt{2x + 3} = 4x$.

Squaring both sides of the equation gives us

$$(\sqrt{2x + 3})^2 = (4x)^2$$
$$2x + 3 = 16x^2$$
$$16x^2 - 2x - 3 = 0$$
$$(2x - 1)(8x + 3) = 0$$
$$x = \frac{1}{2} \quad \text{or} \quad x = -\frac{3}{8}$$

Checking the results in the original equation, we find that $x = \frac{1}{2}$ checks, giving $2 = 2$, but $x = -\frac{3}{8}$ gives $\frac{3}{2} = -\frac{3}{2}$. Therefore, the solution is $x = \frac{1}{2}$. The value $x = -\frac{3}{8}$ is an extraneous root.

Example C

Solve the equation $\sqrt{x - 1} = x - 3$.

Squaring both sides of the equation, we have

$$(\sqrt{x - 1})^2 = (x - 3)^2$$

(Remember, we are squaring each *side* of the equation, not just the terms separately on each side.) Hence,

$$x - 1 = x^2 - 6x + 9$$
$$x^2 - 7x + 10 = 0$$
$$(x - 5)(x - 2) = 0$$
$$x = 5 \quad \text{or} \quad x = 2$$

The solution $x = 5$ checks, but the solution $x = 2$ gives $1 = -1$. Thus, the solution is $x = 5$. The value $x = 2$ is an extraneous root.

Example D

Solve the equation $\sqrt[3]{x - 8} = 2$.

Cubing both sides of the equation, we have $x - 8 = 8$. Thus $x = 16$, which checks.

Example E

Solve the equation $\sqrt{x + 1} + \sqrt{x - 4} = 5$.

This is most easily solved by first placing one of the radicals on the right side and then squaring both sides of the equation:

$$\sqrt{x + 1} = 5 - \sqrt{x - 4}$$
$$(\sqrt{x + 1})^2 = (5 - \sqrt{x - 4})^2$$
$$x + 1 = 25 - 10\sqrt{x - 4} + (x - 4)$$

Now, isolating the radical on one side of the equation and squaring again, we have

$$10\sqrt{x - 4} = 20$$
$$\sqrt{x - 4} = 2$$
$$x - 4 = 4$$
$$x = 8$$

This solution checks.

Example F

Solve the equation $\sqrt{x} - \sqrt[4]{x} = 2$.

We can solve this most easily by handling it as an equation in quadratic form. By letting $y = \sqrt[4]{x}$, we have

$$y^2 - y - 2 = 0$$
$$(y - 2)(y + 1) = 0$$
$$y = 2 \quad \text{or} \quad y = -1$$

Since $y = \sqrt[4]{x}$ we know that a negative value of y does not lead to a solution of the original equation. Therefore, we see that $y = -1$ cannot give us a solution. For the other value, $y = 2$, we have

$$\sqrt[4]{x} = 2 \quad \text{or} \quad x = 16$$

This checks, because $\sqrt{16} - \sqrt[4]{16} = 4 - 2 = 2$. Therefore, the only solution of the original equation is $x = 16$.

Example G

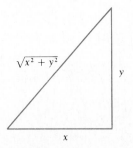

Figure 13–8

The perimeter of a right triangle is 60 ft, and its area is 120 ft^2. Find the lengths of the three sides (see Fig. 13–8).

If we let the two legs of the triangle be x and y, the perimeter can be expressed as $p = x + y + \sqrt{x^2 + y^2}$. Also, the area is $A = \frac{1}{2}xy$. In this way we arrive at the equations $x + y + \sqrt{x^2 + y^2} = 60$ and $xy = 240$. Solving the first of these for the radical, we then have $\sqrt{x^2 + y^2} = 60 - x - y$. Squaring both sides and combining terms, we have

$$0 = 3600 - 120x - 120y + 2xy$$

Solving the second of the original equations for y, we have $y = 240/x$. Substituting, we have

$$0 = 3600 - 120x - 120\left(\frac{240}{x}\right) + 480$$

Multiplying each side by x, dividing through by 120, and rearranging the terms, we have $x^2 - 34x + 240 = 0$, which may be factored into $(x - 10)(x - 24) = 0$. Thus, x can be either 10 or 24. From the equation $xy = 240$, if $x = 10$, then $y = 24$. This means that the two legs are 10 ft and 24 ft, and the hypotenuse is 26 ft. The same solution is found by using the value of 24 ft for x.

Exercises 13–4

In Exercises 1 through 28, solve the given equations.

1. $\sqrt{x - 8} = 2$
2. $\sqrt{x + 4} = 3$
3. $\sqrt{8 - 2x} = x$
4. $\sqrt{3x + 4} = x$
5. $\sqrt{3x + 2} = 3x$
6. $\sqrt{2x + 6} = 2x$
7. $\sqrt{x - 2} = x - 2$
8. $\sqrt{5x - 1} = x - 3$
9. $\sqrt[3]{y - 5} = 3$
10. $\sqrt[4]{5 - x} = 2$
11. $\sqrt{x + 12} = x$
12. $\sqrt{x + 3} = 4x$
13. $5\sqrt{x + 3} = 2x$
14. $4\sqrt{x} = x + 3$
15. $\sqrt{x + 4} = x - 8$
16. $\sqrt{x + 15} = x - 5$
17. $\sqrt{2x + 8} - 2 = 2x$
18. $\sqrt{3x + 3} - 1 = 3x$
19. $2\sqrt{x + 2} - \sqrt{3x + 4} = 1$
20. $\sqrt{x - 1} + \sqrt{x + 2} = 3$
21. $\sqrt{5x + 1} - 1 = 3\sqrt{x}$
22. $\sqrt{2x + 1} + \sqrt{3x} = 11$
23. $\sqrt{2x - 1} - \sqrt{x + 11} = -1$
24. $\sqrt{5x - 4} - \sqrt{x} = 2$
25. $\sqrt[3]{2x - 1} = \sqrt[3]{x + 5}$
26. $\sqrt[4]{x + 10} = \sqrt{x - 2}$
27. $\sqrt{x - 2} = \sqrt[4]{x - 2} + 12$
28. $\sqrt{3x} + \sqrt{3x + 4} = 4$

In Exercises 29 through 32, solve the indicated equations for the indicated letter.

29. The velocity of an object falling under the influence of gravity in terms of its initial velocity v_0, the acceleration due to gravity g, and the height fallen is given by $v = \sqrt{v_0^2 - 2gh}$. Solve this equation for h.

30. In measuring the velocity of water flowing through an opening under given conditions, the equation $v = \sqrt{2g(h_1 - h_2)}$ arises. Solve for h_1.

31. An equation used in analyzing a certain type of concrete beam is $k = \sqrt{2np + (np)^2} - np$. Solve for p.

32. The theory of relativity states that the mass m of an object increases with velocity v according to the relation

$$m = \frac{m_0}{\sqrt{1 - v^2/c^2}}$$

where m_0 is the "rest mass" and c is the velocity of light. Solve for v in terms of m.

In Exercises 33 through 36, set up the proper equations and solve them.

33. Find the dimensions of the rectangle for which the diagonal is 3 in. more than the longer side, which in turn is 3 in. longer than the shorter side.

34. The sides of a certain triangle are $\sqrt{x - 1}$, $\sqrt{5x - 1}$, and 9. Find x when the perimeter of the triangle is 19.

35. An island is 3 mi offshore from the nearest point P on a straight beach. A person in a motorboat travels straight from the island to the beach x mi from P, and then travels 2 mi along the beach away from P. Find x if the person traveled a total of 8 mi.

36. The focal length f of a lens, in terms of its image distance q and object distance p, is given by

$$\frac{1}{f} = \frac{1}{p} + \frac{1}{q}$$

Find p and q if $f = 4$ cm and $p = \sqrt{q}$.

13-5 Exercises for Chapter 13

In Exercises 1 through 10, solve the given systems of equations graphically.

1. $x + 2y = 6$
   $y = 4x^2$

2. $x + y = 3$
   $x^2 + y^2 = 25$

3. $3x + 2y = 6$
   $x^2 + 4y^2 = 4$

4. $x^2 - 2y = 0$
   $y = 3x - 5$

5. $y = x^2 + 1$
   $2x^2 + y^2 = 4$

6. $\dfrac{x^2}{4} + y^2 = 1$
   $x^2 - y^2 = 1$

7. $y = 4 - x^2$
   $y = 2x^2$

8. $xy = -2$
   $y = 1 - 2x^2$

9. $y = x^2 - 2x$
   $y = 1 - e^{-x}$

10. $y = \ln x$
    $y = \sin x$

In Exercises 11 through 20, solve the given systems of equations algebraically.

11. $y = 4x^2$
    $y = 8x$

12. $x + y = 2$
    $xy = 1$

13. $2y = x^2$
    $x^2 + y^2 = 3$

14. $y = x^2$
    $2x^2 - y^2 = 1$

15. $4x^2 + y = 3$
    $2x + 3y = 1$

16. $2x^2 + y^2 = 3$
    $x + 2y = 1$

17. $4x^2 - 7y^2 = 21$
    $x^2 + 2y^2 = 99$

18. $3x^2 + 2y^2 = 11$
    $2x^2 - y^2 = 30$

19. $4x^2 + 3xy = 4$
    $x + 3y = 4$

20. $\dfrac{6}{x} + \dfrac{3}{y} = 4$

    $\dfrac{36}{x^2} + \dfrac{36}{y^2} = 13$

In Exercises 21 through 40, solve the given equations.

21. $x^4 - 20x^2 + 64 = 0$

22. $x^6 - 26x^3 - 27 = 0$

23. $x^{3/2} - 9x^{3/4} + 8 = 0$

24. $x^{1/2} + 3x^{1/4} - 28 = 0$

25. $x^{-2} + 4x^{-1} - 21 = 0$

26. $x^{-4} - 5x^{-2} - 36 = 0$

27. $2x - 3\sqrt{x} - 5 = 0$

28. $x^{-1} + x^{-1/2} = 6$

29. $\left(\dfrac{1}{x+1}\right)^2 - \dfrac{1}{x+1} = 2$

30. $(x^2 + 5x)^2 - 5(x^2 + 5x) = 6$

31. $\sqrt{x + 5} = 4$

32. $\sqrt[3]{x - 2} = 3$

33. $\sqrt{5x - 4} = x$

34. $\sqrt{6x + 8} = 3x$

35. $x - 1 = \sqrt{5x + 9}$

36. $x + 2 = \sqrt{11x - 2}$

37. $\sqrt{x + 1} + \sqrt{x} = 2$

38. $\sqrt{3 + x} + \sqrt{3x - 2} = 1$

39. $\sqrt{3x + 4} + \sqrt{x + 2} = 8$

40. $\sqrt{3x - 2} - \sqrt{x + 7} = 1$

In Exercises 41 through 44 solve for the indicated quantities.

41. The formula for the lateral surface area of a cone can be stated as $S = \pi r \sqrt{r^2 + h^2}$. Solve for r.

42. Under certain conditions, the frequency ω of an *RLC* circuit is given by

$$\omega = \frac{\sqrt{R^2 + 4(L/C)} + R}{2L}$$

Solve for C.

43. In an experiment, an object is allowed to fall, stopped, and then falls for twice the initial time. The total distance the object falls is 45 ft. The equations relating the times t_1 and t_2, in seconds, of fall are $16t_1^2 + 16t_2^2 = 45$, and $t_2 = 2t_1$. Find the times of fall.

44. If two objects collide and the kinetic energy remains constant, the collision is termed perfectly elastic. Under these conditions, if an object of mass m_1 and initial velocity u_1 strikes a second object (initially at rest) of mass m_2, such that the velocities after collision are v_1 and v_2, the following equations are found:

$$m_1 u_1 = m_1 v_1 + m_2 v_2$$
$$\tfrac{1}{2} m_1 u_1^2 = \tfrac{1}{2} m_1 v_1^2 + \tfrac{1}{2} m_2 v_2^2$$

Solve these equations for m_2 in terms of u_1, v_1, and m_1.

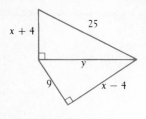

Figure 13-9

In Exercises 45 through 50 set up the appropriate equations and solve them.

45. Find the values of x and y in Fig. 13–9.

46. A rectangular solar panel has an area of 2700 cm^2 and a perimeter of 240 cm. Find the dimensions of the panel.

47. The area of one square metal piece is 9 times the area of a second square metal piece. The sum of the length of one side of one piece and one side of the other piece is 2 cm. What is the area of the smaller piece?

48. The perimeter of a rectangle is 36 cm. The length is x cm and the width is $\sqrt{x+2}$ cm. Find the dimensions of the rectangle.

49. An object is dropped from the top of a building. Six seconds later it is heard to hit the street below. How high is the building? (The velocity of sound is 1100 ft/s and the distance the object falls as a function of time is $s = 16t^2$.)

50. Two trains are approaching the same crossing on tracks which are at right angles to each other. Each is traveling at 60 km/h. If one is 6 km from the crossing when the other is 3 km from it, how much later will they be 4 km apart?

CHAPTER 14

Equations of Higher Degree

In previous chapters we have discussed methods of solving many kinds of equations. Except for special cases, however, we have not solved polynomial equations of degree higher than two (a polynomial equation of the first degree is a linear equation, and a polynomial equation of the second degree is a quadratic equation). In this chapter we shall develop certain methods for solving polynomial equations, especially the higher-degree equations. Since we shall be discussing equations involving only polynomials, in this chapter $f(x)$ will be assumed to be a polynomial.

14–1 The Remainder Theorem and the Factor Theorem

Any polynomial is a function of the form

$$f(x) = a_0x^n + a_1x^{n-1} + \cdots + a_n \tag{14–1}$$

where $a_0 \neq 0$ and n is a positive integer or zero. We will consider only polynomials in which the coefficients $a_0, a_1, \cdots a_n$ are real numbers.

If we divide a polynomial by $x - r$, we find a result of the following form:

$$f(x) = (x - r)q(x) + R \tag{14–2}$$

where $q(x)$ is the quotient and R is the remainder.

Example A

Divide $f(x) = 3x^2 + 5x - 8$ by $x - 2$.

$$
\begin{array}{r}
3x + 11 \\
x - 2 \overline{\smash{)}\, 3x^2 + 5x - 8 } \\
\underline{3x^2 - 6x } \\
11x - 8 \\
\underline{11x - 22 } \\
14
\end{array}
$$

Thus,

$$3x^2 + 5x - 8 = (x - 2)(3x + 11) + 14$$

If we now set $x = r$ in Eq. (14–2), we have $f(r) = q(r)(r - r) + R$, or

$$\underline{f(r) = R} \qquad\qquad\qquad\qquad\qquad (14\text{–}3)$$

The equation above states that the remainder equals the value of the function of x at $x = r$. This leads us to the **remainder theorem,** which states that *if a polynomial $f(x)$ is divided by $x - r$ until a constant remainder (R) is obtained, then $f(r) = R$.*

Example B

In Example A, $f(x) = 3x^2 + 5x - 8$, $R = 14$, $r = 2$.
We find that

$$f(2) = 3(4) + 5(2) - 8 = 14$$

Thus, $f(2) = 14$ verifies that $f(r) = R$.

Example C

Using the remainder theorem, determine the remainder when we divide $3x^3 - x^2 - 20x + 5$ by $x + 4$.

In using the remainder theorem, we determine the remainder when the function is divided by $x - r$ by evaluating the function for $x = r$. To have $x + 4$ in the proper form to identify r, we write it as $x - (-4)$. This means that $r = -4$, and we therefore are to evaluate the function $f(x) = 3x^3 - x^2 - 20x + 5$ for $x = -4$, or find $f(-4)$. Thus,

$$f(-4) = 3(-4)^3 - (-4)^2 - 20(-4) + 5 = -192 - 16 + 80 + 5$$
$$= -123$$

Thus, the remainder when $3x^3 - x^2 - 20x + 5$ is divided by $x + 4$ is -123.

The remainder theorem leads immediately to another important theorem known as the **factor theorem.** The factor theorem states that *if* $f(r) = R = 0$, *then* $x - r$ *is a factor of* $f(x)$. Inspection of Eq. (14–2) justifies this theorem. Recalling material from earlier chapters, for the function $f(x)$ we see that if $f(r) = 0$, then $x = r$ is a *zero* of $f(x)$, $x - r$ is a *factor* of $f(x)$, and $x = r$ is a *solution* of the equation $f(x) = 0$.

Example D

Is $x + 1$ a factor of $f(x) = x^3 + 2x^2 - 5x - 6$?
  Here $r = -1$, and thus

$$f(-1) = -1 + 2 + 5 - 6 = 0$$

Therefore, since $f(-1) = 0$, $x + 1$ is a factor of $f(x)$.

Example E

The expression $x + 2$ is not a factor of the function in Example D, since

$$f(-2) = -8 + 8 + 10 - 6 = 4$$

But $x - 2$ is a factor, since $f(2) = 8 + 8 - 10 - 6 = 0$.

Example F

Determine if $\frac{2}{3}$ is a zero of the function $f(x) = 3x^3 + 4x^2 - 16x + 8$.
  Evaluating $f(\frac{2}{3})$ we have

$$f\left(\frac{2}{3}\right) = 3\left(\frac{2}{3}\right)^3 + 4\left(\frac{2}{3}\right)^2 - 16\left(\frac{2}{3}\right) + 8$$

$$= \frac{8}{9} + \frac{16}{9} - \frac{32}{3} + 8 = \frac{8 + 16 - 96 + 72}{9}$$

$$= 0$$

Since $f(\frac{2}{3}) = 0$, $\frac{2}{3}$ is a zero of the function.

We now have one way of determining whether or not an expression of the form $x - r$ is a factor of a function. By finding $f(r)$, we can determine whether or not $x - r$ is a factor and whether or not r is a zero of the function.

Exercises 14–1

In Exercises 1 through 8, find the remainder R by long division and by the remainder theorem.

1. $(x^3 + 2x^2 - x - 2) \div (x - 1)$ 2. $(x^3 - 3x^2 - x + 2) \div (x - 2)$
3. $(x^3 + 2x + 3) \div (x + 1)$ 4. $(x^4 - 4x^3 - x^2 + x - 100) \div (x + 3)$
5. $(2x^5 - x^2 + 8x + 44) \div (x + 2)$
6. $(x^3 + 4x^2 - 25x - 98) \div (x - 5)$
7. $(2x^4 - 3x^3 - 4x^2 + 2x - 5) \div (x - 3)$
8. $(2x^4 - 10x^2 + 30x - 60) \div (x + 4)$

In Exercises 9 through 16, find the remainder using the remainder theorem.

9. $(x^3 + 2x^2 - 3x + 4) \div (x + 1)$ 10. $(2x^3 - 4x^2 + x - 1) \div (x + 2)$

11. $(x^4 + x^3 - 2x^2 - 5x + 3) \div (x + 4)$ 12. $(2x^4 - x^2 + 5x - 7) \div (x - 3)$

13. $(2x^4 - 7x^3 - x^2 + 8) \div (x - 3)$

14. $(x^4 - 5x^3 + x^2 - 2x + 6) \div (x + 4)$

15. $(x^5 - 3x^3 + 5x^2 - 10x + 6) \div (x - 2)$

16. $(3x^4 - 12x^3 - 60x + 4) \div (x - 5)$

In Exercises 17 through 24, use the factor theorem to determine whether or not the second expression is a factor of the first.

17. $x^2 - 2x - 3, x - 3$ 18. $3x^3 + 2x^2 - 3x - 2, x + 2$

19. $4x^3 + x^2 - 16x - 4, x - 2$ 20. $3x^3 + 14x^2 + 7x - 4, x + 4$

21. $5x^3 - 3x^2 + 4, x - 2$ 22. $x^5 - 2x^4 + 3x^3 - 6x^2 - 4x + 8, x - 2$

23. $x^6 + 1, x + 1$ 24. $x^7 - 128, x + 2$

In Exercises 25 through 28, determine whether or not the given numbers are zeros of the given functions.

25. $f(x) = x^3 - 2x^2 - 9x + 18; \quad 2$

26. $f(x) = 2x^3 + 3x^2 - 8x - 12; \quad -\frac{3}{2}$

27. $f(x) = 4x^4 - 4x^3 + 23x^2 + x - 6; \quad \frac{1}{2}$

28. $f(x) = 2x^4 + 3x^3 - 12x^2 - 7x + 6; \quad -3$

In Exercises 29 through 32, answer the given questions.

29. By division, show that $2x - 1$ is a factor of $f(x) = 4x^3 + 8x^2 - x - 2$. May we therefore conclude that $f(1) = 0$?

30. By division, show that $x^2 + 2$ is a factor of $f(x) = 3x^3 - x^2 + 6x - 2$. May we therefore conclude that $f(-2) = 0$?

31. For what value of k is $x - 2$ a factor of $f(x) = 2x^3 + kx^2 - x + 14$?

32. For what value of k is $x + 1$ a factor of $f(x) = 3x^4 + 3x^3 + 2x^2 + kx - 4$?

14–2 Synthetic Division

We shall now develop a method which greatly simplifies the procedure for dividing a polynomial by an expression of the form $x - r$. Using **synthetic division,** which is an abbreviated form of long division, we can determine the coefficients of the quotient as well as the remainder. Of course, for some values of r, we can easily calculate $f(r)$ directly. However, if the degree of the equation is high, this requires finding and combining high powers of r. Synthetic division therefore allows us to find $f(r)$ easily by finding the remainder. The method is developed in the following example.

Example A

Divide $x^4 + 4x^3 - x^2 - 16x - 14$ by $x - 2$.

We shall first perform this division in the usual manner.

$$
\begin{array}{r}
x^3 + 6x^2 + 11x + 6 \\
x - 2 \overline{\smash{\big)}\, x^4 + 4x^3 - x^2 - 16x - 14} \\
\underline{x^4 - 2x^3} \\
6x^3 - x^2 \\
\underline{6x^3 - 12x^2} \\
11x^2 - 16x \\
\underline{11x^2 - 22x} \\
6x - 14 \\
\underline{6x - 12} \\
-2
\end{array}
$$

We now note that, when we performed this division, we repeated many terms. Also, the only quantities of importance in the function being divided are the coefficients. There is no real need to put in the powers of x all the time. Therefore, we shall now write the above example without any x's and also eliminate the writing of identical terms:

$$
\begin{array}{r}
1 \quad\quad 6 \quad\quad 11 \quad\quad 6 \\
-2 \overline{\smash{\big)}\, 1 \quad\quad 4 \quad -1 \quad -16 \quad -14} \\
\underline{-2} \\
6 \\
\underline{-12} \\
11 \\
\underline{-22} \\
6 \\
\underline{-12} \\
-2
\end{array}
$$

All but the first of the numbers which represent coefficients of the quotient are repeated below. Also, all the numbers below the dividend may be written in two lines. Thus, we have the following form:

$$
\begin{array}{r}
-2 \overline{\smash{\big)}\, 1 \quad\quad 4 \quad -1 \quad -16 \quad -14} \\
\underline{-2 \quad -12 \quad -22 \quad -12} \\
6 \quad\quad 11 \quad\quad 6 \quad -2
\end{array}
$$

All the coefficients of the actual quotient appear except the first, so we shall now repeat the 1 in the bottom line. Also, we shall change -2 to 2, which is the actual value of r. Then, writing r to the right, we have the following form:

$$
\begin{array}{r|}
1 \quad\quad 4 \quad -1 \quad -16 \quad -14 \quad\quad 2 \\
\underline{-2 \quad -12 \quad -22 \quad -12} \\
1 \quad\quad 6 \quad\quad 11 \quad\quad 6 \quad -2
\end{array}
$$

In this form the 1, 6, 11, and 6 represent the coefficients of the x^3, x^2, x, and constant terms of the quotient. The -2 is the remainder. Finally, we find it easier to use addition rather than subtraction in the process, so we change the signs of the numbers in the middle row. Remember that originally the bottom line was found by subtraction. Thus, we have

$$
\begin{array}{rrrrr|r}
1 & 4 & -1 & -16 & -14 & \underline{2} \\
  & 2 & 12 & 22 & 12 & \\
\hline
1 & 6 & 11 & 6 & -2 &
\end{array}
$$

When we inspect this form we find the following: The 1 multiplied by the 2 (r) gives 2, which is the first number of the middle row. The 4 and 2 (of the middle row) added is 6, which is the second number in the bottom row. The 6 multiplied by 2 (r) is 12, which is the second number in the middle row. This 12 and the -1 give 11. The 11 multiplied by 2 is 22. The 22 added to -16 is 6. This 6 multiplied by 2 gives 12. This 12 added to -14 is -2. When this process is followed in general, the method is called *synthetic division*.

Generalizing on this last example, we have the following procedure: We write down the coefficients of $f(x)$, being certain that the powers are in descending order and that zeros are placed in for missing powers. We write the value of r to the right. We carry down the left coefficient, multiply this number by r, and place this product under the second coefficient of the top line. We add the two numbers in this second column and place the result below; then we multiply this number by r and place the result under the third coefficient of the top line. We continue until the bottom row has as many numbers as the top row. The last number in the bottom row is the remainder, the other numbers being the respective coefficients of the quotient. The first term of the quotient is of degree one less than the dividend.

Example B

By synthetic division, divide $x^5 + 2x^4 - 4x^2 + 3x - 4$ by $x + 3$.

In writing down the coefficients of $f(x)$, we must be certain to include a zero for the missing x^3 term. Also, since the divisor is $x + 3$, we must recognize that $r = -3$. The setup and synthetic division follow.

$$
\begin{array}{rrrrrr|r}
1 & 2 & 0 & -4 & 3 & -4 & \underline{-3} \\
  & -3 & 3 & -9 & 39 & -126 & \\
\hline
1 & -1 & 3 & -13 & 42 & -130 &
\end{array}
$$

Thus, the quotient is $x^4 - x^3 + 3x^2 - 13x + 42$ and the remainder is -130. Notice the degree of the dividend is 5 and the degree of the quotient is 4.

Example C

By synthetic division, divide $3x^4 - 5x + 6$ by $x - 4$.

$$
\begin{array}{rrrrr|r}
3 & 0 & 0 & -5 & 6 & \underline{4} \\
  & 12 & 48 & 192 & 748 & \\
\hline
3 & 12 & 48 & 187 & 754 &
\end{array}
$$

Thus, the quotient is $3x^3 + 12x^2 + 48x + 187$ and the remainder is 754.

Example D

By synthetic division, determine whether or not $x - 4$ is a factor of $x^4 + 2x^3 - 15x^2 - 32x - 16$.

$$
\begin{array}{rrrrr|r}
1 & 2 & -15 & -32 & -16 & \underline{4} \\
  & 4 & 24 & 36 & 16 & \\
\hline
1 & 6 & 9 & 4 & 0 &
\end{array}
$$

Since the remainder is zero, $x - 4$ is a factor. We may also conclude that $f(x) = (x - 4)(x^3 + 6x^2 + 9x + 4)$, since the bottom line gives us the coefficients in the quotient.

Example E

By using synthetic division, determine whether $2x - 3$ is a factor of $2x^3 - 3x^2 + 8x - 12$.

▶ We first note that the coefficient of x in the possible factor is not 1. Thus, we cannot use $r = 3$, since the factor is not of the form $x - r$. However, $2x - 3 = 2(x - \frac{3}{2})$, which means that if $2(x - \frac{3}{2})$ is a factor of the function, $2x - 3$ is a factor. If we use $r = \frac{3}{2}$, and find that the remainder is zero, then $x - \frac{3}{2}$ is a factor.

$$
\begin{array}{rrrr|r}
2 & -3 & 8 & -12 & \dfrac{3}{2} \\
  & 3 & 0 & 12 & \\
\hline
2 & 0 & 8 & 0 &
\end{array}
$$

Since the remainder is zero, $x - \frac{3}{2}$ is a factor. Also, the quotient is $2x^2 + 8$, which may be factored into $2(x^2 + 4)$. Thus, 2 is also a factor of the function. This means that $2(x - \frac{3}{2})$ is a factor of the function, and this in turn means that $2x - 3$ is a factor.

Example F

By synthetic division, determine whether or not $\frac{1}{3}$ is a zero of the function $3x^3 + 2x^2 - 4x + 1$.

This problem is equivalent to dividing the function by $x - \frac{1}{3}$. If the remainder is zero, $\frac{1}{3}$ is a zero of the function.

$$
\begin{array}{rrrr|r}
3 & 2 & -4 & 1 & \dfrac{1}{3} \\
  & 1 & 1 & -1 & \\
\hline
3 & 3 & -3 & 0 &
\end{array}
$$

Since the remainder is zero, we conclude that $\frac{1}{3}$ is a zero of the function.

$$3x^3 + 2x^2 - 4x + 1 = \left(x - \frac{1}{3}\right)(3x^2 + 3x - 3)$$

$$= 3\left(x - \frac{1}{3}\right)(x^2 + x - 1)$$

Exercises 14-2

In Exercises 1 through 20, perform the required divisions by synthetic division. Exercises 1 through 16 are the same as those of Section 14-1.

1. $(x^3 + 2x^2 - x - 2) \div (x - 1)$ 2. $(x^3 - 3x^2 - x + 2) \div (x - 2)$

3. $(x^3 + 2x + 3) \div (x + 1)$ 4. $(x^4 - 4x^3 - x^2 + x - 100) \div (x + 3)$

5. $(2x^5 - x^2 + 8x + 44) \div (x + 2)$ 6. $(x^3 + 4x^2 - 25x - 98) \div (x - 5)$

7. $(2x^4 - 3x^3 - 4x^2 + 2x - 5) \div (x - 3)$

8. $(2x^4 - 10x^2 + 30x - 60) \div (x + 4)$

9. $(x^3 + 2x^2 - 3x + 4) \div (x + 1)$ 10. $(2x^3 - 4x^2 + x - 1) \div (x + 2)$

11. $(x^4 + x^3 - 2x^2 - 5x + 3) \div (x + 4)$

12. $(2x^4 - x^2 + 5x - 7) \div (x - 3)$ 13. $(2x^4 - 7x^3 - x^2 + 8) \div (x - 3)$

14. $(x^4 - 5x^3 + x^2 - 2x + 6) \div (x + 4)$

15. $(x^5 - 3x^3 + 5x^2 - 10x + 6) \div (x - 2)$

16. $(3x^4 - 12x^3 - 60x + 4) \div (x - 5)$

17. $(x^6 + 2x^2 - 6) \div (x - 2)$ 18. $(x^5 + 4x^4 - 8) \div (x + 1)$

19. $(x^7 - 128) \div (x - 2)$ 20. $(x^5 + 32) \div (x + 2)$

In Exercises 21 through 32, use the factor theorem and synthetic division to determine whether or not the second expression is a factor of the first.

21. $x^3 + x^2 - x + 2$; $x + 2$ 22. $x^3 + 6x^2 + 10x + 6$; $x + 3$

23. $x^4 - 6x^2 - 3x - 2$; $x - 3$ 24. $2x^4 - 5x^3 - 24x^2 + 5$; $x - 5$

25. $2x^5 - x^3 + 3x^2 - 4$; $x + 1$ 26. $x^5 - 3x^4 - x^2 - 6$; $x - 3$

27. $4x^3 - 6x^2 + 2x - 2$; $x - \frac{1}{2}$ 28. $3x^3 - 5x^2 + x + 1$; $x + \frac{1}{3}$

29. $2x^4 - x^3 + 2x^2 - 3x + 1$; $2x - 1$

30. $6x^4 + 5x^3 - x^2 + 6x - 2$; $3x - 1$

31. $4x^4 + 2x^3 - 8x^2 + 3x + 12$; $2x + 3$

32. $3x^4 - 2x^3 + x^2 + 15x + 4$; $3x + 4$

In Exercises 33 through 36, use synthetic division to determine whether or not the given numbers are zeros of the given functions.

33. $x^4 - 5x^3 - 15x^2 + 5x + 14$; 7 34. $x^4 + 7x^3 + 12x^2 + x + 4$; -4

35. $9x^3 + 9x^2 - x + 2$; $-\frac{2}{3}$ 36. $2x^3 + 13x^2 + 10x - 4$; $\frac{1}{2}$

14-3 The Roots of an Equation

In this section we shall present certain theorems which are useful in determining the number of roots in the equation $f(x) = 0$, and the nature of some of these roots. In dealing with polynomial equations of higher degree, it is helpful to have as much of this kind of information as is readily obtainable before proceeding to solve for the roots.

The first of these theorems is so important that it is called **the fundamental theorem of algebra.** It states that *every polynomial equation has at least one (real or complex) root.* The proof of this theorem is of an advanced nature, and therefore we must accept its validity at this time. However, using the fundamental theorem, we can show the validity of other useful theorems.

Let us now assume that we have a polynomial equation $f(x) = 0$, and that we are looking for its roots. By the fundamental theorem, we know that it has at least one root. Assuming that we can find this root by some means (the factor theorem, for example), we shall call this root r_1. Thus,

$$f(x) = (x - r_1)f_1(x)$$

where $f_1(x)$ is the polynomial quotient found by dividing $f(x)$ by $(x - r_1)$. However, since the fundamental theorem states that any polynomial equation has at least one root, this must apply to $f_1(x) = 0$ as well. Let us assume that $f_1(x) = 0$ has the root r_2. Therefore, this means that $f(x) = (x - r_1)(x - r_2)f_2(x)$. Continuing this process until one of the quotients is a constant a, we have

$$f(x) = a(x - r_1)(x - r_2) \cdots (x - r_n)$$

Note that one linear factor appears each time a root is found, and that the degree of the quotient is one less each time. Thus there are n factors, if the degree of $f(x)$ is n. This leads us to two theorems. The first of these states that *each polynomial of the nth degree can be factored into n linear factors.* The second theorem states that *each polynomial equation of degree n has exactly n roots.*

Example A

Consider the equation $f(x) = 2x^4 - 3x^3 - 12x^2 + 7x + 6 = 0$.

$$2x^4 - 3x^3 - 12x^2 + 7x + 6 = (x - 3)(2x^3 + 3x^2 - 3x - 2)$$
$$2x^3 + 3x^2 - 3x - 2 = (x + 2)(2x^2 - x - 1)$$
$$2x^2 - x - 1 = (x - 1)(2x + 1)$$
$$2x + 1 = 2\left(x + \frac{1}{2}\right)$$

Therefore,

$$2x^4 - 3x^3 - 12x^2 + 7x + 6 = 2(x - 3)(x + 2)(x - 1)\left(x + \frac{1}{2}\right) = 0$$

The degree of $f(x)$ is 4. There are 4 linear factors: $(x - 3)$, $(x + 2)$, $(x - 1)$, and $(x + \frac{1}{2})$. There are 4 roots of the equation: 3, -2, 1, and $-\frac{1}{2}$. Thus, we have verified each of the theorems above for this example.

It is not necessary for each root of an equation to be different from the others. For example, the equation $(x - 1)^2 = 0$ has two roots, both of which are 1. Such roots are referred to as multiple roots.

When we solve the equation $x^2 + 1 = 0$, we get two roots, j and $-j$. In fact, if we have any equation for which the roots are complex, for every root of the form $a + bj$ $(b \neq 0)$, there is also a root of the form $a - bj$. This is so because any quadratic equation can be solved by the quadratic formula. The solutions from the quadratic formula (for an equation of the form $ax^2 + bx + c = 0$) are

$$\frac{-b + \sqrt{b^2 - 4ac}}{2a} \quad \text{and} \quad \frac{-b - \sqrt{b^2 - 4ac}}{2a}$$

and the only difference between these roots is the sign before the radical. Thus we have the following theorem. *If a complex number* $a + bj$ *is the root of* $f(x) = 0$, *its conjugate* $a - bj$ *is also a root.*

Example B

Consider the equation $f(x) = (x - 1)^3(x^2 + x + 1) = 0$.

We observe directly (since three factors of $x - 1$ are already indicated) that there is a triple root of 1. To find the other two roots, we use the quadratic formula on the *factor* $(x^2 + x + 1)$. This is permissible, because what we are actually finding are those values of x which make

$$x^2 + x + 1 = 0$$

For this we have

$$x = \frac{-1 \pm \sqrt{1 - 4}}{2}$$

Thus,

$$x = \frac{-1 + j\sqrt{3}}{2} \quad \text{and} \quad x = \frac{-1 - j\sqrt{3}}{2}$$

Therefore, the roots of $f(x)$ are

$$1, \quad 1, \quad 1, \quad \frac{-1 + j\sqrt{3}}{2}, \quad \frac{-1 - j\sqrt{3}}{2}$$

One further observation can be made from Example B. *Whenever enough roots are known so that the remaining factor is quadratic, it is always possible to find the remaining roots from the quadratic formula.* This is true for finding real or complex roots. If there are n roots and if we find $n - 2$ of these roots, the solution may be completed by using the quadratic formula.

Example C

Solve the equation $3x^3 + 10x^2 - 16x - 32 = 0$ given that $-\frac{4}{3}$ is a root.

Using synthetic division and the given root we have

$$
\begin{array}{rrrr|}
3 & 10 & -16 & -32 \\
 & -4 & -8 & 32 \\
\hline
3 & 6 & -24 & 0 \\
\end{array}
\quad -\frac{4}{3}
$$

Thus, $3x^3 + 10x^2 - 16x - 32 = (x + \frac{4}{3})(3x^2 + 6x - 24)$. We know that $x + \frac{4}{3}$ is a factor from the given root, and that $3x^2 + 6x - 24$ is a factor from the synthetic division. This second factor has a common factor of 3, which means that $3x^2 + 6x - 24 = 3(x^2 + 2x - 8)$. We now see that $x^2 + 2x - 8 = (x + 4)(x - 2)$. Thus,

$$3x^3 + 10x^2 - 16x - 32 = 3\left(x + \frac{4}{3}\right)(x + 4)(x - 2)$$

This means the roots are $-\frac{4}{3}$, -4, and 2.

Example D

Solve the equation $x^4 + 3x^3 - 4x^2 - 10x - 4 = 0$, given that -1 and 2 are roots.

Using synthetic division and the root -1, we have

$$
\begin{array}{rrrrr}
1 & 3 & -4 & -10 & -4 \\
 & -1 & -2 & 6 & 4 \\
\hline
1 & 2 & -6 & -4 & 0 \\
\end{array}
\quad \underline{|-1}
$$

Therefore, we now know that

$$x^4 + 3x^3 - 4x^2 - 10x - 4 = (x + 1)(x^3 + 2x^2 - 6x - 4)$$

We now know that $x - 2$ must be a factor of $x^3 + 2x^2 - 6x - 4$, since it is a factor of the original function. Again, using synthetic division and this time the root 2, we have the following:

$$
\begin{array}{rrrr}
1 & 2 & -6 & -4 \\
 & 2 & 8 & 4 \\
\hline
1 & 4 & 2 & 0 \\
\end{array}
\quad \underline{|2}
$$

Thus,

$$x^4 + 3x^3 - 4x^2 - 10x - 4 = (x + 1)(x - 2)(x^2 + 4x + 2)$$

The roots from this last factor are now found by the quadratic formula:

$$x = \frac{-4 \pm \sqrt{16 - 8}}{2} = \frac{-4 \pm 2\sqrt{2}}{2} = -2 \pm \sqrt{2}$$

Therefore, the roots are -1, 2, $-2 + \sqrt{2}$, $-2 - \sqrt{2}$.

Example E

Solve the equation $3x^4 - 26x^3 + 63x^2 - 36x - 20 = 0$, given that 2 is a double root.

Using synthetic division, we have

$$\begin{array}{rrrrr|l}
3 & -26 & 63 & -36 & -20 & \underline{2} \\
  & 6 & -40 & 46 & 20 & \\
\hline
3 & -20 & 23 & 10 & 0 &
\end{array}$$

Therefore, we know that

$$3x^4 - 26x^3 + 63x^2 - 36x - 20 = (x - 2)(3x^3 - 20x^2 + 23x + 10)$$

Also, since 2 is a double root, it must be a root of the quotient $3x^3 - 20x^2 + 23x + 10$. Using synthetic division again, we have

$$\begin{array}{rrrr|l}
3 & -20 & 23 & 10 & \underline{2} \\
  & 6 & -28 & -10 & \\
\hline
3 & -14 & -5 & 0 &
\end{array}$$

The quotient $3x^2 - 14x - 5$ factors into $(3x + 1)(x - 5)$. Therefore, the roots of the equation are 2, 2, $-\frac{1}{3}$, and 5.

Example F

Solve the equation $2x^4 - x^3 + 7x^2 - 4x - 4 = 0$, given that $2j$ is a root.

Since $2j$ is a root, we know that $-2j$ is also a root. Using synthetic division twice, we can then reduce the remaining factor to a quadratic function.

$$\begin{array}{rrrrr|l}
2 & -1 & 7 & -4 & -4 & \underline{2j} \\
  & 4j & -8 - 2j & 4 - 2j & 4 & \\
\hline
2 & 4j - 1 & -1 - 2j & -2j & 0 & \quad\underline{-2j} \\
  & -4j & & 2j & 2j & \\
\hline
2 & -1 & -1 & & &
\end{array}$$

The quadratic factor $2x^2 - x - 1$ factors into $(2x + 1)(x - 1)$. Therefore, the roots of the equation are $2j$, $-2j$, 1, and $-\frac{1}{2}$.

Exercises 14–3

In Exercises 1 through 24, solve the given equations using synthetic division, given the roots indicated.

1. $x^3 + 2x^2 - x - 2 = 0$ $(r_1 = 1)$
2. $x^3 + 2x^2 + x + 2 = 0$ $(r_1 = -2)$
3. $x^3 + x^2 - 8x - 12 = 0$ $(r_1 = -2)$
4. $x^3 - 1 = 0$ $(r_1 = 1)$
5. $2x^3 + 11x^2 + 20x + 12 = 0$ $(r_1 = -\frac{3}{2})$
6. $2x^3 + 5x^2 - 11x + 4 = 0$ $(r_1 = \frac{1}{2})$
7. $3x^3 + 2x^2 + 3x + 2 = 0$ $(r_1 = j)$

8. $x^3 + 5x^2 + 9x + 5 = 0$ $(r_1 = -2 + j)$

9. $x^4 + x^3 - 2x^2 + 4x - 24 = 0$ $(r_1 = 2, r_2 = -3)$

10. $x^4 + 2x^3 - 4x^2 - 5x + 6 = 0$ $(r_1 = 1, r_2 = -2)$

11. $x^4 - 6x^2 - 8x - 3 = 0$ (−1 is a double root)

12. $4x^4 + 28x^3 + 61x^2 + 42x + 9 = 0$ (−3 is a double root)

13. $6x^4 + 5x^3 - 15x^2 + 4 = 0$ $(r_1 = -\frac{1}{2}, r_2 = \frac{2}{3})$

14. $6x^4 - 5x^3 - 14x^2 + 14x - 3 = 0$ $(r_1 = \frac{1}{3}, r_2 = \frac{3}{2})$

15. $2x^4 - x^3 - 4x^2 + 10x - 4 = 0$ $(r_1 = 1 + j)$

16. $x^4 - 8x^3 - 72x - 81 = 0$ $(r_1 = 3j)$

17. $2x^5 + 11x^4 + 16x^3 - 8x^2 - 32x - 16 = 0$ (−2 is a triple root)

18. $x^5 - 3x^4 + 4x^3 - 4x^2 + 3x - 1 = 0$ (1 is a triple root)

19. $2x^5 + x^4 - 15x^3 + 5x^2 + 13x - 6 = 0$ $(r_1 = 1, r_2 = -1, r_3 = \frac{1}{2})$

20. $12x^5 - 7x^4 + 41x^3 - 26x^2 - 28x + 8 = 0$ $(r_1 = 1, r_2 = \frac{1}{4}, r_3 = -\frac{2}{3})$

21. $x^5 - 3x^4 - x + 3 = 0$ $(r_1 = 3, r_2 = j)$

22. $4x^5 + x^3 - 4x^2 - 1 = 0$ $(r_1 = 1, r_2 = \frac{1}{2}j)$

23. $x^6 + 2x^5 - 4x^4 - 10x^3 - 41x^2 - 72x - 36 = 0$ (−1 is a double root, 2j is a root)

24. $x^6 - x^5 - 2x^3 - 3x^2 - x - 2 = 0$ (j is a double root)

14–4 Rational and Irrational Roots

If we form the product of the factors $(x + 2)(x - 4)(x + 3)$, we obtain $x^3 + x^2 - 14x - 24$. In forming this product, we find that the constant 24 which results is determined only by the numbers 2, 4, and 3. We note that these numbers represent the roots of the equation if the given function is set equal to zero. In fact, if we found all the integral roots of an equation, and represented the equation in the form

$$f(x) = (x - r_1)(x - r_2) \cdots (x - r_k)f_{k+1}(x) = 0$$

where all the roots indicated are integers, the constant term of $f(x)$ must have factors of $r_1, r_2, \ldots, r_k$. This leads us to the theorem which states that *if the coefficient of the highest power of x is 1*, then any integral roots are factors of the constant term of the function.

Example A

The equation $f(x) = x^5 - 4x^4 - 7x^3 + 14x^2 - 44x + 120 = 0$ can be written as

$$(x - 5)(x + 3)(x - 2)(x^2 + 4) = 0$$

We now note that $5(3)(2)(4) = 120$. Thus, the roots 5, −3, and 2 are numerical factors of $|120|$. The theorem states nothing in regard to the signs involved.

If the coefficient of the highest-power term of $f(x)$ is not 1, then this coefficient can be factored from every term of $f(x)$. Thus any equation of the form $f(x) = a_0x^n + a_1x^{n-1} + \cdots + a_n = 0$ can be written in the form

$$f(x) = a_0\left(x^n + \frac{a_1}{a_0}x^{n-1} + \cdots + \frac{a_n}{a_0}\right) = 0 \qquad (14\text{–}4)$$

This equation, along with the theorem above, now gives us another, more inclusive, theorem.

Any rational roots of a polynomial equation

$$f(x) = a_0x^n + a_1x^{n-1} + \cdots + a_n = 0$$

▶ *must be integral factors of a_n divided by integral factors of a_0.*

The same reasoning as previously stated, applied to the factor within parentheses of Eq. (14–4), leads us to this result.

Example B   If $f(x) = 4x^3 - 3x^2 - 25x - 6 = 0$, any rational roots, if they exist, must be integral factors of 6 divided by integral factors of 4. The integral factors of 6 are 1, 2, 3, and 6 and the integral factors of 4 are 1, 2, and 4. Forming all possible positive and negative quotients, any rational roots that exist will be found in the following list: ± 1, $\pm\frac{1}{2}$, $\pm\frac{1}{4}$, ± 2, ± 3, $\pm\frac{3}{2}$, $\pm\frac{3}{4}$, ± 6.

The roots of this equation are -2, 3, and $-\frac{1}{4}$.

There are 16 different possible rational roots in Example B. Since we have no way of telling which of these are the actual roots, we now present a rule which will help us to find these roots. This rule is known as **Descartes' rule of signs.** It states that *the number of positive roots of a polynomial equation $f(x) = 0$ cannot exceed the number of changes in sign in $f(x)$ in going from one term to the next in $f(x)$. The number of negative roots cannot exceed the number of sign changes in $f(-x)$.*

We can reason this way: If $f(x)$ has all positive terms, then any positive number substituted in $f(x)$ must give a positive value for the function. This indicates that the number substituted in the function is not a root. Thus, there must be at least one negative and one positive term in the function for any positive number to be a root. This is not a proof, but does indicate the type of reasoning which is used in developing the theorem.

Example C

By Descartes' rule of signs, determine the maximum number of positive and negative roots of $3x^3 - x^2 - x + 4 = 0$.

Here $f(x) = 3x^3 - x^2 - x + 4$. The first term is positive and the second negative, which indicates a change of sign. The third term is also negative; there is no change of sign from the second to the third term. The fourth term is positive, thus giving us a second change of sign, from the third to the fourth term. Hence there are two changes in sign, and therefore no more than two positive roots of $f(x) = 0$. Then we write

$$f(-x) = 3(-x)^3 - (-x)^2 - (-x) + 4 = -3x^3 - x^2 + x + 4$$

There is only one change of sign in $f(-x)$; therefore, there is one negative root. *When there is just one change of sign in $f(x)$, there is a positive root, and when there is just one change of sign in $f(-x)$, there is a negative root.*

Example D

For the equation $4x^5 - x^4 - 4x^3 + x^2 - 5x - 6 = 0$, we write

$$f(x) = 4x^5 - x^4 - 4x^3 + x^2 - 5x - 6$$

and

$$f(-x) = -4x^5 - x^4 + 4x^3 + x^2 + 5x - 6$$

Thus, there are no more than three positive and two negative roots.

At this point let us summarize the information we can determine about the roots of a polynomial equation $f(x) = 0$ of degree n:

1. There are n roots.
2. Complex roots appear in conjugate pairs.
3. Any rational roots must be factors of the constant term divided by factors of the coefficient of the highest-power term.
4. The maximum number of positive roots is the number of sign changes in $f(x)$, and the maximum number of negative roots is the number of sign changes in $f(-x)$.
5. Once we determine $n - 2$ of the roots, the remaining roots can be found by the quadratic formula.

Synthetic division is normally used to try possible roots. This is because synthetic division is relatively easy to perform, and when a root is found we have the quotient factor, which is of degree one less than the degree of the dividend. Each root we find makes the ensuing work simpler. The following examples indicate the complete method, as well as two other helpful rules.

Example E

Determine the roots of the equation $2x^3 + x^2 + 5x - 3 = 0$.

Since $n = 3$, there are three roots. If we can find one of these roots, we can use the quadratic formula to find the other two. We have $f(x) = 2x^3 + x^2 + 5x - 3$, and therefore there is one positive root. We also have $f(-x) = -2x^3 + x^2 - 5x - 3$, and therefore there are no more than two negative roots. The *possible* rational roots are ± 1, $\pm \frac{1}{2}$, $\pm \frac{3}{2}$, ± 3. Thus, using synthetic division, we shall try these. We first try the root 1 (always a possibility if there are positive roots).

$$
\begin{array}{rrrr|l}
2 & 1 & 5 & -3 & \underline{1} \\
  & 2 & 3 & 8 & \\
\hline
2 & 3 & 8 & 5 &
\end{array}
$$

Thus we see that 1 is not a root, but we have gained some information, if we observe closely. If we try any positive number larger than 1, the results in the last row will be larger positive numbers than we now have. The products will be larger, and therefore the sums will also be larger. Thus there is no positive root larger than 1. This leads to the following rule: *When we are trying a positive root, if the bottom row contains all positive numbers, then there are no roots larger than the value tried.* This rule tells us that there is no reason to try $+\frac{3}{2}$ and $+3$ as roots. Therefore, let us now try $+\frac{1}{2}$.

$$
\begin{array}{rrrr|l}
2 & 1 & 5 & -3 & \dfrac{1}{2} \\
  & 1 & 1 & 3 & \\
\hline
2 & 2 & 6 & 0 &
\end{array}
$$

Hence $+\frac{1}{2}$ is a root, and the remaining factor is $2x^2 + 2x + 6$, which itself factors to $2(x^2 + x + 3)$. By the quadratic formula we find the remaining roots.

$$
x = \frac{-1 \pm \sqrt{1 - 12}}{2} = \frac{-1 \pm j\sqrt{11}}{2}
$$

The three roots are

$$
\frac{1}{2}, \qquad \frac{-1 + j\sqrt{11}}{2} \quad \text{and} \quad \frac{-1 - j\sqrt{11}}{2}
$$

We note that there were actually no negative roots, because the nonpositive roots are complex. Also, in proceeding in this way, we never found it necessary to try any negative roots. It must be admitted, however, that the solutions to all problems may not be so easily determined.

Example F

Determine the roots of the equation $x^4 - 7x^3 + 12x^2 + 4x - 16 = 0$.
We write

$$f(x) = x^4 - 7x^3 + 12x^2 + 4x - 16$$
$$f(-x) = x^4 + 7x^3 + 12x^2 - 4x - 16$$

We see that there are four roots; there are no more than three positive roots, and there is one negative root. The possible rational roots are ± 1, ± 2, ± 4, ± 8, ± 16. Since there is only one negative root, we shall look for this one first. Trying -2, we have

$$
\begin{array}{rrrrr|r}
1 & -7 & +12 & +4 & -16 & \underline{\,-2} \\
 & -2 & +18 & -60 & +112 & \\
\hline
1 & -9 & +30 & -56 & +96 &
\end{array}
$$

If we were to try any negative roots less than -2 (remember, -3 is less than -2), we would find that the numbers would still alternate from term to term in the quotient. Thus, we have this rule: *When we are trying a negative root, if the signs alternate in the bottom row, then there are no roots less than the value tried.* So we next try -1.

$$
\begin{array}{rrrrr|r}
1 & -7 & +12 & +4 & -16 & \underline{\,-1} \\
 & -1 & 8 & -20 & 16 & \\
\hline
1 & -8 & 20 & -16 & 0 &
\end{array}
$$

Thus, -1 is the negative root. Next we shall try $+1$.

$$
\begin{array}{rrrr|r}
1 & -8 & 20 & -16 & \underline{\,1} \\
 & 1 & -7 & 13 & \\
\hline
1 & -7 & 13 & -3 &
\end{array}
$$

Since $+1$ is not a root, we next try $+2$.

$$
\begin{array}{rrrr|r}
1 & -8 & 20 & -16 & \underline{\,2} \\
 & 2 & -12 & 16 & \\
\hline
1 & -6 & 8 & 0 &
\end{array}
$$

Thus $+2$ is a root. It is not necessary to find any more roots by trial and error. We now may use the quadratic formula on the remaining factor $x^2 - 6x + 8$, and find that the roots are 2 and 4. Thus, the roots are -1, 2, 2, and 4. (Note that 2 is a double root.)

In concluding this discussion of the roots of an equation, we note that when a polynomial equation has more than two irrational roots, we cannot generally find these roots by the methods just presented. These irrational roots can be approximated by graphical methods, such as those discussed in Section 2–5. Also, a calculator can be used to determine the values of these roots to the desired accuracy. The use of a calculator is illustrated in the following example.

Example G

Find the irrational root of the equation $x^3 + 2x^2 + 8x - 2 = 0$ which lies between 0 and 1.

We know that whenever a curve crosses the x-axis, that value of x is a root. We are told here that the root we want is between 0 and 1. We would then expect to find the function either positive or negative when $x = 0$, and to have the opposite sign when $x = 1$. To check this, the remainder theorem may be used. We find that $f(0) = -2$ and $f(1) = 9$.

We shall now approximate the value of the root by using a calculator to evaluate $f(x)$ at values approximately halfway between those values where $f(x)$ has different signs. At this point we will try $x = 0.4$. The halfway point between $x = 0$ and $x = 1$ is 0.5, but it is likely that $f(x) = 0$ is nearer $x = 0$ than $x = 1$ since $f(0) = -2$ and $f(1) = 9$. Using the calculator sequence

$$.4 \quad \boxed{x^y} \quad 3 \quad \boxed{+} \quad 2 \quad \boxed{\times} \quad .4 \quad \boxed{x^2}$$

$$\boxed{+} \quad 8 \quad \boxed{\times} \quad .4 \quad \boxed{-} \quad 2 \quad \boxed{=}$$

we find that $f(0.4) = 1.584$. Since $f(0.4)$ is positive we know that the root is between 0 and 0.4.

Since $f(0)$ and $f(0.4)$ are nearly equal in absolute value, we now evaluate $f(0.2)$. This value is -0.312.

We now know that the root is between 0.2 and 0.4. Therefore, we find that $f(0.3) = 0.607$. This tells us that the root is between 0.2 and 0.3, and that it is likely to be closer to 0.2.

Next we find that $f(0.24) = 0.049024$, $f(0.22) = -0.132552$, and $f(0.23) = -0.042033$. Therefore, the root is between 0.23 and 0.24. Since $f(0.235) = 0.0034279$, we know that the value of the root is $x = 0.23$ to two decimal places.

Exercises 14–4

In Exercises 1 through 20, solve the given equations.

1. $x^3 + 2x^2 - x - 2 = 0$ 2. $x^3 + x^2 - 5x + 3 = 0$

3. $x^3 + 2x^2 - 5x - 6 = 0$ 4. $x^3 + 1 = 0$

5. $2x^3 - 5x^2 - 28x + 15 = 0$ 6. $2x^3 - x^2 - 3x - 1 = 0$

7. $3x^3 + 11x^2 + 5x - 3 = 0$ 8. $4x^3 - 5x^2 - 23x + 6 = 0$

9. $x^4 - 11x^2 - 12x + 4 = 0$ 10. $x^4 + x^3 - 2x^2 - 4x - 8 = 0$

11. $x^4 - 2x^3 - 13x^2 + 14x + 24 = 0$ 12. $x^4 - x^3 + 2x^2 - 4x - 8 = 0$

13. $2x^4 - 5x^3 - 3x^2 + 4x + 2 = 0$ 14. $2x^4 + 7x^3 + 9x^2 + 5x + 1 = 0$

15. $12x^4 + 44x^3 + 21x^2 - 11x - 6 = 0$

16. $9x^4 - 3x^3 + 34x^2 - 12x - 8 = 0$

17. $x^5 + x^4 - 9x^3 - 5x^2 + 16x + 12 = 0$

18. $x^6 - x^4 - 14x^2 + 24 = 0$

19. $2x^5 - 5x^4 + 6x^3 - 6x^2 + 4x - 1 = 0$

20. $2x^5 + 5x^4 - 4x^3 - 19x^2 - 16x - 4 = 0$

In Exercises 21 through 28, determine the required quantities. In Exercises 27 and 28, it is necessary to set up equations of higher degree.

21. Under certain conditions, the velocity of an object as a function of time is given by $v = 2t^3 - 11t^2 - 28t - 15$. For what values of t is $v = 0$?

22. The deflection y of a beam at a horizontal distance x from one end is given by $y = k(x^4 - 2Lx^3 + L^3x)$, where L is the length of the beam and k is a constant. For what values of x is the deflection zero?

23. In the theory of the motion of a sphere moving through a fluid, the expression $4r^3 - 3ar^2 - a^3$ is found. In terms of a, solve for r if this expression is zero.

24. The area of a segment of a circle is given approximately by the equation

$$A = \frac{h^3}{2L} + \frac{2Lh}{3}$$

where L is the length of the chord and h is the altitude of the segment. Find h when $L = \frac{3}{2}$ and $A = \frac{13}{24}$.

25. In determining one of the dimensions d of the support columns of a building, the equation $3d^3 + 5d^2 - 400d - 18000 = 0$ is found. Determine this dimension (in inches).

26. Three electric resistors are connected in parallel. The second resistor is $1\ \Omega$ more than the first, and the third resistor is $4\ \Omega$ more than the first. The total resistance of the combination is $1\ \Omega$. To find the first resistance R, we must solve the equation

$$\frac{1}{R} + \frac{1}{R + 1} + \frac{1}{R + 4} = 1$$

Find the values of the resistances.

27. A rectangular box is made from a piece of cardboard 8 in. by 12 in., by cutting a square from each corner and bending up the sides. How large is the side of the square cut out, if the volume of the box is 64 in.^{3}?

28. A slice 2 cm thick is cut off the side of a cube, leaving 75 cm^3 in the remaining volume. Find the length of the edge of the cube.

In Exercises 29 and 30, use a calculator to find the irrational root, to two decimal places, which lies between the listed values. See Example G.

29. $x^3 - 6x^2 + 10x - 4 = 0$ (0 and 1)

30. $x^4 - x^3 - 3x^2 - x - 4 = 0$ (2 and 3)

In Exercises 31 and 32, find all the real roots of the equations to two decimal places. Graph the functions to locate the roots approximately, and use a calculator to complete the solution. See Example G.

31. $x^3 - 2x^2 - 5x + 4 = 0$

32. $x^4 - x^3 - 2x^2 - x - 3 = 0$

14–5 Exercises for Chapter 14

In Exercises 1 through 4, find the remainder of the indicated division by the remainder theorem.

1. $(2x^3 - 4x^2 - x + 4) \div (x - 1)$
2. $(x^3 - 2x^2 + 9) \div (x + 2)$
3. $(4x^3 + x + 4) \div (x + 3)$
4. $(x^4 - 5x^3 + 8x^2 + 15x - 2) \div (x - 3)$

In Exercises 5 through 8, use the factor theorem to determine whether or not the second expression is a factor of the first.

5. $x^4 + x^3 + x^2 - 2x - 3; \quad x + 1$ 6. $2x^3 - 2x^2 - 3x - 2; \quad x - 2$
7. $x^4 + 4x^3 + 5x^2 + 5x - 6; \quad x + 3$ 8. $9x^3 + 6x^2 + 4x + 2; \quad 3x + 1$

In Exercises 9 through 16, use synthetic division to perform the indicated divisions.

9. $(x^3 + 3x^2 + 6x + 1) \div (x - 1)$ 10. $(3x^3 - 2x^2 + 7) \div (x - 3)$
11. $(2x^3 - 3x^2 - 4x + 3) \div (x + 2)$ 12. $(3x^3 - 5x^2 + 7x - 6) \div (x + 4)$
13. $(x^4 - 2x^3 - 3x^2 - 4x - 8) \div (x + 1)$
14. $(x^4 - 6x^3 + x - 8) \div (x - 3)$ 15. $(2x^5 - 46x^3 + x^2 - 9) \div (x - 5)$
16. $(x^6 + 63x^3 + 5x^2 - 9x - 8) \div (x + 4)$

In Exercises 17 through 20, use synthetic division to determine whether or not the given numbers are zeros of the given functions.

17. $x^3 + 8x^2 + 17x - 6; \quad -3$ 18. $2x^3 + x^2 - 4x + 4; \quad -2$
19. $2x^4 - x^3 + 2x^2 + x - 1; \quad \frac{1}{2}$ 20. $6x^4 - 7x^3 + 2x^2 - 9x - 6; \quad -\frac{2}{3}$

In Exercises 21 through 32, find all of the solutions of the given equations, with the aid of synthetic division and with the roots indicated.

21. $x^3 + 8x^2 + 17x + 6 = 0 \quad (r_1 = -3)$
22. $2x^3 + 7x^2 - 6x - 8 = 0 \quad (r_1 = -4)$
23. $3x^4 + 5x^3 + x^2 + x - 10 = 0 \quad (r_1 = 1, r_2 = -2)$
24. $x^4 - x^3 - 5x^2 - x - 6 = 0 \quad (r_1 = 3, r_2 = -2)$
25. $2x^4 + x^3 - 29x^2 - 34x + 24 = 0 \quad (r_1 = -2, r_2 = \frac{1}{2})$
26. $x^4 + x^3 - 11x^2 - 9x + 18 = 0 \quad (r_1 = -3, r_2 = 1)$
27. $4x^4 + 4x^3 + x^2 + 4x - 3 = 0 \quad (r_1 = j)$
28. $x^4 + 2x^3 - 4x - 4 = 0 \quad (r_1 = -1 + j)$
29. $x^5 + 3x^4 - x^3 - 11x^2 - 12x - 4 = 0 \quad (-1 \text{ is a triple root})$
30. $24x^5 + 10x^4 + 7x^2 - 6x + 1 = 0 \quad (r_1 = -1, r_2 = \frac{1}{4}, r_3 = \frac{1}{3})$
31. $x^5 + 4x^4 + 5x^3 - x^2 - 4x - 5 = 0 \quad (r_1 = 1, r_2 = -2 + j)$
32. $2x^5 - x^4 + 8x - 4 = 0 \quad (r_1 = \frac{1}{2}, r_2 = 1 + j)$

In Exercises 33 through 40, solve the given equations.

33. $x^3 + x^2 - 10x + 8 = 0$ 34. $x^3 - 8x^2 + 20x - 16 = 0$

35. $2x^3 - x^2 - 8x - 5 = 0$ 36. $2x^3 - 3x^2 - 11x + 6 = 0$

37. $6x^3 - x^2 - 12x - 5 = 0$ 38. $6x^3 + 19x^2 + 2x - 3 = 0$

39. $2x^4 + x^3 + 3x^2 + 2x - 2 = 0$

40. $2x^4 + 5x^3 - 14x^2 - 23x + 30 = 0$

In Exercises 41 through 52, determine the required quantities. Where appropriate, set up the required equations.

41. For what value of k is $x + 2$ a factor of $f(x) = 3x^3 + kx^2 - 8x - 8$?

42. For what value of k is $x - 3$ a factor of $f(x) = kx^4 - 15x^2 - 5x - 12$?

43. Where does the graph of the function $f(x) = 6x^4 - 14x^3 + 5x^2 + 5x - 2$ cross the x-axis?

44. Where does the graph of the function $f(x) = 2x^4 - 7x^3 + 11x^2 - 28x + 12$ cross the x-axis?

45. The total profit P a manufacturer makes in producing x units of a commodity is given by $P = 2x^3 - 3x^2 - 5x - 150$. Less than how many units of production result in a loss?

46. A certain sphere of radius 2.5 cm sinks to a depth y, which is given by the equation $y^3 - 6y^2 + 16 = 0$. Find the depth to which the sphere sinks.

47. The ends of a 10-ft beam are supported at different levels. The deflection y of the beam is given by $y = kx^2(x^3 + 436x - 4000)$, where x is the horizontal distance from one end and k is a constant. Determine the values of x for which the deflection is zero.

48. Three electric capacitors are connected in series. The capacitance of the second is 1 μF (microfarad) more than the first, and the third is 2 μF more than the second. The capacitance of the combination is 1.33 μF. The equation used to determine C, the capacitance of the first capacitor, is

$$\frac{1}{C} + \frac{1}{C+1} + \frac{1}{C+3} = \frac{3}{4}$$

Find the values of the capacitances.

49. The width of a rectangular box equals the depth, and the length is 4 ft more than the width. The volume of the box is 539 ft^3. What are the dimensions of the box?

50. The hypotenuse of a right triangle is 2 m longer than one of the legs. The area of the triangle is 24 m^2. Find the sides of the triangle.

51. The edges of a rectangular box are 3, 5, and 6 ft. If each edge is increased by the same amount, the volume is doubled. By how much should each edge be increased to accomplish this? (The answer is an irrational number. Use a calculator to determine the answer to the nearest tenth.)

52. If the edge of a cube is increased by 1 mm, its volume is doubled. Find the edge of the cube to two decimal places. (The answer is an irrational number. Use a calculator.)

CHAPTER 15

Determinants and Matrices

In Chapter 4 we first met the concept of a determinant and saw how it is used to solve systems of linear equations. However, at that time we limited our discussion to second- and third-order determinants. In the first two sections of this chapter we shall show methods of evaluating higher-order determinants. In the remainder of the chapter we shall develop the related concept of a matrix and its use in solving systems of linear equations.

15–1 Determinants: Expansion by Minors

From Section 4–7 we recall that *a third-order* **determinant** *is defined by the equation*

$$\begin{vmatrix} a_1 & b_1 & c_1 \\ a_2 & b_2 & c_2 \\ a_3 & b_3 & c_3 \end{vmatrix} = a_1b_2c_3 + a_3b_1c_2 + a_2b_3c_1 - a_3b_2c_1 - a_1b_3c_2 - a_2b_1c_3 \quad (15\text{--}1)$$

If we rearrange the terms on the right and factor a_1, a_2, and a_3 from the terms in which they are contained, we have

$$\begin{vmatrix} a_1 & b_1 & c_1 \\ a_2 & b_2 & c_2 \\ a_3 & b_3 & c_3 \end{vmatrix} = a_1(b_2c_3 - b_3c_2) - a_2(b_1c_3 - b_3c_1) + a_3(b_1c_2 - b_2c_1) \quad (15\text{--}2)$$

Recalling the definition of a second-order determinant, we have

$$
\begin{vmatrix} a_1 & b_1 & c_1 \\ a_2 & b_2 & c_2 \\ a_3 & b_3 & c_3 \end{vmatrix} = a_1 \begin{vmatrix} b_2 & c_2 \\ b_3 & c_3 \end{vmatrix} - a_2 \begin{vmatrix} b_1 & c_1 \\ b_3 & c_3 \end{vmatrix} + a_3 \begin{vmatrix} b_1 & c_1 \\ b_2 & c_2 \end{vmatrix}
\tag{15-3}
$$

In Eq. (15–3) we note that the third-order determinant is expanded with the terms of the expansion as products of the elements of the first column and specific second-order determinants. In each case the elements of the second-order determinant are those elements which are in neither the same row nor the same column as the element from the first column. *These determinants are called* **minors.**

In general, *the minor of a given element of a determinant is the determinant which results by deleting the row and the column in which the element lies.* Consider the following example.

Example A

Consider the determinant

$$
\begin{vmatrix} 1 & 2 & 3 \\ 4 & 5 & 6 \\ 7 & 8 & 9 \end{vmatrix}
$$

We find the minor of the element 1 by deleting the elements in the first row and first column because the element 1 is located in the first row and in the first column. This minor is the determinant

$$
\begin{vmatrix} 5 & 6 \\ 8 & 9 \end{vmatrix}
$$

The minor for the element 2 is formed by deleting the elements in the first row and second column, for this is the location of the 2. The minor of 2 is the determinant

$$
\begin{vmatrix} 4 & 6 \\ 7 & 9 \end{vmatrix}
$$

The minor for the element 6 is the determinant

$$
\begin{vmatrix} 1 & 2 \\ 7 & 8 \end{vmatrix}
$$

We now see that Eq. (15–3) expresses the expansion of a third-order determinant as the sum of the products of the elements of the first column and their minors, with the second term assigned a minus sign. Actually this is only one of several ways of expressing the expansion. However, it does lead to a general theorem regarding the expansion of a determinant of any order. The foregoing provides a basis for this theorem, although it cannot be considered as a proof. The theorem is as follows:

> *The value of a determinant of order n may be found by forming the n products of the elements of any column (or row) and their minors. A product is given a plus sign if the sum of the number of the column and the number of the row in which the element lies is even, and a minus sign if this sum is odd. The algebraic sum of the terms thus obtained is the value of the determinant.*

The following two examples illustrate the expansion of a third-order determinant and a fourth-order determinant by minors using the theorem above.

Example B

Evaluate $\begin{vmatrix} 1 & -3 & -2 \\ 4 & -1 & 0 \\ 4 & 3 & -5 \end{vmatrix}$ by expansion by minors.

Since we may expand by any column or row, let us select the first row. The expansion is as follows:

$$\begin{vmatrix} 1 & -3 & -2 \\ 4 & -1 & 0 \\ 4 & 3 & -5 \end{vmatrix} = +(1)\begin{vmatrix} -1 & 0 \\ 3 & -5 \end{vmatrix} - (-3)\begin{vmatrix} 4 & 0 \\ 4 & -5 \end{vmatrix} + (-2)\begin{vmatrix} 4 & -1 \\ 4 & 3 \end{vmatrix}$$

The first term of the expansion is assigned a plus sign since the element 1 is in column 1, row 1 and $1 + 1 = 2$ (even). The second term is assigned a minus sign since the element -3 is in column 2 and row 1, and $2 + 1 = 3$ (odd). The third term is assigned a plus sign since the element -2 is in column 3 and row 1, and $3 + 1 = 4$ (even). Actually, once the first sign has been properly determined, the others are known since the signs alternate from term to term. Using the definition of a second-order determinant, we complete the evaluation

$$\begin{vmatrix} 1 & -3 & -2 \\ 4 & -1 & 0 \\ 4 & 3 & -5 \end{vmatrix} = +(1)(5 - 0) - (-3)(-20 - 0) + (-2)[12 - (-4)]$$

$$= 1(5) + 3(-20) - 2(16) = 5 - 60 - 32 = -87$$

Expansion of this same determinant by the third column is as follows:

$$\begin{vmatrix} 1 & -3 & -2 \\ 4 & -1 & 0 \\ 4 & 3 & -5 \end{vmatrix} = +(-2)\begin{vmatrix} 4 & -1 \\ 4 & 3 \end{vmatrix} - (0)\begin{vmatrix} 1 & -3 \\ 4 & 3 \end{vmatrix} + (-5)\begin{vmatrix} 1 & -3 \\ 4 & -1 \end{vmatrix}$$

$$= -2(12 + 4) + 0 - 5(-1 + 12) = -87$$

This expansion has one advantage: since one of the elements of the third column is zero, its minor does not have to be evaluated because zero times whatever the value of the determinant will give the product of zero.

Example C

Evaluate

$$\begin{vmatrix} 3 & -2 & 0 & 2 \\ 1 & 0 & -1 & 4 \\ -3 & 1 & 2 & -2 \\ 2 & -1 & 0 & -1 \end{vmatrix}$$

Expanding by the third column, we have

$$\begin{vmatrix} 3 & -2 & 0 & 2 \\ 1 & 0 & -1 & 4 \\ -3 & 1 & 2 & -2 \\ 2 & -1 & 0 & -1 \end{vmatrix} = +(0)\begin{vmatrix} 1 & 0 & 4 \\ -3 & 1 & -2 \\ 2 & -1 & -1 \end{vmatrix} - (-1)\begin{vmatrix} 3 & -2 & 2 \\ -3 & 1 & -2 \\ 2 & -1 & -1 \end{vmatrix}$$

$$+(2)\begin{vmatrix} 3 & -2 & 2 \\ 1 & 0 & 4 \\ 2 & -1 & -1 \end{vmatrix} - (0)\begin{vmatrix} 3 & -2 & 2 \\ 1 & 0 & 4 \\ -3 & 1 & -2 \end{vmatrix}$$

It is not necessary to expand the minors in the first and fourth terms, since the element in each case is zero. The minors in the second and third terms can be expanded as third-order determinants or by minors. This illustrates well that expansion by minors effectively reduces the order of the determinant to be evaluated by one. Completing the evaluation by minors, we have

$$\begin{vmatrix} 3 & -2 & 0 & 2 \\ 1 & 0 & -1 & 4 \\ -3 & 1 & 2 & -2 \\ 2 & -1 & 0 & -1 \end{vmatrix} = \begin{vmatrix} 3 & -2 & 2 \\ -3 & 1 & -2 \\ 2 & -1 & -1 \end{vmatrix} + 2\begin{vmatrix} 3 & -2 & 2 \\ 1 & 0 & 4 \\ 2 & -1 & -1 \end{vmatrix}$$

$$= \left[3\begin{vmatrix} 1 & -2 \\ -1 & -1 \end{vmatrix} - (-3)\begin{vmatrix} -2 & 2 \\ -1 & -1 \end{vmatrix} + 2\begin{vmatrix} -2 & 2 \\ 1 & -2 \end{vmatrix} \right]$$

$$+ 2\left[-(-2)\begin{vmatrix} 1 & 4 \\ 2 & -1 \end{vmatrix} + 0\begin{vmatrix} 3 & 2 \\ 2 & -1 \end{vmatrix} - (-1)\begin{vmatrix} 3 & 2 \\ 1 & 4 \end{vmatrix} \right]$$

$$= [3(-1 - 2) + 3(2 + 2) + 2(4 - 2)]$$
$$\quad + 2[2(-1 - 8) + (12 - 2)]$$
$$= [-9 + 12 + 4] + 2[-18 + 10]$$
$$= +7 + 2(-8) = -9$$

We can use the expansion of determinants by minors to solve systems of linear equations. **Cramer's rule** *for solving systems of equations, as stated in Section 4–7, is valid for any system of n equations in n unknowns.* The following examples illustrate the solution of systems of equations.

Example D

The production of a particular computer component is done in three stages, taking a total of 7 h. The second stage is 1 h less than the first, and the third stage is twice as long as the second stage. How long is each stage of production?

First, we let a = the number of hours of the first production stage, b = the number of hours of the second stage, and c = the number of hours of the third stage.

The total production time of 7 h gives us $a + b + c = 7$. Since the second stage is 1 h less than the first, we then have $b = a - 1$. The fact that the third stage is twice as long as the second gives us $c = 2b$. Stating these equations in standard form for solution, we have

$$a + b + c = 7$$
$$a - b \quad = 1$$
$$2b - c = 0$$

Using Cramer's rule, we have

$$a = \frac{\begin{vmatrix} 7 & 1 & 1 \\ 1 & -1 & 0 \\ 0 & 2 & -1 \end{vmatrix}}{\begin{vmatrix} 1 & 1 & 1 \\ 1 & -1 & 0 \\ 0 & 2 & -1 \end{vmatrix}} = \frac{-(1)\begin{vmatrix} 1 & 1 \\ 2 & -1 \end{vmatrix} + (-1)\begin{vmatrix} 7 & 1 \\ 0 & -1 \end{vmatrix} - 0\begin{vmatrix} 7 & 1 \\ 0 & 2 \end{vmatrix}}{-(1)\begin{vmatrix} 1 & 1 \\ 2 & -1 \end{vmatrix} + (-1)\begin{vmatrix} 1 & 1 \\ 0 & -1 \end{vmatrix} - 0\begin{vmatrix} 1 & 1 \\ 0 & 2 \end{vmatrix}}$$

$$= \frac{-(-1 - 2) - (-7)}{-(-1 - 2) - (-1)} = \frac{3 + 7}{3 + 1} = \frac{10}{4} = \frac{5}{2}$$

Here we expanded each determinant by minors of the second row. Now using the second equation we have $\frac{5}{2} - b = 1$, or $b = \frac{3}{2}$. Using the third equation we have $2(\frac{3}{2}) - c = 0$, or $c = 3$. Thus,

$$a = \frac{5}{2} \text{h}, \qquad b = \frac{3}{2} \text{h}, \quad \text{and} \quad c = 3 \text{ h}$$

This means that the first stage takes 2.5 h, the second stage takes 1.5 h, and the third stage takes 3 h. We see that these times agree with the given information.

Example E   Solve the following system of equations.

$$
\begin{aligned}
x + 2y + z &= 5 \\
2x + z + 2t &= 1 \\
x - y + 3z + 4t &= -6 \\
4x - y - 2t &= 0
\end{aligned}
$$

$$
x = \dfrac{\begin{vmatrix} 5 & 2 & 1 & 0 \\ 1 & 0 & 1 & 2 \\ -6 & -1 & 3 & 4 \\ 0 & -1 & 0 & -2 \end{vmatrix}}{\begin{vmatrix} 1 & 2 & 1 & 0 \\ 2 & 0 & 1 & 2 \\ 1 & -1 & 3 & 4 \\ 4 & -1 & 0 & -2 \end{vmatrix}}
$$

$$
= \dfrac{-(0)\begin{vmatrix} 2 & 1 & 0 \\ 0 & 1 & 2 \\ -1 & 3 & 4 \end{vmatrix} + (-1)\begin{vmatrix} 5 & 1 & 0 \\ 1 & 1 & 2 \\ -6 & 3 & 4 \end{vmatrix} - (0)\begin{vmatrix} 5 & 2 & 0 \\ 1 & 0 & 2 \\ -6 & -1 & 4 \end{vmatrix} + (-2)\begin{vmatrix} 5 & 2 & 1 \\ 1 & 0 & 1 \\ -6 & -1 & 3 \end{vmatrix}}{(1)\begin{vmatrix} 0 & 1 & 2 \\ -1 & 3 & 4 \\ 0 & -2 \end{vmatrix} - 2\begin{vmatrix} 2 & 1 & 2 \\ 1 & 3 & 4 \\ 4 & 0 & -2 \end{vmatrix} + (1)\begin{vmatrix} 2 & 0 & 2 \\ 1 & -1 & 4 \\ 4 & -1 & -2 \end{vmatrix} - (0)\begin{vmatrix} 2 & 0 & 1 \\ 1 & -1 & 3 \\ 4 & -1 & 0 \end{vmatrix}}
$$

$$
= \dfrac{-(-26) - 2(-14)}{1(0) - 2(-18) + 1(18)} = \dfrac{26 + 28}{36 + 18} = \dfrac{54}{54} = 1
$$

In solving for x the determinant in the numerator was evaluated by expanding by the minors of the fourth row, since it contained two zeros. The determinant in the denominator was evaluated by expanding by the minors of the first row. Now we solve for y, and we again note two zeros in the fourth row of the determinant of the numerator.

$$
y = \dfrac{\begin{vmatrix} 1 & 5 & 1 & 0 \\ 2 & 1 & 1 & 2 \\ 1 & -6 & 3 & 4 \\ 4 & 0 & 0 & -2 \end{vmatrix}}{54} = \dfrac{-4\begin{vmatrix} 5 & 1 & 0 \\ 1 & 1 & 2 \\ -6 & 3 & 4 \end{vmatrix} + (-2)\begin{vmatrix} 1 & 5 & 1 \\ 2 & 1 & 1 \\ 1 & -6 & 3 \end{vmatrix}}{54}
$$

$$
= \dfrac{-4(-26) - 2(-29)}{54} = \dfrac{104 + 58}{54} = \dfrac{162}{54} = 3
$$

Substituting these values for x and y into the first equation, we can solve for z. This gives $z = -2$. Again, substituting the values for x and y into the fourth equation, we find $t = \frac{1}{2}$. Thus the required solution is $x = 1$, $y = 3$, $z = -2$, $t = \frac{1}{2}$. The solution can be checked by substituting these values into either the second or third equation.

Exercises 15–1

In Exercises 1 through 16, evaluate the given determinants by expansion by minors.

1. $\begin{vmatrix} 3 & 0 & 0 \\ -2 & 1 & 4 \\ 4 & -2 & 5 \end{vmatrix}$

2. $\begin{vmatrix} 10 & 0 & -3 \\ -2 & -4 & 1 \\ 3 & 0 & 2 \end{vmatrix}$

3. $\begin{vmatrix} -2 & -4 & 2 \\ 1 & 3 & 0 \\ -4 & 5 & 2 \end{vmatrix}$

4. $\begin{vmatrix} 5 & -1 & 2 \\ 8 & 3 & -4 \\ 0 & 2 & -6 \end{vmatrix}$

5. $\begin{vmatrix} -6 & -1 & 3 \\ 2 & -2 & -3 \\ 10 & 1 & -2 \end{vmatrix}$

6. $\begin{vmatrix} 9 & -3 & 1 \\ -1 & 2 & -1 \\ 2 & -1 & 3 \end{vmatrix}$

7. $\begin{vmatrix} -3 & -2 & 5 \\ 1 & 2 & -1 \\ 3 & -4 & 2 \end{vmatrix}$

8. $\begin{vmatrix} 4 & -3 & 3 \\ -3 & 5 & 6 \\ 2 & -1 & 2 \end{vmatrix}$

9. $\begin{vmatrix} 1 & 0 & 1 & 0 \\ 2 & 4 & -3 & 1 \\ 1 & 1 & 1 & 1 \\ 3 & 5 & 0 & 2 \end{vmatrix}$

10. $\begin{vmatrix} 2 & 0 & 3 & 1 \\ -1 & -1 & 4 & 0 \\ 1 & 2 & 1 & 2 \\ 3 & 3 & -2 & -1 \end{vmatrix}$

11. $\begin{vmatrix} 2 & -1 & 1 & -4 \\ 2 & 1 & 3 & -5 \\ 3 & -1 & -1 & 0 \\ 1 & 2 & 2 & 6 \end{vmatrix}$

12. $\begin{vmatrix} 3 & 6 & -2 & 4 \\ 2 & -5 & 2 & 6 \\ 5 & 3 & 4 & 0 \\ 1 & 2 & 0 & -1 \end{vmatrix}$

13. $\begin{vmatrix} 1 & 2 & -1 & -2 \\ 3 & 1 & 2 & 1 \\ -1 & 3 & -1 & 2 \\ 2 & 1 & 3 & -3 \end{vmatrix}$

14. $\begin{vmatrix} 3 & -1 & 2 & -5 \\ 1 & 4 & 2 & 5 \\ -1 & 1 & 1 & 3 \\ 1 & 2 & -1 & -2 \end{vmatrix}$

15. $\begin{vmatrix} 1 & 2 & 1 & 2 & 1 \\ 1 & 0 & 0 & 1 & 0 \\ 0 & 1 & 1 & 0 & 1 \\ 1 & 1 & 2 & 2 & 1 \\ 0 & 1 & 1 & 0 & 2 \end{vmatrix}$

16. $\begin{vmatrix} 3 & 1 & 1 & 1 & 2 \\ 1 & 1 & 0 & 0 & 1 \\ 1 & 1 & 2 & 2 & 3 \\ 0 & 2 & 1 & 0 & 3 \\ 1 & 1 & 0 & 1 & 0 \end{vmatrix}$

In Exercises 17 through 24, solve the given systems of equations by determinants. Evaluate the determinants by expansion by minors.

17. $2x + y + z = 6$
    $x - 2y + 2z = 10$
    $3x - y - z = 4$

18. $2x + y = -1$
    $4x - 2y - z = 5$
    $2x + 3y + 3z = -2$

19. $3x + 6y + 2z = -2$
    $x + 3y - 4z = 2$
    $2x - 3y - 2z = -2$

20. $x + 3y + z = 4$
    $2x - 6y - 3z = 10$
    $4x - 9y + 3z = 4$

21. $x + t = 0$
    $3x + y + z = -1$
    $2y - z + 3t = 1$
    $2z - 3t = 1$

22. $2x + y + z = 4$
    $2y - 2z - t = 3$
    $3y - 3z + 2t = 1$
    $6x - y + t = 0$

23. $x + 2y - z = 6$
    $y - 2z - 3t = -5$
    $3x - 2y + t = 2$
    $2x + y + z - t = 0$

24. $2x + 3y + z = 4$
    $x - 2y - 3z + 4t = -1$
    $3x + y + z - 5t = 3$
    $-x + 2y + z + 3t = 2$

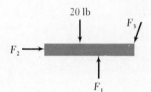

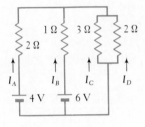

Figure 15-1

In Exercises 25 through 28, solve the given problems by use of determinants, using methods of this section.

25. In applying Kirchhoff's laws (see Exercise 19 of Section 4–6) to the given electric circuit, the following equations are found. Determine the indicated currents in amperes (see Fig. 15–1).

$$I_A + I_B + I_C + I_D = 0$$
$$2I_A - I_B = -2$$
$$3I_C - 2I_D = 0$$
$$I_B - 3I_C = 6$$

26. In analyzing the forces shown in Fig. 15–2, the following equations are derived:

$$F_1 = 20 + 0.8F_3$$
$$F_2 = 0.6F_3$$
$$3F_1 = 40 + 4F_3$$

Find forces F_1, F_2, and F_3.

Figure 15-2

27. A 1% solution, a 5% solution, and a 10% solution of sulfuric acid are to be mixed in order to get 600 mL of a 6% solution. If the volume of the 5% solution equals the volume of the other two solutions together, how much of each is needed?

28. A company budgets $100,000 to buy a fleet of 20 cars, made up of four models costing $4000, $5000, $6000, and $9000 each, respectively. The budget calls for 15 of the $4000 and $5000 models. The total cost of the $5000 and $6000 models is $59,000. How many of each model are in the budget?

15–2 Some Properties of Determinants

Expansion of determinants by minors allows us to evaluate a determinant of any order. However, even for a fourth-order determinant, the amount of work necessary for the evaluation is usually excessive. There are a number of basic properties of determinants which allow us to perform the evaluation with considerably less work. We will present these properties here without proof, although each will be illustrated.

1. *If each element below the principal diagonal of a determinant is zero, then the product of the elements of the principal diagonal is the value of the determinant.*

Example A

The value of the determinant

$$\begin{vmatrix} 2 & 1 & 5 & 8 \\ 0 & -5 & 7 & 9 \\ 0 & 0 & 4 & -6 \\ 0 & 0 & 0 & 3 \end{vmatrix}$$

equals the product $2(-5)(4)(3) = -120$. Since all of the elements below the principal diagonal are zero, there is no need to expand the determinant. It will be noted, however, that if the determinant is expanded by the first column, and successive determinants are expanded by their first columns, the same value is found. Performing the expansion we have

$$\begin{vmatrix} 2 & 1 & 5 & 8 \\ 0 & -5 & 7 & 9 \\ 0 & 0 & 4 & -6 \\ 0 & 0 & 0 & 3 \end{vmatrix} = 2 \begin{vmatrix} -5 & 7 & 9 \\ 0 & 4 & -6 \\ 0 & 0 & 3 \end{vmatrix}$$

$$= 2(-5)\begin{vmatrix} 4 & -6 \\ 0 & 3 \end{vmatrix} = 2(-5)(4)(3) = -120$$

2. *If the corresponding rows and columns of a determinant are interchanged, the value of the determinant is unchanged.*

Example B

For the determinant

$$\begin{vmatrix} 1 & 3 & -1 \\ 2 & 0 & 4 \\ -2 & 5 & -6 \end{vmatrix}$$

if we interchange the first row and first column, the second row and second column, and the third row and third column, we obtain the determinant

$$\begin{vmatrix} 1 & 2 & -2 \\ 3 & 0 & 5 \\ -1 & 4 & -6 \end{vmatrix}$$

By expanding, we can show that the value of each is the same. We obtain very similar expansions if we expand the first by the first column and the second by the first row. These expansions are

$$\begin{vmatrix} 1 & 3 & -1 \\ 2 & 0 & 4 \\ -2 & 5 & -6 \end{vmatrix} = (1)\begin{vmatrix} 0 & 4 \\ 5 & -6 \end{vmatrix} - 2\begin{vmatrix} 3 & -1 \\ 5 & -6 \end{vmatrix} + (-2)\begin{vmatrix} 3 & -1 \\ 0 & 4 \end{vmatrix}$$

$$= (-20) - 2(-13) - 2(12) = -18$$

$$\begin{vmatrix} 1 & 2 & -2 \\ 3 & 0 & 5 \\ -1 & 4 & -6 \end{vmatrix} = (1)\begin{vmatrix} 0 & 5 \\ 4 & -6 \end{vmatrix} - 2\begin{vmatrix} 3 & 5 \\ -1 & -6 \end{vmatrix} + (-2)\begin{vmatrix} 3 & 0 \\ -1 & 4 \end{vmatrix}$$

$$= (-20) - 2(-13) - 2(12) = -18$$

Therefore, we see that

$$\begin{vmatrix} 1 & 3 & -1 \\ 2 & 0 & 4 \\ -2 & 5 & -6 \end{vmatrix} = \begin{vmatrix} 1 & 2 & -2 \\ 3 & 0 & 5 \\ -1 & 4 & -6 \end{vmatrix}$$

3. *If two columns (or rows) of a determinant are identical, the value of the determinant is zero.*

Example C

The value of the determinant

$$\begin{vmatrix} 3 & 5 & 2 \\ -4 & 6 & 9 \\ -4 & 6 & 9 \end{vmatrix}$$

is zero, since the second and third rows are identical. This is easily verified by expanding by the first row.

$$\begin{vmatrix} 3 & 5 & 2 \\ -4 & 6 & 9 \\ -4 & 6 & 9 \end{vmatrix} = 3\begin{vmatrix} 6 & 9 \\ 6 & 9 \end{vmatrix} - 5\begin{vmatrix} -4 & 9 \\ -4 & 9 \end{vmatrix} + 2\begin{vmatrix} -4 & 6 \\ -4 & 6 \end{vmatrix}$$

$$= 3(0) - 5(0) + 2(0) = 0$$

4. *If two columns (or rows) of a determinant are interchanged, the value of the determinant is changed in sign.*

Example D

The values of the determinants

$$
\begin{vmatrix} 3 & 0 & 2 \\ 1 & 1 & 5 \\ 2 & 1 & 3 \end{vmatrix} \quad \text{and} \quad \begin{vmatrix} 2 & 0 & 3 \\ 5 & 1 & 1 \\ 3 & 1 & 2 \end{vmatrix}
$$

differ in sign, since the first and third columns are interchanged. We shall verify this by expanding each by the second column.

$$
\begin{vmatrix} 3 & 0 & 2 \\ 1 & 1 & 5 \\ 2 & 1 & 3 \end{vmatrix} = -(0)\begin{vmatrix} 1 & 5 \\ 2 & 3 \end{vmatrix} + (1)\begin{vmatrix} 3 & 2 \\ 2 & 3 \end{vmatrix} - (1)\begin{vmatrix} 3 & 2 \\ 1 & 5 \end{vmatrix}
$$

$$
= 0 + (9 - 4) - (15 - 2) = 5 - 13 = -8
$$

$$
\begin{vmatrix} 2 & 0 & 3 \\ 5 & 1 & 1 \\ 3 & 1 & 2 \end{vmatrix} = -(0)\begin{vmatrix} 5 & 1 \\ 3 & 2 \end{vmatrix} + (1)\begin{vmatrix} 2 & 3 \\ 3 & 2 \end{vmatrix} - (1)\begin{vmatrix} 2 & 3 \\ 5 & 1 \end{vmatrix}
$$

$$
= 0 + (4 - 9) - (2 - 15) = -5 + 13 = 8
$$

Therefore,

$$
\begin{vmatrix} 3 & 0 & 2 \\ 1 & 1 & 5 \\ 2 & 1 & 3 \end{vmatrix} = -\begin{vmatrix} 2 & 0 & 3 \\ 5 & 1 & 1 \\ 3 & 1 & 2 \end{vmatrix}
$$

5. *If all elements of a column (or row) are multiplied by the same number k, the value of the determinant is multiplied by k.*

Example E

The value of the determinant

$$
\begin{vmatrix} -1 & 0 & 6 \\ 2 & 1 & -2 \\ 0 & 5 & 3 \end{vmatrix}
$$

is multiplied by 3 if the elements of the second row are multiplied by 3. That is,

$$
\begin{vmatrix} -1 & 0 & 6 \\ 6 & 3 & -6 \\ 0 & 5 & 3 \end{vmatrix} = 3 \begin{vmatrix} -1 & 0 & 6 \\ 2 & 1 & -2 \\ 0 & 5 & 3 \end{vmatrix}
$$

By expansion, we can show that

$$\begin{vmatrix} -1 & 0 & 6 \\ 2 & 1 & -2 \\ 0 & 5 & 3 \end{vmatrix} = 47 \quad \text{and} \quad \begin{vmatrix} -1 & 0 & 6 \\ 6 & 3 & -6 \\ 0 & 5 & 3 \end{vmatrix} = 141$$

The validity of this property can be seen by expanding each determinant by the second row. In each case one element is three times the other corresponding element, but the minors are the same. Look at the element in the second row, first column and its minor for each determinant.

$$2 \begin{vmatrix} 0 & 6 \\ 5 & 3 \end{vmatrix} \quad \text{and} \quad 6 \begin{vmatrix} 0 & 6 \\ 5 & 3 \end{vmatrix}$$

The element 6 in the second determinant is three times the corresponding element 2 in the first determinant, and the minors are the same.

---

6. *If all the elements of any column (or row) are multiplied by the same number k, and the resulting numbers are added to the corresponding elements of another column (or row), the value of the determinant is unchanged.*

Example F

The value of the determinant

$$\begin{vmatrix} 4 & -1 & 3 \\ 2 & 2 & 1 \\ 1 & 0 & -3 \end{vmatrix}$$

is unchanged if we multiply each element of the first row by 2, and add these numbers to the corresponding elements of the second row. This gives

$$\begin{vmatrix} 4 & -1 & 3 \\ 2+8 & 2+(-2) & 1+6 \\ 1 & 0 & -3 \end{vmatrix} = \begin{vmatrix} 4 & -1 & 3 \\ 10 & 0 & 7 \\ 1 & 0 & -3 \end{vmatrix}$$

or

$$\begin{vmatrix} 4 & -1 & 3 \\ 10 & 0 & 7 \\ 1 & 0 & -3 \end{vmatrix} = \begin{vmatrix} 4 & -1 & 3 \\ 2 & 2 & 1 \\ 1 & 0 & -3 \end{vmatrix}$$

When each determinant is expanded, the value -37 is obtained. The great value in property 6 is that by its use zeros can be purposely placed in the resulting determinant.

With the use of the properties above, determinants of higher order can be evaluated much more readily. *The technique is to obtain zeros in a given column (or row) in all positions except one. We can then expand by this column (or row), thereby reducing the order of the determinant.* The following example illustrates the method.

Example G

Evaluate

$$\begin{vmatrix} 3 & 2 & -1 & 1 \\ -1 & 1 & 2 & 3 \\ 2 & 2 & 1 & 4 \\ 0 & -1 & -2 & 2 \end{vmatrix}$$

The evaluation is as follows. The small circled numbers above the equals signs refer to the explanations given below the setup.

$$\begin{vmatrix} 3 & 2 & -1 & 1 \\ -1 & 1 & 2 & 3 \\ 2 & 2 & 1 & 4 \\ 0 & -1 & -2 & 2 \end{vmatrix} \stackrel{①}{=} \begin{vmatrix} 0 & 5 & 5 & 10 \\ -1 & 1 & 2 & 3 \\ 2 & 2 & 1 & 4 \\ 0 & -1 & -2 & 2 \end{vmatrix}$$

$$\stackrel{②}{=} \begin{vmatrix} 0 & 5 & 5 & 10 \\ -1 & 1 & 2 & 3 \\ 0 & 4 & 5 & 10 \\ 0 & -1 & -2 & 2 \end{vmatrix}$$

$$\stackrel{③}{=} -(-1) \begin{vmatrix} 5 & 5 & 10 \\ 4 & 5 & 10 \\ -1 & -2 & 2 \end{vmatrix}$$

$$\stackrel{④}{=} 5 \begin{vmatrix} 1 & 1 & 2 \\ 4 & 5 & 10 \\ -1 & -2 & 2 \end{vmatrix}$$

$$\stackrel{⑤}{=} 5(2) \begin{vmatrix} 1 & 1 & 1 \\ 4 & 5 & 5 \\ -1 & -2 & 1 \end{vmatrix} \stackrel{⑥}{=} 10 \begin{vmatrix} 1 & 1 & 1 \\ 0 & 1 & 1 \\ -1 & -2 & 1 \end{vmatrix}$$

$$\overset{⑦}{=} 10\begin{vmatrix} 1 & 1 & 1 \\ 0 & 1 & 1 \\ 0 & -1 & 2 \end{vmatrix} \overset{⑧}{=} 10(1)\begin{vmatrix} 1 & 1 \\ -1 & 2 \end{vmatrix}$$

$$\overset{⑨}{=} 10(2 + 1) = 30$$

① Each element of the second row is multiplied by 3, and the resulting numbers are added to the corresponding elements of the first row. Here we have used property 6. In this way a zero has been placed in column 1, row 1.

② Each element of the second row is multiplied by 2, and the resulting numbers are added to the corresponding elements of the third row. Again, we have used property 6. Also, a zero has been placed in the first column, third row. We now have three zeros in the first column.

③ Expand the determinant by the first column. We have now reduced the determinant to a third-order determinant.

④ Factor 5 from each element of the first row. Here we are using property 5.

⑤ Factor 2 from each element of the third column. Again we are using property 5. Also, by doing this we have reduced the size of the numbers, and the resulting numbers are somewhat easier to work with.

⑥ Each element of the first row is multiplied by -4, and the resulting numbers are added to the corresponding elements of the second row. Here we are using property 6. We have placed a zero in the first column, second row.

⑦ Each element of the first row is added to the corresponding element of the third row. Again, we have used property 6. A zero has been placed in the first column, third row. We now have two zeros in the first column.

⑧ Expand the determinant by the first column.

⑨ Expand the second-order determinant.

A somewhat more systematic method is to place zeros below the principal diagonal and then use property 1. However, all such techniques are essentially equivalent.

Exercises 15–2

In Exercises 1 through 8, evaluate each of the determinants by inspection. Careful observation will allow evaluation by the use of one or more of the basic properties of this section.

1. $\begin{vmatrix} 4 & -5 & 8 \\ 0 & 3 & -8 \\ 0 & 0 & -5 \end{vmatrix}$

2. $\begin{vmatrix} 6 & 4 & 0 \\ 0 & -2 & 3 \\ 0 & 0 & -6 \end{vmatrix}$

3. $\begin{vmatrix} -2 & 0 & 0 \\ 15 & 4 & 0 \\ 2 & -7 & 7 \end{vmatrix}$

4.
$$\begin{vmatrix} 3 & 0 & 0 \\ 0 & 10 & 0 \\ -9 & -1 & -5 \end{vmatrix}$$

5.
$$\begin{vmatrix} -2 & 0 & -1 \\ 5 & 0 & 3 \\ 3 & 0 & -4 \end{vmatrix}$$

6.
$$\begin{vmatrix} -6 & -3 & 1 \\ 1 & 2 & -5 \\ 0 & 0 & 0 \end{vmatrix}$$

7.
$$\begin{vmatrix} 3 & -2 & 4 & 2 \\ 5 & -1 & 2 & -1 \\ 3 & -2 & 4 & 2 \\ 0 & 3 & -6 & 0 \end{vmatrix}$$

8.
$$\begin{vmatrix} -1 & -2 & -2 & 1 \\ 1 & 3 & 3 & -3 \\ -2 & 1 & 1 & 1 \\ 4 & 0 & 0 & -2 \end{vmatrix}$$

In Exercises 9 through 20, evaluate the determinants using the properties given in this section. Do not evaluate directly more than one second-order determinant for each.

9.
$$\begin{vmatrix} 3 & 1 & 0 \\ -2 & 3 & -1 \\ 4 & 2 & 5 \end{vmatrix}$$

10.
$$\begin{vmatrix} 6 & -1 & 3 \\ 0 & 2 & -2 \\ -1 & 4 & 3 \end{vmatrix}$$

11.
$$\begin{vmatrix} 5 & -1 & -2 \\ 3 & -5 & -2 \\ 1 & 4 & 6 \end{vmatrix}$$

12.
$$\begin{vmatrix} -4 & 3 & -2 \\ -2 & 2 & 4 \\ -1 & 5 & -3 \end{vmatrix}$$

13.
$$\begin{vmatrix} 4 & 3 & 6 & 0 \\ 3 & 0 & 0 & 4 \\ 5 & 0 & 1 & 2 \\ 2 & 1 & 1 & 7 \end{vmatrix}$$

14.
$$\begin{vmatrix} -2 & 1 & 3 & 0 \\ 1 & 3 & 0 & 0 \\ 0 & 2 & -3 & -1 \\ 4 & -1 & 2 & 1 \end{vmatrix}$$

15.
$$\begin{vmatrix} 3 & 1 & 2 & -1 \\ 2 & -1 & 3 & -1 \\ 1 & 2 & 1 & 3 \\ 1 & -2 & -3 & 2 \end{vmatrix}$$

16.
$$\begin{vmatrix} 6 & -3 & -6 & 3 \\ -2 & 1 & 2 & -1 \\ 18 & 7 & -1 & 5 \\ 0 & -1 & 10 & 10 \end{vmatrix}$$

17.
$$\begin{vmatrix} 1 & 3 & -3 & 5 \\ 4 & 2 & 1 & 2 \\ 3 & 2 & -2 & 2 \\ 0 & 1 & 2 & -1 \end{vmatrix}$$

18.
$$\begin{vmatrix} -2 & 2 & 1 & 3 \\ 1 & 4 & 3 & 1 \\ 4 & 3 & -2 & -2 \\ 3 & -2 & 1 & 5 \end{vmatrix}$$

19.
$$\begin{vmatrix} 1 & 2 & 0 & 1 & 0 \\ 0 & 2 & 1 & 0 & 1 \\ 1 & 0 & -1 & 1 & -1 \\ -2 & 0 & -1 & 2 & 1 \\ 1 & 0 & 2 & -1 & -2 \end{vmatrix}$$

20.
$$\begin{vmatrix} -1 & 3 & 5 & 0 & -5 \\ 0 & 1 & 7 & 3 & -2 \\ 5 & -2 & -1 & 0 & 3 \\ -3 & 0 & 2 & -1 & 3 \\ 6 & 2 & 1 & -4 & 2 \end{vmatrix}$$

In Exercises 21 through 28, solve the given systems of equations by determinants. Evaluate the determinants by the properties of determinants given in this section.

21. $2x - y + z = 5$
$x + 2y + 3z = 10$
$3x + 3y + 2z = 5$

22. $2x + y + z = 5$
$x + 3y - 3z = -13$
$3x + 2y - z = -1$

23. $3x + 2y + z = 1$
$9x + 2z = 5$
$6x - 4y - z = 3$

24. $3x + y + 2z = 4$
$x - y + 4z = 2$
$6x + 3y - 2z = 10$

25. $2x + y + z = 2$
$3y - z + 2t = 4$
$y + 2z + t = 0$
$3x + 2z = 4$

26. $2x + y + z = 0$
$x - y + 2t = 2$
$2y + z + 4t = 2$
$5x + 2z + 2t = 4$

27. $x + y + 2z = 1$
$2x - y + t = -2$
$x - y - z - 2t = 4$
$2x - y + 2z - t = 0$

28. $3x + y + t = 0$
$3z + 2t = 8$
$6x + 2y + 2z + t = 3$
$3x - y - z - t = 0$

In Exercises 29 through 32, solve the given problems by determinants. In Exercises 30 through 32, the necessary equations must be set up.

29. In applying Kirchhoff's laws (see Exercise 19 of Section 4–6) to the circuit shown in Fig. 15–3, the following equations are found. Determine the indicated currents, in amperes.

$$I_A + I_B + I_C + I_D + I_E = 0$$
$$-2I_A + 3I_B = 0$$
$$3I_B - 3I_C = 6$$
$$-3I_C + I_D = 0$$
$$-I_D + 2I_E = 0$$

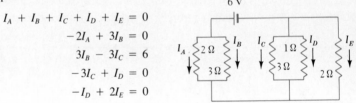

Figure 15-3

30. A land developer subdivides a tract of 100 acres into 120 building lots of 3 types. They have areas of $\frac{1}{2}$ acre, 1 acre, and 2 acres, to sell at $3000, $5000, and $8000, respectively. If all lots are sold, the gross income is $510,000. How many of each type are there in the development?

31. In testing for air pollution, a given air sample contained a total of 6 parts per million (ppm) of four pollutants, sulfur dioxide (SO_2), nitric oxide (NO), nitrogen dioxide (NO_2), and carbon monoxide (CO). The ppm of CO was ten times that of SO_2, which in turn equaled those of NO and NO_2. There was a total of 0.8 ppm of SO_2 and NO. How many ppm of each were present in the air sample?

32. A firm sells four types of appliances. Appliances A, B, C, and D, respectively, sell for $2, $3, $1, and $4 each. On a certain day it sold a total of 33 appliances, with receipts of $91. It sold twice as many of type B as type C, and twice as many of type D as type C. How many of each were sold on this day?

15–3 Matrices: Definitions and Basic Operations

Systems of linear equations occur in several areas of important technical and scientific applications. We indicated a few of these in Chapter 4 and in the first two sections of this chapter. Since a considerable amount of work is generally required to solve a system of equations, numerous methods have been developed for their solution.

Since the use of computers has been rapidly increasing in importance over the last several years, another mathematical concept which can be used to solve systems of equations is becoming used much more widely than in previous years. It is also used in numerous applications other than systems of equations, in such fields as business, economics, and psychology, as well as the scientific and technical areas. Since it is readily adaptable to use on a computer and is applicable to numerous areas, its importance will increase for some time to come. At this point, however, we shall only be able to introduce its definitions and basic operations.

A **matrix** *is an ordered rectangular array of numbers.* To distinguish such an array from a determinant, we shall enclose it within parentheses. As with a determinant, *the individual numbers are called* **elements** *of the matrix.*

Example A

Some examples of matrices are as follows:

$$\begin{pmatrix} 2 & 8 \\ 1 & 0 \end{pmatrix} \quad \begin{pmatrix} 2 & -4 & 6 \\ -1 & 0 & 5 \end{pmatrix} \quad \begin{pmatrix} 4 & 6 \\ 0 & -1 \\ -2 & 5 \\ 3 & 0 \end{pmatrix}$$

$$(-1 \quad 2 \quad 0 \quad 9) \quad \begin{pmatrix} -1 & 8 & 6 & 7 & 9 \\ 2 & 6 & 0 & 4 & 3 \\ 5 & -1 & 8 & 10 & 2 \end{pmatrix}$$

As we can see, it is not necessary for the number of columns and number of rows to be the same, although such is the case for a determinant. However, *if the number of rows does equal the number of columns, the matrix is called a* **square matrix.** We shall find that square matrices are of some special importance. *If all the elements of a matrix are zero, the matrix is called a* **zero matrix.** We shall find it convenient to designate a given matrix by a capital letter.

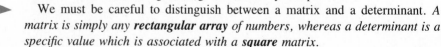
We must be careful to distinguish between a matrix and a determinant. *A matrix is simply any* **rectangular array** *of numbers, whereas a determinant is a specific value which is associated with a* **square** *matrix.*

Example B

Consider the following matrices:

$$A = \begin{pmatrix} 5 & 0 & -1 \\ 1 & 2 & 6 \\ 0 & -4 & -5 \end{pmatrix}, \quad B = \begin{pmatrix} 9 \\ 8 \\ 1 \\ 5 \end{pmatrix}, \quad C = (-1 \quad 6 \quad 8 \quad 9), \quad O = \begin{pmatrix} 0 & 0 \\ 0 & 0 \end{pmatrix}$$

Matrix A is an example of a square matrix, matrix B is an example of a matrix with four rows and one column, matrix C is an example of a matrix with one row and four columns, and matrix O is an example of a zero matrix.

To be able to refer to specific elements of a matrix and to give a general representation, a double-subscript notation is usually employed. That is,

$$A = \begin{pmatrix} a_{11} & a_{12} & a_{13} \\ a_{21} & a_{22} & a_{23} \\ a_{31} & a_{32} & a_{33} \end{pmatrix}$$

We see that the first subscript refers to the row in which the element lies, and the second subscript refers to the column in which the element lies.

Two matrices are said to be **equal** *if and only if they are identical.* That is, they must have the same number of columns, the same number of rows, and the elements must respectively be equal. If these conditions are not satisfied, the matrices are not equal.

Example C

$$\begin{pmatrix} a_{11} & a_{12} & a_{13} \\ a_{21} & a_{22} & a_{23} \end{pmatrix} = \begin{pmatrix} 1 & -5 & 0 \\ 4 & 6 & -3 \end{pmatrix}$$

if and only if $a_{11} = 1$, $a_{12} = -5$, $a_{13} = 0$, $a_{21} = 4$, $a_{22} = 6$, and $a_{23} = -3$.
The matrices

$$\begin{pmatrix} 1 & 2 & 3 \\ -1 & -2 & -5 \end{pmatrix} \quad \text{and} \quad \begin{pmatrix} 1 & 2 & -5 \\ -1 & -2 & 3 \end{pmatrix}$$

are not equal, since the elements in the third column are reversed.
The matrices

$$\begin{pmatrix} 2 & 3 \\ -1 & 5 \end{pmatrix} \quad \text{and} \quad \begin{pmatrix} 2 & 3 \\ -1 & 5 \\ 0 & 0 \end{pmatrix}$$

are not equal, since the number of rows is different.

Example D

If

$$\begin{pmatrix} 2x - y \\ 3x + 2y \end{pmatrix} = \begin{pmatrix} 4 \\ 13 \end{pmatrix}$$

we then know that $2x - y = 4$ and $3x + 2y = 13$. Therefore, to find the values of x and y, we must solve the system of equations

$$2x - y = 4$$
$$3x + 2y = 13$$

Using any of the methods previously developed, we find that $x = 3$ and $y = 2$.

If two matrices have the same number of rows and the same number of columns, their **sum** *is defined as the matrix consisting of the sums of the corresponding elements.* If the number of rows or the number of columns of the two matrices is not equal, they cannot be added.

Example E

$$\begin{pmatrix} 8 & 1 & -5 & 9 \\ 0 & -2 & 3 & 7 \end{pmatrix} + \begin{pmatrix} -3 & 4 & 6 & 0 \\ 6 & -2 & 6 & 5 \end{pmatrix}$$

$$= \begin{pmatrix} 8 + (-3) & 1 + 4 & -5 + 6 & 9 + 0 \\ 0 + 6 & -2 + (-2) & 3 + 6 & 7 + 5 \end{pmatrix}$$

$$= \begin{pmatrix} 5 & 5 & 1 & 9 \\ 6 & -4 & 9 & 12 \end{pmatrix}$$

The matrices

$$\begin{pmatrix} 3 & -5 & 8 \\ 2 & 9 & 0 \\ 4 & -2 & 3 \end{pmatrix} \text{ and } \begin{pmatrix} 3 & -5 & 8 & 0 \\ 2 & 9 & 0 & 0 \\ 4 & -2 & 3 & 0 \end{pmatrix}$$

cannot be added since the second matrix has one more column than the first matrix. The fact that the extra column contains only zeros does not matter.

The product of a number and a matrix (known as **scalar multiplication** *of a matrix) is defined as the matrix whose elements are obtained by multiplying each element of the given matrix by the given number.* That is, kA is the matrix obtained by multiplying the elements of matrix A by k. In this way $A + A$ and $2A$ will result in the same matrix.

Example F

For the matrix

$$A = \begin{pmatrix} -5 & 7 \\ 3 & 0 \end{pmatrix}$$

we have

$$2A = \begin{pmatrix} 2(-5) & 2(7) \\ 2(3) & 2(0) \end{pmatrix} = \begin{pmatrix} -10 & 14 \\ 6 & 0 \end{pmatrix}$$

Also,

$$5A = \begin{pmatrix} -25 & 35 \\ 15 & 0 \end{pmatrix} \quad \text{and} \quad -A = \begin{pmatrix} 5 & -7 \\ -3 & 0 \end{pmatrix}$$

By combining the definitions for the addition of matrices and for the scalar multiplication of a matrix, we can define the subtraction of matrices. That is, *the* **difference** *of matrices* A *and* B *is given by* $A - B = A + (-B)$. Therefore, we would change the sign of each element of B, and proceed as in addition.

By the preceding definitions we can see that the operations of addition, subtraction, and multiplication by a number of matrices are like those for real numbers. For these operations, we say that the algebra of matrices is like the algebra of real numbers. Although it is not our primary purpose to develop the algebra of matrices, we can see that the following laws hold for matrices.

$A + B = B + A$	(commutative law)	(15–4)
$A + (B + C) = (A + B) + C$	(associative law)	(15–5)
$k(A + B) = kA + kB$		(15–6)
$A + O = A$		(15–7)

Here we have let O represent the zero matrix. We shall find in the next section that not all laws for matrix operations are similar to those for real numbers.

Exercises 15–3

In Exercises 1 through 8, determine the value of the literal symbols for the following, in which the equality is properly defined.

1. $\begin{pmatrix} a & b \\ c & d \end{pmatrix} = \begin{pmatrix} 1 & -3 \\ 4 & 7 \end{pmatrix}$

2. $\begin{pmatrix} x & y & z \\ r & -s & -t \end{pmatrix} = \begin{pmatrix} -2 & 7 & -9 \\ 4 & -4 & 5 \end{pmatrix}$

3. $\begin{pmatrix} x \\ x + y \end{pmatrix} = \begin{pmatrix} 2 \\ 5 \end{pmatrix}$

4. $(x \quad x + y \quad x + y + z) = (5 \quad 6 \quad 8)$

5. $\begin{pmatrix} x & x + y \\ x - z & y + z \\ x + t & y - t \end{pmatrix} = \begin{pmatrix} 2 & 3 \\ 4 & -1 \end{pmatrix}$

6. $\begin{pmatrix} 2x - 3y \\ x + 4y \end{pmatrix} = \begin{pmatrix} 13 \\ 1 \end{pmatrix}$

7. $\begin{pmatrix} x \\ x+2 \\ 2y-3 \end{pmatrix} = \begin{pmatrix} 4 \\ y \\ z \end{pmatrix}$

8. $\begin{pmatrix} x & y & z \\ x+y & 2x-y & x+2 \end{pmatrix} = \begin{pmatrix} 2 & -3 \\ z & t \end{pmatrix}$

In Exercises 9 through 12, find the indicated sums of matrices.

9. $\begin{pmatrix} 2 & 3 \\ -5 & 4 \end{pmatrix} + \begin{pmatrix} -1 & 7 \\ 5 & -2 \end{pmatrix}$

10. $\begin{pmatrix} 1 & 0 & 9 \\ 3 & -5 & -2 \end{pmatrix} + \begin{pmatrix} 4 & -1 & 7 \\ 2 & 0 & -3 \end{pmatrix}$

11. $\begin{pmatrix} 5 & -8 \\ -3 & 5 \\ -1 & 6 \end{pmatrix} + \begin{pmatrix} -5 & 8 \\ 4 & 1 \\ 2 & -6 \end{pmatrix}$

12. $\begin{pmatrix} 4 & 2 & -9 \\ -6 & 4 & 7 \\ -1 & 0 & 5 \end{pmatrix} + \begin{pmatrix} -4 & -9 & -2 \\ 3 & 0 & 0 \\ 5 & 10 & -1 \end{pmatrix}$

In Exercises 13 through 20, use the following matrices to determine the indicated matrices.

$$A = \begin{pmatrix} -1 & 4 & -7 & 0 \\ 2 & -6 & -1 & 2 \end{pmatrix}, \quad B = \begin{pmatrix} 1 & 5 & -6 & 3 \\ 4 & -1 & 8 & -2 \end{pmatrix}, \quad C = \begin{pmatrix} 3 & -6 & 9 \\ -4 & 1 & 2 \end{pmatrix}$$

13. $A + B$ 14. $A - B$ 15. $A + C$ 16. $B + C$

17. $2A + B$ 18. $2B + A$ 19. $A - 2B$ 20. $3A - B$

In Exercises 21 through 24, use the given matrices to verify the indicated laws.

$$A = \begin{pmatrix} -1 & 2 & 3 & 7 \\ 0 & -3 & -1 & 4 \\ 9 & -1 & 0 & -2 \end{pmatrix}, \quad B = \begin{pmatrix} 4 & -1 & -3 & 0 \\ 5 & 0 & -1 & 1 \\ 1 & 11 & 8 & 2 \end{pmatrix}$$

21. $A + B = B + A$ 22. $A + 0 = A$

23. $-(A - B) = B - A$ 24. $3(A + B) = 3A + 3B$

In Exercises 25 and 26, find the unknown quantities in the given matrix equalities.

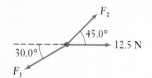

25. The forces shown in Fig. 15–4 can be found by solving for them in the following matrix equation.

$$\begin{pmatrix} F_1 \cos 30.0° - F_2 \cos 45.0° \\ F_1 \sin 30.0° - F_2 \sin 45.0° \end{pmatrix} = \begin{pmatrix} 12.5 \\ 0 \end{pmatrix}$$

Find F_1 and F_2.

Figure 15-4

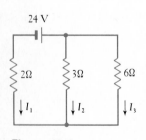

Figure 15-5

26. The electric currents shown in Fig. 15–5 can be found by solving for them in the following matrix equation.

$$\begin{pmatrix} I_1 + I_2 + I_3 \\ -2I_1 + 3I_2 \\ -3I_2 + 6I_3 \end{pmatrix} = \begin{pmatrix} 0 \\ 24 \\ 0 \end{pmatrix}$$

In Exercises 27 and 28, perform the indicated matrix operations.

27. The contractor of a housing development constructs four different types of houses, with either a carport, one-car garage, or a two-car garage. The following matrix

shows the number of houses of each type, and the type of garage.

$$\begin{array}{cccc} & \text{Type A} & \text{Type B} & \text{Type C} & \text{Type D} \end{array}$$

$$\begin{array}{l} \text{Carport} \\ \text{1-car garage} \\ \text{2-car garage} \end{array} \begin{pmatrix} 8 & 6 & 0 & 0 \\ 5 & 4 & 3 & 0 \\ 0 & 3 & 5 & 6 \end{pmatrix}$$

If the contractor builds two additional identical developments, find the matrix showing the total number of each house-garage type built.

28. In taking inventory, a firm finds that it has in one warehouse 6 pieces of 20-ft brass pipe, 8 pieces of 30-ft brass pipe, 11 pieces of 40-ft brass pipe, 5 pieces of 20-ft steel pipe, 10 pieces of 30-ft steel pipe, and 15 pieces of 40-ft steel pipe. This inventory can be represented by the matrix A, below. In each of two other warehouses, the inventory of the same items is represented by the matrix B, below.

$$A = \begin{pmatrix} 6 & 8 & 11 \\ 5 & 10 & 15 \end{pmatrix} \qquad B = \begin{pmatrix} 8 & 3 & 4 \\ 6 & 10 & 5 \end{pmatrix}$$

By matrix addition and scalar multiplication, find the matrix which represents the total number of each item in the three warehouses.

15–4 Multiplication of Matrices

The definition for the multiplication of matrices does not have an intuitive basis. However, through the solution of a system of linear equations we can, at least in part, show why multiplication is defined as it is. Consider Example A.

Example A

If we solve the system of equations

$$2x + y = 1$$
$$7x + 3y = 5$$

we obtain the solution $x = 2$, $y = -3$. In checking this solution in each of the equations, we obtain

$$2(2) + 1(-3) = 1$$
$$7(2) + 3(-3) = 5$$

Let us represent the coefficients of the equations by the matrix $\begin{pmatrix} 2 & 1 \\ 7 & 3 \end{pmatrix}$ and the solutions by the matrix $\begin{pmatrix} 2 \\ -3 \end{pmatrix}$. If we now indicate the multiplication of these matrices, and perform it as shown

$$\begin{pmatrix} 2 & 1 \\ 7 & 3 \end{pmatrix} \begin{pmatrix} 2 \\ -3 \end{pmatrix} = \begin{pmatrix} 2(2) + 1(-3) \\ 7(2) + 3(-3) \end{pmatrix} = \begin{pmatrix} 1 \\ 5 \end{pmatrix}$$

we note that we obtain a matrix which properly represents the right-side values of the equations. (Note carefully the products and sums in the resulting matrix.)

Following reasons along the lines indicated in Example A, we shall now define the **multiplication of matrices.** If the number of columns in a first matrix equals the number of rows in a second matrix, the product of these matrices is formed as follows: *The element in a specified row and a specified column of the product matrix is the sum of the products formed by multiplying each element in the specified row of the first matrix by the corresponding element in the specific column of the second matrix.* The product matrix will have the same number of rows as the first matrix and the same number of columns as the second matrix. Consider the following examples.

Example B

Find the product AB, where

$$A = \begin{pmatrix} 2 & 1 \\ -3 & 0 \\ 1 & 2 \end{pmatrix} \quad \text{and} \quad B = \begin{pmatrix} -1 & 6 & 5 & -2 \\ 3 & 0 & 1 & -4 \end{pmatrix}$$

To find the element in the first row and first column of the product, we find the sum of the products of corresponding elements of the first row of A and first column of B. To find the element in the first row and second column of the product, we find the sum of the products of corresponding elements in the first row of A and the second column of B. We continue this process until we have found the three rows (the number of rows in A) and the four columns (the number of columns in B) of the product. The product is formed as follows.

$$\begin{pmatrix} 2 & 1 \\ -3 & 0 \\ 1 & 2 \end{pmatrix} \begin{pmatrix} -1 & 6 & 5 & -2 \\ 3 & 0 & 1 & -4 \end{pmatrix}$$

$$= \begin{pmatrix} 2(-1) + 1(3) & 2(6) + 1(0) & 2(5) + 1(1) & 2(-2) + 1(-4) \\ -3(-1) + 0(3) & -3(6) + 0(0) & -3(5) + 0(1) & -3(-2) + 0(-4) \\ 1(-1) + 2(3) & 1(6) + 2(0) & 1(5) + 2(1) & 1(-2) + 2(-4) \end{pmatrix}$$

$$= \begin{pmatrix} 1 & 12 & 11 & -8 \\ 3 & -18 & -15 & 6 \\ 5 & 6 & 7 & -10 \end{pmatrix}$$

If we attempt to form the product BA, we find that B has four columns and A has three rows. Since the number of columns in B does not equal the number of rows in A, the product BA cannot be formed. In this way we see that $AB \neq BA$, which means that *matrix multiplication is not commutative (except in special cases)*. Therefore, matrix multiplication differs from the multiplication of real numbers.

Example C

Find the product

$$\begin{pmatrix} -1 & 9 & 3 & -2 \\ 2 & 0 & -7 & 1 \end{pmatrix} \begin{pmatrix} 6 & -2 \\ 1 & 0 \\ 3 & -5 \\ 3 & 9 \end{pmatrix}$$

We can find the product because the first matrix has four columns and the second matrix has four rows. The product is found as follows.

$$\begin{pmatrix} -1 & 9 & 3 & -2 \\ 2 & 0 & -7 & 1 \end{pmatrix} \begin{pmatrix} 6 & -2 \\ 1 & 0 \\ 3 & -5 \\ 3 & 9 \end{pmatrix}$$

$$= \begin{pmatrix} -1(6) + 9(1) + 3(3) + (-2)(3) & -1(-2) + 9(0) + 3(-5) + (-2)(9) \\ 2(6) + 0(1) + (-7)(3) + 1(3) & 2(-2) + 0(0) + (-7)(-5) + 1(9) \end{pmatrix}$$

$$= \begin{pmatrix} -6 + 9 + 9 - 6 & 2 + 0 - 15 - 18 \\ 12 + 0 - 21 + 3 & -4 + 0 + 35 + 9 \end{pmatrix} = \begin{pmatrix} 6 & -31 \\ -6 & 40 \end{pmatrix}$$

There are two special matrices of particular importance in the multiplication of matrices. The first of these is the **identity matrix I,** *which is a square matrix with* 1's *for elements on the principal diagonal, with all other elements zero.* (The principal diagonal starts with element a_{11}.) It has the property that if it is multiplied by another square matrix with the same number of rows and columns, then the second matrix equals the product matrix.

Example D

Show that $AI = IA = A$ for the matrix

$$A = \begin{pmatrix} 2 & -3 \\ 4 & 1 \end{pmatrix}$$

Since A has two rows and two columns, we choose I with two rows and two columns. Therefore, for this case

$$I = \begin{pmatrix} 1 & 0 \\ 0 & 1 \end{pmatrix}$$

(continued on next page)

Forming the indicated products, we have results as follows.

$$AI = \begin{pmatrix} 2 & -3 \\ 4 & 1 \end{pmatrix} \begin{pmatrix} 1 & 0 \\ 0 & 1 \end{pmatrix}$$

$$= \begin{pmatrix} 2(1) + (-3)(0) & 2(0) + (-3)(1) \\ 4(1) + 1(0) & 4(0) + 1(1) \end{pmatrix} = \begin{pmatrix} 2 & -3 \\ 4 & 1 \end{pmatrix}$$

$$IA = \begin{pmatrix} 1 & 0 \\ 0 & 1 \end{pmatrix} \begin{pmatrix} 2 & -3 \\ 4 & 1 \end{pmatrix}$$

$$= \begin{pmatrix} 1(2) + 0(4) & 1(-3) + 0(1) \\ 0(2) + 1(4) & 0(-3) + 1(1) \end{pmatrix} = \begin{pmatrix} 2 & -3 \\ 4 & 1 \end{pmatrix}$$

Therefore, we see that $AI = IA = A$.

For a given square matrix A, its **inverse** A^{-1} is the other important special matrix. *The matrix A and its inverse have the property that*

$$AA^{-1} = A^{-1}A = I \tag{15–8}$$

If the product of two matrices equals the identity matrix, the matrices are called inverses of each other. Under certain conditions the inverse of a given square matrix may not exist, although for most square matrices the inverse does exist. In the next section we shall develop the procedure for finding the inverse of a square matrix, and the section which follows shows how the inverse is used in the solution of systems of equations. At this point we shall simply show that the product of certain matrices equals the identity matrix, and that therefore these matrices are inverses of each other.

Example E

For the given matrices A and B, show that $AB = BA = I$, and therefore that $B = A^{-1}$.

$$A = \begin{pmatrix} 1 & -3 \\ -2 & 7 \end{pmatrix}, \quad B = \begin{pmatrix} 7 & 3 \\ 2 & 1 \end{pmatrix}$$

Forming the products AB and BA, we have the following:

$$AB = \begin{pmatrix} 1 & -3 \\ -2 & 7 \end{pmatrix} \begin{pmatrix} 7 & 3 \\ 2 & 1 \end{pmatrix} = \begin{pmatrix} 7 - 6 & 3 - 3 \\ -14 + 14 & -6 + 7 \end{pmatrix} = \begin{pmatrix} 1 & 0 \\ 0 & 1 \end{pmatrix}$$

$$BA = \begin{pmatrix} 7 & 3 \\ 2 & 1 \end{pmatrix} \begin{pmatrix} 1 & -3 \\ -2 & 7 \end{pmatrix} = \begin{pmatrix} 7 - 6 & -21 + 21 \\ 2 - 2 & -6 + 7 \end{pmatrix} = \begin{pmatrix} 1 & 0 \\ 0 & 1 \end{pmatrix}$$

Since $AB = I$ and $BA = I$, $B = A^{-1}$ and $A = B^{-1}$.

The following example illustrates one kind of application of the multiplication of matrices.

Example F

A particular firm produces three types of machines parts. On a given day it produces 40 of type X, 50 of type Y, and 80 of type Z. Each of type X requires 4 units of material and 1 worker-hour to produce; each of type Y requires 5 units of material and 2 worker-hours to produce; each of type Z requires 3 units of material and 2 worker-hours to produce. By representing the number of each type produced as the matrix $A = (40 \quad 50 \quad 80)$ and the material and time requirements by the matrix

$$B = \begin{pmatrix} 4 & 1 \\ 5 & 2 \\ 3 & 2 \end{pmatrix}$$

the product AB gives the total number of units of material and the total number of worker-hours needed for the day's production in a one-row, two-column matrix.

$$AB = (40 \quad 50 \quad 80) \begin{pmatrix} 4 & 1 \\ 5 & 2 \\ 3 & 2 \end{pmatrix}$$

$$= (160 + 250 + 240 \quad 40 + 100 + 160) = (650 \quad 300)$$

Therefore, 650 units of material and 300 worker-hours are required.

We now have seen how multiplication is defined for matrices. We see that *matrix multiplication is not commutative;* that is, $AB \neq BA$ in general. This is a major difference from the multiplication of real numbers. Another difference is that it is possible that $AB = 0$, even though neither A nor B is 0 (see Exercise 8, below). There are some similarities, however, in that $AI = A$, where we make I and the number 1 equivalent for the two types of multiplication. Also, *the distributive property $A(B + C) = AB + AC$ holds for matrix multiplication.* This points out some more of the properties of the algebra of matrices.

Exercises 15–4

In Exercises 1 through 12, perform the indicated matrix multiplications.

1.
$$(4 \quad -2) \begin{pmatrix} -1 & 0 \\ 2 & 6 \end{pmatrix}$$

2.
$$(-1 \quad 5 \quad -2) \begin{pmatrix} 6 & 3 \\ 2 & -1 \\ 0 & 2 \end{pmatrix}$$

3.
$$\begin{pmatrix} 2 & -3 \\ 5 & -1 \end{pmatrix} \begin{pmatrix} 3 & 0 & -1 \\ 7 & -5 & 8 \end{pmatrix}$$

4.
$$\begin{pmatrix} -7 & 8 \\ 5 & 0 \end{pmatrix} \begin{pmatrix} -9 & 10 \\ 1 & 4 \end{pmatrix}$$

5.
$$\begin{pmatrix} 2 & -3 & 1 \\ 0 & 7 & -3 \end{pmatrix} \begin{pmatrix} 9 \\ -2 \\ 5 \end{pmatrix}$$

6.
$$\begin{pmatrix} 0 & -1 & 2 \\ 4 & 11 & 2 \end{pmatrix} \begin{pmatrix} 3 & -1 \\ 1 & 2 \\ 6 & 1 \end{pmatrix}$$

7.
$$\begin{pmatrix} -1 & -5 \\ 4 & 0 \\ 2 & 10 \end{pmatrix} \begin{pmatrix} 2 & 0 \\ 1 & 1 \end{pmatrix}$$

8.
$$\begin{pmatrix} 12 & -4 \\ 3 & -1 \\ 6 & -2 \end{pmatrix} \begin{pmatrix} 2 & -1 & 3 \\ 6 & -3 & 9 \end{pmatrix}$$

9.
$$\begin{pmatrix} -1 & 7 \\ 3 & 5 \\ 10 & -1 \\ -5 & 12 \end{pmatrix} \begin{pmatrix} 2 & 1 & 0 \\ 5 & -3 & 1 \end{pmatrix}$$

10.
$$\begin{pmatrix} 3 & -1 & 8 \\ 0 & 2 & -4 \\ -1 & 6 & 7 \end{pmatrix} \begin{pmatrix} 7 & -1 \\ 0 & 3 \\ 1 & -2 \end{pmatrix}$$

11.
$$\begin{pmatrix} -9 & -1 & 4 \\ 6 & 9 & -1 \end{pmatrix} \begin{pmatrix} 6 & -5 \\ 4 & 1 \\ -1 & 6 \end{pmatrix}$$

12.
$$\begin{pmatrix} 1 & 2 & -6 & 6 & 1 \\ -2 & 4 & 0 & 1 & 2 \end{pmatrix} \begin{pmatrix} 1 \\ -1 \\ 0 \\ 5 \\ 2 \end{pmatrix}$$

In Exercises 13 through 16, find, if possible, *AB* and *BA*.

13.
$$A = (1 \quad -3 \quad 8), \quad B = \begin{pmatrix} -1 \\ 5 \\ 7 \end{pmatrix}$$

14.
$$A = \begin{pmatrix} -3 & 2 & 0 \\ 1 & -4 & 5 \end{pmatrix}, \quad B = \begin{pmatrix} -2 & 0 \\ 4 & -6 \\ 5 & 1 \end{pmatrix}$$

15.
$$A = \begin{pmatrix} -1 & 2 & 3 \\ 5 & -1 & 0 \end{pmatrix}, \quad B = \begin{pmatrix} 1 \\ -5 \\ 2 \end{pmatrix}$$

16.
$$A = \begin{pmatrix} -2 & 1 & 7 \\ 3 & -1 & 0 \\ 0 & 2 & -1 \end{pmatrix}, \quad B = (4 \quad -1 \quad 5)$$

In Exercises 17 through 20, show that $AI = IA = A$.

17.
$$A = \begin{pmatrix} 1 & 8 \\ -2 & 2 \end{pmatrix}$$

18.
$$A = \begin{pmatrix} -3 & 4 \\ 1 & 2 \end{pmatrix}$$

19.
$$A = \begin{pmatrix} 1 & 3 & -5 \\ 2 & 0 & 1 \\ 1 & -2 & 4 \end{pmatrix}$$

20.
$$A = \begin{pmatrix} -1 & 2 & 0 \\ 4 & -3 & 1 \\ 2 & 1 & 3 \end{pmatrix}$$

In Exercises 21 through 24, determine whether or not $B = A^{-1}$.

21. $A = \begin{pmatrix} 5 & -2 \\ -2 & 1 \end{pmatrix}, \quad B = \begin{pmatrix} 1 & 2 \\ 2 & 5 \end{pmatrix}$

22. $A = \begin{pmatrix} 3 & -4 \\ 5 & -7 \end{pmatrix}, \quad B = \begin{pmatrix} 7 & -4 \\ 5 & -2 \end{pmatrix}$

23. $A = \begin{pmatrix} 1 & -2 & 3 \\ 2 & -5 & 7 \\ -1 & 3 & -5 \end{pmatrix}, \quad B = \begin{pmatrix} 4 & -1 & 1 \\ 3 & -2 & -1 \\ 1 & -1 & -1 \end{pmatrix}$

24. $A = \begin{pmatrix} 1 & -1 & 3 \\ 3 & -4 & 8 \\ -2 & 3 & -4 \end{pmatrix}, \quad B = \begin{pmatrix} 8 & -5 & -4 \\ 4 & -2 & -1 \\ -1 & 1 & 1 \end{pmatrix}$

In Exercises 25 through 28, determine by matrix multiplication whether or not A is the proper matrix of solution values.

25. $\begin{aligned} 3x - 2y &= -1 \\ 4x + y &= 6 \end{aligned} \quad A = \begin{pmatrix} 1 \\ 2 \end{pmatrix}$

26. $\begin{aligned} 4x + y &= -5 \\ 3x + 4y &= 6 \end{aligned} \quad A = \begin{pmatrix} -2 \\ 3 \end{pmatrix}$

27. $\begin{aligned} 3x + y + 2z &= 1 \\ x - 3y + 4z &= -3 \\ 2x + 2y + z &= 1 \end{aligned} \quad A = \begin{pmatrix} -1 \\ 2 \\ 1 \end{pmatrix}$

28. $\begin{aligned} 2x - y + z &= 7 \\ x - 3y + 2z &= 6 \\ 3x + y - z &= 8 \end{aligned} \quad A = \begin{pmatrix} 3 \\ -2 \\ -1 \end{pmatrix}$

In Exercises 29 through 32, perform the indicated matrix multiplications.

29. For the identity matrix with two rows and two columns, show that $(-I)^2 = I$.

30. For the matrix J, where

$$J = \begin{pmatrix} 0 & -1 \\ 1 & 0 \end{pmatrix}$$

show that $J^2 = -I$. (Note the similarity with $j^2 = -1$.)

31. The firm referred to in Exercise 28 of Section 15–3 can determine the total number of feet of brass pipe and of steel pipe in each warehouse by multiplying the matrices of that exercise by the matrix

$$C = \begin{pmatrix} 20 \\ 30 \\ 40 \end{pmatrix}$$

Determine the total number of feet of each type of pipe (a) in the first warehouse and (b) in all three warehouses by matrix multiplication.

32. In the theory related to the reproduction of color photography, the equations

$$\begin{pmatrix} X \\ Y \\ Z \end{pmatrix} = \begin{pmatrix} 1.0 & 0.1 & 0 \\ 0.5 & 1.0 & 0.1 \\ 0.3 & 0.4 & 1.0 \end{pmatrix} \begin{pmatrix} x \\ y \\ z \end{pmatrix}$$

are found. The X, Y, and Z represent the red, green, and blue densities of the reproductions, respectively, and the x, y, and z represent the red, green, and blue densities, respectively, of the subject. Give the equations relating X, Y, and Z and x, y, and z.

15–5 Finding the Inverse of a Matrix

In the last section we introduced the concept of the inverse of a matrix. In this section we shall show how the inverse is found, and in the following section we shall show how this inverse is used in the solution of a system of linear equations.

We shall first show two methods of finding the inverse of a two-row, two-column (2×2) matrix. The first method is as follows:

1. *Interchange the elements on the principal diagonal.*
2. *Change the signs of the off-diagonal elements.*
3. *Divide each resulting element by the determinant of the given matrix.*

This method, which can be used with second-order square matrices but not higher-order matrices, is illustrated in the following example.

Example A

Find the inverse of the matrix

$$A = \begin{pmatrix} 2 & -3 \\ 4 & -7 \end{pmatrix}$$

First we interchange the elements on the principal diagonal and change the signs of the off-diagonal elements. This gives us the matrix

$$\begin{pmatrix} -7 & 3 \\ -4 & 2 \end{pmatrix}$$

Now we find the determinant of the original matrix, which means we evaluate

$$\begin{vmatrix} 2 & -3 \\ 4 & -7 \end{vmatrix} = -2$$

(Note again that the matrix is the array of numbers, whereas the determinant of the matrix has a value associated with it.) We now divide each element of the second matrix by -2. This gives

$$\frac{1}{-2}\begin{pmatrix} -7 & 3 \\ -4 & 2 \end{pmatrix} = \begin{pmatrix} \dfrac{-7}{-2} & \dfrac{3}{-2} \\ \dfrac{-4}{-2} & \dfrac{2}{-2} \end{pmatrix} = \begin{pmatrix} \dfrac{7}{2} & -\dfrac{3}{2} \\ 2 & -1 \end{pmatrix}$$

This last matrix is the inverse of matrix A. Therefore,

$$A^{-1} = \begin{pmatrix} \dfrac{7}{2} & -\dfrac{3}{2} \\ 2 & -1 \end{pmatrix}$$

Check by multiplication gives

$$AA^{-1} = \begin{pmatrix} 2 & -3 \\ 4 & -7 \end{pmatrix}\begin{pmatrix} \dfrac{7}{2} & -\dfrac{3}{2} \\ 2 & -1 \end{pmatrix} = \begin{pmatrix} 7-6 & -3+3 \\ 14-14 & -6+7 \end{pmatrix} = \begin{pmatrix} 1 & 0 \\ 0 & 1 \end{pmatrix} = I$$

The second method involves transforming the given matrix into the identity matrix, while at the same time transforming the identity matrix into the inverse. There are two types of steps allowable in making these transformations.

1. *Every element in any row may be multiplied by any given number other than zero.*
2. *Any row may be replaced by a row whose elements are the sum of a nonzero multiple of itself and a nonzero multiple of another row.*

Some reflection shows that these operations are those which are performed in solving a system of equations by addition or subtraction. The following example illustrates the method.

Example B

Find the inverse of the matrix

$$A = \begin{pmatrix} 2 & -3 \\ 4 & -7 \end{pmatrix}$$

First we set up the given matrix along with the identity matrix in the following manner.

$$\begin{pmatrix} 2 & -3 & | & 1 & 0 \\ 4 & -7 & | & 0 & 1 \end{pmatrix}$$

The vertical line simply shows the separation of the two matrices.

We wish to transform the left matrix into the identity matrix. Therefore, the first requirement is a 1 for element a_{11}. Therefore, we divide all elements of the first row by 2. This gives the following setup.

$$\begin{pmatrix} 1 & -\frac{3}{2} & | & \frac{1}{2} & 0 \\ 4 & -7 & | & 0 & 1 \end{pmatrix}$$

Next we want to have a zero for element a_{21}. Therefore, we shall subtract 4 times each element of row 1 from the corresponding element in row 2, replacing the elements of row 2. This gives us the following setup.

$$\begin{pmatrix} 1 & -\frac{3}{2} & | & \frac{1}{2} & 0 \\ 4-4(1) & -7-4(-\frac{3}{2}) & | & 0-4(\frac{1}{2}) & 1-4(0) \end{pmatrix}$$

or

$$\begin{pmatrix} 1 & -\frac{3}{2} & | & \frac{1}{2} & 0 \\ 0 & -1 & | & -2 & 1 \end{pmatrix}$$

Next, we want to have 1, not -1, for element a_{22}. Therefore, we multiply each element of row 2 by -1. This gives

$$\begin{pmatrix} 1 & -\frac{3}{2} & | & \frac{1}{2} & 0 \\ 0 & 1 & | & 2 & -1 \end{pmatrix}$$

(*continued on next page*)

Finally, we want zero for element a_{12}. Therefore, we add $\frac{3}{2}$ times each element of row 2 to the corresponding elements of row 1, replacing row 1. This gives

$$\begin{pmatrix} 1 + \frac{3}{2}(0) & -\frac{3}{2} + \frac{3}{2}(1) \left| \frac{1}{2} + \frac{3}{2}(2) \right. & 0 + \frac{3}{2}(-1) \\ 0 & 1 & 2 & -1 \end{pmatrix} \quad \text{or} \quad \begin{pmatrix} 1 & 0 \left| \frac{7}{2} \right. & -\frac{3}{2} \\ 0 & 1 \left| 2 \right. & -1 \end{pmatrix}$$

At this point, we have transformed the given matrix into the identity matrix, and the identity matrix into the inverse. Therefore, the matrix to the right of the vertical bar in the last setup is the required inverse. Thus,

$$A^{-1} = \begin{pmatrix} \frac{7}{2} & -\frac{3}{2} \\ 2 & -1 \end{pmatrix}$$

This is the same matrix and inverse as illustrated in Example A.

The idea to be noted most carefully in Example B is the order in which the zeros and ones were placed in transforming the given matrix to the identity matrix. We shall now give another example of finding the inverse for a 2×2 matrix, and then we shall find the inverse for a 3×3 matrix with the same method. This method is applicable for a square matrix of any number of rows or columns.

Example C   Find the inverse of the matrix $\begin{pmatrix} -3 & 6 \\ 4 & 5 \end{pmatrix}$.

$$\begin{pmatrix} -3 & 6 \left| 1 \right. & 0 \\ 4 & 5 \left| 0 \right. & 1 \end{pmatrix} \qquad \text{(original setup)}$$

$$\begin{pmatrix} 1 & -2 \left| -\frac{1}{3} \right. & 0 \\ 4 & 5 \left| 0 \right. & 1 \end{pmatrix} \qquad \text{(row 1 divided by } -3)$$

$$\begin{pmatrix} 1 & -2 \left| -\frac{1}{3} \right. & 0 \\ 0 & 13 \left| \frac{4}{3} \right. & 1 \end{pmatrix} \qquad (-4 \text{ times row 1 added to row 2)}$$

$$\begin{pmatrix} 1 & -2 \left| -\frac{1}{3} \right. & 0 \\ 0 & 1 \left| \frac{4}{39} \right. & \frac{1}{13} \end{pmatrix} \qquad \text{(row 2 divided by 13)}$$

$$\begin{pmatrix} 1 & 0 \left| -\frac{5}{39} \right. & \frac{2}{13} \\ 0 & 1 \left| \frac{4}{39} \right. & \frac{1}{13} \end{pmatrix} \qquad (2 \text{ times row 2 added to row 1)}$$

Therefore, $A^{-1} = \begin{pmatrix} -\frac{5}{39} & \frac{2}{13} \\ \frac{4}{39} & \frac{1}{13} \end{pmatrix}$, which can be checked by multiplication.

Example D   Find the inverse of the matrix $\begin{pmatrix} 1 & 2 & -1 \\ 3 & 5 & -1 \\ -2 & -1 & -2 \end{pmatrix}$.

$$\left(\begin{array}{ccc|ccc} 1 & 2 & -1 & 1 & 0 & 0 \\ 3 & 5 & -1 & 0 & 1 & 0 \\ -2 & -1 & -2 & 0 & 0 & 1 \end{array}\right) \qquad \text{(original setup)}$$

$$\left(\begin{array}{ccc|ccc} 1 & 2 & -1 & 1 & 0 & 0 \\ 0 & -1 & 2 & -3 & 1 & 0 \\ -2 & -1 & -2 & 0 & 0 & 1 \end{array}\right) \qquad \begin{array}{l}(-3 \text{ times row 1 added to} \\ \text{row 2)}\end{array}$$

$$\left(\begin{array}{ccc|ccc} 1 & 2 & -1 & 1 & 0 & 0 \\ 0 & -1 & 2 & -3 & 1 & 0 \\ 0 & 3 & -4 & 2 & 0 & 1 \end{array}\right) \qquad \text{(2 times row 1 added to row 3)}$$

$$\left(\begin{array}{ccc|ccc} 1 & 2 & -1 & 1 & 0 & 0 \\ 0 & 1 & -2 & 3 & -1 & 0 \\ 0 & 3 & -4 & 2 & 0 & 1 \end{array}\right) \qquad \text{(row 2 multiplied by } -1)$$

$$\left(\begin{array}{ccc|ccc} 1 & 0 & 3 & -5 & 2 & 0 \\ 0 & 1 & -2 & 3 & -1 & 0 \\ 0 & 3 & -4 & 2 & 0 & 1 \end{array}\right) \qquad \begin{array}{l}(-2 \text{ times row 2 added to} \\ \text{row 1)}\end{array}$$

$$\left(\begin{array}{ccc|ccc} 1 & 0 & 3 & -5 & 2 & 0 \\ 0 & 1 & -2 & 3 & -1 & 0 \\ 0 & 0 & 2 & -7 & 3 & 1 \end{array}\right) \qquad \begin{array}{l}(-3 \text{ times row 2 added to} \\ \text{row 3)}\end{array}$$

$$\left(\begin{array}{ccc|ccc} 1 & 0 & 3 & -5 & 2 & 0 \\ 0 & 1 & -2 & 3 & -1 & 0 \\ 0 & 0 & 1 & -\frac{7}{2} & \frac{3}{2} & \frac{1}{2} \end{array}\right) \qquad \text{(row 3 divided by 2)}$$

$$\left(\begin{array}{ccc|ccc} 1 & 0 & 3 & -5 & 2 & 0 \\ 0 & 1 & 0 & -4 & 2 & 1 \\ 0 & 0 & 1 & -\frac{7}{2} & \frac{3}{2} & \frac{1}{2} \end{array}\right) \qquad \text{(2 times row 3 added to row 2)}$$

$$\left(\begin{array}{ccc|ccc} 1 & 0 & 0 & \frac{11}{2} & -\frac{5}{2} & -\frac{3}{2} \\ 0 & 1 & 0 & -4 & 2 & 1 \\ 0 & 0 & 1 & -\frac{7}{2} & \frac{3}{2} & \frac{1}{2} \end{array}\right) \qquad \begin{array}{l}(-3 \text{ times row 3 added to} \\ \text{row 1)}\end{array}$$

Therefore, the required inverse matrix is

$$
\begin{pmatrix}
\frac{11}{2} & -\frac{5}{2} & -\frac{3}{2} \\
-4 & 2 & 1 \\
-\frac{7}{2} & \frac{3}{2} & \frac{1}{2}
\end{pmatrix}
$$

which may be checked by multiplication.

In transforming a matrix into the identity matrix, we work on one column at a time, transforming the columns in order from left to right. It is generally wisest to make the element on the principal diagonal for the column 1 first, and then to make all the other elements in the column 0. Looking back to Example D, we see that this procedure has been systematically followed, first on column 1, then on column 2, and finally on column 3.

There are other methods of finding the inverse of a matrix. One of these other methods is shown in Exercises 25 through 28, which follow.

Exercises 15–5

See Appendix E for a computer program for finding the inverse of a 2 × 2 matrix.

In Exercises 1 through 8, find the inverse of each of the given matrices by the method of Example A of this section.

1. $\begin{pmatrix} 2 & -5 \\ -2 & 4 \end{pmatrix}$ 2. $\begin{pmatrix} -6 & 3 \\ 3 & -2 \end{pmatrix}$ 3. $\begin{pmatrix} -1 & 5 \\ 4 & 10 \end{pmatrix}$ 4. $\begin{pmatrix} 8 & -1 \\ -4 & -5 \end{pmatrix}$

5. $\begin{pmatrix} 0 & -4 \\ 2 & 6 \end{pmatrix}$ 6. $\begin{pmatrix} 7 & -2 \\ -6 & 2 \end{pmatrix}$ 7. $\begin{pmatrix} -5 & -4 \\ 2 & 8 \end{pmatrix}$ 8. $\begin{pmatrix} 7 & -3 \\ -1 & -5 \end{pmatrix}$

In Exercises 9 through 24, find the inverse of each of the given matrices by transforming the identity matrix, as in Examples B through D.

9. $\begin{pmatrix} 1 & 2 \\ 2 & 3 \end{pmatrix}$ 10. $\begin{pmatrix} 1 & 5 \\ -1 & -4 \end{pmatrix}$ 11. $\begin{pmatrix} 2 & 4 \\ -1 & -1 \end{pmatrix}$ 12. $\begin{pmatrix} -2 & 6 \\ 3 & -4 \end{pmatrix}$

13. $\begin{pmatrix} 2 & 5 \\ -1 & 2 \end{pmatrix}$ 14. $\begin{pmatrix} -2 & 3 \\ -3 & 5 \end{pmatrix}$ 15. $\begin{pmatrix} 2 & -1 \\ 4 & 6 \end{pmatrix}$ 16. $\begin{pmatrix} 1 & -3 \\ 7 & -5 \end{pmatrix}$

17. $\begin{pmatrix} 1 & -3 & -2 \\ -2 & 7 & 3 \\ 1 & -1 & -3 \end{pmatrix}$ 18. $\begin{pmatrix} 1 & 2 & -1 \\ 3 & 7 & -5 \\ -1 & -2 & 0 \end{pmatrix}$ 19. $\begin{pmatrix} 1 & -1 & -3 \\ 0 & -1 & -2 \\ 2 & 1 & -1 \end{pmatrix}$

20. $\begin{pmatrix} 1 & 4 & 1 \\ -3 & -13 & -1 \\ 0 & -2 & 5 \end{pmatrix}$ 21. $\begin{pmatrix} 1 & 3 & 2 \\ -2 & -5 & -1 \\ 2 & 4 & 0 \end{pmatrix}$ 22. $\begin{pmatrix} 1 & 3 & 4 \\ -1 & -4 & -2 \\ 4 & 9 & 20 \end{pmatrix}$

23. $\begin{pmatrix} 2 & 4 & 0 \\ 3 & 4 & -2 \\ -1 & 1 & 2 \end{pmatrix}$ 24. $\begin{pmatrix} -2 & 6 & 1 \\ 0 & 3 & -3 \\ 4 & -7 & 3 \end{pmatrix}$

In Exercises 25 through 28, find the inverse of each of the given matrices (same as those for Exercises 21 through 24) by use of the following information. For matrix A, its inverse A^{-1} is found from

$$A = \begin{pmatrix} a_{11} & a_{12} & a_{13} \\ a_{21} & a_{22} & a_{23} \\ a_{31} & a_{32} & a_{33} \end{pmatrix}, \quad A^{-1} = \frac{1}{|A|} \begin{pmatrix} \begin{vmatrix} a_{22} & a_{23} \\ a_{32} & a_{33} \end{vmatrix} & -\begin{vmatrix} a_{12} & a_{13} \\ a_{32} & a_{33} \end{vmatrix} & \begin{vmatrix} a_{12} & a_{13} \\ a_{22} & a_{23} \end{vmatrix} \\ -\begin{vmatrix} a_{21} & a_{23} \\ a_{31} & a_{33} \end{vmatrix} & \begin{vmatrix} a_{11} & a_{13} \\ a_{31} & a_{33} \end{vmatrix} & -\begin{vmatrix} a_{11} & a_{13} \\ a_{21} & a_{23} \end{vmatrix} \\ \begin{vmatrix} a_{21} & a_{22} \\ a_{31} & a_{32} \end{vmatrix} & -\begin{vmatrix} a_{11} & a_{12} \\ a_{31} & a_{32} \end{vmatrix} & \begin{vmatrix} a_{11} & a_{12} \\ a_{21} & a_{22} \end{vmatrix} \end{pmatrix}$$

25. $\begin{pmatrix} 1 & 3 & 2 \\ -2 & -5 & -1 \\ 2 & 4 & 0 \end{pmatrix}$ 26. $\begin{pmatrix} 1 & 3 & 4 \\ -1 & -4 & -2 \\ 4 & 9 & 20 \end{pmatrix}$

27. $\begin{pmatrix} 2 & 4 & 0 \\ 3 & 4 & -2 \\ -1 & 1 & 2 \end{pmatrix}$ 28. $\begin{pmatrix} -2 & 6 & 1 \\ 0 & 3 & -3 \\ 4 & -7 & 3 \end{pmatrix}$

In Exercises 29 and 30, perform the indicated matrix operations. They verify the validity of the method of Example A.

29. For the matrix

$$A = \begin{pmatrix} a & b \\ c & d \end{pmatrix}$$

show that

$$\frac{1}{ad - bc} \begin{pmatrix} a & b \\ c & d \end{pmatrix} \begin{pmatrix} d & -b \\ -c & a \end{pmatrix} = \begin{pmatrix} 1 & 0 \\ 0 & 1 \end{pmatrix}$$

30. Find the inverse of matrix A of Exercise 29 by the method of Examples B through D.

15–6 *Matrices and Linear Equations*

As we stated earlier, matrices can be used to solve systems of equations. In this section we shall show one of the methods of how this is done.

Let us consider the system of equations

$$a_1x + b_1y = c_1$$
$$a_2x + b_2y = c_2$$

Recalling the definition of equality of matrices, we can write this system directly in terms of matrices as

$$\begin{pmatrix} a_1x + b_1y \\ a_2x + b_2y \end{pmatrix} = \begin{pmatrix} c_1 \\ c_2 \end{pmatrix}$$

The left side of this equation can be written as the product of two matrices. If we let

$$A = \begin{pmatrix} a_1 & b_1 \\ a_2 & b_2 \end{pmatrix} \quad \text{and} \quad X = \begin{pmatrix} x \\ y \end{pmatrix} \qquad (15\text{--}9)$$

then we have

$$AX = \begin{pmatrix} a_1x + b_1y \\ a_2x + b_2y \end{pmatrix} \qquad (15\text{--}10)$$

Therefore, the system of equations in Eq. (15–10) can be written in terms of matrices as

$$AX = C \qquad (15\text{--}11)$$

where $C = \begin{pmatrix} c_1 \\ c_2 \end{pmatrix}$.

If we now multiply each side of this matrix equation by A^{-1}, we have

$$A^{-1}AX = A^{-1}C$$

Since $A^{-1}A = I$, we have

$$IX = A^{-1}C$$

However, $IX = X$. Therefore,

$$X = A^{-1}C \qquad (15\text{--}12)$$

Equation (15–12) states that *we can solve a system of linear equations by multiplying the one-column matrix of the constants on the right by the inverse of the matrix of the coefficients*. The result is a one-column matrix whose elements are the required values. The following examples illustrate the method.

Example A

Use matrices to solve the system of equations

$$2x - y = 7$$
$$5x - 3y = 18$$

We set up the matrix of the coefficients as

$$A = \begin{pmatrix} 2 & -1 \\ 5 & -3 \end{pmatrix}$$

By either of the methods of the previous section, we can determine the inverse of this matrix to be

$$A^{-1} = \begin{pmatrix} 3 & -1 \\ 5 & -2 \end{pmatrix}$$

We now form the matrix product $A^{-1}C$, where $C = \begin{pmatrix} 7 \\ 18 \end{pmatrix}$. This gives

$$A^{-1}C = \begin{pmatrix} 3 & -1 \\ 5 & -2 \end{pmatrix}\begin{pmatrix} 7 \\ 18 \end{pmatrix} = \begin{pmatrix} 21 - 18 \\ 35 - 36 \end{pmatrix} = \begin{pmatrix} 3 \\ -1 \end{pmatrix}$$

Since $X = A^{-1}C$, this means that

$$\begin{pmatrix} x \\ y \end{pmatrix} = \begin{pmatrix} 3 \\ -1 \end{pmatrix}$$

Therefore, the required solution is $x = 3$ and $y = -1$.

Example B

Use matrices to solve the system of equations

$$2x - y = 3$$
$$6x + 4y = -5$$

Setting up matrices A and C, we have

$$A = \begin{pmatrix} 2 & -1 \\ 6 & 4 \end{pmatrix} \quad \text{and} \quad C = \begin{pmatrix} 3 \\ -5 \end{pmatrix}$$

We now find the inverse of A to be

$$A^{-1} = \begin{pmatrix} \frac{2}{7} & \frac{1}{14} \\ -\frac{3}{7} & \frac{1}{7} \end{pmatrix}$$

Therefore,

$$A^{-1}C = \begin{pmatrix} \frac{2}{7} & \frac{1}{14} \\ -\frac{3}{7} & \frac{1}{7} \end{pmatrix}\begin{pmatrix} 3 \\ -5 \end{pmatrix} = \begin{pmatrix} \frac{6}{7} - \frac{5}{14} \\ -\frac{9}{7} - \frac{5}{7} \end{pmatrix} = \begin{pmatrix} \frac{1}{2} \\ -2 \end{pmatrix}$$

Therefore, the required solution is $x = \frac{1}{2}$ and $y = -2$.

Example C

Use matrices to solve the system of equations

$$
\begin{aligned}
x + 4y - z &= 4 \\
x + 3y + z &= 8 \\
2x + 6y + z &= 13
\end{aligned}
$$

Setting up matrices A and C, we have

$$
A = \begin{pmatrix} 1 & 4 & -1 \\ 1 & 3 & 1 \\ 2 & 6 & 1 \end{pmatrix} \quad \text{and} \quad C = \begin{pmatrix} 4 \\ 8 \\ 13 \end{pmatrix}
$$

To give another example of finding the inverse of a 3×3 matrix, we shall briefly show the steps for finding A^{-1}.

$$
\left(\begin{array}{ccc|ccc} 1 & 4 & -1 & 1 & 0 & 0 \\ 1 & 3 & 1 & 0 & 1 & 0 \\ 2 & 6 & 1 & 0 & 0 & 1 \end{array}\right)
\qquad
\left(\begin{array}{ccc|ccc} 1 & 4 & -1 & 1 & 0 & 0 \\ 0 & -1 & 2 & -1 & 1 & 0 \\ 2 & 6 & 1 & 0 & 0 & 1 \end{array}\right)
$$

$$
\left(\begin{array}{ccc|ccc} 1 & 4 & -1 & 1 & 0 & 0 \\ 0 & -1 & 2 & -1 & 1 & 0 \\ 0 & -2 & 3 & -2 & 0 & 1 \end{array}\right)
\qquad
\left(\begin{array}{ccc|ccc} 1 & 4 & -1 & 1 & 0 & 0 \\ 0 & 1 & -2 & 1 & -1 & 0 \\ 0 & -2 & 3 & -2 & 0 & 1 \end{array}\right)
$$

$$
\left(\begin{array}{ccc|ccc} 1 & 0 & 7 & -3 & 4 & 0 \\ 0 & 1 & -2 & 1 & -1 & 0 \\ 0 & -2 & 3 & -2 & 0 & 1 \end{array}\right)
\qquad
\left(\begin{array}{ccc|ccc} 1 & 0 & 7 & -3 & 4 & 0 \\ 0 & 1 & -2 & 1 & -1 & 0 \\ 0 & 0 & -1 & 0 & -2 & 1 \end{array}\right)
$$

$$
\left(\begin{array}{ccc|ccc} 1 & 0 & 7 & -3 & 4 & 0 \\ 0 & 1 & -2 & 1 & -1 & 0 \\ 0 & 0 & 1 & 0 & 2 & -1 \end{array}\right)
\qquad
\left(\begin{array}{ccc|ccc} 1 & 0 & 7 & -3 & 4 & 0 \\ 0 & 1 & 0 & 1 & 3 & -2 \\ 0 & 0 & 1 & 0 & 2 & -1 \end{array}\right)
$$

$$
\left(\begin{array}{ccc|ccc} 1 & 0 & 0 & -3 & -10 & 7 \\ 0 & 1 & 0 & 1 & 3 & -2 \\ 0 & 0 & 1 & 0 & 2 & -1 \end{array}\right)
\qquad
\text{Thus, } A^{-1} = \begin{pmatrix} -3 & -10 & 7 \\ 1 & 3 & -2 \\ 0 & 2 & -1 \end{pmatrix}
$$

Therefore,

$$
A^{-1}C = \begin{pmatrix} -3 & -10 & 7 \\ 1 & 3 & -2 \\ 0 & 2 & -1 \end{pmatrix}\begin{pmatrix} 4 \\ 8 \\ 13 \end{pmatrix} = \begin{pmatrix} -1 \\ 2 \\ 3 \end{pmatrix}
$$

This means that $x = -1$, $y = 2$, $z = 3$.

Example D

Use matrices to solve the system of equations

$$x + 2y - z = -4$$
$$3x + 5y - z = -5$$
$$-2x - y - 2z = -5$$

Setting up matrices A and C, we have

$$A = \begin{pmatrix} 1 & 2 & -1 \\ 3 & 5 & -1 \\ -2 & -1 & -2 \end{pmatrix} \quad \text{and} \quad C = \begin{pmatrix} -4 \\ -5 \\ -5 \end{pmatrix}$$

We now find the inverse of A to be

$$A^{-1} = \begin{pmatrix} \frac{11}{2} & -\frac{5}{2} & -\frac{3}{2} \\ -4 & 2 & 1 \\ -\frac{7}{2} & \frac{3}{2} & \frac{1}{2} \end{pmatrix}$$

(see Example D of Section 15–5). Therefore,

$$A^{-1}C = \begin{pmatrix} \frac{11}{2} & -\frac{5}{2} & -\frac{3}{2} \\ -4 & 2 & 1 \\ -\frac{7}{2} & \frac{3}{2} & \frac{1}{2} \end{pmatrix} \begin{pmatrix} -4 \\ -5 \\ -5 \end{pmatrix} = \begin{pmatrix} -2 \\ 1 \\ 4 \end{pmatrix}$$

This means that the solution is $x = -2$, $y = 1$, $z = 4$.

After having solved systems of equations in this manner, the reader may feel that the method is much longer and more tedious than previously developed techniques. The principal problem with this method is that a great deal of numerical computation is generally required. However, methods such as this one are easily programmed for use on a computer, which can do the arithmetic work very rapidly. Therefore it is the *method* of solving the system of equations which is of primary importance here.

Exercises 15–6

In Exercises 1 through 8, solve the given systems of equations by using the inverse of the coefficient matrix. The numbers in parentheses refer to exercises from Section 15–5 where the inverses may be checked.

1. $2x - 5y = -14$ (1)
   $-2x + 4y = 11$

2. $-x + 5y = 4$ (3)
   $4x + 10y = -4$

3. $2x + 4y = -9$ (11)
   $-x - y = 2$

4. $2x + 5y = -6$ (13)
   $-x + 2y = -6$

5. $x - 3y - 2z = -8$ (17)
   $-2x + 7y + 3z = 19$
   $x - y - 3z = -3$

6. $x - y - 3z = -1$ (19)
   $-y - 2z = -2$
   $2x + y - z = 2$

7. $x + 3y + 2z = 5$ (21)
   $-2x - 5y - z = -1$
   $2x + 4y = -2$

8. $2x + 4y = -2$ (23)
   $3x + 4y - 2z = -6$
   $-x + y + 2z = 5$

In Exercises 9 through 20, solve the given systems of equations by using the inverse of the coefficient matrix.

9. $2x + 7y = 16$
   $x + 4y = 9$

10. $4x - 3y = -13$
    $-3x + 2y = 9$

11. $2x - 3y = 3$
    $4x - 5y = 4$

12. $x + 2y = 3$
    $3x + 4y = 11$

13. $5x - 2y = -14$
    $3x + 4y = -11$

14. $4x - 3y = -1$
    $8x + 3y = 4$

15. $2x - y = 6$
    $4x + 3y = -10$

16. $4x - y = 3$
    $6x - 3y = 5$

17. $x + 2y + 2z = -4$
    $4x + 9y + 10z = -18$
    $-x + 3y + 7z = -7$

18. $x - 4y - 2z = -7$
    $-x + 5y + 5z = 18$
    $3x - 7y + 10z = 38$

19. $2x + 4y + z = 5$
    $-2x - 2y - z = -6$
    $-x + 2y + z = 0$

20. $4x + y = 2$
    $-2x - y + 3z = -18$
    $2x + y - z = 8$

In Exercises 21 through 24, solve the indicated systems of equations by using the inverse of the coefficient matrix. In Exercises 23 and 24, it is necessary to set up the appropriate equations.

21. Three forces F_1, F_2, and F_3 are acting on a certain beam. The forces (in pounds) can be found by solving the following equations.

$$F_1 + F_2 + F_3 = 30$$
$$4F_1 + F_2 - 4F_3 = 0$$
$$5F_2 - 3F_3 = 4$$

Determine these forces.

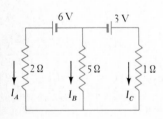

Figure 15-6

22. In applying Kirchhoff's laws (see Exercise 19 of Section 4–6) to the circuit shown in Fig. 15–6, the following equations are found. Determine the indicated currents, in amperes.

$$I_A + I_B + I_C = 0, \qquad 2I_A - 5I_B = 6, \qquad 5I_B - I_C = -3$$

23. Type A doors cost $10 each and type B doors cost $14 each. A builder was billed $220 for a shipment of these doors. She found that had she reversed the number of each on her order, she would have been billed $212. How many of each did she actually receive?

24. Fifty shares of stock A and 30 shares of stock B cost $2600. Thirty shares of stock A and 40 shares of stock B cost $2000. What is the price per share of each stock?

15–7 Exercises for Chapter 15

In Exercises 1 through 8, evaluate the given determinants by expansion by minors.

1. $\begin{vmatrix} 1 & 2 & -1 \\ 4 & 1 & -3 \\ -3 & -5 & 2 \end{vmatrix}$

2. $\begin{vmatrix} 3 & -1 & 2 \\ 7 & -1 & 4 \\ 2 & 1 & -3 \end{vmatrix}$

3. $\begin{vmatrix} -1 & 3 & -7 \\ 0 & 5 & 4 \\ 4 & -3 & -2 \end{vmatrix}$

4. $\begin{vmatrix} 6 & -5 & -7 \\ -1 & 2 & 4 \\ 2 & -3 & 1 \end{vmatrix}$

5. $\begin{vmatrix} 2 & 6 & 2 & 5 \\ 2 & 0 & 4 & -1 \\ 4 & -3 & 6 & 1 \\ 3 & -1 & 0 & -2 \end{vmatrix}$

6. $\begin{vmatrix} 1 & -2 & 2 & 4 \\ 0 & 1 & 2 & 3 \\ 3 & 2 & 2 & 5 \\ 2 & 1 & -2 & 0 \end{vmatrix}$

7. $\begin{vmatrix} 1 & 3 & -2 & 4 \\ 2 & 0 & 3 & -2 \\ 5 & -1 & 5 & -3 \\ -6 & 4 & -1 & 2 \end{vmatrix}$

8. $\begin{vmatrix} 2 & 3 & -1 & -1 \\ -3 & -2 & 5 & -6 \\ 2 & 1 & -3 & 2 \\ 4 & 0 & -2 & 1 \end{vmatrix}$

In Exercises 9 through 16, evaluate the determinants of Exercises 1 through 8 by using the basic properties of determinants.

In Exercises 17 through 20, evaluate the given determinants by using the basic properties of determinants.

17. $\begin{vmatrix} 1 & 0 & -3 & -2 \\ 1 & -1 & 2 & 0 \\ -1 & 1 & 1 & 1 \\ 5 & -1 & 2 & -1 \end{vmatrix}$

18. $\begin{vmatrix} 2 & 6 & -2 & 4 \\ -2 & 2 & -3 & 3 \\ 3 & 2 & 2 & -2 \\ 2 & -6 & 4 & 1 \end{vmatrix}$

19. $\begin{vmatrix} 1 & -1 & 3 & 0 & 2 \\ 4 & 0 & 4 & -2 & 2 \\ 0 & 4 & 0 & -1 & -1 \\ -2 & 2 & -1 & 4 & 0 \\ 1 & -1 & 2 & 0 & 1 \end{vmatrix}$

20. $\begin{vmatrix} 1 & 4 & -3 & 3 & 0 \\ 3 & 1 & -1 & 2 & 2 \\ 1 & 2 & 1 & 1 & 1 \\ -3 & -5 & -5 & 0 & -6 \\ 2 & 2 & -2 & 3 & -2 \end{vmatrix}$

In Exercises 21 through 24, determine the values of the literal symbols.

21. $\begin{pmatrix} a \\ a - b \end{pmatrix} = \begin{pmatrix} 3 \\ 4 \end{pmatrix}$

22. $\begin{pmatrix} r & s \\ s + t & r - u \end{pmatrix} = \begin{pmatrix} -3 & 2 \\ 5 & 1 \end{pmatrix}$

23. $\begin{pmatrix} 2x & 3y & 2z \\ x + y & 2y + z & z - x \end{pmatrix} = \begin{pmatrix} 4 & -9 & 5 \\ a & b & c \end{pmatrix}$

24. $\begin{pmatrix} x - y \\ 2x + 2z \\ 4y + z \end{pmatrix} = \begin{pmatrix} 1 \\ 3 \\ -1 \end{pmatrix}$

In Exercises 25 through 32, use the given matrices and perform the indicated operations.

$$A = \begin{pmatrix} 2 & -3 \\ 4 & 1 \\ -5 & 0 \\ 2 & -3 \end{pmatrix}, \qquad B = \begin{pmatrix} -1 & 0 \\ 4 & -6 \\ -3 & -2 \\ 1 & -7 \end{pmatrix}, \qquad C = \begin{pmatrix} 5 & -6 \\ 2 & 8 \\ 0 & -2 \end{pmatrix}$$

25. $A + B$

26. $2C$

27. $-3B$

28. $B - A$

29. $A - C$

30. $2C - B$

31. $2A - 3B$

32. $2(A - B)$

In Exercises 33 through 36, perform the indicated matrix multiplications.

33. $\begin{pmatrix} 5 & -1 \\ 3 & 2 \end{pmatrix}\begin{pmatrix} 1 \\ -8 \end{pmatrix}$

34.

$$\begin{pmatrix} 6 & -4 & 1 & 0 \\ 2 & 0 & -4 & 3 \end{pmatrix}\begin{pmatrix} 7 & -1 & 6 \\ 4 & 0 & 1 \\ 3 & -2 & 5 \\ 9 & 1 & 0 \end{pmatrix}$$

35. $\begin{pmatrix} -1 & 7 \\ 2 & 0 \\ 4 & -1 \end{pmatrix}\begin{pmatrix} 1 & -4 & 5 \\ 5 & 1 & 0 \end{pmatrix}$

36. $\begin{pmatrix} 0 & -1 & 6 \\ 8 & 1 & 4 \\ 7 & -2 & -1 \end{pmatrix}\begin{pmatrix} 5 & -1 & 7 & 1 & 5 \\ 0 & 1 & 0 & 4 & 1 \\ 1 & -2 & 3 & 0 & 1 \end{pmatrix}$

In Exercises 37 through 44, find the inverses of the given matrices.

37. $\begin{pmatrix} 2 & -5 \\ 2 & -4 \end{pmatrix}$

38. $\begin{pmatrix} -1 & -6 \\ 2 & 10 \end{pmatrix}$

39. $\begin{pmatrix} 7 & -1 \\ 4 & 8 \end{pmatrix}$

40. $\begin{pmatrix} 5 & -1 \\ 4 & -8 \end{pmatrix}$

41. $\begin{pmatrix} 1 & 1 & -2 \\ -1 & -2 & 1 \\ 0 & 3 & 4 \end{pmatrix}$

42. $\begin{pmatrix} -1 & -1 & 2 \\ 2 & 3 & 0 \\ 1 & 4 & 1 \end{pmatrix}$

43. $\begin{pmatrix} 2 & -4 & 3 \\ 4 & -6 & 5 \\ -2 & 1 & -1 \end{pmatrix}$

44. $\begin{pmatrix} 3 & 1 & -4 \\ -3 & 1 & -2 \\ -6 & 0 & 3 \end{pmatrix}$

In Exercises 45 through 52, solve the given systems of equations using the inverse of the coefficient matrix.

45. $2x - 3y = -9$
    $4x - y = -13$

46. $5x - 7y = 62$
    $6x + 5y = -6$

47. $3x + 5y = 29$
    $4x - 7y = -57$

48. $4x - 2y = 1$
    $8x + 2y = 5$

49. $2x - 3y + 2z = 7$
    $3x + y - 3z = -6$
    $x + 4y + z = -13$

50. $2x + 2y - z = 8$
    $x + 4y + 2z = 5$
    $3x - 2y + z = 17$

51. $x + 2y + 3z = 1$
    $3x - 4y - 3z = 2$
    $7x - 6y + 6z = 2$

52. $3x + 2y + z = 2$
    $2x + 3y - 6z = 3$
    $x + 3y + 3z = 1$

In Exercises 53 through 56, solve the given systems of equations by determinants. Use the basic properties of determinants.

53. $3x - 2y + z = 6$
    $2x + 3z = 3$
    $4x - y + 5z = 6$

54. $7x + y + 2z = 3$
    $4x - 2y + 4z = -2$
    $2x + 3y - 6z = 3$

55. $2x - 3y + z - t = -8$
    $4x + 3z + 2t = -3$
    $2y - 3z - t = 12$
    $x - y - z + t = 3$

56. $3x + 2y - 2z - 2t = 0$
    $5y + 3z + 4t = 3$
    $6y - 3z + 4t = 9$
    $6x - y + 2z - 2t = -3$

In Exercises 57 and 58, use the matrix

$$J = \begin{pmatrix} 0 & -1 \\ 1 & 0 \end{pmatrix}$$

57. Show that $J^{-1} = -J$.

58. Show that $J^3 = -J$.

In Exercises 59 and 60, use the matrices

$$A = \begin{pmatrix} 1 & -2 \\ 0 & 3 \end{pmatrix} \quad \text{and} \quad B = \begin{pmatrix} -3 & 1 \\ 2 & -1 \end{pmatrix}$$

59. Show that $(A + B)(A - B) \neq A^2 - B^2$.

60. Show that $(A + B)^2 \neq A^2 + 2AB + B^2$.

In Exercises 61 through 64, solve the given systems of equations by any appropriate method of this chapter.

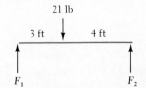

21 lb

3 ft   4 ft

F_1 F_2

Figure 15-7

61. To find the forces F_1 and F_2 shown in Fig. 15–7, it is necessary to solve the following equations.

$$F_1 + F_2 = 21 \qquad 3F_1 - 4F_2 = 0$$

Find F_1 and F_2.

62. Two electric resistors, R_1 and R_2, are tested with currents and voltages such that the following equations are found.

$$2R_1 + 3R_2 = 26 \qquad 3R_1 + 2R_2 = 24$$

Find the resistances R_1 and R_2 (in ohms).

63. A business executive, in pricing three different products, determines that a certain total income should be made on given combinations of sales of these products. Thus, if he sets prices p_1, p_2, and p_3, each respectively, he will arrive at the following equations.

$$p_1 + p_2 + p_3 = 200$$
$$p_1 + 2p_2 + 3p_3 = 430$$
$$4p_1 + 2p_2 + p_3 = 430$$

What are the prices necessary to meet his goals?

64. To find the electric currents (in amperes) indicated in Fig. 15–8, it is necessary to solve the following equations.

$$I_A + I_B + I_C = 0$$
$$5I_A - 2I_B = -4$$
$$2I_B - I_C = 0$$

Find I_A, I_B, and I_C.

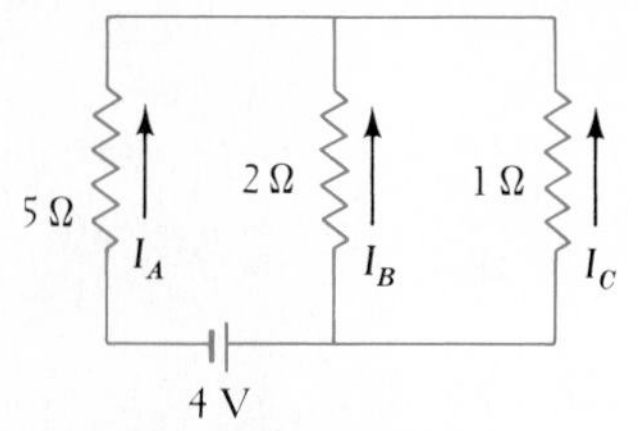

Figure 15-8

In Exercises 65 through 68, set up systems of linear equations and solve by any appropriate method illustrated in this chapter.

65. By mass, three alloys have the following percentages of lead, zinc, and copper.

	Lead	Zinc	Copper
Alloy A	60%	30%	10%
Alloy B	40%	30%	30%
Alloy C	30%	70%	

How many grams of each of alloys A, B, and C must be mixed to get 100 g of an alloy which is 44% lead, 38% zinc, and 18% copper?

66. A sum of $9000 is invested, part at 6%, part at 5%, and part at 4%. The annual income from these investments is $460. The 5% investment yields $50 more than the 4% investment. How much is invested at each rate?

67. The angles A, B, C, and D of a quadrilateral are related in the following way. Angles B and C are supplementary; four times angle A equals 10° less than the sum of angles B, C, and D; six times angle A plus three times angle C equals twice the sum of all four angles. Find the angles.

68. In setting up salaries for personnel in a certain company, the following criteria were used. The total annual budget for salaries of the 3 managers, 15 research persons, 25 technicians, and 40 production workers was $1,230,000. Each manager receives $5000 more than each research person, $10,000 more than each technician, and $13,000 more than each production worker. What salary does each position pay?

In Exercises 69 and 70, perform the indicated operations on the required matrices.

69. An automobile maker has two assembly plants at which cars with 4 cylinders, 6 cylinders, or 8 cylinders, and with either standard transmission or automatic transmission are assembled. The annual production at the first plant of cars with number of cylinders-transmission type (standard-automatic) is as follows: 4—15,000, 10,000; 6—20,000, 18,000; 8—8000, 30,000. At the second plant the production is 4—18,000, 12,000; 6—30,000, 22,000; 8—12,000, 40,000. Set up matrices for this information, and by matrix addition find the matrix for total production by the number of cylinders and type of transmission.

70. Set up a matrix representing the information given in Exercise 65. A given shipment contains 500 g of alloy A, 800 g of alloy B, and 700 g of alloy C. Set up a matrix for this information. By multiplying these matrices, obtain a matrix which gives the total weight of lead, zinc, and copper in the shipment.

CHAPTER 16

Inequalities

Until now we have devoted a great deal of time to the solution of equations. Equation-solving does play an extremely important role in mathematics, but there are also times when we wish to solve inequalities. For example, we have often been faced with the problem of whether or not a given number is real or complex. We determine this from the quadratic formula, by observing the sign of the expression $b^2 - 4ac$. If this expression is positive, the resulting number is real, and if it is negative, the resulting number is complex. This is one of the important uses of inequalities. In this chapter we shall discuss some of the important properties of inequalities and methods of solving inequalities, and illustrate some of their applications in various fields of technology.

16–1 Properties of Inequalities

In Chapter 1 we first came across the signs of inequality. The expression $a < b$ is read as "a is less than b," and the expression $a > b$ is read as "a is greater than b." *These signs define what is known as the* **sense** *(indicated by the direction of the sign) of the inequality.* Two inequalities are said to have the same sense if the signs of inequality point in the same direction. They are said to have the opposite sense if the signs of inequality point in opposite directions. *The two sides of the inequality are called* **members** *of the inequality.*

Example A

The inequalities $x + 3 > 2$ and $x + 1 > 0$ have the same sense, as do the inequalities $3x - 1 < 4$ and $x^2 - 1 < 3$.

The inequalities $x - 4 < 0$ and $x > -4$ have the opposite sense, as do the inequalities $2x + 4 > 1$ and $3x^2 - 7 < 1$.

The **solution** *of an inequality consists of those values of the variable for which the inequality is satisfied.* Most inequalities with which we shall deal are known as **conditional inequalities.** That is, *there are some values of the variable which satisfy the inequality, and also there are some values which do not satisfy it.* Also, *some inequalities are satisfied for all values of the variable. These are called* **absolute inequalities.** A solution of an inequality may consist of only real numbers, as the terms "greater than" and "less than" have not been defined for complex numbers.

Example B

The inequality $x + 1 > 0$ is satisfied by all values of x greater than -1. Thus, the values of x which satisfy this inequality are written as $x > -1$. This illustrates the difference between the solution of an equation and the solution of an inequality. The solution to an equation normally consists of a few specific numbers, whereas *the solution to an inequality normally consists of a range of values of the variable*. Any and all values within this range are termed the solution of the inequality.

Example C

The inequality $x^2 + 1 > 0$ is true for all values of x, since x^2 is never negative. This is an absolute inequality. The inequality shown in Example B is a conditional inequality.

There are occasions when it is convenient to combine an inequality with an equality. For such purposes, the symbols $\leq$, read "less than or equal to," and $\geq$, read "greater than or equal to," are used.

Example D

If we wish to state that x is positive, we would write $x > 0$. However, the value zero is not included in the solution. If we wished to state that x is not negative, that is, that zero is included as part of the solution, we can write $x \geq 0$. In order to state that x is less than or equal to -5, we write $x \leq -5$.

We shall now present the basic operations performed on inequalities. These operations are the same as those performed on equations, but in certain cases the results take on a different form. *The following are referred to as the* **properties of inequalities:**

1. *The sense of an inequality is not changed when the same number is added to—or subtracted from—both members of the inequality.* Symbolically this may be stated as "if $a > b$, then $a + c > b + c$, or $a - c > b - c$."

Example E

9 > 6; thus, 9 + 4 > 6 + 4, or 13 > 10. Also, 9 − 12 > 6 − 12 or −3 > −6.

2. *The sense of an inequality is not changed if both members are multiplied or divided by the same positive number. Symbolically this is stated as,* "if $a > b$, then $ac > bc$, or $a/c > b/c$, provided that $c > 0$."

Example F

8 < 15; thus, 8(2) < 15(2) or 16 < 30. Also, $\frac{8}{2} < \frac{15}{2}$ or $4 < \frac{15}{2}$.

 3. *The sense of an inequality is **reversed** if both members are multiplied or divided by the same negative number. Symbolically this is stated as,* "if $a > b$, then $ac < bc$, or $a/c < b/c$, provided that $c < 0$." *Be very careful to note that we obtain different results, depending on whether both members are multiplied by a positive or by a negative number.*

Example G

4 > −2; thus, 4(−3) < (−2)(−3), or −12 < 6. Also,

$$\frac{4}{-2} < \frac{-2}{-2} \quad \text{or} \quad -2 < 1$$

Example H

If $-\frac{x}{2} \le -4$, then

$$(-2)\left(-\frac{x}{2}\right) \ge (-2)(-4)$$

or

$$x \ge 8$$

Here we note that the sense of the inequality is reversed, but the equality remains.

4. *If both members of an inequality are positive numbers and n is a positive integer, then the inequality formed by taking the nth power of each member, or the nth root of each member, is in the same sense as the given inequality. Symbolically this is stated as,* "if $a > b$, then $a^n > b^n$, or $\sqrt[n]{a} > \sqrt[n]{b}$, provided that $n > 0$, $a > 0$, $b > 0$."

Example I $16 > 9$; thus, $16^2 > 9^2$ or $256 > 81$; also $\sqrt{16} > \sqrt{9}$ or $4 > 3$.

Many inequalities have more than two members. In fact, inequalities with three members are very common. All the operations stated above hold for inequalities with more than two members. Some care must be used, however, in stating inequalities with more than two members.

Example J   In order to state that 5 is less than 6, and also greater than 2, which says that 5 is between 2 and 6, we may write $2 < 5 < 6$, or $6 > 5 > 2$. However, generally the form with the *less than* inequality signs is preferred.

In order to state that a number x may be equal to or greater than 2, **and** also less than 6, we write $2 \leq x < 6$.

By writing $2 \leq x \leq 6$ we are stating that x is greater than or equal to 2, and at the same time less than or equal to 6.

By writing $x \leq -5$, $x > 7$ we are stating that x is less than or equal to -5, *or* greater than 7. This may not be stated as $7 < x \leq -5$, for this shows x as being less than -5, while at the same time greater than 7, and no such numbers exist.

Example K   The inequality $x^2 - 3x + 2 > 0$ is satisfied if x is either greater than 2 or less than 1. This would be written as $x > 2$ or $x < 1$, but it would be incorrect to state it as $1 > x > 2$. (If we wrote it this way, we would be saying that the same value of x is less than 1 and at the same time greater than 2. Of course, as we noted for this type of situation in Example J, no such number exists.) Any inequality must be valid for all values satisfying it. However, we could say that the inequality is not satisfied for $1 \leq x \leq 2$, which means those values of x between or equal to 1 and 2.

We shall now present two examples of other kinds of problems in which the basic properties of inequalities are used.

Example L   If $0 < x < 1$, prove that $x^2 < x$.

From the given inequality we see that x is a positive number less than 1. Thus, if we multiply the members of the given inequality by x, we have $0 < x^2 < x$, which gives the desired result if we consider the middle and right members. Note the meaning of this inequality. The square of any positive number less than 1 is less than the number itself.

Example M

State, by means of an inequality, the conditions that x must satisfy if a point in the xy-plane lies between the lines $x = 1$ and $x = 5$.

  The x-coordinate of any point in this part of the plane is greater than 1, but at the same time less than 5. Thus we have $1 < x < 5$. See Fig. 16–1.

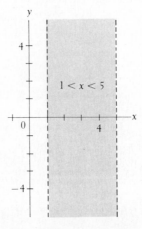

Figure 16–1

  On occasion it is necessary to define a function in a different way for some values of the independent variable than for other values. Inequalities can then be used to denote the intervals over which the different definitions of the function are valid. The following example illustrates this use of inequalities, and includes a graphical representation.

Example N

The electrical intensity within a charged spherical conductor is zero. The intensity on the surface and outside of the sphere is equal to a constant divided by the square of the distance from the center of the sphere. State these relations by using inequalities, and make a graphical representation.

  Let a = the radius of the sphere, r = the distance from the center of the sphere, and E = the electrical intensity.

  The first statement may be written as $E = 0$ if $0 \le r < a$, since this would be read as "the electric intensity is 0 if the distance from the center is less than the radius." Negative values of r are meaningless, which is the reason for saying that r is greater than or equal to zero. The second statement may be written as

(continued on next page)

$E = k/r^2$ if $a \le r$. Making a table of values for this equation, we have the following points:

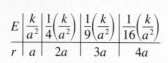

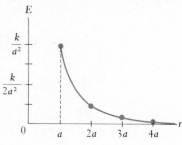

Figure 16–2

The graph of E versus r is shown in Fig. 16–2.

Exercises 16–1

In Exercises 1 through 8, for the inequality $4 < 9$, state the inequality resulting when the operations given are performed on both members.

1. Add 3.
2. Subtract 6.
3. Multiply by 5.
4. Multiply by -2.
5. Divide by -1.
6. Divide by 2.
7. Square both.
8. Take square roots.

In Exercises 9 through 12, for the given inequalities, state the inequality that results when the given operations are performed on both members.

9. $\dfrac{-x}{3} \ge -4$, multiply by -3.
10. $\dfrac{-x}{3} \ge -4$, multiply by 3.
11. $-2x \le -6$, divide by 2.
12. $-2x \ge -6$, divide by -2.

In Exercises 13 through 20, give the inequalities which are equivalent to the following statements about a number x.

13. Greater than -2.
14. Less than 7.
15. Less than or equal to 4.
16. Greater than or equal to -6.
17. Greater than 1 and less than 7.
18. Greater than or equal to -2 and less than 6.
19. Less than -9, or greater than or equal to -4.
20. Less than or equal to 8, or greater than or equal to 12.

In Exercises 21 through 28, state the condition in terms of an inequality that x must satisfy to describe the location of the given point.

21. The point (x, y) lies to the right of the y-axis.
22. The point (x, y) lies to the right of the line $x = 1$.
23. The point (x, y) lies on or to the left of the y-axis.

24. The point (x, y) lies on or to the right of the line $x = -2$.

25. The point (x, y) lies outside of the region between the lines $x = -1$ and $x = 1$.

26. The point (x, y) lies in the region between the lines $x = -1$ and $x = 1$.

27. The point (x, y) lies on or to the right of the line $x = 2$ and to the left of the line $x = 6$.

28. The point (x, y) lies to the left of the line $x = -4$ or to the right of the line $x = 3$.

In Exercises 29 through 32, state the conditions in terms of inequalities that x, or y, or both, must satisfy to describe the location of the given point.

29. The point (x, y) lies in the first quadrant.

30. The point (x, y) lies in the region bounded by the lines $x = 1$, $x = 4$, $y = -3$, and $y = -1$.

31. The point (x, y) lies above the line $x = y$.

32. The point (x, y) lies within three units of the origin. (*Hint:* Use the Pythagorean theorem.)

In Exercises 33 through 36, prove the given inequalities.

33. If $x > 1$, prove that $x^2 > x$.   34. If $x > y > 0$, prove that $\dfrac{1}{y} > \dfrac{1}{x}$.

35. If $0 < x < y$, prove that $\sqrt{xy} < y$.

36. If $x > x^2$, prove that $\sqrt{3x + 1} > x + 1$.

In Exercises 37 through 44, some applications of inequalities are shown.

37. A certain projectile is at a height h of more than 200 m for a certain interval of time after it is launched. Express this statement as an inequality in terms of the height h.

38. The force F which is required to lift one end of a beam must be at least 60 N. Write this as an inequality.

39. The temperature T within a certain refrigeration unit must be at least 36°F and no more than 40°F. Express the temperatures which should not exist in this refrigeration unit by use of inequalities.

40. An earth satellite put into orbit near the earth's surface will have an elliptic orbit if its velocity v is between 18,000 mi/h and 25,000 mi/h. State this as an inequality.

41. A geologist reported that the layers of soil in a certain region were formed between 25,000 years ago and 40,000 years ago. Write this as an inequality, with t representing past time in years.

42. A certain solar panel can produce an electric current i of no more than 5 mA. The current cannot be negative. Write this as an inequality.

43. The electric potential V inside a charged spherical conductor equals a constant k divided by the radius a of the sphere. The potential on the surface of and outside the sphere equals the same constant k divided by the distance r from the center of the sphere. State these relations by the use of inequalities and make a graphical representation of V versus r.

44. A semiconductor diode, an electronic device, has the property that an electric current can flow through it in only one direction. Thus, if a diode is in a circuit with an alternating-current source, the current in the circuit exists only during the half-cycle when the direction is correct for the diode. If a source of current given by $i = 2 \sin 120\pi t$ milliamperes is connected in series with a diode, write the inequalities which are appropriate for the first four half-cycles and graph the resulting current versus the time. Assume that the diode allows a positive current to flow.

16–2 Graphical Solution of Inequalities

Equations can be solved by graphical and by algebraic means. This is also true of inequalities. In this section we shall take up graphical solutions, and in the following section we shall develop algebraic methods. The graphical method is shown in the following examples.

Example A

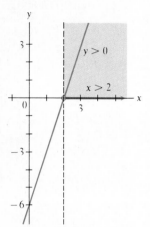

Figure 16–3

Graphically solve the inequality $3x - 2 > 4$.

This means that we want to locate all those values of x which make the left member of this inequality greater than the right member. By subtracting 4 from each member, we have the equivalent inequality $3x - 6 > 0$.

If we find the graph of the function $y = 3x - 6$, all those values of x for which y is positive would satisfy the inequality $3x - 6 > 0$. Thus, we graph the equation $y = 3x - 6$, and find those values of x for which y is positive. From the graph in Fig. 16–3, we see that values of $x > 2$ correspond to $y > 0$. Thus, the values of x which satisfy the inequality are given by $x > 2$. In the figure we have shaded the area where the line is above the x-axis. Then we show the solution by placing an open circle on the x-axis at $(2, 0)$ and drawing an arrow to the right. The open circle means that 2 is not included in the solution.

Example B

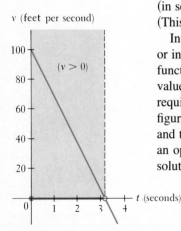

Figure 16–4

The velocity v (in feet per second) of a certain projectile in terms of the time t (in seconds) is given by $v = 100 - 32t$. For how long is the velocity positive? (This can be interpreted as "how long is the projectile moving upward?")

In terms of inequalities, we would like to know for what values of t is $v > 0$, or in other terms, solve the inequality $100 - 32t > 0$. Therefore, we graph the function $v = 100 - 32t$ as shown in Fig. 16–4. From the graph, we see that the values of t which correspond to $v > 0$ are $0 \le t < 3.1$ s, which are therefore the required values of t. The value of 3.1 is approximated from the graph. In the figure we have shaded the area above the t-axis through which the line passes, and then indicated the solution by drawing a line from a solid circle at $(0, 0)$ to an open circle at $(3.1, 0)$. The solid circle shows that $t = 0$ is included in the solution.

Example C

Graphically solve the inequality $2x^2 < x + 3$.

Finding the equivalent inequality, with 0 for a right-hand member, we have $2x^2 - x - 3 < 0$. Thus, those values of x for which y is negative for the function $y = 2x^2 - x - 3$ will satisfy the inequality. So we graph the equation $y = 2x^2 - x - 3$, from the values given in the following table.

x	-2	0	$\frac{1}{4}$	2
y	7	-3	$-\frac{25}{8}$	3

From the graph in Fig. 16–5, we can see that the inequality is satisfied for the values $-1 < x < 1.5$. Note the shaded area below the x-axis, through which the parabola passes, and the interval marked from $(-1, 0)$ to $(1.5, 0)$.

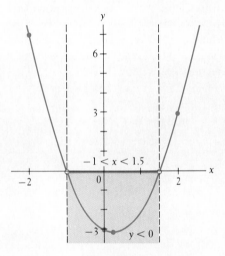

Figure 16–5

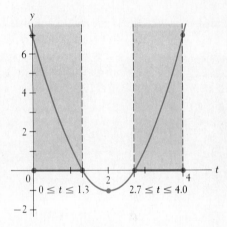

Figure 16–6

Example D

The force F, in newtons, acting on a certain object varies according to the time, in seconds, and is given by the function $F = 2t^2 - 8t + 10$. For what values of t, $0 \le t \le 4$ s, is the force at least 3 N?

To have a force of at least 3 N, we know that $F \ge 3$ N, or $2t^2 - 8t + 10 \ge 3$. This means we are to solve the inequality $2t^2 - 8t + 7 \ge 0$. Using the values in the following table and graphing $y = 2t^2 - 8t + 7$, we see from Fig. 16–6 that the solution is $0 \le t \le 1.3$ s or $2.7 \le t \le 4.0$ s.

t	0	2	4
y	7	-1	7

Summarizing the method for the graphical solution of an inequality, we see that we first write the inequality in an equivalent form with zero on the right. Next we set y equal to the left member and graph the resulting function. Those values of x corresponding to the proper values of y (either above or below the x-axis) are those values which satisfy the inequality.

Example E

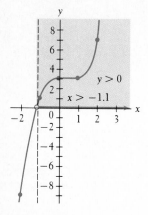

Figure 16–7

Graphically solve the inequality $x^3 > x^2 - 3$.

Finding the equivalent inequality with 0 on the right, we then have $x^3 - x^2 + 3 > 0$. By letting $y = x^3 - x^2 + 3$, we may solve the inequality by finding those values of x for which y is positive.

x	-2	-1	0	1	2
y	-9	1	3	3	7

Approximating the root of the function to be -1.2, the values which satisfy the inequality are given by $x > -1.2$ (Fig. 16–7). If greater accuracy is required, a calculator may be used to find the root. This example also points out the usefulness of graphical methods, since this inequality would prove to be beyond elementary methods if an algebraic solution were required.

Exercises 16–2

In Exercises 1 through 24, solve the given inequalities graphically. Show the solution along the x-axis as in the examples.

1. $2x > 4$
2. $3x < -6$
3. $5x - 1 > 3x$
4. $2x - 3 < x$
5. $7x - 5 < 4x + 3$
6. $2x \geq 6x - 2$
7. $5x \leq 6x - 1$
8. $6 - x < x$
9. $x^2 > 2x$
10. $x^2 < x$
11. $2x + 3 < x^2$
12. $x - 1 < x^2$
13. $x^2 - 5x \leq -4$
14. $7x - 3 \leq -6x^2$
15. $3x^2 - 2x - 8 > 0$
16. $4x^2 - 2x > 5$
17. $x^3 > 1$
18. $x^3 \geq x + 4$
19. $x^4 \leq x^2 - 2x - 1$
20. $x^4 - 6x^3 + 7x^2 > 18 - 12x$
21. $2^x > 0$
22. $\log x > 1$
23. $\sin x < 0$ (limit the graph to the values $0 \leq x \leq 2\pi$)
24. $\cos 2x > 0$ (limit the graph as in Exercise 23)

In Exercises 25 through 32, answer the given questions by solving appropriate inequalities graphically.

25. A salesperson receives $300 monthly, plus a 10% commission on sales. Therefore, his monthly income I in terms of his sales s is $I = 300 + 0.1s$. How much must his sales for the month be in order that his income is at least $500?

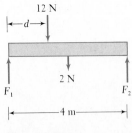

12 N

$\leftarrow d \rightarrow$

2 N

F_1 F_2

$\leftarrow$—4 m—$\rightarrow$

Figure 16–8

26. The basic maintenance on a certain machine is $10 per day. It also costs $2 per hour while in operation. Therefore, the total cost C to operate this machine in terms of the number of hours t that it operates in a given day is $C = 10 + 2t$. How many hours can it operate without the cost exceeding $30 for the day?

27. A beam is supported at each end, as shown in Fig. 16–8. Analyzing the forces on the beam leads to the equation $F_1 = 13 - 3d$. For what values of d is F_1 more than 6 N?

28. An oil company plans to install eight storage tanks, each with a capacity of x liters, and five additional tanks, each with a capacity of y liters. The total capacity of all tanks is 440,000 L, which means that $8x + 5y = 440,000$. If capacity y will be at least 40,000 L, what are the necessary values of capacity x?

29. The electrical resistance R of a certain material depends on the temperature, according to the relation $R = 40 + 0.1T + 0.01T^2$. For what values of the temperature ($T > 0°C$) is the resistance over 42 Ω?

30. The height s of a certain object is given by $s = 140 + 60t - 16t^2$, where t is the time (in seconds). For what values of t is $s > 180$ ft?

31. The cost of producing three of one type of calculator and five of a second type is $50. If the cost of producing each of the second type is less than $10, what are the possible costs of producing each of the first type?

32. A rectangular solar panel is designed so that its length is 20 cm more than its width. For what values of the width is the area more than 2400 cm^2?

16–3 Algebraic Solution of Inequalities

In this section we will show how the algebraic solution of a linear inequality is found in essentially the same way as that of a linear equation. Then we will develop a more general algebraic method of solving inequalities which can be used with polynomial and fractional expressions.

As with an equation, we can perform the same operation on each member when solving an inequality. In doing so, we must be careful when we multiply or divide by a negative number, since the sense of the inequality is reversed. The following examples illustrate the solution of linear inequalities in this way.

Example A

Solve the inequality $3x + 11 \leq 2$.

Here we will isolate x in the left member in the same way we would solve the equation $3x + 11 = 2$.

$$3x + 11 \leq 2 \qquad \text{(original inequality)}$$
$$3x \leq -9 \qquad \text{(subtract 11 from each member)}$$
$$x \leq -3 \qquad \text{(divide each member by 3)}$$

Therefore our solution is $x \leq -3$, which means that any value less than or equal to -3 is a solution.

Example B

Solve the inequality $\frac{3}{2}(1 - x) > \frac{1}{4} - x$.

$$\frac{3}{2}(1 - x) > \frac{1}{4} - x \qquad \text{(original inequality)}$$

$$6(1 - x) > 1 - 4x \qquad \text{(multiply each member by 4)}$$

$$6 - 6x > 1 - 4x \qquad \text{(remove parentheses)}$$

$$-6x > -5 - 4x \qquad \text{(subtract 6 from each member)}$$

$$-2x > -5 \qquad \text{(add $4x$ to each member)}$$

$$x < \frac{5}{2} \qquad \text{(divide each member by -2)}$$

Note that the sense of the inequality was reversed when we divided by -2.

We now recall that one important step in solving inequalities by graphical means was to set up an equivalent inequality with zero as the right member. *This enabled us to find the values which satisfied the inequality by simply determining the **sign** of the function which was set up.* A similar procedure is useful when we solve inequalities algebraically.

A linear function $y = ax + b$ $(a \neq 0)$ is either negative or positive for *all* values to the left of a particular value of x, and has the opposite sign for *all* values to the right of the same value of x. *This value of x which divides the positive and negative intervals is called the **critical value.** This critical value is found where the function is zero.* Thus, to determine where a linear function is positive and where it is negative, we need to find only this critical value, and then determine the *sign* of the function to the left and to the right of this value.

Example C

Solve the inequality $2x - 5 > 1$.

We first find the equivalent inequality with zero on the right. This is done by subtracting 1 from each member. Thus, we have $2x - 6 > 0$. We now set the left member *equal* to zero to find the critical value. Thus, the critical value is 3. We know that the function $2x - 6$ is of one sign for $x < 3$ and has the opposite sign for $x > 3$. Testing values in these intervals, we find that for $x > 3$, $2x - 6 > 0$. Thus, the values which satisfy the inequality are those for which $x > 3$.

Of course, we could have solved this inequality by the method of Examples A and B. However, the idea of using the sign of the function on the left, with zero on the right, is the important concept here.

Example D

Solve the inequality $\frac{1}{2}x - 3 \leq \frac{1}{3} - x$.

We note that this is an inequality combined with an equality. However, the solution proceeds essentially the same as with an inequality.

Multiplying by 6, we have $3x - 18 \leq 2 - 6x$. Subtracting $2 - 6x$ from each member, we have $9x - 20 \leq 0$. The critical value is $\frac{20}{9}$. If $x \leq \frac{20}{9}$, $9x - 20 \leq 0$. Thus, the values which satisfy the inequality are given by $x \leq \frac{20}{9}$.

The preceding analysis is especially useful for solving inequalities involving higher-degree functions or involving fractions with x in the denominator as well as in the numerator. When the equivalent inequality with zero on the right has been found, this inequality is then factored into linear factors (and those quadratic factors which lead to complex roots). Each linear factor can change sign only at its critical value, as all possible values of x are considered. Thus, the function on the left can change sign only where one of its factors changes sign. The function will have the same sign for all values of x less than the leftmost critical value. The sign of the function will also be the same within any given interval between two critical values. All values to the right of the rightmost critical value will also give the function the same sign. *Therefore, we must find all of the critical values and then determine the sign of the function to the left of the leftmost critical value, between the critical values, and to the right of the rightmost critical value. Those intervals in which we have the proper sign will satisfy the given inequality.*

Example E

Solve the inequality $x^2 - 3 > 2x$.

We first find the equivalent inequality with zero on the right. Thus, we have $x^2 - 2x - 3 > 0$. We then factor the left member, and have

$$(x - 3)(x + 1) > 0$$

We find the critical value for each of the factors, for these are the only values for which the function $x^2 - 2x - 3$ is zero. The left critical value is -1, and the right critical value is 3. All values of x to the left of -1 give the same sign for the function. All values of x between -1 and 3 give the function the same sign. All values of x to the right of 3 give the same sign to the function. Therefore, we must determine the sign for $x < -1$, $-1 < x < 3$, and $x > 3$. For the interval $x < -1$, we find that each of the factors is negative. However, the product of two negative numbers gives a positive number. Therefore, if $x < -1$, then $(x - 3)(x + 1) > 0$.

For the interval $-1 < x < 3$, we find that the left factor is negative, but the right factor is positive. The product of a negative and positive number gives a negative number. Thus, for the interval $-1 < x < 3$, $(x - 3)(x + 1) < 0$. For the interval $x > 3$, both factors are positive, making $(x - 3)(x + 1) > 0$. We tabulate the results.

If $\quad\quad x < -1 \quad\quad (x - 3)(x + 1) > 0$

If $-1 < x < 3 \quad\quad (x - 3)(x + 1) < 0$

If $\quad\quad x > 3 \quad\quad (x - 3)(x + 1) > 0$

Thus, the inequality is satisfied for $x < -1$ or $x > 3$.

Example F

Solve the inequality $x^3 - 4x^2 + x + 6 < 0$.

By methods developed in Chapter 14, we factor the function of the left and obtain $(x + 1)(x - 2)(x - 3) < 0$. The critical values are $-1, 2, 3$. We wish to determine the sign of the left member for the intervals $x < -1$, $-1 < x < 2$, $2 < x < 3$, and $x > 3$. This is tabulated here, with the sign of the factors in each case indicated.

Interval	$(x + 1)(x - 2)(x - 3)$			Sign of $(x + 1)(x - 2)(x - 3)$
$x < -1$	$-$	$-$	$-$	$-$
$-1 < x < 2$	$+$	$-$	$-$	$+$
$2 < x < 3$	$+$	$+$	$-$	$-$
$x > 3$	$+$	$+$	$+$	$+$

Thus, the inequality is satisfied for $x < -1$ or $2 < x < 3$.

Example G

Solve the inequality $\dfrac{x - 3}{x + 4} \geq 0$.

The critical values are found from the factors, whether they are in the numerator or in the denominator. Thus, the critical values are -4 and 3. Considering now the *greater than* part of the $\geq$ sign, we set up the following table.

Interval	$\dfrac{x - 3}{x + 4}$	Sign of $\dfrac{x - 3}{x + 4}$
$x < -4$	$\dfrac{-}{-}$	$+$
$-4 < x < 3$	$\dfrac{-}{+}$	$-$
$x > 3$	$\dfrac{+}{+}$	$+$

Thus, the values which satisfy the greater than part of the problem are those for which $x < -4$ or for which $x > 3$. Now considering the equality part of the $\geq$ sign, we note that $x = 3$ is valid, for the fraction is zero. However, if $x = -4$, we have division by zero, and thus x may not equal -4. Therefore, the inequality is satisfied for $x < -4$ or $x \geq 3$.

Example H

Solve the inequality $x^3 - x^2 + x - 1 > 0$.

This leads to $(x^2 + 1)(x - 1) > 0$. There is only one linear factor with a critical value. The factor $x^2 + 1$ is never negative. The inequality is satisfied for $x > 1$.

Example I

Solve the inequality

$$\frac{(x - 2)^2(x + 3)}{4 - x} < 0$$

The critical values are -3, 2, and 4. Thus, we have the following table.

Interval	$\dfrac{(x - 2)^2(x + 3)}{4 - x}$	Sign of $\dfrac{(x - 2)^2(x + 3)}{4 - x}$
$x < -3$	$\dfrac{+ \quad -}{+}$	$-$
$-3 < x < 2$	$\dfrac{+ \quad +}{+}$	$+$
$2 < x < 4$	$\dfrac{+ \quad +}{+}$	$+$
$x > 4$	$\dfrac{+ \quad +}{-}$	$-$

The inequality is satisfied for $x < -3$ or $x > 4$.

Exercises 16–3

In Exercises 1 through 12, solve the given inequalities by the method of Examples A and B.

1. $x - 3 > 2$
2. $x + 5 < 3$
3. $3x + 1 \le -14$
4. $4x \ge 18 + x$
5. $9 - 2x \ge 3$
6. $1 - 5x < 17$
7. $3(x - 2) < x + 5$
8. $2(x - 4) > 3x + 7$
9. $x + 4 \ge 3(x - 3)$
10. $2x - 7 \le 4 - (x + 2)$
11. $\dfrac{1}{3} - \dfrac{x}{2} < x + \dfrac{3}{2}$
12. $\dfrac{x}{5} - 2 > \dfrac{2}{3}(x + 3)$

In Exercises 13 through 40, solve the given inequalities by the method of Examples C through I.

13. $6x - 4 < 8 - x$
14. $2x - 6 \le x + 4$
15. $\frac{1}{3}(x - 6) \ge 4 - x$
16. $7 - x > x - 1$
17. $x^2 - 1 < 0$
18. $x^2 - 4x - 5 > 0$
19. $3x^2 + 5x \ge 2$
20. $2x^2 - 12 \le -5x$
21. $6x^2 + 1 < 5x$
22. $9x^2 + 6x > -1$
23. $x^2 + 4 > 0$
24. $x^4 + 2 < 1$
25. $x^3 + x^2 - 2x > 0$
26. $x^3 - 2x^2 + x > 0$
27. $x^3 + 2x^2 - x - 2 > 0$
28. $x^4 - 2x^3 - 7x^2 + 8x + 12 < 0$

29. $\dfrac{x-8}{3-x} < 0$

30. $\dfrac{x+5}{x-1} > 0$

31. $\dfrac{2x-3}{x+6} \le 0$

32. $\dfrac{3x+1}{x-3} \ge 0$

33. $\dfrac{2}{x^2-x-2} < 0$

34. $\dfrac{-5}{2x^2+3x-2} < 0$

35. $\dfrac{x^2-6x-7}{x+5} > 0$

36. $\dfrac{4-x}{3+2x-x^2} > 0$

37. $\dfrac{6-x}{3-x-4x^2} > 0$

38. $\dfrac{(x-2)^2(5-x)}{(4-x)^3} < 0$

39. $\dfrac{x^4(9-x)(x-5)(2-x)}{(4-x)^5} > 0$

40. $\dfrac{x^3(1-x)(x-2)(3-x)(4-x)}{(5-x)^2(x-6)^3} < 0$

In Exercises 41 through 44, determine the values of x for which the given radicals represent real numbers.

41. $\sqrt{(x-1)(x+2)}$

42. $\sqrt{x^2-3x}$

43. $\sqrt{-x-x^2}$

44. $\sqrt{\dfrac{x^3+6x^2+8x}{3-x}}$

In Exercises 45 through 50, answer the given questions by solving the appropriate inequalities.

45. The velocity (in feet per second) of a certain object in terms of the time t (in seconds) is given by $v = 120 - 32t$. For what values of t is the object ascending ($v > 0$)?

46. The relationship between Fahrenheit degrees and Celsius degrees is $F = \frac{9}{5}C + 32$. For what values of C is $F \ge 98.6$ (normal body temperature)?

47. Determine the values of T for which the resistor of Exercise 29 of Section 16–2 has a resistance between 41 and 42 Ω.

48. The deflection y of a certain beam 9 ft long is given by the equation $y = k(x^3 - 243x + 1458)$. For which values of x (the distance from one end of the beam) is the quantity y/k greater than 216 units?

49. One type of machine part costs $10 each, and a second type costs $20 each. How many of the second type can be purchased if at least 30 of the first type are purchased and a total of $1000 is spent?

50. The length of a microprocessor chip is 2 mm more than its width. If its area is less than 35 mm^2, what values are possible for the width if it must be at least 3 mm?

16–4 Inequalities Involving Absolute Values

If we wish to write the inequality $|x| > 1$ without absolute-value signs, we must note that we are considering values of x which are *numerically* larger than 1. Thus we may write this inequality in the equivalent form $x < -1$ or $x > 1$. We now note that *the original inequality, with an absolute-value sign, can be written in terms of two equivalent inequalities, neither involving absolute values.* If we are asked to write the inequality $|x| < 1$ without the absolute-value signs, we write $-1 < x < 1$ since we are considering values of x which are numerically less than 1.

Following reasoning similar to the above, whenever absolute values are involved in inequalities, the following two relations allow us to write equivalent inequalities without absolute values.

If $|f(x)| > n$, then $f(x) < -n$ or $f(x) > n$.   (16–1)

If $|f(x)| < n$, then $-n < f(x) < n$.   (16–2)

The use of these relations is indicated in the following examples.

Example A

Solve the inequality $|x - 3| < 2$.

Inspection of this inequality shows that we wish to find the values of x which are within 2 units of $x = 3$. Of course, such values are given by $1 < x < 5$. Let us now see how Eq. (16–2) gives us this result.

By using Eq. (16–2), we have

$$-2 < x - 3 < 2$$

By adding 3 to all three members of this inequality, we have

$$1 < x < 5$$

which is the proper interval.

Example B

Solve the inequality $|2x - 1| > 5$.

By using Eq. (16–1), we have

$$2x - 1 < -5 \quad \text{or} \quad 2x - 1 > 5$$

Completing the solution, we have

$$2x < -4 \quad \text{or} \quad 2x > 6$$
$$x < -2 \quad \text{or} \quad x > 3$$

This means that the given inequality is satisfied for $x < -2$ or for $x > 3$.

Example C

Solve the inequality $|3x + 2| \leq 4$.

Although there is a sign of equality involved, we may solve this inequality in the same way as indicated in Eq. (16–2). Since $f(x)$ is linear (and therefore no division by a factor containing x is involved), we may include the equals sign throughout the solution. Therefore, we have

$$-4 \leq 3x + 2 \leq 4$$
$$-6 \leq 3x \leq 2$$
$$-2 \leq x \leq \frac{2}{3}$$

We note that for $x = -2$ and for $x = \frac{2}{3}$, $|3x + 2| = 4$. Thus, the last inequality shown gives the values of x which satisfy the given inequality.

Example D

Solve the inequality $|x^2 + x - 4| > 2$.

By Eq. (16–1), we have $x^2 + x - 4 < -2$ or $x^2 + x - 4 > 2$, which means we want values of x which satisfy *either* of these inequalities. The first inequality becomes

$$x^2 + x - 2 < 0$$
$$(x + 2)(x - 1) < 0$$

which is satisfied for $-2 < x < 1$. The second inequality becomes

$$x^2 + x - 6 > 0$$
$$(x + 3)(x - 2) > 0$$

which is satisfied for $x < -3$ or for $x > 2$. Thus, the original inequality is satisfied for $x < -3$, or for $-2 < x < 1$, or for $x > 2$.

Example E

Solve the inequality $|x^2 + x - 4| < 2$.

By Eq. (16–2), we have $-2 < x^2 + x - 4 < 2$, which we can write as $x^2 + x - 4 > -2$ and $x^2 + x - 4 < 2$, as long as we remember that we want values of x which satisfy *both* of these at the same time. The first inequality can be written as

$$x^2 + x - 4 > -2$$
$$x^2 + x - 2 > 0$$
$$(x + 2)(x - 1) > 0$$

which is satisfied for $x < -2$ or for $x > 1$. The second inequality is

$$x^2 + x - 4 < 2$$
$$x^2 + x - 6 < 0$$

or

$$(x + 3)(x - 2) < 0$$

which is satisfied for $-3 < x < 2$. The values of x which satisfy both of the inequalities are those between -3 and -2 and those between 1 and 2. Thus, the original inequality is satisfied for $-3 < x < -2$ or for $1 < x < 2$.

Exercises 16–4

In Exercises 1 through 16, solve the given inequalities.

1. $|x - 4| < 1$ 2. $|x + 1| < 3$ 3. $|3x - 5| > 2$

4. $|2x - 1| > 1$ 5. $|6x - 5| \le 4$ 6. $|3 - x| \le 2$

7. $|4x + 3| > 3$ 8. $|3x + 1| > 2$ 9. $|x + 4| < 6$

10. $|5x - 10| < 2$ 11. $|2 - 3x| \ge 5$ 12. $|3 - 2x| \ge 6$

13. $|x^2 + 3x - 1| < 3$ 14. $|x^2 - 5x - 1| < 5$

15. $|x^2 + 3x - 1| > 3$ 16. $|x^2 - 5x - 1| > 5$

In Exercises 17 through 20, use inequalities involving absolute values to solve the given problems.

17. The deflection y at a horizontal distance x from the left end of a beam is less than $\frac{1}{4}$ ft within 2 ft of a point 6 ft from the left end. State this with the use of an inequality involving absolute values.

18. A given projectile is at an altitude h between 40 and 60 m for the time between $t = 3$ s and $t = 5$ s. Write this using two inequalities involving absolute values.

19. An object is oscillating at the end of a spring which is suspended from a support. The distance x of the object from the support is given by the equation $|x - 8| < 3$. By solving this inequality, determine the distances (in inches) from the support of the extreme positions of the object.

20. The production p (in barrels) of an oil refinery for the coming month is estimated at $|p - 2,000,000| < 200,000$. By solving this inequality, determine the production which is anticipated.

16–5 Graphical Solution of Inequalities with Two Variables

To this point we have considered inequalities with one variable and certain methods of solving them. We may also graphically solve inequalities involving two variables, such as x and y. In this section we consider the solution of such inequalities, as well as one important type of application.

Let us consider the function $y = f(x)$. We know that the coordinates of points on the graph satisfy the equation $y = f(x)$. However, for points above the graph of the function, we have $y > f(x)$, and for points below the graph of the function we have $y < f(x)$. Consider the following example.

Example A

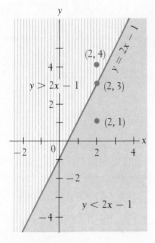

Consider the linear function $y = 2x - 1$. This equation is satisfied for points on the line. For example, the point $(2, 3)$ is on the line and we have $3 = 2(2) - 1 = 3$. Therefore, for points on the line we have $y = 2x - 1$, or $y - 2x + 1 = 0$. The point $(2, 4)$ is above the line, since we have $4 > 2(2) - 1$, or $4 > 3$. Therefore, for points above the line we have $y > 2x - 1$, or $y - 2x + 1 > 0$. In the same way, for points below the line, $y < 2x - 1$ or $y - 2x + 1 < 0$. We note this is true for the point $(2, 1)$, since $1 < 2(2) - 1$, or $1 < 3$. The line for which $y = 2x - 1$, and the regions for which $y > 2x - 1$, and for which $y < 2x - 1$ are shown in Fig. 16–9.

Summarizing,

$$y > 2x - 1 \quad \text{for points } above \text{ the line}$$
$$y = 2x - 1 \quad \text{for points } on \text{ the line}$$
$$y < 2x - 1 \quad \text{for points } below \text{ the line}$$

Figure 16–9

The illustration of Example A leads us to the graphical method of indicating the points which satisfy an inequality with two variables. First we solve the inequality for y and then determine the graph of the function $y = f(x)$. *If we wish to solve the inequality $y > f(x)$, we indicate the appropriate points by shading in the region above the curve. For the inequality $y < f(x)$, we indicate the appropriate points by shading in the region below the curve.* We note that the complete solution to the inequality consists of all points in an entire region of the plane.

Example B

Draw a sketch of the graph of the inequality $y < x + 3$.

First we graph the function $y = x + 3$, as shown in Fig. 16–10. Since we wish to find the points which satisfy the inequality $y < x + 3$, we show these points by shading in the region below the line. We show the line as a *dashed line* to indicate that points on it do not satisfy the inequality.

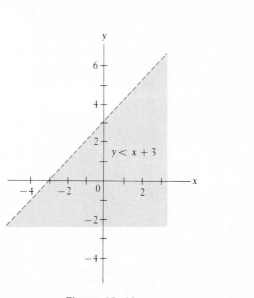

Figure 16–10

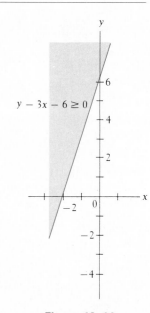

Figure 16–11

Example C

Draw a sketch of the inequality $y - 3x - 6 \geq 0$.

First we state the inequality as $y \geq 3x + 6$. Next the graph of the line $y = 3x + 6$ is drawn, as in Fig. 16–11. The points which satisfy the inequality consist of all points above the line and those points which are on the line. Therefore, we show the line as a *solid line* to indicate that points on it do satisfy the inequality.

Example D

Draw a sketch of the graph of the inequality $y > x^2 - 4$.

    Although the graph of $y = x^2 - 4$ is not a straight line, the method of solution is the same. We graph the function $y = x^2 - 4$ as shown in Fig. 16–12. We then shade in the region above the curve to indicate the points which satisfy the inequality.

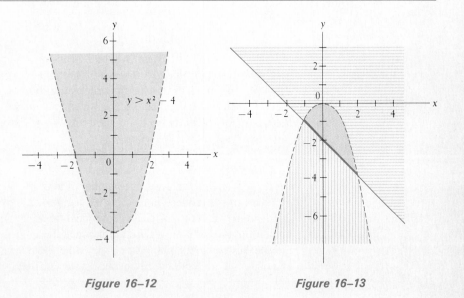

Figure 16–12   *Figure 16–13*

Example E

Draw a sketch of the region which is defined by the system of inequalities $y \geq -x - 2$ and $y + x^2 < 0$.

    In this case we sketch the graph of both inequalities, and then determine the region common to both graphs. First we draw the graph of $y = -x - 2$ (the straight line), and shade in the region above the line. See Fig. 16–13. Next we have $y < -x^2$ and draw the graph of $y = -x^2$, shading the region below it. The sketch of the region which is defined by this system of inequalities is the darkly shaded region below the parabola which is above and on the line.

    An important area in which graphs of inequalities with two or more variables are used is the branch of mathematics known as **linear programming** (in this context, "programming" has no relation to computer programming). This subject is widely applied in industry, business, economics, and technology. The analysis of many social problems can also be made by the use of linear programming.

    Linear programming is used to analyze problems such as those related to maximizing profit, minimizing costs, or the use of materials, with certain constraints of production. The following serves as an example of the use of linear programming.

Example F

A company makes two types of stereo speaker systems, their good-quality system and their highest-quality system. The production of the systems requires assembly of the speaker system itself and the production of the cabinets in which they are installed. The good quality system requires 3 worker-hours for speaker assembly and 2 worker-hours for cabinet production for each complete system. The highest quality system requires 4 worker-hours for speaker assembly and 6 worker-hours for cabinet production for each complete system. Available skilled labor allows for a maximum of 480 worker-hours per week for speaker assembly and a maximum of 540 worker-hours per week for cabinet production. It is anticipated that all systems will be sold and that the profit will be $10 for each good quality system and $25 for each highest quality system. How many of each should be produced to provide the greatest profit?

First, let x = the number of good quality systems and y = the number of highest quality systems made in one week. Thus, the profit p is given by

$$p = 10x + 25y$$

We know that negative numbers are not valid for either x or y, and therefore we have $x \geq 0$ and $y \geq 0$. Also, the number of available worker-hours per week for each part of the production restricts the number of systems which can be made. Both speaker assembly and cabinet production are required for all systems. The number of worker-hours needed to produce the x good quality systems is $3x$ in the speaker assembly shop. Also, $4y$ worker-hours are required in the speaker assembly shop for the highest quality systems. Thus,

$$3x + 4y \leq 480$$

since no more than 480 worker-hours are available in the speaker assembly shop. In the cabinet shop, we have

$$2x + 6y \leq 540$$

since no more than 540 worker-hours are available in the cabinet shop.

Therefore, we wish to maximize the profit p under the **constraints**

$$x \geq 0, \qquad y \geq 0$$
$$3x + 4y \leq 480$$
$$2x + 6y \leq 540$$

In order to do this we sketch the region of points which satisfy this system of inequalities. From the previous examples, we see that the appropriate region is in the first quadrant (since $x \geq 0$ and $y \geq 0$) and under both lines. See Fig. 16–14.

Any point in the shaded region which is defined by the preceding system of inequalities is known as a **feasible point.** In this case it means that it is possible to produce the number of systems of each type according to the coordinates of the point. For example, the point (50, 25) is in the region, which means that it is possible to produce 50 good quality systems and 25 highest quality systems under the given constraints of available skilled labor. However, we wish to find the point which indicates the number of each kind of system which produces the greatest profit.

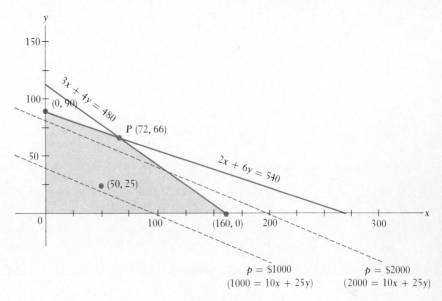

Figure 16–14

If we assume values for the profit, the resulting equations are straight lines. Thus, by finding the greatest value of p for which the line passes through a feasible point, we may solve the given problem. If $p = \$1000$, or if $p = \$2000$, we have the lines shown. Both are possible with various combinations of speaker systems being produced. However, we note the line for $p = \$2000$ passes through feasible points farther from the origin. It is also clear, since these lines are parallel, that the greatest profit attainable is given by the line passing through P, where $3x + 4y = 480$ and $2x + 6y = 540$ intersect. The coordinates of P are $(72, 66)$. Thus, the production should be 72 good quality systems and 66 highest quality systems to produce a weekly profit of $p = 10(72) + 25(66) = \$2370$.

For this type of problem, the solution will be given by one of the vertices of the region. However, it could be any one of them, which means it is possible that only one type of product should be produced. (See Exercise 35.) Thus, we can solve the problem by finding the appropriate region and then testing the coordinates of the vertex points. Here the vertex points are $(160, 0)$ which indicates a profit of $1600, $(0, 90)$ which indicates a profit of $2250, and $(72, 66)$ which indicates a profit of $2370.

Exercises 16–5

In Exercises 1 through 24, draw a sketch of the graph of the given inequality.

1. $y > x - 1$ 2. $y < 3x - 2$

3. $y \geq 2x + 5$ 4. $y \leq 3 - x$

5. $2x + y < 5$ 6. $4x - y > 1$

7. $3x + 2y + 6 > 0$ 8. $x + 4y - 8 < 0$

9. $y < x^2$ 10. $y > -2x^2$

11. $y \geq 1 - x^2$ 12. $y \leq 2x^2 - 3$

13. $x^2 + 2x + y < 0$ 14. $2x^2 - 4x - y > 0$

15. $4x^2 - x - 2y > 0$ 16. $3x^2 + 6x + 2y < 0$

17. $y \leq x^3$ 18. $y \geq 3x - x^3$

19. $y > x^4 - 8$ 20. $y < 32x - x^4$

21. $y < \ln x$ 22. $y > \sin 2x$

23. $y \geq 2 \cos \pi x$ 24. $y \leq 1 - e^{-x}$

In Exercises 25 through 32, draw a sketch of the graph of the region in which the points satisfy the given system of inequalities.

25. $y > x$ 26. $y \leq 2x$ 27. $y \leq 2x^2$ 28. $y > x^2$
     $y > 1 - x$ $y \geq x - 1$ $y > x - 2$ $y < x + 4$

29. $y > \frac{1}{2}x^2$ 30. $y > 4 - x$ 31. $y \geq 0$ 32. $y > 0$
     $y \leq 4x - x^2$ $y < \sqrt{16 - x^2}$ $y \leq \sin x$ $y > 1 - x$
                                          $0 \leq x \leq 3\pi$ $y < e^x$

In Exercises 33 through 36, solve the given linear programming problems.

33. A manufacturer makes two types of calculators, a business model and a scientific model. Each model is assembled in two sets of operations, where each operation is in production 8 h each day. The average time required for a business model in the first operation is 3 min, and 6 min is required in the second operation. The scientific model averages 6 min in the first operation and 4 min in the second operation. All calculators can be sold; the profit for a business model is $8, and the profit for a scientific model is $10. How many of each model should be made each day in order to maximize profit?

34. A company makes brands A and B of breakfast cereal, both of which are enriched with vitamins P and Q. The necessary information about these cereals is given in the following table.

	Cereal A	Cereal B	Min. daily requirement
Vitamin P (units/oz)	1	2	10
Vitamin Q (units/oz)	5	3	30
Cost per ounce	2¢	3¢	

Find the number of ounces of each cereal which together satisfies the minimum daily requirement of vitamins P and Q at the lowest cost. (*Note:* We wish to *minimize* cost; be careful in determining the feasible region.)

35. Using the information of Example F, with the single exception that the profit on each good quality system is $20, how many of each system should be made?

36. A company in competition with the company of Exercise 33 uses identical production methods. However, the profit it receives on its business calculators is only $4, although its profit on its scientific calculators is $10. How many of each type should this company make in order to maximize profit?

16–6 Exercises for Chapter 16

In Exercises 1 through 24, solve the given inequalities algebraically.

1. $2x - 12 > 0$
2. $5 - 3x < 0$
3. $3x + 5 \leq 0$
4. $\frac{1}{4}x - 2 \geq 3x$
5. $3(x - 7) \geq 5x + 8$
6. $2x + 6 < 7(x - 3)$
7. $5x^2 + 9x < 2$
8. $x^2 - 7x \geq 8$
9. $x^2 + 2x > 63$
10. $6x^2 - x > 35$
11. $x^3 + 4x^2 - x > 4$
12. $2x^3 + 4 \leq x^2 + 8x$
13. $\frac{x - 8}{2x + 1} \leq 0$
14. $\frac{3x + 2}{x - 3} > 0$
15. $\frac{(2x - 1)(3 - x)}{x + 4} > 0$
16. $\frac{(3 - x)^2}{2x + 7} \leq 0$
17. $x^4 + x^2 \leq 0$
18. $3x^3 + 7x^2 - 20x < 0$
19. $\frac{1}{x} < 2$
20. $\frac{1}{x - 2} < \frac{1}{4}$
21. $|x - 2| > 3$
22. $|2x - 1| > 5$
23. $|3x + 2| \leq 4$
24. $|4 - 3x| \leq 1$

In Exercises 25 through 32, solve the given inequalities graphically.

25. $6x - 3 < 0$
26. $5 - 8x > 0$
27. $3x - 2 > x$
28. $4 - 5x < 2$
29. $x^2 + 2x + 4 > 0$
30. $3x^2 + 5 > 16x$
31. $x^3 + x + 1 < 0$
32. $\frac{1}{x} > 2$

In Exercises 33 through 36, determine the values of x for which the given radicals represent real numbers.

33. $\sqrt{3 - x}$
34. $\sqrt{x + 5}$
35. $\sqrt{x^2 + 4x}$
36. $\sqrt{\dfrac{x - 1}{x + 2}}$

In Exercises 37 through 44, draw a sketch of the graph of the given inequality.

37. $y > 4 - x$
38. $y < \frac{1}{2}x + 2$
39. $2y - 3x - 4 \leq 0$
40. $3y - x + 6 \geq 0$
41. $y > x^2 + 1$
42. $y < 4x - x^2$
43. $y - x^3 + 1 < 0$
44. $2y + 2x^3 + 6x - 3 > 0$

In Exercises 45 through 48, draw a sketch of the region in which the points satisfy the given systems of inequalities.

45. $y > x + 1$
    $y < 4 - x^2$
46. $y > 2x - x^2$
    $y \geq -2$
47. $y \leq \dfrac{1}{x^2 + 1}$
    $y < x - 1$
48. $y < \cos \frac{1}{2}x$
    $y > \frac{1}{2}e^x$
    $-\pi < x < \pi$

In Exercises 49 through 52, prove the given inequalities.

49. If $x > 0$ and $y < 0$, prove that $\dfrac{1}{x} > \dfrac{1}{y}$.

50. If $x < -1$, prove that $\dfrac{1 - x}{x^2 + 1} > 0$.

51. If $y > 0$ and $x > y + 1$, prove that $\dfrac{x}{y} > \dfrac{y + 1}{x - 1}$.

52. If $x \neq y$, prove that $x^2 + y^2 > 2xy$.

In Exercises 53 through 62, solve the given problems using inequalities.

53. The length L, in centimeters, of a certain metal bar is given by $L = 150 + 0.00025T$, where T is the temperature in degrees Celsius. For what values of T is $L \geq 151$ cm?

54. After conducting tests, it was determined that the stopping distance x (in feet) of a car traveling 60 mi/h was $|x - 290| \leq 35$. Express this inequality without absolute values, and determine the interval of stopping distances which were found in the tests.

55. A piece of wire 100 ft long is to have a piece at least 20 ft cut from it. What lengths may the remaining piece be?

56. What average velocity will allow an object to move more than 150 m in 3 s?

57. One leg of a right triangle is 14 in. longer than the other leg. If the hypotenuse is to be greater than 34 in., what values of the other side are permissible?

58. The length of a rectangular lot is 20 m more than its width. If the area is to be at least 4800 m^2, what values may the width be?

59. Two resistors have a combined resistance of 8 Ω when connected in series. What are the permissible values if they are to have a combined resistance of at least $\frac{3}{2} \Omega$ when connected in parallel? (See Exercise 45 of Section 6–1.)

60. City A is 300 mi from city B. One car starts from A for B one hour before a second car. The first car averages 45 mi/h and the second car averages 60 mi/h for the trip. For what times after the first car starts is the second car ahead of the first car?

61. For 1 g of ice to be melted, then heated to boiling water, and finally vaporized into steam, the relation between the temperature T (in °C) and the number of joules absorbed Q is given by $T = 0$ if $Q < 335$, $T = (Q - 335)/4.19$ if $335 < Q < 750$, and $T = 100$ if $750 < Q < 3010$. Plot the graph of T (along the y-axis) versus Q (along the x-axis) if 3010 J of heat are absorbed by 1 g of ice originally at 0°C.

62. A company produces two types of cameras, the regular model and the deluxe model. For each regular model produced there is a profit of $8, and for each deluxe model the profit is $15. The same amount of materials is used to make each model, but the supply is sufficient only for 450 cameras per day. The deluxe model requires twice the time to produce as the regular model. If only regular models were made, there would be time enough to produce 600 per day. Assuming all models will be sold, how many of each model should be produced if the profit is to be a maximum?

CHAPTER 17

Variation

In many kinds of scientific and technical applications, it is possible to find important relationships among the quantities being measured through experimentation and observation. In studying measured values, it is often possible to see how one quantity changes as other related quantities change. In this chapter we shall see how such information can be used to set up functions which relate the quantities being measured. We shall begin this chapter by studying the meanings of ratio and proportion, which were first introduced in Chapter 3, in more detail.

17–1 Ratio and Proportion

When we first introduced the trigonometric functions in Chapter 3, we used ratios which had a specific meaning. Considering now the general definition, *the **ratio** of a number a to a number b (b ≠ 0) is the quotient a/b*. Thus, a fraction is a ratio.

Any measurement made is the ratio of the measured magnitude to an accepted unit of measurement. For example, when we say that an object is 5 ft long, we are saying that the length of that object is five times as long as an accepted unit of length, the foot. Other examples of ratios are density (weight/volume), relative density (density of object/density of water), and pressure (force/area). As these examples illustrate, ratios may compare quantities of the same kind, or they may express a division of magnitudes of different quantities (such a ratio is also called a **rate**).

Example A

The approximate airline distance from New York to San Francisco is 2500 mi, and the approximate airline distance from New York to Minneapolis is 1000 mi. The ratio of these distances is

$$\frac{2500 \text{ mi}}{1000 \text{ mi}} = \frac{5}{2}$$

Since the units in both are miles, the resulting ratio is a dimensionless number. If a jet travels from New York to San Francisco in 4 h, its average speed is

$$\frac{2500 \text{ mi}}{4 \text{ h}} = 625 \text{ mi/h}$$

In this case we must attach the proper units to the resulting ratio.

As we noted in Example A, we must be careful to attach the proper units to the resulting ratio. Generally, the ratio of measurements of the same kind should be expressed as a dimensionless number. Consider the following example.

Example B

The length of a certain room is 24 ft, and the width of the room is 18 ft. Therefore, the ratio of the length to the width is $\frac{24}{18}$, or $\frac{4}{3}$.

If the width of the room were expressed as 6 yd, we should not express the ratio as 24 ft/6 yd = 4 ft/1 yd. It is much better and more meaningful to first change the units of one of the measurements. Changing the length from 6 yd to 18 ft, we express the ratio as $\frac{4}{3}$, as we saw above. From this ratio we can easily see that the length is $\frac{4}{3}$ as long as the width.

A statement of equality between two ratios is called a **proportion.** One way of denoting a proportion is $a:b = c:d$, which is read "*a* is to *b* as *c* is to *d*." (Another notation used for a proportion is $a:b::c:d$.) Of course, by the definition, $a/b = c/d$, which means that a proportion is an equation. Any operation applicable to an equation is also applicable to a proportion.

Example C

On a certain map 1 in. represents 10 mi. Thus on this map we have a ratio of 1 in./10 mi. To find the distance represented by 3.5 in., we can set up the proportion

$$\frac{3.5 \text{ in.}}{x} = \frac{1 \text{ in.}}{10 \text{ mi}}$$

From this proportion we find the value of $x = 35$ mi.

Example D

If 1 in. = 2.54 cm, what length in centimeters is 15.0 in.?

If we equate the ratio of known lengths to the ratio of the given length to the required length, we can find the required length. This gives us

$$\frac{1 \text{ in.}}{2.54 \text{ cm}} = \frac{15.0 \text{ in.}}{x \text{ cm}}$$

$$x = (15.0)(2.54)$$

$$= 38.1 \text{ cm}$$

Additional illustrations of changing units can be found in Appendix B.

Example E

The magnitude of an electric field E is defined as the ratio between the force F on a charge q and the magnitude of q. This can be written as $E = F/q$. If we know the force exerted on a particular charge at some point in the field, we can determine the force which would be exerted on another charge placed at the same point. For example, if we know that a force of 10 nN is exerted on a charge of 4 nC, we can then determine the force which would be exerted on a charge of 6 nC by the proportion

$$\frac{10 \times 10^{-9}}{4 \times 10^{-9}} = \frac{F}{6 \times 10^{-9}}$$

or

$$F = 1.5 \times 10^{-8} \text{ N} = 15 \text{ nN}$$

Example F

A certain alloy is 5 parts tin and 3 parts lead. How many grams of each are there in 40 g of the alloy?

First, we let x = the number of grams of tin in the given amount of the alloy. Next we note that there are 8 total parts of alloy, of which 5 are tin. Thus, 5 is to 8 as x is to 40. This gives the equation

$$\frac{5}{8} = \frac{x}{40}$$

Multiplying each side by 40, we then find that $x = 25$ g. Therefore, there are 25 g of tin and 15 g of lead, and the ratio of 25 to 15 is the same as 5 to 3.

Exercises 17–1

In Exercises 1 through 8, express the ratios in the simplest form.

1. 18 V to 3 V
2. 27 ft to 18 ft
3. 48 in. to 3 ft
4. 120 s to 4 min
5. 20 qt to 25 gal
6. 4 lb to 8 oz
7. 14 kg to 350 g
8. 200 mm to 5 cm

In Exercises 9 through 16, find the required ratios.

9. A virus 3.0×10^{-5} cm long appears to be 1.2 cm long through a microscope. What is the *magnification* (ratio of image length to object length) of the microscope?

10. The *efficiency* of an engine is defined as the ratio of output to input. Find the efficiency, in percent, of an engine for which the input is 6000 W and output is 4500 W.

11. The ratio of the density of an object to the density of water is known as the *relative density* of the object. If the density of gold is 1200 lb/ft^3 and the density of water is 62.4 lb/ft^3, what is the relative density of gold?

12. The *atomic mass* of an atom of carbon is defined to be 12 u. The ratio of the atomic mass of an atom of oxygen to that of an atom of carbon is $\frac{4}{3}$. What is the atomic mass of an atom of oxygen? (The symbol u represents the *unified atomic mass unit*, where $1\ u = 1.66 \times 10^{-27}$ kg.)

13. A certain body of water exerts a force of 18,000 lb on an area of 20 in.^{2}. Find the pressure (force per unit area) on this area.

14. The electric current in a given circuit is the ratio of the voltage to the resistance. What is the current $(1\ V / 1\ \Omega = 1\ A)$ for a circuit where the voltage is 24.0 V and the resistance is 10.0 Ω?

15. A student finds that 8 divisions can be done on a calculator in 20.0 s. What is the student's rate, in calculations per minute, in doing such division problems?

16. The *heat of vaporization* of a substance is the amount of heat required to change one unit amount of the substance from liquid to vapor. Experimentation shows that 7910 J are needed to change 3.50 g of water to steam. What is the heat of vaporization of water?

In Exercises 17 through 20, find the required quantities from the given proportions.

17. In an electric instrument called a "Wheatstone bridge," electric resistances are related by

$$\frac{R_1}{R_2} = \frac{R_3}{R_4}$$

Find R_2 if $R_1 = 6.00\ \Omega$, $R_3 = 62.5\ \Omega$, and $R_4 = 15.0\ \Omega$.

18. For two connected gears, the relation

$$\frac{d_1}{d_2} = \frac{N_1}{N_2}$$

holds, where d is the diameter of the gear and N is the number of teeth. Find N_1 if $d_1 = 2.60$ in., $d_2 = 11.7$ in., and $N_2 = 45$.

19. For two pulleys connected by a belt, the relation

$$\frac{d_1}{d_2} = \frac{n_1}{n_2}$$

holds, where d is the diameter of the pulley and n is the number of revolutions per unit time it makes. Find n_2 if $d_1 = 4.60$ in., $d_2 = 8.30$ in., and $n_1 = 18.0$ r/min.

20. In a transformer, an electric current in one coil of wire induces a current in a second coil. For a transformer

$$\frac{i_1}{i_2} = \frac{t_2}{t_1}$$

where i is the current and t is the number of windings in each coil, find i_2 for $i_1 = 0.0350$ A, $t_1 = 560$, and $t_2 = 1500$.

In Exercises 21 through 36, answer the given questions by setting up and solving the appropriate proportions.

21. If 1 lb = 454 g, what weight in grams is 20.0 lb?

22. If 9 ft^2 = 1 yd^2, what area in square yards is 45.0 ft^2?

23. If 1.00 hp = 746 W, what power in horsepower is 250 W?

24. If 1.50 L = 1.59 qt, what capacity in quarts is 2.75 L?

25. If 2.00 km = 1.24 mi, what distance in kilometers is 5.00 mi?

26. If 10^4 cm^2 = 10^6 mm^2, what area in square centimeters is 2.50×10^5 mm^2?

27. How many meters per second are equivalent to 45.0 km/h?

28. How many gallons per hour are equivalent to 500 qt/min?

29. The length of a picture is 48.0 in., and its width is 36.0 in. In a reproduction of the picture the length is 36.0 in. What is the width of the reproduction?

30. A car will travel 93.0 mi on 6.00 gal of gasoline. How far will it travel on 10.0 gal of gasoline?

31. In physics, power is defined as the time rate at which work is done. If a given motor does 6000 ft-lb of work in 20.0 s, what work will it do in 4.00 min?

32. It is known that 98.0 kg of sulfuric acid are required to neutralize 80.0 kg of sodium hydroxide. Given 37.0 kg of sulfuric acid, determine the amount of sodium hydroxide which will be neutralized.

33. A physician has 220 mg of medication, which is sufficient for just 2 dosages. The amount given to a patient should be proportional to the patient's weight. For persons of weights of 60 kg and 72 kg, what should be the dosages?

34. A person pays $4500 in state and federal income taxes. Find the amount paid for each if the state taxes were 30% of the taxes paid.

35. A board 10.0 ft long is cut into two pieces, the lengths of which are in the ratio 2:3. Find the lengths of the pieces.

36. A total of 322 bolts are in two containers. The ratio of the number in one container to the number in the other container is 5:9. How many are in each container?

17–2 Variation

Scientific laws are often stated in terms of ratios and proportions. For example, Charles' law can be stated as "for a perfect gas under constant pressure, the ratio of any two volumes this gas may occupy equals the ratio of the absolute temperatures." Symbolically this could be stated as $V_1/V_2 = T_1/T_2$. Thus, if the ratio of the volumes and one of the values of the temperature are known, we can easily find the other temperature.

By multiplying both sides of the proportion of Charles' law by V_2/T_1, we can change the form of the proportion to $V_1/T_1 = V_2/T_2$. This statement says that the ratio of the volume to the temperature (for constant pressure) is constant. Thus, if any pair of values of volume and temperature is known, this ratio of V_1/T_1 can be calculated. This ratio of V_1/T_1 can be called a constant k, which means that Charles' law can be written as $V/T = k$. We now have the statement that the ratio of the volume to temperature is always constant; or, as it is normally stated, "the volume is proportional to the temperature." Therefore, we write $V = kT$, the clearest and most informative statement of Charles' law.

Thus, *for any two quantities always in the same proportion, we say that one is* **proportional to** *(or* **varies directly as***) the second,* and in general this is written as $y = kx$, *where k is the* **constant of proportionality.** This type of relationship is known as **direct variation.**

Example A

The circumference of a circle is proportional to (varies directly as) the radius. We write this symbolically as $C = kr$, where (in this case) $k = 2\pi$.

Example B

The fact that the electric resistance of a wire varies directly as (is proportional to) its length is stated as $R = kL$.

It is very common that, when two quantities are related, the product of the two quantities remains constant. In such a case $yx = k$, or $y = k/x$. This is stated as *"y* **varies inversely as** *x,"* or *"y is* **inversely proportional** *to x."* This type of relationship is known as **inverse variation.**

Example C

Boyle's law states that "at a given temperature, the pressure of a gas varies inversely as the volume." This we write symbolically as $P = k/V$.

For many relationships, one quantity varies as a specified power of another. The terms "varies directly" and "varies inversely" are used in the following examples with the specified power of the relation.

Example D

The statement that the volume of a sphere varies directly as the cube of its radius is written as $V = kr^3$. In this case we know that $k = \frac{4}{3}\pi$.

Example E

The fact that the gravitational force of attraction between two bodies varies inversely as the square of the distance between them is written as $F = k/d^2$.

Finally, *one quantity may vary as the product of two or more other quantities. Such variation is termed* **joint variation.** Also, some quantities are related such that a combination of variations is involved.

Example F

The cost of sheet metal varies jointly as the area of the sheet and the cost per unit area of the metal. This we write as $C = kAc$.

Example G

Newton's law of gravitation states that "the force of gravitation between two objects varies jointly as the product of the masses of the objects, and inversely as the square of the distance between their centers." We write this symbolically as $F = km_1m_2/d^2$.

Once we know how to express the given statement in terms of the variables and the constant of proportionality, we may compute the value of k if one set of values of the variables is known. This value of k can then be used to compute values of one variable, when given the others.

Example H

If y varies inversely as x, and $x = 15$ when $y = 4$, find the value of y when $x = 12$.

First we write $y = k/x$ to denote that y varies inversely as x. Next we substitute $x = 15$ and $y = 4$ into the equation. This leads to

$$4 = \frac{k}{15}$$

or $k = 60$. Thus, for our present discussion the constant of proportionality is 60, and this may be substituted into $y = k/x$, giving

$$y = \frac{60}{x}$$

as the equation between y and x. Now, for any given value of x, we may find the value of y. For $x = 12$, we have

$$y = \frac{60}{12} = 5$$

Example I

The distance an object falls under the influence of gravity varies directly as the square of the time of fall. If an object falls 64.0 ft in 2.00 s, how far will it fall in 3.00 s?

We first write the relation $d = kt^2$. Then we use the given fact that $d = 64.0$ ft for $t = 2.00$ s. This gives $64.0 = 4.00k$. In this way we find that $k = 16.0$ ft/s^2. We now know a general relation between d and t, which is $d = 16.0t^2$. Now we put in the value 3.00 s for t, so that we can find d for this particular value for the time. This gives us $d = 16.0(9.00) = 144$ ft. We also note in this problem that k will normally have a certain set of units associated with it.

Example J

The kinetic energy of a moving object varies jointly as the mass of the object and the square of its velocity. If a 5.00-kg object, traveling at 10.0 m/s, has a kinetic energy of 250 J, find the kinetic energy of an 8.00-kg object traveling at 50.0 m/s.

We first write the relation KE $= kmv^2$. Then we use the known set of values to find k. We write $250 = k(5.00)(100)$, giving

$$k = 0.500\frac{\text{J}}{\text{kg(m/s)}^2}$$

(Actually 1 J $= 1$ kg $(\text{m/s})^2$, which means that $k = 0.500$ with no physical units.) Next we find the desired value for KE by substituting the other given values, and the value of k, in the original relation:

$$\text{KE} = 0.500 \; mv^2 = 0.500(8.00)(2500) = 10{,}000 \text{ J}$$

Example K

The heat developed in a resistor varies jointly as the time and the square of the current in the resistor. If the heat developed in t_0 seconds with a current i_0 passing through the resistor is H_0, how much heat is developed if both the time and current are doubled?

First we set up the relation $H = kti^2$, where t is the time and i is the current. From the given information we can write $H_0 = kt_0i_0^2$. Thus $k = H_0/t_0i_0^2$, and the original equation becomes $H = H_0ti^2/t_0i_0^2$. We now let $t = 2t_0$ and $i = 2i_0$, so that we can find H when the time and current are doubled. With these values we obtain

$$H = \frac{H_0(2t_0)(2i_0)^2}{t_0i_0^2} = \frac{8H_0t_0i_0^2}{t_0i_0^2} = 8H_0$$

This tells us that the heat developed is eight times as much as for the original values of i and t.

Exercises 17–2

In Exercises 1 through 8, express the given statements as equations.

1. y varies directly as z.
2. p varies inversely as q.
3. s varies inversely as the square of t.
4. w is proportional to the cube of L.
5. f is proportional to the square root of x.
6. n is inversely proportional to the $\frac{3}{2}$ power of s.
7. w varies jointly as x and the cube of y.
8. q varies as the square of r and inversely as the fourth power of t.

In Exercises 9 through 16, give the equation relating the variables after evaluating the constant of proportionality for the given set of values.

9. r varies inversely as y, and $r = 2$ when $y = 8$.

10. y varies directly as x, and $y = 18$ when $x = 2$.

11. y varies directly as the square root of x, and $y = 2$ when $x = 64$.

12. n is inversely proportional to the square of p, and $n = \frac{1}{27}$ when $p = 3$.

13. s is inversely proportional to the square root of t, and $s = \frac{1}{2}$ when $t = 49$.

14. f varies inversely as the product xy, and $f = 8$ when $x = 2$ and $y = 5$.

15. p is proportional to q and inversely proportional to the cube of r, and $p = 6$ when $q = 3$ and $r = 2$.

16. v is proportional to t and the square of s, and $v = 80$ when $s = 2$ and $t = 5$.

In Exercises 17 through 24, find the required value by setting up the general equation and then evaluating.

17. Find y when $x = 10$ if y varies directly as x and $y = 20$ when $x = 8$.

18. Find y when $x = 5$ if y varies directly as the square of x and $y = 6$ when $x = 8$.

19. Find s when $t = 10$ if s is inversely proportional to t and $s = 100$ when $t = 5$.

20. Find p for $q = 0.8$ if p is inversely proportional to the square of q and $p = 18$ when $q = 0.2$.

21. Find y for $x = 6$ and $z = 5$ if y varies directly as x and inversely as z and $y = 60$ when $x = 4$ and $z = 10$.

22. Find r when $n = 16$ if r varies directly as the square root of n and $r = 4$ when $n = 25$.

23. Find f when $p = 2$ and $c = 4$ if f varies jointly as p and the cube of c and $f = 8$ when $p = 4$ and $c = 0.1$.

24. Find v when $r = 2$, $s = 3$, and $t = 4$ if v varies jointly as r and s and inversely as the square of t and $v = 8$ when $r = 2$, $s = 6$, and $t = 6$.

In Exercises 25 through 48, solve the given applied problems.

25. Hooke's law states that the force needed to stretch a spring is proportional to the amount the spring is stretched. If 10.0 lb stretches a certain spring 4.00 in., how much will the spring be stretched by a force of 6.00 lb?

26. The amount of heat H required to melt ice is proportional to the mass m of ice which is melted. If it takes 3.35×10^5 J to melt 1000 g of ice, how much heat is required to melt 625 g?

27. The change L_d in length of a copper rod varies directly as the change T_d in temperature of the rod. Set up the equation for this relationship if $L_d = 2.7$ cm when $T_d = 150°C$.

28. In electroplating, the mass m of the material deposited varies directly as the time t during which the electric current is on. Set up the equation for this relationship if 2.50 g are deposited in 5.25 h.

29. A particular type of automobile engine produces p cm^3 of carbon monoxide proportional to the time t that it idles. Find the equation relating p and t if $p = 60,000$ cm^3 for $t = 2.00$ min.

30. In biology it is found that under certain circumstances the rate of increase v of bacteria is proportional to the number N of bacteria present. If $v = 800$ bacteria/h when $N = 4000$ bacteria, what is v when $N = 7500$ bacteria?

31. The heat loss through rockwool insulation is inversely proportional to the thickness of the rockwool. If the loss through 6.00 in. of rockwool is 3200 BTU/h, find the loss through 2.50 in. of rockwool.

32. The index of refraction n of a medium is inversely proportional to the velocity of light v within it. The index of refraction of quartz is 1.46 and the velocity of light in quartz is 2.05×10^8 m/s. What is the index of refraction of a diamond, in which the velocity of light is 1.24×10^8 m/s?

33. The illuminance of a light source varies inversely as the square of the distance from the source. Given that the illuminance is 25.0 units at a distance of 200 cm, find the general relation between the illuminance and distance for this light source.

34. The power P required to propel a ship varies directly as the cube of the speed s of the ship. If 5200 hp will propel a ship at 12.0 mi/h, what power is required to propel it at 15 mi/h?

35. The lift L of a model airplane wing is directly proportional to the square of its width w. If the lift is 50 N for a wing width of 8.0 cm, what is the lift for a width of 6.0 cm?

36. The rate R at which water flows through a pipe is directly proportional to the diameter d of the pipe. If 30 gal/min flow through a pipe of diameter 0.50 in., what is the rate of flow through a pipe of 0.75-in. diameter?

37. The electric resistance of a wire varies directly as its length and inversely as its cross-sectional area. Find the relation between resistance, length, and area for a wire which has a resistance of 0.200 Ω for a length of 200 ft and cross-sectional area of 0.0500 in.^{2}.

38. The general gas law states that the pressure of an ideal gas varies directly as the thermodynamic temperature and inversely as the volume. By first finding the relation between P, T, and V, find V for $P = 400$ kPa and $T = 400$ K, given that $P = 600$ kPa for $V = 10.0$ cm^3 and $T = 300$ K.

39. The average speed s of oxygen molecules in the air is directly proportional to the square root of the absolute temperature T. If the speed of the molecules is 460 m/s at 273 K, what is the speed at 300 K?

40. The period of a pendulum is directly proportional to the square root of its length. Given that a pendulum 2.00 ft long has a period of $\pi/2$ s, what is the period of a pendulum 4.00 ft long?

41. The distance s that an object falls due to gravity varies jointly as the acceleration due to gravity g and the square of the time t of fall. On Earth $g = 32.0$ ft/s^2 and on the moon $g = 5.50$ ft/s^2. On Earth an object falls 144 ft in 3.00 s. How far does an object fall in 7.00 s on the moon?

42. The power of an electric current varies jointly as the resistance and the square of the current. Given that the power is 10.0 W when the current is 0.500 A and the resistance is 40.0 Ω, find the power if the current is 2.00 A and the resistance is 20.0 Ω.

43. Under certain conditions the velocity of an object is proportional to the logarithm of the square root of the time elapsed. Given that $v = 18.0$ cm/s after 4.00 s, what is the velocity after 6.00 s?

44. The level of intensity of a sound wave is proportional to the logarithm of the ratio of the sound intensity to an arbitrary reference intensity, normally defined as 10^{-12} W/m². Assuming that the intensity level is 100 dB for $I = 10^{-2}$ W/m², what is the constant of proportionality?

45. When the volume of a gas changes very rapidly, an approximate relation is that the pressure varies inversely as the $\frac{3}{2}$ power of the volume. Express P as a function of V. Given that $P = 300$ kPa when $V = 100$ cm³, find P when $V = 25.0$ cm³.

46. The x-component of the velocity of an object moving around a circle with constant angular velocity ω varies jointly as ω and $\sin \omega t$. Given that ω is constant at $\pi/6$ rad/s, and the x-component of the velocity is -4π ft/s when $t = 1.00$ s, find the x-component of the velocity when $t = 9.00$ s.

47. The acceleration in the x-direction of the object referred to in Exercise 46 varies jointly as $\cos \omega t$ and the square of ω. Under the same conditions as stated in Exercise 46, calculate the x-component of the acceleration.

48. Under certain conditions, the natural logarithm of the ratio of an electric current at time t to the current at time $t = 0$ is proportional to the time. Given that the current is $1/e$ of its initial value after 0.100 s, what is its value after 0.200 s?

17–3 Exercises for Chapter 17

In Exercises 1 through 8, find the indicated ratios.

1. 1 km to 1 mm   2. 1 mL to 1 L   3. 1 ks to 1 h   4. 1 kW to 1 MW

5. The force required to lift an object is 1500 N if it is applied at an angle of 30° to the horizontal, whereas a force of 2.10 kN is required if it is applied at an angle of 21°. What is the ratio of the first force to the second?

6. In making cement, 6 ft³ of cement were mixed with 1 yd³ of sand. What is the ratio of cement to sand?

7. The area of the world needed for food production is about 5×10^{13} m², and the area needed for housing and industry is about 7×10^{12} m². Find the ratio of the food production area to that for housing and industry.

8. The capacitance C of a capacitor is defined as the ratio of its charge to the voltage. What is the capacitance of a capacitor (in farads) for which the charge is 5.00×10^{-6} C and the voltage is 200 V? (1 F = 1 C/1 V.)

In Exercises 9 through 20, answer the given questions by setting up and solving the appropriate proportions.

9. On a certain map, 1.00 in. represents 16.0 mi. What distance on the map represents 52.0 mi?

10. Given that 1.00 m equals 39.4 in., what length in inches is 2.45 m?

11. Given 1.00 L equals 61.0 in.³, how many liters is 105 in.³?

12. Given that 1.00 BTU equals 1060 J, how many BTU equal 8190 J?

13. A microcomputer printer can print 3600 characters in 30 s. How many characters can it print in 5 min?

14. A given machine can produce 80 bolts in 5.00 min. How many bolts can it produce in an hour?

15. Assuming that 195 g of zinc sulfide are used to produce 128 g of sulfur dioxide, how many grams of sulfur dioxide are produced by the use of 750 g of zinc sulfide?

16. Given that 32.0 lb of oxygen are required to burn 20.0 lb of a certain fuel gas, how much air (which can be assumed to be 21% oxygen) is required to burn 100 lb of the fuel gas?

17. On a certain blueprint a measurement of 25.0 ft is represented by 2.00 in. What is the actual distance between two points if they are 5.75 in. apart on the blueprint?

18. A woman invests $50,000 and a man invests $20,000 in a partnership. If profits are to be shared in the ratio that each invested in the partnership, how much does each receive from $10,500 in profits?

19. The perimeter of a rectangular machine part is 210 cm. The ratio of the length to the width is 7 to 3. What are the dimensions of the part?

20. A certain gasoline company sells regular gas and lead-free gas in the ratio of 4 to 5. If, in a month, the company sells a total of 18,000,000 gal, how many of each type were sold?

In Exercises 21 through 24, give the equation relating the variables after evaluating the constant of proportionality for the given set of values.

21. y varies directly as the square of x, and $y = 27$ when $x = 3$.

22. f varies inversely as l, and $f = 5$ when $l = 8$.

23. v is directly proportional to x and inversely proportional to the cube of y, and $v = 10$ when $x = 5$ and $y = 4$.

24. r varies jointly as u, v, and the square of w, and $r = 8$ when $u = 2$, $v = 4$, and $w = 3$.

In Exercises 25 through 50, solve the given applied problems.

25. For a lever balanced at the fulcrum, the relation

$$\frac{F_1}{F_2} = \frac{L_2}{L_1}$$

holds, where F_1 and F_2 are forces on opposite sides of the fulcrum at distances L_1 and L_2, respectively. If $F_1 = 4.50$ lb, $F_2 = 6.75$ lb, and $L_1 = 17.5$ in., find L_2.

26. A company finds that the volume V of sales of a certain item and the price P of the item are related by

$$\frac{P_1}{P_2} = \frac{V_2}{V_1}$$

Find V_2 if $P_1 = \$8.00$, $P_2 = \$6.00$, and $V_1 = 3000$ per week.

27. The work done by a given force varies directly as the distance through which it acts. If a given force does 180 N · m of work through 27.0 m, how much work does it do through 45.0 m?

28. In modern physics we learn that the energy of a photon (a "particle" of light) is directly proportional to its frequency f. Given that the constant of proportionality is 6.63×10^{-34} J · s (this is known as *Planck's constant*), what is the energy of a photon whose frequency is 3.00×10^{15} Hz?

29. The power of a gas engine is proportional to the area of the piston. If an engine with a piston area of 8.00 in.^{2} can develop 30.0 hp, what power is developed by an engine with a piston area of 6.00 in.^{2}?

30. Ohm's law states that the voltage across a given resistor is proportional to the current i in the resistor. Given that 18.0 V are across a certain resistor in which a current of 8.00 A is flowing, express the general relationship between voltage and current for this resistor.

31. An apartment owner charges rent R proportional to the floor area A of the apartment. Find the equation used relating R and A if an apartment of 900 ft^2 rents for $250 per month.

32. A manufacturer determines that the number r of aluminum cans that can be made by recycling n used cans is proportional to n. How many cans can be made from 50,000 used cans if $r = 115$ cans for $n = 125$ cans?

33. The surface area of a sphere varies directly as the square of its radius. The surface area of a certain sphere is 36π square units when the radius is 3.00 units. What is the surface area when the radius is 4.00 units?

34. The exposure time t required to make a photographic negative varies directly as the square of the f-number of the lens. If 0.0048 s is required at $f/8$, what time is required at $f/10$?

35. The rate of emission of radiant energy from the surface of a body is proportional to the fourth power of the thermodynamic temperature. Given that a 25.0 W (the rate of emission) lamp has an operating temperature of 2500 K, what is the operating temperature of a similar 40.0 W lamp?

36. In economics it is often found that the demand D for a product varies inversely as the price P. If the demand is 500 units per week at a cost of $8.00, what would the demand be if the price were $10.00 each?

37. The force F between two parallel wires carrying electric currents is inversely proportional to the distance between the wires. If a force of 0.750 N exists between wires which are 1.25 cm apart, what is the force between them if they are separated by 1.75 cm?

38. The acceleration of gravity g on a satellite in orbit around the earth varies inversely as the square of its distance from the center of the earth. If $g = 8.7$ m/s^2 for a satellite at an altitude of 400 km above the surface of the earth, find g if it is 1000 km above the surface. The radius of the earth is 6.4×10^6 m.

39. The velocity of a pulse traveling in a string varies directly as the square root of the tension of the string. Given that the velocity in a certain string is 450 ft/s when the tension is 20.0 lb, determine the velocity if the tension were 30.0 lb.

40. The frequency of vibration of a wire varies directly as the square root of the tension on the wire. Express the relation between f and T. If $f = 400$ Hz when $T = 1.00$ N, find f when $T = 3.40$ N.

41. The velocity of a jet of fluid flowing from an opening in the side of a container is proportional to the square root of the depth of the opening. If the velocity of the jet from an opening at a depth of 4.00 ft is 16.0 ft/s, what is the velocity of a jet from an opening at a depth of 25.0 ft?

42. The difference in pressure in a fluid between that at the surface and that at a point below varies jointly as the density of the fluid and the depth of the point. The density of water is 1000 kg/m^3, and the density of alcohol is 800 kg/m^3. This difference in pressure at a point 0.200 m below the surface of water is 1.96 kPa. What is the difference in pressure at a point 0.300 m below the surface of alcohol?

43. The crushing load of a pillar varies as the fourth power of its radius and inversely as the square of its length. Express L in terms of r and l for a pillar 20.0 ft tall and 1.00 ft in diameter which is crushed by a load of 20.0 tons.

44. The acoustical intensity of a sound wave is proportional to the square of the pressure amplitude and inversely proportional to the velocity of the wave. Given that the intensity is 0.474 W/m^2 for a pressure amplitude of 20.0 Pa and a velocity of 346 m/s, what is the intensity if the pressure amplitude is 15.0 Pa and the velocity is 320 m/s?

45. In any given electric circuit containing an inductance and capacitance, the resonant frequency is inversely proportional to the square root of the capacitance. If the resonant frequency in a circuit is 25.0 Hz and the capacitance is 100 μF, what is the resonant frequency of this circuit if the capacitance is 25.0 μF?

46. The safe uniformly distributed load on a horizontal beam, supported at both ends, varies jointly as the width and the square of the depth and inversely as the distance between supports. Given that one beam has double the dimensions of another, how many times heavier is the safe load it can support than the first can support?

47. Kepler's third law of planetary motion states that the square of the period of any planet is proportional to the cube of the mean radius (about the sun) of that planet, with the constant of proportionality being the same for all planets. Using the fact that the period of the earth is one year and its mean radius is 93.0 million miles, calculate the mean radius for Venus, given that its period is 7.38 months.

48. The amount of heat per unit of time passing through a wall t units thick is proportional to the temperature difference ΔT and the area A, and is inversely proportional to t. The constant of proportionality is called the coefficient of conductivity. Calculate the coefficient of conductivity of a 0.210-m-thick concrete wall if 0.419 J/s passes through an area of 0.0105 m^2 when the temperature difference is 10.5°C.

49. The percentage error in determining an electric current due to a small error in reading a galvanometer is proportional to $\tan \theta + \cot \theta$, where θ is the angular deflection of the galvanometer. Given that the percentage error is 2.00% when $\theta = 4.0°$, determine the percentage error as a function of θ.

50. Newton's law of gravitation is stated in Example G of Section 17–2. Here k is the same for any two objects. A spacecraft is traveling the 240,000 mi from the earth to the moon, whose mass is $\frac{1}{81}$ that of the earth. How far from the earth is the gravitational force of the earth on the spacecraft equal to the gravitational force of the moon on the spacecraft?

CHAPTER 18

Progressions and the Binomial Theorem

In this chapter we are going to consider briefly the properties of certain sequences of numbers. In itself a sequence is a set of numbers arranged in some specified manner. The kinds of sequences we shall consider are those which form what are known as arithmetic progressions, those which form geometric progressions, and those which are used in the expansion of a binomial to a power. Applications of progressions can be found in many areas, including interest calculations and certain areas in physics. Also, progressions and binomial expansions are of importance in developing mathematical topics which in themselves have wide technical application.

18-1 Arithmetic Progressions

An **arithmetic progression** *(AP) is a sequence of numbers in which each number after the first can be obtained from the preceding one by adding to it a fixed number called the* **common difference.**

Example A

The sequence 2, 5, 8, 11, 14, . . . is an AP with a common difference of 3. The sequence 7, 2, -3, -8, . . . is an AP with a common difference of -5.

If we know the first term of an AP, we can find any other term in the progression by successively adding the common difference enough times for the desired term to be obtained. This, however, is a very inefficient method, and we can learn more about the progression if we establish a general way of finding any particular term.

In general, if a is the first term and d the common difference, the second term is $a + d$, the third term is $a + 2d$, and so forth. If we are looking for the nth term, we note that we need only add d to the first term $n - 1$ times. Thus, *the nth term l of an AP is given by*

$$l = a + (n - 1)d \qquad\qquad (18–1)$$

Occasionally l is referred to as the last term of an AP, but this is somewhat misleading. In reality there is no actual limit to the possible number of terms in an AP, although we may be interested only in a particular number of them. For this reason it is clearer to call l the nth term, rather than the last term. Also, this is the reason for writing three dots after the last indicated term of an AP. These dots indicate that the progression may continue.

Example B

Find the tenth term of the progression 2, 5, 8, ...

By subtracting any given term from the following term, we find that the common difference is $d = 3$. From the terms given, we know that the first term is $a = 2$. From the statement of the problem, the desired term is the tenth, or $n = 10$. Thus, we may find the tenth term l by

$$l = 2 + (10 - 1)3 = 2 + 9 \cdot 3 = 29$$

Example C

Find the number of terms in the progression for which $a = 5$, $l = -119$, and $d = -4$.

Substitution into Eq. (18–1) gives

$$-119 = 5 + (n - 1)(-4)$$

which leads to

$$-124 = -4n + 4$$
$$4n = 128$$
$$n = 32$$

Example D

How many numbers between 10 and 1000 are divisible by 6?

We must first find the smallest and the largest numbers in this range which are divisible by 6. These numbers are 12 and 996. Obviously the common difference between all multiples of 6 is 6. Thus, $a = 12$, $l = 996$, and $d = 6$. Therefore,

$$996 = 12 + (n - 1)6$$
$$6n = 990$$
$$n = 165$$

Thus there are 165 numbers between 10 and 1000 which are divisible by 6.

Example E

A block sliding down an inclined plane was given an initial velocity of 8 cm/s. If it accelerates such that it gains velocity by 3 cm/s during each second, after how many seconds is its velocity 35 cm/s?

Here we see that the velocity (in centimeters per second) after each second is

$$11, 14, 17, \ldots, 35, \ldots$$

Therefore, $a = 11$ (the 8 cm/s was at the beginning—after 0 s), $d = 3, l = 35$, and we are to find n.

$$35 = 11 + (n - 1)(3)$$
$$24 = 3n - 3$$
$$3n = 27$$
$$n = 9$$

Therefore, the velocity of the block is 35 cm/s after 9 s.

Another important quantity concerning an AP is the sum s of the first n terms. We can indicate this sum by either of the two equations,

$$s = a + (a + d) + (a + 2d) + \cdots + (l - d) + l$$

or

$$s = l + (l - d) + (l - 2d) + \cdots + (a + d) + a$$

If we now add these equations, we have

$$2s = (a + l) + (a + l) + (a + l) + \cdots + (a + l) + (a + l)$$

There is one factor $(a + l)$ for each term, and there are n terms. Thus, *the sum of the first n terms is given by*

$$s = \frac{n}{2}(a + l) \tag{18–2}$$

Example F

Find the sum of the first 1000 positive integers.

Here $a = 1, l = 1000$, and $n = 1000$. Thus,

$$s = \frac{1000}{2}(1 + 1000) = 500(1001) = 500{,}500$$

Example G

Find the sum of the AP for which $n = 10$, $a = 4$, and $d = -5$.

We first find the nth term. We write

$$l = 4 + (10 - 1)(-5) = 4 - 45 = -41$$

Now we can solve for s:

$$s = \frac{10}{2}(4 - 41) = 5(-37) = -185$$

Example H

For an AP, given that $a = 2$, $d = \frac{3}{2}$, and $s = 72$, find n and l.

First we substitute the given values into Eqs. (18–1) and (18–2) in order to identify what is known and how we may proceed. Substituting $a = 2$ and $d = \frac{3}{2}$ in Eq. (18–1), we obtain

$$l = 2 + (n - 1)\left(\frac{3}{2}\right)$$

Substituting $s = 72$ and $a = 2$ in Eq. (18–2), we obtain

$$72 = \frac{n}{2}(2 + l)$$

We note that n and l appear in both equations, which means that we must solve them simultaneously. Substituting the expression for l from the first equation into the second equation we proceed with the solution.

$$72 = \frac{n}{2}\left[2 + 2 + (n - 1)\left(\frac{3}{2}\right)\right]$$

$$72 = 2n + \frac{3n(n - 1)}{4}$$

$$288 = 8n + 3n^2 - 3n$$

$$3n^2 + 5n - 288 = 0$$

$$n = \frac{-5 \pm \sqrt{25 - 4(3)(-288)}}{6} = \frac{-5 \pm \sqrt{3481}}{6} = \frac{-5 \pm 59}{6}$$

Since n must be a positive integer we find that $n = \dfrac{-5 + 59}{6} = 9$. Using this value in the expression for l, we find

$$l = 2 + (9 - 1)\left(\frac{3}{2}\right) = 14$$

Therefore, $n = 9$ and $l = 14$.

Example 1

Each swing of a pendulum is measured to be 3.00 in. shorter than the preceding swing. If the first swing is 10.0 ft, determine the total distance traveled by the pendulum bob in the first five swings.

In this problem each of the terms of the progression represents the distance traveled in each respective swing. This means that $a = 10.0$. Also, since 3.00 in. $= 0.25$ ft, we see that $d = -0.25$. Using these values, we find the distance traversed in the fifth swing to be

$$l = 10.0 + 4(-0.25) = 9.00$$

The total distance traversed is

$$s = \frac{5}{2}(10.0 + 9.00) = 47.5 \text{ ft}$$

Exercises 18–1

In Exercises 1 through 4, write five terms of the AP with the given values.

1. $a = 4, d = 2$ 2. $a = 6, d = -\frac{1}{2}$

3. Third term $= 5$, fifth term $= -3$

4. Second term $= -2$, fifth term $= 7$

In Exercises 5 through 12, find the nth term of the AP with the given values.

5. $1, 4, 7, \ldots n = 8$ 6. $-6, -4, -2, \ldots n = 10$

7. $18, 13, 8, \ldots n = 17$ 8. $2, \frac{1}{2}, -1, \ldots n = 25$

9. $a = -7, d = 4, n = 12$ 10. $a = \frac{3}{2}, d = \frac{1}{6}, n = 50$

11. $a = b, d = 2b, n = 25$ 12. $a = -c, d = 3c, n = 30$

In Exercises 13 through 16 find the indicated sum of the terms of the AP.

13. $n = 20, a = 4, l = 40$ 14. $n = 8, a = -12, l = -26$

15. $n = 10, a = -2, d = -\frac{1}{2}$ 16. $n = 40, a = 3, d = \frac{1}{3}$

In Exercises 17 through 28, find any of the values of $a, d, l, n,$ or s that are missing.

17. $a = 5, d = 8, l = 45$ 18. $a = -2, n = 60, l = 28$

19. $a = \frac{5}{3}, n = 20, s = \frac{40}{3}$ 20. $a = 0.1, l = -5.9, s = -8.7$

21. $d = 3, n = 30, s = 1875$ 22. $d = 9, l = 86, s = 455$

23. $a = 74, d = -5, l = -231$ 24. $a = -\frac{9}{7}, n = 19, l = -\frac{36}{7}$

25. $a = -5, d = \frac{1}{2}, s = \frac{23}{2}$ 26. $d = -2, n = 50, s = 0$

27. $a = -c, l = \frac{b}{2}, s = 2b - 4c$ 28. $a = 3b, n = 7, d = \frac{b}{3}$

In Exercises 29 through 44, find the indicated quantities.

29. Sixth term $= 56$, tenth term $= 72$ (find a, d, s for $n = 10$).

30. Seventeenth term $= -91$, second term $= -73$ (find a, d, s for $n = 40$).

31. Fourth term = 2, tenth term = 0 (find a, d, s for $n = 10$).

32. Third term = 1, sixth term = -8 (find a, d, s for $n = 12$).

33. Find the sum of the first 100 integers.

34. Find the sum of the first 100 odd integers.

35. Find the sum of the first 200 multiples of 5.

36. Find the number of multiples of 8 between 99 and 999.

37. A man accepts a position which pays $18,200 per year, and receives a raise of $1250 each year. During what year of his association with the firm will his salary be $28,200?

38. A body falls 16.0 ft during the first second, 48.0 ft during the second second, 80.0 ft during the third second, and so on. How far will it fall in the 20th second?

39. For the object in Exercise 38, what is the total distance fallen in the first 20 s?

40. A well-driller charges $3 for drilling the first foot of a well, and for every foot thereafter she charges 1¢ more than the preceding foot. How much does she charge for drilling a 500-ft well?

41. A car depreciates by $650 each year. If its present value is $5200, after how many years will it be considered to have no value?

42. A clock strikes the number of times of the hour. How many strikes does it make in one day?

43. Show that the sum of the first n positive integers is $\frac{1}{2}n(n + 1)$.

44. Show that the sum of the first n positive odd integers is n^2.

18–2 Geometric Progressions

A second type of sequence is the **geometric progression.** *A geometric progression (GP) is a sequence of numbers in which each number after the first can be obtained from the preceding one by multiplying it by a fixed number called the* **common ratio.** One important application of geometric progressions is in computing interest on savings accounts. Other applications can be found in biology and physics.

Example A

The sequence 2, 4, 8, 16, . . . forms a GP with a common ratio of 2. The sequence 9, 3, 1, $\frac{1}{3}$, . . . forms a GP with a common ratio of $\frac{1}{3}$.

If we know the first term, we can then find any other desired term by multiplying by the common ratio a sufficient number of times. When we do this for a general GP, we can determine the nth term in terms of the first term a, the common ratio r, and n. Thus, the second term is ra, the third term is r^2a, and so forth. In general, *the expression for the nth term is*

$$l = ar^{n-1}$$

(18–3)

Example B

Find the eighth term of the GP 8, 4, 2, ...
Here $a = 8$, $r = \frac{1}{2}$, and $n = 8$. The eighth term is given by

$$l = 8\left(\frac{1}{2}\right)^{8-1} = \frac{8}{2^7} = \frac{1}{16}$$

Example C

Find the tenth term of a GP when the second term is 3, the fourth term is 9, and $r > 0$.
We can find r, if we let $a = 3$, $l = 9$, and $n = 3$ (we are at this time considering the progression made up of 3, the next number, and 9). Thus,

$$9 = 3r^2 \quad \text{or} \quad r = \sqrt{3}$$

We now can find a, by using just two terms (a and 3) for a progression:

$$3 = a(\sqrt{3})^{2-1} \quad \text{or} \quad a = \sqrt{3}$$

We now can find the tenth term directly:

$$l = \sqrt{3}(\sqrt{3})^{10-1} = \sqrt{3}(\sqrt{3})^9 = \sqrt{3}(3^4\sqrt{3}) = 3^5 = 243$$

We could have shortened this procedure one step by letting the second term be the first term of a new progression of nine terms. If the first term is of no importance in itself, this is perfectly acceptable.

Example D

Under certain circumstances, 20% of a substance changes chemically each 10.0 min. If there are originally 100 g of a substance, how much will remain after an hour?
Let P = the portion of the substance remaining after each minute. From the statement of the problem, $r = 0.8$ (80% remains after each 10-min period), $a = 100$, and n represents the number of minutes elapsed. This means $P = 100(0.8)^{n/10}$. It is necessary to divide n by 10 because the ratio is given for a 10-min period. In order to find P when $n = 60$, we write

$$\begin{aligned} P &= 100(0.8)^6 \\ &= 100(0.262) \\ &= 26.2 \text{ g} \end{aligned}$$

This means that 26.2 g are left after an hour.

A general expression for the sum of the first n terms of a geometric progression may be found by directly forming the sum and multiplying this equation by r. By doing this, we have

$$s = a + ar + ar^2 + \cdots + ar^{n-1}$$
$$rs = ar + ar^2 + ar^3 + \cdots + ar^n$$

If we now subtract the first of these equations from the second, we have $rs - s = ar^n - a$. All other terms cancel by subtraction. Solving this equation for s and writing both the numerator and the denominator in the final solution in the form generally used, we obtain for *the sum of the first n terms of a geometric series*

$$s = \frac{a(1 - r^n)}{1 - r} \qquad (r \neq 1)$$

(18–4)

Example E

Find the sum of the first seven terms of the GP $2, 1, \frac{1}{2}, \ldots$.
Here $a = 2$, $r = \frac{1}{2}$, and $n = 7$. Hence

$$s = \frac{2(1 - (\frac{1}{2})^7)}{1 - \frac{1}{2}} = \frac{2(1 - \frac{1}{128})}{\frac{1}{2}} = 4\left(\frac{127}{128}\right) = \frac{127}{32}$$

Example F

If \$100 is invested each year at 5% interest compounded annually, what would be the total amount of the investment after 10 years (before the eleventh deposit is made)?

After one year the amount invested will have added to it the interest for the year. Thus, for the last \$100 invested, its value will become $\$100(1 + 0.05) = \$100(1.05) = \$105$. The next to last \$100 will have interest added twice. After one year its value becomes \$100(1.05), and after two years its value becomes $[\$100(1.05)](1.05) = \$100(1.05)^2$. In the same way, the value of the first \$100 becomes $\$100(1.05)^{10}$, since it will have interest added 10 times. This means we are asked to sum the progression

$$100(1.05) + 100(1.05)^2 + 100(1.05)^3 + \cdots + 100(1.05)^{10}$$

or

$$100[1.05 + (1.05)^2 + (1.05)^3 + \cdots + (1.05)^{10}]$$

For the progression in the brackets we have $a = 1.05$, $r = 1.05$, and $n = 10$. Thus,

$$s = \frac{1.05[1 - (1.05)^{10}]}{1 - 1.05} = \frac{1.05}{-0.05}(1 - 1.628895) = 13.2068$$

This value is easily obtained by the use of the $\boxed{x^y}$ key on a calculator. Therefore, the total value of these \$100 investments is $100(13.2068) = \$1320.68$. We see that \$320.68 of interest has been earned.

Exercises 18–2

In Exercises 1 through 4, write down the first five terms of the GP with the given values.

1. $a = 45$, $r = \frac{1}{3}$ 2. $a = 9$, $r = -\frac{2}{3}$

3. $a = 2$, $r = 3$ 4. $a = -3$, $r = 2$

In Exercises 5 through 12, find the nth term of the GP which has the given values.

5. $\frac{1}{2}, 1, 2, \ldots$ $(n = 6)$ 6. $10, 1, 0.1, \ldots$ $(n = 8)$

7. $125, -25, 5, \ldots$ $(n = 7)$ 8. $0.1, 0.3, 0.9, \ldots$ $(n = 5)$

9. $a = -27$, $r = -\frac{1}{3}$, $n = 6$ 10. $a = 48$, $r = \frac{1}{2}$, $n = 9$

11. $a = 2$, $r = 10$, $n = 7$ 12. $a = -2$, $r = 2$, $n = 6$

In Exercises 13 through 16, find the sum of the n terms of the GP with the given values.

13. $a = 8$, $r = 2$, $n = 5$ 14. $a = 162$, $r = -\frac{1}{3}$, $n = 6$

15. $a = 192$, $l = 3$, $n = 4$ 16. $a = 9$, $l = -243$, $n = 4$

In Exercises 17 through 24, find any of the values of a, r, l, n, or s that are missing.

17. $a = \frac{1}{16}$, $r = 4$, $n = 6$ 18. $r = 0.2$, $l = 0.00032$, $n = 7$

19. $r = \frac{3}{2}$, $n = 5$, $s = 211$ 20. $r = -\frac{1}{2}$, $l = \frac{1}{8}$, $n = 7$

21. $l = 27$, $n = 4$, $s = 40$ 22. $a = 3$, $n = 5$, $l = 48$

23. $a = 75$, $r = \frac{1}{5}$, $l = \frac{3}{25}$ 24. $r = -2$, $n = 6$, $s = 42$

In Exercises 25 through 40, find the indicated quantities.

25. Find the tenth term of a GP if the fourth term is 8 and the seventh term 16.

26. Find the sum of the first 8 terms of the geometric progression for which the fifth term is 5, the seventh term is 10, and $r > 0$.

27. A woman accepts a managerial position at a beginning salary of $24,000 and a 10% increase each year for 8 years. What will be her salary during her eighth year in the position?

28. What is the value of an investment of $10,000 after 10 years if it earns 6% annual interest, compounded semiannually. (6% annual interest compounded semiannually means that 3% interest is added each six months.)

29. A person invests $100 each year for 5 years. How much is the investment worth if the interest is 6% compounded annually?

30. A person invests $1000 each year for 8 years. How much is the investment worth if the interest is 8% compounded quarterly? (See Exercise 28.)

31. The number of bacteria in a particular culture increases by 50% each hour. How long will it take the number of bacteria to increase by a factor of 10?

32. If you decided to save money by putting away 1¢ on a given day, 2¢ one week later, 4¢ a week later, and so on, how much would you have to put aside 1 year later?

33. A ball is dropped from a height of 8.00 ft, and on each rebound it rises $\frac{1}{2}$ of the height it last fell. What is the total distance the ball has traveled when it hits the ground for the fourth time?

34. How many direct ancestors (parents, grandparents, and so on) does a person have in the ten generations which preceded him or her?

35. The American Wire Gauge standard of wire diameters is based on a geometric progression. The ratio of one diameter to the next is the 39th root of 92. If the diameter of No. 30 wire is 0.0100 in., what is the diameter of the wire which is ten sizes larger (No. 20 wire)?

36. A tank with a 22% acid solution has 25% of its contents removed and replaced with water. If this is done four times, what is the percent of acid in the final solution in the tank?

37. The half-life of tungsten 176 is 80.0 min. This means that half of a given amount will disintegrate in 80.0 min. After 160 min three-fourths will have disintegrated. How much will disintegrate in 120 min?

38. The power on a satellite is supplied by a radioactive isotope. On a given satellite the power decreases by 0.2% each day. What percent of the initial power remains after one year?

39. Derive a formula for s in terms of a, r, and l.

40. Write down several terms of a general GP. Then verify the statement that, if the logarithm of each term is taken, the resulting sequence is an AP.

18–3 Geometric Progressions with Infinitely Many Terms

If we consider the sum of the first n terms of the GP with terms $1, \frac{1}{2}, \frac{1}{4}, \ldots$, we find that we get the values in the following table.

n	2	3	4	5	6	7	8	9	10
s	$\frac{3}{2}$	$\frac{7}{4}$	$\frac{15}{8}$	$\frac{31}{16}$	$\frac{63}{32}$	$\frac{127}{64}$	$\frac{255}{128}$	$\frac{511}{256}$	$\frac{1023}{512}$

We see that as n gets larger, the numerator of each fraction becomes more nearly twice the denominator. In fact, we would find that if we continued to compute s as n becomes larger, s can be found as close to the value 2 as desired, although it will never actually reach the value 2. For example, if $n = 100$, $s = 2 - 1.6 \times 10^{-30}$, which could be written as

$$1.99999999999999999999999999999984$$

to 32 significant figures. In the formula for the sum of n terms of a GP,

$$s = a\frac{1 - r^n}{1 - r}$$

the term r^n becomes exceedingly small, and if we consider n as being sufficiently large, we can see that this term is effectively zero. *If this term were exactly zero,* then the sum would be

$$s = 1\frac{1 - 0}{1 - \frac{1}{2}} = 2$$

The only problem is that we cannot find any number large enough for n to make $(\frac{1}{2})^n$ zero. There is, however, an accepted notation for this. This notation is

$$\lim_{n \to \infty} r^n = 0 \qquad (\text{if } |r| < 1)$$

and it is read as "the limit, as n *approaches* infinity, of r to the nth power is zero."

The symbol ∞ is read as **infinity,** *but it must not be thought of as a number.* It is simply a symbol which stands for a *process* of considering numbers which become large without bound. The number which is called the **limit** of the sums is simply the number which the sums get closer and closer to, as *n* is considered to approach infinity. This notation and terminology are of particular importance in the calculus.

If we consider values of *r* such that $|r| < 1$, and let the values of *n* become unbounded, we find that $\lim_{n \to \infty} r^n = 0$. The formula for the sum of a geometric progression with infinitely many terms then becomes

$$s = \frac{a}{1 - r} \qquad\qquad (18–5)$$

(If $r \geq 1$, *s* is unbounded in value.)

Example A

Find the sum of the geometric progression for which $a = 4$, $r = \frac{1}{8}$, and for which *n* increases without bound.

$$s = \frac{4}{1 - \frac{1}{8}} = \frac{4}{1} \cdot \frac{8}{7} = \frac{32}{7}$$

Example B

Find the fraction which has as its decimal form 0.44444444 . . . .
This decimal form can be thought of as being

$$0.4 + 0.04 + 0.004 + 0.0004 + \cdots$$

which means that it can also be thought of as the sum of a GP with infinitely many terms, where $a = 0.4$ and $r = 0.1$. With these considerations, we have

$$s = \frac{0.4}{1 - 0.1} = \frac{0.4}{0.9} = \frac{4}{9}$$

Thus, the fraction $\frac{4}{9}$ and the decimal 0.4444 . . . represent the same number.

Example C

Find the fraction which has as its decimal form 0.121212 . . . .
This decimal form can be considered as being

$$0.12 + 0.0012 + 0.000012 + \cdots$$

which means that we have a GP with infinitely many terms, and that $a = 0.12$ and $r = 0.01$. Thus,

$$s = \frac{0.12}{1 - 0.01} = \frac{0.12}{0.99} = \frac{4}{33}$$

Therefore, the decimal 0.121212 . . . and the fraction $\frac{4}{33}$ represent the same number.

The decimals in Examples B and C are called **repeating decimals,** because *the numbers in the decimal form appear endlessly in a particular order*. These two examples verify the theorem that any repeating decimal represents a rational number. However, all repeating decimals do not necessarily start repeating immediately. If numbers never do repeat, the decimal represents an irrational number. For example, there are no repeating decimals which represent π or $\sqrt{2}$.

Example D

Find the fraction which has as its decimal form the repeating decimal 0.50345345345 . . . .

We first separate the decimal into the beginning, nonrepeating part, and the infinite repeating decimal, which follows. Thus we have 0.50 + 0.00345345345 . . . . This means that we are to add $\frac{50}{100}$ to the fraction which represents the sum of the terms of the GP 0.00345 + 0.00000345 + $\cdots$. For this GP, $a = 0.00345$ and $r = 0.001$. We find the sum of this GP to be

$$s = \frac{0.00345}{1 - 0.001} = \frac{0.00345}{0.999} = \frac{115}{33,300} = \frac{23}{6660}$$

Therefore,

$$0.50345345 \ldots = \frac{5}{10} + \frac{23}{6660} = \frac{5(666) + 23}{6660} = \frac{3353}{6660}$$

Example E

Each swing of a certain pendulum bob is 95% as long as the preceding swing. How far does the bob travel in coming to rest if the first swing is 40.0 in. long?

We are to find the sum of a geometric progression with infinitely many terms, for which $a = 40.0$ and $r = 95\% = \frac{19}{20}$. Substituting these values into Eq. (18–5), we obtain

$$s = \frac{40.0}{1 - \frac{19}{20}} = \frac{40.0}{\frac{1}{20}} = (40.0)(20) = 800 \text{ in.}$$

Therefore, the pendulum bob travels 800 in. (about 67 ft) in coming to rest.

Exercises 18–3

In Exercises 1 through 12, find the sum of the given geometric progressions.

1. $4, 2, 1, \frac{1}{2}, \ldots$

2. $6, -2, \frac{2}{3}, \ldots$

3. $5, 1, 0.2, 0.04, \ldots$

4. $2, \sqrt{2}, 1, \ldots$

5. $20, -1, 0.05, \ldots$

6. $9, 8.1, 7.29, \ldots$

7. $1, \frac{7}{8}, \frac{49}{64}, \ldots$

8. $6, -4, \frac{8}{3}, \ldots$

9. $1, 0.0001, 0.00000001, \ldots$

10. $30, -9, 2.7, \ldots$

11. $2 + \sqrt{3}, 1, 2 - \sqrt{3}, \ldots$

12. $1 + \sqrt{2}, -1, \sqrt{2} - 1, \ldots$

In Exercises 13 through 24, find the fractions equal to the given decimals.

13. 0.33333 . . .
14. 0.55555 . . .
15. 0.404040 . . .
16. 0.070707 . . .
17. 0.181818 . . .
18. 0.272727 . . .
19. 0.273273273 . . .
20. 0.792792792 . . .
21. 0.366666 . . .
22. 0.66424242 . . .
23. 0.100841841841 . . .
24. 0.184561845618456 . . .

In Exercises 25 through 28, solve the given problems by use of the sum of a geometric progression with infinitely many terms.

25. If the ball in Exercise 33 of Section 18–2 is allowed to bounce indefinitely, what is the total distance it will travel?

26. An object suspended on a spring is oscillating up and down. If the first oscillation is 10.0 cm and each oscillation thereafter is nine-tenths of the preceding one, find the total distance the object travels.

27. In coming to rest, a record turntable makes $\frac{8}{10}$ as many revolutions in a second as in the previous second. How many revolutions does the turntable make in coming to rest if it makes 0.50 r in the first second after the power is turned off?

28. A square has sides of 20 cm each. Another square is inscribed in the original square by joining the midpoints of the sides. Assuming that such inscribed squares can be formed endlessly, what is the sum of the areas of all of the squares? See Fig. 18–1.

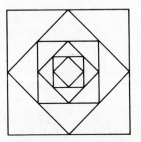

Figure 18-1

18–4 The Binomial Theorem

If we wished to find the roots of the equation $(x + 2)^5 = 0$, we note that there are five factors of $x + 2$, which in turn tells us the only root is $x = -2$. However, if we wished to expand the expression $(x + 2)^5$, a number of repeated multiplications would be needed. This would be a relatively tedious operation. In this section we shall develop *the* **binomial theorem,** *by which it is possible to expand binomials to any given power without direct multiplication.* Such direct expansion can be helpful and labor-saving in developing certain mathematical topics. We may also expand certain expressions where direct multiplication is not actually possible. Also, the binomial theorem is used to develop the necessary expressions for use in certain technical applications.

By direct multiplication, we may obtain the following expansions of the binomial $a + b$.

$$(a + b)^0 = 1$$
$$(a + b)^1 = a + b$$
$$(a + b)^2 = a^2 + 2ab + b^2$$
$$(a + b)^3 = a^3 + 3a^2b + 3ab^2 + b^3$$
$$(a + b)^4 = a^4 + 4a^3b + 6a^2b^2 + 4ab^3 + b^4$$
$$(a + b)^5 = a^5 + 5a^4b + 10a^3b^2 + 10a^2b^3 + 5ab^4 + b^5$$

An inspection indicates certain properties which these expansions have, and which we shall assume are valid for the expansion of $(a + b)^n$, where n is any positive integer. These properties are as follows:

1. There are $n + 1$ terms.
2. The first term is a^n and the final term is b^n.
3. Progressing from the first term to the last, the exponent of a decreases by 1 from term to term, the exponent of b increases by 1 from term to term, and the sum of the exponents of a and b in each term is n.
4. The coefficients of terms equidistant from the ends are equal.
5. If the coefficient of any term is multiplied by the exponent of a in that term, and this product is divided by the number of that term, we obtain the coefficient of the next term.

Example A

Using the above properties we shall develop the expansion for $(a + b)^5$.

From Property 1 we know that there are six terms. From Property 2 we know that the first term is a^5 and the final term is b^5. From Property 3 we know that the factors of a and b in terms 2, 3, 4, and 5 are $a^4 b$, $a^3 b^2$, $a^2 b^3$, and ab^4, respectively.

From Property 5 we obtain the coefficients of terms 2, 3, 4, and 5. In the first term, a^5, the coefficient is 1. Multiplying by 5, the power of a, and dividing by 1, the number of the term, we obtain 5, which is the coefficient of the second term. Thus, the second term is $5a^4 b$. Again using Property 5 we obtain the coefficient of the third term. The coefficient of the second term is 5. Multiplying by 4, and dividing by 2, we obtain 10. This means that the third term is $10a^3 b^2$.

From Property 4 we know that the coefficient of the fifth term is the same as the second, and the coefficient of the fourth term is the same as the third. Thus,

$$(a + b)^5 = a^5 + 5a^4 b + 10a^3 b^2 + 10a^2 b^3 + 5ab^4 + b^5$$

It is not necessary to use the above properties directly to expand a given binomial. If they are applied to $(a + b)^n$ we may obtain a general formula for the expansion of a binomial. Thus, *the binomial theorem states that the following* **binomial formula** *is valid for all values of n* (the binomial theorem is proven through advanced methods).

$$(a + b)^n = a^n + na^{n-1}b + \frac{n(n-1)}{2!}a^{n-2}b^2 + \frac{n(n-1)(n-2)}{3!}a^{n-3}b^3 + \cdots + b^n \qquad (18\text{--}6)$$

*The notation n! is read as "**n factorial**."* It denotes the product of the first n integers. Thus, $2! = 1 \cdot 2$, $3! = 1 \cdot 2 \cdot 3$, $4! = 1 \cdot 2 \cdot 3 \cdot 4$, and so on. Evaluating these products we see that $2! = 2$, $3! = 6$, $4! = 24$, and so on.

Example B

By use of the binomial formula expand $(2x + 3)^6$.

In using the binomial formula for $(2x + 3)^6$ we use $2x$ for a, 3 for b, and 6 for n. Thus,

$$(2x + 3)^6 = (2x)^6 + 6(2x)^5(3) + \frac{(6)(5)}{2}(2x)^4(3^2) + \frac{(6)(5)(4)}{(2)(3)}(2x)^3(3^3)$$

$$+ \frac{(6)(5)(4)(3)}{(2)(3)(4)}(2x)^2(3^4) + \frac{(6)(5)(4)(3)(2)}{(2)(3)(4)(5)}(2x)(3^5) + 3^6$$

$$= 64x^6 + 576x^5 + 2160x^4 + 4320x^3 + 4860x^2 + 2916x + 729$$

For the first few integral powers of a binomial, the coefficients can be obtained by setting them up in the following pattern, known as **Pascal's triangle.**

$n = 0$						1						
$n = 1$					1		1					
$n = 2$				1		2		1				
$n = 3$			1		3		3		1			
$n = 4$		1		4		6		4		1		
$n = 5$	1		5		10		10		5		1	
$n = 6$	1	6		15		20		15		6		1

We note that the first and last coefficient shown in each row is 1, and the second and next-to-last coefficients are equal to n. Other coefficients are obtained by adding the two nearest coefficients in the row above. This pattern may be continued indefinitely, although the use of Pascal's triangle is cumbersome for high values of n.

Example C

By use of Pascal's triangle, expand $(5s - 2t)^4$.

Here we note that $n = 4$. Thus, the coefficients of the five terms are 1, 4, 6, 4, and 1, respectively. Also, here we use $5s$ for a and $-2t$ for b. We are expanding this as $[(5s) + (-2t)]^4$. Therefore,

$$(5s - 2t)^4 = (5s)^4 + 4(5s)^3(-2t) + 6(5s)^2(-2t)^2 + 4(5s)(-2t)^3 + (-2t)^4$$

$$= 625s^4 - 1000s^3t + 600s^2t^2 - 160st^3 + 16t^4$$

In certain uses of a binomial expansion it is not necessary to obtain all terms. Only the first few terms are required. The following example illustrates finding the first four terms of an expansion.

Example D

Find the first four terms of the expansion of $(x + 7)^{12}$.

Here we use x for a, 7 for b, and 12 for n. Thus, from the binomial formula we have

$$(x + 7)^{12} = x^{12} + 12x^{11}(7) + \frac{(12)(11)}{2}x^{10}(7^2) + \frac{(12)(11)(10)}{(2)(3)}x^9(7^3) + \cdots$$

$$= x^{12} + 84x^{11} + 3234x^{10} + 75{,}460x^9 + \cdots$$

If we let $a = 1$ and $b = x$ in the binomial formula, we obtain the **binomial series**

$$(1 + x)^n = 1 + nx + \frac{n(n - 1)}{2!}x^2 + \frac{n(n - 1)(n - 2)}{3!}x^3 + \cdots \quad (18\text{--}7)$$

which through advanced methods can be shown to be valid for any real number n if $|x| < 1$. When n is either negative or a fraction, we obtain an infinite series. In such a case, we calculate as many terms as may be needed although such a series is not obtainable through direct multiplication. The binomial series may be used to develop important expressions which are used in applications and more advanced mathematics topics.

Example E

Using the binomial series, find the first four terms of the expansion for $\dfrac{1}{\sqrt{1 - x}}$.

In order to use the binomial series, we write

$$\frac{1}{\sqrt{1 - x}} = [1 + (-x)]^{-1/2}$$

Thus,

$$[1 + (-x)]^{-1/2} = 1 + \left(-\frac{1}{2}\right)(-x) + \frac{(-\frac{1}{2})(-\frac{1}{2} - 1)}{2!}(-x)^2$$

$$+ \frac{(-\frac{1}{2})(-\frac{1}{2} - 1)(-\frac{1}{2} - 2)}{3!}(-x)^3 + \cdots$$

or

$$\frac{1}{\sqrt{1 - x}} = 1 + \frac{1}{2}x + \frac{3}{8}x^2 + \frac{5}{16}x^3 + \cdots$$

The binomial series can also be used to make certain numerical calculations. These calculations can also be done on a calculator, but the binomial series method was often used before calculators became generally available.

Example F

Approximate the value of $\sqrt[3]{1006}$ to five decimal places.

This approximation can be made by use of the first three terms of the binomial expansion if we write $\sqrt[3]{1006} = \sqrt[3]{1000(1.006)} = 10\sqrt[3]{1.006} = 10\sqrt[3]{1 + 0.006}$. By expanding $(1 + 0.006)^{1/3}$, we have

$$(1 + 0.006)^{1/3} = 1 + \frac{1}{3}(0.006) + \frac{(\frac{1}{3})(-\frac{2}{3})}{2}(0.006)^2 + \cdots$$

$$= 1 + 0.002 - 0.000004$$

$$= 1.001996$$

Therefore, $\sqrt[3]{1006} = 10(1.001996) = 10.01996$. The use of additional terms of the expansion will not affect the first five decimal places. The use of this type of approximation is helpful if a calculator is not available, or if accuracy beyond that available on the calculator is required.

Exercises 18–4

In Exercises 1 through 8, expand and simplify the given expressions by use of the binomial formula.

1. $(t + 1)^3$
2. $(3x - 2)^3$
3. $(2x - 1)^4$
4. $(x^2 + 3)^4$
5. $(x + 2)^5$
6. $(xy - z)^5$
7. $(2a - b^2)^6$
8. $(\frac{a}{x} + x)^6$

In Exercises 9 through 12, expand and simplify the given expressions by use of Pascal's triangle.

9. $(5x - 3)^4$
10. $(b + 4)^5$
11. $(2a + 1)^6$
12. $(x - 3)^7$

In Exercises 13 through 20, find the first four terms of the indicated expansions.

13. $(x + 2)^{10}$
14. $(x - 3)^8$
15. $(2a - 1)^7$
16. $(3b + 2)^9$
17. $\left(x^2 - \frac{y}{2}\right)^{12}$
18. $\left(2a - \frac{1}{x}\right)^{11}$
19. $\left(a^2 + \frac{1}{2b}\right)^{20}$
20. $\left(2x^2 + \frac{y}{3}\right)^{15}$

In Exercises 21 through 28, find the first four terms of the indicated expansions by use of the binomial series.

21. $(1 + x)^8$
22. $(1 + x)^{-1/3}$
23. $(1 - x)^{-2}$
24. $(1 - 2x)^9$
25. $\sqrt{1 + x}$
26. $\dfrac{1}{\sqrt{1 + x}}$
27. $\dfrac{1}{\sqrt{9 - 9x}}$
28. $\sqrt{4 + x^2}$

In Exercises 29 through 36, approximate the values of the given expressions to three decimal places by use of three terms of the appropriate binomial series.

29. $\sqrt{1.1}$
30. $\sqrt{0.9}$
31. $\sqrt[3]{994}$ (see Example F)
32. $\sqrt[3]{9}$ ($\sqrt[3]{9} = \sqrt[3]{8 + 1} = \sqrt[3]{8(1 + \frac{1}{8})} = 2\sqrt[3]{1 + \frac{1}{8}}$)
33. $\sqrt[3]{82}$
34. $\sqrt[4]{15}$
35. $(1.02)^{-4}$
36. $(0.97)^{-2}$

See Appendix E for a computer program for finding coefficients of a binomial series.

In Exercises 37 through 40, find the indicated terms by use of the following information. The $r + 1$ term of the expansion of $(a + b)^n$ is given by

$$\frac{n(n - 1)(n - 2) \cdots (n - r + 1)}{r!} a^{n-r} b^r$$

37. The term involving b^5 in $(a + b)^8$.

38. The term involving y^6 in $(x + y)^{10}$.

39. The fifth term of $(2x - 3b)^{12}$.

40. The sixth term of $(a - b)^{14}$.

In Exercises 41 through 44, find the indicated expansions.

41. In determining the change of the rate of emission of energy from the surface of a body at temperature T, the expression $(T + h)^4$ is used. Expand this expression.

42. In determining the probability of a given number of heads or tails when 8 coins are tossed, the expression $(H + T)^8$ can be used. The various coefficients give the number of chances in 256 that a certain number of heads and tails will result. For example, the coefficient of the $H^2 T^6$ term gives the chances in 256 that 2 heads and 6 tails will result. Expand this expression.

43. In the theory associated with the magnetic field due to an electric current, the expression $1 - \dfrac{x}{\sqrt{a^2 + x^2}}$ is found. By expanding $(a^2 + x^2)^{-1/2}$ find the first three nonzero terms which could be used to approximate the given expression.

44. Find the first four terms of the expansion of $(1 + x)^{-1}$ and then divide $1 + x$ into 1. Compare the results.

18–5 Exercises for Chapter 18

In Exercises 1 through 8, find the indicated term of each progression.

1. $1, 6, 11, \ldots$ (17th)　　　　　　　2. $1, -3, -7, \ldots$ (21st)

3. $\frac{1}{2}, 0.1, 0.02, \ldots$ (9th)　　　　　4. $0.025, 0.01, 0.004, \ldots$ (7th)

5. $8, \frac{7}{2}, -1, \ldots$ (16th)　　　　　　6. $-1, -\frac{5}{3}, -\frac{7}{3}, \ldots$ (25th)

7. $\frac{3}{4}, \frac{1}{2}, \frac{1}{3}, \ldots$ (7th)　　　　　　8. $\frac{2}{3}, 1, \frac{3}{2}, \ldots$ (7th)

In Exercises 9 through 12, find the sum of each progression with the indicated values.

9. $a = -4, n = 15, l = 17$ (AP)

10. $a = 3, d = -\frac{2}{3}, n = 10$

11. $a = 16, r = -\frac{1}{2}, n = 10$

12. $a = 64, l = 729, n = 7$ (GP, $r > 0$)

In Exercises 13 through 24, find the indicated quantities for the appropriate progressions.

13. $a = 17, d = -2, n = 9, s = ?$　　14. $d = \frac{4}{3}, a = -3, l = 17, n = ?$

15. $a = 18, r = \frac{1}{2}, n = 6, l = ?$　　16. $l = \frac{49}{8}, r = -\frac{2}{7}, s = \frac{17199}{288}, a = ?$

17. $a = 8, l = -2.5, s = 22, d = ?$

18. $a = 2, d = 0.2, n = 11, s = ?$

19. $n = 6, r = -0.25, s = 204.75, l = ?$

20. $a = 10, r = 0.1, s = 11.111, n = ?$

21. $a = -1, l = 32, n = 12, s = ?$ (AP)

22. $a = 1, l = 64, s = 325, n = ?$ (AP)

23. $a = 1, n = 7, l = 64, s = ?$ 24. $a = \frac{1}{4}, n = 6, l = 8, s = ?$

In Exercises 25 through 32, find the fractions equal to the given decimals.

25. $0.77777\ldots$	26. $0.030303\ldots$	27. $0.757575\ldots$
28. $0.484848\ldots$	29. $0.123123123\ldots$	30. $0.0727272\ldots$
31. $0.166666\ldots$	32. $0.25399399399\ldots$	

In Exercises 33 through 40, expand and simplify the given expressions. In Exercises 37 through 40, find the first four terms of the appropriate expansion.

33. $(x - 2)^4$	34. $(s + 2t)^4$	35. $(x^2 + 1)^5$	36. $(3n - a)^6$

37. $(a + 2b^2)^{10}$

38. $\left(\dfrac{x}{4} - y\right)^{12}$

39. $\left(p^2 - \dfrac{q}{6}\right)^9$

40. $\left(2s^2 - \dfrac{3}{2t}\right)^{14}$

In Exercises 41 through 48, find the first four terms of the indicated expansions by use of the binomial series.

41. $(1 + x)^{12}$	42. $(1 + x)^{10}$	43. $\sqrt{1 + x^2}$	44. $(4 - 4x)^{-1}$
45. $\sqrt{1 - a^2}$	46. $\sqrt{1 + b^4}$	47. $(1 - 2x)^{-3}$	48. $(1 + 4x)^{-1/4}$

In Exercises 49 through 52, approximate the values of the given expressions to three decimal places by use of three terms of the appropriate binomial series.

49. $\sqrt{908}$	50. $\sqrt[3]{61}$	51. $(8.04)^{-1}$	52. $(4.06)^{-1/2}$

In Exercises 53 through 68 solve the given problems by use of an appropriate progression or expansion.

53. Find the sum of the first 1000 positive even integers.

54. How many numbers divisible by 4 lie between 23 and 121?

55. Fifteen layers of logs are so piled that there are 20 logs in the bottom layer, and each layer contains one log less than the layer below it. How many logs are in the pile?

56. A contractor employed in the construction of a building was penalized for taking more time than the contract allowed. He forfeited $150 for the first day late, $225 for the second day, $300 for the third day, and so forth. If he forfeited a total of $6750, how many additional days did he require to complete the building?

57. What is the value after 20 years of an investment of $2500, if it draws interest at 5% compounded annually?

58. A person invests $1000 each year for 20 years. How much is the total investment worth if the annual interest is 6% and it is compounded quarterly?

59. A business estimates that the salvage value of a piece of machinery decreases by 20% each year. If the machinery is purchased for $80,000, what is its value after 10 years?

60. A tank contains 100 L of a given chemical. Thirty liters are drawn off and replaced with water. Then 30 L of the resulting solution are drawn off and replaced with water. If this operation is performed a total of five times, how much of the original chemical remains?

61. A certain object, after being heated, cools at such a rate that its temperature decreases 10% each minute. If the object is originally heated to 100°C, what is its temperature 10.0 min later?

62. If the population of a certain town increases 20% each year, how long will it take for the population to double?

63. A piece of paper 0.004 in. thick is cut in half. These two pieces are then placed one on the other and they are cut in half. If this is repeated such that the paper is cut in half 40 times, how high will the pile be?

64. After the power is turned off, an object on a nearly frictionless surface slows down such that it travels 99.9% as far during one second as during the previous second. If it travels 100 cm during the first second after the power is turned off, how far does it travel while stopping?

65. A ball, starting from rest, rolls down a uniform incline so that it covers 10 in. during the first second, 30 in. during the second second, 50 in. during the third second, and so on. How long will it take to cover $333\frac{1}{3}$ ft?

66. The successive distances traveled by a pendulum bob are 90 cm, 60 cm, 40 cm, ... Find the total distance the bob travels before it comes to rest.

67. In finding the partial pressure P_F of fluorine gas under certain conditions, the equation

$$P_F = \frac{(1 + 2 \times 10^{-10}) - \sqrt{1 + 4 \times 10^{-10}}}{2} \text{ atm}$$

is found. By using three terms of the expansion for $\sqrt{1 + x}$ approximate the value of this expression.

68. The terms $a, a + 12, a + 24$ form an AP, and the terms $a, a + 24, a + 12$ form a GP. Find these progressions.

CHAPTER 19

Additional Topics in Trigonometry

The definitions of the trigonometric functions were first introduced in Section 3–2, and were again summarized in Section 7–1. If we take a close look at these definitions, we find that there are many relationships among the various functions. In this chapter we shall study some of these basic relationships, as well as develop certain others. Also, we shall show how equations with trigonometric functions are solved. Finally, we shall develop the concept of the inverse trigonometric functions, which were first introduced in Chapter 3.

19–1 Fundamental Trigonometric Identities

The definition of the sine of an angle is $\sin \theta = y/r$, and the definition of the cosecant of an angle is $\csc \theta = r/y$. But we know that $1/(r/y) = y/r$, which means that $\sin \theta = 1/\csc \theta$. In writing this down we made no reference to any particular angle, and since the definitions hold for *any* angle, this relation between the $\sin \theta$ and $\csc \theta$ also holds for any angle. *A relation such as this, which holds for any value of the variable, is called an* **identity.** Of course, specific values where division by zero would be indicated are excluded.

Such trigonometric identities are important for a number of reasons. We have already actually made limited use of some of them in Section 9–4 when we graphed certain trigonometric functions. We also used an important identity in deriving the law of cosines in Chapter 8. When we consider equations with trigonometric functions later in this chapter, we will find that the solution of such equations depends on the proper use of identities. In the study of calculus, there are certain types of problems which require the use of trigonometric identities for solution (even problems in which trigonometric functions do not appear). Also, they are used in developing expressions and solving equations in certain technical areas.

Several important identities exist among the six trigonometric functions, and we shall develop these identities in this section. We shall also show how we can use the basic identities to verify other identities among the functions.

By the definitions, we have

$$\sin \theta \csc \theta = \frac{y}{r} \cdot \frac{r}{y} = 1 \quad \text{or} \quad \sin \theta = \frac{1}{\csc \theta} \quad \text{or} \quad \csc \theta = \frac{1}{\sin \theta}$$

$$\cos \theta \sec \theta = \frac{x}{r} \cdot \frac{r}{x} = 1 \quad \text{or} \quad \cos \theta = \frac{1}{\sec \theta} \quad \text{or} \quad \sec \theta = \frac{1}{\cos \theta}$$

$$\tan \theta \cot \theta = \frac{y}{x} \cdot \frac{x}{y} = 1 \quad \text{or} \quad \tan \theta = \frac{1}{\cot \theta} \quad \text{or} \quad \cot \theta = \frac{1}{\tan \theta}$$

$$\frac{\sin \theta}{\cos \theta} = \frac{y/r}{x/r} = \frac{y}{x} = \tan \theta; \qquad \frac{\cos \theta}{\sin \theta} = \frac{x/r}{y/r} = \frac{x}{y} = \cot \theta$$

Also, by the definitions and the Pythagorean theorem in the form of $x^2 + y^2 = r^2$, we arrive at the following identities.

By dividing the Pythagorean relation through by r^2, we have

$$\left(\frac{x}{r}\right)^2 + \left(\frac{y}{r}\right)^2 = 1 \quad \text{which leads us to} \quad \cos^2\theta + \sin^2\theta = 1$$

By dividing the Pythagorean relation by x^2, we have

$$1 + \left(\frac{y}{x}\right)^2 = \left(\frac{r}{x}\right)^2 \quad \text{which leads us to} \quad 1 + \tan^2\theta = \sec^2\theta$$

By dividing the Pythagorean relation by y^2, we have

$$\left(\frac{x}{y}\right)^2 + 1 = \left(\frac{r}{y}\right)^2 \quad \text{which leads us to} \quad \cot^2\theta + 1 = \csc^2\theta$$

The term $\cos^2 \theta$ is the common way of writing $(\cos \theta)^2$, and thus it means to square the value of the cosine of the angle. Obviously the same holds true for the other functions.

Summarizing these results, we have the following important identities among the trigonometric functions:

(19–1) $\quad \sin \theta = \dfrac{1}{\csc \theta}$	$\cos \theta = \dfrac{1}{\sec \theta} \quad$ (19–2)
(19–3) $\quad \tan \theta = \dfrac{1}{\cot \theta}$	$\tan \theta = \dfrac{\sin \theta}{\cos \theta} \quad$ (19–4)
(19–5) $\quad \cot \theta = \dfrac{\cos \theta}{\sin \theta}$	$\sin^2\theta + \cos^2\theta = 1 \quad$ (19–6)
(19–7) $\quad 1 + \tan^2\theta = \sec^2\theta$	$1 + \cot^2 \theta = \csc^2 \theta \quad$ (19–8)

In using these identities, θ may stand for any angle or number or expression representing an angle or number.

Example A

$$\sin(x + 1) = \frac{1}{\csc(x + 1)}$$

$$\tan 157° = \frac{\sin 157°}{\cos 157°}, \qquad 1 + \tan^2\left(\frac{\pi}{6}\right) = \sec^2\left(\frac{\pi}{6}\right)$$

Example B

We shall verify three of the identities for particular values of θ.

From Table 3, we find that $\tan 53° = 1.327$ and $\cot 53° = 0.7536$. Considering Eq. (19–3) and comparing $\tan 53°$ with the reciprocal of $\cot 53°$, we have

$$\frac{1}{\cot 53°} = \frac{1}{0.7536} = 1.327 \quad \text{and} \quad \tan 53° = 1.327$$

This verifies Eq. (19–3) for these values.

Using Table 3 or a calculator, we find that $\sin 157° = 0.3907$ and $\cos 157° = -0.9205$. Considering Eq. (19–4) and dividing, we find that

$$\frac{\sin 157°}{\cos 157°} = \frac{0.3907}{-0.9205} = -0.4245$$

Checking, we see that $\tan 157° = -0.4245$, which verifies that

$$\tan 157° = \frac{\sin 157°}{\cos 157°}$$

From Section 3–3, we recall that

$$\sin 45° = \frac{1}{\sqrt{2}} = \frac{\sqrt{2}}{2} \quad \text{and} \quad \cos 45° = \frac{\sqrt{2}}{2}$$

Using Eq. (19–6), we have

$$\sin^2 45° + \cos^2 45° = \left(\frac{\sqrt{2}}{2}\right)^2 + \left(\frac{\sqrt{2}}{2}\right)^2 = \frac{1}{2} + \frac{1}{2} = 1$$

We see that this identity checks for these values.

A great many identities exist among the trigonometric functions. We are going to use the basic identities already developed in Eqs. (19–1) through (19–8), along with a few additional ones developed in later sections, to prove the validity of still other identities. *The ability to prove such identities depends to a large extent on being very familiar with the basic identities,* so that you can recognize them in somewhat different forms. If you do not learn these basic identities and learn them well, you will have difficulty in following the examples and doing the exercises. The more readily you recognize these forms, the more easily you will be able to prove such identities.

In proving identities, we should look for combinations which appear in, or are very similar to, those in the basic identities. Consider the following examples.

Example C

In proving the identity

$$\sin x = \frac{\cos x}{\cot x}$$

we know that $\cot x = \dfrac{\cos x}{\sin x}$. Since $\sin x$ appears on the left, substituting for $\cot x$ on the right will eliminate $\cot x$ and introduce $\sin x$. This should help us proceed in proving the identity. Thus,

$$\sin x = \frac{\cos x}{\cot x}$$

$$= \frac{\cos x}{\dfrac{\cos x}{\sin x}} = \frac{\cos x}{1} \cdot \frac{\sin x}{\cos x}$$

$$= \sin x$$

By showing that the right side may be changed exactly to $\sin x$, the expression on the left side, we have proved the identity.

Some important points should be made in relation to the proof of the identity of Example C. We must recognize what basic identities may be useful. The proof of an identity requires the use of basic algebraic operations, and these must be done carefully and correctly. Although in Example C we changed the right side to the form on the left, we could have changed the left to the form on the right. From this, and the fact that various substitutions are possible, we see that there is a variety of procedures which can be used to prove any given identity.

Example D

Prove that $\tan \theta \csc \theta = \sec \theta$.

In proving this identity we know that $\tan \theta = \dfrac{\sin \theta}{\cos \theta}$ and also that $\dfrac{1}{\cos \theta} = \sec \theta$. Thus, by substituting for $\tan \theta$ we introduce $\cos \theta$ in the denominator, which is equivalent to introducing $\sec \theta$ in the numerator. Therefore, changing only the left side, we have

$$\tan \theta \csc \theta = \frac{\sin \theta}{\cos \theta} \csc \theta = \frac{1}{\cos \theta} \sin \theta \csc \theta$$

$$= \sec \theta \sin \theta \frac{1}{\sin \theta} = \sec \theta$$

or

$$\tan \theta \csc \theta = \sec \theta$$

Many variations of the preceding steps are possible. Also, we could have changed only the right side to obtain the form on the left. For example,

$$\tan\theta\,\csc\theta = \sec\theta$$

$$= \frac{1}{\cos\theta} = \frac{\sin\theta}{\cos\theta\,\sin\theta} = \frac{\sin\theta}{\cos\theta}\frac{1}{\sin\theta}$$

$$= \tan\theta\,\csc\theta$$

In proving the identities of Examples C and D we have shown that the expression on one side of the equals sign can be changed into the expression on the other side. Although making the restriction that we change only one side is not entirely necessary, *we shall restrict the method of proof to changing only one side into the same form as the other side.* In this way, we know precisely to which form we are to change, and therefore by looking ahead we are better able to make the proper changes.

▶   There is no set procedure which can be stated for working with identities. The most important factors are to be able to *recognize the proper forms,* to be able to *see what effect any change may have* before we actually perform it, and then *perform it correctly.* Normally *it is easier to change the more complicated side of an identity to the same form as the less complicated side.* If the two sides are of approximately the same complexity, a close look at each side usually suggests steps which will lead to the solution.

Example E

Prove the identity $\dfrac{\cos x\,\csc x}{\cot^2 x} = \tan x$.

First, we note that the left-hand side has several factors and the right-hand side has only one. Therefore, let us transform the left-hand side. Next, we note that we want $\tan x$ as the final result. We know that $\cot x = 1/\tan x$. Thus

$$\frac{\cos x\,\csc x}{\cot^2 x} = \frac{\cos x\,\csc x}{1/\tan^2 x} = \cos x\,\csc x\,\tan^2 x$$

At this point, we have two factors of $\tan x$ on the left. Since we want only one, let us factor out one. Therefore,

$$\cos x\,\csc x\,\tan^2 x = \tan x(\cos x\,\csc x\,\tan x)$$

Now, replacing $\tan x$ within the parentheses by $\sin x/\cos x$, we have

$$\tan x(\cos x\,\csc x\,\tan x) = \frac{\tan x(\cos x\,\csc x\,\sin x)}{\cos x}$$

Now we may cancel $\cos x$. Also, $\csc x\,\sin x = 1$ from Eq. (19-1). Finally,

$$\frac{\tan x(\cos x\,\csc x\,\sin x)}{\cos x} = \tan x\left(\frac{\cos x}{\cos x}\right)(\csc x\,\sin x)$$

$$= \tan x(1)(1) = \tan x$$

Since we have transformed the left-hand side into $\tan x$, we have proven the identity. Of course, it is not necessary to rewrite expressions as we did in this example. This was done here only to include the explanations between steps.

Example F

Prove the identity $\sin^4 x - \cos^4 x = \sin^2 x - \cos^2 x$.

We note that either side of this identity may be factored, but that one of the factors of the left side is the expression which appears on the right. Therefore, by factoring, and the use of Eq. (19–6), we have (changing only the form of the left side)

$$\sin^4 x - \cos^4 x = (\sin^2 x - \cos^2 x)(\sin^2 x + \cos^2 x)$$
$$= (\sin^2 x - \cos^2 x)(1)$$

or

$$\sin^4 x - \cos^4 x = \sin^2 x - \cos^2 x$$

Example G

Prove the identity $\dfrac{\sec^2 y}{\cot y} - \tan^3 y = \tan y$.

Here we shall simplify the left side. We note that we can remove cot y from the denominator since tan $y = 1/\cot y$. Also, the presence of $\sec^2 y$ suggests the use of Eq. (19–7). Therefore, we have

$$\frac{\sec^2 y}{\cot y} - \tan^3 y = \frac{\sec^2 y}{1/\tan y} - \tan^3 y = \sec^2 y \tan y - \tan^3 y$$
$$= \tan y(\sec^2 y - \tan^2 y) = \tan y(1) = \tan y$$

or

$$\frac{\sec^2 y}{\cot y} - \tan^3 y = \tan y$$

Here we have used Eq. (19–7) in the form $\sec^2 y - \tan^2 y = 1$.

Example H

Prove the identity $\dfrac{1 - \sin x}{\sin x \cot x} = \dfrac{\cos x}{1 + \sin x}$.

The combination $1 - \sin x$ also suggests $1 - \sin^2 x$, since multiplying $(1 - \sin x)$ by $(1 + \sin x)$ gives $1 - \sin^2 x$, which can then be replaced by $\cos^2 x$. Thus, changing only the left side we have

$$\frac{1 - \sin x}{\sin x \cot x} = \frac{(1 - \sin x)(1 + \sin x)}{\sin x \cot x (1 + \sin x)}$$

$$= \frac{1 - \sin^2 x}{\sin x \left(\dfrac{\cos x}{\sin x}\right)(1 + \sin x)} = \frac{\cos^2 x}{\cos x (1 + \sin x)}$$

or

$$\frac{1 - \sin x}{\sin x \cot x} = \frac{\cos x}{1 + \sin x}$$

Example I

Prove the identity $\sec^2 x + \csc^2 x = \sec^2 x \csc^2 x$.

Here we note the presence of $\sec^2 x$ and $\csc^2 x$ on each side. This suggests the possible use of the square relationships. By replacing the $\sec^2 x$ on the right-hand side by $1 + \tan^2 x$, we can create $\csc^2 x$ plus another term. The left-hand side is the $\csc^2 x$ plus another term, so this procedure should help. Thus, changing only the right side,

$$\sec^2 x + \csc^2 x = \sec^2 x \csc^2 x$$
$$= (1 + \tan^2 x)(\csc^2 x)$$
$$= \csc^2 x + \tan^2 x \csc^2 x$$

Now we note that $\tan x = \sin x / \cos x$ and $\csc x = 1/\sin x$. Thus,

$$\sec^2 x + \csc^2 x = \csc^2 x + \left(\frac{\sin^2 x}{\cos^2 x}\right)\left(\frac{1}{\sin^2 x}\right)$$

$$= \csc^2 x + \frac{1}{\cos^2 x}$$

$$= \csc^2 x + \sec^2 x$$

We could have used many other variations of this procedure, and they would have been perfectly valid.

Example J

Prove the identity $\dfrac{\csc x}{\tan x + \cot x} = \cos x$.

Here we shall simplify the left-hand side until we have the expression which appears on the right-hand side.

$$\frac{\csc x}{\tan x + \cot x} = \frac{\csc x}{\tan x + \dfrac{1}{\tan x}} = \frac{\csc x}{\dfrac{\tan^2 x + 1}{\tan x}}$$

$$= \frac{\csc x \tan x}{\tan^2 x + 1} = \frac{\csc x \tan x}{\sec^2 x}$$

$$= \frac{\dfrac{1}{\sin x} \cdot \dfrac{\sin x}{\cos x}}{\dfrac{1}{\cos^2 x}} = \frac{1}{\sin x} \cdot \frac{\sin x}{\cos x} \cdot \frac{\cos^2 x}{1}$$

$$= \cos x$$

Therefore, we have shown that $\dfrac{\csc x}{\tan x + \cot x} = \cos x$, which proves the identity.

Exercises 19–1

In Exercises 1 through 4, verify the indicated basic identities for the given angles.

1. Verify Eq. (19–3) for $\theta = 56°$.

2. Verify Eq. (19–5) for $\theta = 80°$.

3. Verify Eq. (19–6) for $\theta = \dfrac{2\pi}{3}$.

4. Verify Eq. (19–7) for $\theta = \dfrac{7\pi}{6}$.

In Exercises 5 through 56, prove the given identities.

See Appendix E for a computer program for verifying a trigonometric identity.

5. $\dfrac{\cot \theta}{\cos \theta} = \csc \theta$ 6. $\dfrac{\tan y}{\sin y} = \sec y$ 7. $\dfrac{\sin x}{\tan x} = \cos x$ 8. $\dfrac{\csc \theta}{\sec \theta} = \cot \theta$

9. $\sin y \cot y = \cos y$ 10. $\cos x \tan x = \sin x$ 11. $\sin x \sec x = \tan x$

12. $\cot \theta \sec \theta = \csc \theta$ 13. $\csc^2 x(1 - \cos^2 x) = 1$ 14. $\cos^2 x(1 + \tan^2 x) = 1$

15. $\sin x(1 + \cot^2 x) = \csc x$ 16. $\sec \theta(1 - \sin^2 \theta) = \cos \theta$

17. $\sin x(\csc x - \sin x) = \cos^2 x$ 18. $\cos y(\sec y - \cos y) = \sin^2 y$

19. $\tan y(\cot y + \tan y) = \sec^2 y$ 20. $\csc x(\csc x - \sin x) = \cot^2 x$

21. $\sin x \tan x + \cos x = \sec x$ 22. $\sec x \csc x - \cot x = \tan x$

23. $\cos \theta \cot \theta + \sin \theta = \csc \theta$ 24. $\csc x \sec x - \tan x = \cot x$

25. $\sec \theta \tan \theta \csc \theta = \tan^2 \theta + 1$ 26. $\sin x \cos x \tan x = 1 - \cos^2 x$

27. $\cot \theta \sec^2 \theta - \cot \theta = \tan \theta$ 28. $\sin y + \sin y \cot^2 y = \csc y$

29. $\tan x + \cot x = \sec x \csc x$ 30. $\tan x + \cot x = \tan x \csc^2 x$

31. $\cos^2 x - \sin^2 x = 1 - 2 \sin^2 x$ 32. $\tan^2 y \sec^2 y - \tan^4 y = \tan^2 y$

33. $\dfrac{\sin x}{1 - \cos x} = \csc x + \cot x$ 34. $\dfrac{1 + \cos x}{\sin x} = \dfrac{\sin x}{1 - \cos x}$

35. $\dfrac{\sec x + \csc x}{1 + \tan x} = \csc x$ 36. $\dfrac{\cot x + 1}{\cot x} = 1 + \tan x$

37. $\tan^2 x \cos^2 x + \cot^2 x \sin^2 x = 1$ 38. $\dfrac{\sin \theta}{\csc \theta} + \dfrac{\cos \theta}{\sec \theta} = 1$

39. $\dfrac{\sec \theta}{\cos \theta} - \dfrac{\tan \theta}{\cot \theta} = 1$ 40. $\dfrac{\csc \theta}{\sin \theta} - \dfrac{\cot \theta}{\tan \theta} = 1$

41. $\dfrac{1 - 2 \cos^2 x}{\sin x \cos x} = \tan x - \cot x$ 42. $\cos^3 x \csc^3 x \tan^3 x = \csc^2 x - \cot^2 x$

43. $4 \sin x + \tan x = \dfrac{4 + \sec x}{\csc x}$ 44. $\dfrac{1 + \tan x}{\sin x} - \sec x = \csc x$

45. $\sec x + \tan x + \cot x = \dfrac{1 + \sin x}{\cos x \sin x}$ 46. $\csc \theta(\csc \theta - \cot \theta) = \dfrac{1}{1 + \cos \theta}$

47. $\dfrac{\cos \theta + \sin \theta}{1 + \tan \theta} = \cos \theta$ 48. $\dfrac{\sec x - \cos x}{\tan x} = \sin x$

49. $(\tan x + \cot x)\sin x \cos x = 1$

50. $\sec x(\sec x - \cos x) + \dfrac{\cos x - \sin x}{\cos x} + \tan x = \sec^2 x$

51. $2 \sin^4 x - 3 \sin^2 x + 1 = \cos^2 x(1 - 2 \sin^2 x)$

52. $\dfrac{\sin^4 x - \cos^4 x}{1 - \cot^4 x} = \sin^4 x$

53. $\dfrac{\cot 2y}{\sec 2y - \tan 2y} - \dfrac{\cos 2y}{\sec 2y + \tan 2y} = \sin 2y + \csc 2y$

54. $\dfrac{1}{2}\sin 5y\left(\dfrac{\sin 5y}{1 - \cos 5y} + \dfrac{1 - \cos 5y}{\sin 5y}\right) = 1$

55. $1 + \sin^2 x + \sin^4 x + \cdots = \sec^2 x$ 56. $1 - \tan^2 x + \tan^4 x - \cdots = \cos^2 x$

In Exercises 57 through 60, prove the identities which arise in the given area of application.

57. For an object of weight w on an inclined plane the coefficient of friction μ, related to the frictional force between the plane and object, can be found from the equation $\mu w \cos \theta = w \sin \theta$, where θ is the angle between the plane and horizontal. Solve for μ and show that $\mu = \tan \theta$.

58. In finding the change in $\tan^2 x$ for a given change in x, the expression $2 \tan x \sec^2 x$ is used. Show that this expression is equal to $2 \sin x \sec^3 x$.

59. In determining the rate of radiation by an accelerated electric charge, it is necessary to show that $\sin^3 \theta = \sin \theta - \cos^2 \theta \sin \theta$. Show that this is valid, by transforming the left-hand side.

60. In determining the path of least time between two points under certain circumstances, it is necessary to show that

$$\sqrt{\dfrac{1 + \cos \theta}{1 - \cos \theta}}\,\sin \theta = 1 + \cos \theta$$

Show this by transforming the left-hand side.

In Exercises 61 and 62, use the given substitutions to show that the given equations are valid. $\left(\text{In each, } 0 < \theta < \dfrac{\pi}{2}.\right)$

61. If $x = 2 \tan \theta$, show that $\sqrt{4 + x^2} = 2 \sec \theta$.

62. If $x = 3 \sin \theta$, show that $\sqrt{9 - x^2} = 3 \cos \theta$.

19–2 Sine and Cosine of the Sum and Difference of Two Angles

There are other important relations among the trigonometric functions. The most important and useful relations are those which involve twice an angle and half an angle. To obtain these relations, we shall first derive the expressions for the sine and cosine of the sum and difference of two angles. These expressions will lead directly to the desired relations of double and half angles.

In Fig. 19–1, the angle α is in standard position, and the angle β has as its initial side the terminal side of α. Thus the angle of interest, $\alpha + \beta$, is in standard position. From a point P, on the terminal side of $\alpha + \beta$, perpendiculars are dropped to the x-axis and to the terminal side of α, at the points M and Q respectively. Then perpendiculars are dropped from Q to the x-axis and to the line MP at points N and R respectively. By this construction, $\angle RPQ$ is equal to $\angle \alpha$. ($\angle RQO = \angle \alpha$ by alternate interior angles; $\angle RQO + \angle RQP = 90°$ by construction; $\angle RQP + \angle RPQ = 90°$ by the sum of the angles of a triangle being 180°.

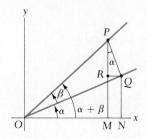

Figure 19–1

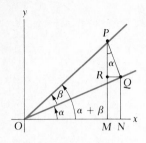

Figure 19–1

Thus, $\angle\alpha = \angle RPQ$.) By definition,

$$\sin(\alpha + \beta) = \frac{MP}{OP} = \frac{MR + RP}{OP} = \frac{NQ}{OP} + \frac{RP}{OP}$$

These last two fractions do not define any function of either α or β, and therefore we multiply the first fraction (numerator and denominator) by OQ and the second fraction by QP. When we do this and rearrange the fractions, we have functions of α and β:

$$\sin(\alpha + \beta) = \frac{NQ}{OP} + \frac{RP}{OP} = \frac{NQ}{OQ} \cdot \frac{OQ}{OP} + \frac{RP}{QP} \cdot \frac{QP}{OP}$$

$$= \sin\alpha\cos\beta + \cos\alpha\sin\beta$$

Using the same figure, we can also obtain the expression for $\cos(\alpha + \beta)$. Thus we have the relations

$$\sin(\alpha + \beta) = \sin\alpha\cos\beta + \cos\alpha\sin\beta \qquad (19\text{–}9)$$

and

$$\cos(\alpha + \beta) = \cos\alpha\cos\beta - \sin\alpha\sin\beta \qquad (19\text{–}10)$$

Example A

Find $\sin 75°$ from $\sin 75° = \sin(45° + 30°)$.

$$\sin 75° = \sin(45° + 30°) = \sin 45°\cos 30° + \cos 45°\sin 30°$$

$$= \frac{\sqrt{2}}{2}\cdot\frac{\sqrt{3}}{2} + \frac{\sqrt{2}}{2}\cdot\frac{1}{2} = \frac{\sqrt{6}}{4} + \frac{\sqrt{2}}{4} = \frac{\sqrt{6} + \sqrt{2}}{4}$$

$$= 0.9659$$

Example B

Verify that $\sin 90° = 1$, by finding $\sin(60° + 30°)$.

$$\sin 90° = \sin(60° + 30°) = \sin 60°\cos 30° + \cos 60°\sin 30°$$

$$= \frac{\sqrt{3}}{2}\cdot\frac{\sqrt{3}}{2} + \frac{1}{2}\cdot\frac{1}{2} = \frac{3}{4} + \frac{1}{4} = 1$$

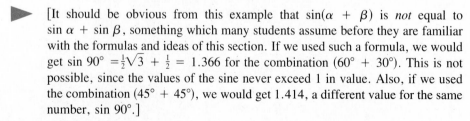

[It should be obvious from this example that $\sin(\alpha + \beta)$ is *not* equal to $\sin\alpha + \sin\beta$, something which many students assume before they are familiar with the formulas and ideas of this section. If we used such a formula, we would get $\sin 90° = \frac{1}{2}\sqrt{3} + \frac{1}{2} = 1.366$ for the combination $(60° + 30°)$. This is not possible, since the values of the sine never exceed 1 in value. Also, if we used the combination $(45° + 45°)$, we would get 1.414, a different value for the same number, $\sin 90°$.]

From Eqs. (19–9) and (19–10), we can easily find expressions for $\sin(\alpha - \beta)$ and $\cos(\alpha - \beta)$. This is done by finding $\sin(\alpha + (-\beta))$ and $\cos(\alpha + (-\beta))$. Thus, we have

$$\sin(\alpha - \beta) = \sin(\alpha + (-\beta)) = \sin \alpha \cos(-\beta) + \cos \alpha \sin(-\beta)$$

Since $\cos(-\beta) = \cos \beta$ and $\sin(-\beta) = -\sin \beta$ (see Exercise 53 of Section 7–2), we have

$$\sin(\alpha - \beta) = \sin \alpha \cos \beta - \cos \alpha \sin \beta \qquad (19\text{–}11)$$

In the same manner we find that

$$\cos(\alpha - \beta) = \cos \alpha \cos \beta + \sin \alpha \sin \beta \qquad (19\text{–}12)$$

Example C

Find $\cos 15°$ from $\cos(45° - 30°)$.

$$\cos 15° = \cos(45° - 30°) = \cos 45° \cos 30° + \sin 45° \sin 30°$$
$$= \frac{\sqrt{2}}{2} \cdot \frac{\sqrt{3}}{2} + \frac{\sqrt{2}}{2} \cdot \frac{1}{2} = \frac{\sqrt{6} + \sqrt{2}}{4} = 0.9659$$

We get the same result as in Example A, which should be the case since $\sin 75° = \cos 15°$. (See Section 3–4.)

Example D

Evaluate $\cos 23° \cos 67° - \sin 23° \sin 67°$.

We note that this expression fits the form of the right side of Eq. (19–10), and so

$$\cos 23° \cos 67° - \sin 23° \sin 67° = \cos(23° + 67°)$$
$$= \cos 90°$$
$$= 0$$

Thus, we are able to evaluate this by recognizing the form of the given expression. Evaluation by a calculator will verify the result.

By using Eqs. (19–9) and (19–10), we can determine expressions for $\tan(\alpha + \beta)$, $\cot(\alpha + \beta)$, $\sec(\alpha + \beta)$, and $\csc(\alpha + \beta)$. These expressions are less applicable than those for the sine and cosine, and therefore we shall not derive them here, although the expression for $\tan(\alpha + \beta)$ is found in the exercises at the end of this section. By using Eqs. (19–11) and (19–12), we can find similar expressions for the functions of $(\alpha - \beta)$.

Certain trigonometric identities can also be worked out by using the formulas derived in this section. The following examples illustrate the use of these formulas in identities.

Example E

Prove that $\sin(180° + x) = -\sin x$.

By using Eq. (19–9) we have

$$\sin(180° + x) = \sin 180° \cos x + \cos 180° \sin x$$

Since $\sin 180° = 0$ and $\cos 180° = -1$, we have

$$\sin 180° \cos x + \cos 180° \sin x = (0)\cos x + (-1)\sin x$$

or

$$\sin(180° + x) = -\sin x$$

Example F

Show that $\dfrac{\sin(\alpha - \beta)}{\sin \alpha \sin \beta} = \cot \beta - \cot \alpha$.

By using Eq. (19–11), we have

$$\frac{\sin(\alpha - \beta)}{\sin \alpha \sin \beta} = \frac{\sin \alpha \cos \beta - \cos \alpha \sin \beta}{\sin \alpha \sin \beta} = \frac{\sin \alpha \cos \beta}{\sin \alpha \sin \beta} - \frac{\cos \alpha \sin \beta}{\sin \alpha \sin \beta}$$

$$= \frac{\cos \beta}{\sin \beta} - \frac{\cos \alpha}{\sin \alpha} = \cot \beta - \cot \alpha$$

Example G

Show that

$$\sin\left(\frac{\pi}{4} + x\right)\cos\left(\frac{\pi}{4} + x\right) = \frac{1}{2}(\cos^2 x - \sin^2 x)$$

$$\sin\left(\frac{\pi}{4} + x\right)\cos\left(\frac{\pi}{4} + x\right)$$

$$= \left(\sin\frac{\pi}{4}\cos x + \cos\frac{\pi}{4}\sin x\right)\left(\cos\frac{\pi}{4}\cos x - \sin\frac{\pi}{4}\sin x\right)$$

$$= \sin\frac{\pi}{4}\cos\frac{\pi}{4}\cos^2 x - \sin^2\frac{\pi}{4}\sin x \cos x + \cos^2\frac{\pi}{4}\sin x \cos x - \sin^2 x \sin\frac{\pi}{4}\cos\frac{\pi}{4}$$

$$= \frac{\sqrt{2}}{2}\frac{\sqrt{2}}{2}\cos^2 x - \left(\frac{\sqrt{2}}{2}\right)^2 \sin x \cos x + \left(\frac{\sqrt{2}}{2}\right)^2 \sin x \cos x - \frac{\sqrt{2}}{2}\frac{\sqrt{2}}{2}\sin^2 x$$

$$= \frac{1}{2}\cos^2 x - \frac{1}{2}\sin^2 x = \frac{1}{2}(\cos^2 x - \sin^2 x)$$

Example H

Show that $\sin(x + y)\cos y - \cos(x + y)\sin y = \sin x$.

If we let $x + y = z$, we note that the left-hand side of the above expression becomes $\sin z \cos y - \cos z \sin y$, which is the proper form for $\sin(z - y)$. By replacing z with $x + y$, we obtain $\sin(x + y - y)$, which is $\sin x$. Therefore, the above expression has been shown to be true. We again see that proper recognition of a basic form leads to the solution.

Exercises 19–2

In Exercises 1 through 4, determine the values of the given functions as indicated.

1. Find $\sin 105°$ by using $105° = 60° + 45°$.

2. Find $\cos 75°$ by using $75° = 30° + 45°$.

3. Find $\cos 15°$ by using $15° = 60° - 45°$.

4. Find $\sin 15°$ by using $15° = 45° - 30°$.

In Exercises 5 through 8, evaluate the indicated functions with the following given information: $\sin \alpha = \frac{4}{5}$ (in first quadrant), and $\cos \beta = -\frac{12}{13}$ (in second quadrant).

5. $\sin(\alpha + \beta)$ 6. $\cos(\beta - \alpha)$ 7. $\cos(\alpha + \beta)$ 8. $\sin(\alpha - \beta)$

In Exercises 9 through 12, reduce each of the given expressions to a single term. Expansion of any term is not necessary; proper recognition of the form of the expression leads to the proper result.

9. $\sin x \cos 2x + \sin 2x \cos x$ 10. $\sin 3x \cos x - \sin x \cos 3x$

11. $\cos(x + y)\cos y + \sin(x + y)\sin y$ 12. $\cos(2x - y)\cos y - \sin(2x - y)\sin y$

In Exercises 13 through 16, evaluate each of the given expressions. Proper recognition of the given form leads to the result. Verify each result by use of a calculator.

13. $\sin 122° \cos 32° - \cos 122° \sin 32°$ 14. $\cos 250° \cos 70° + \sin 250° \sin 70°$

15. $\cos 312° \cos 48° - \sin 312° \sin 48°$ 16. $\sin 56° \cos 124° + \cos 56° \sin 124°$

In Exercises 17 through 28, prove the given identities.

17. $\sin(270° - x) = -\cos x$ 18. $\sin(90° + x) = \cos x$

19. $\cos\left(\frac{\pi}{2} - x\right) = \sin x$ 20. $\cos\left(\frac{3\pi}{2} + x\right) = \sin x$

21. $\cos(30° + x) = \dfrac{\sqrt{3} \cos x - \sin x}{2}$ 22. $\sin(120° - x) = \dfrac{\sqrt{3} \cos x + \sin x}{2}$

23. $\sin\left(\frac{\pi}{4} + x\right) = \dfrac{\sin x + \cos x}{\sqrt{2}}$ 24. $\cos\left(\frac{\pi}{3} + x\right) = \dfrac{\cos x - \sqrt{3} \sin x}{2}$

25. $\sin(x + y)\sin(x - y) = \sin^2 x - \sin^2 y$

26. $\cos(x + y)\cos(x - y) = \cos^2 x - \sin^2 y$

27. $\cos(\alpha + \beta) + \cos(\alpha - \beta) = 2 \cos \alpha \cos \beta$

28. $\cos(x - y) + \sin(x + y) = (\cos x + \sin x)(\cos y + \sin y)$

In Exercises 29 through 32, additional trigonometric identities are shown. Derive these in the indicated manner. Equations (19–13), (19–14), and (19–15) are known as the product formulas.

29. By dividing Eq. (19–9) by Eq. (19–10), show that

$$\tan(\alpha + \beta) = \frac{\tan \alpha + \tan \beta}{1 - \tan \alpha \tan \beta}$$

(*Hint:* Divide numerator and denominator by $\cos \alpha \cos \beta$.)

30. By adding Eqs. (19–9) and (19–11), derive the equation

$$\sin \alpha \cos \beta = \tfrac{1}{2}[\sin (\alpha + \beta) + \sin (\alpha - \beta)] \qquad (19\text{–}13)$$

31. By adding Eqs. (19–10) and (19–12), derive the equation

$$\cos \alpha \cos \beta = \tfrac{1}{2}[\cos (\alpha + \beta) + \cos (\alpha - \beta)] \qquad (19\text{–}14)$$

32. By subtracting Eq. (19–10) from Eq. (19–12), derive

$$\sin \alpha \sin \beta = \tfrac{1}{2}[\cos (\alpha - \beta) - \cos (\alpha + \beta)] \qquad (19\text{–}15)$$

In Exercises 33 through 36, additional trigonometric identities are shown. Derive them by letting $\alpha + \beta = x$ and $\alpha - \beta = y$, which leads to $\alpha = \tfrac{1}{2}(x + y)$ and $\beta = \tfrac{1}{2}(x - y)$. The resulting equations are known as the factor formulas.

33. Use Eq. (19–13) and the substitutions above to derive the equation

$$\sin x + \sin y = 2 \sin \tfrac{1}{2}(x + y) \cos \tfrac{1}{2}(x - y) \qquad (19\text{–}16)$$

34. Use Eqs. (19–9) and (19–11) and the substitutions above to derive the equation

$$\sin x - \sin y = 2 \sin \tfrac{1}{2}(x - y) \cos \tfrac{1}{2}(x + y) \qquad (19\text{–}17)$$

35. Use Eq. (19–14) and the substitutions above to derive the equation

$$\cos x + \cos y = 2 \cos \tfrac{1}{2}(x + y) \cos \tfrac{1}{2}(x - y) \qquad (19\text{–}18)$$

36. Use Eq. (19–15) and the substitutions above to derive the equation

$$\cos x - \cos y = -2 \sin \tfrac{1}{2}(x + y) \sin \tfrac{1}{2}(x - y) \qquad (19\text{–}19)$$

In Exercises 37 through 40, use the equations of this section to solve the given problems.

37. In determining the motion of an object, the expression

$$\cos \alpha \sin (\omega t + \phi) - \sin \alpha \cos (\omega t + \phi)$$

is found. Simplify this expression.

38. Under certain conditions, the current in an electric circuit is given by

$$i = I_m[\sin(\omega t + \alpha) \cos \phi + \cos(\omega t + \alpha) \sin \phi]$$

Simplify the expression on the right.

39. The displacements y_1 and y_2 of two waves traveling through the same medium are given by the equations $y_1 = A \sin 2\pi(t/T - x/\lambda)$ and $y_2 = A \sin 2\pi(t/T + x/\lambda)$. Find an expression for the displacement $y_1 + y_2$ of the combination of the waves.

40. In the analysis of the angles of incidence i and reflection r of a light ray subject to certain conditions, the following expression is found:

$$E_2\left(\frac{\tan r}{\tan i} + 1\right) = E_1\left(\frac{\tan r}{\tan i} - 1\right)$$

Show that an equivalent expression is

$$E_2 = E_1 \frac{\sin(r - i)}{\sin(r + i)}$$

19–3 Double-Angle Formulas

If we let $\beta = \alpha$ in Eqs. (19–9) and (19–10), we can derive the important double-angle formulas. Thus, by making this substitution in Eq. (19–9), we have

$$\sin(\alpha + \alpha) = \sin(2\alpha) = \sin \alpha \cos \alpha + \cos \alpha \sin \alpha = 2 \sin \alpha \cos \alpha$$

Using the same substitution in Eq. (19–10), we have

$$\cos(\alpha + \alpha) = \cos \alpha \cos \alpha - \sin \alpha \sin \alpha = \cos^2\alpha - \sin^2\alpha$$

By using the basic identity (19–6), other forms of this last equation may be derived. Thus, summarizing these formulas, we have

$$\sin 2\alpha = 2 \sin \alpha \cos \alpha \qquad (19\text{–}20)$$
$$\cos 2\alpha = \cos^2\alpha - \sin^2\alpha \qquad (19\text{–}21)$$
$$= 2 \cos^2\alpha - 1 \qquad (19\text{–}22)$$
$$= 1 - 2 \sin^2\alpha \qquad (19\text{–}23)$$

We should note carefully that these equations give expressions for the sine and cosine of twice an angle in terms of functions of the angle. They can be used any time we have expressed one angle as twice another. These double-angle formulas are widely used in applications of trigonometry, especially in the calculus. They should be known and recognized quickly in any of the various forms.

Example A

If $\alpha = 30°$, we have $\cos 2(30°) = \cos 60° = \cos^2 30° - \sin^2 30°$.
If $\alpha = 3x$, we have $\sin 2(3x) = \sin 6x = 2 \sin 3x \cos 3x$.
If $2\alpha = x$, we may write $\alpha = x/2$, which means that

$$\sin 2\left(\frac{x}{2}\right) = \sin x = 2 \sin \frac{x}{2} \cos \frac{x}{2}$$

Example B

Using the double-angle formulas, simplify the expression

$$\cos^2 2x - \sin^2 2x$$

By using Eq. (19–21) and letting $\alpha = 2x$, we have

$$\cos^2 2x - \sin^2 2x = \cos 2(2x) = \cos 4x$$

Example C

Verify the values of $\sin 90°$ and $\cos 90°$ by use of the functions of $45°$.

$$\sin 90° = \sin 2(45°) = 2 \sin 45° \cos 45° = 2\left(\frac{\sqrt{2}}{2}\right)\left(\frac{\sqrt{2}}{2}\right) = 1$$

$$\cos 90° = \cos 2(45°) = \cos^2 45° - \sin^2 45° = \left(\frac{\sqrt{2}}{2}\right)^2 - \left(\frac{\sqrt{2}}{2}\right)^2 = 0$$

Example D

Using a calculator, verify the values of sin 142° and cos 142° by use of the functions of 71°.

$$\sin 142° = 2 \sin 71° \cos 71° = 0.6156615$$

By using the calculator sequence

71 | SIN | | × | 71 | COS | | × | 2 | = |

we obtain the above result. Also, it is found by use of the sequence 142 | SIN |. Therefore, the value by use of the double-angle formula is verified.

In the same way,

$$\cos 142° = \cos^2 71° - \sin^2 71° = -0.7880108$$

is found by the calculator sequence

71 | COS | | x^2 | | − | 71 | SIN | | x^2 | | = |

and by the calculator sequence 142 | COS |. Therefore, this double-angle formula is verified.

Example E

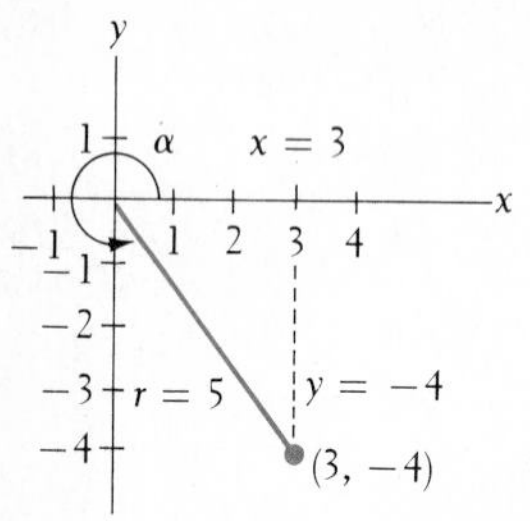

Figure 19–2

Given that $\cos \alpha = \frac{3}{5}$ (in the fourth quadrant), find $\sin 2\alpha$.

Knowing that $\cos \alpha = \frac{3}{5}$ for an angle in the fourth quadrant, we then determine that

$$\sin \alpha = -\frac{4}{5}$$

(see Fig. 19–2). Thus,

$$\sin 2\alpha = 2\left(-\frac{4}{5}\right)\left(\frac{3}{5}\right) = -\frac{24}{25}$$

Example F

Prove the identity $\dfrac{2}{1 + \cos 2x} = \sec^2 x$.

$$\frac{2}{1 + \cos 2x} = \frac{2}{1 + (2\cos^2 x - 1)} = \frac{2}{2\cos^2 x} = \sec^2 x$$

Example G

Show that $\dfrac{\sin 3x}{\sin x} + \dfrac{\cos 3x}{\cos x} = 4 \cos 2x$.

The first step is to combine the two fractions on the left, so that we can see if any usable forms will emerge. Thus, the left side becomes

$$\frac{\sin 3x \cos x + \cos 3x \sin x}{\sin x \cos x}$$

We now note that the numerator is of the form $\sin(A + x)$, where $A = 3x$. Also, the denominator is $\frac{1}{2} \sin 2x$. Making these substitutions, we have

$$\frac{\sin(3x + x)}{\frac{1}{2} \sin 2x} = \frac{2 \sin 4x}{\sin 2x}$$

By expanding $\sin 4x$ into $2 \sin 2x \cos 2x$, we obtain

$$\frac{2(2 \sin 2x \cos 2x)}{\sin 2x} = 4 \cos 2x$$

Therefore, the expression is shown to be valid.

Exercises 19–3

In Exercises 1 through 4, determine the values of the indicated functions in the given manner.

1. Find $\sin 60°$ by using the functions of $30°$.

2. Find $\sin 120°$ by using the functions of $60°$.

3. Find $\cos 120°$ by using the functions of $60°$.

4. Find $\cos 60°$ by using the functions of $30°$.

In Exercises 5 through 8, use a calculator to verify the values found by use of the double-angle formulas as in Example D.

5. Find $\sin 158°$ directly and by use of the functions of $79°$.

6. Find $\sin 84°$ directly and by use of the functions of $42°$.

7. Find $\cos 96°$ directly and by use of the functions of $48°$.

8. Find $\cos 276°$ directly and by use of the functions of $138°$.

In Exercises 9 through 12, evaluate the indicated functions with the given information.

9. Find $\sin 2x$ if $\cos x = \frac{4}{5}$ (in first quadrant).

10. Find $\cos 2x$ if $\sin x = -\frac{12}{13}$ (in third quadrant).

11. Find $\cos 2x$ if $\tan x = \frac{1}{2}$ (in third quadrant).

12. Find $\sin 4x$ if $\sin x = \frac{3}{5}$ (in first quadrant) $[4x = 2(2x)]$.

In Exercises 13 through 16, simplify the given expressions. Expansion of any term is not necessary; proper recognition of the form of the expression leads to the proper result.

13. $4 \sin 4x \cos 4x$

14. $4 \sin^2 x \cos^2 x$

15. $1 - 2 \sin^2 4x$

16. $\sin^2 4x - \cos^2 4x$

In Exercises 17 through 28, prove the given identities.

17. $\cos^2 \alpha - \sin^2 \alpha = 2 \cos^2 \alpha - 1$

18. $\cos^2 \alpha - \sin^2 \alpha = 1 - 2 \sin^2 \alpha$

19. $\cos^4 x - \sin^4 x = \cos 2x$

20. $(\sin x + \cos x)^2 = 1 + \sin 2x$

21. $\dfrac{\sin 2\theta}{1 + \cos 2\theta} = \tan \theta$

22. $\dfrac{2 \tan \alpha}{1 + \tan^2 \alpha} = \sin 2\alpha$

23. $\dfrac{1 - \tan^2 x}{\sec^2 x} = \cos 2x$

24. $1 - \cos 2\theta = \dfrac{2}{1 + \cot^2 \theta}$

25. $2 \csc 2x \tan x = \sec^2 x$

26. $2 \sin x + \sin 2x = \dfrac{2 \sin^3 x}{1 - \cos x}$

27. $\dfrac{\sin 3x}{\sin x} - \dfrac{\cos 3x}{\cos x} = 2$

28. $\dfrac{\sin 3x}{\sin x} + \dfrac{\cos 3x}{\cos x} = 4 \cos 2x$

In Exercises 29 and 30, prove the given identities by letting $3x = 2x + x$.

29. $\sin 3x = 3 \cos^2 x \sin x - \sin^3 x$

30. $\cos 3x = \cos^3 x - 3 \sin^2 x \cos x$

In Exercises 31 through 34, solve the given problems.

31. In Exercise 29 of Section 19–2, let $\beta = \alpha$, and show that

$$\tan 2\alpha = \frac{2 \tan \alpha}{1 - \tan^2 \alpha}$$

32. Given that $x = \cos 2\theta$ and $y = \sin \theta$, find the relation between x and y by eliminating θ.

33. The equation for the displacement of a certain object at the end of a spring is $y = A \sin 2t + B \cos 2t$. Show that this equation may be written as $y = C \sin(2t + \alpha)$ where $C = \sqrt{A^2 + B^2}$ and $\tan \alpha = B/A$. (*Hint:* Let $A/C = \cos \alpha$ and $B/C = \sin \alpha$.)

34. To find the horizontal range R of a projectile, the equation $R = vt \cos \alpha$ is used, where α is the angle between the line of fire and the horizontal, v is the initial velocity of the projectile, and t is the time of flight. It can be shown that $t = (2v \sin \alpha)/g$, where g is the acceleration due to gravity. Show that $R = (v^2 \sin 2\alpha)/g$.

19–4 Half-Angle Formulas

If we let $\theta = \alpha/2$ in the identity $\cos 2\theta = 1 - 2 \sin^2 \theta$ and then solve for $\sin(\alpha/2)$, we obtain

$$\sin \frac{\alpha}{2} = \pm \sqrt{\frac{1 - \cos \alpha}{2}} \qquad (19\text{–}24)$$

Also, with the same substitution in the identity $\cos 2\theta = 2 \cos^2 \theta - 1$, which is then solved for $\cos(\alpha/2)$, we have

$$\cos \frac{\alpha}{2} = \pm \sqrt{\frac{1 + \cos \alpha}{2}} \qquad (19\text{–}25)$$

In each of Eqs. (19–24) and (19–25), the sign chosen depends on the quadrant in which $\alpha/2$ lies.

We can use these half-angle formulas to find values of the functions of angles which are half of those for which the functions are known. Examples A through F illustrate how these identities are used in evaluations and in identities.

Example A

We can find sin 15° by using the relation

$$\sin 15° = \sqrt{\frac{1 - \cos 30°}{2}} = \sqrt{\frac{1 - 0.8660}{2}} = 0.2588$$

Here the plus sign is used, since 15° is in the first quadrant.

Example B

We can find cos 165° by use of the relation

$$\cos 165° = -\sqrt{\frac{1 + \cos 330°}{2}} = -\sqrt{\frac{1 + 0.8660}{2}} = -0.9659$$

Here the minus sign is used, since 165° is in the second quadrant, and the cosine of a second-quadrant angle is negative.

Example C

Simplify $\sqrt{\dfrac{1 - \cos 114°}{2}}$ by expressing the result in terms of one-half the given angle. Then using a calculator, show that the values are equal.

We note that the given expression fits the form of the right side of Eq. (19–24), which means that

$$\sqrt{\frac{1 - \cos 114°}{2}} = \sin \tfrac{1}{2}(114°) = \sin 57°$$

By using the calculator sequence

1 $\boxed{-}$ 114 $\boxed{\text{COS}}$ $\boxed{=}$ $\boxed{\div}$ 2 $\boxed{=}$ $\boxed{\sqrt{x}}$

we obtain the result 0.8386706. The same result is obtained from the sequence 57 $\boxed{\text{SIN}}$, which verifies the answer.

Example D

Simplify the expression $\sqrt{18 - 18 \cos 4x}$.

First we factor the 18 from each of the terms under the radical, and note that $18 = 9(2)$ and 9 is a perfect square. This leads to

$$\sqrt{18 - 18 \cos 4x} = \sqrt{9(2)(1 - \cos 4x)} = 3\sqrt{2(1 - \cos 4x)}$$

This last expression is very similar to that for $\sin (\alpha /2)$, except that no 2 appears in the denominator. Therefore, multiplying the numerator and the denominator under the radical by 2 leads to the solution.

$$3\sqrt{2(1 - \cos 4x)} = 3\sqrt{\frac{4(1 - \cos 4x)}{2}} = 6\sqrt{\frac{1 - \cos 4x}{2}} = 6 \sin \frac{4x}{2} = 6 \sin 2x$$

Example E

Prove the identity $\sec \dfrac{\alpha}{2} + \csc \dfrac{\alpha}{2} = \dfrac{2\left(\sin \dfrac{\alpha}{2} + \cos \dfrac{\alpha}{2}\right)}{\sin \alpha}$

By expressing $\sin \alpha$ as $2 \sin \dfrac{\alpha}{2} \cos \dfrac{\alpha}{2}$, we have

$$\dfrac{2\left(\sin \dfrac{\alpha}{2} + \cos \dfrac{\alpha}{2}\right)}{2 \sin \dfrac{\alpha}{2} \cos \dfrac{\alpha}{2}} = \dfrac{\sin \dfrac{\alpha}{2}}{\sin \dfrac{\alpha}{2} \cos \dfrac{\alpha}{2}} + \dfrac{\cos \dfrac{\alpha}{2}}{\sin \dfrac{\alpha}{2} \cos \dfrac{\alpha}{2}}$$

$$= \dfrac{1}{\cos \dfrac{\alpha}{2}} + \dfrac{1}{\sin \dfrac{\alpha}{2}} = \sec \dfrac{\alpha}{2} + \csc \dfrac{\alpha}{2}$$

Example F

We can find relations for the other functions of $\alpha/2$ by expressing these functions in terms of $\sin(\alpha/2)$ and $\cos(\alpha/2)$. For example:

$$\sec \dfrac{\alpha}{2} = \dfrac{1}{\cos \dfrac{\alpha}{2}} = \pm \dfrac{1}{\sqrt{\dfrac{1 + \cos \alpha}{2}}} = \pm \sqrt{\dfrac{2}{1 + \cos \alpha}}$$

Example G

Show that $2 \cos^2 \dfrac{x}{2} - \cos x = 1$.

The first step is to substitute for $\cos(x/2)$, which will result in each term containing x on the left being in terms of x, and no $x/2$ terms will exist. This might allow us to combine terms. So we perform this operation, and we have for the left side

$$2\left(\dfrac{1 + \cos x}{2}\right) - \cos x$$

Combining terms, we can complete the proof:

$$1 + \cos x - \cos x = 1$$

Exercises 19–4

In Exercises 1 through 4, use the half-angle formulas to evaluate the given functions.

1. $\cos 15°$ 2. $\sin 22.5°$ 3. $\sin 75°$ 4. $\cos 112.5°$

In Exercises 5 through 8, simplify the given expressions by giving the results in terms of one-half the given angle. Then use a calculator to verify the results, as in Example C.

5. $\sqrt{\dfrac{1 - \cos 236°}{2}}$ 6. $\sqrt{\dfrac{1 + \cos 98°}{2}}$

7. $\sqrt{1 + \cos 164°}$ 8. $\sqrt{2 - 2 \cos 328°}$

In Exercises 9 through 12, use the half-angle formulas to simplify the given expressions.

9. $\sqrt{\dfrac{1 - \cos 6\alpha}{2}}$ 10. $\sqrt{\dfrac{4 + 4 \cos 8\beta}{2}}$

11. $\sqrt{8 + 8 \cos 4x}$ 12. $\sqrt{2 - 2 \cos 16x}$

In Exercises 13 through 16, evaluate the indicated functions with the information given.

13. Find the value of $\sin(\alpha / 2)$, if $\cos \alpha = \frac{12}{13}$ (in first quadrant).

14. Find the value of $\cos(\alpha / 2)$, if $\sin \alpha = -\frac{4}{5}$ (in third quadrant).

15. Find the value of $\cos(\alpha / 2)$, if $\tan \alpha = -\frac{7}{24}$ (in second quadrant).

16. Find the value of $\sin(\alpha / 2)$, if $\cos \alpha = \frac{8}{17}$ (in fourth quadrant).

In Exercises 17 through 20, derive the required expressions.

17. Derive an expression for $\csc(\alpha / 2)$ in terms of $\cos \alpha$.

18. Derive an expression for $\sec(\alpha / 2)$ in terms of $\sec \alpha$.

19. Derive an expression for $\tan(\alpha / 2)$ in terms of $\sin \alpha$ and $\cos \alpha$.

20. Derive an expression for $\cot(\alpha / 2)$ in terms of $\sin \alpha$ and $\cos \alpha$.

In Exercises 21 through 28, prove the given identities.

21. $\sin \dfrac{\alpha}{2} = \dfrac{1 - \cos \alpha}{2 \sin \dfrac{\alpha}{2}}$ 22. $2 \cos \dfrac{x}{2} = (1 + \cos x)\sec \dfrac{x}{2}$

23. $2 \sin^2 \dfrac{x}{2} + \cos x = 1$ 24. $2 \cos^2 \dfrac{\theta}{2} \sec \theta = \sec \theta + 1$

25. $\cos \dfrac{\theta}{2} = \dfrac{\sin \theta}{2 \sin \dfrac{\theta}{2}}$ 26. $\cos^2 \dfrac{x}{2}\left[1 + \left(\dfrac{\sin x}{1 + \cos x}\right)^2\right] = 1$

27. $2 \sin^2 \dfrac{\alpha}{2} - \cos^2 \dfrac{\alpha}{2} = \dfrac{1 - 3 \cos \alpha}{2}$ 28. $\tan \dfrac{\alpha}{2} = \dfrac{\sin \alpha}{1 + \cos \alpha}$

In Exercises 29 and 30, use the half-angle formulas to solve the given problems.

29. In the kinetic theory of gases, the expression

$$\sqrt{(1 - \cos \alpha)^2 + \sin^2\alpha \cos^2\beta + \sin^2\alpha \sin^2\beta}$$

is found. Show that this expression equals $2 \sin \dfrac{\alpha}{2}$.

30. The index of refraction n, the angle of a prism A, and the minimum angle of refraction ϕ are related by

$$n = \dfrac{\sin \dfrac{A + \phi}{2}}{\sin \dfrac{A}{2}}$$

Show that an equivalent expression is

$$n = \sqrt{\dfrac{1 + \cos \phi}{2}} + \left(\cot \dfrac{A}{2}\right)\sqrt{\dfrac{1 - \cos \phi}{2}}$$

19–5 Trigonometric Equations

One of the most important uses of the trigonometric identities is in the solution of equations involving the trigonometric functions. When equations are written in terms of more than one function, the identities provide a way of transforming many of them to equations or factors involving only one function of the same angle. If we can accomplish this we can employ algebraic methods from then on to complete the solution. No general methods exist for the solution of such equations, but the following examples illustrate methods which prove to be useful.

Example A

Solve the equation $2 \cos \theta - 1 = 0$ for all values of θ such that $0 \le \theta < 2\pi$.

Solving the equation for $\cos \theta$, we obtain $\cos \theta = \frac{1}{2}$. The problem asks for all values of θ between 0 and 2π that satisfy the equation. We know that the cosines of angles in the first and fourth quadrants are positive. Also, we know that

$$\cos \frac{\pi}{3} = \frac{1}{2}$$

Therefore,

$$\theta = \frac{\pi}{3} \quad \text{and} \quad \theta = \frac{5\pi}{3}$$

Example B

Solve the equation $5(1 + \sin x) = 2 \sin x + 3$ for $0 \le x < 2\pi$.

Solving for $\sin x$, we have

$$5(1 + \sin x) = 2 \sin x + 3$$
$$5 + 5 \sin x = 2 \sin x + 3$$
$$3 \sin x = -2$$
$$\sin x = -\frac{2}{3}$$

Since x cannot be found from known values by inspection, we use a calculator. We must remember that the answer is to be expressed in radians. Therefore, with the calculator in radian mode, we use the sequence

$$2 \quad \boxed{\div} \quad 3 \quad \boxed{=} \quad \boxed{+/-} \quad \boxed{\text{ARCSIN}}$$

and we obtain $x = -0.7297$. However, this is not in the range of values $0 \le x < 2\pi$. Therefore, recalling that $\sin x$ is negative in the third and fourth quadrants, we add 0.7297 to π and subtract it from 2π. This gives the solutions $x = 3.871$ and $x = 5.553$.

Example C

Solve the equation $2 \cos^2 x - \sin x - 1 = 0$ $(0 \leq x < 2\pi)$.

By use of the identity $\sin^2 x + \cos^2 x = 1$, this equation may be put in terms of $\sin x$ only. Thus, we have

$$2(1 - \sin^2 x) - \sin x - 1 = 0$$
$$-2 \sin^2 x - \sin x + 1 = 0$$
$$2 \sin^2 x + \sin x - 1 = 0$$

or

$$(2 \sin x - 1)(\sin x + 1) = 0$$

Just as in solving algebraic equations, we can set each factor equal to zero to find valid solutions. Thus, $\sin x = \frac{1}{2}$ and $\sin x = -1$. For the range between 0 and 2π, the value $\sin x = \frac{1}{2}$ gives values of x as $\pi/6$ and $5\pi/6$, and $\sin x = -1$ gives the value $x = 3\pi/2$. Thus, the complete solution is $x = \pi/6$, $x = 5\pi/6$, and $x = 3\pi/2$.

Example D

Solve the equation $\sec^2 x + 2 \tan x - 6 = 0$ $(0 \leq x < 2\pi)$.

By use of the identity $1 + \tan^2 x = \sec^2 x$ we may express this equation in terms of $\tan x$ only. Therefore,

$$\tan^2 x + 1 + 2 \tan x - 6 = 0$$
$$\tan^2 x + 2 \tan x - 5 = 0$$

We note that this expression is not factorable. Therefore, using the quadratic formula, we obtain the following solution:

$$\tan x = \frac{-2 \pm \sqrt{4 + 20}}{2} = -1 \pm 2.4495$$

Therefore, $\tan x = 1.4495$ and $\tan x = -3.4495$. In radians, we find that $\tan x = 1.4495$ for $x = 0.9669$. Since $\tan x$ is also positive in the third quadrant, we have $x = 4.108$ as well. In the same way, using $\tan x = -3.4495$, we obtain $x = 1.853$ and $x = 4.995$. Therefore, the correct solutions are $x = 0.9669$, $x = 1.853$, $x = 4.108$, and $x = 4.995$.

Example E

Solve the equation $\sin 2x + \sin x = 0$ $(0 \leq x < 2\pi)$.

By using the double-angle formula for $\sin 2x$, we can write the equation in the form

$$2 \sin x \cos x + \sin x = 0 \quad \text{or} \quad \sin x(2 \cos x + 1) = 0$$

The first factor gives $x = 0$ or $x = \pi$. The second factor, for which $\cos x = -\frac{1}{2}$, gives $x = 2\pi/3$ and $x = 4\pi/3$. Thus, the complete solution is $x = 0$, $x = 2\pi/3$, $x = \pi$, and $x = 4\pi/3$.

Example F

Solve the equation $\cos \frac{x}{2} = 1 + \cos x \ (0 \le x < 2\pi)$.

By using the half-angle formula for $\cos (x/2)$ and then squaring both sides of the resulting equation, this equation can be solved.

$$\pm \sqrt{\frac{1 + \cos x}{2}} = 1 + \cos x$$

$$\frac{1 + \cos x}{2} = 1 + 2 \cos x + \cos^2 x$$

Simplifying this last equation, we have

$$2 \cos^2 x + 3 \cos x + 1 = 0$$
$$(2 \cos x + 1)(\cos x + 1) = 0$$

The values of the cosine which come from these factors are $\cos x = -\frac{1}{2}$ and $\cos x = -1$. Thus, the values of x which satisfy the last equation are $x = 2\pi/3$, $x = 4\pi/3$, and $x = \pi$. However, when we solved this equation, we squared both sides of it. In doing this we may have introduced extraneous solutions (see Section 13–3). Thus, we must check each solution in the original equation to see if it is valid. Hence,

$$\cos \frac{\pi}{3} \stackrel{?}{=} 1 + \cos \frac{2\pi}{3} \quad \text{or} \quad \frac{1}{2} \stackrel{?}{=} 1 + \left(-\frac{1}{2}\right) \quad \text{or} \quad \frac{1}{2} = \frac{1}{2}$$

$$\cos \frac{2\pi}{3} \stackrel{?}{=} 1 + \cos \frac{4\pi}{3} \quad \text{or} \quad -\frac{1}{2} \stackrel{?}{=} 1 + \left(-\frac{1}{2}\right) \quad \text{or} \quad -\frac{1}{2} \ne \frac{1}{2}$$

$$\cos \frac{\pi}{2} \stackrel{?}{=} 1 + \cos \pi \quad \text{or} \quad 0 \stackrel{?}{=} 1 - 1 \quad \text{or} \quad 0 = 0$$

Thus, the apparent solution $x = 4\pi/3$ is not a solution of the original equation. The correct solutions are $x = 2\pi/3$ and $x = \pi$.

Example G

Solve the equation $\tan 3\theta - \cot 3\theta = 0 \ (0 \le \theta < 2\pi)$.

$$\tan 3\theta - \frac{1}{\tan 3\theta} = 0 \quad \text{or} \quad \tan^2 3\theta = 1 \quad \text{or} \quad \tan 3\theta = \pm 1$$

Thus,

$$3\theta = \frac{\pi}{4}, \frac{3\pi}{4}, \frac{5\pi}{4}, \frac{7\pi}{4}, \frac{9\pi}{4}, \frac{11\pi}{4}, \frac{13\pi}{4}, \frac{15\pi}{4}, \frac{17\pi}{4}, \frac{19\pi}{4}, \frac{21\pi}{4}, \frac{23\pi}{4}$$

Here we must include values of angles which when divided by 3 give angles between 0 and 2π. Thus, we need values of 3θ from 0 to 6π. The solutions are

$$\theta = \frac{\pi}{12}, \frac{\pi}{4}, \frac{5\pi}{12}, \frac{7\pi}{12}, \frac{3\pi}{4}, \frac{11\pi}{12}, \frac{13\pi}{12}, \frac{5\pi}{4}, \frac{17\pi}{12}, \frac{19\pi}{12}, \frac{7\pi}{4}, \frac{23\pi}{12}$$

It is noted that these values satisfy the original equation. Since we multiplied through by $\tan 3\theta$ in the solution, any value of θ which leads to $\tan 3\theta = 0$ would not be valid, since this would indicate division by zero in the original equation.

Example H

Solve the equation $\cos 3x \cos x + \sin 3x \sin x = 1$ $(0 \leq x < 2\pi)$.

   The left side of this equation is of the general form $\cos(A - x)$, where $A = 3x$. Therefore,

$$\cos 3x \cos x + \sin 3x \sin x = \cos(3x - x) = \cos 2x$$

The original equation becomes

$$\cos 2x = 1$$

This equation is satisfied if $2x = 0$ and $2x = 2\pi$. The solutions are $x = 0$ and $x = \pi$. Only through recognition of the proper trigonometric form can we readily solve this equation.

Exercises 19–5

In Exercises 1 through 32, solve the given trigonometric equations for values of x so that $0 \leq x < 2\pi$.

1. $\sin x - 1 = 0$
2. $2 \sin x + 1 = 0$
3. $\tan x + 1 = 0$
4. $2 \cos x + 1 = 0$
5. $2(2 + \cos x) = 3 + \cos x$
6. $4 \tan x + 2 = 3(1 + \tan x)$
7. $7 \sin x - 2 = 3(2 - \sin x)$
8. $3 - 4 \cos x = 7 - (2 - \cos x)$
9. $4 \cos^2 x - 1 = 0$
10. $\sin^2 x - 1 = 0$
11. $4 \sin^2 x - 3 = 0$
12. $3 \tan^2 x - 1 = 0$
13. $2 \sin^2 x - \sin x = 0$
14. $3 \cos x - 4 \cos^2 x = 0$
15. $\sin 4x - \cos 2x = 0$
16. $\sin 4x - \sin 2x = 0$
17. $\sin 2x \sin x + \cos x = 0$
18. $\cos 2x + \sin^2 x = 0$
19. $2 \sin x - \tan x = 0$
20. $\sin x - \sin \dfrac{x}{2} = 0$
21. $2 \cos^2 x - 2 \cos 2x - 1 = 0$
22. $2 \cos^2 2x + 1 = 3 \cos 2x$
23. $\sin^2 x - 2 \sin x - 1 = 0$
24. $\tan^2 x - 5 \tan x + 6 = 0$
25. $4 \tan x - \sec^2 x = 0$
26. $\tan^2 x - 2 \sec^2 x + 4 = 0$
27. $\sin 2x \cos x - \cos 2x \sin x = 0$
28. $\cos 3x \cos x - \sin 3x \sin x = 0$
29. $\sin 2x + \cos 2x = 0$
30. $2 \sin 4x + \csc 4x = 3$
31. $\tan x + 3 \cot x = 4$
32. $\sin x \sin \tfrac{1}{2}x = 1 - \cos x$

In Exercises 33 through 36, solve the indicated equations.

33. In finding the dimensions of the largest cylinder which can be inscribed in a sphere, the equation $\sin \theta - 3 \sin \theta \cos^2 \theta = 0$ must be solved for θ, for $0 < \theta < \pi/2$. Solve for the value of θ which satisfies the equation.

34. Vectors of magnitudes 3 and 2 are directed at an angle θ, such that the vertical component of the first vector equals the horizontal component of the second vector. Find the angle θ, such that $0 < \theta < \pi$.

35. The angular displacement θ of a certain pendulum in terms of the time t is given by $\theta = e^{-0.1t}(\cos 2t + 3\sin 2t)$. What is the smallest value of t for which the displacement is zero?

36. The vertical displacement y of an object at the end of a spring, which itself is being moved up and down, is given by $y = 2\cos 4t + \sin 2t$. Find the smallest value of t (in seconds) for which $y = 0$.

In Exercises 37 through 40, solve the given equations for $0 \le x < 2\pi$. Then solve each graphically and compare results.

37. $3\sin x - 1 = 0$

38. $4\cos x + 3 = 0$

39. $5\cos 2x + 1 = 0$

40. $4\sin 3x + 1 = 0$

In Exercises 41 through 44, solve the given equations graphically.

41. $\sin 2x = x$

42. $\cos 2x = 4 - x^2$

43. In the study of light diffraction, the equation $\tan \theta = \theta$ is found. Solve this equation for $0 \le \theta < 2\pi$.

44. An equation used in astronomy is $\theta - e\sin\theta = M$. Solve for θ for $e = 0.25$ and $M = 0.75$.

19–6 Introduction to the Inverse Trigonometric Functions

When we studied logarithms, we found that we often wished to change a given expression from exponential to logarithmic form, or from logarithmic to exponential form. Each of these forms has its advantages for particular purposes. We found that the exponential function $y = b^x$ can also be written in logarithmic form with x as a function of y, or $x = \log_b y$. We then represented both of these functions as y in terms of x, saying that the letter used for the dependent and independent variables did not matter, when we wished to express a functional relationship. Since it is standard to use y as the dependent variable and x as the independent variable, we wrote the logarithmic function as $y = \log_b x$.

These two functions, the exponential function $y = b^x$ and the logarithmic function $y = \log_b x$, are called **inverse functions.** *This means that if we solve for the independent variable in terms of the dependent variable in one, we will arrive at the functional relationship expressed by the other.* It also means that, for every value of x, there is only one corresponding value of y.

Just as we are able to solve $y = b^x$ for the exponent by writing it in logarithmic form, there are times when it is necessary to solve for the independent variable (the angle) in trigonometric functions. Therefore, we define the **inverse sine of x** by the relation

$$y = \arcsin x \quad \text{(the notation } y = \sin^{-1} x \text{ is also used)} \quad (19\text{--}26)$$

Similar relations exist for the other inverse trigonometric relations. In Eq. (19–26), x is the value of the sine of the angle y, and therefore the most meaningful way of reading it is ''y is the angle whose sine is x.''

Example A

$y = \arccos x$ is read as ''y is the angle whose cosine is x.''

The equation $y = \arctan 2x$ is read as ''y is the angle whose tangent is $2x$.''

It is important to emphasize that $y = \arcsin x$ and $x = \sin y$ express the same relationship between x and y. The advantage of having both forms is that a trigonometric relation may be expressed in terms of a function of an angle or in terms of the angle itself.

If we consider closely the equation $y = \arcsin x$ and possible values of x, we note that there are an unlimited number of possible values of y for a given value of x. Consider the following example.

Example B

For $y = \arcsin x$, if $x = \frac{1}{2}$, we have $y = \arcsin \frac{1}{2}$. This means that we are to find an angle whose sine is $\frac{1}{2}$. We know that $\sin (\pi/6) = \frac{1}{2}$. Therefore, $y = \pi/6$.

However, we also know that $\sin (5\pi/6) = \frac{1}{2}$. Therefore, $y = 5\pi/6$ is also a proper value. If we consider negative angles, such as $-7\pi/6$, or angles generated by additional rotations, such as $13\pi/6$, we conclude that there are an unlimited number of possible values for y.

To have a properly defined **function** *in mathematics, there must be only one value of the dependent variable for a given value of the independent variable. A* **relation,** *on the other hand, may have more than one such value.* Therefore, we see that $y = \arcsin x$ is not really a function, although it is properly a relation. It is necessary to restrict the values of y in order to define the **inverse trigonometric functions,** and this is done in the following section. It is the purpose of this section to introduce the necessary notation and to develop an understanding of the basic concept. The following examples further illustrate the meaning of the notation.

Example C

If $y = \arccos 0$, y is the angle whose cosine is zero. The smallest positive angle for which this is true is $\pi/2$. Therefore, $y = \pi/2$ is an acceptable value.

If $y = \arctan 1$, an acceptable value for y is $\pi/4$. This is the same as saying $\tan(\pi/4) = 1$.

Example D

If $y = \arcsin(-0.3279)$, a calculator in radian mode gives the value -0.3341, which is a fourth-quadrant angle expressed as a negative angle. The smallest positive angle for which $y = \arcsin(-0.3279)$ is found by adding 0.3341 to π, which gives the value 3.4757.

For $y = \arccos(-0.8249)$, a calculator in radian mode gives the value 2.5408, which is the smallest positive value. It is a second-quadrant angle.

Example E

Given that $y = \sec 2x$, solve for x.

We first express the inverse relation as $2x = \operatorname{arcsec} y$. Then we solve for x by dividing through by 2. Thus, we have $x = \frac{1}{2} \operatorname{arcsec} y$. Note that we first wrote the inverse relation by writing the expression for the angle, which in this case was

$2x$. Just as $\sec 2x$ and $2 \sec x$ are different relations, so are the arcsec $2x$ and $2 \operatorname{arcsec} x$.

Example F

Given that $4y = \operatorname{arccot} 2x$, solve for x.

Writing this as the cotangent of $4y$ (since the given expression means "$4y$ is the angle whose cotangent is $2x$"), we have

$$2x = \cot 4y \quad \text{or} \quad x = \frac{1}{2} \cot 4y$$

Example G

Given that $y = \pi - \operatorname{arccsc} \frac{1}{3}x$, solve for x.

Before solving for x, we must find the expression for $\operatorname{arccsc} \frac{1}{3}x$, which is

$$\operatorname{arccsc} \tfrac{1}{3}x = \pi - y$$

Then,

$$\frac{1}{3}x = \csc(\pi - y) \quad \text{or} \quad x = 3 \csc y$$

[since $\csc(\pi - y) = \csc y$].

Exercises 19–6

In Exercises 1 through 8, write down the meaning of each of the given equations. See Example A.

1. $y = \arctan x$
2. $y = \operatorname{arcsec} x$
3. $y = \operatorname{arccot} 3x$
4. $y = \operatorname{arccsc} 4x$
5. $y = 2 \arcsin x$
6. $y = 3 \arctan x$
7. $y = 5 \arccos 2x$
8. $y = 4 \arcsin 3x$

In Exercises 9 through 20, find the smallest positive angle (in terms of π) for each of the given expressions.

9. $\arccos \dfrac{1}{2}$
10. $\arcsin 1$
11. $\operatorname{arcsec}(-\sqrt{2})$

12. $\arccos \dfrac{\sqrt{2}}{2}$
13. $\arctan(-1)$
14. $\operatorname{arccsc}(-1)$

15. $\arctan \sqrt{3}$
16. $\operatorname{arcsec} 2$
17. $\operatorname{arccot}(-\sqrt{3})$

18. $\arcsin\left(-\dfrac{\sqrt{3}}{2}\right)$
19. $\operatorname{arccsc} \sqrt{2}$
20. $\operatorname{arccot} \dfrac{\sqrt{3}}{3}$

In Exercises 21 through 28, find the angle given by a calculator in radian mode and the smallest positive angle (if it is different) for each of the given expressions.

21. arccos 0.1942   22. arcsin 0.6308   23. arctan(-1.7652)

24. arccos(-0.7075)   25. arcsin(-0.0874)   26. arctan 6.8764

27. arcsin 0.2195   28. arccos(-0.1869)

In Exercises 29 through 36, solve the given equations for x.

29. $y = \sin 3x$

30. $y = \cos(x - \pi)$

31. $y = \arctan\left(\dfrac{x}{4}\right)$

32. $y = 2 \arcsin\left(\dfrac{x}{6}\right)$

33. $y = 1 + \sec 3x$

34. $4y = 5 - \csc 8x$

35. $1 - y = \arccos(1 - x)$

36. $2y = \text{arccot } 3x - 5$

In Exercises 37 through 40, determine the required quadrants.

37. In which quadrants is arcsin x if $0 < x < 1$?

38. In which quadrants is arctan x if $0 < x < 1$?

39. In which quadrants is arccos x if $-1 < x < 0$?

40. In which quadrants is arcsin x if $-1 < x < 0$?

In Exercises 41 through 44, solve the given problems with the use of the inverse trigonometric relations.

41. A body is moving along a straight line in such a way that its acceleration is directed toward a fixed point, and is proportional to its distance from that point. Its position is given by $x = A \cos (t \sqrt{k/m})$. Solve for k.

42. Under certain conditions the magnetic potential V at a distance r from a circuit with a current I is given by

$$V = \frac{kI \cos \theta}{r^2}$$

where k is a constant and θ is the angle between the radius vector r and the direction perpendicular to the plane of the circuit. Solve for θ.

43. The magnitude of a certain ray of polarized light is given by the equation $E = E_0 \cos \theta \sin \theta$. Solve for θ.

44. The equation for the displacement of a certain object oscillating at the end of a spring is given by $y = A \sin 2t + B \cos 2t$. Solve for t. (*Hint:* See Exercise 33 of Section 19–3.)

19–7 *The Inverse Trigonometric Functions*

We noted in the preceding section that we could find many values of y if we assumed some value for x in the relation $y = \arcsin x$. As these relations were defined, any one of the various possibilities would be considered correct. This, however, does not meet a basic requirement for a function, and it also leads to ambiguity. *In order to define the* **inverse trigonometric functions** *properly, so* *this ambiguity does not exist, there must be only a single value of y for any given value of x.* Therefore, the following values are defined for the given functions.

$$-\frac{\pi}{2} \le \text{Arcsin } x \le \frac{\pi}{2}, \; 0 \le \text{Arccos } x \le \pi, \; -\frac{\pi}{2} < \text{Arctan } x < \frac{\pi}{2}$$

$$0 < \text{Arccot } x < \pi, \; 0 \le \text{Arcsec } x \le \pi, \; -\frac{\pi}{2} \le \text{Arccsc } x \le \frac{\pi}{2}$$

(19–27)

This means that when we are looking for a value of y to correspond to a given value for x, we must use a value of y as defined in Eqs. (19–27). The capital letter designates the use of the inverse trigonometric *function*.

Example A

$$\text{Arcsin}\left(\frac{1}{2}\right) = \frac{\pi}{6}$$

This is the only value of the function which lies within the defined range. The value $5\pi/6$ is not correct, since it lies outside the defined range of values.

Example B

$$\text{Arccos}\left(-\frac{1}{2}\right) = \frac{2\pi}{3}$$

Other values such as $4\pi/3$ and $-2\pi/3$ are not correct, since they are not within the defined range of values for the function Arccos x.

Example C

$$\text{Arctan}(-1) = -\frac{\pi}{4}$$

This is the only value within the defined range for the function Arctan x. We must remember that *when x is negative for Arcsin x and Arctan x, the value of y is a fourth-quadrant angle, expressed as a **negative angle**.* This is a direct result of the definition.

Example D

$$\text{Arcsin}\left(-\frac{\sqrt{3}}{2}\right) = -\frac{\pi}{3} \qquad \text{Arccos}(-1) = \pi$$

$$\text{Arctan } 0 = 0 \qquad \text{Arctan}(\sqrt{3}) = \frac{\pi}{3}$$

Example E

By use of a calculator in radian mode, we find the following values.

$$\text{Arcsin } 0.6294 = 0.6808 \qquad \text{Arcsin}(-0.1568) = -0.1574$$

$$\text{Arccos}(-0.8026) = 2.5024 \qquad \text{Arctan}(-1.9268) = -1.0921$$

We note that the calculator gives values which are in the defined ranges of values for each function.

One might logically ask why these values are chosen when there are so many different possibilities. The values are so chosen that, if x is positive, the resulting answer gives an angle in the first quadrant. We must, however, account for the possibility that x might be negative. We could not choose second-quadrant angles for Arcsin x. Since the sine of a second-quadrant angle is also positive, this then would lead to ambiguity. The sine is negative for fourth-quadrant angles, and to have a continuous range of values we must express the fourth-quadrant angles in the form of negative angles. This range is also chosen for Arctan x, for similar reasons. However, Arccos x cannot be chosen in this way, since the cosines of fourth-quadrant angles are also positive. Thus, again to keep a continuous range of values for Arccos x, the second-quadrant angles are chosen for negative values of x.

As for the values for the other functions, we chose values such that if x is positive, the result is also an angle in the first quadrant. As for negative values of x, it rarely makes any difference, since either positive values of x arise, or we can use one of the other functions. Our definitions, however, are those which are generally used.

The graphs of the inverse trigonometric relations can be used to show the fact that many values of y correspond to a given value of x. We can also show the ranges used in defining the inverse trigonometric functions, and that these ranges are specific sections of the curves.

Since $y = \arcsin x$ and $x = \sin y$ are equivalent equations, we can obtain the graph of the inverse sine by sketching the sine curve *along the y-axis*. In Figs. (19–3), (19–4), and (19–5), the graphs of three inverse trigonometric relations are shown, with the heavier portions indicating the graphs of the inverse trigonometric functions. The graphs of the other inverse relations are found in the same way.

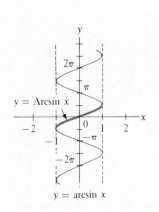

Figure 19–3

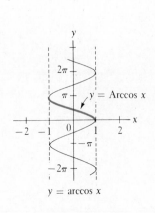

Figure 19–4

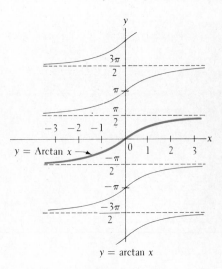

Figure 19–5

If we know the value of x for one of the inverse functions, we can find the trigonometric functions of the angle. If general relations are desired, a representative triangle is very useful. The following examples illustrate these methods.

Example F

Find cos(Arcsin 0.5). (Remember again: The inverse functions yield *angles.*)
     We know Arcsin 0.5 is a first-quadrant angle, since 0.5 is positive. Thus we find Arcsin $0.5 = \pi/6$. The problem now becomes one of finding $\cos(\pi/6)$. This is, of course, $\sqrt{3}/2$, or 0.8660.

Example G

$$\sin(\text{Arccot } 1) = \sin\frac{\pi}{4} = \frac{\sqrt{2}}{2} = 0.7071$$

$$\tan[\text{Arccos}(-1)] = \tan \pi = 0$$

Example H

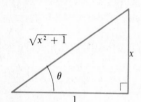

Figure 19–6

Find sin(Arctan x).
     We know that Arctan x is another way of stating "the angle whose tangent is x." Thus, let us draw a right triangle (as in Fig. 19–6) and label one of the acute angles θ, the side opposite θ as x, and the side adjacent to θ as 1. In this way we see that, by definition, $\tan \theta = x/1$, or $\theta = $ Arctan x, which means θ is the desired angle. By the Pythagorean theorem, the hypotenuse of this triangle is $\sqrt{x^2 + 1}$. Now we find that the sin θ, which is the same as sin (Arctan x), is $x/\sqrt{x^2 + 1}$, from the definition of the sine. Thus,

$$\sin(\text{Arctan } x) = \frac{x}{\sqrt{x^2 + 1}}$$

Example I

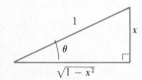

Figure 19–7

Find cos(2 Arcsin x).
     From Fig. 19–7, we see that $\theta = $ Arcsin x. From the double-angle formulas, we have

$$\cos 2\theta = 1 - 2 \sin^2\theta$$

Thus, since $\sin \theta = x$, we have

$$\cos(2 \text{ Arcsin } x) = 1 - 2x^2$$

Example J

For the triangle shown in Fig. 19–8, solve for angle A in terms of the given sides a, b, and c.

From the law of cosines we have

$$a^2 = b^2 + c^2 - 2bc \cos A$$

Solving for angle A, we have

$$2bc \cos A = b^2 + c^2 - a^2$$

$$\cos A = \frac{b^2 + c^2 - a^2}{2bc}$$

$$A = \text{Arccos}\left(\frac{b^2 + c^2 - a^2}{2bc}\right)$$

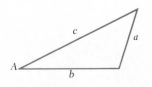

Figure 19–8

Exercises 19–7

In Exercises 1 through 20, evaluate the given expressions.

1. $\text{Arccos}(\frac{1}{2})$
2. $\text{Arcsin}(1)$
3. $\text{Arcsin } 0$

4. $\text{Arccos } 0$
5. $\text{Arctan}(-\sqrt{3})$
6. $\text{Arcsin}(-\frac{1}{2})$

7. $\text{Arcsec } 2$
8. $\text{Arccot } \sqrt{3}$
9. $\text{Arctan}\left(\dfrac{\sqrt{3}}{3}\right)$

10. $\text{Arctan } 1$
11. $\text{Arcsin}\left(-\dfrac{\sqrt{2}}{2}\right)$
12. $\text{Arccos}\left(-\dfrac{\sqrt{3}}{2}\right)$

13. $\text{Arccsc } \sqrt{2}$
14. $\text{Arccot } 1$
15. $\sin(\text{Arctan } \sqrt{3})$

16. $\tan\left(\text{Arcsin }\dfrac{\sqrt{2}}{2}\right)$
17. $\cos[\text{Arctan}(-1)]$
18. $\sec[\text{Arccos}(-\frac{1}{2})]$

19. $\cos(2 \text{ Arcsin } 1)$
20. $\sin(2 \text{ Arctan } 2)$

In Exercises 21 through 32, use a calculator to evaluate the given expressions.

21. $\text{Arctan}(-3.7321)$
22. $\text{Arccos}(-0.6561)$
23. $\text{Arcsin}(-0.8326)$

24. $\text{Arctan } 0.2846$
25. $\text{Arccos } 0.1291$
26. $\text{Arcsin } 0.2119$

27. $\text{Arctan } 8.2614$
28. $\text{Arcsin}(-0.8881)$
29. $\tan[\text{Arccos}(-0.6281)]$

30. $\cos[\text{Arctan}(-1.2256)]$
31. $\sin[\text{Arctan}(-0.2297)]$

32. $\tan[\text{Arcsin}(-0.3019)]$

In Exercises 33 through 40, find an algebraic expression for each of the expressions given.

33. $\tan(\text{Arcsin } x)$
34. $\sin(\text{Arccos } x)$
35. $\cos(\text{Arcsec } x)$

36. $\cot(\text{Arccot } x)$
37. $\sec(\text{Arccsc } 3x)$
38. $\tan(\text{Arcsin } 2x)$

39. $\sin(2 \text{ Arcsin } x)$
40. $\cos(2 \text{ Arctan } x)$

In Exercises 41 and 42, prove that the given expressions are equal. This can be done by use of the relation for sin $(\alpha + \beta)$ and by showing that the sine of the sum of angles on the left equals the sine of the angle on the right.

41. $\text{Arcsin} \dfrac{3}{5} + \text{Arcsin} \dfrac{5}{13} = \text{Arcsin} \dfrac{56}{65}$ 42. $\text{Arctan} \dfrac{1}{3} + \text{Arctan} \dfrac{1}{2} = \dfrac{\pi}{4}$

In Exercises 43 and 44, verify the given expressions.

43. $\text{Arcsin } 0.5 + \text{Arccos } 0.5 = \dfrac{\pi}{2}$

44. $\text{Arctan } \sqrt{3} + \text{Arccot } \sqrt{3} = \dfrac{\pi}{2}$

In Exercises 45 and 46, solve for angle A for the triangles in the given figures in terms of the given sides and angles.

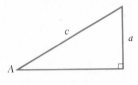

Figure 19–9

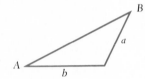

Figure 19–10

45. Figure 19–9   46. Figure 19–10

In Exercises 47 and 48, derive the given expressions.

47. For the triangle in Fig. 19–11 show that $\alpha = \text{Arctan}(a + \tan \beta)$.

48. Show that the length of the pulley belt indicated in Fig. 19–12 is given by the expression

$$L = 24 + 11\pi + 10 \, \text{Arcsin}\!\left(\dfrac{5}{13}\right)$$

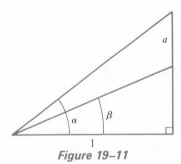

Figure 19–11

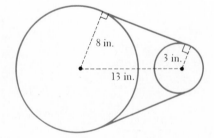

Figure 19–12

19–8 Exercises for Chapter 19

In Exercises 1 through 8, determine the values of the indicated functions in the given manner.

1. Find sin 120° by using 120° = 90° + 30°.
2. Find cos 30° by using 30° = 90° − 60°.
3. Find sin 135° by using 135° = 180° − 45°.
4. Find cos 225° by using 225° = 180° + 45°.
5. Find cos 180° by using 180° = 2(90°).
6. Find sin 180° by using 180° = 2(90°).
7. Find sin 45° by using 45° = $\frac{1}{2}$(90°).
8. Find cos 45° by using 45° = $\frac{1}{2}$(90°).

In Exercises 9 through 16, simplify the given expressions by use of one of the basic formulas of the chapter. Then use a calculator to verify the result by finding the value of the original expression and the value of the simplified expression.

9. sin 14° cos 38° + cos 14° sin 38°
10. $\cos^2 148° − \sin^2 148°$
11. 2 sin 46° cos 46°
12. cos 73° cos 142° + sin 73° sin 142°
13. $\sqrt{\dfrac{1 + \cos 12°}{2}}$
14. cos 3° cos 215° − sin 3° sin 215°
15. $1 − 2 \sin^2 82°$
16. $\sqrt{\dfrac{1 − \cos 166°}{2}}$

In Exercises 17 through 24, simplify each of the given expressions. Expansion of any term is not necessary; proper recognition of the form of the expression leads to the proper result.

17. sin 2x cos 3x + cos 2x sin 3x
18. cos 7x cos 3x + sin 7x sin 3x
19. 8 sin 6x cos 6x
20. 10 sin 5x cos 5x
21. $2 − 4 \sin^2 6x$
22. $\cos^2 2x − \sin^2 2x$
23. $\sqrt{2 + 2 \cos 2x}$
24. $\sqrt{32 − 32 \cos 4x}$

In Exercises 25 through 32, evaluate the given expressions.

25. Arcsin(-1)
26. Arcsec $\sqrt{2}$
27. Arccos(0.9659)
28. Arctan(-0.6249)
29. tan[Arcsin($-\frac{1}{2}$)]
30. cos[Arctan($-\sqrt{3}$)]
31. Arcsin(tan π)
32. Arccos$\left[\tan\left(-\dfrac{\pi}{4} \right) \right]$

In Exercises 33 through 60, prove the given identities.

33. $\dfrac{\sec y}{\csc y} = \tan y$
34. cos θ csc θ = cot θ
35. sin x(csc x − sin x) = $\cos^2 x$
36. cos y(sec y − cos y) = $\sin^2 y$
37. $\dfrac{1}{\sin \theta} − \sin \theta = \cot \theta \cos \theta$
38. sin θ sec θ csc θ cos θ = 1

39. $\cos\theta\cot\theta + \sin\theta = \csc\theta$

40. $\dfrac{\sin x\cot x + \cos x}{\cot x} = 2\sin x$

41. $\dfrac{\sec^4 x - 1}{\tan^2 x} = 2 + \tan^2 x$

42. $\cos^2 y - \sin^2 y = \dfrac{1 - \tan^2 y}{1 + \tan^2 y}$

43. $2\csc 2x\cot x = 1 + \cot^2 x$

44. $\cos^8 x - \sin^8 x = (\cos^4 x + \sin^4 x)\cos 2x$

45. $\dfrac{1 - \sin^2\theta}{1 - \cos^2\theta} = \cot^2\theta$

46. $\dfrac{1 + \cos 2\theta}{\cos^2\theta} = 2$

47. $\dfrac{\cos 2\theta}{\cos^2\theta} = 1 - \tan^2\theta$

48. $\dfrac{\sin 2\theta\sec\theta}{2} = \sin\theta$

49. $\sin\dfrac{\theta}{2}\cos\dfrac{\theta}{2} = \dfrac{\sin\theta}{2}$

50. $\sin\dfrac{x}{2} = \dfrac{\sec x - 1}{2\sec x\sin\left(\dfrac{x}{2}\right)}$

51. $\sec x + \tan x = \dfrac{\cos x}{1 - \sin x}$

52. $\dfrac{\cos\theta - \sin\theta}{\cos\theta + \sin\theta} = \dfrac{\cot\theta - 1}{\cot\theta + 1}$

53. $\cos(x - y)\cos y - \sin(x - y)\sin y = \cos x$

54. $\sin 3y\cos 2y - \cos 3y\sin 2y = \sin y$

55. $\sin 4x(\cos^2 2x - \sin^2 2x) = \dfrac{\sin 8x}{2}$

56. $\csc 2x + \cot 2x = \cot x$

57. $\dfrac{\sin x}{\csc x - \cot x} = 1 + \cos x$

58. $\cos x - \sin\dfrac{x}{2} = \left(1 - 2\sin\dfrac{x}{2}\right)\left(1 + \sin\dfrac{x}{2}\right)$

59. $\dfrac{\sin(x + y) + \sin(x - y)}{\cos(x + y) + \cos(x - y)} = \tan x$

60. $\sec\dfrac{x}{2} + \csc\dfrac{x}{2} = \dfrac{2\left(\sin\dfrac{x}{2} + \cos\dfrac{x}{2}\right)}{\sin x}$

In Exercises 61 through 64, solve for x.

61. $y = 2\cos 2x$

62. $y - 2 = 2\tan\left(x - \dfrac{\pi}{2}\right)$

63. $y = \dfrac{\pi}{4} - 3\arcsin 5x$

64. $2y = \text{arcsec } 4x - 2$

In Exercises 65 through 76, solve the given equations for x such that $0 \le x < 2\pi$.

65. $3(\tan x - 2) = 1 + \tan x$

66. $5\sin x = 3 - (\sin x + 2)$

67. $2(1 - 2\sin^2 x) = 1$

68. $\sec^2 x = 2\tan x$

69. $\cos^2 2x - 1 = 0$

70. $2\sin 2x + 1 = 0$

71. $4\cos^2 x - 3 = 0$

72. $\cos 2x = \sin x$

73. $\sin^2 x - \cos^2 x + 1 = 0$

74. $\cos 3x\cos x + \sin 3x\sin x = 0$

75. $\sin^2\left(\dfrac{x}{2}\right) - \cos x + 1 = 0$

76. $\sin x + \cos x = 1$

In Exercises 77 through 80, find an algebraic expression for each of the expressions.

77. $\tan(\text{Arccot } x)$

78. $\cos(\text{Arccsc } x)$

79. $\sin(2 \text{ Arccos } x)$

80. $\cos(\pi - \text{Arctan } x)$

In Exercises 81 through 84, use the given substitutions to show that the given equations are valid. (In each, $0 \le \theta < \pi/2$.)

81. If $x = \cos \theta$, show that $\sqrt{1 - x^2} = \sin \theta$

82. If $x = 2 \sec \theta$, show that $\sqrt{x^2 - 4} = 2 \tan \theta$

83. If $x = \tan \theta$, show that $\dfrac{x}{\sqrt{1 + x^2}} = \sin \theta$

84. If $x = \cos \theta$, show that $\dfrac{\sqrt{1 - x^2}}{x} = \tan \theta$

In Exercises 85 through 96, use the formulas and methods of this chapter to solve the given problems.

85. In finding the area between a section of the curve of $y = \cos^3 x$ and the x-axis, it is necessary to transform $\cos^3 x$ to $\cos x - \sin^2 x \cos x$. Show that this is valid.

86. In the theory dealing with the motion of fluid in cylinders, the expression

$$4 \sin^2\alpha \cos^2\alpha + (\cos^2\alpha - \sin^2\alpha)^2$$

is found. Simplify this expression.

87. If the area bounded by $y = \sin x$ between 0 and π and the x-axis is rotated about the x-axis, a volume is generated. In finding this volume it is necessary to change $\sin^2 x$ into $\frac{1}{2}(1 - \cos 2x)$. Show that this change is valid.

88. In surveying, when determining an azimuth (a measure used for reference purposes), it might be necessary to simplify the expression

$$\frac{1}{2 \cos \alpha \cos \beta} - \tan \alpha \tan \beta$$

Perform this operation by expressing it in the simplest possible form when $\alpha = \beta$.

89. An object is under the influence of a central force (an example of one type of central force is the attraction of the sun for the earth). The y-coordinate of its path is given by $y = 20 \sin \frac{1}{3}t$. Solve for t.

90. In finding the pressure exerted by soil on a retaining wall, the expression $1 - \cos 2\phi - \sin 2\phi \tan \alpha$ is found. Show that this expression can also be written as $2 \sin \phi(\sin \phi - \cos \phi \tan \alpha)$.

91. In the theory dealing with the reflection of light, the expression

$$\frac{\cos (\phi + \alpha)}{\cos (\phi - \alpha)}$$

is found. Express this in terms of functions of α if $\phi = 2\alpha$.

92. In developing an expression for the power in an alternating-current circuit, the expression $\sin \omega t \sin(\omega t + \phi)$ is found. Show that this expression can be written as $\frac{1}{2}[\cos \phi - \cos(2\omega t + \phi)]$.

93. In the theory of interference of light, the expression $1 - 2r^2\cos \beta + r^4$ is found. Show that this expression can be written as $(1 - r^2)^2 + 4r^2\sin^2(\beta/2)$.

94. In the theory of diffraction of light, an equation found is

$$y = R \sin 2\pi\left(\frac{t}{T} - \frac{a}{\lambda}\right)\cos \alpha - R \cos 2\pi\left(\frac{t}{T} - \frac{a}{\lambda}\right)\sin \alpha$$

Show that

$$y = R \sin 2\pi\left(\frac{t}{T} - \frac{a}{\lambda} - \frac{\alpha}{2\pi}\right)$$

95. If a plane surface inclined at angle θ moves horizontally, the angle for which the lifting force of the air is a maximum is found by solving the equation $2 \sin \theta \cos^2\theta - \sin^3\theta = 0$, where $0 < \theta < 90°$. Solve for θ.

96. The electric current as a function of the time for a particular circuit is given by $i = 8e^{-20t}(\sqrt{3} \cos 10t - \sin 10t)$. Find the time in seconds when the current is first zero.

CHAPTER 20

Plane Analytic Geometry

We first introduced the graph of a function in Chapter 2, and since that time we have made extensive use of graphs for representing functions. In Chapter 4 we showed the graph of the linear function and in Chapter 6 we graphed the quadratic function. Also, we have used graphs to represent the trigonometric functions, the exponential and logarithmic functions, and the inverse trigonometric functions. We have also seen how graphs may be used in solving equations and systems of equations. In this chapter we shall consider certain basic principles relating to the graphs of functions. We shall also further show how certain graphs can be constructed by recognizing the form of the equation.

The underlying principle of analytic geometry is the relationship of geometry to algebra. A great deal can be learned about a geometric figure if we can find the function which represents its graph. Also, by analyzing certain characteristics of a function, we can obtain useful information as to the nature of its graph.

It is necessary when studying many of the concepts of the calculus to have the ability to recognize certain curves and their basic characteristics. Also, curves have technical applications, many of which are illustrated or indicated in the examples and exercises in this chapter.

20–1 Basic Definitions

In this section we shall develop certain basic concepts which will be needed for future use in establishing the proper relationships between an equation and a curve.

The first of these concepts involves the distance between any two points in the coordinate plane. If these two points lie on a line parallel to the x-axis, the **directed distance** from the first point $A(x_1, y)$ to the second point $B(x_2, y)$ is denoted as $\overline{AB}$ and is defined as $x_2 - x_1$. We can see that $\overline{AB}$ is *positive if B is to the right of A, and it is negative if B is to the left of A.* Similarly, the directed distance between two points $C(x, y_1)$ and $D(x, y_2)$ on a line parallel to the y-axis is $y_2 - y_1$. We see that $\overline{CD}$ *is positive if D is above C, and it is negative if D is below C.*

Example A

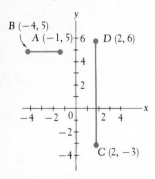

Figure 20–1

The line segment joining $A(-1, 5)$ and $B(-4, 5)$ in Fig. 20–1 is parallel to the x-axis. Therefore, the directed distance

$$\overline{AB} = -4 - (-1) = -3$$

and the directed distance

$$\overline{BA} = -1 - (-4) = 3$$

Also in Fig. 20–1, the line segment joining $C(2, -3)$ and $D(2, 6)$ is parallel to the y-axis. The directed distance

$$\overline{CD} = 6 - (-3) = 9$$

and the directed distance

$$\overline{DC} = -3 - 6 = -9$$

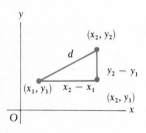

Figure 20–2

We now wish to find the length of a line segment joining any two points in the plane. If these points are on a line which is not parallel to either of the axes (Fig. 20–2), we must use the Pythagorean theorem to find the distance between them. By making a right triangle with the line segment joining the two points as the hypotenuse, and line segments parallel to the axes as the legs, we have the formula which gives the distance between any two points in the plane. This formula, called *the* **distance formula,** *is*

$$d = \sqrt{(x_2 - x_1)^2 + (y_2 - y_1)^2} \tag{20–1}$$

Here we choose the positive square root since we are concerned only with the magnitude of the length of the line segment.

Example B

The distance between $(3, -1)$ and $(-2, -5)$ is given by

$$d = \sqrt{[(-2) - 3]^2 + [(-5) - (-1)]^2}$$
$$= \sqrt{(-5)^2 + (-4)^2} = \sqrt{25 + 16} = \sqrt{41}$$

It makes no difference which point is chosen as (x_1, y_1) and which is chosen as (x_2, y_2), since the difference in the x-coordinates (and y-coordinates) is squared. We also obtain $\sqrt{41}$ if we set up the distance as

$$d = \sqrt{[3 - (-2)]^2 + [(-1) - (-5)]^2}$$

Another important quantity for a line is its *slope,* which we first defined in Chapter 4. Here we shall give it a somewhat more general definition and develop its meaning in more detail, as well as review the basic meaning.

The **slope** *gives a measure of the direction of a line, and is defined as the vertical directed distance from one point to another on the same straight line, divided by the horizontal directed distance from the first point to the second.* Thus, the slope, *m,* is given by

$$m = \frac{y_2 - y_1}{x_2 - x_1} \qquad\qquad (20\text{--}2)$$

Example C

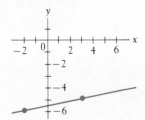

Figure 20–3

The slope of the line joining $(3, -5)$ and $(-2, -6)$ is

$$m = \frac{-6 - (-5)}{-2 - 3} = \frac{-6 + 5}{-5} = \frac{1}{5}$$

See Fig. 20–3. Again we may interpret either of the points as (x_1, y_1) and the other as (x_2, y_2). We can also obtain the slope of this same line from

$$m = \frac{-5 - (-6)}{3 - (-2)} = \frac{1}{5}$$

The larger the numerical value of the slope of a line, the more nearly vertical is the line. Also, *a line rising to the right has a positive slope, and a line falling to the right has a negative slope.*

Example D

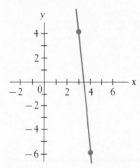

Figure 20–4

The line through the two points in Example C has a positive slope, which is numerically small. From Fig. 20–3 it can be seen that the line rises slightly to the right.

The line joining $(3, 4)$ and $(4, -6)$ has a slope of

$$m = \frac{4 - (-6)}{3 - 4} = -10$$

This line falls sharply to the right (see Fig. 20–4).

From the definition of slope, we may conclude that the slope of a line parallel to the *y*-axis cannot be defined (the *x*-coordinates of any two points on the line

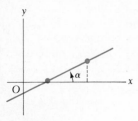

Figure 20-5

would be the same, which would then necessitate division by zero). This, however, does not prove to be of any trouble.

If a given line is extended indefinitely in either direction, it must cross the x-axis at some point unless it is parallel to the x-axis. *The angle measured from the x-axis in a positive direction to the line is called the* **inclination** *of the line* (see Fig. 20–5). The inclination of a line parallel to the x-axis is defined to be zero. *An alternative definition of slope, in terms of the inclination, is*

$$m = \tan \alpha, \qquad 0° \leq \alpha < 180° \tag{20-3}$$

where α is the inclination. This can be seen from the fact that the slope can be defined in terms of any two points on the line. Thus, if we choose as one of these points that point where the line crosses the x-axis and any other point, we see from the definition of the tangent of an angle that Eq. (20–3) is in agreement with Eq. (20–2).

Example E

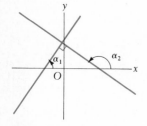

Figure 20-6

The slope of a line with an inclination of 45° is

$$m = \tan 45° = 1.000$$

The slope of a line having an inclination of 120° is

$$m = \tan 120° = -\sqrt{3} = -1.732$$

See Fig. 20–6.

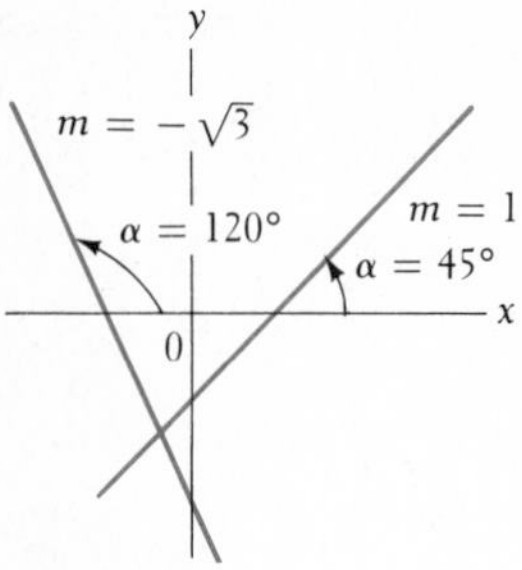

Any two parallel lines crossing the x-axis will have the same inclination. Therefore, *the slopes of parallel lines are equal.* This can be stated as

$$m_1 = m_2 \qquad \text{for } \| \text{ lines} \tag{20-4}$$

If two lines are perpendicular, this means that there must be 90° between their inclinations (Fig. 20–7). The relation between their inclinations is

$$\alpha_2 = \alpha_1 + 90°$$

which can be written as

$$90° - \alpha_2 = -\alpha_1$$

Taking the tangent of each of the angles in this last relation, we have

$$\tan(90° - \alpha_2) = \tan(-\alpha_1)$$

Figure 20-7

or

$$\cot \alpha_2 = -\tan \alpha_1$$

since a function of the complement of an angle is equal to the cofunction of that angle (see Section 3–4), and since $\tan(-\alpha) = -\tan \alpha$ (see Exercise 53 of Section 7–2). But $\cot \alpha = 1/\tan \alpha$, which means that $1/\tan \alpha_2 = -\tan \alpha_1$. Using the inclination definition of slope, we may write, as *the relation between slopes of perpendicular lines,*

$$m_2 = -\frac{1}{m_1} \quad \text{or} \quad m_1 m_2 = -1 \qquad \text{for } \perp \text{ lines} \qquad (20\text{–}5)$$

Example F

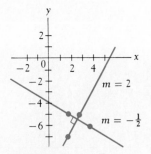

Figure 20–8

The line through $(3, -5)$ and $(2, -7)$ has a slope of

$$m_1 = \frac{-5 + 7}{3 - 2} = 2$$

The line through $(4, -6)$ and $(2, -5)$ has a slope of

$$m_2 = \frac{-6 - (-5)}{4 - 2} = -\frac{1}{2}$$

These lines are perpendicular. See Fig. 20–8.

Using the formulas for distance and slope, we can show certain basic geometric relations. The following examples illustrate the use of the formulas, and thus show the use of algebra in solving problems which are basically geometric. This illustrates the methods of analytic geometry.

Example G

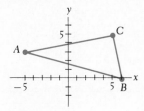

Figure 20–9

Show that line segments joining $A(-5, 3)$, $B(6, 0)$, and $C(5, 5)$ form a right triangle (see Fig. 20–9).

If these points are vertices of a right triangle, the slopes of two of the sides must be negative reciprocals. This would show perpendicularity. Thus, we find the slopes of the three lines to be

$$m_{AB} = \frac{3 - 0}{-5 - 6} = -\frac{3}{11}, \qquad m_{AC} = \frac{3 - 5}{-5 - 5} = \frac{1}{5}, \qquad m_{BC} = \frac{0 - 5}{6 - 5} = -5$$

We see that the slopes of AC and BC are negative reciprocals, which means that $AC \perp BC$. From this we can conclude that the triangle is a right triangle.

Example H

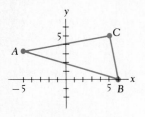

Figure 20–9

Find the area of the triangle in Example G.

Since the right angle is at C, the legs of the triangle are AC and BC. The area is one-half the product of the lengths of the legs of a right triangle. The lengths of the legs are

$$d_{AC} = \sqrt{(-5-5)^2 + (3-5)^2} = \sqrt{104} = 2\sqrt{26}$$

and

$$d_{BC} = \sqrt{(6-5)^2 + (0-5)^2} = \sqrt{26}$$

Therefore, the area is

$$A = \frac{1}{2}(2\sqrt{26})(\sqrt{26}) = 26$$

Exercises 20–1

In Exercises 1 through 10, find the distances between the given pairs of points.

1. $(3, 8)$ and $(-1, -2)$
2. $(-1, 3)$ and $(-8, -4)$
3. $(4, -5)$ and $(4, -8)$
4. $(-3, 7)$ and $(2, 10)$
5. $(-1, 0)$ and $(5, -7)$
6. $(15, -1)$ and $(-11, 1)$
7. $(-4, -3)$ and $(3, -3)$
8. $(-2, 5)$ and $(-2, -2)$
9. $(1.22, -3.45)$ and $(-1.07, -5.16)$
10. $(-5.6, 2.3)$ and $(8.2, -7.5)$

In Exercises 11 through 20, find the slopes of the lines through the points in Exercises 1 through 10.

In Exercises 21 through 24, find the slopes of the lines with the given inclinations.

21. $30°$
22. $60°$
23. $150°$
24. $135°$

In Exercises 25 through 28, find the inclinations of the lines with the given slopes.

25. 0.3640
26. 0.8243
27. -6.691
28. -1.428

In Exercises 29 through 32, determine whether the lines through the two pairs of points are parallel or perpendicular.

29. $(6, -1)$ and $(4, 3)$, and $(-5, 2)$ and $(-7, 6)$
30. $(-3, 9)$ and $(4, 4)$, and $(9, -1)$ and $(4, -8)$
31. $(-1, -4)$ and $(2, 3)$, and $(-5, 2)$ and $(-19, 8)$
32. $(-1, -2)$ and $(3, 6)$, and $(2, -6)$ and $(5, 0)$

In Exercises 33 through 36, determine the value of k.

33. The distance between $(-1, 3)$ and $(11, k)$ is 13.
34. The distance between $(k, 0)$ and $(0, 2k)$ is 10.
35. The points $(6, -1)$, $(3, k)$, and $(-3, -7)$ are all on the same line.
36. The points in Exercise 35 are the vertices of a right triangle, with the right angle at $(3, k)$.

In Exercises 37 through 40, show that the given points are vertices of the indicated geometric figures.

37. Show that the points (2, 3), (4, 9), and (−2, 7) are the vertices of an isosceles triangle.

38. Show that (−1, 3), (3, 5), and (5, 1) are the vertices of a right triangle.

39. Show that (3, 2), (7, 3), (−1, −3), and (3, −2) are the vertices of a parallelogram.

40. Show that (−5, 6), (0, 8), (−3, 1), and (2, 3) are the vertices of a square.

In Exercises 41 and 42, find the indicated areas.

41. Find the area of the triangle of Exercise 38.

42. Find the area of the square of Exercise 40.

In Exercises 43 and 44, use the following information to find the midpoints between the given points on a straight line.

The *midpoint* between points (x_1, y_1) and (x_2, y_2) on a straight line is the point

$$\left(\frac{x_1 + x_2}{2}, \frac{y_1 + y_2}{2} \right)$$

43. (−4, 9) and (6, 1)

44. (−1, 6) and (−13, −8)

20-2 The Straight Line

In Chapter 4 we derived the *slope-intercept form* of the equation of the straight line. Here we shall extend the development to include other forms of the equation of the straight line. Also, other methods of finding and applying these equations will be shown. For completeness, we shall review some of the material in Chapter 4.

Using the definition of slope, we can derive the general type of equation which always represents a straight line. This is another basic method of analytic geometry. That is, equations of a particular form can be shown to represent a particular type of curve. When we recognize the form of the equation, we know the kind of curve it represents. This can be of great assistance in sketching the graph.

A straight line can be defined as a "curve" with constant slope. By this we mean that for any two different points chosen on a given line, if the slope is calculated, the same value is always found. Thus, if we consider one point (x_1, y_1) on a line to be fixed (Fig. 20–10), and another point $P(x, y)$ which can *represent* any other point on the line, we have

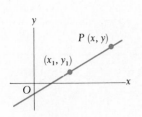

Figure 20–10

$$m = \frac{y - y_1}{x - x_1}$$

which can be written as

$$y - y_1 = m(x - x_1) \tag{20-6}$$

Equation (20–6) is known as the **point-slope form** *of the equation of a straight line.* It is useful when we know the slope of a line and some point through which the line passes. Direct substitutions can then give us the equation of the line. Such information is often available about a given line.

Example A

Find the equation of the line which passes through $(-4, 1)$ with a slope of -2. By using Eq. (20–6), we find that

$$y - 1 = (-2)(x + 4)$$

which can be simplified to

$$y + 2x + 7 = 0$$

This line is shown in Fig. 20–11.

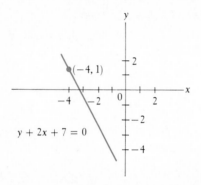

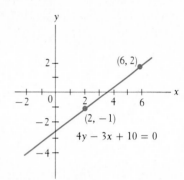

Figure 20–11   *Figure 20–12*

Example B

Find the equation of the line through $(2, -1)$ and $(6, 2)$. We first find the slope of the line through these points:

$$m = \frac{2 + 1}{6 - 2} = \frac{3}{4}$$

Then, by using either of the two known points and Eq. (20–6), we can find the equation of the line:

$$y + 1 = \frac{3}{4}(x - 2)$$

or

$$4y + 4 = 3x - 6$$

or

$$4y - 3x + 10 = 0$$

This line is shown in Fig. 20–12.

Equation (20–6) can be used for any line except one parallel to the y-axis. Such a line has an undefined slope. However, it does have the property that all points have the same x-coordinate, regardless of the y-coordinate. *We represent a line parallel to the y-axis as*

$$x = a \qquad\qquad (20\text{–}7)$$

A line parallel to the x-axis has a slope of 0. From Eq. (20–6), we can find its equation to be $y = y_1$. To keep the same form as Eq. (20–7), we normally write this as

$$y = b \qquad\qquad (20\text{–}8)$$

Example C

The line $x = 2$ is a line parallel to the y-axis and two units to the right of it. This line is shown in Fig. 20–13.

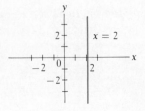

Figure 20–13   Figure 20–14

Example D

The line $y = -4$ is a line parallel to the x-axis and 4 units below it. This line is shown in Fig. 20–14.

If we choose the special point $(0, b)$, which is the y-intercept of the line, as the point to use in Eq. (20–6), we have

$$y - b = m(x - 0)$$

or

$$y = mx + b \qquad\qquad (20\text{–}9)$$

Equation (20–9) is the **slope-intercept form** *of the equation of a straight line*, and we first derived it in Chapter 4. Its primary usefulness lies in the fact that once we find the equation of a line and then write it in slope-intercept form, we know that the slope of the line is the coefficient of the x-term and that it crosses the y-axis with the coordinate indicated by the constant term.

Example E

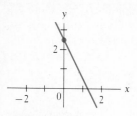

Figure 20–15

Find the slope and the y-intercept of the straight line whose equation is $2y + 4x - 5 = 0$.

We write this equation in slope-intercept form:

$$2y = -4x + 5$$

$$y = -2x + \frac{5}{2}$$

Since the coefficient of x in this form is -2, the slope of the line is -2. The constant on the right is $\frac{5}{2}$, which means that the y-intercept is $\frac{5}{2}$. See Fig. 20–15.

From Eqs. (20–6) and (20–9), and from the examples of this section, we see that the equation of the straight line has certain characteristics: we have a term in y, a term in x, and a constant term if we simplify as much as possible. *This form is represented by the equation*

$$\underline{Ax + By + C = 0}$$ (20–10)

which is known as the **general form** *of the equation of the straight line.* We have seen this form before in Chapter 4. Now we have shown why it represents a straight line.

Example F

Find the general form of the equation of the line which is parallel to the line $3x + 2y - 6 = 0$ and which passes through the point $(-1, 2)$.

Since the line whose equation we want is parallel to the line $3x + 2y - 6 = 0$, it must have the same slope. Therefore, writing $3x + 2y - 6 = 0$ in slope-intercept form, we have

$$3x + 2y - 6 = 0$$

$$2y = -3x + 6$$

$$y = -\frac{3}{2}x + 3$$

Therefore, the slope of $3x + 2y - 6 = 0$ is $-\frac{3}{2}$, which means that the slope of the required line is also $-\frac{3}{2}$. We use $m = -\frac{3}{2}$, the given point $(-1, 2)$, and the point-slope form to find the equation.

$$y - 2 = -\frac{3}{2}(x + 1)$$

$$2y - 4 = -3(x + 1)$$

$$2y - 4 = -3x - 3$$

$$3x + 2y - 1 = 0$$

This last form is the general form of the equation of the line. This line and the line $3x + 2y - 6 = 0$ are shown in Fig. 20–16.

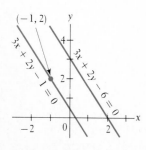

Figure 20–16

In many physical situations a linear relationship exists between variables. A few examples of situations where such relationships exist are between (1) the distance traveled by an object and the elapsed time, when the velocity is constant, (2) the amount a spring stretches and the force applied, (3) the change in electric resistance and the change in temperature, (4) the force applied to an object and the resulting acceleration, and (5) the pressure at a certain point within a liquid and the depth of the point. The following example illustrates the use of a straight line in dealing with an applied problem.

Example G

Under the condition of constant acceleration, the velocity of an object varies linearly with the time. If after 1 s a certain object has a velocity of 40 ft/s, and 3 s later it has a velocity of 55 ft/s, find the equation relating the velocity and time, and graph this equation. From the graph determine the initial velocity (the velocity when $t = 0$) and the velocity after 6 s.

If we treat the velocity v as the dependent variable, and the time t as the independent variable, the slope of the straight line is

$$m = \frac{v_2 - v_1}{t_2 - t_1}$$

Using the information given in the problem, we have

$$m = \frac{55 - 40}{4 - 1} = 5$$

Then, using the point-slope form of the equation of a straight line, we have $v - 40 = 5(t - 1)$, or $v = 5t + 35$, which is the required equation (see Fig. 20–17). For purposes of graphing the line, the values given are sufficient. Of course, there is no need to include negative values of t, since these have no physical meaning. From the graph we see that the line crosses the v-axis at 35. This means that the initial velocity is 35 ft/s. Also, when $t = 6$ we see that $v = 65$ ft/s.

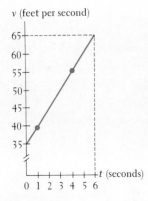

Figure 20–17

Exercises 20–2

In Exercises 1 through 20, find the equation of each of the lines with the given properties.

1. Passes through $(-3, 8)$ with a slope of 4.
2. Passes through $(-2, -1)$ with a slope of -2.
3. Passes through $(-2, -5)$ and $(4, 2)$.
4. Passes through $(-3, -5)$ and $(-2, 3)$.
5. Passes through $(1, 3)$ and has an inclination of $45°$.
6. Has a y-intercept of -2 and an inclination of $120°$.
7. Passes through $(6, -3)$ and is parallel to the x-axis.
8. Passes through $(-4, -2)$ and is perpendicular to the x-axis.
9. Is parallel to the y-axis and is 3 units to the left of it.
10. Is parallel to the x-axis and is 5 units below it.

11. Has an x-intercept of 4 and a y-intercept of -6.

12. Has an x-intercept of -3 and a slope of 2.

13. Is perpendicular to a line with a slope of 3 and passes through $(1, -2)$.

14. Is perpendicular to a line with a slope of -4 and has a y-intercept of 3.

15. Is parallel to a line with a slope of $-\frac{1}{2}$ and has an x-intercept of 4.

16. Is parallel to a line through $(-1, 2)$ and $(3, 1)$ and passes through $(1, 2)$.

17. Is parallel to a line through $(7, -1)$ and $(4, 3)$ and has a y-intercept of -2.

18. Is perpendicular to the line joining $(4, 2)$ and $(3, -5)$ and passes through $(4, 2)$.

19. Is perpendicular to the line $5x - 2y - 3 = 0$ and passes through $(3, -4)$.

20. Is parallel to the line $2y - 6x - 5 = 0$ and passes through $(-4, -5)$.

In Exercises 21 through 24, draw the lines with the given equations.

21. $4x - y = 8$ 22. $2x - 3y - 6 = 0$

23. $3x + 5y - 10 = 0$ 24. $4y = 6x - 9$

See Appendix E for a computer program for graphing a straight line.

In Exercises 25 through 28, reduce the equations to slope-intercept form and determine the slope and y-intercept.

25. $3x - 2y - 1 = 0$ 26. $4x + 2y - 5 = 0$

27. $5x - 2y + 5 = 0$ 28. $6x - 3y - 4 = 0$

In Exercises 29 through 32, determine the value of k.

29. What is the value of k if the lines $4x - ky = 6$ and $6x + 3y + 2 = 0$ are to be parallel?

30. What must k equal in Exercise 29 if the given lines are to be perpendicular?

31. What must be the value of k if the lines $3x - y = 9$ and $kx + 3y = 5$ are to be perpendicular?

32. What must k equal in Exercise 31 if the given lines are to be parallel?

In Exercises 33 and 34, show that the given lines are parallel. In Exercises 35 and 36, show that the given lines are perpendicular.

33. $3x - 2y + 5 = 0$ and $4y = 6x - 1$

34. $3y - 2x = 4$ and $6x - 9y = 5$

35. $6x - 3y - 2 = 0$ and $x + 2y - 4 = 0$

36. $4x - y + 2 = 0$ and $2x + 8y - 1 = 0$

In Exercises 37 through 40, find the equations of the given lines.

37. The line which has an x-intercept of 4 and a y-intercept the same as the line $2y - 3x - 4 = 0$.

38. The line which is perpendicular to the line $8x + 2y - 3 = 0$ and has the same x-intercept as this line.

39. The line with a slope of -3 which also passes through the intersection of the lines $5x - y = 6$ and $x + y = 12$.

40. The line which passes through the point of intersection of $2x + y - 3 = 0$ and $x - y - 3 = 0$ and through the point $(4, -3)$.

In Exercises 41 through 52, some applications and methods involving straight lines are shown.

41. The length l of a pulley belt is 8 cm longer than twice the circumference c of one of the pulley wheels. Express l as a function of c.

42. An acid solution is made from x liters of a 20% solution and y liters of a 30% solution. If a total of 20 L is made, find the equation relating x and y.

43. One microcomputer printer can print x characters per second and a second printer can print y characters per second. If the first prints for 50 s and the second for 60 s and they print a total of 12,200 characters, find the equation relating x and y.

44. A light ray travels (from the left) along the line $x + y = 1$ and reflects off the x-axis. Write the equation for the reflected ray.

45. The average velocity of an object is defined as the change in displacement s divided by the corresponding change in time t. Find the equation relating the displacement s and time t for an object for which the average velocity is 50 m/s and $s = 10$ m when $t = 0$ s.

46. A wall is 15 cm thick. At the outside, the temperature is 3°C and at the inside, it is 23°C. If the temperature changes at a constant rate through the wall, write an equation of the temperature T in the wall as a function of the distance x from the outside to the inside of the wall.

47. Within certain limits, the amount which a spring stretches varies linearly with the amount of force applied. If a spring whose natural length is 15 in. stretches 2.0 in. when 3.0 lb of force are applied, find the equation relating the length of the spring and the applied force.

48. The electric resistance of a certain resistor increases by 0.005 Ω for every increase of 1°C. Given that its resistance is 2.000 Ω at 0°C, find the equation relating the resistance and temperature. From the equation find the resistance when the temperature is 50°C.

49. The amount of heat required to raise the temperature of water varies linearly with the increase in temperature. However, a certain quantity of heat is required to change ice (at 0°C) into water without a change in temperature. An experiment is performed with 10 g of ice at 0°C. It is found that ice requires 4.19 kJ to change it into water at 20°C. Another 1.26 kJ is required to warm the water to 50°C. How many kilojoules are required to melt the ice at 0°C into water at 0°C?

50. The pressure at a point below the surface of a body of water varies linearly with the depth of the water. If the pressure at a depth of 10.0 m is 199 kPa and the pressure at a depth of 30.0 m is 395 kPa, what is the pressure at the surface? (This is the atmospheric pressure of the air above the water.)

51. A survey of the traffic on a particular highway showed that the number of cars passing a particular point each minute varied linearly from 6:30 A.M. to 8:30 A.M. on workday mornings. The study showed that an average of 45 cars passed the point in one minute at 7 A.M. and that 115 cars passed in one minute at 8 A.M. If n is the number of cars passing the point in one minute and t is the number of minutes after 6:30 A.M., find the equation relating n and t, and graph the equation. From the graph, determine n at 6:30 A.M. and at 8:30 A.M.

52. The total fixed cost for a company to operate a certain plant is $400 per day. It also costs $2 for each unit produced in the plant. Find the equation relating the total cost C of operating the plant and the number of units n produced each day. Graph the equation, assuming that $n \le 300$.

x	x^2	y
0	0	2
1	1	5
2	4	14
3	9	29
4	16	50
5	25	77

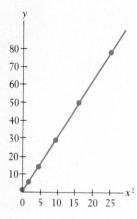

Figure 20–18

In Exercises 53 through 56, treat the given nonlinear functions as linear functions in order to sketch their graphs. As an example, $y = 2 + 3x^2$ can be sketched as a straight line by graphing y as a function of x^2. A table of values for this graph is shown along with the corresponding graph in Fig. 20–18.

53. Sketch the graph of the function $y = 1 + \sqrt{x}$ by graphing y as a function of $\sqrt{x}$.

54. Using this method, sketch the function $C = 10/E$ as a linear function. This is the equation between the voltage across a capacitor and the capacitance, assuming the charge on the capacitor is constant.

55. A spacecraft is launched such that its altitude h, in kilometers, is given by $h = 300 + 2t^{3/2}$ for $0 \leq t < 100$ s. Sketch this as a linear function.

56. The current i, in amperes, in a certain electric circuit is given by $i = 6(1 - e^{-t})$. Sketch this as a linear function.

In Exercises 57 through 60, plot the given nonlinear functions on semilogarithmic or logarithmic paper to show that they are linear when plotted. In Section 12–8 we noted that curves plotted on logarithmic and semilogarithmic paper often become straight lines. Since distances corresponding to log y and log x are plotted automatically if the graph is logarithmic in the respective directions, many nonlinear functions are linear when plotted on this paper.

57. A function of the form $y = ax^n$ is straight when plotted on logarithmic paper, since log y = log a + n log x is in the form of a straight line. The variables are log y and log x; the slope can be found from (log y − log a)/log x = n, and the intercept is a. (To get the slope from the graph, it is necessary to measure vertical and horizontal distances between two points. The log y intercept is found where log x = 0, and this occurs when x = 1.) Plot $y = 3x^4$ on logarithmic paper to verify this analysis.

58. A function of the form $y = a(b^x)$ is a straight line on semilogarithmic paper, since log y = log a + x log b is in the form of a straight line. The variables are log y and x, the slope is log b, and the intercept is a. [To get the slope from the graph, we must calculate (log y − log a)/x for some set of values x and y. The intercept is read directly off the graph where x = 0.] Plot $y = 3(2^x)$ on semilogarithmic paper to verify this analysis.

59. If experimental data are plotted on logarithmic paper and the points lie on a straight line, it is possible to determine the function (see Exercise 57). The following data come from an experiment to determine the functional relationship between the tension in a string and the velocity of a wave in the string. From the graph on logarithmic paper, determine v as a function of T.

v (meters per second)	16.0	32.0	45.3	55.4	64.0
T (newtons)	0.100	0.400	0.800	1.20	1.60

60. If experimental data are plotted on semilogarithmic paper, and the points lie on a straight line, it is possible to determine the function (see Exercise 58). The following data come from an experiment designed to determine the relationship between the voltage across an inductor and the time, after the switch is opened. Determine v as a function of t.

v (volts)	40	15	5.6	2.2	0.8
t (milliseconds)	0.0	20	40	60	80

20–3 The Circle

We have found that we can obtain a general equation which represents a straight line by considering a fixed point on the line and then a general point $P(x, y)$ which can represent any other point on the same line. Mathematically we can state this as "the line is the **locus** of a point $P(x, y)$ which *moves* along the line." That is, the point $P(x, y)$ can be considered as a variable point which moves along the line.

In this way we can define a number of important curves. A **circle** is defined as *the locus of a point P(x, y) which moves so that it is always equidistant from a fixed point.* We call this fixed distance the **radius,** *and we call the fixed point the* **center** *of the circle.* Thus, using this definition, calling the fixed point (h, k) and the radius r, we have

$$\sqrt{(x - h)^2 + (y - k)^2} = r$$

or, by squaring both sides, we have

$$(x - h)^2 + (y - k)^2 = r^2 \qquad (20–11)$$

Equation (20–11) is called the **standard equation** *of a circle with center at (h, k) and radius r* (Fig. 20–19).

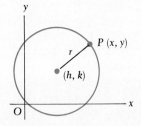

Figure 20–19

Example A

The equation $(x - 1)^2 + (y + 2)^2 = 16$ represents a circle with center at $(1, -2)$ and a radius of 4. We determine these values by considering the equation of this circle to be in the form of Eq. (20–11) as

$$(x - 1)^2 + [y - (-2)]^2 = 4^2$$

Note carefully the way in which we find the y-coordinate of the center. This circle is shown in Fig. 20–20.

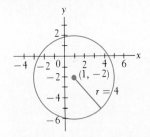

Figure 20–20

If the center of the circle is at the origin, which means that the coordinates of the center are $(0, 0)$, the equation of the circle becomes

$$x^2 + y^2 = r^2 \qquad (20–12)$$

A circle of this type clearly exhibits an important property of the graphs of many equations. *It is* **symmetrical** *to the x-axis and also to the y-axis.* Symmetry to the x-axis can be thought of as meaning that the lower half of the curve is a reflection of the upper half, and conversely. It can be shown that *if $-y$ can replace y in an* equation without changing the equation, the graph of the equation is **symmetrical to the x-axis.** Symmetry to the y-axis is similar. *If* $-x$ *can replace x in the equation without changing the equation, the graph is* **symmetrical to the y-axis.**

This type of circle is symmetrical to the origin as well as being symmetrical to both axes. The meaning of symmetry to the origin is that the origin is the midpoint of any two points (x, y) and $(-x, -y)$ which are on the curve. Thus, *if* $-x$ *can replace x, and* $-y$ *replace y at the same time, without changing the equation, the graph of the equation is* **symmetrical to the origin.**

Example B

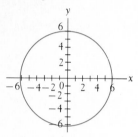

Figure 20–21

The equation of the circle with its center at the origin and with a radius of 6 is $x^2 + y^2 = 36$.

The symmetry of this circle can be shown analytically by the substitutions mentioned above. Replacing x by $-x$, we obtain $(-x)^2 + y^2 = 36$. Since $(-x)^2 = x^2$, this equation can be rewritten as $x^2 + y^2 = 36$. Since this substitution did not change the equation, the graph is symmetrical to the y-axis.

Replacing y by $-y$, we obtain $x^2 + (-y)^2 = 36$, which is the same as $x^2 + y^2 = 36$. This means that the curve is symmetrical to the x-axis.

Replacing x by $-x$, and simultaneously replacing y by $-y$, we obtain $(-x)^2 + (-y)^2 = 36$, which is the same as $x^2 + y^2 = 36$. This means that the curve is symmetrical to the origin. The circle is shown in Fig. 20–21.

Example C

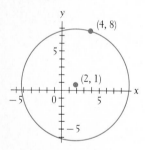

Figure 20–22

Find the equation of the circle with center at (2, 1) and which passes through (4, 8).

In Eq. (20–11), we can determine the equation if we can find h, k, and r for this circle. From the given information, $h = 2$ and $k = 1$. To find r, we use the fact that *all points on the circle must satisfy the equation of the circle.* The point (4, 8) must satisfy Eq. (20–11), with $h = 2$ and $k = 1$. Thus, $(4 - 2)^2 + (8 - 1)^2 = r^2$. From this relation we find $r^2 = 53$. The equation of the circle is

$$(x - 2)^2 + (y - 1)^2 = 53$$

This circle is shown in Fig. 20–22.

Example D

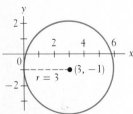

Figure 20–23

Find the equation of the circle with center at (3, -1) and which is tangent to the y-axis.

We know that $h = 3$ and $k = -1$. Since the circle is tangent to the y-axis, the radius is the distance from the center to the y-axis. Therefore, $r = 3$, since the center is 3 units to the right of the y-axis ($h = 3$). Therefore the equation of the circle is

$$(x - 3)^2 + (y + 1)^2 = 9$$

The circle is shown in Fig. 20–23.

If we multiply out each of the terms in Eq. (20–11), we may combine the resulting terms to obtain

$$x^2 - 2hx + h^2 + y^2 - 2ky + k^2 = r^2$$

$$x^2 + y^2 - 2hx - 2ky + (h^2 + k^2 - r^2) = 0 \qquad \textbf{(20–13)}$$

Since each of h, k, and r is constant for any given circle, the coefficients of x and y and the term within parentheses in Eq. (20–13) are constants. Equation (20–13) can then be written as

$$x^2 + y^2 + Dx + Ey + F = 0 \qquad \textbf{(20–14)}$$

Equation (20–14) is called the **general equation** *of the circle.* It tells us that any equation which can be written in that form will represent a circle.

Example E

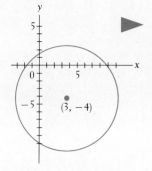

Figure 20–24

Find the center and radius of the circle

$$x^2 + y^2 - 6x + 8y - 24 = 0$$

We can find this information if we write the given equation in standard form. To do so, we must complete the square in the x-terms and also in the y-terms. This is done by first writing the equation in the form

$$(x^2 - 6x \quad\ ) + (y^2 + 8y \quad\ ) = 24$$

To complete the square of the x-terms, we take half of 6, which is 3, square it, and add the result, 9, to each side of the equation. In the same way, we complete the square of the y-terms by adding 16 to each side of the equation, which gives

$$(x^2 - 6x + 9) + (y^2 + 8y + 16) = 24 + 9 + 16$$
$$(x - 3)^2 + (y + 4)^2 = 49$$

Thus, the center is $(3, -4)$, and the radius is 7 (see Fig. 20–24).

Example F

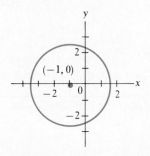

Figure 20–25

The equation $3x^2 + 3y^2 + 6x - 20 = 0$ can be seen to represent a circle by writing it in general form. This is done by dividing through by 3. In this way we have $x^2 + y^2 + 2x - \frac{20}{3} = 0$. To determine the center and radius, we write the equation in standard form and complete the necessary squares. This leads to

$$(x^2 + 2x + 1) + y^2 = \frac{20}{3} + 1$$

$$(x + 1)^2 + y^2 = \frac{23}{3}$$

Thus, the center is $(-1, 0)$; the radius is $\sqrt{\frac{23}{3}} = 2.77$ (see Fig. 20–25).

We can also see that this circle is symmetrical to the x-axis, but it is not symmetrical to the y-axis or to the origin. If we replace y by $-y$, the equation does not change, but if we replace x by $-x$, the term $6x$ in the original equation becomes negative, and the equation *does* change.

Exercises 20–3

In Exercises 1 through 4, determine the center and radius of each circle.

1. $(x - 2)^2 + (y - 1)^2 = 25$ 2. $(x - 3)^2 + (y + 4)^2 = 49$

3. $(x + 1)^2 + y^2 = 4$ 4. $x^2 + (y + 6)^2 = 64$

In Exercises 5 through 20, find the equation of each of the circles from the given information.

5. Center at $(0, 0)$, radius 3   6. Center at $(0, 0)$, radius 1

7. Center at $(2, 2)$, radius 4   8. Center at $(0, 2)$, radius 2

9. Center at $(-2, 5)$, radius $\sqrt{5}$ 10. Center at $(-3, -5)$, radius 8

11. Center at $(-3, 0)$, radius $\frac{1}{2}$ 12. Center at $(\frac{3}{2}, -2)$, radius $\frac{5}{2}$

13. Center at $(2, 1)$, passes through $(4, -1)$.

14. Center at $(-1, 4)$, passes through $(-2, 3)$.

15. Center at $(-3, 5)$, tangent to the x-axis.

16. Center at $(2, -4)$, tangent to the y-axis.

17. Tangent to both axes and the lines $y = 4$ and $x = 4$.

18. Tangent to both axes, radius 4, and in the second quadrant.

19. Center on the line $5x = 2y$, radius 5, tangent to the x-axis.

20. The points $(3, 8)$ and $(-3, 0)$ are the ends of a diameter.

In Exercises 21 through 32, determine the center and radius of each circle. Sketch each circle.

21. $x^2 + (y - 3)^2 = 4$ 22. $(x - 2)^2 + (y + 3)^2 = 49$

23. $4(x + 1)^2 + 4(y - 5)^2 = 81$ 24. $2(x + 4)^2 + 2(y + 3)^2 = 25$

25. $x^2 + y^2 - 25 = 0$ 26. $x^2 + y^2 - 9 = 0$

27. $x^2 + y^2 - 2x - 8 = 0$ 28. $x^2 + y^2 - 4x - 6y - 12 = 0$

29. $x^2 + y^2 + 8x - 10y - 8 = 0$ 30. $x^2 + y^2 + 8x + 6y = 0$

31. $2x^2 + 2y^2 - 4x - 8y - 1 = 0$ 32. $3x^2 + 3y^2 - 12x + 4 = 0$

In Exercises 33 through 36, determine whether the circles with the given equations are symmetrical to either axis or the origin.

33. $x^2 + y^2 = 100$ 34. $x^2 + y^2 - 4x - 5 = 0$

35. $x^2 + y^2 + 8y - 9 = 0$ 36. $x^2 + y^2 - 2x + 4y - 3 = 0$

In Exercises 37 through 44, find the indicated quantities.

37. Determine where the circle $x^2 - 6x + y^2 - 7 = 0$ crosses the x-axis.

38. Find the points of intersection of the circle $x^2 + y^2 - x - 3y = 0$ and the line $y = x - 1$.

39. Find the locus of a point $P(x, y)$ which moves so that its distance from $(2, 4)$ is twice its distance from $(0, 0)$. Describe the locus.

40. Find the equation of the locus of a point $P(x, y)$ which moves so that the line joining it and $(2, 0)$ is always perpendicular to the line joining it and $(-2, 0)$. Describe the locus.

41. If an electrically charged particle enters a magnetic field of flux density B with a velocity v at right angles to the field, the path of the particle is a circle. The radius of the path is given by $R = mv/Bq$, where m is the mass of the particle and q is its charge. If a proton ($m = 1.67 \times 10^{-27}$ kg, $q = 1.60 \times 10^{-19}$ C) enters a magnetic field of flux density 1.50 T (tesla: $V \cdot s/m^2$) with a velocity of 2.40×10^8 m/s, find the equation of the path of the proton. (The fact that a charged particle travels in a circular path in a magnetic field is an important basis for the construction of a cyclotron.)

42. For a constant (magnitude) impedance and capacitive reactance, sketch the graph of resistance versus inductive reactance (see Section 11–7).

43. A cylindrical oil tank, 4.00 ft in diameter, is on its side. A circular hole, with center 18.0 in. below the center of the end of the tank, has been drilled in the end of the tank. If the radius of the hole is 3.00 in., what is the equation of the circle of the end of the tank and the equation of the circle of the hole? Use the center of the end of the tank as the origin.

44. A student is drawing a friction drive in which two circular disks are in contact with each other. They are represented by circles in the drawing. The first has a radius of 10.0 cm and the second has a radius of 12.0 cm. What is the equation of each circle if the origin is at the center of the first circle and the x-axis passes through the center of the second circle?

20–4 The Parabola

Another important curve is the parabola. We have come across this curve several times in the earlier chapters. In Chapter 6 we showed that the graph of a quadratic function is a parabola. In this section we define the parabola more generally, and thereby find the general form of its equation.

A **parabola** *is defined as the locus of a point P(x, y) which moves so that it is always equidistant from a given line and a given point. The given line is called the* **directrix,** *and the given point is called the* **focus.** The line through the focus which is perpendicular to the directrix is called the **axis** of the parabola. The point midway between the directrix and focus is the **vertex** of the parabola. Using the definition, we shall find the equation of the parabola for which the focus is the point $(p, 0)$ and the directrix is the line $x = -p$. By choosing the focus and directrix in this manner, we shall be able to find a general representation of the equation of a parabola with its vertex at the origin.

According to the definition of the parabola, the distance from a point $P(x, y)$ on the parabola to the focus $(p, 0)$ must equal the distance from $P(x, y)$ to the directrix $x = -p$. The distance from P to the focus can be found by use of the distance formula. The distance from P to the directrix is the perpendicular distance, and this can be found as the distance between two points on a line parallel to the x-axis. These distances are indicated in Fig. 20–26. Thus, we have

$$\sqrt{(x - p)^2 + (y - 0)^2} = x + p$$

Squaring both sides of this equation, we have

$$(x - p)^2 + y^2 = (x + p)^2$$

or

$$x^2 - 2px + p^2 + y^2 = x^2 + 2px + p^2$$

Simplifying, we obtain

$$\underline{y^2 = 4px}$$ (20–15)

Equation (20–15) is called the **standard form** *of the equation of a parabola with its axis along the x-axis and the vertex at the origin.* Its symmetry to the x-axis can be proven since $(-y)^2 = 4px$ is the same as $y^2 = 4px$.

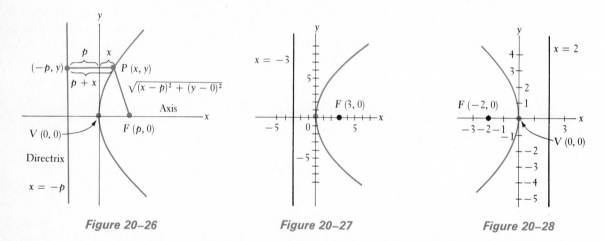

Figure 20–26   Figure 20–27   Figure 20–28

Example A   Find the coordinates of the focus, the equation of the directrix, and sketch the graph of the parabola $y^2 = 12x$.

From the form of the equation, we know that the vertex is at the origin (Fig. 20–27). The coefficient 12 tells us that the focus is $(3, 0)$, since $p = \frac{12}{4}$ [note that the coefficient of x in Eq. (20–15) is $4p$]. Also, this means that the directrix is the line $x = -3$.

Example B

If the focus is to the left of the origin, with the directrix an equal distance to the right, the coefficient of the x-term will be negative. This tells us that the parabola opens to the left, rather than to the right, as is the case when the focus is to the right of the origin. For example, the parabola $y^2 = -8x$ has its vertex at the origin, its focus at $(-2, 0)$, and the line $x = 2$ as its directrix. This is consistent with Eq. (20–15), in that $4p = -8$, or $p = -2$. The parabola opens to the left as shown in Fig. 20–28.

If we chose the focus as the point $(0, p)$ and the directrix as the line $y = -p$, we would find that the resulting equation is

$$x^2 = 4py \tag{20–16}$$

This is the standard form of the equation of a parabola with the y-axis as its axis and the vertex at the origin. Its symmetry to the y-axis can be proven, since $(-x)^2 = 4py$ is the same as $x^2 = 4py$. We note that the difference between this equation and Eq. (20–15) is that x is squared and y appears to the first power in Eq. (20–16), rather than the reverse, as in Eq. (20–15).

Example C

The parabola $x^2 = 4y$ has its vertex at the origin, focus at the point $(0, 1)$, and its directrix the line $y = -1$. The graph is shown in Fig. 20–29, and we see in this case that the parabola opens up.

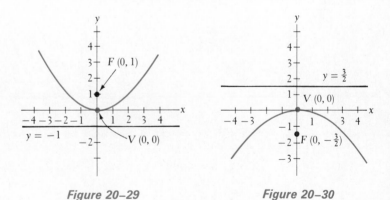

Figure 20–29   Figure 20–30

Example D

The parabola $x^2 = -6y$ has its vertex at the origin, its focus at the point $(0, -\frac{3}{2})$, and its directrix the line $y = \frac{3}{2}$. This parabola opens down, as Fig. 20–30 shows.

Example E

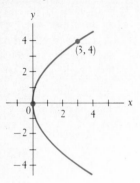

Figure 20–31

Find the equation of the parabola with vertex at the origin, axis along the x-axis, and which passes through $(3, 4)$.

Since the parabola has its vertex at the origin and axis along the x-axis, its general form is the equation given by Eq. (20–15), or $y^2 = 4px$. We can find the value of p by using the fact that it passes through $(3, 4)$. These coordinates must satisfy the equation. This means that

$$4^2 = 4p(3), \quad \text{or} \quad p = \frac{4}{3}$$

Therefore, the equation is $y^2 = \frac{16}{3}x$. The parabola is shown in Fig. 20–31.

Equations (20–15) and (20–16) give us the general form of the equation of a parabola with its vertex at the origin and its focus on one of the coordinate axes. The following example illustrates the use of the definition of the parabola to find the equation of a parabola which has its vertex at a point other than at the origin.

Example F

Using the definition of the parabola, determine the equation of the parabola with its focus at $(2, 3)$ and its directrix the line $y = -1$ (see Fig. 20–32).

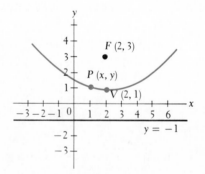

Figure 20–32

Choosing a general point $P(x, y)$ on the parabola, and equating the distances from this point to $(2, 3)$ and to the line $y = -1$, we have

$$\sqrt{(x - 2)^2 + (y - 3)^2} = y + 1$$

Squaring both sides of this equation and simplifying, we have

$$(x - 2)^2 + (y - 3)^2 = (y + 1)^2$$
$$x^2 - 4x + 4 + y^2 - 6y + 9 = y^2 + 2y + 1$$

or

$$8y = 12 - 4x + x^2$$

When we see it in this form, we note that this type of equation has appeared numerous times in earlier chapters. The x-term and the constant (12 in this case) are characteristic of a parabola which does not have its vertex at the origin if the directrix is parallel to the x-axis.

Thus we can conclude that *a parabola is characterized by the presence of the square of either (but not both) x or y, and a first-power term in the other.* In this way we can recognize a parabola by inspecting the equation.

Exercises 20–4

In Exercises 1 through 12, determine the coordinates of the given parabolas. Sketch each curve.

1. $y^2 = 4x$ 2. $y^2 = 16x$ 3. $y^2 = -4x$ 4. $y^2 = -16x$

5. $x^2 = 8y$ 6. $x^2 = 10y$ 7. $x^2 = -4y$ 8. $x^2 = -12y$

9. $y^2 = 2x$ 10. $x^2 = 14y$ 11. $y = x^2$ 12. $x = 3y^2$

In Exercises 13 through 20, find the equations of the parabolas satisfying the given conditions.

13. Focus $(3, 0)$, directrix $x = -3$ 14. Focus $(-2, 0)$, directrix $x = 2$

15. Focus $(0, 4)$, vertex $(0, 0)$ 16. Focus $(-3, 0)$, vertex $(0, 0)$

17. Vertex $(0, 0)$, directrix $y = -1$ 18. Vertex $(0, 0)$, directrix $y = \frac{1}{2}$

19. Vertex at $(0, 0)$, axis along the y-axis, passes through $(-1, 8)$.

20. Vertex at $(0, 0)$, axis along the x-axis, passes through $(2, -1)$.

In Exercises 21 through 24, use the definition of the parabola to find the equations of the parabolas satisfying the given conditions. Sketch each curve.

21. Focus $(6, 1)$, directrix $x = 0$ 22. Focus $(1, -4)$, directrix $x = 2$

23. Focus $(1, 1)$, vertex $(1, 3)$ 24. Vertex $(-2, -4)$, directrix $x = 3$

In Exercises 25 through 34, find the indicated quantities.

25. In calculus it can be shown that a light ray coming from the focus of a parabola will be reflected off parallel to the axis of the parabola. Suppose that a light ray from the focus of a parabolic reflector described by the equation $y^2 = 16x$ strikes the reflector at the point $(1, 4)$. Along what line does the incident ray move, and along what line does the reflected ray move?

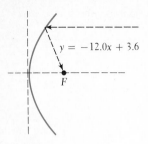

$$y = -12.0x + 3.6$$

F

Figure 20–33

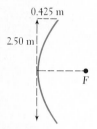

0.425 m

2.50 m

F

Figure 20–34

26. A radio wave reflector has a parabolic cross section, and a wave that comes in parallel to its axis will be reflected through the focus. A particular reflector is designed for the incoming and reflected waves, as shown in Fig. 20–33 (measurements in meters). What is the equation of the parabola?

27. A linear solar reflector has a parabolic cross section, which is shown in Fig. 20–34. From the given dimensions, what is the focal length (vertex to focus) of the reflector?

28. The rate of development of heat H (measured in watts) in a resistor of resistance R (measured in ohms) of an electric circuit is given by $H = Ri^2$, where i is the current (measured in amperes) in the resistor. Sketch the graph of H versus i, if $R = 6.0 \ \Omega$.

29. Under certain conditions, a cable which hangs between two supports can be closely approximated as being parabolic. Assuming that this cable hangs in the shape of a parabola, find its equation if a point 10 ft horizontally from its lowest point is 1.0 ft above its lowest point. Choose the lowest point as the origin of the coordinate system.

30. A wire is fastened 36.0 ft up on each of two telephone poles which are 200 ft apart. Halfway between the poles the wire is 30.0 ft above the ground. Assuming the wire is parabolic, find the height of the wire 50.0 ft from either pole.

31. The period T (measured in seconds) of the oscillation for resonance in an electric circuit is given by $T = 2\pi \sqrt{LC}$, where L is the inductance (in henrys) and C is the capacitance (in farads). Sketch the graph of the period versus capacitance for a constant inductance of 1 H. Assume values of C from 1 μF to 250 μF for this is a common range of values.

32. The velocity v of a jet of water flowing from an opening in the side of a certain container is given by $v = 8\sqrt{h}$, where h is the depth of the opening. Sketch a graph of v (in feet per second) versus h (in feet).

33. A small island is 4 km from a straight shoreline. A ship channel is equidistant between the island and the shoreline. Write an equation for the channel.

34. Under certain circumstances, the maximum power P in an electric circuit varies as the square of the voltage of the source E_0 and inversely as the internal resistance R_i of the source. If 10 W is the maximum power for a source of 2.0 V and internal resistance of 0.10 Ω, sketch the graph of P versus E_0, if R_i remains constant.

20–5 The Ellipse

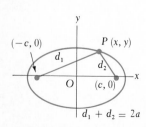

$(-c, 0)$

$P(x, y)$

d_1

d_2

O $(c, 0)$

x

y

$d_1 + d_2 = 2a$

Figure 20–35

The next important curve we shall discuss is the ellipse. *The **ellipse** is defined as the locus of a point $P(x, y)$ which moves so that the sum of its distances from two fixed points is constant. These two fixed points are called the **foci** of the ellipse.* We will let this sum of distances be $2a$ and the foci be the points $(c, 0)$ and $(-c, 0)$. These choices set up the ellipse with its center at the origin such that c is the length of the line segment from the center to a focus. Also, by choosing $2a$ as the sum of distances, we will see that the length a has a special meaning.

From the definition of an ellipse, we have (see Fig. 20–35)

$$\sqrt{(x - c)^2 + y^2} + \sqrt{(x + c)^2 + y^2} = 2a$$

From the section on solving equations involving radicals (Section 13–4), we find that we should remove one of the radicals to the right side, and then square each side. Thus we have the following steps:

$$\sqrt{(x + c)^2 + y^2} = 2a - \sqrt{(x - c)^2 + y^2}$$
$$(x + c)^2 + y^2 = 4a^2 - 4a\sqrt{(x - c)^2 + y^2} + (\sqrt{(x - c)^2 + y^2})^2$$
$$x^2 + 2cx + c^2 + y^2 = 4a^2 - 4a\sqrt{(x - c)^2 + y^2} + x^2 - 2cx + c^2 + y^2$$
$$4a\sqrt{(x - c)^2 + y^2} = 4a^2 - 4cx$$
$$a\sqrt{(x - c)^2 + y^2} = a^2 - cx$$
$$a^2(x^2 - 2cx + c^2 + y^2) = a^4 - 2a^2cx + c^2x^2$$
$$(a^2 - c^2)x^2 + a^2y^2 = a^2(a^2 - c^2)$$

At this point we define (and the usefulness of this definition will be shown presently) $a^2 - c^2 = b^2$. This substitution gives us

$$b^2x^2 + a^2y^2 = a^2b^2$$

Dividing through by a^2b^2, we have

$$\frac{x^2}{a^2} + \frac{y^2}{b^2} = 1 \qquad (20\text{–}17)$$

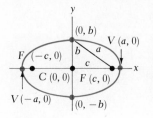

Figure 20–36

If we let $y = 0$, we find that the x-intercepts are $(-a, 0)$ and $(a, 0)$. We see that the distance $2a$, originally chosen as the sum of the distances in the derivation of Eq. (20–17), is the distance between the x-intercepts. *These two points are known as the* **vertices** *of the ellipse, and the line between them is known as the* **major axis** *of the ellipse. Thus, a is called the* **semimajor axis.**

If we now let $x = 0$, we find the y-intercepts to be $(0, -b)$ and $(0, b)$. *The line joining these two intercepts is called the* **minor axis** *of the ellipse (Fig. 20–36) which means b is called the* **semiminor axis.** The point $(0, b)$ is on the ellipse, and is also equidistant from $(-c, 0)$ and $(c, 0)$. Since the sum of the distances from these points to $(0, b)$ equals $2a$, the distance from $(c, 0)$ to $(0, b)$ must be a. Thus, we have a right triangle formed by line segments of lengths a, b, and c, with a as the hypotenuse. From this *we have the relation*

$$a^2 = b^2 + c^2 \qquad (20\text{–}18)$$

between the distances a, b, and c. This equation also shows us why b was defined as it was in the derivation.

Equation (20–17) is called the **standard equation** *of the ellipse with its major axis along the x-axis and its center at the origin.*

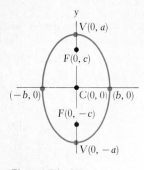

Figure 20-37

If we choose points on the y-axis as the foci, *the standard equation of the ellipse, with its center at the origin and its major axis along the y-axis, is*

$$\frac{y^2}{a^2} + \frac{x^2}{b^2} = 1 \qquad (20\text{--}19)$$

In this case the vertices are $(0, a)$ and $(0, -a)$, the foci are $(0, c)$ and $(0, -c)$, and the ends of the minor axis are $(b, 0)$ and $(-b, 0)$. See Fig. 20-37.

The symmetry of the ellipses given by Eqs. (20-17) and (20-19) to each of the axes and the origin can be proven, since each of x and y can be replaced by its negative in these equations without changing the equations.

Example A

The ellipse

$$\frac{x^2}{25} + \frac{y^2}{9} = 1$$

has vertices at $(5, 0)$ and $(-5, 0)$. Its minor axis extends from $(0, 3)$ to $(0, -3)$, as we see in Fig. 20-38. This information is directly obtainable from this form of the equation, since the 25 tells us that $a^2 = 25$, or $a = 5$. In the same way we have $b = 3$. Since we know both a^2 and b^2, we can find c^2 from the relation $c^2 = a^2 - b^2$. Thus, $c^2 = 16$, or $c = 4$. Thus, the foci are $(4, 0)$ and $(-4, 0)$.

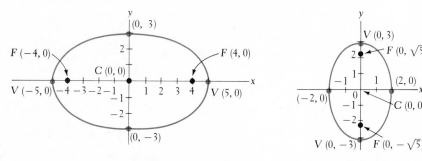

Figure 20-38   **Figure 20-39**

Example B

The ellipse

$$\frac{x^2}{4} + \frac{y^2}{9} = 1$$

has vertices at $(0, 3)$ and $(0, -3)$. The minor axis extends from $(2, 0)$ to $(-2, 0)$. This we can find directly from the equation, since $a^2 = 9$ and $b^2 = 4$. One might ask how we chose $a^2 = 9$ in this example, and $a^2 = 25$ in the preceding example. Since $a^2 = b^2 + c^2$, a is always larger than b. Thus, we can tell which axis (x or y) the major axis is along by seeing which number (in the denominator) is larger, when the equation is written in standard form. The larger one stands for a^2. In this example, the foci are $(0, \sqrt{5})$ and $(0, -\sqrt{5})$. See Fig. 20-39.

Example C

Find the coordinates of the vertices, the ends of the minor axis, and the foci of the ellipse $4x^2 + 16y^2 = 64$.

This equation must be put in standard form first, which we do by dividing through by 64. When this is done, we obtain

$$\frac{x^2}{16} + \frac{y^2}{4} = 1$$

Thus, $a = 4$, $b = 2$, and $c = 2\sqrt{3}$. The vertices are at $(4, 0)$ and $(-4, 0)$. The ends of the minor axis are $(0, 2)$ and $(0, -2)$, and the foci are $(2\sqrt{3}, 0)$ and $(-2\sqrt{3}, 0)$. See Fig. 20–40.

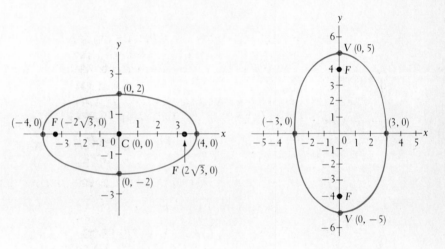

Figure 20–40   Figure 20–41

Example D

Find the equation of the ellipse for which the center is at the origin, the major axis is along the y-axis, the minor axis is 6 units long, and there are 8 units between foci.

Directly we know that $2b = 6$, or $b = 3$. Also $2c = 8$, or $c = 4$. Thus, we find that $a = 5$. Since the major axis is along the y-axis, we have the equation

$$\frac{y^2}{25} + \frac{x^2}{9} = 1$$

This ellipse is shown in Fig. 20–41.

Example E

Find the equation of the ellipse with its center at the origin, an end of its minor axis at $(2, 0)$, and which passes through $(-1, \sqrt{6})$.

Since the center is at the origin and an end of the minor axis is at $(2, 0)$, we know that the ellipse is of the form of Eq. (20–19), and that $b = 2$. Thus, we have

$$\frac{y^2}{a^2} + \frac{x^2}{2^2} = 1$$

In order to find a^2 we use the fact that the ellipse passes through $(-1, \sqrt{6})$. This means that these coordinates satisfy the equation of the ellipse. This gives us

$$\frac{(\sqrt{6})^2}{a^2} + \frac{(-1)^2}{4} = 1$$

$$\frac{6}{a^2} + \frac{1}{4} = 1$$

$$\frac{6}{a^2} = \frac{3}{4}$$

$$a^2 = 8$$

Thus, the equation of the ellipse is

$$\frac{y^2}{8} + \frac{x^2}{4} = 1$$

The ellipse is shown in Fig. 20–42.

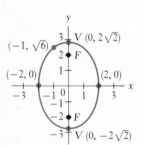

Figure 20–42

Equations (20–17) and (20–19) give us the standard form of the equation of an ellipse with its center at the origin and its foci on one of the coordinate axes. The following example illustrates the use of the definition of the ellipse to find the equation of an ellipse with its center at a point other than the origin.

Example F

Using the definition, find the equation of the ellipse with foci at $(1, 3)$ and $(9, 3)$, with a major axis of 10.

Using the same method as in the derivation of Eq. (20–17), we have the following steps. (Remember, the sum of distances in the definition equals the length of the major axis.)

$$\sqrt{(x-1)^2 + (y-3)^2} + \sqrt{(x-9)^2 + (y-3)^2} = 10$$
$$\sqrt{(x-1)^2 + (y-3)^2} = 10 - \sqrt{(x-9)^2 + (y-3)^2}$$
$$(x-1)^2 + (y-3)^2 = 100 - 20\sqrt{(x-9)^2 + (y-3)^2}$$
$$+ (\sqrt{(x-9)^2 + (y-3)^2})^2$$
$$x^2 - 2x + 1 + y^2 - 6y + 9 = 100 - 20\sqrt{(x-9)^2 + (y-3)^2}$$
$$+ x^2 - 18x + 81 + y^2 - 6y + 9$$

$$20\sqrt{(x-9)^2 + (y-3)^2} = 180 - 16x$$
$$5\sqrt{(x-9)^2 + (y-3)^2} = 45 - 4x$$
$$25(x^2 - 18x + 81 + y^2 - 6y + 9) = 2025 - 360x + 16x^2$$
$$25x^2 - 450x + 2250 + 25y^2 - 150y = 2025 - 360x + 16x^2$$
$$9x^2 - 90x + 25y^2 - 150y + 225 = 0$$

The additional x- and y-terms are characteristic of the equation of an ellipse whose center is not at the origin (see Fig. 20–43).

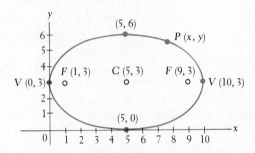

Figure 20–43

We can conclude that *the equation of an ellipse is characterized by the presence of both an x^2- and a y^2-term, having different coefficients (in value but not in sign)*. The difference between the equation of an ellipse and that of a circle is that the coefficients of the squared terms in the equation of the circle are the same, whereas those of the ellipse differ.

Exercises 20–5

In Exercises 1 through 12, find the coordinates of the vertices and foci of the given ellipses. Sketch each curve.

1. $\dfrac{x^2}{4} + \dfrac{y^2}{1} = 1$ 2. $\dfrac{x^2}{100} + \dfrac{y^2}{64} = 1$ 3. $\dfrac{x^2}{25} + \dfrac{y^2}{36} = 1$

4. $\dfrac{x^2}{49} + \dfrac{y^2}{81} = 1$ 5. $4x^2 + 9y^2 = 36$ 6. $x^2 + 36y^2 = 144$

7. $49x^2 + 4y^2 = 196$ 8. $25x^2 + y^2 = 25$ 9. $8x^2 + y^2 = 16$

10. $2x^2 + 3y^2 = 6$ 11. $4x^2 + 25y^2 = 25$ 12. $9x^2 + 4y^2 = 9$

In Exercises 13 through 20, find the equations of the ellipses satisfying the given conditions. The center of each is at the origin.

13. Vertex $(15, 0)$, focus $(9, 0)$ 14. Minor axis 8, vertex $(0, -5)$

15. Focus $(0, 2)$, major axis 6 16. Semiminor axis 2, focus $(3, 0)$

17. Vertex $(8, 0)$, passes through $(2, 3)$. 18. Focus $(0, 2)$, passes through $(-1, \sqrt{3})$.

19. Passes through $(2, 2)$ and $(1, 4)$. 20. Passes through $(-2, 2)$ and $(1, \sqrt{6})$.

In Exercises 21 through 24, find the equations of the ellipses with the given properties by use of the definition of an ellipse.

21. Foci at $(-2, 1)$ and $(4, 1)$, a major axis of 10.

22. Foci at $(-3, -2)$ and $(-3, 8)$, major axis 26.

23. Vertices at $(1, 5)$ and $(1, -1)$, foci at $(1, 4)$ and $(1, 0)$.

24. Vertices at $(-2, 1)$ and $(-2, 5)$, foci at $(-2, 2)$ and $(-2, 4)$.

In Exercises 25 through 34, solve the given problems.

25. Show that the ellipse $2x^2 + 3y^2 - 8x - 4 = 0$ is symmetrical to the x-axis.

26. Show that the ellipse $5x^2 + y^2 - 3y - 7 = 0$ is symmetrical to the y-axis.

27. The arch of a bridge across a stream is in the form of half an ellipse above the water level. If the span of the arch at water level is 100 ft and the maximum height of the arch above water level is 30 ft, what is the equation of the arch? Choose the origin of the coordinate system at the most convenient point.

Figure 20–44

28. An elliptical gear (Fig. 20–44) which rotates about its center is kept continually in mesh with a circular gear which is free to move horizontally. If the equation of the ellipse of the gear (with the origin of the coordinate system at its center) in its present position is $3x^2 + 7y^2 = 20$, how far does the center of the circular gear move going from one extreme position to the other? (Assume the units are centimeters.)

29. A satellite to study the earth's atmosphere has a minimum altitude of 500 mi and a maximum altitude of 2000 mi. If the path of the satellite about the earth is an ellipse with the center of the earth at one focus, what is the equation of its path? (Assume the radius of the earth is 4000 mi.)

30. An electric circuit is caused to flow in a loop of wire rotating in a magnetic field. In a study of this phenomenon, a piece of wire is cut into two pieces, one of which is bent into a circle and the other into a square. If the sum of areas of the circle and square is always π units, find the relation between the radius r of the circle and the side x of the square. Sketch the graph of r versus x.

31. A vertical pipe 6.0 inches in diameter is to pass through a roof inclined at 45°. What are the dimensions of the elliptical hole which must be cut in the roof for the pipe?

32. The ends of a horizontal tank 20.0 ft long are ellipses, which can be described by the equation $9x^2 + 20y^2 = 180$, where x and y are measured in feet. The area of an ellipse is $A = \pi ab$. Find the volume of the tank.

33. A draftsman draws a series of triangles with a base from $(-3, 0)$ to $(3, 0)$ and a perimeter of 14 cm (all measurements in centimeters). Find the equation of the curve on which all of the third vertices of the triangles are located.

34. A drainage trough 1.50 m wide, 0.50 m deep, and 8.00 m long has a semielliptical cross section. At capacity what volume does it hold? (See Exercise 32.)

20–6 The Hyperbola

The final curve we shall discuss in detail is the hyperbola. *The **hyperbola** is defined as the locus of a point $P(x, y)$ which moves so that the difference of the distances from two fixed points is a constant. These fixed points are the **foci** of the hyperbola.* We choose the foci of the hyperbola as the points $(c, 0)$ and

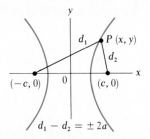

Figure 20–45

(−c, 0) (see Fig. 20–45) and the constant difference to be $2a$. As with the ellipse, these choices make c the length of the line segment from the center to a focus and a (as we will see) the length of the line segment from the center to a vertex. Therefore, we have

$$\sqrt{(x + c)^2 + y^2} - \sqrt{(x - c)^2 + y^2} = 2a$$

Following this same procedure as in the preceding section, we find the equation of the hyperbola to be

$$\frac{x^2}{a^2} - \frac{y^2}{b^2} = 1 \tag{20–20}$$

When we derive this equation, we have a definition of the relation between a, b, and c which is different from that for the ellipse. This relation is

$$c^2 = a^2 + b^2 \tag{20–21}$$

An analysis of Eq. (20–20) will reveal the significance of a and b. First, by letting $y = 0$, we find that the x-intercepts are $(a, 0)$ and $(-a, 0)$, just as they are for the ellipse. *These points are called the* **vertices** *of the hyperbola.* By letting $x = 0$, we find that we have imaginary solutions for y, which means that there are no points on the curve which correspond to a value of $x = 0$.

To find the significance of b, we shall solve Eq. (20–20) for y in a particular form. First we have

$$\frac{y^2}{b^2} = \frac{x^2}{a^2} - 1$$
$$= \frac{x^2}{a^2} - \frac{a^2 x^2}{a^2 x^2}$$
$$= \frac{x^2}{a^2}\left(1 - \frac{a^2}{x^2}\right)$$

Multiplying through by b^2 and then taking the square root of each side, we have

$$y^2 = \frac{b^2 x^2}{a^2}\left(1 - \frac{a^2}{x^2}\right)$$
$$y = \pm\frac{bx}{a}\sqrt{1 - \frac{a^2}{x^2}} \tag{20–22}$$

We note that, if large values of x are assumed in Eq. (20–22), the quantity under the radical becomes approximately 1. In fact, the larger x becomes, the nearer 1 this expression becomes, since the x^2 in the denominator of a^2/x^2 makes this term nearly zero. Thus, for large values of x, Eq. (20–22) is approximately

$$y = \pm\frac{bx}{a} \qquad\qquad (20\text{–}23)$$

Equation (20–23) can be seen to represent the equations for two straight lines, each of which passes through the origin. One has a slope of b/a and the other a slope of $-b/a$. *These lines are called the* **asymptotes** *of the hyperbola. An asymptote is a line which a curve approaches as one of the variables approaches some particular value.* The tangent curve also has asymptotes, as we can see in Fig. 9–18. We can designate this limiting procedure with notation introduced in Chapter 18 by saying that

$$y \to \frac{bx}{a} \quad \text{as} \quad x \to \infty$$

Since straight lines are easily sketched, the easiest way to sketch a hyperbola is to draw its asymptotes and then to draw the hyperbola so that it comes closer and closer to these lines as x becomes larger numerically. To draw in the asymptotes, the usual procedure is to first draw a small rectangle, $2a$ by $2b$, with the origin in the center. Then straight lines are drawn through opposite vertices. These lines are the asymptotes (see Fig. 20–46). Thus we see that the significance of the value of b lies in the slope of the asymptotes of the hyperbola.

Equation (20–20) is called the **standard equation** *of the hyperbola with its center at the origin. It has a* **transverse axis** *of length $2a$ along the x-axis and a* **conjugate axis** *of length $2b$ along the y-axis.* This means that a represents the length of the semitransverse axis, and b represents the length of the semiconjugate axis (see Fig. 20–47). From the definition of c we know that c is the length of the line segment from the center to a focus. Also, c is the length of the semidiagonal as shown in Fig. 20–47. This shows us the geometric meaning of the relation among a, b, and c which is given in Eq. (20–21).

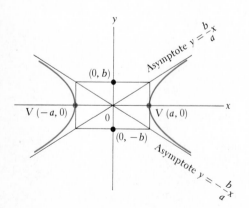

Figure 20–46

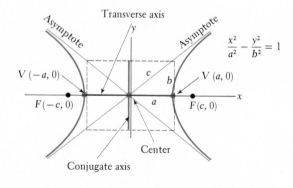

Figure 20–47

If the transverse axis is along the y-axis and the conjugate axis is along the x-axis, the equation of a hyperbola with its center at the origin is

$$\frac{y^2}{a^2} - \frac{x^2}{b^2} = 1 \qquad (20\text{–}24)$$

The symmetry of the hyperbolas given by Eqs. (20–20) and (20–24) to each of the axes and to the origin can be proven since each of x and y can be replaced by its negative in these equations without changing the equations.

Example A

The hyperbola

$$\frac{x^2}{16} - \frac{y^2}{9} = 1$$

has vertices at $(4, 0)$ and $(-4, 0)$. Its transverse axis extends from one vertex to the other. Its conjugate axis extends from $(0, 3)$ to $(0, -3)$. Since $c^2 = a^2 + b^2$, we find that $c = 5$, which means the foci are the points $(5, 0)$ and $(-5, 0)$. Drawing in the rectangle and then the asymptotes (Fig. 20–48), we draw the hyperbola from each vertex toward the asymptotes. Thus, the curve is sketched.

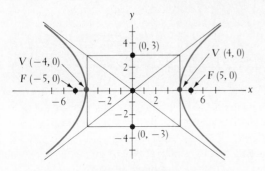

Figure 20–48

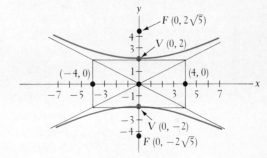

Figure 20–49

Example B

The hyperbola

$$\frac{y^2}{4} - \frac{x^2}{16} = 1$$

has vertices at $(0, 2)$ and $(0, -2)$. Its conjugate axis extends from $(4, 0)$ to $(-4, 0)$. The foci are $(0, 2\sqrt{5})$ and $(0, -2\sqrt{5})$. Since, with 1 on the right side of the equation, the y^2-term is the positive term, this hyperbola is in the form of Eq. (20–24). [Example A illustrates a hyperbola of the type of Eq. (20–20), since the x^2-term is the positive term.] Since $2a$ extends along the y-axis, we see that the equations of the asymptotes are $y = \pm(a/b)x$. This is not a contradiction of Eq. (20–23), but an extension of it for the case of a hyperbola with its transverse axis along the y-axis. The ratio a/b simply expresses the slope of the asymptote (see Fig. 20–49).

Example C

Determine the coordinates of the vertices and foci of the hyperbola $4x^2 - 9y^2 = 36$.

First, by dividing through by 36, we can put this equation in standard form. Thus, we have

$$\frac{x^2}{9} - \frac{y^2}{4} = 1$$

The transverse axis is along the x-axis with vertices at $(3, 0)$ and $(-3, 0)$. Since $c^2 = a^2 + b^2$, $c = \sqrt{13}$. This means that the foci are at $(\sqrt{13}, 0)$ and $(-\sqrt{13}, 0)$ (see Fig. 20–50).

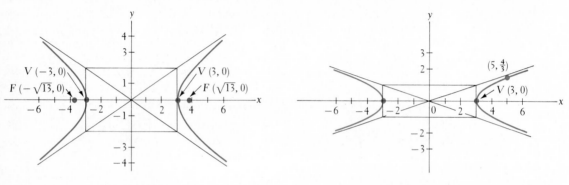

Figure 20–50   Figure 20–51

Example D

Find the equation of the hyperbola with its center at the origin, a vertex at $(3, 0)$, and which passes through $(5, \frac{4}{3})$.

When we know that the hyperbola has its center at the origin and a vertex at $(3, 0)$, we know that the standard form of its equation is given by Eq. (20–20). This also tells us that $a = 3$. By using the fact that the coordinates of the point $(5, \frac{4}{3})$ must satisfy the equation, we are able to find the value of b^2. Substituting, we have

$$\frac{25}{9} - \frac{\left(\frac{16}{9}\right)}{b^2} = 1$$

from which we find $b^2 = 1$. Therefore, the equation of the hyperbola is

$$\frac{x^2}{9} - \frac{y^2}{1} = 1$$

or

$$x^2 - 9y^2 = 9$$

This hyperbola is shown in Fig. 20–51.

Equations (20–20) and (20–24) give us the standard form of the equation of the hyperbola with its center at the origin and its foci on one of the coordinate axes. There is one other important equation form which represents a hyperbola, and that is

$$xy = c \qquad\qquad (20\text{–}25)$$

The asymptotes of this hyperbola are the coordinate axes, and the foci are on the line $y = x$, or on the line $y = -x$, if c is negative. This hyperbola is symmetrical to the origin, for if $-x$ replaces x, and $-y$ replaces y at the same time, we have $(-x)(-y) = c$, or $xy = c$. The equation remains unchanged. However, if either $-x$ replaces x, or if $-y$ replaces y, but not both, the sign on the left is changed. Therefore, it is not symmetrical to either axis. The c here represents a constant and is not related to the focus. The following example illustrates this type of hyperbola.

Example E

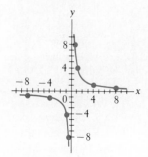

Figure 20–52

Plot the graph of the equation $xy = 4$.

We find the values of the table below, and then plot the appropriate points. Here it is permissible to use a limited number of points, since we know that the equation represents a hyperbola (Fig. 20–52). Thus, using $y = 4/x$, we obtain the values in the table.

x	-8	-4	-1	$-\frac{1}{2}$	$\frac{1}{2}$	1	4	8
y	$-\frac{1}{2}$	-1	-4	-8	8	4	1	$\frac{1}{2}$

We conclude that the equation of a hyperbola is characterized by the presence of both an x^2- and a y^2-term, having different signs, or by the presence of an xy-term with no squared terms.

Exercises 20–6

In Exercises 1 through 12, find the coordinates of the vertices and the foci of the given hyperbolas. Sketch each curve.

1. $\dfrac{x^2}{25} - \dfrac{y^2}{144} = 1$

2. $\dfrac{x^2}{16} - \dfrac{y^2}{4} = 1$

3. $\dfrac{y^2}{9} - \dfrac{x^2}{1} = 1$

4. $\dfrac{y^2}{2} - \dfrac{x^2}{2} = 1$

5. $2x^2 - y^2 = 4$

6. $3x^2 - y^2 = 9$

7. $2y^2 - 5x^2 = 10$

8. $3y^2 - 2x^2 = 6$

9. $4x^2 - y^2 + 4 = 0$

10. $9x^2 - y^2 - 9 = 0$

11. $4x^2 - 9y^2 = 16$

12. $y^2 - 9x^2 = 25$

In Exercises 13 through 20, find the equations of the hyperbolas satisfying the given conditions. The center of each is at the origin.

13. Vertex (3, 0), focus (5, 0)

14. Vertex (0, 1), focus (0, $\sqrt{3}$)

15. Conjugate axis = 12, vertex (0, 10)

16. Focus (8, 0), transverse axis = 4

17. Passes through (2, 3), focus (2, 0).

18. Passes through (8, $\sqrt{3}$), vertex (4, 0).

19. Passes through (5, 4) and (3, $\frac{4}{5}\sqrt{5}$).

20. Passes through (1, 2) and (2, $2\sqrt{2}$).

In Exercises 21 through 24, sketch the graphs of the hyperbolas given.

21. $xy = 2$ 22. $xy = 10$ 23. $xy = -2$ 24. $xy = -4$

In Exercises 25 through 28, find the equations of the hyperbolas with the given properties by use of the definition of the hyperbola.

25. Foci at (1, 2) and (11, 2), with a transverse axis of 8.

26. Vertices (-2, 4) and (-2, -2), with conjugate axis of 4.

27. Center at (2, 0), vertex at (3, 0), conjugate axis of 6.

28. Center at (1, -1), focus at (1, 4), vertex at (1, 2).

In Exercises 29 through 34, solve the given problems.

29. Sketch the graph of impedance versus resistance if the reactance $X_L - X_C$ of a given electric circuit is constant at 60 Ω (see Section 11–7).

30. One statement of Boyle's law is that the product of the pressure and volume, for constant temperature, remains a constant for a perfect gas. If one set of values for a perfect gas under the condition of constant temperature is that the pressure is 300 kPa for a volume of 8.0 L, sketch a graph of pressure versus volume.

31. The relationship between the frequency f, wavelength λ, and the velocity v of a wave is given by $v = f\lambda$. The velocity of light is a constant, being 3.0×10^{10} cm/s. The visible spectrum ranges in wavelength from about 4.0×10^{-5} cm (violet) to about 7.0×10^{-5} cm (red). Sketch a graph of frequency f (in Hertz) as a function of wavelength for the visible spectrum.

32. Wavelengths of gamma rays vary from about 10^{-8} cm to about 10^{-13} cm, but the gamma rays have the same velocity as light. On logarithmic paper, plot the graph of frequency versus wavelength for gamma rays. What type of curve results when this type of hyperbola is plotted on logarithmic paper?

33. An electronic instrument located at point P records the sound of a rifle shot and the impact of the bullet striking the target at the same instant. Show that P lies on a branch of a hyperbola.

34. Two concentric hyperbolas are called conjugate hyperbolas if the transverse and conjugate axes of one are respectively the conjugate and transverse axes of the other. What is the equation of the hyperbola conjugate to the hyperbola of Exercise 14?

20–7 Translation of Axes

Until now, except by direct use of the definition, the equations considered for the parabola, the ellipse, and the hyperbola have been restricted to the particular cases in which the vertex of the parabola is at the origin and the center of the ellipse or hyperbola is at the origin. In this section we shall consider, without specific use of the definition, the equations of these curves for the cases in which the axis of the curve is parallel to one of the coordinate axes. This is done by **translation of axes.**

We choose a point (h, k) in the xy-coordinate plane and let this point be the origin of another coordinate system, the $x'y'$-coordinate system. The x'-axis is parallel to the x-axis and the y'-axis is parallel to the y-axis. Every point in the plane now has two sets of coordinates associated with it, (x, y) and (x', y'). From Fig. 20–53 we see that

$$x = x' + h \quad \text{and} \quad y = y' + k \tag{20–26}$$

Equations (20–26) can also be written in the form

$$x' = x - h \quad \text{and} \quad y' = y - k \tag{20–27}$$

The following examples illustrate the use of Eqs. (20–27) in the analysis of equations of the parabola, ellipse, and hyperbola.

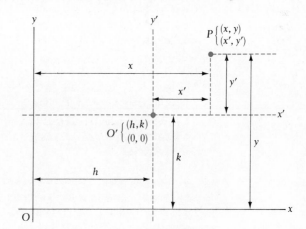

Figure 20–53

Example A

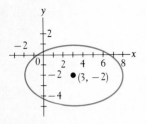

Figure 20–54

Describe the locus of the equation

$$\frac{(x-3)^2}{25} + \frac{(y+2)^2}{9} = 1$$

In this equation, given that $h = 3$ and $k = -2$, we have $x' = x - 3$ and $y' = y + 2$. In terms of x' and y', the equation is

$$\frac{(x')^2}{25} + \frac{(y')^2}{9} = 1$$

We recognize this equation as that of an ellipse (see Fig. 20–54) with a semi-major axis of 5 and a semiminor axis of 3. The center of the ellipse is at $(3, -2)$ since this was the choice of h and k to make the equation fit a standard form.

Example B

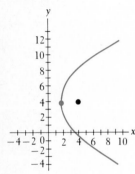

Figure 20–55

Find the equation of the parabola with vertex at $(2, 4)$ and focus at $(4, 4)$.

If we let the origin of the $x'y'$-coordinate system be the point $(2, 4)$, the point $(4, 4)$ would be the point $(2, 0)$ in the $x'y'$-system. This means that $p = 2$ and $4p = 8$ (Fig. 20–55). In the $x'y'$-system, the equation is

$$(y')^2 = 8(x')$$

Since $(2, 4)$ is the origin of the $x'y'$-system, this means that $h = 2$ and $k = 4$. Using Eq. (20–27), we have

$$(y - 4)^2 = 8(x - 2)$$

as the equation of the parabola in the xy-coordinate system. If this equation is multiplied out, and like terms are combined, we obtain

$$y^2 - 8x - 8y + 32 = 0$$

Example C

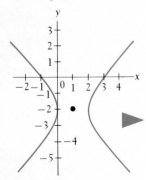

Figure 20–56

Find the center of the hyperbola $2x^2 - y^2 - 4x - 4y - 4 = 0$.

To analyze this curve, we first complete the square in the x-terms and in the y-terms. This will allow us to recognize properly the choice of h and k.

$$2x^2 - 4x - y^2 - 4y = 4$$
$$2(x^2 - 2x \quad) - (y^2 + 4y \quad) = 4$$
$$2(x^2 - 2x + 1) - (y^2 + 4y + 4) = 4 + 2 - 4$$

We note here that when we added 1 to complete the square of the x-terms within the parentheses, we were actually adding 2 to the left side. Thus, we added 2 to the right side. Similarly, when we added 4 to the y-terms within the parentheses, we were actually subtracting 4 from the left side. Continuing, we have

$$2(x - 1)^2 - (y + 2)^2 = 2$$
$$\frac{(x - 1)^2}{1} - \frac{(y + 2)^2}{2} = 1$$

Thus, if we let $h = 1$ and $k = -2$, the equation in the $x'y'$-system becomes

$$\frac{(x')^2}{1} - \frac{(y')^2}{2} = 1$$

This means that the center is at $(1, -2)$, since this point corresponds to the origin of the $x'y'$-coordinate system (see Fig. 20–56).

Example D

Find the vertex of the parabola $2x^2 - 12x - 3y + 15 = 0$.

First, we complete the square in the x-terms, placing all other resulting terms on the right. By factoring the resulting expression on the right, we may determine the values of h and k.

$$2x^2 - 12x = 3y - 15$$
$$2(x^2 - 6x \quad\;\;) = 3y - 15$$
$$2(x^2 - 6x + 9) = 3y - 15 + 18$$
$$2(x - 3)^2 = 3(y + 1)$$
$$(x - 3)^2 = \frac{3}{2}(y + 1)$$
$$x'^2 = \frac{3}{2}y'$$

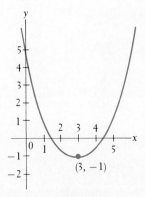

Figure 20–57

Therefore, the vertex is at $(3, -1)$, since this point corresponds to the origin of the $x'y'$-coordinate system. See Fig. 20–57.

Example E

Glass beakers are to be made with a height of 3 in. Express the surface area of the beakers in terms of the radius of the base. Sketch the graph of area versus radius.

The total surface area of a beaker is the sum of the area of the base and the lateral surface area of the side. In general, this surface area S in terms of the radius r of the base and the height h of the side is

$$S = \pi r^2 + 2\pi r h$$

Since h is constant at 3 in., we have

$$S = \pi r^2 + 6\pi r$$

which is the desired relationship.

For the purposes of sketching the graph of S and r, we now complete the square of the r terms:

$$S = \pi(r^2 + 6r)$$
$$S + 9\pi = \pi(r^2 + 6r + 9)$$
$$S + 9\pi = \pi(r + 3)^2$$
$$(r + 3)^2 = \frac{1}{\pi}(S + 9\pi)$$

(continued on next page)

We note that this equation represents a parabola with vertex at $(-3, -9\pi)$ for its coordinates (r, S). Since $4p = \frac{1}{\pi}$, $p = \frac{1}{4\pi}$ and the focus of this parabola is at $(-3, \frac{1}{4\pi} - 9\pi)$. Only positive values for S and r have meaning, and therefore the part of the graph for negative r is shown as a dashed curve (see Fig. 20–58).

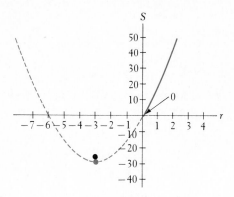

Figure 20–58

Exercises 20–7

In Exercises 1 through 8, describe the locus of each of the given equations. Identify the type of curve and its center (vertex if it is a parabola). Sketch each curve.

1. $(y - 2)^2 = 4(x + 1)$

2. $\dfrac{(x + 4)^2}{4} + \dfrac{(y - 1)^2}{1} = 1$

3. $\dfrac{(x - 1)^2}{4} - \dfrac{(y - 2)^2}{9} = 1$

4. $(y + 5)^2 = -8(x - 2)$

5. $\dfrac{(x + 1)^2}{1} + \dfrac{y^2}{9} = 1$

6. $\dfrac{(y - 4)^2}{16} - \dfrac{(x + 2)^2}{4} = 1$

7. $(x + 3)^2 = -12(y - 1)$

8. $\dfrac{x^2}{16} + \dfrac{(y + 1)^2}{1} = 1$

In Exercises 9 through 20, find the equation of each of the curves described by the given information.

9. Parabola: vertex $(-1, 3)$, $p = 4$, axis parallel to x-axis.

10. Parabola: vertex $(2, -1)$, directrix $y = 3$.

11. Parabola: vertex $(-3, 2)$, focus $(-3, 3)$.

12. Parabola: focus $(2, 4)$, directrix $x = 6$.

13. Ellipse: center $(-2, 2)$, focus $(-5, 2)$, vertex $(-7, 2)$.

14. Ellipse: center $(0, 3)$, focus $(12, 3)$, major axis 26 units.

15. Ellipse: vertices $(-2, -3)$ and $(-2, 5)$, end of minor axis $(0, 1)$.

16. Ellipse: foci $(1, -2)$ and $(1, 10)$, minor axis 5 units.

17. Hyperbola: vertex $(-1, 1)$, focus $(-1, 4)$, center $(-1, 2)$.

18. Hyperbola: foci $(2, 1)$ and $(8, 1)$, conjugate axis 6 units.

19. Hyperbola: vertices $(2, 1)$ and $(-4, 1)$, focus $(-6, 1)$.

20. Hyperbola: center $(1, -4)$, focus $(1, 1)$, transverse axis 8 units.

In Exercises 21 through 28, determine the center (or vertex if the curve is a parabola) of the given curves. Sketch each curve.

21. $x^2 + 2x - 4y - 3 = 0$ 22. $y^2 - 2x - 2y - 9 = 0$

23. $4x^2 + 9y^2 + 24x = 0$ 24. $2x^2 + 9y^2 + 8x - 72y + 134 = 0$

25. $9x^2 - y^2 + 8y - 7 = 0$ 26. $5x^2 - 4y^2 + 20x + 8y = 4$

27. $2x^2 - 4x = 9y - 2$ 28. $4x^2 + 16x = y - 20$

In Exercises 29 through 32, find the required equations.

29. Find the equation of the hyperbola with asymptotes $x - y = -1$ and $x + y = -3$, and vertex $(3, -1)$.

30. The circle $x^2 + y^2 + 4x - 5 = 0$ passes through the foci and the ends of the minor axis of an ellipse which has its major axis along the x-axis. Find the equation of the ellipse.

31. A first parabola has its vertex at the focus of a second parabola, and its focus at the vertex of the second parabola. If the equation of the second parabola is $y^2 = 4x$, find the equation of the first parabola.

32. Verify each of the following equations as being the standard form as indicated. Parabola, vertex at (h, k), axis parallel to the x-axis:

$$(y - k)^2 = 4p(x - h)$$

Parabola, vertex at (h, k), axis parallel to the y-axis:

$$(x - h)^2 = 4p(y - k)$$

Ellipse, center at (h, k), major axis parallel to the x-axis:

$$\frac{(x - h)^2}{a^2} + \frac{(y - k)^2}{b^2} = 1$$

Ellipse, center at (h, k), major axis parallel to the y-axis:

$$\frac{(y - k)^2}{a^2} + \frac{(x - h)^2}{b^2} = 1$$

Hyperbola, center at (h, k), transverse axis parallel to the x-axis:

$$\frac{(x - h)^2}{a^2} - \frac{(y - k)^2}{b^2} = 1$$

Hyperbola, center at (h, k), transverse axis parallel to the y-axis:

$$\frac{(y - k)^2}{a^2} - \frac{(x - h)^2}{b^2} = 1$$

In Exercises 33 through 36, solve the given problems.

33. The power supplied to a circuit by a battery with a voltage E and an internal resistance r is given by $P = EI - rI^2$, where P is the power (in watts) and I is the current (in amperes). Sketch the graph of P vs. I for a 6.0 V battery with an internal resistance of 0.30 Ω.

34. A calculator company determined that its total income I from the sale of a particular type of calculator is given by $I = 100x - x^2$, where x is the selling price of the calculator. Sketch a graph of I vs. x.

35. The planet Pluto moves about the sun in an elliptical orbit, with the sun at one focus. The closest that Pluto approaches the sun is 2.8 billion miles, and the farthest it gets from the sun is 4.6 billion miles. If the sun is at the origin of a coordinate system and the other focus is on the positive x-axis, what is the equation of the path of Pluto?

36. A rectangular tract of land is to have a perimeter of 800 m. Express the area in terms of its width and sketch the graph.

20–8 The Second-Degree Equation

The equations of the circle, parabola, ellipse, and hyperbola are all special cases of the same general equation. In this section we discuss this equation and how to identify the particular form it takes when it represents a specific type of curve.

Each of these curves can be represented by a **second-degree equation** *of the form*

$$Ax^2 + Bxy + Cy^2 + Dx + Ey + F = 0 \qquad (20\text{--}28)$$

[This equation is the same as Eq. (13–1).] The coefficients of the second-degree terms determine the type of curve which results. Recalling the discussions of the general forms of the equations of the circle, parabola, ellipse, and hyperbola from the previous sections of this chapter, we have the following results.

Equation (20–28) represents the indicated curve for the given conditions for A, B, and C.

1. If $A = C$, $B = 0$, a circle.
2. If $A \neq C$ (but they have the same sign), $B = 0$, an ellipse.
3. If A and C have different signs, $B = 0$, a hyperbola.
4. If $A = 0$, $C = 0$, $B \neq 0$, a hyperbola.
5. If either $A = 0$ or $C = 0$ (but not both), $B = 0$, a parabola.
   (Special cases, such as a single point or no real locus, can also result.)

Another conclusion about Eq. (20–28) is that, if either $D \neq 0$ or $E \neq 0$ (or both), the center (or vertex of a parabola) of the curve is not at the origin. If $B \neq 0$, the axis of the curve has been rotated. We have considered only one such case (the hyperbola $xy = c$) in this chapter.

Example A

The equation $2x^2 = 3 - 2y^2$ represents a circle. This can be seen by putting it in the form of Eq. (20–28). This form is

$$2x^2 + 2y^2 - 3 = 0$$

Here $A = C$ and, since there is no xy-term, $B = 0$. This means that it is a circle. If we write the equation as $x^2 + y^2 = \frac{3}{2}$ it fits the form of Eq. (20–12).

Example B

The equation $3x^2 = 6x - y^2 + 3$ represents an ellipse. Before we analyze the equation, we should put it in the form of Eq. (20–28). For the given equation, this form is

$$3x^2 + y^2 - 6x - 3 = 0$$

Here we see that $B = 0$, A and C have the same sign, and $A \neq C$. Therefore, it is an ellipse. The $-6x$ term indicates that the center of the ellipse is not at the origin.

Example C

The equation $2(x + 3)^2 = y^2 + 2x^2$ represents a parabola. Putting it in the form of Eq. (20–28), we have

$$2(x^2 + 6x + 9) = y^2 + 2x^2$$
$$2x^2 + 12x + 18 = y^2 + 2x^2$$
$$y^2 - 12x - 18 = 0$$

We now note that $A = 0$, $B = 0$, and $C \neq 0$. This indicates that the equation represents a parabola. Here, the -18 term indicates that the vertex is not at the origin.

Example D

Identify the curve represented by the equation $2x^2 + 12x = y^2 - 14$. Determine the appropriate important quantities associated with the curve, and sketch the graph.

Writing this equation in the form of Eq. (20–28), we have

$$2x^2 - y^2 + 12x + 14 = 0$$

In this form we identify the equation as representing a hyperbola, since A and C have different signs, and $B = 0$. We now write it in the standard form of a hyperbola.

$$2x^2 + 12x - y^2 = -14$$
$$2(x^2 + 6x \quad) - y^2 = -14$$
$$2(x^2 + 6x + 9) - y^2 = -14 + 18$$
$$2(x + 3)^2 - y^2 = 4$$
$$\frac{(x + 3)^2}{2} - \frac{y^2}{4} = 1$$
$$\frac{x'^2}{2} - \frac{y'^2}{4} = 1$$

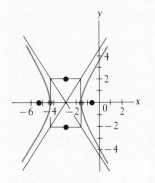

Figure 20–59

Thus, we see that the center (h, k) of the hyperbola is the point $(-3, 0)$. Also, $a = \sqrt{2}$ and $b = 2$. This means that the vertices are $(-3 + \sqrt{2}, 0)$ and $(-3 - \sqrt{2}, 0)$, and the conjugate axis extends from $(-3, 2)$ to $(-3, -2)$. Also, $c^2 = 2 + 4 = 6$, which means that $c = \sqrt{6}$. The foci are $(-3 + \sqrt{6}, 0)$ and $(-3 - \sqrt{6}, 0)$. The graph is shown in Fig. 20–59.

Example E

Identify the curve represented by the equation $4y^2 - 23 = 4(4x + 3y)$. Determine the appropriate important quantities associated with the curve, and sketch the graph.

Writing this equation in the form of Eq. (20–28), we have

$$4y^2 - 23 = 16x + 12y$$
$$4y^2 - 16x - 12y - 23 = 0$$

Therefore, we recognize the equation as representing a parabola, since $A = 0$ and $B = 0$. Now writing the equation in the standard form of a parabola, we have

$$4y^2 - 12y = 16x + 23$$
$$4(y^2 - 3y \quad) = 16x + 23$$
$$4\left(y^2 - 3y + \frac{9}{4}\right) = 16x + 23 + 9$$
$$4\left(y - \frac{3}{2}\right)^2 = 16(x + 2)$$
$$\left(y - \frac{3}{2}\right)^2 = 4(x + 2)$$
$$y'^2 = 4x'$$

We now note that the vertex is the point $(-2, \frac{3}{2})$ and that $p = 1$. Also, it is symmetric to the x'-axis. Therefore, the focus is $(-1, \frac{3}{2})$ and the directrix is $x = -3$. The graph is shown in Fig. 20–60.

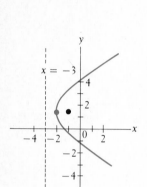

Figure 20–60

In Chapter 13, when these curves were first introduced, they were referred to as **conic sections.** If a plane is passed through a cone, the intersection of the plane and the cone results in one of these curves, the curve formed depends on the angle of the plane with respect to the axis of the cone. This is indicated in Fig. 20–61.

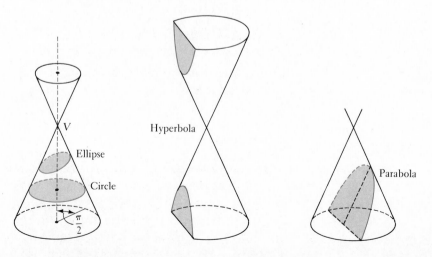

Figure 20–61

Exercises 20–8

In Exercises 1 through 20, identify each of the equations as representing either a circle, parabola, ellipse, or hyperbola.

1. $x^2 + 2y^2 - 2 = 0$ 2. $x^2 - y = 0$

3. $2x^2 - y^2 - 1 = 0$ 4. $3x^2 + 3y^2 - 1 = 0$

5. $2x^2 + 2y^2 - 3y - 1 = 0$ 6. $x^2 - 2y^2 - 3x - 1 = 0$

7. $2x^2 - x - y = 1$ 8. $2x^2 + 4y^2 - y - 2x = 4$

9. $x^2 = y^2 - 1$ 10. $3x^2 = 2y - 4y^2$

11. $x^2 = y - y^2$ 12. $y = 3 - 6x^2$

13. $x(y + 3x) = x^2 + xy - y^2 + 1$ 14. $x(2 - x) = y^2$

15. $2xy + x - 3y = 6$ 16. $(y + 1)^2 = x^2 + y^2 - 1$

17. $2x(x - y) = y(3 - y - 2x)$ 18. $2x^2 = x(x - 1) + 4y^2$

19. $y(3 - 2y) = 2(x^2 - y^2)$ 20. $4x(x - 1) = 2x^2 - 2y^2 + 3$

In Exercises 21 through 28, identify the curve represented by each of the given equations. Determine the appropriate important quantities associated with the curve, and sketch the graph.

21. $x^2 = 8(y - x - 2)$ 22. $x^2 = 6x - 4y^2 - 1$

23. $y^2 = 2(x^2 - 2x - 2y)$ 24. $4x^2 + 4 = 9 - 8x - 4y^2$

25. $y^2 + 42 = 2x(10 - x)$ 26. $x^2 - 4y = y^2 + 4(1 - x)$

27. $4(y^2 - 4x - 2) = 5(4y - 5)$ 28. $2(2x^2 - y) = 8 - y^2$

In Exercises 29 through 32, use the given values to determine the type of curve represented.

29. For the equation $x^2 + ky^2 = a^2$, what type of curve is represented if (a) $k = 1$, (b) $k < 0$, and (c) if $k > 0$ $(k \neq 1)$?

30. For the equation $\dfrac{x^2}{4 - C} - \dfrac{y^2}{C} = 1$, what type of curve is represented if (a) $C < 0$, (b) $0 < C < 4$? (For $C > 4$, see Exercise 32.)

31. In Eq. (20–28), if $A = C \neq 0$ and $B = D = E = F = 0$, what locus is described?

32. For the equation in Exercise 30, what type of locus is described if $C > 4$?

In Exercises 33 through 36, set up the necessary equation and then determine the type of curve it represents.

33. For a given alternating-current circuit the resistance and capacitive reactance are constant. What type of curve is represented by the equation relating impedance and inductive reactance? (See Section 11–7.)

34. A room is 8.0 ft high, and the length is 6.5 ft longer than the width. Express the volume V of the (rectangular) room in terms of the width w. What type of curve is represented by the equation?

35. The sides of a rectangle are $2x$ and y, and the diagonal is $x + 5$. What type of curve is represented by the equation relating x and y?

36. The legs of a right triangle are x and y and the hypotenuse is $x + 2$. What type of curve is represented by the equation relating x and y?

20–9 Polar Coordinates

Thus far we have graphed all curves in one coordinate system. This system, the rectangular coordinate system, is probably the most useful and widely applicable system. However, for certain types of curves, other coordinate systems prove to be better adapted. These coordinate systems are widely used, especially when certain applications of higher mathematics are involved. We shall discuss one of these systems here.

Instead of designating a point by its x- and y-coordinates, we can specify its location by its radius vector and the angle which the radius vector makes with the x-axis. Thus, the r and θ that are used in the definitions of the trigonometric functions can also be used as the coordinates of points in the plane. The important aspect of choosing coordinates is that, for each set of values, there must be only one point which corresponds to this set. We can see that this condition is satisfied by the use of r and θ as coordinates. *In* **polar coordinates** *the origin is called the* **pole,** *and the half-line for which the angle is zero (equivalent to the positive x-axis) is called the* **polar axis** (see Fig. 20–62).

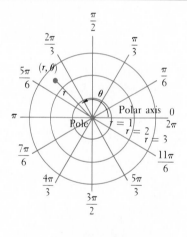

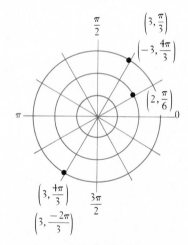

Figure 20–62   Figure 20–63

Example A

If $r = 2$ and $\theta = \pi/6$, we have the point as indicated in Fig. 20–63. The coordinates (r, θ) of this point are written as $(2, \pi/6)$ when polar coordinates are used. This point corresponds to $(\sqrt{3}, 1)$ in rectangular coordinates.

One difference between rectangular coordinates and polar coordinates is that, for each point in the plane, there are limitless possibilities for the polar coordinates of that point. For example, the point $(2, \pi/6)$ can also be represented by $(2, 13\pi/6)$ since the angles $\pi/6$ and $13\pi/6$ are coterminal. We also remove one restriction on r that we imposed in the definition of the trigonometric functions. That is, r is allowed to take on positive and negative values. If r is considered negative, then the point is found on the opposite side of the pole from that on which it is positive.

Example B

The coordinates $(3, 4\pi/3)$ and $(3, -2\pi/3)$ represent the same point. However, the point $(-3, 4\pi/3)$ is on the opposite side of the pole, three units from the pole. Another possible set of coordinates for $(-3, 4\pi/3)$ is $(3, \pi/3)$ (see Fig. 20–63).

The relationships between the polar coordinates of a point and the rectangular coordinates of the same point come directly from the definitions of the trigonometric functions. Those most commonly used are

$$x = r \cos \theta, \qquad y = r \sin \theta \qquad\qquad (20\text{--}29)$$

and

$$\tan \theta = \frac{y}{x}, \qquad r = \sqrt{x^2 + y^2} \qquad\qquad (20\text{--}30)$$

The following examples show the use of Eqs. (20–29) and (20–30) in changing coordinates in one system to coordinates in the other system. Also, they are used to transform equations from one system to the other.

Example C

Using Eqs. (20–29), we can transform the polar coordinates of $(4, \pi/4)$ into the rectangular coordinates $(2\sqrt{2}, 2\sqrt{2})$, since

$$x = 4 \cos \frac{\pi}{4} = 4\left(\frac{\sqrt{2}}{2}\right) = 2\sqrt{2}$$

and

$$y = 4 \sin \frac{\pi}{4} = 4\left(\frac{\sqrt{2}}{2}\right) = 2\sqrt{2}$$

Example D

Using Eqs. (20–30), we can transform the rectangular coordinates $(3, -5)$ into polar coordinates.

$$\tan \theta = -\frac{5}{3}, \qquad \theta = 5.25 \quad (\text{or } -1.03)$$

$$r = \sqrt{3^2 + (-5)^2} = 5.83$$

We know that θ is a fourth-quadrant angle since x is positive and y is negative. Therefore, the point $(3, -5)$ in rectangular coordinates can be expressed as the point $(5.83, 5.25)$ in polar coordinates. Other polar coordinates for the point are also possible.

Example E

Find the polar equation of the circle $x^2 + y^2 = 2x$.

Here we are given the equation of a circle in rectangular coordinates x and y and are to change it into an equation expressed in polar coordinates r and θ. Since

$$r^2 = x^2 + y^2 \quad \text{and} \quad x = r\cos\theta$$

we have

$$r^2 = 2r\cos\theta$$

or

$$r = 2\cos\theta$$

as the appropriate polar equation.

Example F

Find the rectangular equation of the *rose* $r = 4\sin 2\theta$.

Using the relation $2\sin\theta\cos\theta = \sin 2\theta$, we have $r = 8\sin\theta\cos\theta$. Then, using Eqs. (20–29) and (20–30), we have

$$\sqrt{x^2 + y^2} = 8\left(\frac{y}{r}\right)\left(\frac{x}{r}\right) = \frac{8xy}{r^2} = \frac{8xy}{x^2 + y^2}$$

Squaring both sides, we obtain

$$x^2 + y^2 = \frac{64x^2y^2}{(x^2 + y^2)^2} \quad \text{or} \quad (x^2 + y^2)^3 = 64x^2y^2$$

From this example we can see that plotting the graph from the rectangular equation would be complicated. However, as we will see in the following section, plotting this graph in polar coordinates is quite simple.

Exercises 20–9

In Exercises 1 through 12, plot the given polar coordinate points on polar coordinate paper.

1. $\left(3, \dfrac{\pi}{6}\right)$ 2. $(2, \pi)$ 3. $\left(\dfrac{5}{2}, -\dfrac{2\pi}{5}\right)$ 4. $\left(5, -\dfrac{\pi}{3}\right)$

5. $\left(-2, \dfrac{7\pi}{6}\right)$ 6. $\left(-5, \dfrac{\pi}{4}\right)$ 7. $\left(-3, -\dfrac{5\pi}{4}\right)$ 8. $\left(-4, -\dfrac{5\pi}{3}\right)$

9. $\left(0.5, -\dfrac{8\pi}{3}\right)$ 10. $(2.2, -6\pi)$ 11. $(2, 2)$ 12. $(-1, -1)$

In Exercises 13 through 16, find a set of polar coordinates for each of the given points expressed in rectangular coordinates.

13. $(\sqrt{3}, 1)$ 14. $(-1, -1)$ 15. $\left(-\dfrac{\sqrt{3}}{2}, -\dfrac{1}{2}\right)$ 16. $(-5, 4)$

In Exercises 17 through 20, find the rectangular coordinates corresponding to the points for which the polar coordinates are given.

17. $\left(8, \dfrac{4\pi}{3}\right)$ 18. $(-4, -\pi)$ 19. $\left(3, -\dfrac{\pi}{8}\right)$ 20. $(-1, -1)$

In Exercises 21 through 28, find the polar equation of the given rectangular equations.

21. $x = 3$ 22. $y = 2$ 23. $x^2 + y^2 = a^2$
24. $x^2 + y^2 = 4y$ 25. $y^2 = 4x$ 26. $x^2 - y^2 = a^2$
27. $x^2 + 4y^2 = 4$ 28. $y = x^2$

In Exercises 29 through 36, find the rectangular equation of each of the given polar equations.

29. $r = \sin \theta$ 30. $r = 4 \cos \theta$ 31. $r \cos \theta = 4$
32. $r \sin \theta = -2$ 33. $r = 2(1 + \cos \theta)$ 34. $r = 1 - \sin \theta$
35. $r^2 = \sin 2\theta$ 36. $r^2 = 16 \cos 2\theta$

In Exercises 37 through 40, find the required equations.

37. If we refer back to Eqs. (7–10) and (7–11), we see that the length along the arc of a circle and the area within a circular sector vary with two variables. These variables are those which we refer to as the polar coordinates. Express the arc length s and the area A in terms of rectangular coordinates. (How must we express θ?)

38. Under certain conditions, the x- and y-components of a magnetic field B are given by the equations

$$B_x = \frac{-ky}{x^2 + y^2} \quad \text{and} \quad B_y = \frac{kx}{x^2 + y^2}$$

Write these equations in terms of polar coordinates.

39. Express the equation of the cable (see Exercise 29 of Section 20–4) in polar coordinates.

40. The polar equation of the path of a weather satellite of the earth is

$$r = \frac{4800}{1 + 0.14 \cos \theta}$$

where r is measured in miles. Find the rectangular equation of the path of this satellite. The path is an ellipse, with the earth at one of the foci.

20–10 Curves in Polar Coordinates

The basic method for finding a curve in polar coordinates is the same as in rectangular coordinates. We assume values of θ and then find the corresponding values of r. These points are plotted and joined, thus forming the curve which represents the function. However, there are certain basic curves which can be sketched directly from the equation. The following examples illustrate these methods.

Example A

The graph of the polar equation $r = 3$ is a circle of radius 3, with center at the pole. This can be seen to be the case, since $r = 3$ for all possible values of θ. It is not really necessary to find specific points for this circle. See Fig. 20–64.

Example B

The graph of $\theta = \pi/6$ is a straight line through the pole. It represents all points for which $\theta = \pi/6$ for all possible values of r, positive or negative. See Fig. 20–64.

Example C

Plot the graph of $r = 1 + \cos\theta$.

We find the following table of values of r corresponding to the assumed values of θ.

θ	0	$\frac{\pi}{4}$	$\frac{\pi}{2}$	$\frac{3\pi}{4}$	π	$\frac{5\pi}{4}$	$\frac{3\pi}{2}$	$\frac{7\pi}{4}$	2π
r	2	1.7	1	0.3	0	0.3	1	1.7	2

We now see that the points on the curve start repeating, and it is unnecessary to find additional points. This curve is called a **cardioid** and is shown in Fig. 20–65.

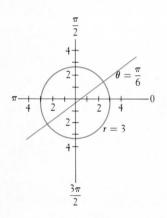

Figure 20–64

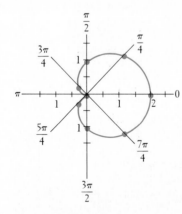

Figure 20–65

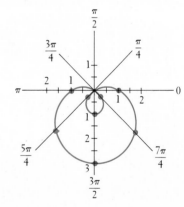

Figure 20–66

Example D

Plot the graph of $r = 1 - 2\sin\theta$.

θ	0	$\frac{\pi}{4}$	$\frac{\pi}{2}$	$\frac{3\pi}{4}$	π	$\frac{5\pi}{4}$	$\frac{3\pi}{2}$	$\frac{7\pi}{4}$	2π
r	1	-0.4	-1	-0.4	1	2.4	3	2.4	1

Particular care should be taken in plotting the points for which r is negative. This curve is known as a **limaçon** and is shown in Fig. 20–66.

Example E

Plot the graph of $r = 2 \cos 2\theta$.

θ	0	$\frac{\pi}{12}$	$\frac{\pi}{6}$	$\frac{\pi}{4}$	$\frac{\pi}{3}$	$\frac{5\pi}{12}$	$\frac{\pi}{2}$	$\frac{7\pi}{12}$	$\frac{2\pi}{3}$	$\frac{3\pi}{4}$	$\frac{5\pi}{6}$	$\frac{11\pi}{12}$	π
r	2	1.7	1	0	-1	-1.7	-2	-1.7	-1	0	1	1.7	2

For the values of θ from π to 2π, the values of r repeat. We have a four-leaf **rose** (Fig. 20–67).

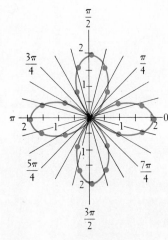

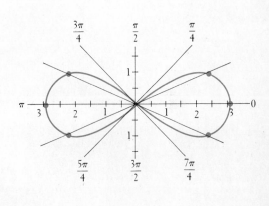

Figure 20–67　　　　　　　　　　*Figure 20–68*

Example F

Plot the graph of $r^2 = 9 \cos 2\theta$.

θ	0	$\frac{\pi}{8}$	$\frac{\pi}{4}$	$\cdots$	$\frac{3\pi}{4}$	$\frac{7\pi}{8}$	π
r	± 3	± 2.5	0		0	± 2.5	± 3

There are no values of r corresponding to values of θ in the range $\pi/4 < \theta < 3\pi/4$, since twice these angles are in the second and third quadrants, and the cosine is negative for such angles. The value of r^2 cannot be negative. Also, the values of r repeat for $\theta > \pi$. The figure is called a **lemniscate** (Fig. 20–68).

Exercises 20–10

In Exercises 1 through 28, plot the given curves in polar coordinates.

1. $r = 4$　　　　2. $r = 2$　　　　3. $\theta = \frac{3\pi}{4}$　　　　4. $\theta = \frac{5\pi}{3}$

5. $r = 4 \sec \theta$　　6. $r = 4 \csc \theta$　　7. $r = 2 \sin \theta$　　8. $r = 3 \cos \theta$

9. $r = 1 - \cos \theta$　(cardioid)　　　　10. $r = \sin \theta - 1$　(cardioid)

11. $r = 2 - \cos\theta$ (limaçon)

12. $r = 2 + 3\sin\theta$ (limaçon)

13. $r = 4\sin 2\theta$ (rose)

14. $r = 2\sin 3\theta$ (rose)

15. $r^2 = 4\sin 2\theta$ (lemniscate)

16. $r^2 = 2\sin\theta$

17. $r = 2^\theta$ (spiral)

18. $r = 10^\theta$ (spiral)

19. $r = -4\sin 4\theta$ (rose)

20. $r = -\cos 3\theta$ (rose)

21. $r = \dfrac{1}{2 - \cos\theta}$ (ellipse)

22. $r = \dfrac{1}{1 - \cos\theta}$ (parabola)

23. $r = \dfrac{6}{1 - 2\cos\theta}$ (hyperbola)

24. $r = \dfrac{6}{3 - 2\sin\theta}$ (ellipse)

25. $r = 4\cos\frac{1}{2}\theta$

26. $r = 2 + \cos 3\theta$

27. $r = 3 - \sin 3\theta$

28. $r = 4\tan\theta$

In Exercises 29 through 32, sketch the indicated graphs.

29. The charged particle (see Exercise 41 of Section 20–3) is originally traveling along the line $\theta = \pi$ toward the pole. When it reaches the pole it enters the magnetic field. Its path is then described by the polar equation $r = 2a\sin\theta$, where a is the radius of the circle. Sketch the graph of this path.

30. A cam is shaped such that the equation of the edge of the upper "half" is given by $r = 2 + \cos\theta$, and the equation of the edge of the lower "half" by $r = 3/(2 - \cos\theta)$. Plot the curve which represents the shape of the cam.

31. A satellite at a height proper to make one revolution per day around the earth will have for an excellent approximation of its projection on the earth of its path the curve

$$r^2 = R^2 \cos 2\left(\theta + \frac{\pi}{2}\right)$$

where R is the radius of the earth. Sketch the path of the projection.

32. Sketch the graph of the rectangular equation

$$4(x^6 + 3x^4y^2 + 3x^2y^4 + y^6 - x^4 - 2x^2y^2 - y^4) + y^2 = 0$$

[*Hint:* The equation can be written as $4(x^2 + y^2)^3 - 4(x^2 + y^2)^2 + y^2 = 0$. Transform to polar coordinates, and then sketch the curve.]

20–11 Exercises for Chapter 20

In Exercises 1 through 12, find the equation of the indicated curve subject to the given conditions. Sketch each curve.

1. Straight line: passes through $(1, -7)$ with a slope of 4.

2. Straight line: passes through $(-1, 5)$ and $(-2, -3)$.

3. Straight line: perpendicular to $3x - 2y + 8 = 0$ and has a y-intercept of -1.

4. Straight line: parallel to $2x - 5y + 1 = 0$ and has an x-intercept of 2.

5. Circle: center at $(1, -2)$, passes through $(4, -3)$.

6. Circle: tangent to the line $x = 3$, center at $(5, 1)$.

7. Parabola: focus $(3, 0)$, vertex $(0, 0)$.

8. Parabola: directrix $y = -5$, vertex $(0, 0)$.

9. Ellipse: vertex (10, 0), focus (8, 0), center (0, 0).

10. Ellipse: center (0, 0), passes through (0, 3) and (2, 1).

11. Hyperbola: vertex (0, 13), center (0, 0), conjugate axis of 24.

12. Hyperbola: foci (0, 10) and (0, −10), vertex (0, 8).

In Exercises 13 through 24, find the indicated quantities for each of the given equations. Sketch each curve.

13. $x^2 + y^2 + 6x - 7 = 0$, center and radius

14. $x^2 + y^2 - 4x + 2y - 20 = 0$, center and radius

15. $x^2 = -20y$, focus and directrix

16. $y^2 = 24x$, focus and directrix

17. $16x^2 + y^2 = 16$, vertices and foci

18. $2y^2 - 9x^2 = 18$, vertices and foci

19. $2x^2 - 5y^2 = 8$, vertices and foci

20. $2x^2 + 25y^2 = 50$, vertices and foci

21. $x^2 - 8x - 4y - 16 = 0$, vertex and focus

22. $y^2 - 4x + 4y + 24 = 0$, vertex and directrix

23. $4x^2 + y^2 - 16x + 2y + 13 = 0$, center

24. $x^2 - 2y^2 + 4x + 4y + 6 = 0$, center

In Exercises 25 through 32, plot the given curves in polar coordinates.

25. $r = 4(1 + \sin\theta)$

26. $r = 1 + 2\cos\theta$

27. $r = 4\cos 3\theta$

28. $r = -3\sin\theta$

29. $r = \cot\theta$

30. $r = \dfrac{1}{2(\sin\theta - 1)}$

31. $r = 2\sin\left(\dfrac{\theta}{2}\right)$

32. $r = \theta$

In Exercises 33 through 36, find the polar equation of each of the given rectangular equations.

33. $y = 2x$ 34. $2xy = 1$ 35. $x^2 - y^2 = 16$ 36. $x^2 + y^2 = 7 - 6y$

In Exercises 37 through 40, find the rectangular equation of each of the given polar equations.

37. $r = 2\sin 2\theta$

38. $r^2 = \sin\theta$

39. $r = \dfrac{4}{2 - \cos\theta}$

40. $r = \dfrac{2}{1 - \sin\theta}$

In Exercises 41 through 44, determine the number of real solutions to the given systems of equations by sketching the curves. (See Section 13–1.)

41. $x^2 + y^2 = 9$
    $4x^2 + y^2 = 16$

42. $y = e^x$
    $x^2 - y^2 = 1$

43. $x^2 + y^2 - 4y - 5 = 0$
    $y^2 - 4x^2 - 4 = 0$

44. $x^2 - 4y^2 + 2x - 3 = 0$
    $y^2 - 4x - 4 = 0$

In Exercises 45 through 76, solve the given problems.

45. Find the points of intersection of the ellipses $25x^2 + 4y^2 = 100$ and $4x^2 + 9y^2 = 36$.

46. Find the points of intersection of the hyperbola $y^2 - x^2 = 1$ and the ellipse $x^2 + 25y^2 = 25$.

47. In two ways show that the line segments joining $(-3, 11)$, $(2, -1)$, and $(14, 4)$ form a right triangle.

48. Show that the altitudes of the triangle with vertices $(2, -4)$, $(3, -1)$, and $(-2, 5)$ meet at a single point.

49. By means of the definition of a parabola, find the equation of the parabola with focus at $(3, 1)$ and directrix the line $y = -3$. Find the same equation by the method of translation of axes.

50. Repeat the instructions of Exercise 49 for the parabola with focus at $(0, 2)$ and directrix the line $y = 4$.

51. A piece x centimeters long is cut from a board 250 cm long. Sketch the graph of the remaining length y as a function of x.

52. The acceleration of an object is defined as the change in velocity v divided by the corresponding change in time t. Find the equation relating the velocity v and time t for an object for which the acceleration is 20 ft/s^2 and $v = 5.0$ ft/s when $t = 0$ s.

53. In an electric circuit the voltage V equals the product of the current I and the resistance R. Sketch the graph of V (in volts) vs. I (in amperes) for a circuit in which $R = 3\ \Omega$.

54. An airplane touches down when landing at 100 mi/h. Its velocity v while coming to a stop is given by $v = 100 - 20{,}000t$, where t is the time in hours. Sketch the graph of v vs. t.

55. In a certain electric circuit, two resistors having resistances R_1 and R_2, respectively, act as a voltage divider. The relation between them is $\alpha = R_1/(R_1 + R_2)$, where α is a constant less than 1. Sketch the curve of R_1 vs. R_2 (both in ohms), if $\alpha = \frac{1}{2}$.

56. Let C and F denote corresponding Celsius and Fahrenheit temperature readings. If the equation relating the two is linear, determine this equation given that $C = 0$ when $F = 32$ and $C = 100$ when $F = 212$.

57. A solar panel reflector is circular and has a circumference of 4.50 m. Taking the origin of a coordinate system at the center of the reflector, what is the equation of the circumference?

58. A friction disk drive wheel is tangent to another wheel, as shown in Fig. 20–69. The radii of the wheels are 2 in. and 6 in. What is the equation of the circumference of each wheel if the coordinate system is located at the center of the drive wheel as shown?

59. The arch of a small bridge across a stream is parabolic. If, at water level, the span of the arch is 80 ft and the maximum height of the span above water level is 20 ft, what is the equation of the arch? Choose the most convenient point for the origin of the coordinate system.

60. If a source of light is placed at the focus of a parabolic reflector, the reflected rays are parallel. Where is the focus of a parabolic reflector which is 8 cm across and 6 cm deep?

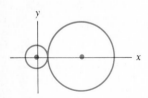

Figure 20–69

61. Sketch the curve of the total surface area of a right circular cylinder as a function of its radius, if the height is always 10 units.

62. At very low temperatures certain metals have an electric resistance of zero. This phenomenon is called superconductivity. A magnetic field also affects the super-conductivity. A certain level of magnetic field, the threshold field, is related to the temperature T by

$$\frac{H_T}{H_0} = 1 - \left(\frac{T}{T_0}\right)^2$$

where H_0 and T_0 are specifically defined values of magnetic field and temperature. Sketch H_T/H_0 vs. T/T_0.

63. The vertical position of a projectile is given by $y = 120t - 16t^2$, and its horizontal position is given by $x = 60t$. By eliminating the time t, determine the path of the projectile. Sketch this path for the length of time it would need to strike the ground, assuming level terrain.

64. The rate r in grams per second at which a substance is formed in a chemical reaction is given by $r = 10.0 - 0.010t^2$. Sketch the graph of r vs. t.

65. The inside of the top of an arch is a semiellipse with an equation of $x^2 + 4y^2 = 100$. The outside of the arch is also a semiellipse with an equation $2x^2 + 5y^2 = 500$. If measurements are in meters, what is the thickness of the arch (a) at each base and (b) at the center?

66. The earth moves about the sun in an elliptical path. If the closest the earth gets to the sun is 90.5 million miles, and the farthest it gets from the sun is 93.5 million miles, find the equation of the path of the earth. The sun is at one focus of the ellipse. Assume the major axis to be along the x-axis and that the center of the ellipse is at the origin.

67. Soon after reaching the vicinity of the moon, Apollo 11 (the first spacecraft to land a man on the moon) went into an elliptical lunar orbit. The closest the craft was to the moon in this orbit was 70 mi, and the farthest it was from the moon was 190 mi. What was the equation of the path if the moon was at one of the foci of the ellipse? Assume the major axis to be along the x-axis and that the center of the ellipse is at the origin. The radius of the moon is 1080 mi.

68. A machine-part designer wishes to make a model for an elliptical cam by placing two pins in her design board, putting a loop of string over the pins, and marking off the outline by keeping the string taut. (Note that she is using the definition of the ellipse.) If the cam is to measure 10 cm by 6 cm, how long should the loop of string be and how far apart should the pins be?

69. Under certain conditions the work W done on a wire by increasing the tension from T_1 to T_2 is given by $W = k(T_2^2 - T_1^2)$, where k is a constant depending on the properties of the wire. If the work is constant for various sets of initial and final tensions, sketch a graph of T_2 vs. T_1.

70. In order to study the relationship between velocity and pressure of a stream of water, a pipe is constructed such that its cross section is a hyperbola. If the pipe is 10 ft long, 1 ft in diameter at the narrow point (middle), and 2 ft in diameter at the ends, what is the equation of the cross section of the pipe? (In physics it is shown that where the velocity of a fluid is greatest, the pressure is the least.)

71. A hallway 16 ft wide has a ceiling whose cross section is a semiellipse. The ceiling is 10 ft high at the walls and 14 ft high at the center. Find the height of the ceiling 4 ft from each wall.

72. A 60-ft rope passes over a pulley 10 ft above the ground and an object on the ground is attached at one end. A man holds the other end at a level of 4 ft above the ground. If the man walks away from the pulley, express the height of the object above the ground in terms of the distance the man is from directly below the object. Sketch the graph of distance vs. height.

73. Express the equation of the artificial satellite (see Exercise 29 of Section 20–5) in polar coordinates.

74. The x- and y-coordinates of the position of a certain moving object as functions of time are given by $x = 2t$ and $y = \sqrt{8t^2 + 1}$. Find the polar equation of the path of the object.

75. Under a force which varies inversely as the square of the distance from an attracting object (such as the force the sun exerts on the earth), it can be shown that the equation that an object follows is given in general by

$$\frac{1}{r} = a + b \cos \theta$$

where a and b are constant for a particular path. By transforming this equation to rectangular coordinates, show that this equation represents one of the conic sections, the particular section depending on the values of a and b. It is through this kind of analysis that we know the paths of the planets and comets are conic sections.

76. The sound produced by a jet engine was measured at a distance of 100 m in all directions. The loudness of the sound (in decibels) was found to be $d = 115 + 10 \cos \theta$, where the $0°$ line for the angle θ is directed in front of the engine. Sketch the graph of d vs. θ in polar coordinates (use d as r).

CHAPTER 21

Introduction to Statistics and Empirical Curve Fitting

When a mathematician wishes to state the relation between the area of a circle and its radius, he or she writes $A = \pi r^2$ and knows that this relation is true for all circles. When an engineer wishes to find the safe load which a particular cable is able to support, he or she cannot so simply write an equation for the relation, because the load which a cable may support depends on the diameter of the cable, the material of which it is made, the quality of material in the particular cable, any possible defects in its manufacture, and anything else which could make this cable different from all other cables. Thus, to determine the load which a cable can support, an engineer may test many cables of particular specifications. He or she will find that most of the cables can support approximately the same load, but that there is some variation, and occasionally perhaps a great variation for some reason or other.

The engineer, in testing a number of cables and thereby selecting a certain type of cable, is making use of the basic methods of statistics. That is, the engineer (1) collects data, (2) analyzes this data, and (3) interprets this data. In the first three sections of this chapter, we shall discuss some of the basic methods of tabulating and analyzing this type of statistical information. The sections which follow then show how to establish a function in equation form from appropriate statistical data.

21–1 Frequency Distributions

In statistics we deal with various sets of numbers. These could be measurements of length, test scores, weights of objects, or numerous other possibilities. We could deal with the entire set of numbers, but this is often too large to be practical. In such a case we deal with a selected (assumed to be representative) sample.

Example A

A company produces a machine part which is designed to be 3.80 cm long. If 10,000 are produced daily, it could be impractical to test each one to determine if it meets specifications. Therefore, it might be decided to test every tenth, or every hundredth, part.

If only ten such parts were produced daily, it might be possible to test each one for specifications.

One way of organizing data in order to develop some understanding of it is to *tabulate the number of occurrences for each particular value within the set. This is called a* **frequency distribution.**

Example B

A test station measured the loudness of the sound of jet aircraft taking off from a certain airport. The decibel readings of the first twenty jets were as follows: 110, 95, 100, 115, 105, 110, 120, 110, 115, 105, 90, 95, 105, 110, 100, 115, 105, 120, 95, 110.

We can see that it is difficult to determine any pattern to the readings. Therefore, we set up the following table to show the frequency distribution.

Decibel reading	90	95	100	105	110	115	120
Frequency	1	3	2	4	5	3	2

We note that the distribution and pattern of readings is clearer in this form.

Just as graphs are useful in representing algebraic functions, so also are they a very convenient method of representing frequency distributions. There are several useful methods of graphing such distributions, among which *the most important are the* **histogram** *and the* **frequency polygon.** The following examples illustrate these graphical representations.

Example C

A histogram represents a particular set of data by use of rectangles. The width of the base of each rectangle is the same, and is labeled for one of the readings. The height of the rectangle represents the number of values of a given reading. In Fig. 21–1 a histogram representing the data of the frequency distribution of Example B is shown.

Example D

The frequency polygon is used to represent a set of data by plotting as abscissas (x-values) the values of the readings, and as ordinates (y-values) the number of occurrences of each reading (the frequency). The resulting points are joined by straight-line segments. Figure 21–2 shows a frequency polygon of the data of Example B.

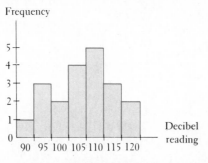

Figure 21–1

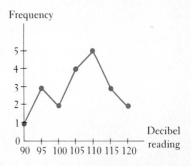

Figure 21–2

Often the number of measurements is sufficiently large that it is not feasible to tabulate the number of occurrences for a particular value. In other cases such a table does not give a clear idea of the distribution. In these cases we may designate certain intervals, and tabulate the number of values within each interval. Frequency distributions, histograms, and frequency polygons can be used to represent data tabulated in this way.

Example E

A certain mathematics course had 80 students enrolled in it. After all the tests and exams were recorded, the instructors determined a numerical average (based on 100) for each student. The following is a list of the numerical grades, with the number of students receiving each.

22—1, 37—1, 40—2, 44—1, 47—1, 53—1, 55—3, 56—1, 60—3,
61—1, 63—4, 65—2, 66—1, 67—5, 68—2, 70—2, 71—4, 72—3,
74—5, 75—4, 77—7, 78—4, 79—1, 81—2, 82—4, 84—1, 85—1,
86—3, 87—2, 88—1, 90—2, 92—1, 93—2, 95—1, 97—1

We can observe from this listing that it is difficult to see just how the grades were distributed. Therefore the instructors grouped the grades into intervals, which included five possible grades in each interval. This led to the following table.

Interval	20–24	25–29	30–34	35–39	40–44	45–49	50–54	55–59
Number in interval	1	0	0	1	3	1	1	4
Interval	60–64	65–69	70–74	75–79	80–84	85–89	90–94	95–99
Number in interval	8	10	14	16	7	7	5	2

We can see that the distribution of grades becomes much clearer in the above table. Finally, since the school graded students only with the letters A, B, C, D, and F, the instructors then grouped the grades in intervals below 60, from 60 to 69, from 70 to 79, from 80 to 89, and from 90 to 100, and assigned the

(continued on next page)

appropriate letter grade to each numerical grade within each interval. This led to the following table.

Grade	A	B	C	D	F
Number receiving this grade	7	14	30	18	11

Thus, we can see the distribution according to letter grades. We can also note that, if the number of intervals were reduced much further, it would be difficult to draw any reasonable conclusions regarding the distribution of grades.

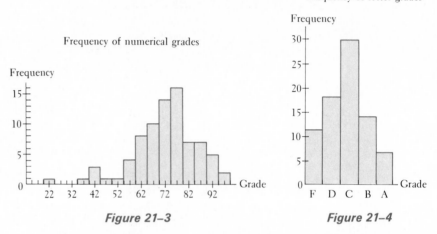

Figure 21–3   *Figure 21–4*

Histograms showing the frequency of numerical grades and letter grades are shown in Figs. 21–3 and 21–4. In Fig. 21–3, each interval is represented by one number, the middle number. In Fig. 21–5 a frequency polygon of numerical grades is shown, also with each interval represented by the middle number.

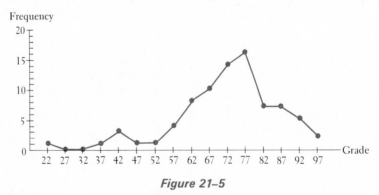

Figure 21–5

We can see from Example E that care must be taken in grouping data. If the number of intervals is too small, important characteristics of the data may not be apparent. If the number of intervals is too large, it may be difficult to analyze the data to determine the distribution and patterns.

Exercises 21–1

In Exercises 1 through 16, use the following sets of numbers.

A: 3, 6, 4, 2, 5, 4, 7, 6, 3, 4, 6, 4, 5, 7, 3

B: 25, 26, 23, 24, 25, 28, 26, 27, 23, 28, 25

C: 48, 53, 49, 45, 55, 49, 47, 55, 48, 57, 51, 46

D: 105, 108, 103, 108, 106, 104, 109, 104, 110, 108, 108, 104, 113, 106, 107, 106, 107, 109, 105, 111, 109, 108

In Exercises 1 through 4, set up a frequency distribution table, indicating the frequency of each number given in the indicated set.

1. Set A
2. Set B
3. Set C
4. Set D

In Exercises 5 through 8, set up a frequency distribution table, indicating the frequency of numbers for the given intervals of the given sets.

5. Intervals 2–3, 4–5, and 6–7 for set A.
6. Intervals 22–24, 25–27, and 28–30 for set B.
7. Intervals 43–45, 46–48, 49–51, 52–54, and 55–57 for set C.
8. Intervals 101–105, 106–110, and 111–115 for set D.

In Exercises 9 through 12, draw histograms for the data in the given exercise.

9. Exercise 1
10. Exercise 4
11. Exercise 7
12. Exercise 8

In Exercises 13 through 16, draw frequency polygons for the data in the given exercise.

13. Exercise 1
14. Exercise 4
15. Exercise 7
16. Exercise 8

In Exercises 17 through 30, find the indicated quantities.

17. A bottling machine is designed to pour 750 mL of soda into a bottle. During the day sample bottles were tested, and the following volumes were recorded (in milliliters).

    751, 754, 748, 750, 752, 751, 747, 748, 753, 750, 746, 751, 753, 749, 753

    Form a frequency distribution table indicating the frequency of each volume recorded in this data.

18. For the data of Exercise 17, draw a histogram.

19. For the data of Exercise 17, draw a frequency polygon.

20. For the data of Exercise 17, form a frequency distribution table for intervals 746–748, 749–751, and 752–754. Then draw a histogram to represent these data.

21. A strobe light is designed to flash every 2.25 s at a certain setting. Sample bulbs were tested with the results in the following table.

Number of bulbs	2	7	18	41	56	32	8	3	3
Time between flashes (seconds)	2.21	2.22	2.23	2.24	2.25	2.26	2.27	2.28	2.29

    Draw a histogram for these data.

22. Draw a frequency polygon for the data of Exercise 21.

23. The data in the following table were compiled when a particular type of cable was being tested for its breaking load.

Load interval (pounds)	830–839	840–849	850–859	860–869	870–879	880–889	890–899
Number of cables breaking	4	14	48	531	85	22	2

Draw a frequency polygon for these data.

24. Draw a histogram for the data of Exercise 23.

25. A researcher, testing an electric circuit, found the following values for the current (in milliamperes) in the circuit on successive trials:

    3.44, 3.48, 3.40, 3.29, 3.46, 3.41, 3.37, 3.45, 3.47, 3.43, 3.38, 3.50, 3.41, 3.42, 3.44, 3.46, 3.39, 3.47

    Form a histogram for the intervals 3.25–3.29, 3.30–3.34, and so on.

26. Form a histogram for the data given in Exercise 25 for the intervals 3.21–3.30, 3.31–3.40, and so on.

27. One hundred motorists were asked to maintain a miles per gallon record for their cars, all the same model. The records which were reported are shown in the following table.

Miles per gallon	19.0–19.4	19.5–19.9	20.0–20.4	20.5–20.9
Number reporting	4	9	14	24

Miles per gallon	21.0–21.4	21.5–21.9	22.0–22.4	22.5–22.9
Number reporting	28	16	4	1

Draw a histogram for this data.

28. Draw a frequency polygon for the data of Exercise 27.

29. Take two dice and toss them 100 times, recording at each toss the sum which appears. Draw a frequency polygon of the sum and the frequency with which it occurred.

30. From the financial section of a newspaper, record (to the nearest dollar) the closing price of the first 50 stocks listed. Form a frequency table with intervals 0–9, 10–19, and so forth. Form a histogram of these data.

21–2 Measures of Central Tendency

Tables and graphical representations give a general description of data. However, it is often profitable and convenient to find representative values for the location of the center of distribution, and other numbers to give a measure of the deviation from this central value. In this way we can obtain an arithmetical description of the data. We shall now discuss the values commonly used to measure the location of the center of the distribution. These are referred to as *measures of central tendency*.

The first of these measures of central tendency is the **median.** *The median is the middle number, that number for which there are as many above it as below it in the distribution.* If there is no middle number, the median is that number halfway between the two numbers nearest the middle of the distribution.

Example A

Given the numbers 5, 2, 6, 4, 7, 4, 7, 2, 8, 9, 4, 11, 9, 1, 3, we first arrange them in numerical order. This arrangement is

$$1, 2, 2, 3, 4, 4, 4, 5, 6, 7, 7, 8, 9, 9, 11$$

Since there are 15 numbers, the middle number is the eighth. Since the eighth number is 5, the median is 5.

If the number 11 is not included in this set of numbers, and there are only 14 numbers in all, the median is that number halfway between the seventh and eighth numbers. Since the seventh is 4 and the eighth is 5, the median is 4.5.

Example B

In the distribution of grades given in Example E of Section 21–1, the median is 74. There are 80 grades in all, and if they are listed in numerical order we find that the 39th through the 43rd grade is 74. This means that the 40th and 41st grades are both 74. The number halfway between the 40th and 41st grades would be the median. The fact that both are 74 means that the median is also 74.

Another very widely applied measure of central tendency is the **arithmetic mean.** *The mean is calculated by finding the sum of all the values and then dividing by the number of values.* (The arithmetic mean is the number most people call the "average." However, in statistics the word *average* has the more general meaning of a measure of central tendency.)

Example C

The arithmetic mean of the numbers given in Example A is found by finding the sum of all the numbers and dividing by 15. Thus, letting $\bar{x}$ (read as "*x* bar") represent the mean, we have

$$\bar{x} = \frac{5 + 2 + 6 + 4 + 7 + 4 + 7 + 2 + 8 + 9 + 4 + 11 + 9 + 1 + 3}{15}$$

$$= \frac{82}{15} = 5.5$$

Thus, the mean is 5.5.

If we wish to find the arithmetic mean of a large number of values, and if some of them appear more than once, the calculation can be somewhat simplified. By multiplying each number by its frequency, adding these results, and then dividing by the total number of values considered, the mean can be calculated. Letting $\bar{x}$

represent the mean of the values $x_1, x_2, \ldots, x_n$, which occur $f_1, f_2, \ldots, f_n$ times, respectively, we have

$$\bar{x} = \frac{x_1 f_1 + x_2 f_2 + \cdots + x_n f_n}{f_1 + f_2 + \cdots + f_n} \tag{21-1}$$

(This notation can be simplified by letting

$$x_1 f_1 + x_2 f_2 + \cdots + x_n f_n = \sum_{i=1}^{n} x_i f_i \qquad \text{and} \qquad f_1 + f_2 + \cdots + f_n = \sum_{i=1}^{n} f_i$$

This notation is often used to indicate a sum; Σ *is known as the* **summation sign.**)

Example D

Using Eq. (21–1) to find the arithmetic mean of the numbers of Example A, we have

$$\bar{x} = \frac{1(1) + 2(2) + 3(1) + 4(3) + 5(1) + 6(1) + 7(2) + 8(1) + 9(2) + 11(1)}{15}$$

$$= \frac{82}{15} = 5.5$$

Example E

We find the arithmetic mean of the grades in Example E of Section 21–1 by

$$\bar{x} = \frac{22(1) + (37)(1) + (40)(2) + \cdots + (67)(5) + \cdots + (97)(1)}{80}$$

$$= \frac{5733}{80} = 71.7$$

Another measure of central tendency is *the* **mode,** *which is the value which appears most frequently.* If two or more values appear with the same greatest frequency, each is a mode. If no value is repeated, there is no mode.

Example F

The mode of the numbers in Example A is 4, since it appears three times and no other value appears more than twice.

The modes of the numbers

     1, 2, 2, 4, 5, 5, 6, 7

are 2 and 5, since each appears twice and no other number is repeated.

There is no mode for the values

     1, 2, 5, 6, 7, 9

since none of the values is repeated.

Example G

In order to determine the frictional force between two specially designed surfaces, the force to move a block with one surface along an inclined plane with the other surface is measured 10 times. The results, with forces in newtons, were:

$$2.2, 2.4, 2.1, 2.2, 2.5, 2.2, 2.4, 2.7, 2.1, 2.5$$

Find the mean, median, and mode of these forces.

To find the mean we sum the values of the forces and divide this total by 10. This gives

$$F_{mean} = \frac{2.2 + 2.4 + 2.1 + 2.2 + 2.5 + 2.2 + 2.4 + 2.7 + 2.1 + 2.5}{10}$$

$$= \frac{23.3}{10}$$

$$= 2.3 \text{ N}$$

The median is found by arranging the values in order and finding the middle value. The values in order are

$$2.1, 2.1, 2.2, 2.2, 2.2, 2.4, 2.4, 2.5, 2.5, 2.7$$

Since there are 10 values, we see that the fifth value is 2.2 and the sixth is 2.4. The value midway between these is 2.3, which is the median. Therefore, the median force is 2.3 N.

The mode is 2.2 N, since this value appears three times, which is more than any other value.

Exercises 21-2

See Appendix E for a computer program for finding the arithmetic mean.

In Exercises 1 through 12, use the following sets of numbers. They are the same as those used in Exercises 21-1.

A: 3, 6, 4, 2, 5, 4, 7, 6, 3, 4, 6, 4, 5, 7, 3

B: 25, 26, 23, 24, 25, 28, 26, 27, 23, 28, 25

C: 48, 53, 49, 45, 55, 49, 47, 55, 48, 57, 51, 46

D: 105, 108, 103, 108, 106, 104, 109, 104, 110, 108, 108, 104, 113, 106, 107, 106, 107, 109, 105, 111, 109, 108

In Exercises 1 through 4, determine the median of the numbers of the given set.

1. Set A 2. Set B 3. Set C 4. Set D

In Exercises 5 through 8, determine the arithmetic mean of the numbers of the given set.

5. Set A 6. Set B 7. Set C 8. Set D

In Exercises 9 through 12, determine the mode of the numbers of the given set.

9. Set A 10. Set B 11. Set C 12. Set D

In Exercises 13 through 24, the required sets of numbers are those in Section 21–1. Find the indicated measures of central tendency.

13. The median of the volumes in Exercise 17.

14. The mean of the volumes in Exercise 17.

15. The mode of the volumes in Exercise 17.

16. The median of the times in Exercise 21.

17. The mean of the times in Exercise 21.

18. The mode of the times in Exercise 21.

19. The median of the currents in Exercise 25.

20. The mean of the currents in Exercise 25.

21. The mode of the currents in Exercise 25.

22. The median of the miles per gallon in Exercise 27. (Use the middle value for each interval.)

23. The mean of the miles per gallon in Exercise 27. (Use the middle value for each interval.)

24. The median of the prices in Exercise 30.

In Exercises 25 through 28, find the indicated measures of central tendency.

25. The weekly salaries (in dollars) for the workers in a small factory are as follows:

    250, 350, 275, 225, 175, 300, 200, 350, 275, 400, 300, 225, 250, 300

    Find the median and the mode of the salaries.

26. Find the mean salary for the salaries in Exercise 25.

27. In a particular month the electrical usage, rounded to the nearest 100 kWh (kilowatt hours), of 1000 homes in a certain city was summarized as follows:

Number of homes	22	80	106	185	380	122	90	15
Usage	500	600	700	800	900	1000	1100	1200

    Find the mean of the electrical usage.

28. Find the median and the mode of the electrical usage in Exercise 27.

In Exercises 29 through 32, use the following information. The **midrange,** another measure of central tendency of a frequency distribution, is found by finding the sum of the highest and the lowest values and dividing this sum by two. Using this definition, find the midrange of the numbers in the indicated sets of numbers at the beginning of this exercise set.

29. Set A　　　30. Set B　　　31. Set C　　　32. Set D

21–3 Standard Deviation

In the preceding section we discussed measures of central tendency of sets of data. However, regardless of the measure which may be used, it does not tell us whether the data are grouped closely together or spread over a large range of values. Therefore, we also need some measure of the deviation, or dispersion, of the values from the median or mean. If the dispersion is small and the numbers are grouped closely together, the measure of central tendency is more reliable and descriptive of the data than the case in which the spread is greater.

There are several measures of dispersion, and we discuss in this section one which is very widely used. It is called the **standard deviation.**

The standard deviation of a set of numbers is given by the equation

$$s = \sqrt{\overline{(x - \bar{x})^2}} \tag{21–2}$$

The definition of s indicates that the following steps are to be taken in computing its value.

1. Find the arithmetic mean $\bar{x}$ of the set of numbers.
2. Subtract the mean from each of the numbers of the set.
3. Square these differences.
4. Find the arithmetic mean of these squares. (Note that a bar over any quantity signifies the mean of the set of values of that quantity.)
5. Find the square root of this last arithmetic mean.

Defined in this way, s must be a positive number, and thus indicates a deviation from the mean, regardless of whether or not individual numbers are greater or less than the mean.

Example A

Find the standard deviation of the numbers 1, 5, 4, 2, 6, 2, 1, 1, 5, 3. The most efficient method of finding s is to make a table in which Steps 1–4 are indicated.

x	$x - \bar{x}$	$(x - \bar{x})^2$
1	-2	4
5	2	4
4	1	1
2	-1	1
6	3	9
2	-1	1
1	-2	4
1	-2	4
5	2	4
3	0	0
30		32

$$\bar{x} = \frac{30}{10} = 3$$

$$\overline{(x - \bar{x})^2} = \frac{32}{10} = 3.2$$

$$s = \sqrt{3.2} = 1.8$$

Example B

Find the standard deviation of the numbers given in Example A of Section 21–2.

Since several of the numbers appear more than once, it is helpful to use the frequency of each number in the table. Therefore, we have the following table and calculations.

x	f	fx	$x - \bar{x}$	$(x - \bar{x})^2$	$f(x - \bar{x})^2$
1	1	1	-4.5	20.25	20.25
2	2	4	-3.5	12.25	24.50
3	1	3	-2.5	6.25	6.25
4	3	12	-1.5	2.25	6.75
5	1	5	-0.5	0.25	0.25
6	1	6	0.5	0.25	0.25
7	2	14	1.5	2.25	4.50
8	1	8	2.5	6.25	6.25
9	2	18	3.5	12.25	24.50
11	1	11	5.5	30.25	30.25
	15	82			123.75

$$\bar{x} = \frac{82}{15} = 5.5$$

$$\overline{f(x - \bar{x})^2} = \frac{123.75}{15} = 8.25$$

$$s = \sqrt{8.25} = 2.9$$

We can see from Examples A and B that there is a great deal of calculational work required to find the standard deviation of a set of numbers. It is possible to reduce this work somewhat, and we now show how this may be done.

To find the mean of the sum of two sets of data, x_i and y_i, where the number of values in each set is the same, we can determine the mean of the x's and the mean of the y's and add the results. That is

$$\overline{x + y} = \bar{x} + \bar{y} \qquad (21-3)$$

This is true by the meaning of the definition of the arithmetic mean.

$$\overline{x + y} = \frac{x_1 + x_2 + \cdots + x_n + y_1 + y_2 + \cdots + y_n}{n}$$

$$= \frac{x_1 + x_2 + \cdots x_n}{n} + \frac{y_1 + y_2 + \cdots + y_n}{n} = \bar{x} + \bar{y}$$

Also, when we find the arithmetic mean, if all the numbers contain a constant factor, this number can be factored before the mean is found. That is,

$$\overline{kx} = k\bar{x} \qquad (21-4)$$

If we multiply out the quantity under the radical in Eq. (21–2), and then apply Eq. (21–3), we obtain

$$\overline{x^2 - 2x\bar{x} + \bar{x}^2} = \overline{x^2} - \overline{2x\bar{x}} + \overline{\bar{x}^2}$$

In this equation the number 2 and $\bar{x}$ are constants for any given problem. Thus, by using Eq. (21–4), we have

$$\overline{x^2} - \overline{2x\bar{x}} + \overline{\bar{x}^2} = \overline{x^2} - 2\bar{x}(\bar{x}) + \bar{x}^2(1) = \overline{x^2} - \bar{x}^2$$

Substituting this last result into Eq. (21–2), we have

$$s = \sqrt{\overline{x^2} - \overline{x}^2} \tag{21-5}$$

Equation (21–5) shows us that the standard deviation s may be found by finding the mean of the squares of the x's, subtracting the square of $\overline{x}$, and then finding the square root. This eliminates the step of subtracting $\overline{x}$ from each x, a step required by the original definition. Obviously, a calculator is very useful in making these calculations.

Example C

By using Eq. (21–5), find s for the numbers in Example A. Again, a table is the most convenient form.

x	x^2
1	1
5	25
4	16
2	4
6	36
2	4
1	1
1	1
5	25
3	9
30	122

$$\overline{x} = \frac{30}{10} = 3, \qquad \overline{x}^2 = 9$$

$$\overline{x^2} = \frac{122}{10} = 12.2$$

$$\overline{x^2} - \overline{x}^2 = 12.2 - 9 = 3.2$$

$$s = \sqrt{3.2} = 1.8$$

Example D

In an ammeter, two resistances are connected in parallel. Most of the current passing through the meter goes through the one called the shunt. In order to determine the accuracy of the resistance of shunts being made for ammeters, a manufacturer tested a sample of 100 shunts. The resistance of each, to the nearest hundredth of an ohm, is indicated in the following table. Calculate the standard deviation of the resistances of the shunts.

R (ohms)	f	fR	fR^2
0.200	1	0.200	0.0400
0.210	3	0.630	0.1323
0.220	5	1.100	0.2420
0.230	10	2.300	0.5290
0.240	17	4.080	0.9792
0.250	40	10.000	2.5000
0.260	13	3.380	0.8788
0.270	6	1.620	0.4374
0.280	3	0.840	0.2352
0.290	2	0.580	0.1682
	100	24.730	6.1421

$$\overline{R} = \frac{\Sigma fR}{\Sigma f} = \frac{24.73}{100} = 0.2473$$

$$\overline{R}^2 = 0.06115729$$

$$\overline{R^2} = \frac{\Sigma fR^2}{\Sigma f} = \frac{6.142}{100} = 0.06142$$

$$s = \sqrt{\overline{R^2} - \overline{R}^2} = 0.016$$

The arithmetic mean of the resistances is 0.247 Ω, with a standard deviation of 0.016 Ω.

Example E

Find the standard deviation of the grades in Example E of Section 21–1. Use the frequency distribution as grouped in intervals 20–24, 25–29, and so forth, and then assume that each value in the interval is equal to the representative value (middle value) of the interval. (This method is not exact, but when a problem involves a large number of values, the method provides a very good approximation and eliminates a great deal of arithmetic work.)

x	f	fx	fx^2
22	1	22	484
27	0	0	0
32	0	0	0
37	1	37	1,369
42	3	126	5,292
47	1	47	2,209
52	1	52	2,704
57	4	228	12,996
62	8	496	30,752
67	10	670	44,890
72	14	1,008	72,576
77	16	1,232	94,864
82	7	574	47,068
87	7	609	52,983
92	5	460	42,320
97	2	194	18,818
	80	5,755	429,325

$$\bar{x} = \frac{\Sigma fx}{\Sigma f} = \frac{5,755}{80} = 71.9375$$

$$\bar{x}^2 = 5,175.0039$$

$$\overline{x^2} = \frac{\Sigma fx^2}{80} = \frac{429,325}{80} = 5,366.5625$$

$$s = \sqrt{\overline{x^2} - \bar{x}^2} = 13.8$$

We have seen that the standard deviation is a measure of the dispersion of a set of data. Generally, if we make a great many measurements of a given type, we would expect to find a majority of them near the arithmetic mean. If this is the case, the distribution probably follows, at least to a reasonable extent, the **standard normal distribution curve.** See Fig. 21–6. This curve gives a theoretical distribution about the arithmetic mean, assuming that the number of values in the distribution becomes infinite, and its equation is

$$y = \frac{1}{\sqrt{2\pi}} e^{-x^2/2} \tag{21-6}$$

An important feature of the standard deviation is that *in a normal distribution, about 68% of the values are found within the interval $\bar{x} - s$ to $\bar{x} + s$.*

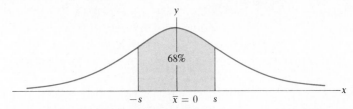

The standard normal distribution curve

Figure 21–6

Example F

In Example B, 9 of the 15 values, or 60%, are between 2.6 and 8.4. Here $\bar{x} = 5.5$ and $s = 2.9$. Thus, $\bar{x} - s = 2.6$ and $\bar{x} + s = 8.4$. Therefore, we see that the percentage in this interval is slightly less than in a normal distribution.

In Example E, using $\bar{x} = 72$ and $s = 14$, we have $\bar{x} - s = 58$ and $\bar{x} + s = 86$. We also note that 59 of the 80 values are in the interval from 58 to 86, which means that 74% of the values are in this interval.

Exercises 21–3

In Exercises 1 through 20, use the following sets of numbers. They are the same as those used in Exercises 21–1 and 21–2.

A: 3, 6, 4, 2, 5, 4, 7, 6, 3, 4, 6, 4, 5, 7, 3

B: 25, 26, 23, 24, 25, 28, 26, 27, 23, 28, 25

C: 48, 53, 49, 45, 55, 49, 47, 55, 48, 57, 51, 46

D: 105, 108, 103, 108, 106, 104, 109, 104, 110, 108, 108, 104, 113, 106, 107, 106, 107, 109, 105, 111, 109, 108

In Exercises 1 through 4, use Eq. (21–2) to find the standard deviation s for the indicated sets of numbers.

1. Set A 2. Set B 3. Set C 4. Set D

In Exercises 5 through 8, use Eq. (21–5) to find the standard deviation s for the indicated sets of numbers.

5. Set A 6. Set B 7. Set C 8. Set D

In Exercises 9 through 16, find the standard deviation s for the indicated sets of numbers.

9. The salaries in Exercise 25 of Section 21–2.

10. The measurements of electric current in Exercise 25 of Section 21–1.

11. The decibel readings in Example B of Section 21–1.

12. The measurements of volume in Exercise 17 of Section 21–1.

13. The measurements of cable-breaking loads in Exercise 23 of Section 21–1.

14. The measurements of strobe light times in Exercise 21 of Section 21–1.

15. The measurements of electric power usage in Exercise 27 of Section 21–2.

16. The measurements of miles per gallon in Exercise 27 of Section 21–1.

See Appendix E for a computer program for finding the standard deviation.

In Exercises 17 through 24, find the percentage of values in the interval $\bar{x} - s$ to $\bar{x} + s$ for the indicated sets of numbers.

17. Set A 18. Set B 19. Set C 20. Set D

21. The salaries in Exercise 25 of Exercises 21–2.

22. The measurements of electric current in Exercise 25 of Exercises 21–1.

23. The measurements of cable-breaking loads in Exercise 23 of Exercises 21–1.

24. The measurements of miles per gallon in Exercise 27 of Exercises 21–1.

21–4 Fitting a Straight Line to a Set of Points

We have considered statistical methods for dealing with one variable. We have discussed methods of tabulating, graphing, measuring the central tendency and the deviations from this value for one variable. We now shall discuss how to obtain a relationship between two variables for which a set of points is known.

In this section we shall show a method of "fitting" a straight line to a given set of points. In the following section fitting suitable nonlinear curves to given sets of points will be discussed. Some of the reasons for doing this are (1) to express a concise relationship between the variables, (2) to use the equation to predict certain fundamental results, (3) to determine the reliability of certain sets of data, and (4) to use the data for testing theoretical concepts.

We shall assume for the examples and exercises of the remainder of this chapter that there is some relationship between the variables. Often when we are analyzing statistics for variables between which we think a relationship might exist, the points are so scattered as to give no reasonable idea regarding the possible functional relationship. We shall assume here that such combinations of variables have been discarded in the analysis.

Example A

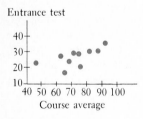

Figure 21–7

All the students enrolled in the mathematics course referred to in Example E of Section 21–1 took an entrance test in mathematics. To study the reliability of this test, an instructor tabulated the test scores of 10 students (selected at random), along with their course average, and made a graph of these figures (see table below and Fig. 21–7).

Student	Entrance test score, based on 40	Course average, based on 100
A	29	63
B	33	88
C	22	77
D	17	67
E	26	70
F	37	93
G	30	72
H	32	81
I	23	47
J	30	74

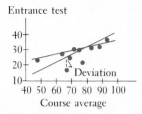

Entrance test

Course average

Figure 21–8

We now ask whether or not there is a functional relationship between the test scores and the course grades. Certainly no clear-cut relationship exists, but in general we see that the higher the test score, the higher the course grade. This leads to the possibility that there might be some straight line, from which none of the points would vary too significantly. If such a line could be found, then it could be the basis of predictions as to the possible success a student might have in the course, on the basis of his or her grade on the entrance test. Assuming that such a straight line exists, the problem is to find the equation of this line. Figure 21–8 shows two such possible lines.

There are various methods of determining the line which best fits the given points. We shall employ the one which is most widely used: the **method of least squares.** *The basic principle of this method is that the sum of the squares of the deviation of all points from the best line (in accordance with this method) is the least it can be. By* **deviation** *we mean the difference between the y-value of the line and the y-value for the point (of original data) for a particular value of x.*

Example B

In Fig. 21–8 the deviation of one point is indicated. The point (67, 17) (student D of Example A) has a deviation of 8 from the indicated line of Fig. 21–8. Thus we square the value of this deviation to obtain 64. The method of least squares requires that the sum of all such squares be a minimum in order to determine the line which fits best.

Therefore, in applying this method, it is necessary to use the equation of a straight line and the coordinates of the points of the data. The deviation of these is indicated and squared. The problem then arises as to the method of determining the constants m and b in the equation of the line $y = mx + b$ for which these squares are a minimum. Since we are dealing with squared quantities, the problem is one of finding the minimum of a quadratic-type function. The following example indicates the method employed.

Example C

For what value of x is the quadratic function $y = x^2 - 8x + 19$ a minimum?

By inspection we see that this equation represents a parabola. From the properties of this type of parabola, we know that the minimum value of y will occur at the vertex. Thus, we are to find the x-coordinate of the vertex. To do this, we complete the square of the x-terms. And so we have

$$y = (x^2 - 8x + 16) + 3 \quad \text{or} \quad y = (x - 4)^2 + 3$$

We now see that if $x = 4$, $y = 3$. If x is anything other than 4, $y > 3$, since $(x - 4)$ is squared and is always positive for values other than $x = 4$. Therefore, the minimum value of this function is 3, and it occurs for $x = 4$.

The previous discussion indicates the type of reasoning used in developing the equations for the best straight line to fit a set of data. However, it is not a derivation, as a complete derivation requires more advanced mathematical methods.

By finding the deviations, squaring, and then applying the method above for finding the minimum of the sum of the squares, *the equation of the* **least-squares line,**

$$y = mx + b \qquad\qquad (21\text{–}7)$$

can be found by the values

$$m = \frac{\overline{xy} - \overline{x}\,\overline{y}}{s_x^2} \qquad\qquad (21\text{–}8)$$

and

$$b = \overline{y} - m\overline{x} \qquad\qquad (21\text{–}9)$$

In Eqs. (21–8) and (21–9), the x's and y's are those of the points in the data and s_x is the standard deviation of the x-values.

Example D

Find the least-squares line of the data of Example A.

Here the y-values will be the entrance-test scores, and the x-values the course averages. The results are best obtained by tabulating the necessary quantities, as we do in the following table.

x	y	xy	x^2
63	29	1,827	3,969
88	33	2,904	7,744
77	22	1,694	5,929
67	17	1,139	4,489
70	26	1,820	4,900
93	37	3,441	8,649
72	30	2,160	5,184
81	32	2,592	6,561
47	23	1,081	2,209
74	30	2,220	5,476
732	279	20,878	55,110

$\overline{x} = \dfrac{732}{10} = 73.2, \quad \overline{x^2} = 5358.24$

$\overline{y} = \dfrac{279}{10} = 27.9, \quad \overline{x}\,\overline{y} = 2042.28$

$\overline{xy} = \dfrac{20,878}{10} = 2087.8$

$\overline{x^2} = \dfrac{55,110}{10} = 5511.0$

$s_x^2 = \overline{x^2} - \overline{x}^2 = 152.76$

$m = \dfrac{\overline{xy} - \overline{x}\,\overline{y}}{s_x^2} = \dfrac{2087.8 - 2042.28}{152.76}$

$\qquad = 0.2979838$

$b = \overline{y} - m\overline{x} = 27.9 - (0.2979838)(73.2) = 6.09$

$y = 0.298x + 6.09$

Thus, the equation of the line is $y = 0.298x + 6.09$. (See Fig. 21–9.)

We note that when plotting the least-squares line, we know two points on it. These are $(0, b)$ and $(\bar{x}, \bar{y})$. We can see that $(\bar{x}, \bar{y})$ satisfies the equation from the solution for b [Eq. (21–9)].

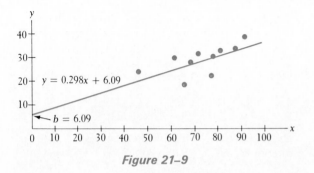

Figure 21–9

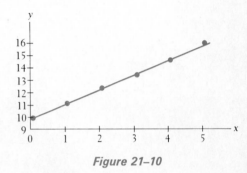

Figure 21–10

Example E

In an experiment to determine the relation between the load on a spring and the length of the spring, the following data were found.

Load (pounds)	0.0	1.0	2.0	3.0	4.0	5.0
Length (inches)	10.0	11.2	12.3	13.4	14.6	15.9

Find the least-squares line for this data, which express the length as a function of the load.

x	y	xy	x^2
0.0	10.0	0.0	0.0
1.0	11.2	11.2	1.0
2.0	12.3	24.6	4.0
3.0	13.4	40.2	9.0
4.0	14.6	58.4	16.0
5.0	15.9	79.5	25.0
15.0	77.4	213.9	55.0

$$\bar{x} = \frac{15.0}{6} = 2.50, \quad \bar{x^2} = 6.25$$

$$\bar{y} = \frac{77.4}{6} = 12.9, \quad \bar{x}\,\bar{y} = 32.25$$

$$\bar{xy} = \frac{213.9}{6} = 35.65$$

$$\bar{x^2} = \frac{55.0}{6} = 9.1666667$$

$$s_x^2 = \bar{x^2} - \bar{x}^2 = 9.1666667 - 6.25 = 2.9166667$$

$$m = \frac{35.65 - 32.25}{2.9166667} = 1.1657143$$

$$b = 12.9 - (1.1657143)(2.50) = 9.985714$$

Therefore, the least-squares line in this case is

$$y = 1.17x + 9.99$$

where y is the length of the spring and x is the load. In Fig. 21–10, the line shown is the least-squares line, and the points are the experimental data points.

Exercises 21-4

In Exercises 1 through 8, find the equation of the least-squares line for the given data. In each case graph the line and the data points on the same graph.

1. The points in the following table:

x	4	6	8	10	12
y	1	4	5	8	9

2. The points in the following table:

x	1	2	3	4	5	6	7
y	10	17	28	37	49	56	72

3. The points in the following table:

x	20	26	30	38	48	60
y	160	145	135	120	100	90

4. The points in the following table:

x	1	3	6	5	8	10	4	7	3	8
y	15	12	10	8	9	2	11	9	11	7

5. In an electrical experiment, the following data were found for the values of current and voltage for a particular element of the circuit. Find the voltage as a function of the current.

Voltage (volts)	3.00	4.10	5.60	8.00	10.50
Current (milliamperes)	15.0	10.8	9.30	3.55	4.60

6. The velocity of a falling object was found each second by use of an electric device as shown. Find the velocity as a function of time.

Velocity (meters per second)	9.70	19.5	29.5	39.4	49.2	58.9	68.6
Time (seconds)	1.00	2.00	3.00	4.00	5.00	6.00	7.00

7. In an experiment on the photoelectric effect, the frequency of the light being used was measured as a function of the stopping potential (the voltage just sufficient to stop the photoelectric current) with the results given below. Find the least-squares line for V as a function of f. The frequency for $V = 0$ is known as the *threshold frequency*. From the graph, determine the threshold frequency.

V (volts)	0.350	0.600	0.850	1.10	1.45	1.80
f (petahertz)	0.550	0.605	0.660	0.735	0.805	0.880

8. If gas is cooled under conditions of constant volume, it is noted that the pressure falls nearly proportionally as the temperature. If this were to happen until there was no pressure, the theoretical temperature for this case is referred to as *absolute zero*. In an elementary experiment, the following data were found for pressure and temperature for a gas under constant volume.

P (kilopascals)	133	143	153	162	172	183
T (degrees Celsius)	0.0	20	40	60	80	100

Find the least-squares line for P as a function of T, and, from the graph, determine the value of absolute zero found in this experiment.

The linear coefficient of correlation, a measure of the relatedness of two variables, is defined by $r = m(s_x/s_y)$. Due to its definition, the values of r lie in the range $-1 \le r \le 1$. If r is near 1, the correlation is considered good. For values of r between $-\frac{1}{2}$ and $+\frac{1}{2}$, the correlation is poor. If r is near -1, the variables are said to be negatively correlated; that is, one increases while the other decreases.

In Exercises 9 through 12, compute r for the given sets of data.

9. Exercise 1   10. Exercise 2   11. Exercise 4   12. Example A

21–5 Fitting Nonlinear Curves to Data

If the experimental points do not appear to be on a straight line, but we recognize them as being approximately on some other type of curve, the method of least squares can be extended to use on these other curves. For example, if the points are apparently on a parabola, we could use the function $y = a + bx^2$. To use the above method, we shall extend the least-squares line to

$$y = m[f(x)] + b \tag{21–10}$$

Here, $f(x)$ must be calculated first, and then the problem can be treated as a least-squares line to find the values of m and b. Some of the functions $f(x)$ which may be considered for use are x^2, $1/x$, and 10^x.

Example A

Find the least-squares curve $y = mx^2 + b$ for the points in the following table.

y	1	5	12	24	53	76
x	0	1	2	3	4	5

The necessary quantities are tabulated for purposes of calculation.

x	y	$f(x) = x^2$	yx^2	$(x^2)^2$
0	1	0	0	0
1	5	1	5	1
2	12	4	48	16
3	24	9	216	81
4	53	16	848	256
5	76	25	1900	625
	171	55	3017	979

$$\overline{x^2} = \frac{55}{6} = 9.1666667, \qquad \overline{x^2}^2 = 84.027778$$

$$\bar{y} = \frac{171}{6} = 28.5, \qquad \overline{x^2}\,\bar{y} = 261.25$$

$$\overline{yx^2} = \frac{3017}{6} = 502.83333, \qquad \overline{(x^2)^2} = \frac{979}{6} = 163.16667$$

$$s_{(x^2)}^2 = 163.16667 - 84.027778 = 79.138892$$

$$m = \frac{502.83333 - 261.25}{79.138892} = 3.0526499 \qquad b = 28.5 - (3.0526499)(9.1666667)$$

$$= 0.52$$

Therefore, the desired equation is $y = 3.05x^2 + 0.52$. The graph of this equation and the data points are shown in Fig. 21–11.

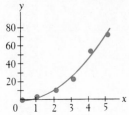

Figure 21–11

Example B

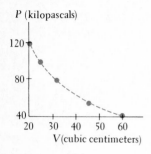

P(kilopascals)

Figure 21–12

In a physics experiment designed to measure the pressure and volume of a gas at constant temperature, the following data were found. When the points were plotted, they were seen to approximate the hyperbola $y = c/x$. Find the least-squares approximation to this hyperbola $[y = m(1/x) + b]$ (see Fig. 21–12).

P(kPa)	$V(cm^3)$	$x(=V)$	$y(=P)$	$f(x) = \dfrac{1}{x}$	$y\left(\dfrac{1}{x}\right)$	$\left(\dfrac{1}{x}\right)^2$
120.0	21.0	21.0	120.0	0.0476190	5.7142857	0.0022676
99.2	25.0	25.0	99.2	0.0400000	3.9680000	0.0016000
81.3	31.8	31.8	81.3	0.0314465	2.5566038	0.0009889
60.6	41.1	41.1	60.6	0.0243309	1.4744526	0.0005920
42.7	60.1	60.1	42.7	0.0166389	0.7104825	0.0002769
			403.8	0.1600353	14.4238246	0.0057254

(*Note on calculator use:* The final digits for the values shown may vary on the type of calculator used, and how values are used. Here all individual values are shown assuming that the calculator displays eight digits, but actually carries more. The value of $1/x$ was found from the value of x, with the eight-digit display indicated. However, the values of $y(1/x)$ and $(1/x)^2$ were found using the extra digits carried by the calculator for $1/x$. The sums were found using the rounded-off values shown.)

$$\overline{\frac{1}{x}} = 0.0320071 \qquad \overline{\left(\frac{1}{x}\right)^2} = 0.0010245 \qquad \overline{y} = 80.76 \qquad \overline{\frac{1}{x}y} = 2.5848902$$

$$\overline{y\left(\frac{1}{x}\right)} = 2.8847649 \qquad \overline{\left(\frac{1}{x}\right)^2} = 0.0011451 \qquad s^2_{(1/x)} = 0.0011451 - 0.0010245$$
$$= 0.0001206$$

$$m = \frac{2.8847649 - 2.5848902}{0.0001206} \qquad b = 80.76 - 2486.5232(0.0320071)$$
$$= 2486.5232 \qquad \qquad = 1.17$$

Thus the equation of the hyperbola $y = m(1/x) + b$ is

$$y = \frac{2490}{x} + 1.17$$

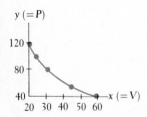

y (=*P*)

Figure 21–13

The graph of this hyperbola and the points representing the data are shown in Fig. 21–13.

Example C

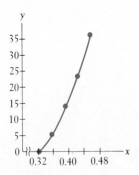

Figure 21-14

It has been found experimentally that the tensile strength of brass (a copper-zinc alloy) increases (within certain limits) with the percentage of zinc. The following table indicates the values which have been found (also see Fig. 21–14).

Tensile strength (10^5 lb/in.^{2})	0.32	0.36	0.40	0.44	0.48
Percentage of zinc	0	5	13	22	34

Fit a curve of the form $y = m(10^x) + b$ to the data. Let x = tensile strength ($\times\ 10^5$) and y = percentage of zinc.

x	y	$f(x) = 10^x$	$y(10^x)$	$(10^x)^2$
0.32	0	2.0892961	0.000000	4.3651583
0.36	5	2.2908677	11.454338	5.2480746
0.40	13	2.5118864	32.654524	6.3095734
0.44	22	2.7542287	60.593031	7.5857758
0.48	34	3.0199517	102.67836	9.1201084
	74	12.6662306	207.38025	32.6286905

(See the note on calculator use in Example B.)

$$\overline{10^x} = 2.5332461 \qquad \overline{10^x y} = 37.492042$$

$$\overline{10^{x2}} = 6.4173359 \qquad \overline{y(10^x)} = 41.47605$$

$$\bar{y} = 14.8 \qquad \overline{(10^x)^2} = 6.5257381$$

$$s^2_{(10^x)} = 6.5257381 - 6.4173359 = 0.1084022$$

$$m = \frac{41.47605 - 37.492042}{0.1084022} = 36.752095$$

$$b = 14.8 - (36.752095)(2.5332461) = -78.3$$

The equation of the curve is $y = 36.8(10^x) - 78.3$. It must be remembered that for practical purposes, y must be positive. The graph of the equation is shown in Fig. 21–15, with the solid portion denoting the meaningful part of the curve. The points of the data are also shown.

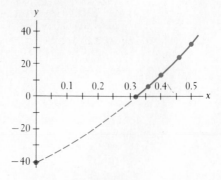

Figure 21-15

Exercises 21–5

In each of the following exercises, find the indicated least-squares curve. Sketch the curve and plot the data points on the same graph.

1. For the points in the following table, find the least-squares curve $y = mx^2 + b$.

x	2	4	6	8	10
y	12	38	72	135	200

2. For the points in the following table, find the least-squares curve $y = m\sqrt{x} + b$.

x	0	4	8	12	16
y	1	9	11	14	15

3. For the points in the following table, find the least-squares curve $y = m(1/x) + b$.

x	1.10	2.45	4.04	5.86	6.90	8.54
y	9.85	4.50	2.90	1.75	1.48	1.30

4. For the points in the following table, find the least-squares curve $y = m(10^x) + b$.

x	0.00	0.200	0.500	0.950	1.325
y	6.00	6.60	8.20	14.0	26.0

5. The following data were found for the distance y that an object rolled down an inclined plane in time t. Determine the least-squares curve $y = mt^2 + b$.

y (centimeters)	6.0	23	55	98	148
t (seconds)	1.0	2.0	3.0	4.0	5.0

6. The increase in length of a certain metallic rod was measured in relation to particular increases in temperature. If y represents the increase in length for the corresponding increase in temperature x, find the least-squares curve $y = m(x^2) + b$ for these data.

x (degrees Celsius)	50.0	100	150	200	250
y (centimeters)	1.00	4.40	9.40	16.4	24.0

7. Use the data of Exercise 6 and determine the least-squares curve $y = m(10^z) + b$ where $z = x/1000$.

8. Measurements were made of the current in an electric circuit as a function of time. The circuit contained a resistance of 5 Ω and an inductance of 10 H. The following data were found.

i (amperes)	0.00	2.52	3.45	3.80	3.92
t (seconds)	0.00	2.00	4.00	6.00	8.00

   Find the least-squares curve $i = m(e^{-0.5t}) + b$ for this data. From the equation, determine the value the current approaches as t approaches infinity.

9. The resonant frequency of an electric circuit containing a 4μF capacitor was measured as a function of an inductance in the circuit. The following data were found.

f (hertz)	490	360	250	200	170
L (henrys)	1.0	2.0	4.0	6.0	9.0

Find the least-squares curve $f = m(1/\sqrt{L}) + b$.

10. The displacement of a pendulum bob from its equilibrium position as a function of time gave the following results.

Displacement (centimeters)	0.00	7.80	10.0	8.10	0.00
Time (seconds)	0.00	0.90	1.60	2.20	3.15

Find the least-squares curve of the form $y = m(\sin x) + b$, expressing the displacement as y and the time as x.

21–6 Exercises for Chapter 21

In Exercises 1 through 4, use the following set of numbers:

2.3, 2.6, 4.2, 3.6, 3.5, 4.1, 4.8, 2.5, 3.0, 4.1, 3.8

1. Determine the median of the set of numbers.

2. Determine the arithmetic mean of the set of numbers.

3. Determine the standard deviation of the set of numbers.

4. Construct a frequency table with intervals 2.0–2.9, 3.0–3.9, and 4.0–4.9.

In Exercises 5 through 12, use the following set of numbers.

109, 103, 113, 110, 113, 106, 114, 101, 108, 101, 112, 106, 115, 102, 107, 110, 102, 115, 106, 105

5. Construct a frequency distribution table with intervals 101–103, 104–106, and so on.

6. Determine the median. 　　　　　　　　7. Determine the mode.

8. Determine the mean. 　　　　　　　　　9. Determine the standard deviation.

10. Draw a frequency polygon for the data for Exercise 5.

11. Draw a histogram for the data of Exercise 5.

12. Construct a frequency distribution table with intervals of 101–105, 106–110, and 111–115. Then draw a histrogram for these data.

In Exercises 13 through 18, use the following data:

An important property of oil is its coefficient of viscosity, which gives a measure of how well it flows. In order to determine the viscosity of a certain motor oil, a refinery took samples from 12 different storage tanks and tested them at 50°C. The results (in pascal-seconds) were 0.24, 0.28, 0.29, 0.26, 0.27, 0.26, 0.25, 0.27, 0.28, 0.26, 0.26, 0.25.

13. Determine the arithmetic mean. 　　　　14. Determine the median.

15. Determine the standard deviation. 　　　16. Make a histogram.

17. Make a frequency polygon. 　　　　　　18. Determine the mode.

In Exercises 19 through 24, use the following data:

Two machine parts are considered satisfactorily assembled if their total thickness (to the nearest 0.01 in.) is between or equal to 0.92 and 0.94 in. One hundred sample assemblies are tested, and the thicknesses, to the nearest 0.01 in., are given in the following table:

Total thickness	0.90	0.91	0.92	0.93	0.94	0.95	0.96
Number	3	9	31	38	12	5	2

19. Determine the arithmetic mean.   20. Determine the median.

21. Determine the mode.   22. Make a histogram.

23. Determine the standard deviation.   24. Make a frequency polygon.

In Exercises 25 through 28, use the following data:

A Geiger counter records the presence of high-energy nuclear particles. Even though no apparent radioactive source is present, a certain number of particles will be recorded. These are primarily cosmic rays, which are caused by very high-energy particles from outer space. In an experiment to measure the amount of cosmic radiation, the number of counts were recorded during 200 5-s intervals. The following table gives the number of counts, and the number of 5-s intervals having this number of counts. Draw a frequency curve for this data.

Counts	0	1	2	3	4	5	6	7	8	9	10
Intervals	3	10	25	45	29	39	26	11	7	2	3

25. Determine the median.   26. Determine the arithmetic mean.

27. Make a histogram.   28. Make a frequency polygon.

In Exercises 29 through 34, find the indicated least-squares curves.

29. In a certain experiment, the resistance of a certain resistor was measured as a function of the temperature. The data found were as follows:

R (ohms)	25.0	26.8	28.9	31.2	32.8	34.7
T (degrees Celsius)	0.0	20.0	40.0	60.0	80.0	100

Find the least-squares line for this data, expressing R as a function of T. Sketch the line and data points on the same graph.

30. The solubility of sodium chloride (table salt) (in kilograms per cubic meter of water) as a function of temperature (degrees Celsius) is measured as follows:

Solubility	357	360	366	373	384
Temperature	0.0	20.0	40.0	60.0	80.0

Find the least-squares straight line for these data.

31. The coefficient of friction of an object sliding down an inclined plane was measured as a function of the angle of incline of the plane. The following results were obtained.

Coefficient of friction	0.16	0.34	0.55	0.85	1.24	1.82	2.80
Angle (degrees)	10.0	20.0	30.0	40.0	50.0	60.0	70.0

Find the least-squares curve of the form $y = m(\tan x) + b$, expressing the coefficient of friction as y and the angle as x.

32. An electric device recorded the total distance (in centimeters) which an object fell at 0.1 s intervals. Following are the data which were found.

Distance	4.90	19.5	44.0	78.1	122.0
Time	0.100	0.200	0.300	0.400	0.500

Find the least-squares curve of the form $y = mx^2 + b$, using y to represent distance and x to represent time.

33. The period of a pendulum as a function of its length was measured, giving the following results.

Period (seconds)	1.10	1.90	2.50	2.90	3.30
Length (feet)	1.00	3.00	5.00	7.00	9.00

Find the least-squares curve of the form $y = m\sqrt{x} + b$, which expresses the period as a function of the length.

34. In an elementary experiment which measured the wavelength of sound as a function of the frequency, the following results were obtained.

Wavelength (centimeters)	140	107	81.0	70.0	60.0
Frequency (hertz)	240	320	400	480	560

Find the least-squares curve of the form $y = m(1/x) + b$ for this data, expressing wavelength as y and frequency as x.

APPENDIX A

Study Aids

A–1 Introduction

The primary objective of this text is to give you an understanding of mathematics so that you can use it effectively as a tool in your technology. Without understanding of the basic methods, knowledge is usually short-lived. However, if you do understand, you will find your work much more enjoyable and rewarding. This is true in any course you may take, be it in mathematics or in any other field.

Mathematics is an indispensable tool in almost all scientific fields of study. You will find it used to a greater and greater degree as you work in your chosen field. Generally, in the introductory portions of allied courses, it is enough to have a grasp of elementary concepts in algebra and geometry. However, as you develop in your field, the need for more mathematics will be apparent. This text is designed to develop these necessary tools so they will be available in your allied courses. You cannot derive the full benefit from your mathematics course unless you devote the necessary amount of time to developing a sound understanding of the subject.

It is assumed in this text that you have a background which includes geometry and some algebra. Therefore, many of the topics covered in this book may seem familiar to you, especially in the earlier chapters. However, it is likely that your background in some of these areas is not complete, either because you have not studied mathematics for a year or two or because you did not understand the topics when you first encountered them. If a topic is familiar, do not reason that there is no sense in studying it again, but take the opportunity to clarify any points on which you are not certain. In almost every topic you will probably find certain points which can use further study. If the topic is new, use your time effectively to develop an understanding of the methods involved, and do not simply memorize problems of a certain type.

There is only one good way to develop the understanding and working knowledge necessary in any course, and that is to *work with it*. Many students consider mathematics difficult. They will tell you that it is their lack of mathematical ability and the complexity of the material itself which make it difficult. Some topics in mathematics, especially in the more advanced areas, do require a certain aptitude for full comprehension. However, a large proportion of poor grades in elementary mathematics courses result from the fact that the student is not willing to put in the necessary time to develop the understanding. The student takes a quick glance through the material, tries a few exercises, is largely unsuccessful, and then decides that the material is "impossible." A detailed reading of the text, a careful following of the illustrative examples, and then solving the exercises would lead to more success and therefore would make the work much more enjoyable and rewarding. No matter what text is used or what methods are used in the course or what other variables may be introduced, if you do not put in an adequate amount of time studying, you will not derive the proper results. More detailed suggestions for study are included in the following section.

If you consider these suggestions carefully, and follow good study habits, you should enjoy a successful learning experience in this course as well as in other courses.

A–2 Suggestions for Study

When you are studying the material presented in this text, the following suggestions may help you to derive full benefit from the time you devote to it.

1. Before attempting to do the exercises, read through the material preceding them.
2. Follow the illustrative examples carefully, being certain that you know how to proceed from step to step. You should then have a good idea of the methods involved.
3. Work through the exercises, spending a reasonable amount of time on each problem. If you cannot solve a certain problem in a reasonable amount of time, leave it and go to the next. Return to this problem later. If you find many problems difficult, you should reread the explanatory material and the examples to determine what point or points you have not understood.
4. When you have completed the exercises, or at least most of them, glance back through the explanatory material to be sure you understand the methods and principles.
5. If you have gone through the first four steps and certain points still elude you, ask to have these points clarified in class. Do not be afraid to ask questions; only be sure that you have made a sincere effort on your own before you ask them.

Some study habits which are useful not only here but in all of your other subjects are the following:

1. Put in the time required to develop the material fully, being certain that you are making effective use of your time. A good place to study helps immeasurably.
2. Learn the *methods and principles* being presented. Memorize as little as possible, for although there are certain basic facts which are more expediently learned by memorization, these should be kept to a minimum.
3. Keep up with the material in all of your courses. Do not let yourself get so behind in your studies that it becomes difficult to make up the time. Usually the time is never really made up. Studying only before tests is a poor way of learning, and is usually rather ineffective.
4. When you are taking examinations, always read each question carefully before attempting the solution. Solve those you find easiest first, and do not spend too much time on any one problem. Also, use all the time available for the examination. If you finish early, use the remainder of the time to check your work.

A-3 Problem Analysis

Drill-type problems require a working knowledge of the methods presented. However, they do not require, in general, much analysis before being put in proper form for solution. Stated problems, on the other hand, do require proper interpretation before they can be put in a form for solution. The remainder of this section is devoted to some suggestions for solving stated problems.

We have to put stated problems in symbolic form before we attempt to solve them. It is this step which most students find difficult. Because such problems require the student to do more than merely go through a certain routine, they demand more analysis and thus appear more "difficult." There are several reasons for the student's difficulty, some of them being: (1) unsuccessful previous attempts at solving such problems, leading the student to believe that all stated problems are "impossible"; (2) failure to read the problem carefully; (3) a poorly organized approach to the solution; and (4) improper and incomplete interpretation of the statements given. The first two of these can be overcome only with the proper attitude and care.

There are over 70 completely worked examples of stated problems (as well as many other problems which indicate a similar analysis) throughout this text, illustrating proper interpretations and approaches to these problems. Therefore, we shall not include specific examples here. However, we shall set forth the method of analysis of any stated problem. Such an analysis generally follows these steps:

1. Read the problem carefully.
2. Carefully identify known and unknown quantities.
3. Draw a figure when appropriate (which is quite often the case).
4. Write, in symbols, the relations given in the statements.
5. Solve for the desired quantities.

If you follow this step-by-step method and write out the solution neatly, you should find that stated problems lend themselves to solution more readily than you had previously found.

Units of Measurement and Approximate Numbers

B-1 Units of Measurement; the Metric System

The solution of most technical problems involves the use of the basic operations on numbers, where many of these numbers represent some sort of measurement or calculation. Therefore, associated with these numbers are **units of measurement,** and for the calculations to be meaningful, we must know these units. For example, if we measure the length of an object to be 12, we must know whether it is being measured in feet, yards, or some other specified unit of length.

Certain universally accepted **base units** *are used to measure fundamental quantities.* The units for numerous other quantities are expressed in terms of the base units. Fundamental quantities for which base units are defined are (1) length, (2) mass or force, depending on the system of units being used, (3) time, (4) electric current, (5) temperature, (6) amount of substance, and (7) luminous intensity. *Other units, referred to as* **derived units,** *are expressible in terms of the units for these quantities.*

Even though all other quantities can be expressed in terms of the fundamental ones, many have units which are given a specified name. This is done primarily for those quantities which are used very commonly, although it is not done for all such quantities. For example, the volt is defined as a meter^2-kilogram/second^3-ampere, which is in terms of (a unit of length)^2(a unit of mass)/(a unit of time)^3(a unit of electric current). The unit for acceleration has no special name, and is left in terms of the base units; for example, feet/second^2. For convenience, special symbols are usually used to designate units. The units for acceleration would be written as ft/s^2.

Two basic systems of units, the **metric system** and the **United States Customary** system, are in use today. (The U.S. Customary system is also known as the *British system,* but the metric system is now used in Great Britain.) Nearly every country in the world now uses the metric system. In the United States both systems are used, and a national committee is now coordinating a voluntary conversion to the use of the metric system. Some major U.S. industrial firms have completed the conversion for their products, and most others will have done so by the end of the 1980s.

Therefore, for the present, both systems are of importance although the metric system will eventually be used almost universally. For that reason, where units are used, some of the exercises and examples have metric units and others have U.S. Customary units. Technicians and engineers need to have some knowledge of both systems.

Although more than one system has been developed in which metric units are used, the system which is now accepted as the metric system is the **International System of Units (SI).** This was established in 1960 and uses some different

Table B–1. Quantities and Their Associated Units

Quantity	Quantity symbol	U.S. Customary Name	U.S. Customary Symbol	Metric (SI) Name	Metric (SI) Symbol	In terms of other SI units
Length	s	foot	ft	**meter**	m	
Mass	m	slug		**kilogram**	kg	
Force	F	pound	lb	newton	N	$m \cdot kg/s^2$
Time	t	second	s	**second**	s	
Area	A		ft^2		m^2	
Volume	V		ft^3		m^3	
Capacity	V	gallon	gal	liter	L	$(1\ L = 1\ dm^3)$
Velocity	v		ft/s		m/s	
Acceleration	a		ft/s^2		m/s^2	
Density	d, ρ		lb/ft^3		kg/m^3	
Pressure	p		lb/ft^2	pascal	Pa	N/m^2
Energy, work	E, W		ft·lb	joule	J	N·m
Power	P	horsepower	hp	watt	W	J/s
Period	T		s		s	
Frequency	f		1/s	hertz	Hz	1/s
Angle	θ	radian	rad	radian	rad	

Special Notes:
1. The SI base units are shown in boldface type.
2. The units symbols shown above are those which are used in the text. Many of them were adopted with the adoption of the SI system. This means, for example, that we use s rather than sec for seconds, and A rather than amp for amperes. Also, other units such as volt are not spelled out, a common practice in the past. When a given unit is used with both systems, we use the SI symbol for the unit.
3. The liter and degree Celsius are not actually SI units. However, they are recognized for use with the SI system due to their practical importance. Also, the symbol for liter has several variations. Presently L is recognized for use in the United States and Canada, 1 is recognized by the International Committee of Weights and Measures, and ℓ is also recognized for use in several countries.

definitions for base units from the previously developed metric units. However, the measurement of the base units in the SI system is more accessible, and the differences are very slight. *Therefore, when we refer to the metric system, we are using SI units.*

As we have stated, in each system the base units are specified, and all other units are then expressible in terms of these. Table B–1 lists the fundamental quantities, as well as many other commonly used quantities, along with the symbols and names of units used to represent each quantity shown.

In the U.S. Customary system, the base unit of length is the *foot,* and that of force is the *pound.* In the metric system, the base unit of length is the *meter,* and that of mass is the *kilogram.* Here, we see a difference in the definition of the systems which causes some difficulty when units are converted from one system

Table B–1. Continued

Quantity	Quantity symbol	U.S. Customary		Metric (SI)		
		Name	Symbol	Name	Symbol	In terms of other SI units
Electric current	I, i	ampere	A	**ampere**	A	
Electric charge	q	coulomb	C	coulomb	C	A·s
Electric potential	V, E	volt	V	volt	V	J/A·s
Capacitance	C	farad	F	farad	F	s/Ω
Inductance	L	henry	H	henry	H	Ω·s
Resistance	R	ohm	Ω	ohm	Ω	V/A
Thermodynamic temperature	T			**kelvin**	K	
Temperature	T	Fahrenheit degree	°F	degree Celsius	°C	(1°C = 1 K)
Quantity of heat	Q	British thermal unit	Btu	joule	J	
Amount of substance	n			**mole**	mol	
Luminous intensity	I	candlepower	cp	**candela**	cd	

4. Other units of time, along with their symbols, which are recognized for use with the SI system and are used in this text are: minute, min; hour, h; day, d.

5. There are many additional specialized units which are used with the SI system. However, most of those which appear in this text are shown in the table. A few of the specialized units are noted when used in the text. One which is frequently used is that for revolution, r.

6. Other common U.S. units which are used in the text are as follows: inch, in.; yard, yd; mile, mi; ounce, oz.; ton; quart, qt; acre.

7. There are a number of units which were used with the metric system prior to the development of the SI system. However, many of these are not to be used with the SI system. Among those which were commonly used are the dyne, erg, and calorie.

to the other. That is, a base unit in the U.S. Customary system is a unit of force, and a base unit in the metric system is a unit of mass.

The distinction between mass and force is very significant in physics, and the weight of an object is the force with which it is attracted to the earth. Weight, which is therefore a force, is different from mass, which is a measure of the inertia an object exhibits. Although they are different quantities, mass and weight are, however, very closely related. In fact, the weight of an object equals its mass multiplied by the acceleration due to gravity. Near the surface of the earth the acceleration due to gravity is nearly constant, although it decreases as the distance from the earth increases. Therefore, near the surface of the earth the weight of an object is directly proportional to its mass. However, at great distances from the earth the weight of an object will be zero, whereas its mass does not change.

Since force and mass are different, it is not strictly correct to convert pounds to kilograms. However, since the pound is the base unit in the U.S. Customary system, and the kilogram is the base unit in the metric system, at the earth's surface 1 kg corresponds to 2.21 lb, in the sense that the force of gravity on a 1 kg mass is 2.21 lb.

When designating units for weight, we use pounds in the U.S. Customary system. In the metric system, although kilograms are used for weight, it is preferable to specify the mass of an object in kilograms. The force of gravity on an object is designated in newtons, and the use of the term weight is avoided unless its meaning is completely clear.

As for the other fundamental quantities, both systems use the *second* as the base unit of time. In the SI system the *ampere* is defined as the base unit of electric current, and this can also be used in the U.S. Customary system. As for temperature, *degrees Fahrenheit* are used with the U.S. Customary system, and *degrees Celsius* (formerly centigrade) are used with the metric system (actually, the *kelvin* is defined as the base unit, where the temperature in kelvins is the temperature in degrees Celsius plus 273.16). In the SI system, the base unit for the amount of a substance is the *mole,* and the base unit of luminous intensity is the *candela.* These last two are of limited importance to our use in this text.

Due to greatly varying sizes of certain quantities, the metric system employs certain prefixes to units to denote different orders of magnitude. These prefixes, with their meanings and symbols, are shown in Table B–2.

Table B–2. Metric Prefixes

Prefix	Factor	Symbol	Prefix	Factor	Symbol
exa	10^{18}	E	deci	10^{-1}	d
peta	10^{15}	P	centi	10^{-2}	c
tera	10^{12}	T	milli	10^{-3}	m
giga	10^{9}	G	micro	10^{-6}	μ
mega	10^{6}	M	nano	10^{-9}	n
kilo	10^{3}	k	pico	10^{-12}	p
hecto	10^{2}	h	femto	10^{-15}	f
deca	10^{1}	da	atto	10^{-18}	a

Example A

Some of the more commonly used units which use the prefixes in Table B–2, along with their meanings, are shown below.

Unit	Symbol	Meaning	Unit	Symbol	Meaning
megohm	MΩ	10^6 ohms	milligram	mg	10^{-3} gram
kilometer	km	10^3 meters	microfarad	μF	10^{-6} farad
centimeter	cm	10^{-2} meter	nanosecond	ns	10^{-9} second

(Mega is shortened to meg when used with "ohm.")

When designating units of area or volume, where square or cubic units are used, we use exponents in the designation. For example, we use m^2 rather than sq m, or $in.^3$ rather than cu in.

When we are working with *numbers that represent units of measurement (referred to as* **denominate numbers**), it is sometimes necessary to change from one set of units to another. *A change within a given system is called a* **reduction,** and *a change from one system to another is called a* **conversion.** Table B–3 gives some basic reduction and conversion factors. It should be noted that some calculators are programmed to do conversions.

Table B–3. Reduction and Conversion Factors

1 in. = 2.54 cm (exact)	$1\ ft^3$ = 28.32 L
1 km = 0.6214 mi	1 L = 1.057 qt
1 lb = 453.6 g	1 Btu = 778.0 ft · lb
1 kg = 2.205 lb	1 hp = 550 ft · lb/s (exact)
1 lb = 4.448 N	1 hp = 746.0 W

The advantages of the metric system should be evident. Reductions within the system are made by use of powers of 10, which amounts to moving the decimal point. Reductions in the British system have numerous different multiples that must be used. Thus, comparisons and changes within the metric system are much simpler than in the British system.

To change a given number of one set of units into another set of units, *we perform algebraic operations with units in the same manner as we do with any algebraic symbol.* Consider the following example.

Example B

If we had a number representing feet per second to be multiplied by another number representing seconds per minute, as far as the units are concerned, we have

$$\frac{ft}{s} \times \frac{s}{min} = \frac{ft \times \cancel{s}}{\cancel{s} \times min} = \frac{ft}{min}$$

This means that the final result would be in feet per minute.

In changing a number of one set of units to another set of units, we use reduction and conversion factors and the principle illustrated in Example B. The convenient way to use the values in the tables is in the form of fractions. Since the given values are equal to each other, their quotient is 1. For example, since 1 in. = 2.54 cm,

$$\frac{1 \text{ in.}}{2.54 \text{ cm}} = 1 \quad \text{or} \quad \frac{2.54 \text{ cm}}{1 \text{ in.}} = 1$$

since each represents the division of a certain length by itself. Multiplying a quantity by 1 does not change its value. The following examples illustrate reduction and conversion of units.

Example C

Reduce 20 kg to milligrams.

$$20 \text{ kg} = 20 \text{ kg}\left(\frac{10^3 \text{ g}}{1 \text{ kg}}\right)\left(\frac{10^3 \text{ mg}}{1 \text{ g}}\right)$$
$$= 20 \times 10^6 \text{ mg}$$
$$= 2.0 \times 10^7 \text{ mg}$$

We note that this result is found essentially by moving the decimal point three places when changing from kilograms to grams, and another three places when changing from grams to milligrams.

Example D

Change 30 mi/h to feet per second.

$$30\frac{\text{mi}}{\text{h}} = \left(30\frac{\text{mi}}{\text{h}}\right)\left(\frac{5280 \text{ ft}}{1 \text{ mi}}\right)\left(\frac{1 \text{ h}}{60 \text{ min}}\right)\left(\frac{1 \text{ min}}{60 \text{ s}}\right) = \frac{(30)(5280) \text{ ft}}{(60)(60) \text{ s}} = 44\frac{\text{ft}}{\text{s}}$$

Note that the only units remaining after the division are those required.

Example E

Change 575 g/cm³ to kilograms per cubic meter.

$$575\frac{\text{g}}{\text{cm}^3} = \left(575\frac{\text{g}}{\text{cm}^3}\right)\left(\frac{100 \text{ cm}}{1 \text{ m}}\right)^3\left(\frac{1 \text{ kg}}{1000 \text{ g}}\right)$$
$$= \left(575\frac{\text{g}}{\text{cm}^3}\right)\left(\frac{10^6 \text{ cm}^3}{1 \text{ m}^3}\right)\left(\frac{1 \text{ kg}}{10^3 \text{ g}}\right)$$
$$= 575 \times 10^3\frac{\text{kg}}{\text{m}^3} = 5.75 \times 10^5\frac{\text{kg}}{\text{m}^3}$$

Example F

Change 62.8 lb/in.^{2} to newtons per square meter.

$$62.8\frac{lb}{in.^2} = \left(62.8\frac{lb}{in.^2}\right)\left(\frac{4.448\ N}{1\ lb}\right)\left(\frac{1\ in.}{2.54\ cm}\right)^2\left(\frac{100\ cm}{1\ m}\right)^2$$

$$= \left(62.8\frac{\cancel{lb}}{\cancel{in.^2}}\right)\left(\frac{4.448\ N}{1\ \cancel{lb}}\right)\left(\frac{1\ \cancel{in.^2}}{2.54^2\ \cancel{cm^2}}\right)\left(\frac{10^4\ \cancel{cm^2}}{1\ m^2}\right)$$

$$= \frac{(62.8)(4.448)(10^4)}{(2.54)^2}\frac{N}{m^2} = 4.33\times 10^5\frac{N}{m^2}$$

Exercises B–1

In Exercises 1 through 4, give the symbol and meaning for the given unit.

1. megahertz   2. kilowatt   3. millimeter   4. picosecond

In Exercises 5 through 8, give the name and meaning for the units whose symbols are given.

5. kV   6. GΩ   7. mA   8. pF

In Exercises 9 through 38, make the indicated reductions or conversions.

9. Reduce 1 km to centimeters.   10. Reduce 1 kg to milligrams.

11. Reduce 1 mi to inches.   12. Reduce 1 gal to pints (2 pt = 1 qt).

13. Convert 5.25 in. to centimeters.   14. Convert 6.50 kg to pounds.

15. Convert 15.7 qt to liters.   16. Convert 100 km to miles.

17. Reduce 1 ft^2 to square inches.   18. Reduce 1 yd^3 to cubic feet.

19. Reduce 250 mm^2 to square meters.   20. Reduce 30.8 kL to milliliters.

21. Convert 4.50 lb to grams.   22. Convert 0.360 in. to meters.

23. Convert 829 in.^{3} to liters.   24. Convert 0.0680 kL to cubic feet.

25. Convert 1 hp to kilogram centimeters per second.

26. Convert 8.75 Btu to joules.   27. Convert 75.0 W to horsepower.

28. Convert 300 mL to quarts.

29. An Atlas rocket weighed 260,000 lb at takeoff. How many tons is this?

30. An airplane is flying at 37,000 ft. What is its altitude in miles?

31. The speedometer of a car is calibrated in kilometers per hour. If the speedometer of such a car reads 60, how fast in miles per hour is the car traveling?

32. The acceleration due to gravity is about 980 cm/s^2. Convert this to feet per second squared.

33. The speed of sound is about 1130 ft/s. Change this speed to miles per hour.

34. The density of water is about 62.4 lb/ft^3. Convert this to kilograms per cubic meter.

35. The average density of the earth is about 5.52 g/cm^3. Convert this to pounds per cubic foot.

36. The moon travels about 1,500,000 miles in about 28 days in one rotation about the earth. Express its velocity in feet per second.

37. At sea level, atmospheric pressure is about 14.7 lb/in.^{2}. Express this pressure in pascals.

38. The earth's surface receives energy from the sun at the rate of 1.35 kW/m^2. Reduce this to joules per second square centimeter.

B–2 Approximate Numbers and Significant Digits

When we perform calculations on numbers, we must consider the accuracy of these numbers, since this affects the accuracy of the results obtained. Most of the numbers involved in technical and scientific work are *approximate,* having been arrived at through some process of measurement. However, certain other numbers are *exact,* having been arrived at through some definition or counting process. We can determine whether or not a number is approximate or exact if we know how the number was determined.

Example A

If we measure the length of a rope to be 15.3 ft, we know that the 15.3 is approximate. A more precise measuring device may cause us to determine the length as 15.28 ft. However, regardless of the method of measurement used, we shall not be able to determine this length exactly.

If a voltage shown on a voltmeter is read as 116 V, the 116 is approximate. A more precise voltmeter may show the voltage as 115.7 V. However, this voltage cannot be determined exactly.

Example B

If a computer counts the names it has processed and prints this number as 837, this 837 is exact. We know the number of names was not 836 or 838. Since 837 was determined through a counting process, it is exact.

When we say that 60 s = 1 min, the 60 is exact, since this is a definition. By this definition there are exactly 60 s in 1 min.

When we are writing approximate numbers we often have to include some zeros so that the decimal point will be properly located. However, *except for these zeros, all other digits are considered to be* **significant digits.** When we make computations with approximate numbers, we must know the number of significant digits. The following example illustrates how we determine this.

Example C

All numbers in this example are assumed to be approximate.

34.7 has three significant digits.

8900 has two significant digits. We assume that the two zeros are place holders (unless we have specific knowledge to the contrary).

0.039 has two significant digits. The zeros are for proper location of the decimal point.

706.1 has four significant digits. The zero is not used for the location of the decimal point. It shows specifically the number of tens in the number.

5.90 has three significant digits. The zero is not necessary as a place holder, and should not be written unless it is significant.

Other approximate numbers with the proper number of significant digits are listed below.

96,000	two	0.0709	three	1.070	four
30,900	three	6.000	four	700.00	five
4.006	four	0.0005	one	20,008	five

Note from the example above that *all nonzero digits are significant. Zeros, other than those used as place holders for proper positioning of the decimal point, are also significant.*

In computations involving approximate numbers, the position of the decimal point as well as the number of significant digits is important. *The* **precision** *of a number refers directly to the decimal position of the last significant digit, whereas the* **accuracy** *of a number refers to the number of significant digits in the number.* Consider the illustrations in the following example.

Example D

Suppose that you are measuring an electric current with two ammeters. One ammeter reads 0.031 A and the second ammeter reads 0.0312 A. The second reading is more precise, in that the last significant digit is the number of ten-thousandths, and the first reading is expressed only to thousandths. The second reading is also more accurate, since it has three significant digits rather than two.

A machine part is measured to be 2.5 cm long. It is coated with a film 0.025 cm thick. The thickness of the film has been measured to a greater precision, although the two measurements have the same accuracy: two significant digits.

A segment of a newly completed highway is 9270 ft long. The concrete surface is 0.8 ft thick. Of these two numbers, 9270 is more accurate, since it contains three significant digits, and 0.8 is more precise, since it is expressed to tenths.

The last significant digit of an approximate number is known not to be completely accurate. It has usually been determined by estimation or **rounding off.** However, we do know that it is at most in error by one-half of a unit in its place value.

Example E

When we measure the length of the rope referred to in Example A to be 15.3 ft, we are saying that the length is at least 15.25 ft and no longer than 15.35 ft. Any value between these two, rounded off to tenths, would be expressed as 15.3 ft.

In converting the fraction $\frac{2}{3}$ to the decimal form 0.667, we are saying that the value is between 0.6665 and 0.6675.

The principle of rounding off a number is to write the closest approximation, with the last significant digit in a specified position, or with a specified number of significant digits. We shall now formalize the process of rounding off as follows: If we want a certain number of significant digits, we examine the digit in the next place to the right. If this digit is less than 5, we accept the digit in the last place. If the next digit is 5 or greater, we increase the digit in the last place by 1, and this resulting digit becomes the final significant digit of the approximation. If necessary, we use zeros to replace other digits in order to locate the decimal point properly. Except when the next digit is a 5, and no other nonzero digits are discarded, we have the closest possible approximation with the desired number of significant digits.

Example F

70,360 rounded off to three significant digits is 70,400.
70,430 rounded off to three significant digits is 70,400.
187.35 rounded off to four significant digits is 187.4.
71,500 rounded off to two significant digits is 72,000.

With the advent of computers and calculators, another method of reducing numbers to a specified number of significant digits is used. This is the process of **truncation,** in which the digits beyond a certain place are discarded. For example, 3.17482 truncated to thousandths is 3.174. For our purposes in this text, when working with approximate numbers, we have used only rounding off.

Exercises B–2

In Exercises 1 through 8, determine whether the numbers given are exact or approximate.

1. There are 24 h in one day.
2. The velocity of light is 186,000 mi/s.
3. The 3-stage rocket took 74.6 h to reach the moon.
4. A person bought 5 lb of nails for $1.56.
5. The melting point of gold is 1063°C.
6. The 21 students had an average test grade of 81.6.
7. A building lot 100 ft by 200 ft cost $3200.
8. In a certain city 5% of the people have their money in a bank that pays 5% interest.

In Exercises 9 through 16, determine the number of significant digits in the given approximate numbers.

9. 37.2; 6844   10. 3600; 730   11. 107; 3004   12. 0.8735; 0.0075
13. 6.80; 6.08   14. 90,050; 105,040   15. 30,000; 30,000.0   16. 1.00; 0.01

In Exercises 17 through 24, determine which of each pair of approximate numbers is (a) more precise, and (b) more accurate.

17. 3.764; 2.81   18. 0.041; 7.673   19. 30.8; 0.01   20. 70,370; 50,400

21. 0.1; 78.0   22. 7040; 37.1   23. 7000; 0.004   24. 50.060; 8.914

In Exercises 25 through 32, round off each of the given approximate numbers (a) to three significant digits, and (b) to two significant digits.

25. 4.933   26. 80.53   27. 57,893   28. 30,490

29. 861.29   30. 9555   31. 0.30505   32. 0.7350

B–3 Arithmetic Operations with Approximate Numbers

 When performing arithmetic operations on approximate numbers, we must be careful not to express the result to a precision or accuracy which is not warranted. The following two examples illustrate how a false indication of the accuracy of a result could be obtained when using approximate numbers.

Example A

A pipe is made in two sections. The first is measured to be 16.3 ft long and the second is measured to be 0.927 ft long. A plumber wants to know what the total length will be when the two sections are put together.

At first, it appears we might simply add the numbers as follows to obtain the necessary result.

$$
\begin{array}{r}
16.3 \ \ \text{ft} \\
\underline{0.927 \ \text{ft}} \\
17.227 \ \text{ft}
\end{array}
$$

However, the first length is precise only to tenths, and the digit in this position was obtained by rounding off. It might have been as small as 16.25 ft or as large as 16.35 ft. If we consider only the precision of this first number, the total length might be as small as 17.177 ft or as large as 17.277 ft. These two values agree when rounded off to two significant digits (17). They vary by 0.1 when rounded off to tenths (17.2 and 17.3). When rounded to hundredths, they do not agree at all, since the third significant digit is different (17.18 and 17.28). Therefore, there is no agreement at all in the digits after the third when these two numbers are rounded off to a precision beyond tenths. This may also be deemed reasonable, since the first length is not expressed beyond tenths. The second number does not further change the precision of the result, since it is expressed to thousandths. Therefore we may conclude that the result must be rounded off at least to tenths, the precision of the first number.

Example B

We can find the area of a rectangular piece of land by multiplying the length, 207.54 ft, by the width, 81.4 ft. Performing the multiplication, we find the area to be (207.54 ft)(81.4 ft) $=$ 16,893.756 ft^2.

However, we know this length and width were found by measurement and that the least each could be is 207.535 ft and 81.35 ft. Multiplying these values, we find the least value for the area to be

$$(207.535 \text{ ft})(81.35 \text{ ft}) = 16,882.97225 \text{ ft}^2$$

The greatest possible value for the area is

$$(207.545 \text{ ft})(81.45 \text{ ft}) = 16,904.54025 \text{ ft}^2$$

We now note that the least possible and greatest possible values of the area agree when rounded off to three significant digits (16,900 ft^2) and there is no agreement in digits beyond this if the two values are rounded off to a greater accuracy. Therefore we can conclude that the accuracy of the result is good to three significant digits, or certainly no more than four. We also note that the width was accurate to three significant digits, and the length to five significant digits.

The following rules are based on reasoning similar to that in Examples A and B; we go by these rules when we perform the basic arithmetic operations on approximate numbers.

1. When approximate numbers are added or subtracted, the result is expressed with the precision of the least precise number.
2. When approximate numbers are multiplied or divided, the result is expressed with the accuracy of the least accurate number.
3. When the root of an approximate number is found, the result is accurate to the accuracy of the number.

If a calculator is not being used, there is another procedure which can be helpful. Before and during the calculation all numbers except the least precise or least accurate may be rounded off to one place beyond that of the least precise or least accurate number.

Example C

Add the approximate numbers 73.2, 8.0627, 93.57, 66.296.

The least precise of these numbers is 73.2. Therefore, before performing the addition we may round off the other number to hundredths. If we were using a calculator, this would not be done. In either case, after the addition we round off the result to tenths. This leads to

$$
\begin{array}{ccc}
73.2 & & 73.2 \\
8.06 & & 8.0627 \\
93.57 & \text{or} & 93.57 \\
\underline{66.30} & & \underline{66.296} \\
241.13 & & 241.1287
\end{array}
$$

Therefore, the final result is 241.1.

Example D

If we multiply 2.4832 by 30.5, we obtain 75.7376. However, since 30.5 has only three significant digits, we express the product and result as $(2.4832)(30.5) = 75.7$.

To find the square root of 3.7, we may express the result only to two significant digits. Thus, $\sqrt{3.7} = 1.9$.

The rules stated in this section are usually sufficiently valid for the computations encountered in technical work. They are intended only as good practical rules for working with approximate numbers. It was recognized in Examples A and B that the last significant digit obtained by these rules is subject to some possible error. Therefore it is possible that the most accurate result is not obtained by their use, although this is not often the case.

If an exact number is included in a calculation, there is no limitation to the number of decimal positions it may take on. *The accuracy of the result is limited only by the approximate numbers involved.*

Exercises B–3

In Exercises 1 through 4, add the given approximate numbers.

1.	3.8	2.	26	3.	0.36294	4.	56.1
	0.154		5.806		0.086		3.0645
	47.26		147.29		0.5056		127.38
					0.74		0.055

In Exercises 5 through 8, subtract the given approximate numbers.

5.	468.14	6.	1.03964	7.	57.348	8.	8.93
	36.7		0.69		26.5		6.8947

In Exercises 9 through 12, multiply the given approximate numbers.

9. $(3.64)(17.06)$ 10. $(0.025)(70.1)$

11. $(704.6)(0.38)$ 12. $(0.003040)(6079.52)$

In Exercises 13 through 16, divide the given approximate numbers.

13. $608 \div 3.9$

14. $0.4962 \div 827$

15. $\dfrac{596,000}{22}$

16. $\dfrac{53.267}{0.3002}$

In Exercises 17 through 20, find the indicated square roots of the given approximate numbers.

17. $\sqrt{32}$ 18. $\sqrt{6.5}$ 19. $\sqrt{19.3}$ 20. $\sqrt{0.0694}$

In Exercises 21 through 24, evaluate the given expression. All numbers are approximate.

21. $3.862 + 14.7 - 8.3276$

22. $(3.2)(0.386) + 6.842$

23. $\dfrac{8.60}{0.46} + (0.9623)(3.86)$

24. $9.6 - 0.1962(7.30)$

In Exercises 25 through 28, perform the indicated operations. The first number given is approximate and the second number is exact.

25. $3.62 + 14$

26. $17.382 - 2.5$

27. $(0.3142)(60)$

28. $8.62 \div 1728$

In Exercises 29 through 38, the solution to some of the problems will require the use of reduction and conversion factors.

29. Two forces, 18.6 lb and 2.382 lb, are acting on an object. What is the sum of these forces?

30. Three sections of a bridge are measured to be 52.3 ft, 36.38 ft, and 38 ft, respectively. What is the total length of these three sections?

31. Two planes are reported to have flown at speeds of 938 mi/h and 1400 km/h, respectively. Which plane is faster, and by how many miles per hour?

32. The density of a certain type of iron is 7.10 g/cm^3. The density of a type of tin is 448 lb/ft^3. Which is greater?

33. If the temperature of water is raised from 4°C to 30°C, its density reduces by 0.420%. If the density of water at 4°C is 62.4 lb/ft^3, what is its density at 30°C?

34. The power (in watts) developed in an electric circuit is found by multiplying the current (in amperes) by the voltage. In a certain circuit the current is 0.0125 A and the voltage is 12.68 V. What is the power that is developed?

35. A certain ore is 5.3% iron. How many tons of ore must be refined to obtain 45,000 lb of iron?

36. A certain computer has a data transfer rate of 1200 kilobytes/s. How many bytes (a *byte* is a unit, usually a single character, of computer memory) can this computer transfer in 3.28 μs?

37. In order to find the velocity (in feet per second) of an object which has fallen a certain height, we calculate the square root of the product of 64.4 (an approximate number) and the height in feet. What is the velocity of an object which has fallen 63 m?

38. A student reports the current in a certain experiment to be 0.02 A at one time and later notes that it is 0.023 A. The student then states that the change in current is 0.003 A. What is wrong with this conclusion?

Review of Geometry

Figure C–1

Figure C–2

Figure C–3

Figure C–4

Figure C–5

In this appendix we shall present geometric terminology and formulas that are related to the basic geometric figures. It is intended only as a brief summary of basic geometry.

Geometry deals with the properties and measurement of angles, lines, surfaces, and volumes, and the basic figures that are formed. We shall restrict our attention in this appendix to the basic figures and concepts which are related to these figures.

Repeating the definition in Section 3–1, an **angle** is generated by rotating a half-line about its endpoint from an initial position to a terminal position. One complete rotation of a line about a point is defined to be an angle of 360 **degrees** (Fig. C–1), written as 360°. A **straight angle** contains 180° (Fig. C–2), and a **right angle** contains 90° (Fig. C–3) and is shown by the symbol ⌐. If two lines meet so that the angle between them is 90°, the lines are said to be **perpendicular.**

An angle less than 90° is an **acute angle** (Fig. C–4). An angle greater than 90° but less than 180° is an **obtuse angle** (Fig. C–5). **Supplementary angles** are two angles whose sum is 180°, and **complementary angles** are two angles whose sum equals 90°.

In a plane, if a line crosses two **parallel** or nonparallel lines, it is called a **transversal.** If a transversal crosses a pair of parallel lines, certain pairs of equal angles result. In Fig. C–6, the **corresponding angles** are equal. (That is,

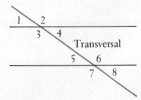

Figure C–6

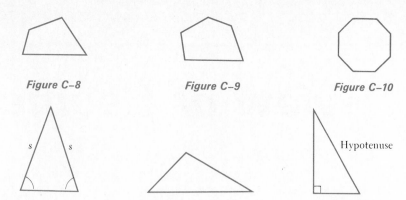

Figure C–7　　　　Figure C–8　　　　Figure C–9　　　　Figure C–10

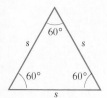

Figure C–11

Figure C–12　　　　Figure C–13　　　　Figure C–14

$\angle 1 = \angle 5$, $\angle 2 = \angle 6$, $\angle 3 = \angle 7$, $\angle 4 = \angle 8$.) Also, the **alternate interior angles** are equal ($\angle 3 = \angle 6$ and $\angle 4 = \angle 5$).

Adjacent angles have a common vertex and a side common to them. For example, in Fig. C–6, ∡ 5 and 6 are adjacent angles. **Vertical angles** are equal angles formed "across" the point of intersection of two intersecting lines. In Fig. C–6, ∡ 5 and 8 are vertical angles.

When a part of the plane is bounded and closed by straight line segments, it is called a **polygon.** In general, polygons are named according to the number of sides they contain. A **triangle** has three sides (Fig. C–7), a **quadrilateral** has four sides (Fig. C–8), a **pentagon** has five sides (Fig. C–9), and so on. In a **regular polygon,** all the sides are of equal length (Fig. C–10), and all the interior angles are equal.

There are several important types of triangles. In an **equilateral triangle,** the three sides are equal and the three angles are also equal, each being 60° (Fig. C–11). In an **isosceles triangle,** two of the sides are equal, as are the two **base angles** (the angles opposite the equal sides) (Fig. C–12). In a **scalene triangle,** no two sides are equal, and none of the angles is a right angle (Fig. C–13). In a **right triangle,** one of the angles is a right angle (Fig. C–14). The side opposite the right angle is called the **hypotenuse.**

There are certain basic properties of triangles which we shall mention here. One very important property is that

the sum of the three angles of any triangle is 180°.

Another property is that the three **medians** (line segments drawn from a vertex to the **midpoint** of the opposite side) meet at a single point (Fig. C–15). This point of intersection is called the **centroid** of the triangle. Also, the three **angle bisectors** meet at a common point (Fig. C–16), as do the three **altitudes** (heights), which are drawn from a vertex perpendicular to the opposite side (or the extension of the opposite side) (Fig. C–17).

Of particular importance is the **Pythagorean theorem,** which states that

in a right triangle, the square of the length of the hypotenuse equals the sum of the squares of the lengths of the other two sides.

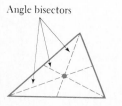

Medians

Centroid

Figure C–15

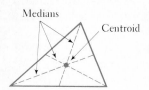

Angle bisectors

Figure C–16

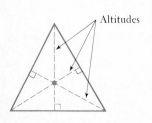

Altitudes

Figure C–17

Two triangles are said to be **congruent** if corresponding angles are equal and if corresponding sides are equal. Also,

> *two* **similar** *triangles have the properties that* (1) *the corresponding angles are equal, and* (2) *the corresponding sides are proportional.*

See Fig. C–18.

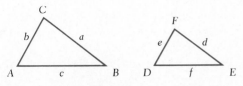

$$\triangle ABC \sim \triangle DEF \qquad (\sim \text{ means ``similar to''})$$

$$\angle A = \angle D, \ \angle B = \angle E, \ \angle C = \angle F$$

$$\frac{a}{d} = \frac{b}{e} = \frac{c}{f}$$

Figure C–18

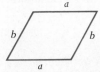

Figure C–19

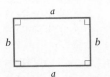

Figure C–20

There are also several important types of quadrilaterals. A **parallelogram** is a quadrilateral with opposite sides **parallel** (extensions of the sides will not intersect) (Fig. C–19). Also, opposite sides and opposite angles of a parallelogram are equal. A **rectangle** is a parallelogram with intersecting sides perpendicular, which means that all four interior angles are right angles (Fig. C–20). It also means that opposite sides of a rectangle are equal and parallel. A **square** is a rectangle, all sides of which are equal (Fig. C–21). A **trapezoid** is a quadrilateral with two parallel sides (Fig. C–22). These parallel sides are called the **bases** of the trapezoid. A **rhombus** is a parallelogram, all four sides of which are equal (Fig. C–23). One important property of all quadrilaterals is that the sum of the four angles is 360°.

All the points on a **circle** are the same distance from a fixed point in the plane. This point is the **center** of the circle. The distance from the center to a point on the circle is the **radius** of the circle. The distance between two points on the circle and on a line passing through the center of the circle is the **diameter** of the circle. Thus the diameter is twice the radius. See Fig. C–24.

Figure C–21

Figure C–22

Figure C–23

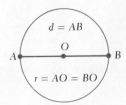

Figure C–24

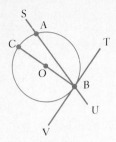

Figure C–25

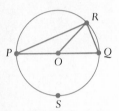

Figure C–26

Figure C–27

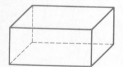

Figure C–28

Figure C–29

Also associated with the circle is the **chord,** which is a line segment having its endpoints on the circle. A **tangent** is a line that touches a circle (does not pass through) at one point. A **secant** is a line that passes through two points of a circle. In Fig. C–25, *AB* is a chord, *TV* is a tangent line, and *SU* is a secant line.

An **arc** is a part of a circle. When two radii form an angle at the center, the angle is called a **central angle.** An **inscribed angle** of an arc is one for which the endpoints of the arc are points on the sides of the angle, and for which the vertex is a point of the arc, although not an endpoint. In Fig. C–26, *RQ* is an arc and ∠*RPQ* is an inscribed angle equal to one-half *RQ*. Angle *ROQ* is a central angle.

There are two important properties of a circle which we shall mention here.

1. *A tangent to a circle is perpendicular to the radius drawn to the point of contact.*
2. *An inscribed angle is one-half of its intercepted arc. Thus, an angle inscribed in a semicircle is a right angle.*

There are two important measures associated with the plane figures we have discussed to this point. They are its **perimeter** and its **area.** The perimeter of a plane figure is the distance around it. Formulas for the areas of the important plane figures are found in the following section of this appendix.

We now note the solid figures of the greatest importance. A **polyhedron** is a solid figure which is bounded by planes. The plane surfaces of the polyhedron are called **faces** (Fig. C–27). A **prism** is a polyhedron whose bases are parallel and equal polygons and whose side faces are parallelograms (Fig. C–28). A prism of particular importance is the **rectangular solid** (Fig. C–29). The faces of a rectangular solid are all rectangles. The base of a **pyramid** is a polygon (Fig. C–30). The other faces, the **lateral faces,** meet at a common point, the **vertex.**

A **right circular cylinder** is generated by revolving a rectangle about one of its sides (Fig. C–31). A **right circular cone** is generated by revolving a right triangle about one of its legs (Fig. C–32). A **sphere** is generated by revolving a circle about a diameter (Fig. C–33).

The **lateral area** of a solid is the area other than that of the bases. The **total area** is the lateral area plus the area of the bases. Formulas for the important areas and those for the **volumes** of solid figures are found in the following section of this appendix.

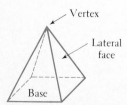

Figure C–30

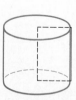

Figure C–31

Figure C–32

Figure C–33

C–2 Basic Geometric Formulas

For the indicated figures, the following symbols are used: A = area, B = area of base, c = circumference, S = lateral area, V = volume.

1. **Triangle.** $A = \frac{1}{2}bh$ (Fig. C–34)
2. **Pythagorean theorem.** $c^2 = a^2 + b^2$ (Fig. C–35)
3. **Parallelogram.** $A = bh$ (Fig. C–36)
4. **Trapezoid.** $A = \frac{1}{2}(a + b)h$ (Fig. C–37)
5. **Circle.** $A = \pi r^2$, $c = 2\pi r$ (Fig. C–38)
6. **Rectangular solid.** $A = 2(lw + lh + wh)$, $V = lwh$ (Fig. C–39)
7. **Cube.** $A = 6e^2$, $V = e^3$ (Fig. C–40)
8. **Any cylinder or prism with parallel bases.** $V = Bh$ (Fig. C–41)
9. **Right circular cylinder.** $S = 2\pi rh$, $V = \pi r^2 h$ (Fig. C–42)
10. **Any cone or pyramid.** $V = \frac{1}{3}Bh$ (Fig. C–43)
11. **Right circular cone.** $S = \pi rs$, $V = \frac{1}{3}\pi r^2 h$ (Fig. C–44)
12. **Sphere.** $A = 4\pi r^2$, $V = \frac{4}{3}\pi r^3$ (Fig. C–45)

As we noted, the **perimeter** of a plane figure is the distance around it. For example, the perimeter p of the triangle in Fig. C–35 is $p = a + b + c$. The perimeter of a circle is its **circumference.**

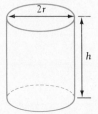

Figure C–34

Figure C–35

Figure C–36

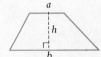

Figure C–37

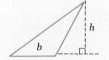

Figure C–38

Figure C–39

Figure C–40

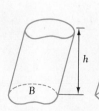

Figure C–41

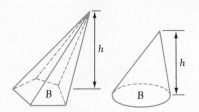

Figure C–42

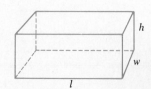

Figure C–43

Figure C–44

Figure C–45

C–3 Exercises for Appendix C

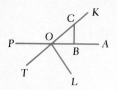

Figure C–46

In Exercises 1 through 4, refer to Fig. C–46, in which AOP and KOT are straight lines. $\angle TOP = 40°$, $\angle OBC = 90°$, and $\angle AOL = 55°$. Determine the indicated angles.

1. $\angle LOT$ 2. $\angle POL$ 3. $\angle KOP$ 4. $\angle KCB$

In Exercises 5 through 8, refer to Figs. C–47 and C–48. In Fig. C–47, $AB \parallel FC$, $AD \perp BE$, and $\angle A = 38°$. In Fig. C–48, $AB \parallel CD$, $AB = AC$, and $AD = CD$. (The symbol $\parallel$ means ''is parallel to,'' and the symbol $\perp$ means ''is perpendicular to.'') Find the indicated angles.

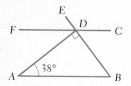

Figure C–47

5. In Fig. C–47, $\angle BDC$ 6. In Fig. C–47, $\angle CDE$
7. In Fig. C–48, $\angle DCA$ 8. In Fig. C–48, $\angle ADC$

In Exercises 9 through 12, refer to Figs. C–49 and C–50. In each, $ABCD$ is a parallelogram. In Fig. C–49, $\triangle BEC$ is isosceles, and $\angle BCE = 40°$. In Fig. C–50, $\angle FED = 34°$. Determine the indicated angles.

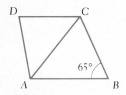

Figure C–48

9. In Fig. C–49, $\angle CEB$ 10. In Fig. C–49, $\angle ADC$
11. In Fig. C–50, $\angle EDF$ 12. In Fig. C–50, $\angle DCB$

In Exercises 13 through 16, refer to Fig. C–51, where AB is a diameter, line TB is tangent to the circle at B, and $\angle ABC = 65°$. Determine the indicated angles.

13. $\angle CBT$ 14. $\angle BCT$ 15. $\angle CAB$ 16. $\angle BTC$

In Exercises 17 through 20, refer to Fig. C–52. Determine the indicated arcs and angles.

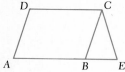

Figure C–49

17. $\overset{\frown}{BC}$ 18. $\overset{\frown}{AB}$ 19. $\angle ABC$ 20. $\angle ACB$

In Exercises 21 through 32, find the indicated sides of the right triangle shown in Fig. C–53. Where necessary, round off the results to four significant digits.

	a	b	c		a	b	c
21.	3	4	?	22.	9	12	?
23.	8	15	?	24.	24	10	?
25.	6	?	10	26.	2	?	4
27.	5.126	?	7.245	28.	3.074	?	9.452
29.	?	12.32	16.15	30.	?	10.49	18.72
31.	?	15.08	32.59	32.	?	5.472	36.54

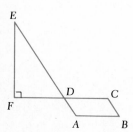

Figure C–50

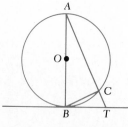

Figure C–51

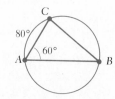

Figure C–52

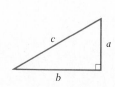

Figure C–53

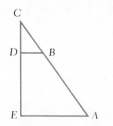

Figure C-54

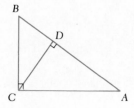

Figure C-55

In Exercises 33 through 36, use Fig. C–54. In this figure $BD \parallel AE$, $AE = 18$, and $DB = 6$.

33. If $DC = 7$, find CE. 34. If $CA = 27$, find CB.

35. If $CB = 10$, find BA. 36. If $CE = 24$, find DE.

In Exercises 37 through 40, determine the required values. In Fig. C–55, $AD = 9$ and $AC = 12$.

37. Two triangles are similar, and the sides of the larger triangle are 3.0 in., 5.0 in., and 6.0 in., and the shortest side of the other triangle is 2.0 in. Find the remaining sides of the smaller triangle.

38. Two triangles are similar. The angles of the smaller triangle are 50°, 100°, and 30°, and the sides of the smaller triangle are 7.00 in., 9.00 in., and 4.57 in. The longest side of the larger triangle is 15.00 in. Find the other two sides and the three angles of the larger triangle.

39. In Fig. C–55, find AB. 40. In Fig. C–55, find BC.

In Exercises 41 through 52, find the perimeters (or circumferences) of the indicated figures.

41. A triangle with sides 6 ft, 8 ft, and 11 ft.

42. A quadrilateral with sides 3 in., 4 in., 6 in., and 9 in.

43. An isosceles triangle whose equal sides are 3 yd long and whose third side is 4 yd long.

44. An equilateral triangle whose sides are 7 ft long.

45. A rectangle 5.14 in. long and 4.09 in. wide.

46. A parallelogram whose longer side is 13.7 mm and whose shorter side is 11.9 mm.

47. A square of side 8.18 cm.

48. A rhombus of side 15.6 m.

49. A trapezoid whose longer base is 74.2 cm, whose shorter base is 46.8 cm, and whose other sides are equal to the shorter base.

50. A trapezoid whose bases are 17.8 in. and 7.4 in., and whose other sides are each 8.1 in.

51. A circle of radius 10.0 ft.

52. A circle of diameter 7.06 cm.

In Exercises 53 through 60, find the areas of the indicated figures.

53. A parallelogram of base 7.0 in. and height 4.0 in.

54. A rectangle of base 8.2 ft and height 2.5 ft.

55. A triangle of base 6.3 yd and height 4.1 yd.

56. A triangle of base 14.2 cm and height 6.83 cm.

57. A trapezoid of bases 3.0 m and 9.0 m and height 4.0 m.

58. A trapezoid of bases 18.5 in. and 26.3 in. and height 10.5 in.

59. A circle of radius 7.00 in.

60. A semicircle of diameter 8.24 cm.

In Exercises 61 through 72, evaluate the volume of the indicated figures.

61. A rectangular solid of length 9.00 ft, width 6.00 ft, and height 4.00 ft.

62. A cube of edge 7.15 ft.

63. Prism, square base of side 2.0 mm, altitude 5.0 mm.

64. Prism, trapezoidal base (bases 4.0 in. and 6.0 in. and height 3.0 in.), altitude 6.0 in.

65. Cylinder, radius of base 7.00 in., altitude 6.00 in.

66. Cylinder, diameter of base 6.36 cm, altitude 18.0 cm.

67. Pyramid, rectangular base 12.5 in. by 8.75 in., altitude 4.20 in.

68. Pyramid, equilateral triangular base of side 3.00 in., altitude 4.25 in.

69. Cone, radius of base 2.66 ft, altitude 1.22 yd.

70. Cone, diameter of base 16.3 cm, altitude 18.4 cm.

71. Sphere, radius 5.48 m.

72. Sphere, diameter 15.7 yd.

In Exercises 73 through 80, find the total surface area of the indicated figure for the given values.

73. Prism, rectangular base 4.00 in. by 8.00 in., altitude 6.00 in.

74. Prism, parallelogram base (base 16.0 in., height 4.25 in., perimeter 42.0 in.), altitude 6.45 in.

75. Cylinder, radius of base 8.58 m, altitude 1.38 m.

76. Cylinder, diameter of base 12.5 ft, altitude 4.60 ft.

77. Sphere, radius 16.0 ft.

78. Sphere, diameter 15.3 in.

79. Cone, radius of base 7.05 cm, slant height 8.45 cm.

80. Cone, diameter of base 76.2 mm, altitude 22.1 mm.

APPENDIX D

The Scientific Calculator

D–1 Introduction

Until the early 1970s the personal calculational device of many technicians and scientists was the slide rule. However, with the development of the microprocessor chip, the pocket scientific electronic calculator is now readily available. Since the calculator is easier to use, and has much greater accuracy and calculational ability than the slide rule, it is now the personal calculational device of technicians and scientists, as well as many others.

In Section 1–4 the calculator was briefly introduced. We noted there that a scientific calculator can be used for any of the calculations which may be found in this text, but that one which performs only the basic arithmetic operations is not sufficient.

There are a great many types and models of calculators. However, the discussions in this appendix are based on the operations which can be performed by use of the basic keys of a scientific calculator and are general enough to apply to most such calculators. Some of the variations and special features which may be found are noted. Nearly all calculators come with a manual that can be used to learn the operations of the calculator, or to supplement this material on the variations and special features which a given calculator may have.

In this appendix we discuss the calculator keys and operations which are used for data entry, arithmetic operations, special functions (squares, square roots, reciprocals, powers and roots), trigonometric and inverse trigonometric functions, exponential and logarithmic functions, and calculator memory. Examples are included here and in many examples throughout the text.

Many models of calculators use individual keys for each number and function. Other models use many of the keys for more than one purpose. However, the difference in operation is generally minor. Also, the labeling of many keys varies from one model to another. Some of these variations which are used are noted.

When a number is entered, or a result calculated, it shows on the display at the top of the calculator. *We will use an eight-digit display,* as well as a possible display of the exponent of 10 for scientific notation, in our discussion of the calculator. Many calculators do display ten or more digits, although the operation is essentially the same. If the result contains more significant digits than the display can show, the result shown will be truncated or rounded off. We will use a rounded off display.

In making entries or calculations, it is possible that a number with too many digits or one which is too large or too small cannot be entered. Also, certain operations do not have defined results. If such an entry or calculation is attempted, the calculator will make an error indication such as E, ⊓, or a flashing display. Operations which can result in an error indication include division by zero, square root of a negative number, logarithm of a negative number, and the inverse trigonometric function of a value outside of the appropriate interval.

Also, some calculators will give a special display when batteries need charging or replacement. The user should become acquainted with any special displays the calculator may use.

As we noted in Section 1–4, the calculator logic determines the order in which entries must be made in order to obtain the desired result. We also noted that *in this text we assume that the calculator uses algebraic logic.* Some discussion of this is found in Section 1–4 but *the user must become acquainted with the logic of the particular calculator being used.*

D–2 Calculator Data Entry and Function Keys

Following is a listing of certain basic keys which may be found on a scientific calculator, and which can be of use in performing calculations for the exercises in this text. There are other keys and other calculational capabilities which are also found on many calculators. Along with each key is a description and an illustration of its basic use. In the examples shown at the right, to perform the indicated entry or calculation, use the given sequence of entries and operations. The final display is also shown.

It should be emphasized that not all the keys which will probably be found on a given calculator are discussed, but that a scientific calculator will have most of the keys which are described. It should be noted that the use of many of the keys may be beyond the scope of the reader until the appropriate text material has been covered.

Keys	Examples
0, 1, . . . , 9 **Digit Keys** These keys are used to enter the digits 0 through 9 in the display, or to enter an exponent of 10 when scientific notation is used.	To enter: 37514 Sequence: 3, 7, 5, 1, 4 Display: 37514.
$\cdot$ **Decimal Point Key** This key is used to enter a decimal point.	To enter: 375.14 Sequence: 3, 7, 5, $\cdot$, 1, 4 Display: 375.14
$+/-$ **Change Sign Key** This key is used to change the sign of the number on the display, or to change the sign of the exponent when scientific notation is used. (May be designated as CHS.)	To enter: -375.14 Sequence: 3, 7, 5, $\cdot$, 1, 4, $+/-$ Display: -375.14 (From here on the entry of a number will be shown as one operation.)
π **Pi Key** This key is used to enter π to the number of digits of the display.	To enter: π Sequence: π Display: 3.1415927 (For calculators with dual-purpose keys, see $2nd$ key.)
EE **Enter Exponent Key** This key is used to enter an exponent when scientific notation is used. After the key is pressed, the exponent is entered. For a negative exponent, the $+/-$ key is pressed after the exponent is entered. (May be designated as $E\ EX$.)	To enter: 2.936×10^8 Sequence: 2.936, EE, 8 Display: 2.936 08 To enter: -2.936×10^{-8} Sequence: 2.936, $+/-$, EE, 8, $+/-$ Display: $-2.936\ -08$
$=$ **Equals Key** This key is used to complete a calculation to give the required result.	See the following examples for illustrations of the use of this key. If a calculator does not have an equals key, it does not use algebraic logic.

Keys	Examples
$+$ **Add Key** This key is used to add the next entry to the previous entry or result.	Evaluate: 37.56 + 241.9 Sequence: 37.56, $+$, 241.9, $=$ Display: 279.46
$-$ **Subtract Key** This key is used to subtract the next entry from the previous entry or result.	Evaluate: 37.56 − 241.9 Sequence: 37.56, $-$, 241.9, $=$ Display: −204.34
$\times$ **Multiply Key** This key is used to multiply the previous entry or result by the next entry.	Evaluate: 8.75 × 30.92 Sequence: 8.75, $\times$, 30.92, $=$ Display: 270.55
$\div$ **Divide Key** This key is used to divide the previous entry or result by the next entry.	Evaluate: 8.75 ÷ 30.92 Sequence: 8.75, $\div$, 30.92, $=$ Display: 0.2829884 (truncated or rounded off)

( ) **Parentheses Keys**
 These keys are used to group specified numbers in a calculation.
	Evaluate: 3.586(30.72 + 47.92)
	 Sequence: 3.586, $\times$, (,
	      30.72, $+$, 47.92
	       ) , $=$
	 Display: 282.00304

CE **Clear Entry Key** This key is used to clear the last entry. Its use will not affect any other part of a calculation. On some calculators one press of the C/CE or CL key is used for this purpose.	Evaluate: 37.56 + 241.9, with an improper entry of 242.9 Sequence: 37.56, $+$, 242.9, CE, 241.9, $=$ Display: 279.46
C **Clear Key** This key is used to clear the display and information being calculated (not including memory), so that a new calculation may be started. For calculators with a C/CE or C key, and no CE key, a second press on these keys is used for this purpose.	To clear previous calculation Sequence: C Display: 0.

Keys	Examples
2nd **Second Function Key** This key is used on calculators on which many of the keys serve dual purposes. It is pressed before the second key functions are activated.	Evaluate: 422π on a calculator where π is a second use of a key Sequence: 422, $\times$, 2nd, π, $=$ Display: 1325.7521
x^2 **Square Key** This key is used to square the number on the display.	Evaluate: 37.4^2 Sequence: 37.4, x^2 Display: 1398.76
$\sqrt{x}$ **Square Root Key** This key is used to find the square root of the number on the display.	Evaluate: $\sqrt{37.4}$ Sequence: 37.4, $\sqrt{x}$ Display: 6.1155539
$1/x$ **Reciprocal Key** This key is used to find the reciprocal of the number on the display.	Evaluate: $\dfrac{1}{37.4}$ Sequence: 37.4, $1/x$ Display: 0.0267380
x^y **x to the y Power Key** This key is used to raise x, the first entry, to the y power, the second entry. (Use only if $x \geq 0$.) (May also be designated as y^x .)	Evaluate: $3.73^{1.5}$ Sequence: 3.73, x^y, 1.5, $=$ Display: 7.2038266
DRG **Degree-Radian-Grad Key** This key is used to designate a displayed angle as being measured in degrees, radians, or grads (100 grads = 90°).	This key should show the appropriate angle measurement before the calculation is started. (On some calculators this key is also used to convert from one angle measure to another.)
SIN **Sine Key** This key is used to find the sine of the angle on the display.	Evaluate: sin 37.4° Sequence: DRG, 37.4, SIN Display: 0.6073758
COS **Cosine Key** This key is used to find the cosine of the angle on the display.	Evaluate: cos 2.475 (rad) Sequence: DRG, 2.475, COS Display: −0.7859330

Keys	Examples
TAN **Tangent Key** This key is used to find the tangent of the angle on the display.	Evaluate: $\tan(-24.9°)$ Sequence: (D) RG , 24.9, +/− , TAN Display: -0.4641845
Cotangent, Secant, Cosecant These functions are found through their reciprocal relation with the tangent, cosine, and sine functions, respectively. See Section 19–1.	Evaluate: $\cot 2.841$ (rad) Sequence: D R G , 2.841, TAN , $1/x$ Display: -3.2259549
INV **Inverse Function Key** This key is used (prior to the appropriate trigonometric function key) to find the angle whose trigonometric function is on the display. Some calculators have ARCSIN or SIN^{-1} (and equivalent) keys for this purpose. The INV key may also be used to evaluate other inverse functions.	Evaluate: Arcsin 0.1758 (the angle whose sine is 0.1758) in degrees Sequence: (D) RG , .1758, INV , SIN Display: 10.125216
LOG **Common Logarithm Key** This key is used to find the logarithm to the base 10 of the number on the display.	Evaluate: log 37.45 Sequence: 37.45, LOG Display: 1.5734518
LN **Natural Logarithm Key** This key is used to find the logarithm to the base e of the number on the display.	Evaluate: ln 0.8421 Sequence: 0.8421, LN Display: -0.1718565
10^x **10 to the x Power Key** This key is used to find antilogarithms (base 10) of the number on the display. (The x^y key can be used for this purpose, if the calculator does not have this key.)	Evaluate: Antilog 0.7265 (or $10^{0.7265}$) Sequence: .7265, 10^x Display: 5.3272122

Keys	Examples
$\boxed{e^x}$ *e* **to the** *x* **Power Key** This key is used to raise *e* to the power on the display.	Evaluate: $e^{-4.05}$ Sequence: 4.05, $\boxed{+/-}$, $\boxed{e^x}$ Display: 0.0174224
$\boxed{\text{STO}}$ **Store in Memory Key** This key is used to store the number on the display in the memory.	Store in memory: 56.02 Sequence: 56.02, $\boxed{\text{STO}}$ Display: 56.02
$\boxed{\text{RCL}}$ **Recall from Memory Key** This key is used to recall the number in the memory to the display. (May be designated as $\boxed{\text{MR}}$.)	Recall from memory: 56.02 Sequence: $\boxed{\text{RCL}}$ Display: 56.02
$\boxed{\text{M}}$ **Other Memory Keys** Some calculators use an $\boxed{\text{M}}$ key to store a number in the memory. It also may add the entry to the number in the memory. There are also keys for other operations on the number in the memory.	

D–3 Combined Operations

Throughout the text there are numerous exercises in which numerical calculations are required. The use of a calculator can save a great deal of time in making these calculations. Many of them require only a basic two- or three-step use of the calculator, and a knowledge of the basic use of the keys is generally sufficient to make such calculations.

There are also many types of problems which require more extensive calculations for their solutions, and for many of these problems a calculator is the only practical way of making the necessary calculations. Throughout the book there are examples of how such calculations are done on a calculator. Therefore, we will not present any examples here.

It can be seen from the examples in the text that a calculator is especially valuable for certain types of problems. These include problems in which special functions such as square roots, exponentials, or logarithms are involved. The calculator is very useful for problems which are primarily arithmetic, especially when there are several significant digits involved. In solving triangles, the calculator is particularly valuable. Also, there are some examples and exercises in which the calculator plays an important role in developing a concept. However, it is still necessary to learn the mathematics required in order to set up and understand a given problem.

The exercises which follow provide an opportunity for additional practice in the use of a calculator.

D–4 Exercises for Appendix D

In Exercises 1 through 36, perform all calculations on a scientific calculator.

1. $47.08 + 8.94$
2. $654.1 + 407.7$
3. $4724 - 561.9$
4. $0.9365 - 8.077$
5. 0.0396×471
6. 26.31×0.9393
7. $76.7 \div 194$
8. $52,060 \div 75.09$
9. 3.76^2
10. 0.986^2
11. $\sqrt{0.2757}$
12. $\sqrt{60.36}$
13. $\dfrac{1}{0.0749}$
14. $\dfrac{1}{607.9}$
15. $(19.66)^{2.3}$
16. $(8.455)^{1.75}$
17. $\sin 47.3°$
18. $\sin 1.15$
19. $\cos 3.85$
20. $\cos 119.1°$
21. $\tan 306.8°$
22. $\tan 0.537$
23. $\sec 6.11$
24. $\csc 242.0°$
25. Arcsin 0.6607 (in degrees)
26. Arccos(-0.8311) (in radians)
27. Arctan(-2.441) (in radians)
28. Arcsin 0.0737 (in degrees)
29. $\log 3.857$
30. $\log 0.9012$
31. $\ln 808$
32. $\ln 70.5$
33. $10^{0.545}$
34. $10^{-0.0915}$
35. $e^{-5.17}$
36. $e^{1.672}$

In Exercises 37 through 90, perform all calculations on a scientific calculator. In these exercises some of the combined operations which are encountered in certain types of problems are given. For specific types of applications, solve problems from the appropriate sections of the text.

37. $(4.38 + 9.07) \div 6.55$
38. $(382 + 964) \div 844$
39. $4.38 + (9.07 \div 6.55)$
40. $382 + (964 \div 844)$
41. $\dfrac{5.73 \times 10^{11}}{20.61 - 7.88}$
42. $\dfrac{7.09 \times 10^{23}}{284 + 839}$
43. $50.38\pi^2$
44. $\dfrac{5\pi}{14.6}$
45. $\sqrt{1.65^2 + 6.44^2}$
46. $\sqrt{0.735^2 + 0.409^2}$
47. $3(3.5)^4 - 4(3.5)^2$
48. $\dfrac{3(-1.86)}{(-1.86)^2 + 1}$

49. $29.4 \cos 72.5°$

50. $\dfrac{477}{\sin 58.7°}$

51. $\dfrac{4 + \sqrt{(-4)^2 - 4(3)(-9)}}{2(3)}$

52. $\dfrac{-5 - \sqrt{5^2 - 4(4)(-7)}}{2(4)}$

53. $\dfrac{0.176(180)}{\pi}$

54. $\dfrac{209.6\pi}{180}$

55. $\dfrac{1}{2}\left(\dfrac{51.4\pi}{180}\right)(7.06)^2$

56. $\dfrac{1}{2}\left(\dfrac{148.2\pi}{180}\right)(49.13)^2$

57. $\text{Arcsin} \dfrac{27.3 \sin 36.5°}{46.8}$

58. $\dfrac{0.684 \sin 76.1°}{\sin 39.5°}$

59. $\sqrt{3924^2 + 1762^2 - 2(3924)(1762)\cos 106.2°}$

60. $\text{Arccos} \dfrac{8.09^2 + 4.91^2 - 9.81^2}{2(8.09)(4.91)}$

61. $\sqrt{5.81 \times 10^8} + \sqrt[3]{7.06 \times 10^{11}}$

62. $(6.074 \times 10^{-7})^{2/5} - (1.447 \times 10^{-5})^{4/9}$

63. $\dfrac{3}{2\sqrt{7} - \sqrt{6}}$

64. $\dfrac{7\sqrt{5}}{4\sqrt{5} - \sqrt{11}}$

65. $\text{Arctan} \dfrac{7.37}{5.06}$

66. $\text{Arctan} \dfrac{46.3}{-25.5}$

67. $2 + \dfrac{\log 12}{\log 7}$

68. $\dfrac{10^{0.4115}}{\pi}$

69. $\dfrac{26}{2}(-1.450 + 2.075)$

70. $\dfrac{4.55(1 - 1.08^{15})}{1 - 1.08}$

71. $\sin^2\left(\dfrac{\pi}{7}\right) + \cos^2\left(\dfrac{\pi}{7}\right)$

72. $\sec^2\left(\dfrac{2}{9}\pi\right) - \tan^2\left(\dfrac{2}{9}\pi\right)$

73. $\sin 31.6° \cos 58.4° + \sin 58.4° \cos 31.6°$

74. $\cos^2 296.7° - \sin^2 296.7°$

75. $\sqrt{(1.54 - 5.06)^2 + (-4.36 - 8.05)^2}$

76. $\sqrt{(7.03 - 2.94)^2 + (3.51 - 6.44)^2}$

77. $\dfrac{(4.001)^2 - 16}{4.001 - 4}$

78. $\dfrac{(2.001)^2 + 3(2.001) - 10}{2.001 - 2}$

79. $\dfrac{4\pi}{3}(8.01^3 - 8.00^3)$

80. $4\pi(76.3^2 - 76.0^2)$

81. $0.01\left(\dfrac{1}{2}\sqrt{2} + \sqrt{2.01} + \sqrt{2.02} + \dfrac{1}{2}\sqrt{2.03}\right)$

82. $0.2\left[\dfrac{1}{2}(3.5)^2 + 3.7^2 + 3.9^2 + \dfrac{1}{2}(4.1)^2\right]$

83. $\dfrac{e^{0.45} - e^{-0.45}}{e^{0.45} + e^{-0.45}}$

84. $\ln \sin 2e^{-0.055}$

85. $\ln\dfrac{2 - \sqrt{2}}{2 - \sqrt{3}}$

86. $\sqrt{\dfrac{9}{2} + \dfrac{9 \sin 0.2\pi}{8\pi}}$

87. $2 + \dfrac{0.3}{4} - \dfrac{(0.3)^2}{64} + \dfrac{(0.3)^3}{512}$

88. $\dfrac{1}{2} + \dfrac{\pi\sqrt{3}}{360} - \dfrac{1}{4}\left(\dfrac{\pi}{180}\right)^2$

89. $160(1 - e^{-1.50})$

90. $e^{-3.60}(\cos 1.20 + 2 \sin 1.20)$

Selected Computer Programs in BASIC

E–1 Introduction

In this appendix there are 14 selected computer programs. They are written in the programming language BASIC. This language is chosen because of its simplicity and widespread availability on computer systems.

It is not the intent here to teach how to write a computer program in BASIC. The programs are intended for those who have some experience with computer programs, or who have resources available to learn the fundamentals of programming in BASIC. However, only a knowledge of the fundamentals is necessary to use and learn from these programs.

Most of the programs in this appendix are reasonably simple, and when you have some experience you will see how they can be modified and expanded to provide additional information. Also, you will see how many additional programs can be developed. These programs illustrate how a few of the calculations and problems in the text can be solved by use of a computer. The number of programs which could be used for the problems in this text is essentially limitless.

There are many special statements, such as INPUT and PRINT, which are used in BASIC. They provide the computer with the instructions which it must follow. You should have a knowledge of the fundamental BASIC statements before attempting to use these programs. Among those which are necessary are REM, INPUT, LET, PRINT, GOTO, FOR (with NEXT), IF (with THEN), and END. Also, the system commands of NEW, LOAD, LIST, and RUN are necessary.

As well as knowing the fundamental statements, it is necessary to know the symbols for the arithmetic operations and the various mathematical functions. The symbols $+$, $-$, $*$, and $/$ are used for addition, subtraction, multiplication, and division, respectively. For exponentiation, $**$, $\uparrow$, or $\wedge$ may be used, depending on the particular computer. A list of the fundamental mathematical functions is shown in the following table.

Table E–1

BASIC Expression	Function	Comment
SIN(X)	sin x	x in radians
COS(X)	cos x	x in radians
TAN(X)	tan x	x in radians
ATN(X)	Arctan x	Result in radians
LOG(X)	ln x	Natural logarithm
EXP(X)	e^x	Exponential function
SQR(X)	$\sqrt{x}$	Square root
ABS(X)	$|x|$	Absolute value
SGN(X)		Result: $+1$ if $x > 0$
		-1 if $x < 0$
		0 if $x = 0$

It should be noted that the trigonometric functions use angles expressed in radians and not degrees. Also, many computers include only the Arctangent and do not have the Arcsine or Arccosine. If a problem involves the use of functions or values which the computer does not have, it may be necessary to define them in the particular program.

There are many variations of BASIC. Therefore, the programs given here (which were written for use on an Apple IIe® computer) may have to be modified to fit the form of BASIC you may be using. However, there are several that may not need modification for use on another computer.

Following are the BASIC programs. There are 14 different programs from 13 different chapters throughout the text. They are identified by the chapter in which the program is appropriate. The program REM statement identifies the type of problem which is involved.

E–2 Programs

Chapter 2

```
10   REM  CURVE PLOTTING: Y=X^2        70   LET Y = X * X
20   GR                                80   COLOR= 12
30   COLOR= 15                         90   PLOT 3 * X + 19,30 - Y
40   HLIN 0,39 AT 30                   100  NEXT X
50   VLIN 0,39 AT 19                   110  END
60   FOR X =  - 5 TO 5
```

Chapter 3

```
10   REM   SOLUTION OF A RIGHT
20   REM   TRIANGLE GIVEN
30   REM   SIDES A AND B
40   PRINT "ENTER SIDES A AND B "
50   PRINT
60   INPUT "SIDE A = ";SA
70   INPUT "SIDE B = ";SB
80   PRINT
90   LET A =  ATN (SA / SB)
100  LET SC = SA /  SIN (A)
110  LET A = A * 180 / 3.14159265
120  LET B = 90 - A
130  PRINT "ANGLE A = ";A
140  PRINT "ANGLE B = ";B
150  PRINT "SIDE C = ";SC
160  END
```

Chapter 4

```
10   REM   SOLVE A 2X2 SYSTEM
20   REM   BY DETERMINANTS
30   PRINT "ENTER COEFFICIENTS"
40   PRINT
50   INPUT "A1 = ";A1: INPUT "B1 =
     ";B1: INPUT "C1 = ";C1
60   PRINT
70   INPUT "A2 = ";A2: INPUT "B2 =
     ";B2: INPUT "C2 = ";C2
80   PRINT
90   LET D = A1 * B2 - A2 * B1
100  IF D = 0 THEN  GOTO 170
110  LET X = (C1 * B2 - C2 * B1)/ D
120  LET Y = (A1 * C2 - A2 * C1)/ D
130  PRINT "X = ";X
140  PRINT
150  PRINT "Y = ";Y
160  GOTO 180
170  PRINT "NO UNIQUE SOLUTION"
180  END
```

Chapter 6

```
10   REM   SOLUTION OF A QUADRATIC
20   REM   EQUATION BY USE OF THE
30   REM   QUADRATIC FORMULA
40   PRINT "ENTER THE COEFFICIENTS
     A,B,C"
50   PRINT
```

```
60   INPUT "A = ";A: INPUT "B = ";
     B: INPUT "C = ";C
70   PRINT
80   IF (B * B - 4 * A * C) < 0 THEN
     GOTO 150
90   LET R =  SQR (B * B - 4 * A *
     C)
100  PRINT
110  PRINT " X1 = ";( - B + R) /
     (2 * A)
120  PRINT
130  PRINT " X2 = ";( - B - R) /
     (2 * A)
140  GOTO 160
150  PRINT "ROOTS ARE IMAGINARY"
160  END
```

Chapter 8: Vectors

```
10   REM   ADDITION OF TWO VECTORS
20   REM   STANDARD POSITION ANGLES
30   PRINT "    ENTER MAGNITUDE AND
     ANGLE OF VECTOR A  AND THEN
     OF VECTOR B"
40   PRINT
50   INPUT " A = ";A: INPUT "TA =
     ";TA: INPUT " B = ";B: INPUT
     "TB = ";TB
60   PRINT
70   LET PI = 3.141592654
80   LET TA = TA * PI / 180
90   LET TB = TB * PI / 180
100  LET RX = A *  COS (TA) + B *
     COS (TB)
110  LET RY = A *  SIN (TA) + B *
     SIN (TB)
120  LET TR = 180 / PI *  ATN (RY
     / RX)
130  LET R =  SQR (RX * RX + RY *
     RY)
140  PRINT "RESULTANT MAGNTIUDE =
     ";R
150  IF RX > 0 AND RY > 0 THEN  PRINT
     "RESULTANT ANGLE = ";TR
160  IF RX < 0 THEN  PRINT "RESUL
     TANT ANGLE = ";TR + 180
170  IF RX > 0 AND RY < 0 THEN  PRINT
     "RESULTANT ANGLE = ";TR + 360
180  END
```

Chapter 8: Triangle

```
10   REM   SOLUTION OF A TRIANGLE
20   REM   GIVEN THREE SIDES
30   PRINT "ENTER THE THREE SIDES"
40   PRINT
50   INPUT "SA = ";SA: INPUT "SB =
     ";SB: INPUT "SC = ";SC
60   PRINT
70   LET PI = 3.141592654
80   DEF  FN ACS(X) =  ATN ( SQR (
     1 - X * X) / X) - ( SGN (X) -
     1) * PI / 2
90   LET A =  FN ACS((SB * SB + SC
     * SC - SA * SA) / (2 * SB *
     SC))
100  LET B =  FN ACS((SA * SA + S
     C * SC - SB * SB) / (2 * SA *
     SC))
110  LET A = A * 180 / PI
120  LET B = B * 180 / PI
130  LET C = 180 - A - B
140  PRINT
150  PRINT " ANGLE A = ";A
160  PRINT " ANGLE B = ";B
170  PRINT " ANGLE C = ";C
180  END
```

Chapter 9

```
10   REM  PLOT Y = A SIN BX
20   GR
30   PRINT " ENTER COEFFICIENTS A
     AND B"
35   COLOR= 15
40   PRINT
50   INPUT "A = ";A: INPUT "B = ";B
60   PRINT
70   COLOR= 15
80   HLIN 0,39 AT 20
90   VLIN 0,39 AT 0
100  FOR X = 0 TO 6.4 STEP .2
110  COLOR= 12
120  PLOT 6 * X,20 - 10 *  SIN (X)
130  COLOR= 9
140  PLOT 6 * X,20 - A * 10 *  SIN
     (B * X)
150  NEXT X
160  END
```

Chapter 11

```
10   REM   COMPLEX NUMBERS
20   REM   RECTANGULAR FORM
30   REM   TO POLAR FORM
40   PRINT "ENTER X AND Y"
50   PRINT
60   INPUT "X = ";X: INPUT "Y = ";Y
70   PRINT
80   LET R =  SQR (X * X + Y * Y)
90   IF X = 0 THEN  GOTO 140
100  LET T =  ATN (Y / X) * 180 /
     3.141592654
110  IF X > 0 AND Y >  = 0 THEN  PRINT
     R;"/";T
120  IF X < 0 THEN  PRINT R;"/";T
     + 180
130  IF X > 0 AND Y < 0 THEN  PRINT
     R;"/";T + 360
140  IF X = 0 AND Y > 0 THEN  PRINT
     R;"/90"
150  IF X = 0 AND Y < 0 THEN  PRINT
     R;"/270"
160  END
```

Chapter 12

```
10   REM   PROPERTIES OF LOGARITHMS
20   PRINT "   ENTER TWO POSITIVE
     NUMBERS"
30   PRINT
40   INPUT "A = ";A: INPUT "B = ";B
50   PRINT
60   PRINT "LOG ";A;" + LOG ";B;"
     = LOG ";A * B
70   PRINT
80   PRINT "LOG ";A;" - LOG ";B;"
     = LOG ";A / B
90   PRINT
100  PRINT A;" LOG ";B;" = LOG ";
     B ^ A
110  END
```

Chapter 15

```
10    REM   INVERSE OF A 2X2 MATRIX
20    PRINT "ENTER THE ELEMENTS"
30    PRINT
40    INPUT "A11 = ";A: INPUT "A12
      = ";B: INPUT "A21 = ";C: INPUT
      "A22 = ";D
50    PRINT
60    PRINT "MATRIX" TAB( 10)A; TAB(
      18)B
70    PRINT
80    PRINT   TAB( 10)C; TAB( 18)D
90    PRINT
100   PRINT
110   LET E = A * D - B * C
120   IF E = 0 THEN  GOTO 170
130   PRINT "INVERSE" TAB( 10)D /
      E; TAB( 18) - B / E
140   PRINT
150   PRINT   TAB( 10) - C / E; TAB(
      18)A / E
160   GOTO 180
170   PRINT "NO INVERSE"
180   END
```

Chapter 18

```
10    REM   BINOMIAL EXPANSION OF
20    REM   FIRST 4 COEFFICIENTS OF
30    REM   (AX + B)^N
40    PRINT "ENTER A,B,N"
50    PRINT
60    INPUT "A = ";A: INPUT "B = ";
      B: INPUT "N = ";N
70    PRINT
80    PRINT "COEFFICIENTS"
90    PRINT
100   LET C1 = A ^ N
110   LET C2 = N * A ^ (N - 1) * B
120   LET C3 = N * (N - 1) * A ^ (
      N - 2) * B * B / 2
130   LET C4 = N * (N - 1) * (N -
      2) * A ^ (N - 3) * B * B * B
      / 6
140   PRINT "C1 = ";C1: PRINT "C2
      = ";C2: PRINT "C3 = ";C3: PRINT
      "C4 = ";C4
150   END
```

Chapter 19

```
10    REM   VERIFY TRIGONOMETRIC
20    REM   IDENTITY
30    PRINT
40    PRINT "SINX TANX + COSX = SECX"
50    PRINT
60    PRINT "ENTER X"
70    PRINT
80    INPUT " X = ";X
90    PRINT
100   LET X = X * 3.14159265 / 180
110   PRINT "SINX TANX + COSX = ";
       SIN (X) *  TAN (X) +  COS (X)
120   PRINT
130   PRINT "SECX = ";1 /  COS (X)
140   END
```

Chapter 20

```
10    REM   GRAPH A STRAIGHT LINE
20    REM   GIVEN TWO POINTS
30    REM   (Y-INT ON GRAPH VALID
40    REM    ONLY IF .4<M<.6)
50    PRINT
60    PRINT "ENTER COORDINATES"
70    PRINT
80    INPUT "X1 = ";X1: INPUT "Y1 =
       ";Y1: INPUT "X2 = ";X2: INPUT
      "Y2 = ";Y2
90    PRINT
100   LET M = (Y2 - Y1) / (X2 - X1)
110   LET B = Y1 - M * X1
120   GR
130   COLOR= 15
140   HLIN 0,39 AT 19
150   VLIN 0,39 AT 38 * (1 - M)
160   FOR X = 0 TO 39
170   LET Y = M * X + B
180   COLOR= 12
190   PLOT X,29 - Y - B
200   NEXT X
210   END
```

Chapter 21

```
10    REM   ARITHMETIC MEAN AND
20    REM   STANDARD DEVIATION
30    REM   OF N NUMBERS
40    PRINT
50    PRINT "ENTER N"
60    LET SQ = O: LET SUM = O: PRINT
70    INPUT "N = ";N
80    PRINT
90    PRINT "ENTER NUMBERS"
100   FOR C = 1 TO N
110   PRINT
120   PRINT
130   INPUT "    ";D
140   LET SUM = SUM + D
150   LET SQ = SQ + D * D
160   NEXT C
170   PRINT
180   LET M = SUM / N
190   LET X = SQ / N
200   LET S =  SQR (X - M * M)
210   PRINT "ARITHMETIC MEAN = ";M
220   PRINT
230   PRINT "STANDARD DEVIATION =
      ";S
240   END
```

Tables

Table 1. Powers and Roots

No.	Sq.	Sq. Root	Cube	Cube Root	No.	Sq.	Sq. Root	Cube	Cube Root
1	1	1.000	1	1.000	51	2 601	7.141	132 651	3.708
2	4	1.414	8	1.260	52	2 704	7.211	140 608	3.733
3	9	1.732	27	1.442	53	2 809	7.280	148 877	3.756
4	16	2.000	64	1.587	54	2 916	7.348	157 464	3.780
5	25	2.236	125	1.710	55	3 025	7.416	166 375	3.803
6	36	2.449	216	1.817	56	3 136	7.483	175 616	3.826
7	49	2.646	343	1.913	57	3 249	7.550	185 193	3.849
8	64	2.828	512	2.000	58	3 364	7.616	195 112	3.871
9	81	3.000	729	2.080	59	3 481	7.681	205 379	3.893
10	100	3.162	1 000	2.154	60	3 600	7.746	216 000	3.915
11	121	3.317	1 331	2.224	61	3 721	7.810	226 981	3.936
12	144	3.464	1 728	2.289	62	3 844	7.874	238 328	3.958
13	169	3.606	2 197	2.351	63	3 969	7.937	250 047	3.979
14	196	3.742	2 744	2.410	64	4 096	8.000	262 144	4.000
15	225	3.873	3 375	2.466	65	4 225	8.062	274 625	4.021
16	256	4.000	4 096	2.520	66	4 356	8.124	287 496	4.041
17	289	4.123	4 913	2.571	67	4 489	8.185	300 763	4.062
18	324	4.243	5 832	2.621	68	4 624	8.246	314 432	4.082
19	361	4.359	6 859	2.668	69	4 761	8.307	328 509	4.102
20	400	4.472	8 000	2.714	70	4 900	8.367	343 000	4.121
21	441	4.583	9 261	2.759	71	5 041	8.426	357 911	4.141
22	484	4.690	10 648	2.802	72	5 184	8.485	373 248	4.160
23	529	4.796	12 167	2.844	73	5 329	8.544	389 017	4.179
24	576	4.899	13 824	2.884	74	5 476	8.602	405 224	4.198
25	625	5.000	15 625	2.924	75	5 625	8.660	421 875	4.217
26	676	5.099	17 576	2.962	76	5 776	8.718	438 976	4.236
27	729	5.196	19 683	3.000	77	5 929	8.775	456 533	4.254
28	784	5.292	21 952	3.037	78	6 084	8.832	474 552	4.273
29	841	5.385	24 389	3.072	79	6 241	8.888	493 039	4.291
30	900	5.477	27 000	3.107	80	6 400	8.944	512 000	4.309
31	961	5.568	29 791	3.141	81	6 561	9.000	531 441	4.327
32	1 024	5.657	32 768	3.175	82	6 724	9.055	551 368	4.344
33	1 089	5.745	35 937	3.208	83	6 889	9.110	571 787	4.362
34	1 156	5.831	39 304	3.240	84	7 056	9.165	592 704	4.380
35	1 225	5.916	42 875	3.271	85	7 225	9.220	614 125	4.397
36	1 296	6.000	46 656	3.302	86	7 396	9.274	636 056	4.414
37	1 369	6.083	50 653	3.332	87	7 569	9.327	658 503	4.431
38	1 444	6.164	54 872	3.362	88	7 744	9.381	681 472	4.448
39	1 521	6.245	59 319	3.391	89	7 921	9.434	704 969	4.465
40	1 600	6.325	64 000	3.420	90	8 100	9.487	729 000	4.481
41	1 681	6.403	68 921	3.448	91	8 281	9.539	753 571	4.498
42	1 764	6.481	74 088	3.476	92	8 464	9.592	778 688	4.514
43	1 849	6.557	79 507	3.503	93	8 649	9.644	804 357	4.531
44	1 936	6.633	85 184	3.530	94	8 836	9.695	830 584	4.547
45	2 025	6.708	91 125	3.557	95	9 025	9.747	857 375	4.563
46	2 116	6.782	97 336	3.583	96	9 216	9.798	884 736	4.579
47	2 209	6.856	103 823	3.609	97	9 409	9.849	912 673	4,595
48	2 304	6.928	110 592	3.634	98	9 604	9.899	941 192	4.610
49	2 401	7.000	117 649	3.659	99	9 801	9.950	970 299	4.626
50	2 500	7.071	125 000	3.684	100	10 000	10.000	1 000 000	4.642

Table 2. Four-Place Logarithms

N	0	1	2	3	4	5	6	7	8	9
10	0000	0043	0086	0128	0170	0212	0253	0294	0334	0374
11	0414	0453	0492	0531	0569	0607	0645	0682	0719	0755
12	0792	0828	0864	0899	0934	0969	1004	1038	1072	1106
13	1139	1173	1206	1239	1271	1303	1335	1367	1399	1430
14	1461	1492	1523	1553	1584	1614	1644	1673	1703	1732
15	1761	1790	1818	1847	1875	1903	1931	1959	1987	2014
16	2041	2068	2095	2122	2148	2175	2201	2227	2253	2279
17	2304	2330	2355	2380	2405	2430	2455	2480	2504	2529
18	2553	2577	2601	2625	2648	2672	2695	2718	2742	2765
19	2788	2810	2833	2856	2878	2900	2923	2945	2967	2989
20	3010	3032	3054	3075	3096	3118	3139	3160	3181	3201
21	3222	3243	3263	3284	3304	3324	3345	3365	3385	3404
22	3424	3444	3464	3483	3502	3522	3541	3560	3579	3598
23	3617	3636	3655	3674	3692	3711	3729	3747	3766	3784
24	3802	3820	3838	3856	3874	3892	3909	3927	3945	3962
25	3979	3997	4014	4031	4048	4065	4082	4099	4116	4133
26	4150	4166	4183	4200	4216	4232	4249	4265	4281	4298
27	4314	4330	4346	4362	4378	4393	4409	4425	4440	4456
28	4472	4487	4502	4518	4533	4548	4564	4579	4594	4609
29	4624	4639	4654	4669	4683	4698	4713	4728	4742	4757
30	4771	4786	4800	4814	4829	4843	4857	4871	4886	4900
31	4914	4928	4942	4955	4969	4983	4997	5011	5024	5038
32	5051	5065	5079	5092	5105	5119	5132	5145	5159	5172
33	5185	5198	5211	5224	5237	5250	5263	5276	5289	5302
34	5315	5328	5340	5353	5366	5378	5391	5403	5416	5428
35	5441	5453	5465	5478	5490	5502	5514	5527	5539	5551
36	5563	5575	5587	5599	5611	5623	5635	5647	5658	5670
37	5682	5694	5705	5717	5729	5740	5752	5763	5775	5786
38	5798	5809	5821	5832	5843	5855	5866	5877	5888	5899
39	5911	5922	5933	5944	5955	5966	5977	5988	5999	6010
40	6021	6031	6042	6053	6064	6075	6085	6096	6107	6117
41	6128	6138	6149	6160	6170	6180	6191	6201	6212	6222
42	6232	6243	6253	6263	6274	6284	6294	6304	6314	6325
43	6335	6345	6355	6365	6375	6385	6395	6405	6415	6425
44	6435	6444	6454	6464	6474	6484	6493	6503	6513	6522
45	6532	6542	6551	6561	6571	6580	6590	6599	6609	6618
46	6628	6637	6646	6656	6665	6675	6684	6693	6702	6712
47	6721	6730	6739	6749	6758	6767	6776	6785	6794	6803
48	6812	6821	6830	6839	6848	6857	6866	6875	6884	6893
49	6902	6911	6920	6928	6937	6946	6955	6964	6972	6981
50	6990	6998	7007	7016	7024	7033	7042	7050	7059	7067
51	7076	7084	7093	7101	7110	7118	7126	7135	7143	7152
52	7160	7168	7177	7185	7193	7202	7210	7218	7226	7235
53	7243	7251	7259	7267	7275	7284	7292	7300	7308	7316
54	7324	7332	7340	7348	7356	7364	7372	7380	7388	7396

Table 2. Continued

N	0	1	2	3	4	5	6	7	8	9
55	7404	7412	7419	7427	7435	7443	7451	7459	7466	7474
56	7482	7490	7497	7505	7513	7520	7528	7536	7543	7551
57	7559	7566	7574	7582	7589	7597	7604	7612	7619	7627
58	7634	7642	7649	7657	7664	7672	7679	7686	7694	7701
59	7709	7716	7723	7731	7738	7745	7752	7760	7767	7774
60	7782	7789	7796	7803	7810	7818	7825	7832	7839	7846
61	7853	7860	7868	7875	7882	7889	7896	7903	7910	7917
62	7924	7931	7938	7945	7952	7959	7966	7973	7980	7987
63	7993	8000	8007	8014	8021	8028	8035	8041	8048	8055
64	8062	8069	8075	8082	8089	8096	8102	8109	8116	8122
65	8129	8136	8142	8149	8156	8162	8169	8176	8182	8189
66	8195	8202	8209	8215	8222	8228	8235	8241	8248	8254
67	8261	8267	8274	8280	8287	8293	8299	8306	8312	8319
68	8325	8331	8338	8344	8351	8357	8363	8370	8376	8382
69	8388	8395	8401	8407	8414	8420	8426	8432	8439	8445
70	8451	8457	8463	8470	8476	8482	8488	8494	8500	8506
71	8513	8519	8525	8531	8537	8543	8549	8555	8561	8567
72	8573	8579	8585	8591	8597	8603	8609	8615	8621	8627
73	8633	8639	8645	8651	8657	8663	8669	8675	8681	8686
74	8692	8698	8704	8710	8716	8722	8727	8733	8739	8745
75	8751	8756	8762	8768	8774	8779	8785	8791	8797	8802
76	8808	8814	8820	8825	8831	8837	8842	8848	8854	8859
77	8865	8871	8876	8882	8887	8893	8899	8904	8910	8915
78	8921	8927	8932	8938	8943	8949	8954	8960	8965	8971
79	8976	8982	8987	8993	8998	9004	9009	9015	9020	9025
80	9031	9036	9042	9047	9053	9058	9063	9069	9074	9079
81	9085	9090	9096	9101	9106	9112	9117	9122	9128	9133
82	9138	9143	9149	9154	9159	9165	9170	9175	9180	9186
83	9191	9196	9201	9206	9212	9217	9222	9227	9232	9238
84	9243	9248	9253	9258	9263	9269	9274	9279	9284	9289
85	9294	9299	9304	9309	9315	9320	9325	9330	9335	9340
86	9345	9350	9355	9360	9365	9370	9375	9380	9385	9390
87	9395	9400	9405	9410	9415	9420	9425	9430	9435	9440
88	9445	9450	9455	9460	9465	9469	9474	9479	9484	9489
89	9494	9499	9504	9509	9513	9518	9523	9528	9533	9538
90	9542	9547	9552	9557	9562	9566	9571	9576	9581	9586
91	9590	9595	9600	9605	9609	9614	9619	9624	9628	9633
92	9638	9643	9647	9652	9657	9661	9666	9671	9675	9680
93	9685	9689	9694	9699	9703	9708	9713	9717	9722	9727
94	9731	9736	9741	9745	9750	9754	9759	9763	9768	9773
95	9777	9782	9786	9791	9795	9800	9805	9809	9814	9818
96	9823	9827	9832	9836	9841	9845	9850	9854	9859	9863
97	9868	9872	9877	9881	9886	9890	9894	9899	9903	9908
98	9912	9917	9921	9926	9930	9934	9939	9943	9948	9952
99	9956	9961	9965	9969	9974	9978	9983	9987	9991	9996

Table 3. Four-Place Values of Trigonometric Functions

Degrees	Sin θ	Cos θ	Tan θ	Cot θ		Degrees	Sin θ	Cos θ	Tan θ	Cot θ	
0°00′	0.0000	1.0000	0.0000	—	90°00′	8°00′	0.1392	0.9903	0.1405	7.115	82°00′
10	0.0029	1.0000	0.0029	343.8	50	10	0.1421	0.9899	0.1435	6.968	50
20	0.0058	1.0000	0.0058	171.9	40	20	0.1449	0.9894	0.1465	6.827	40
30	0.0087	1.0000	0.0087	114.6	30	30	0.1478	0.9890	0.1495	6.691	30
40	0.0116	0.9999	0.0116	85.94	20	40	0.1507	0.9886	0.1524	6.561	20
50	0.0145	0.9999	0.0145	68.75	10	50	0.1536	0.9881	0.1554	6.435	10
1°00′	0.0175	0.9998	0.0175	57.29	89°00′	9°00′	0.1564	0.9877	0.1584	6.314	81°00′
10	0.0204	0.9998	0.0204	49.10	50	10	0.1593	0.9872	0.1614	6.197	50
20	0.0233	0.9997	0.0233	42.96	40	20	0.1622	0.9868	0.1644	6.084	40
30	0.0262	0.9997	0.0262	38.19	30	30	0.1650	0.9863	0.1673	5.976	30
40	0.0291	0.9996	0.0291	34.37	20	40	0.1679	0.9858	0.1703	5.871	20
50	0.0320	0.9995	0.0320	31.24	10	50	0.1708	0.9853	0.1733	5.769	10
2°00′	0.0349	0.9994	0.0349	28.64	88°00′	10°00′	0.1736	0.9848	0.1763	5.671	80°00′
10	0.0378	0.9993	0.0378	26.43	50	10	0.1765	0.9843	0.1793	5.576	50
20	0.0407	0.9992	0.0407	24.54	40	20	0.1794	0.9838	0.1823	5.485	40
30	0.0436	0.9990	0.0437	22.90	30	30	0.1822	0.9833	0.1853	5.396	30
40	0.0465	0.9989	0.0466	21.47	20	40	0.1851	0.9827	0.1883	5.309	20
50	0.0494	0.9988	0.0495	20.21	10	50	0.1880	0.9822	0.1914	5.226	10
3°00′	0.0523	0.9986	0.0524	19.08	87°00′	11°00′	0.1908	0.9816	0.1944	5.145	79°00′
10	0.0552	0.9985	0.0553	18.07	50	10	0.1937	0.9811	0.1974	5.066	50
20	0.0581	0.9983	0.0582	17.17	40	20	0.1965	0.9805	0.2004	4.989	40
30	0.0610	0.9981	0.0612	16.35	30	30	0.1994	0.9799	0.2035	4.915	30
40	0.0640	0.9980	0.0641	15.60	20	40	0.2022	0.9793	0.2065	4.843	20
50	0.0669	0.9978	0.0670	14.92	10	50	0.2051	0.9787	0.2095	4.773	10
4°00′	0.0698	0.9976	0.0699	14.30	86°00′	12°00′	0.2079	0.9781	0.2126	4.705	78°00′
10	0.0727	0.9974	0.0729	13.73	50	10	0.2108	0.9775	0.2156	4.638	50
20	0.0756	0.9971	0.0758	13.20	40	20	0.2136	0.9769	0.2186	4.574	40
30	0.0785	0.9969	0.0787	12.71	30	30	0.2164	0.9763	0.2217	4.511	30
40	0.0814	0.9967	0.0816	12.25	20	40	0.2193	0.9757	0.2247	4.449	20
50	0.0843	0.9964	0.0846	11.83	10	50	0.2221	0.9750	0.2278	4.390	10
5°00′	0.0872	0.9962	0.0875	11.43	85°00′	13°00′	0.2250	0.9744	0.2309	4.331	77°00′
10	0.0901	0.9959	0.0904	11.06	50	10	0.2278	0.9737	0.2339	4.275	50
20	0.0929	0.9957	0.0934	10.71	40	20	0.2306	0.9730	0.2370	4.219	40
30	0.0958	0.9954	0.0963	10.39	30	30	0.2334	0.9724	0.2401	4.165	30
40	0.0987	0.9951	0.0992	10.08	20	40	0.2363	0.9717	0.2432	4.113	20
50	0.1016	0.9948	0.1022	9.788	10	50	0.2391	0.9710	0.2462	4.061	10
6°00′	0.1045	0.9945	0.1051	9.514	84°00′	14°00′	0.2419	0.9703	0.2493	4.011	76°00′
10	0.1074	0.9942	0.1080	9.255	50	10	0.2447	0.9696	0.2524	3.962	50
20	0.1103	0.9939	0.1110	9.010	40	20	0.2476	0.9689	0.2555	3.914	40
30	0.1132	0.9936	0.1139	8.777	30	30	0.2504	0.9681	0.2586	3.867	30
40	0.1161	0.9932	0.1169	8.556	20	40	0.2532	0.9674	0.2617	3.821	20
50	0.1190	0.9929	0.1198	8.345	10	50	0.2560	0.9667	0.2648	3.776	10
7°00′	0.1219	0.9925	0.1228	8.144	83°00′	15°00′	0.2588	0.9659	0.2679	3.732	75°00′
10	0.1248	0.9922	0.1257	7.953	50	10	0.2616	0.9652	0.2711	3.689	50
20	0.1276	0.9918	0.1287	7.770	40	20	0.2644	0.9644	0.2742	3.647	40
30	0.1305	0.9914	0.1317	7.596	30	30	0.2672	0.9636	0.2773	3.606	30
40	0.1334	0.9911	0.1346	7.429	20	40	0.2700	0.9628	0.2805	3.566	20
50	0.1363	0.9907	0.1376	7.269	10	50	0.2728	0.9621	0.2836	3.526	10
8°00′	0.1392	0.9903	0.1405	7.115	82°00′	16°00′	0.2756	0.9613	0.2867	3.487	74°00′
	Cos θ	Sin θ	Cot θ	Tan θ	Degrees		Cos θ	Sin θ	Cot θ	Tan θ	Degrees

Table 3. Continued

Degrees	Sin θ	Cos θ	Tan θ	Cot θ		Degrees	Sin θ	Cos θ	Tan θ	Cot θ	
16°00′	0.2756	0.9613	0.2867	3.487	74°00′	24°00′	0.4067	0.9135	0.4452	2.246	66°00′
10	0.2784	0.9605	0.2899	3.450	50	10	0.4094	0.9124	0.4487	2.229	50
20	0.2812	0.9596	0.2931	3.412	40	20	0.4120	0.9112	0.4522	2.211	40
30	0.2840	0.9588	0.2962	3.376	30	30	0.4147	0.9100	0.4557	2.194	30
40	0.2868	0.9580	0.2994	3.340	20	40	0.4173	0.9088	0.4592	2.177	20
50	0.2896	0.9572	0.3026	3.305	10	50	0.4200	0.9075	0.4628	2.161	10
17°00′	0.2924	0.9563	0.3057	3.271	73°00′	25°00′	0.4226	0.9063	0.4663	2.145	65°00′
10	0.2952	0.9555	0.3089	3.237	50	10	0.4253	0.9051	0.4699	2.128	50
20	0.2979	0.9546	0.3121	3.204	40	20	0.4279	0.9038	0.4734	2.112	40
30	0.3007	0.9537	0.3153	3.172	30	30	0.4305	0.9026	0.4770	2.097	30
40	0.3035	0.9528	0.3185	3.140	20	40	0.4331	0.9013	0.4806	2.081	20
50	0.3062	0.9520	0.3217	3.108	10	50	0.4358	0.9001	0.4841	2.066	10
18°00′	0.3090	0.9511	0.3249	3.078	72°00′	26°00′	0.4384	0.8988	0.4877	2.050	64°00′
10	0.3118	0.9502	0.3281	3.047	50	10	0.4410	0.8975	0.4913	2.035	50
20	0.3145	0.9492	0.3314	3.018	40	20	0.4436	0.8962	0.4950	2.020	40
30	0.3173	0.9483	0.3346	2.989	30	30	0.4462	0.8949	0.4986	2.006	30
40	0.3201	0.9474	0.3378	2.960	20	40	0.4488	0.8936	0.5022	1.991	20
50	0.3228	0.9465	0.3411	2.932	10	50	0.4514	0.8923	0.5059	1.977	10
19°00′	0.3256	0.9455	0.3443	2.904	71°00′	27°00′	0.4540	0.8910	0.5095	1.963	63°00′
10	0.3283	0.9446	0.3476	2.877	50	10	0.4566	0.8897	0.5132	1.949	50
20	0.3311	0.9436	0.3508	2.850	40	20	0.4592	0.8884	0.5169	1.935	40
30	0.3338	0.9426	0.3541	2.824	30	30	0.4617	0.8870	0.5206	1.921	30
40	0.3365	0.9417	0.3574	2.798	20	40	0.4643	0.8857	0.5243	1.907	20
50	0.3393	0.9407	0.3607	2.773	10	50	0.4669	0.8843	0.5280	1.894	10
20°00′	0.3420	0.9397	0.3640	2.747	70°00′	28°00′	0.4695	0.8829	0.5317	1.881	62°00′
10	0.3448	0.9387	0.3673	2.723	50	10	0.4720	0.8816	0.5354	1.868	50
20	0.3475	0.9377	0.3706	2.699	40	20	0.4746	0.8802	0.5392	1.855	40
30	0.3502	0.9367	0.3739	2.675	30	30	0.4772	0.8788	0.5430	1.842	30
40	0.3529	0.9356	0.3772	2.651	20	40	0.4797	0.8774	0.5467	1.829	20
50	0.3557	0.9346	0.3805	2.628	10	50	0.4823	0.8760	0.5505	1.816	10
21°00′	0.3584	0.9336	0.3839	2.605	69°00′	29°00′	0.4848	0.8746	0.5543	1.804	61°00′
10	0.3611	0.9325	0.3872	2.583	50	10	0.4874	0.8732	0.5581	1.792	50
20	0.3638	0.9315	0.3906	2.560	40.	20	0.4899	0.8718	0.5619	1.780	40
30	0.3665	0.9304	0.3939	2.539	30	30	0.4924	0.8704	0.5658	1.767	30
40	0.3692	0.9293	0.3973	2.517	20	40	0.4950	0.8689	0.5696	1.756	20
50	0.3719	0.9283	0.4006	2.496	10	50	0.4975	0.8675	0.5735	1.744	10
22°00′	0.3746	0.9272	0.4040	2.475	68°00′	30°00′	0.5000	0.8660	0.5774	1.732	60°00′
10	0.3773	0.9261	0.4074	2.455	50	10	0.5025	0.8646	0.5812	1.720	50
20	0.3800	0.9250	0.4108	2.434	40	20	0.5050	0.8631	0.5851	1.709	40
30	0.3827	0.9239	0.4142	2.414	30	30	0.5075	0.8616	0.5890	1.698	30
40	0.3854	0.9228	0.4176	2.394	20	40	0.5100	0.8601	0.5930	1.686	20
50	0.3881	0.9216	0.4210	2.375	10	50	0.5125	0.8587	0.5969	1.675	10
23°00′	0.3907	0.9205	0.4245	2.356	67°00′	31°00′	0.5150	0.8572	0.6009	1.664	59°00′
10	0.3934	0.9194	0.4279	2.337	50	10	0.5175	0.8557	0.6048	1.653	50
20	0.3961	0.9182	0.4314	2.318	40	20	0.5200	0.8542	0.6088	1.643	40
30	0.3987	0.9171	0.4348	2.300	30	30	0.5225	0.8526	0.6128	1.632	30
40	0.4014	0.9159	0.4383	2.282	20	40	0.5250	0.8511	0.6168	1.621	20
50	0.4041	0.9147	0.4417	2.264	10	50	0.5275	0.8496	0.6208	1.611	10
24°00′	0.4067	0.9135	0.4452	2.246	66°00′	32°00′	0.5299	0.8480	0.6249	1.600	58°00′
	Cos θ	Sin θ	Cot θ	Tan θ	Degrees		Cos θ	Sin θ	Cot θ	Tan θ	Degrees

Table 3. Continued

Degrees	Sin θ	Cos θ	Tan θ	Cot θ		Degrees	Sin θ	Cos θ	Tan θ	Cot θ	
32°00′	0.5299	0.8480	0.6249	1.600	58°00′	39°00′	0.6293	0.7771	0.8098	1.235	51°00′
10	0.5324	0.8465	0.6289	1.590	50	10	0.6316	0.7753	0.8146	1.228	50
20	0.5348	0.8450	0.6330	1.580	40	20	0.6338	0.7735	0.8195	1.220	40
30	0.5373	0.8434	0.6371	1.570	30	30	0.6361	0.7716	0.8243	1.213	30
40	0.5398	0.8418	0.6412	1.560	20	40	0.6383	0.7698	0.8292	1.206	20
50	0.5422	0.8403	0.6453	1.550	10	50	0.6406	0.7679	0.8342	1.199	10
33°00′	0.5446	0.8387	0.6494	1.540	57°00′	40°00′	0.6428	0.7660	0.8391	1.192	50°00′
10	0.5471	0.8371	0.6536	1.530	50	10	0.6450	0.7642	0.8441	1.185	50
20	0.5495	0.8355	0.6577	1.520	40	20	0.6472	0.7623	0.8491	1.178	40
30	0.5519	0.8339	0.6619	1.511	30	30	0.6494	0.7604	0.8541	1.171	30
40	0.5544	0.8323	0.6661	1.501	20	40	0.6517	0.7585	0.8591	1.164	20
50	0.5568	0.8307	0.6703	1.492	10	50	0.6539	0.7566	0.8642	1.157	10
34°00′	0.5592	0.8290	0.6745	1.483	56°00′	41°00′	0.6561	0.7547	0.8693	1.150	49°00′
10	0.5616	0.8274	0.6787	1.473	50	10	0.6583	0.7528	0.8744	1.144	50
20	0.5640	0.8258	0.6830	1.464	40	20	0.6604	0.7509	0.8796	1.137	40
30	0.5664	0.8241	0.6873	1.455	30	30	0.6626	0.7490	0.8847	1.130	30
40	0.5688	0.8225	0.6916	1.446	20	40	0.6648	0.7470	0.8899	1.124	20
50	0.5712	0.8208	0.6959	1.437	10	50	0.6670	0.7451	0.8952	1.117	10
35°00′	0.5736	0.8192	0.7002	1.428	55°00′	42°00′	0.6691	0.7431	0.9004	1.111	48°00′
10	0.5760	0.8175	0.7046	1.419	50	10	0.6713	0.7412	0.9057	1.104	50
20	0.5783	0.8158	0.7089	1.411	40	20	0.6734	0.7392	0.9110	1.098	40
30	0.5807	0.8141	0.7133	1.402	30	30	0.6756	0.7373	0.9163	1.091	30
40	0.5831	0.8124	0.7177	1.393	20	40	0.6777	0.7353	0.9217	1.085	20
50	0.5854	0.8107	0.7221	1.385	10	50	0.6799	0.7333	0.9271	1.079	10
36°00′	0.5878	0.8090	0.7265	1.376	54°00′	43°00′	0.6820	0.7314	0.9325	1.072	47°00′
10	0.5901	0.8073	0.7310	1.368	50	10	0.6841	0.7294	0.9380	1.066	50
20	0.5925	0.8056	0.7355	1.360	40	20	0.6862	0.7274	0.9435	1.060	40
30	0.5948	0.8039	0.7400	1.351	30	30	0.6884	0.7254	0.9490	1.054	30
40	0.5972	0.8021	0.7445	1.343	20	40	0.6905	0.7234	0.9545	1.048	20
50	0.5995	0.8004	0.7490	1.335	10	50	0.6926	0.7214	0.9601	1.042	10
37°00′	0.6018	0.7986	0.7536	1.327	53°00′	44°00′	0.6947	0.7193	0.9657	1.036	46°00′
10	0.6041	0.7969	0.7581	1.319	50	10	0.6967	0.7173	0.9713	1.030	50
20	0.6065	0.7951	0.7627	1.311	40	20	0.6988	0.7153	0.9770	1.024	40
30	0.6088	0.7934	0.7673	1.303	30	30	0.7009	0.7133	0.9827	1.018	30
40	0.6111	0.7916	0.7720	1.295	20	40	0.7030	0.7112	0.9884	1.012	20
50	0.6134	0.7898	0.7766	1.288	10	50	0.7050	0.7092	0.9942	1.006	10
38°00′	0.6157	0.7880	0.7813	1.280	52°00′	45°00′	0.7071	0.7071	1.000	1.000	45°00′
10	0.6180	0.7862	0.7860	1.272	50						
20	0.6202	0.7844	0.7907	1.265	40		Cos θ	Sin θ	Cot θ	Tan θ	Degrees
30	0.6225	0.7826	0.7954	1.257	30						
40	0.6248	0.7808	0.8002	1.250	20						
50	0.6271	0.7790	0.8050	1.242	10						
39°00′	0.6293	0.7771	0.8098	1.235	51°00′						
	Cos θ	Sin θ	Cot θ	Tan θ	Degrees						

Table 4. Natural Logarithms and Exponential Functions

x	$\ln x$	e^x	e^{-x}	x	$\ln x$	e^x	e^{-x}	x	$\ln x$	e^x	e^{-x}
0.0		1.0000	1.0000	4.5	1.5041	90.017	0.0111	9.0	2.1972	8,103	0.0001
0.1	−2.3026	1.1052	0.9048	4.6	1.5261	99.484	0.0101	9.1	2.2083	8,955	
0.2	−1.6094	1.2214	0.8187	4.7	1.5476	109.95	0.0091	9.2	2.2192	9,897	
0.3	−1.2040	1.3499	0.7408	4.8	1.5686	121.51	0.0082	9.3	2.2300	10,938	
0.4	−0.9163	1.4918	0.6703	4.9	1.5892	134.29	0.0074	9.4	2.2407	12,088	
0.5	−0.6931	1.6487	0.6065	5.0	1.6094	148.41	0.0067	9.5	2.2513	13,360	0.00007
0.6	−0.5108	1.8221	0.5488	5.1	1.6292	164.02	0.0061	9.6	2.2618	14,765	
0.7	−0.3567	2.0138	0.4966	5.2	1.6487	181.27	0.0055	9.7	2.2721	16,318	
0.8	−0.2231	2.2255	0.4493	5.3	1.6677	200.34	0.0050	9.8	2.2824	18,034	
0.9	−0.1054	2.4596	0.4066	5.4	1.6864	221.41	0.0045	9.9	2.2925	19,930	
1.0	0.0000	2.7183	0.3679	5.5	1.7047	244.69	0.0041	10	2.3026	22,026	4.5×10^{-5}
1.1	0.0953	3.0042	0.3329	5.6	1.7228	270.43	0.0037	15	2.7081		
1.2	0.1823	3.3201	0.3012	5.7	1.7405	298.87	0.0033	20	2.9957	4.9×10^8	2.1×10^{-9}
1.3	0.2624	3.6693	0.2725	5.8	1.7579	330.30	0.0030	25	3.2189		
1.4	0.3365	4.0552	0.2466	5.9	1.7750	365.04	0.0027	30	3.4012	1.1×10^{13}	9.4×10^{-14}
1.5	0.4055	4.4817	0.2231	6.0	1.7918	403.43	0.0025	35	3.5553		
1.6	0.4700	4.9530	0.2019	6.1	1.8083	445.86		40	3.6889	2.4×10^{17}	4.2×10^{-18}
1.7	0.5306	5.4739	0.1827	6.2	1.8245	492.75		45	3.8067		
1.8	0.5878	6.0496	0.1653	6.3	1.8405	544.57		50	3.9120	5.2×10^{21}	1.9×10^{-22}
1.9	0.6419	6.6859	0.1496	6.4	1.8563	601.85		55	4.0073		
2.0	0.6931	7.3891	0.1353	6.5	1.8718	665.14	0.0015	60	4.0943	1.1×10^{26}	8.8×10^{-27}
2.1	0.7419	8.1662	0.1225	6.6	1.8871	735.10		70	4.2485		
2.2	0.7885	9.0250	0.1108	6.7	1.9021	812.41		80	4.3820	5.5×10^{34}	1.8×10^{-35}
2.3	0.8329	9.9742	0.1003	6.8	1.9169	897.85		90	4.4998		
2.4	0.8755	11.023	0.0907	6.9	1.9315	992.27		100	4.6052	2.7×10^{43}	3.7×10^{-44}
2.5	0.9163	12.182	0.0821	7.0	1.9459	1097	0.0009				
2.6	0.9555	13.464	0.0743	7.1	1.9601	1212					
2.7	0.9933	14.880	0.0672	7.2	1.9741	1339					
2.8	1.0296	16.445	0.0608	7.3	1.9879	1480					
2.9	1.0647	18.174	0.0550	7.4	2.0015	1636					
3.0	1.0986	20.086	0.0498	7.5	2.0149	1808	0.0006				
3.1	1.1314	22.198	0.0450	7.6	2.0281	1998					
3.2	1.1632	24.533	0.0408	7.7	2.0412	2208					
3.3	1.1939	27.113	0.0369	7.8	2.0541	2441					
3.4	1.2238	29.964	0.0334	7.9	2.0669	2697					
3.5	1.2528	33.115	0.0302	8.0	2.0794	2981	0.0003				
3.6	1.2809	36.598	0.0273	8.1	2.0919	3294					
3.7	1.3083	40.447	0.0247	8.2	2.1041	3641					
3.8	1.3350	44.701	0.0224	8.3	2.1163	4024					
3.9	1.3610	49.402	0.0202	8.4	2.1282	4447					
4.0	1.3863	54.598	0.0183	8.5	2.1401	4915	0.0002				
4.1	1.4110	60.340	0.0166	8.6	2.1518	5432					
4.2	1.4351	66.686	0.0150	8.7	2.1633	6003					
4.3	1.4586	73.700	0.0136	8.8	2.1748	6634					
4.4	1.4816	81.451	0.0123	8.9	2.1861	7332					

ln 10	2.3026
2 ln 10	4.6052
3 ln 10	6.9078
4 ln 10	9.2103
5 ln 10	11.5129

Answers to Odd-Numbered Exercises

Exercises 1–1, p. 5

1. Integer, rational, real; irrational, real   **3.** Imaginary; irrational, real   **5.** $3, \frac{7}{2}$ **7.** $\frac{6}{7}, \sqrt{3}$ **9.** $6 < 8$

11. $\pi > -1$ **13.** $-4 < -3$ **15.** $-\frac{1}{3} > -\frac{1}{2}$ **17.** $\frac{1}{3}, -\frac{1}{2}$ **19.** $-\frac{\pi}{5}, \frac{1}{x}$

21.   **23.**

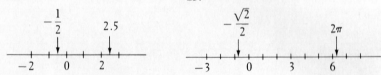

25. $-18, -|-3|, -1, \sqrt{5}, \pi, |-8|, 9$ **27.** (a) Positive integer, (b) negative integer   **29.** (a) To right of origin, (b) to left of -4 **31.** P, V variables; c constant   **33.** Yes

Exercises 1–2, 1–3, p. 10

1. 11   **3.** -11 **5.** 9   **7.** -3 **9.** -24 **11.** 35   **13.** -3 **15.** 20   **17.** 40   **19.** -1
21. 9   **23.** undefined   **25.** 20   **27.** -5 **29.** -9 **31.** 24   **33.** -6 **35.** 3   **37.** Commutative law of multiplication   **39.** Distributive law   **41.** Associative law of addition   **43.** Associative law of multiplication   **45.** Positive   **47.** $\frac{4}{2} = 2; \frac{2}{4} = \frac{1}{2}; 2 \neq \frac{1}{2}$ **49.** $8(2000 + 1000)$; distributive law

Exercises 1–4, p. 13

1. 3.62   **3.** -4.54 **5.** 0.1356   **7.** 6.086   **9.** $43.011 = 43.011$, commutative law of addition
11. $478.30341 = 478.30341$, distributive law   **13.** -8.21 **15.** 0.976   **17.** (a) 0.36, (b) 0.36
19. No, -0.0000073 **21.** 0.2424242, (b) 3.1415927   **23.** Error   **25.** 23.8 points/s

Exercises 1–5, p. 19

1. x^7 **3.** $2b^6$ **5.** m^2 **7.** $\frac{1}{n^4}$ **9.** a^8 **11.** t^{20} **13.** $8n^3$ **15.** a^2x^8 **17.** $\frac{8}{b^3}$ **19.** $\frac{x^8}{16}$ **21.** 1

23. 3   **25.** $\frac{1}{6}$ **27.** s^2 **29.** $-t^{14}$ **31.** $64x^{12}$ **33.** 1   **35.** b^2 **37.** $\frac{a}{x^2}$ **39.** $\frac{x^3}{64a^3}$

41. $64g^2s^6$ **43.** $\frac{5a}{n}$ **45.** -53 **47.** -10 **49.** 6230   **51.** -0.421 **53.** $\frac{-wLx^3}{12EI}$ **55.** 68.9 W

Exercises 1–6, p. 22

1. 45,000 **3.** 0.00201 **5.** 3.23 **7.** 18.6 **9.** 4×10^4 **11.** 8.7×10^{-3} **13.** 6.89×10^0
15. 6.3×10^{-2} **17.** 5.6×10^{13} **19.** 2.2×10^8 **21.** 4.85×10^{10} **23.** 3.67×10^5
25. 9.965×10^{-3} **27.** 5.0350×10^1 **29.** 2.25×10^4 lb/in.^{2} **31.** 0.0000000000016 W
33. 4500 Ω **35.** 1.1×10^{-5} ft **37.** 0.000000000000000000000000167 g **39.** 3.6×10^4 km
41. 2.59×10^{10} cm^2 **43.** $6.03 \times 10^{-3} \; \Omega = 6.03$ mΩ

Exercises 1–7, p. 26

1. 5 **3.** -11 **5.** -7 **7.** 20 **9.** 5 **11.** -6 **13.** 5 **15.** 31 **17.** $3\sqrt{2}$ **19.** $2\sqrt{3}$
21. $2\sqrt{11}$ **23.** $3\sqrt{7}$ **25.** $8\sqrt{3}$ **27.** 7 **29.** 10 **31.** $3\sqrt{10}$ **33.** 9.24 **35.** 0.6877
37. 9.7 s **39.** 40 m **41.** No; true for $a \geq 0$

Exercises 1–8, p. 30

1. $8x$ **3.** $4x + y$ **5.** $a + c - e$ **7.** $-a^2b - a^2b^2$ **9.** $4s + 4$ **11.** $-v + 5x - 4$
13. $5a - 5$ **15.** $-5a + 2$ **17.** $-2t + 5u$ **19.** $7r - 2s$ **21.** $19c - 50$ **23.** $3n - 9$
25. $-2t^2 + 18$ **27.** $6a$ **29.** $2a\sqrt{xy} + 1$ **31.** $4c - 6$ **33.** $8p - 5q$ **35.** $-4x^2 + 22$
37. $4a - 3$ **39.** $-6t + 13$ **41.** $R + 1$ **43.** $3x - 395$

Exercises 1–9, p. 32

1. a^3x **3.** $-a^2c^3x^3$ **5.** $-8a^3x^5$ **7.** $2a^8x^3$ **9.** $a^2x + a^2y$ **11.** $-3s^3 + 15st$ **13.** $5m^3n + 15m^2n$
15. $3x^2 - 3xy + 6x$ **17.** $a^2b^2c^5 - ab^3c^5 - a^2b^3c^4$ **19.** $acx^4 + acx^3y^3$ **21.** $x^2 + 2x - 15$
23. $2x^2 + 9x - 5$ **25.** $6a^2 - 7ab + 2b^2$ **27.** $6s^2 + 11st - 35t^2$ **29.** $2x^3 + 5x^2 - 2x - 5$
31. $x^3 + 2x^2 - 8x$ **33.** $x^3 - 2x^2 - x + 2$ **35.** $x^5 - x^4 - 6x^3 + 4x^2 + 8x$ **37.** $2a^2 - 16a - 18$
39. $2x^3 + 6x^2 - 8x$ **41.** $4x^2 - 20x + 25$ **43.** $x^2 + 6ax + 9a^2$ **45.** $x^2y^2z^2 - 4xyz + 4$
47. $2x^2 + 32x + 128$ **49.** $-x^3 + 2x^2 + 5x - 6$ **51.** $6x^4 + 21x^3 + 12x^2 - 12x$
53. $8000 - 5x^2$ **55.** $wl^4 - 2wl^2x^2 + wx^4$

Exercises 1–10, p. 35

1. $-4x^2y$ **3.** $4t^4/r^2$ **5.** $4x^2$ **7.** $-6a$ **9.** $a^2 + 4y$ **11.** $t - 2rt^2$ **13.** $q + 2p - 4q^3$

15. $\dfrac{2L}{R} - R$ **17.** $\dfrac{1}{3a} - \dfrac{2b}{3a} + \dfrac{a}{b}$ **19.** $x^2 + a$ **21.** $2x + 1$ **23.** $x - 1$ **25.** $4x^2 - x - 1, R = -3$

27. $x + 5$ **29.** $x^2 + x - 6$ **31.** $2x^2 + 4x + 2, R = 4x + 4$ **33.** $x^2 - 2x + 4$ **35.** $x - y$

37. $V^2 - \dfrac{aV}{RT} + \dfrac{ab}{RT}$ **39.** $\dfrac{1}{R_1} + \dfrac{1}{R_2} + \dfrac{1}{R_3}$

Exercises 1–11, p. 39

1. 9 **3.** -1 **5.** 10 **7.** 5 **9.** -3 **11.** 1 **13.** $-\frac{7}{2}$ **15.** 8 **17.** $\frac{10}{3}$ **19.** $-\frac{13}{3}$
21. 2 **23.** 0 **25.** 9.5 **27.** -1.5 **29.** True for all values of x, identity. **31.** Left side 2 greater than
right side for all values of x.

Exercises 1–12, p. 42

1. $\dfrac{b}{a}$ **3.** $\dfrac{m - 1}{4}$ **5.** $\dfrac{c + 6}{a}$ **7.** $2a + 8$ **9.** $\dfrac{E}{I}$ **11.** $\dfrac{P}{2\pi f}$ **13.** $\dfrac{PL^2}{\pi^2 I}$ **15.** $\dfrac{v_0 - v}{t}$

17. $\dfrac{a + PV^2}{V}$ **19.** $\dfrac{Q_1 + PQ_1}{P}$ **21.** $\dfrac{l - a + d}{d}$ **23.** $\dfrac{L - 2d - \pi r_2}{\pi}$ **25.** $\dfrac{V_r A - V_0}{V_0 A}$

27. $\dfrac{wL}{R(w + L)}$ **29.** 28.2 in. **31.** 32.3°C

Exercises 1–13, p. 46

1. \$5200, \$2600 **3.** 2.0 A, 4.2 A, 6.5 A **5.** 20 acres at \$20,000/acre, 50 acres at \$10,000/acre
7. 20 girders **9.** 6.5 ft **11.** 36 in., 54 in. **13.** 6.9 km (main), 9.5 km (others)
15. 360 s (6 min), first car **17.** 1600 mi/h, 2000 mi/h **19.** 6 L

Exercises for Chapter 1, p. 47

1. -10 **3.** -20 **5.** 2 **7.** -25 **9.** -4 **11.** 5 **13.** $4r^2t^4$ **15.** $\dfrac{6m^2}{nt^2}$ **17.** $\dfrac{z^6}{y^2}$ **19.** $\dfrac{8t^3}{s^2}$

21. $3\sqrt{5}$ **23.** $2\sqrt{5}$ **25.** 18.12 **27.** -1.440 **29.** $-a - 2ab$ **31.** $5xy + 3$ **33.** $2x^2 + 9x - 5$
35. $x^2 + 16x + 64$ **37.** $hk - 3h^2k^4$ **39.** $7a - 6b$ **41.** $13xy - 10z$ **43.** $2x^3 - x^2 - 7x - 3$

45. $-3x^2y + 24xy^2 - 48y^3$ **47.** $-9p^2 + 3pq + 18p^2q$ **49.** $\dfrac{6q}{p} - 2 + \dfrac{3q^4}{p^3}$ **51.** $2x - 5$

53. $x^2 - 2x + 3$ **55.** $4x^3 - 2x^2 + 6x,\ R = -1$ **57.** $15r - 3s - 3t$ **59.** $y^2 + 5y - 1,\ R = 4$ **61.** 4
63. -5 **65.** $-\frac{7}{3}$ **67.** 3 **69.** $-\frac{19}{5}$ **71.** 1 **73.** 2.5×10^4 mi/h **75.** $176{,}000{,}000{,}000$ C/kg

77. 2×10^{-7} in. **79.** 0.000018 Pa·s **81.** $\dfrac{5a - 2}{3}$ **83.** $\dfrac{4 - 2n}{3}$ **85.** ϕ / B **87.** $\dfrac{I - P}{Pr}$

89. $\dfrac{nE - Ir}{In}$ **91.** $\dfrac{D_p(N + 2)}{N}$ **93.** $\dfrac{L - L_0}{L_0(t_2 - t_1)}$ **95.** $\dfrac{2S - n^2d + nd}{2n}$ **97.** 1.80×10^{-2} ft **99.** 1.45 N

101. $12x - 2x^2$ **103.** $252T_f - 6250$ **105.** 450 kg to carbon dioxide, 50 kg to carbon monoxide.
107. \$15, \$5 **109.** 66 cm by 56 cm **111.** After 1.4 h **113.** 400 L

Exercises 2–1, p. 57

1. $A = \pi r^2$ **3.** $c = 2\pi r$ **5.** $A = 5l$ **7.** $A = s^2;\ s = \sqrt{A}$ **9.** $d = 55t$ **11.** $w = 3000 - 10t$
13. $I = 8t$ **15.** $C = 25 + 20h$ **17.** $3, -1$ **19.** $11, -7$ **21.** $-18, 70$ **23.** $\frac{5}{2}, -\frac{1}{2}$
25. $\frac{1}{4}a + \frac{1}{2}a^2, 0$ **27.** $3s^2 + s + 6, 12s^2 - 2s + 6$ **29.** $62.9, 260$ **31.** $0.01988, -0.2998$
33. Square the value of the independent variable and add 2. **35.** Cube the value of the independent variable and
subtract this value from 6 times the value of the independent variable. **37.** $y \neq 1$ **39.** $y \geq 1$ **41.** 500 Pa

43. 7.8 **45.** $\frac{1}{16}p^2 + \dfrac{(60 - p)^2}{4\pi}$ **47.** $0.006T^2 + 2.8T + 0.012hT + 2.8h + 0.006h^2$

Exercises 2–2, p. 61

1. $(2, 1), (-1, 2), (-2, -3)$ **3.** **5.** Isosceles triangle **7.** Rectangle

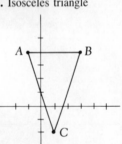

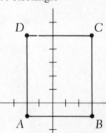

9. $(5, 4)$ **11.** $(3, -2)$ **13.** On a line parallel to the y-axis, one unit to the right. **15.** On a line parallel to the
x-axis, three units above. **17.** On a line bisecting the first and third quadrants. **19.** 0 **21.** To the right of the
y-axis. **23.** To the left of a line which is parallel to the y-axis and one unit to its left. **25.** Fourth quadrant
27. First, third

Exercises 2–3, p. 65

1. **3.** **5.** **7.** **9.**

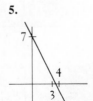
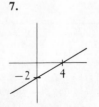
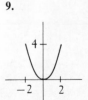

11.

13.

15.

17.

19.

21.

23.

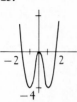

25.

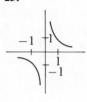

27.

29.

31.

33.

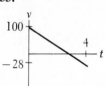

35.

37.

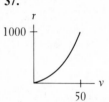

39.

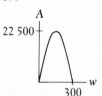

41.

43.

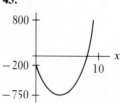

45. $y = x$

$y = |x|$

Exercises 2–4, p. 70

1.

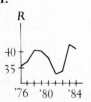

3.

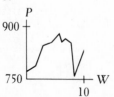

5.

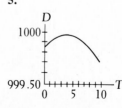

7.

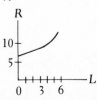

9. (a) 132.1°C, (b) 0.7 min   **11.** (a) 7.8 V, (b) 43°C   **13.** 999.81 kg/m³   **15.** 3.7 kg
17. (a) 138 cm³, (b) 228 kPa   **19.** 264 kPa   **21.** 0.34   **23.** 76 m²   **25.** 130.3°C   **27.** 420 kPa

Exercises 2–5, p. 74
1. 2   **3.** $-\frac{9}{4}$ **5.** 3.5   **7.** 0.5   **9.** 0.0, -1.0 **11.** No (real) zeros   **13.** 0.7   **15.** 2.5
17. $-2.8, 1.8$ **19.** $-1.6, 2.1$ **21.** $-2.0, 0.0, 2.0$ **23.** 0.0, 1.3   **25.** 3.5   **27.** No real zeros
29. After 1.9 s   **31.** 5.1 N   **33.** 13, 77   **35.** 1.8 ft from end

Exercises for Chapter 2, p. 75
1. $V = 8\pi r^2$ **3.** $F = \frac{9}{5}C + 32$ **5.** 16, -47 **7.** $-5, -27$ **9.** 3, $\sqrt{1-4h}$
11. $6hx + 3h^2 - 2h$ **13.** $-3.67, 16.7$ **15.** $0.16572, -0.21566$
17.   **19.**   **21.**   **23.**   **25.**   **27.**

29. -0.5 **31.** 0, 4   **33.** $-1.5, 1.0$ **35.** $-2.4, 0.0, 2.4$ **37.** $-1.2, 1.2$ **39.** 0   **41.** 0.4
43. 0.2, 5.8   **45.** 1.4   **47.** $-0.7, 0.7$ **49.** On a line in the first quadrant, one unit to the right of y-axis.
51. 12.3
53. 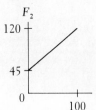   **55.** 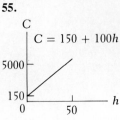   **57.**    **59.** 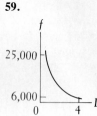   **61.**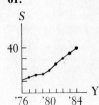

63. 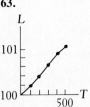   **65.** After 1.3 s and 4.7 s

Exercises 3–1, p. 81
1.   **3.**   **5.** 405°, $-315°$ **7.** 510°, $-210°$ **9.** 430°30′, $-289°30′$
11. 638.1°, $-81.9°$ **13.** 15.2°   **15.** 82.91°   **17.** 15.2°
19. 86.05°   **21.** 301.27°   **23.** 96.13°   **25.** 47°30′   **27.** 19°45′
29. 5°37′   **31.** 24°55′

33. **35.** 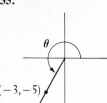 **37.** 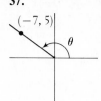 **39.**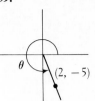

41. $21.710°$

43. $86°16'26''$

Exercises 3–2, p. 85

1. $\sin\theta = \frac{4}{5}$, $\cos\theta = \frac{3}{5}$, $\tan\theta = \frac{4}{3}$, $\cot\theta = \frac{3}{4}$, $\sec\theta = \frac{5}{3}$, $\csc\theta = \frac{5}{4}$

3. $\sin\theta = \frac{8}{17}$, $\cos\theta = \frac{15}{17}$, $\tan\theta = \frac{8}{15}$, $\cot\theta = \frac{15}{8}$, $\sec\theta = \frac{17}{15}$, $\csc\theta = \frac{17}{8}$

5. $\sin\theta = \frac{40}{41}$, $\cos\theta = \frac{9}{41}$, $\tan\theta = \frac{40}{9}$, $\cot\theta = \frac{9}{40}$, $\sec\theta = \frac{41}{9}$, $\csc\theta = \frac{41}{40}$

7. $\sin\theta = \frac{\sqrt{15}}{4}$, $\cos\theta = \frac{1}{4}$, $\tan\theta = \sqrt{15}$, $\cot\theta = \frac{1}{\sqrt{15}}$, $\sec\theta = 4$, $\csc\theta = \frac{4}{\sqrt{15}}$

9. $\sin\theta = \frac{1}{\sqrt{2}}$, $\cos\theta = \frac{1}{\sqrt{2}}$, $\tan\theta = 1$, $\cot\theta = 1$, $\sec\theta = \sqrt{2}$, $\csc\theta = \sqrt{2}$

11. $\sin\theta = \frac{2}{\sqrt{29}}$, $\cos\theta = \frac{5}{\sqrt{29}}$, $\tan\theta = \frac{2}{5}$, $\cot\theta = \frac{5}{2}$, $\sec\theta = \frac{\sqrt{29}}{5}$, $\csc\theta = \frac{\sqrt{29}}{2}$

13. $\sin\theta = 0.846$, $\cos\theta = 0.534$, $\tan\theta = 1.58$, $\cot\theta = 0.631$, $\sec\theta = 1.87$, $\csc\theta = 1.18$

15. $\sin\theta = 0.1521$, $\cos\theta = 0.9884$, $\tan\theta = 0.1539$, $\cot\theta = 6.498$, $\sec\theta = 1.012$, $\csc\theta = 6.575$

17. $\frac{5}{13}$, $\frac{12}{5}$ **19.** $\frac{1}{\sqrt{2}}$, $\sqrt{2}$ **21.** 0.882, 1.33 **23.** 0.246, 3.94 **25.** $\sin\theta = \frac{4}{5}$, $\tan\theta = \frac{4}{3}$

27. $\tan\theta = \frac{1}{2}$, $\sec\theta = \frac{\sqrt{5}}{2}$ **29.** $\sec\theta$ **31.** $\frac{x}{y} \cdot \frac{y}{r} = \frac{x}{r} = \cos\theta$

Exercises 3–3, p. 89

1. $\sin 40° = 0.64$, $\cos 40° = 0.77$, $\tan 40° = 0.84$, $\cot 40° = 1.19$, $\sec 40° = 1.30$, $\csc 40° = 1.56$

3. $\sin 15° = 0.26$, $\cos 15° = 0.97$, $\tan 15° = 0.27$, $\cot 15° = 3.73$, $\sec 15° = 1.04$, $\csc 15° = 3.86$

5. 0.381 **7.** 1.58 **9.** 0.9626 **11.** 0.99414 **13.** 0.4085 **15.** 1.57 **17.** $70.97°$ **19.** $65.70°$

21. $11.718°$ **23.** $49.453°$ **25.** $53.44°$ **27.** $81.787°$ **29.** 0.3256 **31.** 2.394 **33.** $40°10'$

35. $82°50'$ **37.** 0.5528 **39.** 0.8767 **41.** $72°47'$ **43.** $35°8'$ **45.** 182 ft/s **47.** 71.11 V

Exercises 3–4, p. 94

1.

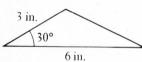

3.

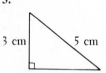

5. $B = 12.2°$, $b = 1450$, $c = 6850$

7. $A = 25.8°$, $B = 64.2°$, $b = 311$

9. $A = 57.9°$, $a = 20.2$, $b = 12.6$

11. $A = 22.1°$, $a = 33.3$, $B = 67.9°$

13. $a = 30.21$, $B = 57.90°$, $b = 48.16$ **15.** $A = 52.15°$, $B = 37.85°$, $c = 71.85$

17. $A = 52.5°$, $b = 0.661$, $c = 1.09$ **19.** $A = 15.82°$, $a = 0.5239$, $c = 1.922$

21. $A = 65.89°$, $B = 24.11°$, $c = 648.46$ **23.** $a = 3.3621$, $B = 77.025°$, $c = 14.974$ **25.** 4.45

27. $40.24°$ **29.** $A = 52°20'$, $b = 0.684$, $c = 1.12$ **31.** $A = 58°40'$, $a = 143$, $B = 31°20'$

33. $a = c \sin A$, $b = c \cos A$, $B = 90° - A$ **35.** $A = 90° - B$, $b = a \tan B$, $c = a/\cos B$

Exercises 3–5, p. 97

1. 384 ft **3.** 9.50 ft **5.** $0.38°$ **7.** 260.6 m **9.** 25.4 ft **11.** $34.3°$ **13.** 5.26 mi **15.** 1.29 m

17. $8.8°$ **19.** 1080 mi **21.** 8.199 cm **23.** $103.34°$ **25.** 29.831 m **27.** 200 m

Exercises for Chapter 3, p. 99

1. $377.0°$, $-343.0°$ **3.** $142.5°$, $-577.5°$ **5.** $31.9°$ **7.** $38.1°$ **9.** $17°30'$ **11.** $49°42'$
13. $\sin \theta = \frac{7}{25}$, $\cos \theta = \frac{24}{25}$, $\tan \theta = \frac{7}{24}$, $\cot \theta = \frac{24}{7}$, $\sec \theta = \frac{25}{24}$, $\csc \theta = \frac{25}{7}$
15. $\sin \theta = \frac{1}{\sqrt{2}}$, $\cos \theta = \frac{1}{\sqrt{2}}$, $\tan \theta = 1$, $\cot \theta = 1$, $\sec \theta = \sqrt{2}$, $\csc \theta = \sqrt{2}$ **17.** 0.923, 2.40
19. 0.447, 1.12 **21.** 0.952 **23.** 1.853 **25.** $18.2°$ **27.** $57.57°$ **29.** 1.686 **31.** 0.5664
33. $32°10'$ **35.** $62°10'$ **37.** $a = 1.83$, $B = 73.0°$, $c = 6.27$ **39.** $A = 51.5°$, $B = 38.5°$, $c = 104$
41. $B = 52.5°$, $b = 15.6$, $c = 19.7$ **43.** $A = 31.61°$, $a = 4.006$, $B = 58.39°$
45. $a = 0.6292$, $B = 40.33°$, $b = 0.5341$ **47.** $A = 48.813°$, $B = 41.187°$, $b = 10.196$
49. 44.4 N   **51.** 0.6120 μm   **53.** $30.6°$ **55.** 1.598 mi   **57.** 4820 ft   **59.** $3.12°$ **61.** 47.2 ft
63. 0.184 km   **65.** $24.94°$ **67.** 155.0 m   **69.** 67.1 ft   **71.** 1.83 km   **73.** 464 m

Exercises 4–1, p. 105

1. Yes, no   **3.** Yes, yes   **5.** -1, -16 **7.** $-\frac{9}{5}$, $-\frac{16}{5}$ **9.** Yes   **11.** No   **13.** No   **15.** Yes
17. Yes   **19.** Yes

Exercises 4–2, p. 110

1. 4 **3.** -5 **5.** $\frac{2}{7}$ **7.** $\frac{1}{2}$

9.
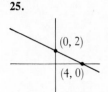

11.
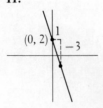

13.
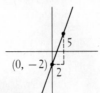

15.
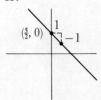

17. $m = -2$, $b = 1$

19. $m = 1$, $b = 4$

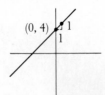

21. $m = \frac{5}{2}$, $b = -2$

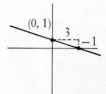

23. $m = -\frac{1}{3}$, $b = 1$

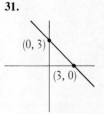

25.
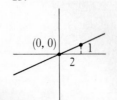

27.
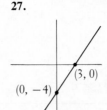

29.
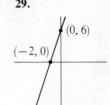

31.

33.

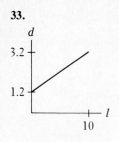

35.

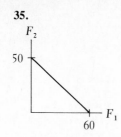

Exercises 4–3, p. 113

1. $x = 3.0, y = 1.0$ **3.** $x = 3.0, y = 0.0$ **5.** $x = 2.2, y = -0.3$ **7.** $x = -0.9, y = -2.3$
9. $x = 6.2, y = -0.5$ **11.** $x = 0.0, y = 3.0$ **13.** $x = -14.0, y = -5.0$ **15.** $x = 4.0,$
$y = 7.5$ **17.** $x = -3.6, y = -1.4$ **19.** $x = 1.1, y = 0.8$ **21.** Dependent   **23.** $x = -1.2,$
$y = -3.6$ **25.** $x = 1.5, y = 4.5$ **27.** Inconsistent   **29.** $l = 9.0$ km, $w = 3.0$ km
31. $I = 0.9$ A, $E = 3.6$ V

Exercises 4–4, p. 119

1. $x = 1, y = -2$ **3.** $x = 7, y = 3$ **5.** $x = -1, y = -4$ **7.** $x = \frac{1}{2}, y = 2$ **9.** $x = -\frac{1}{3}, y = 4$
11. $x = \frac{9}{22}, y = -\frac{16}{11}$ **13.** $x = 3, y = 1$ **15.** $x = -1, y = -2$ **17.** $x = 1, y = 2$ **19.** Inconsistent
21. $x = -\frac{14}{5}, y = -\frac{16}{5}$ **23.** $x = 2.38, y = 0.45$ **25.** $x = \frac{1}{2}, y = -4$ **27.** $x = -\frac{2}{3}, y = 0$
29. Dependent   **31.** $x = -1, y = -2$ **33.** $i_1 = 1.5$ A, $i_2 = 0.5$ A   **35.** 6250 L, 3750 L
37. 7.5 m from one end   **39.** 4,000,000 calc/s, 2,500,000 calc/s   **41.** 105 mL and 75 mL
43. $300 fixed cost, $0.25 per booklet

Exercises 4–5, p. 126

1. -10 **3.** 29   **5.** 32   **7.** 93   **9.** 0.9   **11.** 96   **13.** $x = 3, y = 1$ **15.** $x = -1, y = -2$
17. $x = 1, y = 2$ **19.** Inconsistent   **21.** $x = -\frac{14}{5}, y = -\frac{16}{5}$ **23.** $x = 2.38, y = 0.45$ **25.** $x = \frac{1}{2},$
$y = -4$ **27.** $x = -\frac{2}{3}, y = 0$ **29.** Dependent   **31.** $x = -1, y = -2$ **33.** $s_0 = 15$ ft, $v = 10$ ft/s
35. $3200, $2800   **37.** 58, 31   **39.** 4200, 1400   **41.** $R = \frac{1}{600}T + \frac{11}{30}$

Exercises 4–6, p. 132

1. $x = 2, y = -1, z = 1$ **3.** $x = 4, y = -3, z = 3$ **5.** $x = \frac{1}{2}, y = \frac{2}{3}, z = \frac{1}{6}$ **7.** $x = \frac{2}{3}, y = -\frac{1}{3}, z = 1$
9. $x = \frac{4}{15}, y = -\frac{3}{5}, z = \frac{1}{3}$ **11.** $x = -2, y = \frac{2}{3}, z = \frac{1}{3}$ **13.** $x = \frac{3}{4}, y = 1, z = -\frac{1}{2}$
15. $r = 0, s = 0, t = 0, u = -1$ **17.** Eliminate x and the resulting system with y and z is dependent; choose a
z value, then find y, then x; if $z = 0, y = -6, x = -10$; if $z = 1, y = -11, x = -17$.   **19.** $-\frac{3}{22}$A, $-\frac{39}{110}$A, $\frac{27}{55}$A
21. 9.43 N, 8.33 N, 1.67 N   **23.** 16 parts/h, 20 parts/h, 28 parts/h   **25.** 50 cm³ alloy B, 50 cm³ alloy C

Exercises 4–7, p. 139

1. 122   **3.** 651   **5.** -439 **7.** 202   **9.** 128   **11.** 0.128   **13.** $x = -1, y = 2, z = 0$
15. $x = 2, y = -1, z = 1$ **17.** $x = 4, y = -3, z = 3$ **19.** $x = \frac{1}{2}, y = \frac{2}{3}, z = \frac{1}{6}$
21. $x = \frac{2}{3}, y = -\frac{1}{3}, z = 1$ **23.** $x = \frac{4}{15}, y = -\frac{3}{5}, z = \frac{1}{3}$ **25.** $x = -2, y = \frac{2}{3}, z = \frac{1}{3}$
27. $x = \frac{3}{4}, y = 1, z = -\frac{1}{2}$ **29.** $A = 125$ N, $B = 60$ N, $F = 75$ N   **31.** $12,000, $7000, $1000

Exercises for Chapter 4, p. 141

1. -17 **3.** -35 **5.** -4 **7.** $\frac{2}{7}$

9. $m = -2, b = 4$

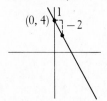

11. $m = 4, b = -\frac{5}{2}$

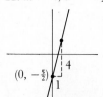

13. $x = 2.0, y = 0.0$ **15.** $x = 2.2, y = 2.7$ **17.** $x = 1.5, y = -1.9$ **19.** $x = 1.5, y = 0.4$
21. $x = 1, y = 2$ **23.** $x = \frac{1}{2}, y = -2$ **25.** $x = -\frac{1}{3}, y = \frac{7}{4}$ **27.** $x = \frac{11}{19}, y = -\frac{26}{19}$
29. $x = -\frac{6}{19}, y = \frac{36}{19}$ **31.** $x = 1.10, y = 0.54$ **33.** $x = 1, y = 2$ **35.** $x = \frac{1}{2}, y = -2$
37. $x = -\frac{1}{3}, y = \frac{7}{4}$ **39.** $x = \frac{11}{19}, y = -\frac{26}{19}$ **41.** $x = -\frac{6}{19}, y = \frac{36}{19}$ **43.** $x = 1.10, y = 0.54$
45. -115 **47.** 220 **49.** $x = 2, y = -1, z = 1$ **51.** $x = \frac{2}{3}, y = -\frac{1}{2}, z = 0$
53. $r = 3, s = -1, t = \frac{3}{2}$ **55.** $x = -\frac{1}{2}, y = \frac{1}{2}, z = 3$ **57.** $x = 2, y = -1, z = 1$
59. $x = \frac{2}{3}, y = -\frac{1}{2}, z = 0$ **61.** $r = 3, s = -1, t = \frac{3}{2}$ **63.** $x = -\frac{1}{2}, y = \frac{1}{2}, z = 3$
65. $x = \frac{8}{3}, y = -8$ **67.** $x = 1, y = 3$ **69.** -6 **71.** $-\frac{4}{3}$ **73.** $R_1 = 5 \ \Omega, R_2 = 2 \ \Omega$
75. 232 lb, 24 lb, 201 lb   **77.** 450 mi/h, 50 mi/h   **79.** 380, 140   **81.** 11 ft
83. 79% nickel, 16% iron, 5% molybdenum

Exercises 5–1, p. 148

1. $40x - 40y$ **3.** $2x^3 - 8x^2$ **5.** $y^2 - 36$ **7.** $9v^2 - 4$ **9.** $16x^2 - 25y^2$ **11.** $144 - 25a^2b^2$
13. $25f^2 + 40f + 16$ **15.** $4x^2 + 28x + 49$ **17.** $x^2 - 2x + 1$ **19.** $16a^2 + 56axy + 49x^2y^2$
21. $x^2 - 4xy + 4y^2$ **23.** $36s^2 - 12st + t^2$ **25.** $x^2 + 6x + 5$ **27.** $c^2 + 9c + 18$ **29.** $6x^2 + 13x - 5$
31. $20x^2 - 21x - 5$ **33.** $20v^2 + 13v - 15$ **35.** $6x^2 - 13xy - 63y^2$ **37.** $2x^2 - 8$ **39.** $8a^3 - 2a$
41. $6ax^2 + 24abx + 24ab^2$ **43.** $20n^4 + 100n^3 + 125n^2$ **45.** $3s^3 - 36s^2 + 108s$
47. $16a^3 - 48a^2 + 36a$ **49.** $x^2 + y^2 + 2xy + 2x + 2y + 1$ **51.** $x^2 + y^2 + 2xy - 6x - 6y + 9$
53. $125 - 75t + 15t^2 - t^3$ **55.** $8x^3 + 60x^2t + 150xt^2 + 125t^3$ **57.** $x^3 + 8$ **59.** $64 - 27x^3$
61. $Ri_1^2 + 2Ri_1i_2 + Ri_2^2$ **63.** $4p^2 + 8pDA + 4D^2A^2$ **65.** $192 - 16t - 16t^2$ **67.** $P + 3Pr + 3Pr^2 + Pr^3$

Exercises 5–2, p. 151

1. $6(x + y)$ **3.** $5(a - 1)$ **5.** $3x(x - 3)$ **7.** $7b(by - 4)$ **9.** $6n(2n + 1)$ **11.** $2(x + 2y - 4z)$
13. $3ab(b - 2 + 4b^2)$ **15.** $4pq(3q - 2 - 7q^2)$ **17.** $2(a^2 - b^2 + 2c^2 - 3d^2)$ **19.** $(x + 2)(x - 2)$
21. $(10 + y)(10 - y)$ **23.** $(6a + 1)(6a - 1)$ **25.** $(9s + 5t)(9s - 5t)$ **27.** $(12n + 13p^2)(12n - 13p^2)$
29. $2(x + 2)(x - 2)$ **31.** $3(x + 3z)(x - 3z)$ **33.** $(x^2 + 4)(x + 2)(x - 2)$
35. $(x^4 + 1)(x^2 + 1)(x + 1)(x - 1)$ **37.** $(3 + b)(x - y)$ **39.** $(2 + a)(r + s)$ **41.** $(a - b)(a + x)$
43. $(x + 2)(x - 2)(x + 3)$ **45.** $i(R_1 + R_2 + r)$ **47.** $m(v_1 + v_2)(v_1 - v_2)$
49. $Pb(L + b)(L - b)$ **51.** $e(e + 2)(e - 2)$

Exercises 5–3, p. 156

1. $(x + 1)(x + 4)$ **3.** $(s - 7)(s + 6)$ **5.** $(t + 8)(t - 3)$ **7.** $(x + 1)^2$ **9.** $(x - 2)^2$
11. $(3x + 1)(x - 2)$ **13.** $(3y + 1)(y - 3)$ **15.** $(2s + 11)(s + 1)$ **17.** $(3f - 1)(f - 5)$
19. $(2t - 3)(t + 5)$ **21.** $(3t - 4u)(t - u)$ **23.** $(4x - 7)(x + 1)$ **25.** $(9x - 2y)(x + y)$
27. $(2m + 5)^2$ **29.** $(2x - 3)^2$ **31.** $(3t - 4)(3t - 1)$ **33.** $(8b - 1)(b + 4)$ **35.** $(4p - q)(p - 6q)$
37. $(12x - y)(x + 4y)$ **39.** $2(x - 1)(x - 6)$ **41.** $2(2x - 1)(x + 4)$ **43.** $x(x + 6)(x - 2)$
45. $(x + 1)^3$ **47.** $(2x + 1)(4x^2 - 2x + 1)$ **49.** $16(t - 4)(t + 2)$ **51.** $(x - 2L)(x - L)$
53. $(s + 4)(s + 12)$

Exercises 5–4, p. 160

1. $\frac{14}{21}$ **3.** $\frac{2ax^2}{2xy}$ **5.** $\frac{2x-4}{x^2+x-6}$ **7.** $\frac{ax^2-ay^2}{x^2-xy-2y^2}$ **9.** $\frac{7}{11}$ **11.** $\frac{2xy}{4y^2}$ **13.** $\frac{2}{x+1}$ **15.** $\frac{x-5}{2x-1}$

17. $\frac{1}{4}$ **19.** $\frac{3}{4}x$ **21.** $\frac{1}{5a}$ **23.** $\frac{3a-2b}{2a-b}$ **25.** $\frac{4x^2+1}{(2x+1)(2x-1)}$ (cannot be reduced)   **27.** $3x$

29. $\frac{1}{2y^2}$ **31.** $\frac{x-4}{x+4}$ **33.** $\frac{2x-1}{x+8}$ **35.** $\frac{5x+4}{x(x+3)}$ **37.** $(x^2+4)(x-2)$ **39.** $\frac{x^2y^2(y+x)}{y-x}$

41. $\frac{x+3}{x-3}$ **43.** $-\frac{1}{2}$ **45.** $-\frac{2x-1}{x}$ **47.** $\frac{(x+5)(x-3)}{(5-x)(x+3)}$ **49.** $\frac{x^2+xy+y^2}{x+y}$

51. $\frac{(x+1)^2}{x^2-x+1}$ **53.** (a)   **55.** (a)

Exercises 5–5, p. 164

1. $\frac{3}{28}$ **3.** $6xy$ **5.** $\frac{7}{18}$ **7.** $\frac{xy^2}{bz^2}$ **9.** $4t$ **11.** $3(u+v)$ **13.** $\frac{10}{3(a+4)}$ **15.** $\frac{x-3}{x(x+3)}$ **17.** $\frac{3x}{5a}$

19. $\frac{(x+1)(x-1)(x-4)}{4(x+2)}$ **21.** $\frac{x^2}{a+x}$ **23.** $\frac{15}{4}$ **25.** $\frac{3}{4x+3}$ **27.** $\frac{7x-1}{3x+5}$ **29.** $\frac{7x^4}{3a^4}$

31. $\frac{4t(2t-1)(t+5)}{(2t+1)^2}$ **33.** $\frac{1}{2}(x+y)$ **35.** $(x+y)(3p+7q)$ **37.** $\frac{12wv^2}{g}$ **39.** $\frac{(T+100)(T-400)}{2(T-40)}$

Exercises 5–6, p. 170

1. $\frac{9}{5}$ **3.** $\frac{8}{x}$ **5.** $\frac{5}{4}$ **7.** $\frac{3+7ax}{4x}$ **9.** $\frac{ax-b}{x^2}$ **11.** $\frac{30+ax^2}{25x^3}$ **13.** $\frac{14-a^2}{10a}$

15. $\frac{-x^2+4x+xy+y-2}{xy}$ **17.** $\frac{7}{2(2x-1)}$ **19.** $\frac{5-3x}{2x(x+1)}$ **21.** $\frac{-3}{4(s-3)}$ **23.** $\frac{x+6}{x^2-9}$

25. $\frac{2x-5}{(x-4)^2}$ **27.** $\frac{x+27}{(x-5)(x+5)(x-6)}$ **29.** $\frac{9x^2+x-2}{(3x-1)(x-4)}$ **31.** $\frac{13t^2+27t}{(t-3)(t+2)(t+3)^2}$

33. $\frac{x+1}{x-1}$ **35.** $-\frac{(x+1)(x^3+x^2-3x-1)}{x^2(x+2)}$ **37.** $-\frac{(3x+4)(x-1)}{2x}$ **39.** $\frac{2s}{r-s}$

41. $\frac{h}{(x+1)(x+h+1)}$ **43.** $\frac{-2hx-h^2}{x^2(x+h)^2}$ **45.** $\frac{y^2-rx+r^2}{r^2}$ **47.** $\frac{2a-1}{a^2}$ **49.** $\frac{3\pi l^3-12cl^2+\pi c^3}{3\pi l^3}$

51. $\frac{b(y^2-x^2)}{(x^2+y^2)^2}$

Exercises 5–7, p. 174

1. 4   **3.** -3 **5.** $\frac{7}{2}$ **7.** $\frac{16}{21}$ **9.** -9 **11.** $-\frac{6}{5}$ **13.** $\frac{5}{3}$ **15.** -2 **17.** $\frac{3}{4}$ **19.** 6   **21.** -5

23. $-\frac{7}{8}$ **25.** No solution   **27.** $\frac{2}{3}$ **29.** $\frac{3b}{1-2b}$ **31.** $\frac{(2b-1)(b+6)}{2(b-1)}$ **33.** $\frac{n_1 V-nV}{n_1}$

35. $\frac{2EI-2IV_0-p^2-ImV^2}{IV^2}$ **37.** $\frac{AIS-AMc}{I}$ **39.** $\frac{rR-rR_2+RR_2}{R_2-R}$ **41.** $\frac{kA_1A_2R-A_1L_2}{A_2}$

43. $\frac{24EID}{x^4-4Lx^3+6L^2x^2}$ **45.** 3.6 s   **47.** 60 m, 36 m

Exercises for Chapter 5, p. 176

1. $12ax+15a^2$ **3.** $4a^2-49b^2$ **5.** $4a^2+4a+1$ **7.** $b^2+3b-28$ **9.** $2x^2-13x-45$
11. $16c^2+6cd-d^2$ **13.** $3(s+3t)$ **15.** $a^2(x^2+1)$ **17.** $(x+12)(x-12)$
19. $(20r+t^2)(20r-t^2)$ **21.** $(3t-1)^2$ **23.** $(5t+1)^2$ **25.** $(x+8)(x-7)$ **27.** $(t-9)(t+4)$
29. $(2x-9)(x+4)$ **31.** $(2x+5)(2x-7)$ **33.** $(5b-1)(2b+5)$ **35.** $4(x+4)(x-4)$
37. $(x+3)^3$ **39.** $(2x+3)(4x^2-6x+9)$ **41.** $(a-3)(b^2+1)$ **43.** $(x+5)(n-x+5)$

45. $\dfrac{16x^2}{3a^2}$　　**47.** $\dfrac{3x + 1}{2x - 1}$　　**49.** $\dfrac{16}{5x(x - y)}$　　**51.** $\dfrac{6}{5 - x}$　　**53.** $\dfrac{x + 2}{2x(7x - 1)}$　　**55.** $\dfrac{1}{x - 1}$　　**57.** $\dfrac{16x - 15}{36x^2}$

59. $\dfrac{5y + 6}{2xy}$　　**61.** $\dfrac{-2(2a + 3)}{a(a + 2)}$　　**63.** $\dfrac{2x^2 - x + 1}{x(x + 3)(x - 1)}$　　**65.** $\dfrac{12x^2 - 7x - 4}{2(x - 1)(x + 1)(4x - 1)}$

67. $\dfrac{x^3 + 6x^2 - 2x + 2}{x(x - 1)(x + 3)}$　　**69.** 2　　**71.** $\dfrac{7}{2c + 4}$　　**73.** $-\dfrac{(a - 1)^2}{2a}$　　**75.** 6

77. $\frac{1}{4}[(x + y)^2 - (x - y)^2] = \frac{1}{4}(x^2 + 2xy + y^2 - x^2 + 2xy - y^2) = \frac{1}{4}(4xy) = xy$　　**79.** $(x - 10)(x + 7)$

81. $x(2x - 5)^2$　　**83.** $\dfrac{w(s - d)(s + d)}{2d}$　　**85.** $\dfrac{1 - t}{(t + 1)^3}$　　**87.** $\dfrac{120T^4 + 20w^2x^2T^2 - 3w^4x^4}{120T^4}$

89. $\dfrac{4(3s + 4)}{s(s + 4)^2}$　　**91.** $\dfrac{\mu R}{r + R + \mu R}$　　**93.** $\dfrac{fp}{p - f}$　　**95.** $\dfrac{wL^3 - 24\theta EI}{4L}$　　**97.** $\frac{48}{7}$ days

99. 15 s　　**101.** 3.1 days

Exercises 6–1, p. 185

1. $a = 1, b = -8, c = 5$　　**3.** $a = 1, b = -2, c = -4$　　**5.** Not quadratic　　**7.** $a = 1, b = -1, c = 0$
9. $2, -2$　　**11.** $\frac{3}{2}, -\frac{3}{2}$　　**13.** $-1, 9$　　**15.** $3, 4$　　**17.** $0, -2$　　**19.** $\frac{1}{3}, -\frac{1}{3}$　　**21.** $\frac{1}{3}, 4$　　**23.** $-4, -4$
25. $\frac{2}{3}, \frac{3}{2}$　　**27.** $\frac{1}{2}, -\frac{3}{2}$　　**29.** $2, -1$　　**31.** $2b, -2b$　　**33.** $0, \frac{5}{2}$　　**35.** $\frac{5}{2}, -\frac{9}{2}$　　**37.** $0, -2$
39. $b - a, -b - a$　　**41.** 10 s　　**43.** $-3, -5$　　**45.** 4 Ω, 12 Ω　　**47.** 12 mm, 8 mm

Exercises 6–2, p. 189

1. $-5, 5$　　**3.** $-\sqrt{7}, \sqrt{7}$　　**5.** $-3, 7$　　**7.** $-3 \pm \sqrt{7}$　　**9.** $2, -4$　　**11.** $-2, -1$　　**13.** $2 \pm \sqrt{2}$
15. $-5, 3$　　**17.** $-3, \frac{1}{2}$　　**19.** $\frac{1}{6}(3 \pm \sqrt{33})$　　**21.** $\frac{1}{4}(1 \pm \sqrt{17})$　　**23.** $-b \pm \sqrt{b^2 - c}$

Exercises 6–3, p. 193

1. $2, -4$　　**3.** $-2, -1$　　**5.** $2 \pm \sqrt{2}$　　**7.** $-5, 3$　　**9.** $-3, \frac{1}{2}$　　**11.** $\frac{1}{6}(3 \pm \sqrt{33})$　　**13.** $\frac{1}{4}(1 \pm \sqrt{17})$
15. $\frac{1}{4}(7 \pm \sqrt{17})$　　**17.** $\frac{1}{2}(-5 \pm \sqrt{-5})$　　**19.** $\frac{1}{6}(1 \pm \sqrt{109})$　　**21.** $\frac{3}{2}, -\frac{3}{2}$　　**23.** $\frac{3}{4}, -\frac{5}{8}$

25. $-0.54, 0.74$　　**27.** $-0.26, 2.43$　　**29.** $-c \pm \sqrt{c^2 + 1}$　　**31.** $\dfrac{b + 1 \pm \sqrt{-3b^2 + 2b + 4b^2a + 1}}{2b^2}$

33. 0.382 Pa　　**35.** 0.915 in.　　**37.** 4.48 ft, 6.48 ft　　**39.** 8 in.

Exercises 6–4, p. 199

1.
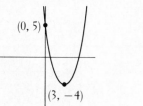

3.
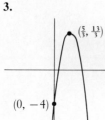

5.
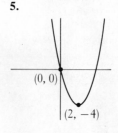

7.
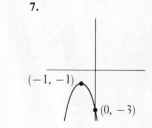

9.
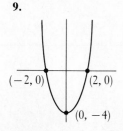

11.
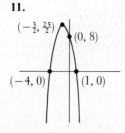

13.
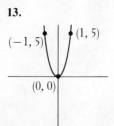

15.
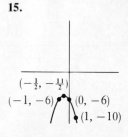

17. $-1.2, 1.2$ **19.** $0.5, 3.1$ **21.** No real roots   **23.** $-2.4, 1.2$

25.

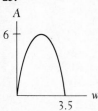

27.

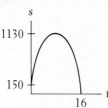

(a) After 16 s
(b) 1130 ft
(c) 3.3 s, 12.3 s

Exercises for Chapter 6, p. 200

1. $-4, 1$ **3.** $2, 8$ **5.** $\frac{1}{3}, -4$ **7.** $\frac{1}{2}, \frac{5}{3}$ **9.** $0, \frac{25}{6}$ **11.** $-\frac{3}{2}, \frac{7}{2}$ **13.** $-10, 11$ **15.** $-1 \pm \sqrt{6}$

17. $-4, \frac{9}{2}$ **19.** $\frac{1}{8}(3 \pm \sqrt{41})$ **21.** $\frac{1}{2}(-1 \pm 3\sqrt{-1})$ **23.** $\frac{1}{6}(-2 \pm \sqrt{58})$ **25.** $-2 \pm 2\sqrt{2}$

27. $\frac{1}{3}(-4 \pm \sqrt{10})$ **29.** $-1, \frac{5}{4}$ **31.** $\frac{1}{4}(-3 \pm \sqrt{-47})$ **33.** $\dfrac{-1 \pm \sqrt{-1}}{a}$ **35.** $\dfrac{-3 \pm \sqrt{9 + 4a^3}}{2a}$

37. $-5, 6$ **39.** $\frac{1}{4}(1 \pm \sqrt{33})$ **41.** $3 \pm \sqrt{7}$ **43.** 0 (3 is not a solution)

45.

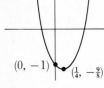

47.

49. $-1.7, 1.2$ **51.** No real roots

53. $-5 \pm 5\sqrt{-79}$ **55.** $\dfrac{E \pm \sqrt{E^2 - 4PR}}{2R}$ **57.** 10 units   **59.** $\dfrac{2.5 \pm \sqrt{6.25 - 50.4n}}{25.2}$

61. 8 V, 12 V   **63.** 3.06 mm   **65.** 0.106 in.   **67.** 25

Exercises 7–1, p. 205

1. $+, -, -$ **3.** $+, +, -$ **5.** $+, +, +$ **7.** $+, -, +$

9. $\sin \theta = \dfrac{1}{\sqrt{5}}$, $\cos \theta = \dfrac{2}{\sqrt{5}}$, $\tan \theta = \frac{1}{2}$, $\cot \theta = 2$, $\sec \theta = \frac{1}{2}\sqrt{5}$, $\csc \theta = \sqrt{5}$

11. $\sin \theta = -\dfrac{3}{\sqrt{13}}$, $\cos \theta = -\dfrac{2}{\sqrt{13}}$, $\tan \theta = \frac{3}{2}$, $\cot \theta = \frac{2}{3}$, $\sec \theta = -\frac{1}{2}\sqrt{13}$, $\csc \theta = -\frac{1}{3}\sqrt{13}$

13. $\sin \theta = \frac{12}{13}$, $\cos \theta = -\frac{5}{13}$, $\tan \theta = -\frac{12}{5}$, $\cot \theta = -\frac{5}{12}$, $\sec \theta = -\frac{13}{5}$, $\csc \theta = \frac{13}{12}$

15. $\sin \theta = -\dfrac{2}{\sqrt{29}}$, $\cos \theta = \dfrac{5}{\sqrt{29}}$, $\tan \theta = -\frac{2}{5}$, $\cot \theta = -\frac{5}{2}$, $\sec \theta = \frac{1}{5}\sqrt{29}$, $\csc \theta = -\frac{1}{2}\sqrt{29}$

17. II   **19.** II   **21.** IV   **23.** III

Exercises 7–2, p. 211

1. $\sin 20°$; $-\cos 40°$ **3.** $-\tan 75°$, $-\csc 58°$ **5.** $-\sin 57°$; $-\cot 6°$ **7.** $\cos 40°$; $-\tan 40°$
9. $-\sin 15.00° = -0.2588$ **11.** $-\cos 73.70° = -0.2807$ **13.** $\tan 39.15° = 0.8141$
15. $\sec 31.67° = 1.175$ **17.** -0.5228 **19.** -0.7620 **21.** -0.3461 **23.** -3.910
25. $237.99°$, $302.01°$ **27.** $66.40°$, $293.60°$ **29.** $15.82°$, $195.82°$ **31.** $102.00°$, $282.00°$
33. $119.45°$ **35.** $263.07°$ **37.** $306.21°$ **39.** $299.24°$ **41.** -0.7003 **43.** -0.7771
45. $<$ **47.** $=$ **49.** 72.7 lb   **51.** 416

53. $\cos(-\theta) = \dfrac{x}{r}$; $\tan(-\theta) = \dfrac{-y}{x}$; $\cot(-\theta) = \dfrac{x}{-y}$; $\sec(-\theta) = \dfrac{r}{x}$; $\csc(-\theta) = \dfrac{r}{-y}$

55. (a) 5.671 (b) -1.428

Exercises 7–3, p. 216

1. $\dfrac{\pi}{12}, \dfrac{5\pi}{6}$ **3.** $\dfrac{5\pi}{12}, \dfrac{11\pi}{6}$ **5.** $\dfrac{7\pi}{6}, \dfrac{3\pi}{2}$ **7.** $\dfrac{8\pi}{9}, \dfrac{13\pi}{9}$ **9.** $72°$, $270°$ **11.** $10°$, $315°$ **13.** $170°$, $300°$

15. $15°$, $27°$ **17.** 0.401 **19.** 4.40 **21.** 5.82 **23.** 3.12 **25.** $43.0°$ **27.** $171.9°$ **29.** $140.4°$
31. $939.7°$ **33.** 0.7071 **35.** 3.732 **37.** -0.8660 **39.** -8.327 **41.** 0.9056 **43.** -0.890
45. -0.479 **47.** -0.149 **49.** 0.3141, 2.827 **51.** 2.932, 6.074 **53.** 0.8309, 5.452
55. 2.442, 3.841 **57.** $1.43\pi = 4.49$ (closest 3-sig.-digit result)   **59.** 15.7 cm/s

Exercises 7–4, p. 221

1. 10.5 in.   **3.** 52.4 in.^{2} **5.** $0.382 = 21.9°$ **7.** 4240 ft^2 **9.** 0.52 rad/s   **11.** 34.73 m^2
13. 0.262 ft   **15.** 129 km   **17.** 5650 cm   **19.** 25.7 ft   **21.** 86.2 rad/s   **23.** 3710 mi/h
25. -2.145 cm   **27.** 5200 ft/min   **29.** 2.60×10^{-6} rad/s   **31.** $25,100$ ft/min   **33.** 69.00 cm^2
35. 4.848×10^{-6} (all three values)   **37.** 7.13×10^{7} mi

Exercises for Chapter 7, p. 224

1. $\sin\theta = \frac{4}{5}$, $\cos\theta = \frac{3}{5}$, $\tan\theta = \frac{4}{3}$, $\cot\theta = \frac{3}{4}$, $\sec\theta = \frac{5}{3}$, $\csc\theta = \frac{5}{4}$

3. $\sin\theta = -\dfrac{2}{\sqrt{53}}$, $\cos\theta = \dfrac{7}{\sqrt{53}}$, $\tan\theta = -\frac{2}{7}$, $\cot\theta = -\frac{7}{2}$, $\sec\theta = \dfrac{\sqrt{53}}{7}$, $\csc\theta = -\dfrac{\sqrt{53}}{2}$

5. $-\cos 48°$; $\tan 14°$ **7.** $-\sin 71°$; $\sec 15°$ **9.** $\dfrac{2\pi}{9}; \dfrac{17\pi}{20}$ **11.** $\dfrac{4\pi}{15}; \dfrac{9\pi}{8}$ **13.** $252°$; $130°$

15. $12°$; $330°$ **17.** $32.1°$ **19.** $206.7°$ **21.** 1.745 **23.** 0.3534 **25.** 4.574 **27.** 2.38
29. -0.4147 **31.** -0.4678 **33.** -1.080 **35.** -0.4264 **37.** -1.638 **39.** 4.140
41. -0.5878 **43.** -0.8660 **45.** 0.5569 **47.** 1.197 **49.** $10.30°$, $190.30°$
51. $118.23°$, $241.77°$ **53.** 0.5759, 5.707 **55.** 4.187, 5.238 **57.** $223.76°$ **59.** $246.78°$
61. -120 V   **63.** 0.4292 ft   **65.** 108 cm^2 **67.** 188 in./s   **69.** 18.98 cm   **71.** 6.702 mm
73. $84,800$ cm/min   **75.** 3.58×10^{5} km

Exercises 8–1, p. 231

1.   **3.**   **5.**   **7.**   **9.**   **11.**   **13.**

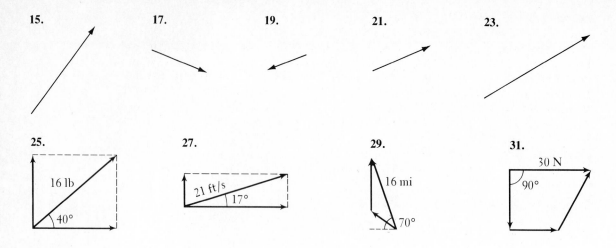

15. **17.** **19.** **21.** **23.**

25. **27.** **29.** **31.**

16 lb 40° · 21 ft/s 17° · 16 mi 70° · 30 N 90°

Exercises 8–2, p. 234

1. 662, 352 **3.** −349, −664 **5.** 3.22, 7.97 **7.** −62.9, 44.1 **9.** 2.08, −8.80
11. −2.53, −0.788 **13.** −0.8088, 0.3296 **15.** 88,920, 12,240 **17.** 2040 km/h, 951 km/h
19. 43.4 lb, −13.7 lb **21.** 115 km, 88.3 km **23.** 18,530 mi/h, −825.3 mi/h

Exercises 8–3, p. 239

1. $R = 24.2$, $\theta = 52.6°$ with A **3.** $R = 7.781$, $\theta = 66.63°$ with A **5.** $R = 10.0$, $\theta = 58.8°$
7. $R = 2.74$, $\theta = 111.0°$ **9.** $R = 2130$, $\theta = 107.7°$ **11.** $R = 1.426$, $\theta = 299.12°$
13. $R = 29.2$, $\theta = 10.8°$ **15.** $R = 47.0$, $\theta = 101.0°$ **17.** $R = 27.27$, $\theta = 33.14°$
19. $R = 12.735$, $\theta = 25.216°$ **21.** $R = 50.1$, $\theta = 50.3°$ **23.** $R = 233$, $\theta = 124.5°$

Exercises 8–4, p. 243

1. 81.5 lb, 56.5° from 45-lb force **3.** 848.0 N, 10.06° from second force **5.** 510 mi, 28.1° S of W
7. 37.79 mi, 77.14° S of W **9.** 92.5 ft/s, 51.6° above horizontal **11.** 31.6 lb, 71.6°
13. 206 m/s, 14.0° from direction of plane **15.** 34.8°, 57.4 lb **17.** 71.3 lb
19. 128 m/s, 20.2° from horizontal **21.** 6.671 rad/s^2 **23.** 4.06 A/m, 11.6° with magnet

Exercises 8–5, p. 249

1. $b = 38.1$, $C = 66.0°$, $c = 46.1$ **3.** $a = 2800$, $b = 2620$, $C = 108.0°$
5. $B = 12.20°$, $C = 149.57°$, $c = 7.448$ **7.** $a = 110.5$, $A = 149.70°$, $C = 9.57°$
9. $A = 125.64°$, $a = 0.0776$, $c = 0.00560$ **11.** $A = 99.4°$, $b = 55.1$, $c = 24.4$
13. $A = 68.01°$, $a = 5520$, $c = 5376$
15. $A_1 = 61.36°$, $C_1 = 70.51°$, $c_1 = 5.628$; $A_2 = 118.64°$, $C_2 = 13.23°$, $c_2 = 1.366$
17. $A_1 = 107.3°$, $a_1 = 5280$, $C_1 = 41.3°$; $A_2 = 9.9°$, $a_2 = 952$, $C_2 = 138.7°$ **19.** No solution
21. 7.83 ft, 10.6 ft **23.** 884 N **25.** 25,200 ft **27.** 23.2 mi **29.** 27,300 km

Exercises 8–6, p. 254

1. $A = 50.3°$, $B = 75.7°$, $c = 6.31$ **3.** $A = 70.9°$, $B = 11.1°$, $c = 4750$
5. $A = 34.72°$, $B = 40.67°$, $C = 104.61°$ **7.** $A = 18.21°$, $B = 22.28°$, $C = 139.51°$
9. $A = 6.0°$, $B = 16.0°$, $c = 1150$ **11.** $A = 82.3°$, $b = 21.6$, $C = 11.4°$
13. $A = 39.09°$, $B = 36.24°$, $b = 97.22$ **15.** $A = 46.94°$, $B = 61.82°$, $C = 71.24°$
17. $A = 137.9°$, $B = 33.7°$, $C = 8.4°$ **19.** $b = 29.5$, $C = 14.4°$, $c = 14.9$ **21.** 1140 ft
23. 0.039 km **25.** 96.93° **27.** 8.88 km/h, 13.4° with bank **29.** 42.53°

Exercises for Chapter 8, p. 255

1. $A_x = 57.4$, $A_y = 30.5$ **3.** $A_x = -0.7485$, $A_y = -0.5357$ **5.** $R = 602$, $\theta = 57.1°$ with A
7. $R = 5960$, $\theta = 33.60°$ with A **9.** $R = 965$, $\theta = 8.6°$ **11.** $R = 26.12$, $\theta = 146.03°$
13. $R = 71.93$, $\theta = 336.50°$ **15.** $R = 99.42$, $\theta = 359.57°$ **17.** $b = 18.1$, $C = 64.0°$, $c = 17.5$
19. $A = 21.2°$, $b = 34.8$, $c = 51.5$ **21.** $a = 17{,}340$, $b = 24{,}660$, $C = 7.99°$
23. $A = 39.88°$, $a = 51.94$, $C = 30.03°$
25. $A_1 = 54.8°$, $a_1 = 12.7$, $B_1 = 68.6°$; $A_2 = 12.0°$, $a_2 = 3.24$, $B_2 = 111.4°$
27. $A = 32.3°$, $b = 267$, $C = 17.7°$ **29.** $A = 148.7°$, $B = 9.3°$, $c = 5.66$
31. $a = 1782$, $b = 1920$, $C = 16.00°$ **33.** $A = 33.95°$, $B = 24.48°$, $C = 121.57°$
35. $A = 20.6°$, $B = 35.6°$, $C = 123.8°$ **37.** -155.7, 81.14 **39.** 0.0291 h
41. 27.0 ft/s, $33.7°$ with horizontal **43.** 269 mi/h, 468 mi/h **45.** 1067 ft **47.** 174.4 m
49. $57.7°$ **51.** 186 mi **53.** 109.6 lb **55.** 310.1 lb, $329.75°$

Exercises 9–1, p. 261

1. $0, -0.7, -1, -0.7, 0, 0.7, 1, 0.7, 0, -0.7, -1, -0.7, 0, 0.7, 1, 0.7, 0$

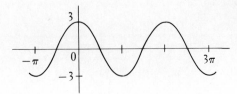

3. $-3, -2.1, 0, 2.1, 3, 2.1, 0, -2.1, -3, -2.1, 0, 2.1, 3, 2.1, 0, -2.1, -3$

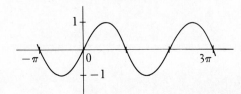

5.

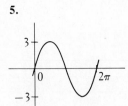

7.

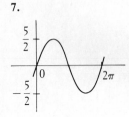

9.

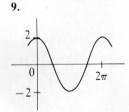

11.

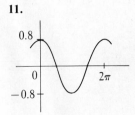

13.

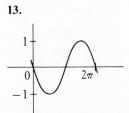

15.

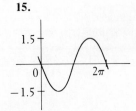

17.

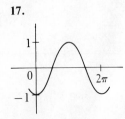

19.

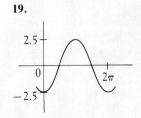

21. 0, 0.84, 0.91, 0.14, −0.76, −0.96, −0.28, 0.66   **23.** 1, 0.54, −0.42, −0.99, −0.65, 0.28, 0.96, 0.75

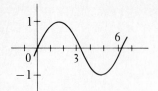   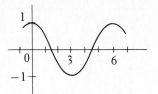

Exercises 9–2, p. 265

1. $\dfrac{\pi}{3}$ **3.** $\dfrac{\pi}{4}$ **5.** $\dfrac{\pi}{6}$ **7.** $\dfrac{\pi}{8}$ **9.** 1   **11.** $\frac{1}{2}$ **13.** 6π **15.** 3π **17.** 3   **19.** $\dfrac{2}{\pi}$

21.

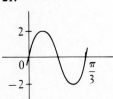

23.

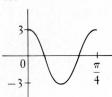

25.

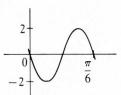

27.

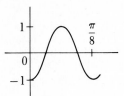

29.

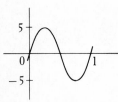

31.

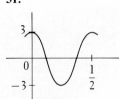

33.

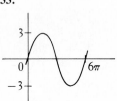

35.

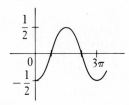

37.

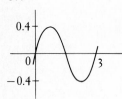

39.

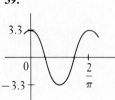

41.

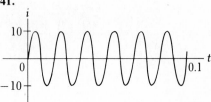

43.

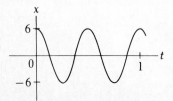

Exercises 9–3, p. 268

1. $1, 2\pi, \dfrac{\pi}{6}$

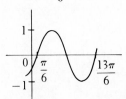

3. $1, 2\pi, -\dfrac{\pi}{6}$

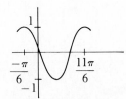

5. $2, \pi, -\dfrac{\pi}{4}$

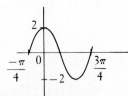

7. $1, \pi, \dfrac{\pi}{2}$

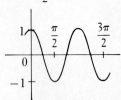

9. $\dfrac{1}{2}, 4\pi, \dfrac{\pi}{2}$

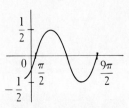

11. $3, 6\pi, -\pi$

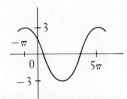

13. $1, 2, -\dfrac{1}{8}$

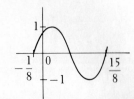

15. $\dfrac{3}{4}, \dfrac{1}{2}, \dfrac{1}{20}$

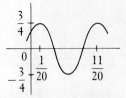

17. $0.6, 1, \dfrac{1}{2\pi}$

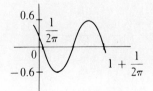

19. $4, \dfrac{2}{3}, -\dfrac{2}{3\pi}$

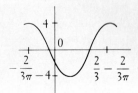

21. $1, \dfrac{2}{\pi}, \dfrac{1}{\pi}$

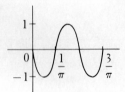

23. $\dfrac{3}{2}, 2, -\dfrac{\pi}{6}$

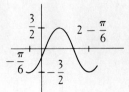

25.

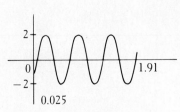

27.

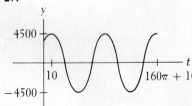

Exercises 9–4, p. 272

1. Undef., -1.7, -1, -0.58, 0, 0.58,
  1, 1.7, undef., -1.7, -1, -0.58, 0

3. Undef., 2, 1.4, 1.2, 1, 1.2, 1.4, 2,
  undef., -2, -1.4, -1.2, -1

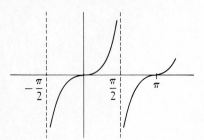

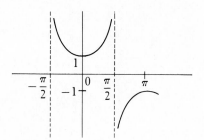

5.

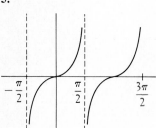

7.

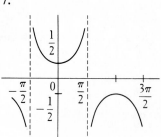

9.

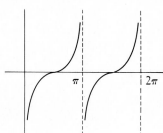

11.

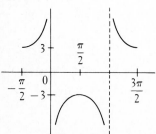

13.

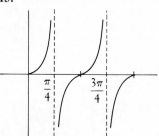

15.

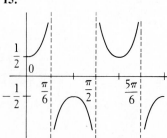

17.

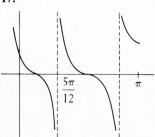

19.

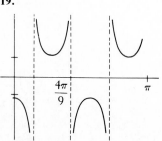

21.

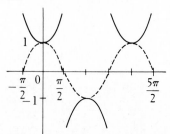

23.

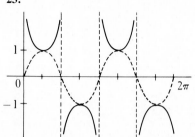

25.

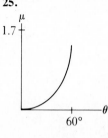

27.

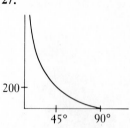

Exercises 9–5, p. 277

1.

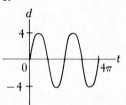

3.

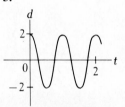

5.

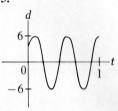

7.

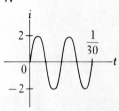

9.

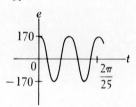

11.

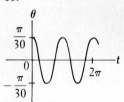

13.

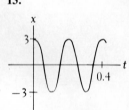

15.

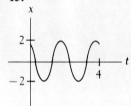

Exercises 9–6, p. 282

1.

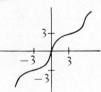

3.

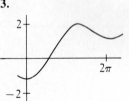

5.

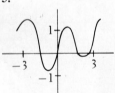

7.

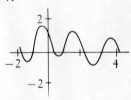

9.

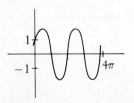

11.

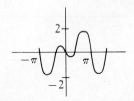

13.

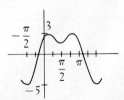

15.

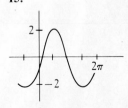

17.

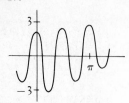

19.

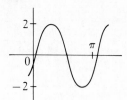

21.

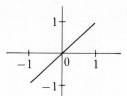

23.

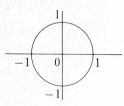

25.

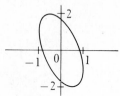

27.

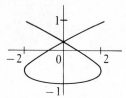

29.

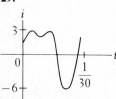

31.

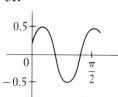

33.

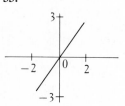

Exercises for Chapter 9, p. 283

1.

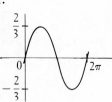

3.

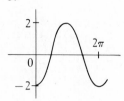

5.

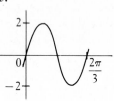

7.

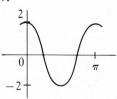

9.

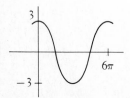

11.

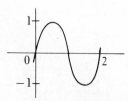

13.

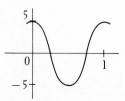

15.

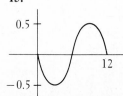

17.

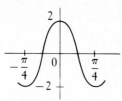

19.

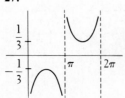

21.

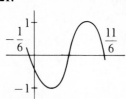

23.

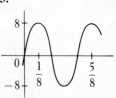

25.

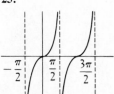

27.

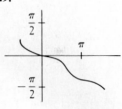

29.

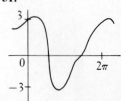

31.

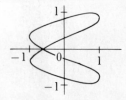

33.

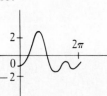

35.

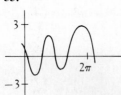

37.

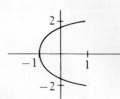

39.

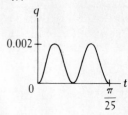

41.

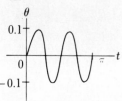

43.

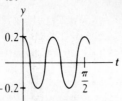

45.

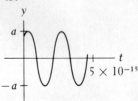

47.

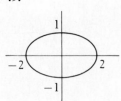

49.

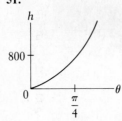

51.

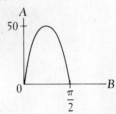

53.

Exercises 10–1, p. 289

1. x^3 **3.** $\dfrac{1}{a^4}$ **5.** $\frac{1}{25}$ **7.** $\frac{25}{4}$ **9.** $\dfrac{4a^2}{x^2}$ **11.** $\dfrac{n^2}{5a}$ **13.** 1   **15.** -7 **17.** $\dfrac{3}{x^2}$ **19.** $\dfrac{1}{7^3 a^3 x^3}$

21. $\dfrac{n^3}{2}$ **23.** $\dfrac{1}{a^3 b^6}$ **25.** $\dfrac{1}{a+b}$ **27.** $\dfrac{2x^2 + 3y^2}{x^2 y^2}$ **29.** $\dfrac{2^3}{3^5}$ **31.** $\dfrac{a^3 b^2}{3}$ **33.** $\dfrac{b^3}{432a}$ **35.** $\dfrac{4}{t^4 v^4}$

37. $\dfrac{x^8 - y^2}{x^4 y^2}$ **39.** $\dfrac{36a^2 + 1}{9a^4}$ **41.** $\frac{10}{9}$ **43.** $\dfrac{ab}{a + b}$ **45.** $\dfrac{4n^2 - 4n + 1}{n^4}$ **47.** $\frac{45}{2}$ **49.** $-\dfrac{x}{y}$

51. $\dfrac{a^2 - ax + x^2}{ax}$ **53.** $\dfrac{t^2 + t + 2}{t^2}$ **55.** $\dfrac{2x}{(x + 1)(x - 1)}$ **57.** $\dfrac{rR}{r + R}$ **59.** $\dfrac{r_1 r_2}{(\mu - 1)(r_2 - r_1)}$

61. $\left(\dfrac{a}{b}\right)^{-n} = \dfrac{1}{\left(\dfrac{a}{b}\right)^n} = \dfrac{1}{\dfrac{a^n}{b^n}} = \dfrac{b^n}{a^n} = \left(\dfrac{b}{a}\right)^n$

Exercises 10–2, p. 293

1. 5   **3.** 3   **5.** 16   **7.** 10^{25} **9.** $\frac{1}{2}$ **11.** $\frac{1}{16}$ **13.** 25   **15.** 4096   **17.** $\frac{1}{110}$ **19.** $\frac{6}{7}$ **21.** $-\frac{1}{2}$

23. 24   **25.** $\frac{39}{1000}$ **27.** $\frac{9}{100}$ **29.** 2.059   **31.** 0.53891   **33.** $a^{7/6}$ **35.** $\dfrac{1}{y^{9/10}}$ **37.** $s^{23/12}$ **39.** $\dfrac{1}{y^{13/12}}$

41. $2ab^2$ **43.** $\dfrac{1}{8a^3 b^{9/4}}$ **45.** $\dfrac{4x}{(4x^2 + 1)^{1/2}}$ **47.** $\dfrac{27}{64t^3}$ **49.** $\dfrac{b^{11/10}}{2a^{1/12}}$ **51.** $\frac{2}{3}x^{1/6} y^{11/12}$ **53.** $\dfrac{x}{(x + 2)^{1/2}}$

55. $\dfrac{a^2 + 1}{a^4}$ **57.** $\dfrac{a + 1}{a^{1/2}}$ **59.** $\dfrac{5x^2 - 2x}{(2x - 1)^{1/2}}$ **61.** 4.72×10^{22} **63.** 55.0%

Exercises 10–3, p. 298

1. $2\sqrt{6}$ **3.** $3\sqrt{5}$ **5.** $xy^2\sqrt{y}$ **7.** $qr^3\sqrt{pr}$ **9.** $x\sqrt{5}$ **11.** $3ac^2\sqrt{2ab}$ **13.** $2\sqrt[3]{2}$ **15.** $2\sqrt[5]{3}$

17. $2\sqrt[3]{a^2}$ **19.** $2st\sqrt[4]{4r^3 t}$ **21.** 2   **23.** $ab\sqrt[3]{b^2}$ **25.** $\frac{1}{2}\sqrt{6}$ **27.** $\dfrac{\sqrt{ab}}{b}$ **29.** $\frac{1}{2}\sqrt[3]{6}$ **31.** $\frac{1}{3}\sqrt[5]{27}$

33. $2\sqrt{5}$ **35.** 2   **37.** 200   **39.** 2000   **41.** $\sqrt{2a}$ **43.** $\frac{1}{2}\sqrt{2}$ **45.** $\sqrt[3]{2}$ **47.** $\sqrt[8]{2}$ **49.** $\frac{1}{6}\sqrt{6}$

51. $\dfrac{\sqrt{b(a^2 + b)}}{ab}$ **53.** $a + b$ **55.** $\frac{1}{2}\sqrt{4x^2 + 1}$ **57.** $\dfrac{\pi\sqrt{6}}{4} = 1.92$ s   **59.** $\sqrt{2ag}$

Exercises 10–4, p. 301

1. $7\sqrt{3}$ **3.** $\sqrt{5} - \sqrt{7}$ **5.** $3\sqrt{5}$ **7.** $-4\sqrt{3}$ **9.** $-2\sqrt{2}$ **11.** $19\sqrt{7}$ **13.** $-4\sqrt{5}$
15. $23\sqrt{3} - 6\sqrt{2}$ **17.** $\frac{7}{3}\sqrt{15}$ **19.** 0   **21.** $13\sqrt[3]{3}$ **23.** $\sqrt[4]{2}$ **25.** $(a - 2b^2)\sqrt{ab}$

27. $(3 - 2a)\sqrt{10}$ **29.** $(2b - a)\sqrt[3]{3a^2 b}$ **31.** $\dfrac{(a^2 - c^3)\sqrt{ac}}{a^2 c^3}$ **33.** $\dfrac{(a - 2b)\sqrt[3]{ab^2}}{ab}$ **35.** $\dfrac{2b\sqrt{a^2 - b^2}}{b^2 - a^2}$

37. $15\sqrt{3} - 11\sqrt{5} = 1.3840144$ **39.** $\frac{1}{6}\sqrt{6} = 0.4082483$ **41.** $-\dfrac{b}{a}$

Exercises 10–5, p. 304

1. $\sqrt{30}$ **3.** $2\sqrt{3}$ **5.** 2   **7.** $2\sqrt[5]{2}$ **9.** 50   **11.** 16   **13.** $\frac{1}{3}\sqrt{30}$ **15.** $\frac{1}{33}\sqrt{165}$
17. $\sqrt{6} - \sqrt{15}$ **19.** $8 - 12\sqrt{3}$ **21.** -1 **23.** $39 - 12\sqrt{3}$ **25.** $48 + 9\sqrt{15}$ **27.** $36 + 13\sqrt{66}$
29. $a\sqrt{b} + \sqrt{ac}$ **31.** $5n\sqrt{3} + 10\sqrt{mn}$ **33.** $2a - 3b + 2\sqrt{2ab}$ **35.** $\sqrt{6} - \sqrt{10} - 2$ **37.** $\sqrt[6]{72}$

39. $\sqrt[12]{a^3 b^7 c^4}$ **41.** 1   **43.** $2x\sqrt{2x} + x\sqrt[6]{8y^4} - 2\sqrt[6]{x^3 y^2} - y$ **45.** $\dfrac{2 - a - a^2}{a}$

47. $4x^2 + x - 2y - 4x\sqrt{x - 2y}$ (valid for $x \geq 2y$)   **49.** $-1 - \sqrt{66} = -9.1240384$

51. $158 + 9\sqrt{182} = 279.41664$ **53.** $\dfrac{5x^2 + 2x}{\sqrt{2x + 1}}$ **55.** $\dfrac{c}{a}$

Exercises 10–6, p. 306

1. $\sqrt{7}$ **3.** $\frac{1}{2}\sqrt{14}$ **5.** $\frac{1}{6}\sqrt[3]{9x^2}$ **7.** $\dfrac{\sqrt[6]{200}}{2}$ **9.** $\frac{1}{2}\sqrt[6]{4a^3}$ **11.** $\dfrac{2 - \sqrt{6}}{2}$ **13.** $\dfrac{a\sqrt{2} - b\sqrt{a}}{a}$

15. $\dfrac{3\sqrt{a} - \sqrt{3b}}{3}$ **17.** $\frac{1}{4}(\sqrt{7} - \sqrt{3})$ **19.** $\frac{1}{3}(\sqrt{35} + \sqrt{14})$ **21.** $-\frac{3}{8}(\sqrt{5} + 3)$ **23.** $\frac{1}{13}(9 + \sqrt{3})$

25. $\frac{1}{11}(\sqrt{7} + 3\sqrt{2} - 6 - \sqrt{14})$ **27.** $\frac{1}{13}(4 - \sqrt{3})$ **29.** $\frac{1}{17}(-56 + 9\sqrt{15})$ **31.** $\frac{1}{14}(4 - \sqrt{2})$

33. $\frac{2x + 2\sqrt{xy}}{x - y}$ **35.** $\frac{8(3\sqrt{a} + 2\sqrt{b})}{9a - 4b}$ **37.** $\frac{2c + 4\sqrt{2cd} + 3d}{2c - d}$ **39.** $-\frac{\sqrt{x^2 - y^2} + \sqrt{x^2 + xy}}{y}$

41. $14 - 2\sqrt{42} = 1.0385186$ **43.** $-\frac{16 + 5\sqrt{30}}{26} = -1.6686972$ **45.** $\frac{2500 - 50\sqrt{V}}{2500 - V}$

Exercises for Chapter 10, p. 307

1. $\frac{2}{a^2}$ **3.** $\frac{2d^3}{c}$ **5.** 375   **7.** $\frac{1}{8000}$ **9.** $\frac{t^4}{9}$ **11.** -28 **13.** $64a^2b^5$ **15.** $-8m^9n^6$

17. $\frac{2y - x^2}{x^2y}$ **19.** $\frac{2y}{x + y}$ **21.** $\frac{b}{ab - 3}$ **23.** $\frac{(x^3y^3 - 1)^{1/3}}{y}$ **25.** $4a(a^2 + 4)^{1/2}$ **27.** $\frac{-2(x + 1)}{(x - 1)^3}$

29. $2\sqrt{17}$ **31.** $b^2c\sqrt{ab}$ **33.** $3ab^2\sqrt{a}$ **35.** $2tu\sqrt{21st}$ **37.** $\frac{5\sqrt{2s}}{2s}$ **39.** $\frac{1}{9}\sqrt{33}$ **41.** $mn^2\sqrt[4]{8m^2n}$

43. $\sqrt{2}$ **45.** $14\sqrt{2}$ **47.** $-7\sqrt{7}$ **49.** $3ax\sqrt{2x}$ **51.** $(2a + b)\sqrt[3]{a}$ **53.** $10 - \sqrt{55}$

55. $4\sqrt{3} - 4\sqrt{5}$ **57.** $-45 - 7\sqrt{17}$ **59.** $33 - 7\sqrt{21}$ **61.** $\frac{6x + \sqrt{3xy}}{12x - y}$ **63.** $-\frac{8 + \sqrt{6}}{29}$

65. $\frac{13 - 2\sqrt{35}}{29}$ **67.** $\frac{6x - 13a\sqrt{x} + 5a^2}{9x - 25a^2}$ **69.** $\frac{\sqrt{a^2b^2 + a}}{a}$ **71.** $\frac{15 - 2\sqrt{15}}{4}$

73. $2\sqrt{13} + 5\sqrt{6} = 19.458551$ **75.** $51 - 7\sqrt{105} = -20.728655$ **77.** $\frac{e^{i\alpha t}}{e^{i\omega t}}$ **79.** 0.011

81. $\frac{\sqrt[3]{4MN^2\rho^2}}{2N\rho}$ **83.** $\frac{\sqrt{3RMT}}{M}$; 483 m/s   **85.** $\frac{\sqrt{LC_1C_2(C_1 + C_2)}}{2\pi LC_1C_2}$

Exercises 11-1, p. 313

1. $9j$ **3.** $-2j$ **5.** $0.6j$ **7.** $2j\sqrt{2}$ **9.** $\frac{1}{2}j\sqrt{7}$ **11.** $-j\sqrt{\frac{2}{5}} = -\frac{1}{5}j\sqrt{10}$ **13.** $-7; 7$

15. $4; -4$ **17.** $-j$ **19.** 1   **21.** 0   **23.** $-2j$ **25.** $2 + 3j$ **27.** $-2 + 3j$
29. $3\sqrt{2} - 2\sqrt{2}j$ **31.** -1 **33.** $6 + 7j$ **35.** $-2j$ **37.** $x = 2, y = -2$ **39.** $x = 10, y = -6$
41. $x = 0, y = -1$ **43.** $x = -2, y = 3$ **45.** Yes   **47.** It is a real number.

Exercises 11-2, p. 316

1. $5 - 8j$ **3.** $-9 + 6j$ **5.** $7 - 5j$ **7.** $-5j$ **9.** -1 **11.** $-8 + 21j$ **13.** $7 + 49j$
15. $-22.4 + 6.4j$ **17.** $22 + 3j$ **19.** $-42 - 6j$ **21.** $-18j\sqrt{2}$ **23.** $25j\sqrt{5}$ **25.** $3j\sqrt{3}$
27. $-4\sqrt{3} + 6j\sqrt{7}$ **29.** $-28j$ **31.** $3\sqrt{7} + 3j$ **33.** $-40 - 42j$ **35.** $-2 - 2j$
37. $\frac{1}{29}(-30 + 12j)$ **39.** $0.075 + 0.025j$ **41.** $-j$ **43.** $\frac{1}{11}(-13 + 8j\sqrt{2})$ **45.** $\frac{1}{3}(2 + 5j\sqrt{2})$
47. $\frac{1}{5}(-1 + 3j)$ **49.** $38 + 4j$ **51.** $(-1 + j)^2 + 2(-1 + j) + 2 = 1 - 2j + j^2 - 2 + 2j + 2 = 0$
53. $(a + bj) + (a - bj) = 2a$ **55.** $(a + bj) - (a - bj) = 2bj$

Exercises 11-3, p. 319

1.   **3.**   **5.** $5 + 4j$ **7.** $8 + j$

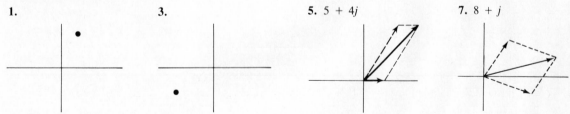

9. $4 + 4j$

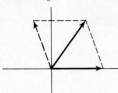

11. $-3j$

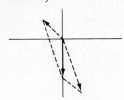

13. $-1 + 4j$

15. $-2 + 3j$

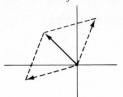

17. $7 + j$

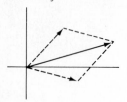

19. $4 - 11j$

21. $-2j$

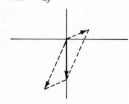

23. $-13 + j$

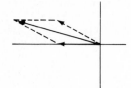

25.

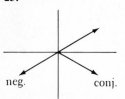

27.

conj.   neg.

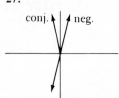

29.

$-6 + 3j$

$-2 + j$ •

$6 - 3j$ •

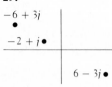

31.

$-9 + 3j$ •

$3 - j$

$9 - 3j$

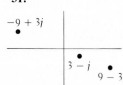

neg.   conj.

Exercises 11–4, p. 323

1. $10(\cos 36.9° + j \sin 36.9°)$

3. $5(\cos 306.9° + j \sin 306.9°)$

5. $3.61(\cos 123.7° + j \sin 123.7°)$

7. $6.00(\cos 203.6° + j \sin 203.6°)$

9. $2(\cos 60° + j \sin 60°)$

11. $8.062(\cos 295.84° + j \sin 295.84°)$

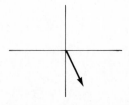

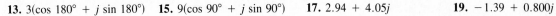

13. $3(\cos 180° + j \sin 180°)$ **15.** $9(\cos 90° + j \sin 90°)$ **17.** $2.94 + 4.05j$ **19.** $-1.39 + 0.800j$

   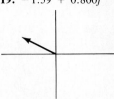

21. -6 **23.** 8   **25.** $11.97 - 3.082j$ **27.** $-0.500 - 0.866j$

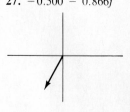

29. $-4.71 + 0.595j$ **31.** $-0.6052 - 0.7096j$ **33.** $7.32j$ **35.** $-6.961 + 86.14j$

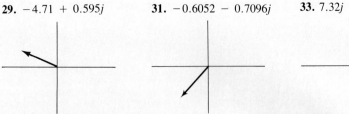 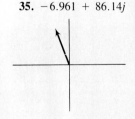

37. $R = 0.200$ mm, $\theta = 53.1°$ **39.** $28.0 + 3.00j$ kV

Exercises 11–5, p. 327
1. $3.00e^{1.05j}$ **3.** $4.50e^{4.93j}$ **5.** $375.5e^{1.666j}$ **7.** $0.515e^{3.46j}$ **9.** $4.06e^{1.77j}$ **11.** $9245e^{5.172j}$ **13.** $5.00e^{5.36j}$
15. $3.61e^{2.55j}$ **17.** $6.37e^{0.386j}$ **19.** $825.7e^{3.836j}$ **21.** $3.00\angle 28.6°$; $2.63 + 1.44j$
23. $4.64\angle 106.0°$; $-1.28 + 4.46j$ **25.** $3.20\angle 310.0°$; $2.06 - 2.45j$
27. $0.1724\angle 136.99°$; $-0.1261 + 0.1176j$ **29.** $0.546e^{0.415j}$; 0.546 A

Exercises 11–6, p. 333
1. $8(\cos 80° + j \sin 80°)$ **3.** $3(\cos 250° + j \sin 250°)$ **5.** $2(\cos 35° + j \sin 35°)$
7. $2.4(\cos 110° + j \sin 110°)$ **9.** $8(\cos 105° + j \sin 105°)$ **11.** $256(\cos 0° + j \sin 0°)$
13. $0.305 + 1.70j = 1.73\angle 79.8°$ **15.** $-5134 + 10{,}570j = 11{,}750 (\cos 115.91° + j \sin 115.91°)$
17. $65.0(\cos 345.7° + j \sin 345.7°) = 63 - 16j$ **19.** $61.4(\cos 343.9° + j \sin 343.9°)$; $59 - 17j$
21. $2.21(\cos 71.6° + j \sin 71.6°)$; $\frac{7}{10} + \frac{21}{10}j$ **23.** $0.385(\cos 120.5° + j \sin 120.5°) = \frac{1}{169}(-33 + 56j)$
25. $625(\cos 212.5° + j \sin 212.5°) = -527 - 336j$ **27.** $609(\cos 281.5° + j \sin 281.5°)$; $122 - 597j$
29. $2(\cos 30° + j \sin 30°)$, $2(\cos 210° + j \sin 210°)$ **31.** $-0.364 + 1.67j$, $-1.26 - 1.15j$, $1.63 - 0.520j$
33. $1, -1, j, -j$ **35.** $j, -\frac{1}{2}(\sqrt{3} + j), \frac{1}{2}(\sqrt{3} - j)$
37. $[\frac{1}{2}(1 - j\sqrt{3})]^3 = \frac{1}{8}[1 - 3(j\sqrt{3}) + 3(j\sqrt{3})^2 - (j\sqrt{3})^3] = \frac{1}{8}[1 - 3j\sqrt{3} - 9 + 3j\sqrt{3}] = \frac{1}{8}(-8) = -1$

Exercises 11–7, p. 340
1. 30.0V   **3.** (a) 18.0 Ω   (b) $-56.3°$ (c) 54.0 V   **5.** (a) 3.00 Ω   (b) $-90.0°$
7. (a) 11.7 Ω   (b) $-59.0°$ **9.** 44.4 V   **11.** 1040 Ω, 6118 Ω   **13.** 6638 Ω, voltage leads by 49.91°
15. 2.00×10^{-4} F   **17.** 1.30×10^{-11} F = 13.0 pF   **19.** 3.88 W

Exercises for Chapter 11, p. 341

1. $10 - j$ **3.** $6 + 2j$ **5.** $9 + 2j$ **7.** $-12 + 66j$ **9.** $\frac{1}{85}(21 + 18j)$ **11.** $\frac{1}{53}(50 - 16j)$

13. $\frac{1}{10}(-12 + 9j)$ **15.** $\frac{1}{5}(13 + 11j)$ **17.** $x = -\frac{2}{3}, y = -2$ **19.** $x = -\frac{1}{2}, y = \frac{1}{2}$

21. $3 + 11j$ **23.** $4 + 8j$

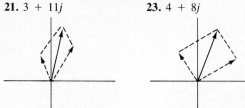

25. $1.41(\cos 315° + j \sin 315°) = 1.41e^{5.50j}$

27. $7.28(\cos 254.1° + j \sin 254.1°) = 7.28e^{4.43j}$

29. $4.67(\cos 76.8° + j \sin 76.8°); 4.67e^{1.34j}$

31. $10(\cos 0° + j \sin 0°); 10e^{0j}$ **33.** $-1.41 - 1.41j$

35. $-2.789 + 4.163j$ **37.** $0.2 + 0.6j$

39. $26.31 - 6.427j$ **41.** $1.94 + 0.495j$ **43.** $-8.346 + 23.96j$ **45.** $15(\cos 84° + j \sin 84°)$

47. $20\underline{/263°}$ **49.** $8(\cos 59° + j \sin 59°)$ **51.** $14.29\underline{/133.61°}$ **53.** $1.26\underline{/59.7°}$ **55.** $9682\underline{/249.52°}$

57. $1024(\cos 160° + j \sin 160°)$ **59.** $27\underline{/331.5°}$ **61.** $32(\cos 270° + j \sin 270°) = -32j$

63. $\dfrac{625}{2}(\cos 270° + j \sin 270°) = -\dfrac{625}{2}j$ **65.** $1.00 + 1.73j, -2, 1.00 - 1.73j$

67. $\cos 67.5° + j \sin 67.5°, \cos 157.5° + j \sin 157.5°, \cos 247.5° + j \sin 247.5°, \cos 337.5° + j \sin 337.5°$

69. $12.5\ \Omega, -53.1°$ **71.** $10.1\ \Omega$, current leads by $8.7°$ **73.** 40.09 V   **75.** 814 lb, $317.5°$

77. $\dfrac{u - j\omega n}{u^2 + \omega^2 n^2}$ **79.** $e^{j\pi} = \cos \pi + j \sin \pi = -1$

Exercises 12–1, p. 346

1. 3 **3.** $\frac{1}{81}$ **5.** $\log_3 27 = 3$ **7.** $\log_4 256 = 4$ **9.** $\log_4(\frac{1}{16}) = -2$ **11.** $\log_2(\frac{1}{64}) = -6$

13. $\log_8 2 = \frac{1}{3}$ **15.** $\log_{1/4}(\frac{1}{16}) = 2$ **17.** $81 = 3^4$ **19.** $9 = 9^1$ **21.** $5 = 25^{1/2}$ **23.** $3 = 243^{1/5}$

25. $0.1 = 10^{-1}$ **27.** $16 = (0.5)^{-4}$ **29.** 2 **31.** -2 **33.** 343 **35.** $\frac{1}{4}$ **37.** 9 **39.** $\frac{1}{64}$

41. 0.2 **43.** -3 **45.** $t = \log_2\left(\dfrac{N}{1000}\right)$ **47.** $N = N_0 e^{-kt}$

Exercises 12–2, p. 350

1.   **5.**   **7.**   **9.**

11. 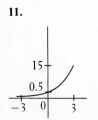   **13.** 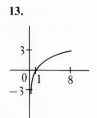   **15.** 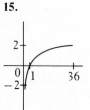   **17.** 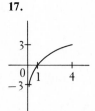   **19.**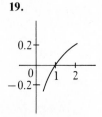

21.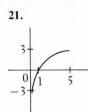

23.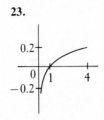

25.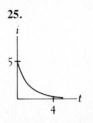

27.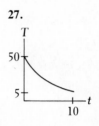

29.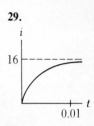

Exercises 12–3, p. 355

1. $\log_5 x + \log_5 y$ **3.** $\log_7 5 - \log_7 a$ **5.** $3 \log_2 a$ **7.** $\log_6 a + \log_6 b + \log_6 c$ **9.** $\frac{1}{4}\log_5 y$
11. $\frac{1}{2}\log_2 x - 2\log_2 a$ **13.** $\log_b ac$ **15.** $\log_5 3$ **17.** $\log_b x^{3/2}$ **19.** $\log_e 4n^3$ **21.** -5 **23.** 2.5
25. $\frac{1}{2}$ **27.** $\frac{3}{4}$ **29.** $2 + \log_3 2$ **31.** $-1 - \log_2 3$ **33.** $\frac{1}{2}(1 + \log_3 2)$ **35.** $4 + 3\log_2 3$

37. $3 + \log_{10} 3$ **39.** $-2 + 3\log_{10} 3$ **41.** $y = 2x$ **43.** $y = \dfrac{x}{5}$ **45.** $y = \dfrac{49}{x^3}$

47. $y = 2(2ax)^{1/5}$ **49.** $y = \dfrac{2}{x}$ **51.** $y = \dfrac{x^2}{25}$ **53.** 0.602   **55.** -0.301 **57.**

59. $S = \log_e\left(\dfrac{T^C}{P^{nR}}\right)$

Exercises 12–4, p. 360

1. 2.754   **3.** -1.194 **5.** 6.966   **7.** -3.9311 **9.** 1.8649   **11.** -0.72948 **13.** 27,400
15. 0.04960   **17.** 2000.4   **19.** 0.7234   **21.** 1.4284   **23.** 0.0057882   **25.** 2.5159   **27.** $0.7325 - 2$
29. 5.9435   **31.** $0.9845 - 2$ **33.** 57.2   **35.** 0.0238   **37.** 5.32×10^8 **39.** 8.28×10^{-10}
41. 9.0607   **43.** $-0.8697 = 0.1303 - 1$ **45.** 5.1761   **47.** -99 **49.** 15.2 dB

Exercises 12–5, p. 363

1. 85.5   **3.** 0.0374   **5.** 98.5   **7.** 1.76   **9.** 94,600   **11.** 3.84   **13.** 1.50   **15.** 25.3   **17.** 3.74
19. 0.00319   **21.** 1.44×10^{341} **23.** 7.37×10^{101} **25.** 79,900 m²   **27.** 753 mg   **29.** 332 m/s
31. 0.232 ft

Exercises 12–6, p. 368

1. 3.258   **3.** 0.4460   **5.** -0.6898 **7.** -4.91623 **9.** 1.92   **11.** 3.418   **13.** 1.933   **15.** 1.795
17. 3.940   **19.** 0.3322   **21.** -0.008335 **23.** -4.34766 **25.** 3.8104   **27.** -0.37793 **29.** 8.94
31. 0.4757   **33.** 11,800   **35.** 8.66   **37.** $y = 3x$ **39.** 8.155%   **41.** 0.384 s   **43.** 257 years

Exercises 12–7, p. 373

1. 4   **3.** 0.861   **5.** 0.587   **7.** 0.6439   **9.** 0.285   **11.** 0.203   **13.** 4.11   **15.** 14.2   **17.** 4
19. $\frac{1}{4}$ **21.** 1.649   **23.** 3   **25.** -0.162 **27.** 0.906   **29.** 4   **31.** 2   **33.** 4   **35.** 1.42
37. 0.00203 s   **39.** 6.84 min   **41.** 10^{11} **43.** $10^{8.25} = 1.78 \times 10^8$ **45.** $n = 20e^{-0.04t}$
47. 3.922×10^{-4} **49.** 3.35   **51.** 3.50

Exercises 12–8, p. 377

1.   **3.**   **5.**   **7.**   **9.**

11.   **13.**   **15.** 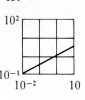 **17.**   **19.**

21.   **23.**   **25.** 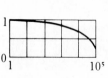 **27.**   **29.**

31.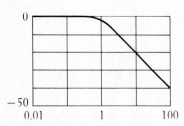

Exercises for Chapter 12, p. 379

1. $10{,}000$ **3.** $\frac{1}{5}$ **5.** 6 **7.** $\frac{5}{3}$ **9.** 6 **11.** 100 **13.** $\log_3 2 + \log_3 x$ **15.** $2\log_3 t$ **17.** $2 + \log_2 7$

19. $2 - \log_3 x$ **21.** $1 + \frac{1}{2}\log_4 3$ **23.** $3 + 4\log_{10} x$ **25.** $y = \dfrac{4}{x}$ **27.** $y = \dfrac{8}{x}$ **29.** $y = \dfrac{15}{x}$ **31.** $y = 2x$

33.

35.

37.

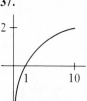

39.

41. 9.42   **43.** 4.77   **45.** 2.18   **47.** 0.728   **49.** 0.805   **51.** 4.30   **53.** 2   **55.** 30

57.

59.

61. $P = 10^{(a + bT)/T}$ **63.** $t = -0.2 \ln\left(\dfrac{i}{i_0}\right)$ **65.** 52.8%

67. $p = kV^{-\gamma}$ **69.** -2.51 V   **71.** 4.68 g·mol/L

73. 1987   **75.**

Exercises 13–1, p. 386

1. $x = 1.8, y = 3.6; x = -1.8, y = -3.6$ **3.** $x = 0.0, y = -2.0; x = 2.7, y = -0.7$

5. $x = 1.5, y = 0.2$ **7.** $x = 1.6, y = 2.5$

9. $x = 1.1, y = 2.8; x = -1.1, y = 2.8; x = 2.4, y = -1.8; x = -2.4, y = -1.8$ **11.** No solution

13. $x = -2.8, y = -1.0; x = 2.8, y = 1.0; x = 2.8, y = -1.0; x = -2.8, y = 1.0$

15. $x = 0.7, y = 0.7; x = -0.7, y = -0.7$ **17.** $x = 0.0, y = 0.0; x = 0.9, y = 0.8$

19. $x = -1.1, y = 3.0; x = 1.8, y = 0.2$ **21.** $x = 1.0, y = 0.0$ **23.** $x = 3.6, y = 1.0$

25. 360 m, 210 m   **27.** 1.3 A, 1.7 A; 2.3 A, 0.7 A

Exercises 13–2, p. 390

1. $x = 0, y = 1; x = 1, y = 2$ **3.** $x = -\frac{19}{5}, y = \frac{17}{5}; x = 5, y = -1$ **5.** $x = 1, y = 0$

7. $x = \frac{2}{7}(3 + \sqrt{2}), y = \frac{2}{7}(-1 + 2\sqrt{2}); x = \frac{2}{7}(3 - \sqrt{2}), y = \frac{2}{7}(-1 - 2\sqrt{2})$ **9.** $x = 1, y = 1$

11. $x = \frac{2}{3}, y = \frac{9}{2}; x = -3, y = -1$ **13.** $x = -2, y = 4; x = 2, y = 4$

15. $x = 1, y = 2; x = -1, y = 2$ **17.** $x = 1, y = 0; x = -1, y = 0; x = \frac{1}{2}\sqrt{6}, y = \frac{1}{2}; x = -\frac{1}{2}\sqrt{6}, y = \frac{1}{2}$

19. $x = \sqrt{19}, y = \sqrt{6}; x = \sqrt{19}, y = -\sqrt{6}; x = -\sqrt{19}; y = \sqrt{6}; x = -\sqrt{19}, y = -\sqrt{6}$

21. $x = \frac{1}{11}\sqrt{22}, y = \frac{1}{11}\sqrt{770}; x = \frac{1}{11}\sqrt{22}, y = -\frac{1}{11}\sqrt{770}; x = -\frac{1}{11}\sqrt{22}, y = \frac{1}{11}\sqrt{770};$

$x = -\frac{1}{11}\sqrt{22}, y = -\frac{1}{11}\sqrt{770}$ **23.** $x = -5, y = -2; x = -5, y = 2; x = 5, y = -2; x = 5, y = 2$

25. 0 ft, 0 s; 50 ft, 1.25 s   **27.** 8, 13   **29.** 4.47 km, 8.94 km   **31.** 8 in., 12 in.

Exercises 13–3, p. 394

1. $-3, -2, 2, 3$ **3.** $2, -2, j, -j$ **5.** $-\frac{1}{2}, \frac{1}{4}$ **7.** $-\frac{1}{2}, \frac{1}{2}, \frac{1}{6}j\sqrt{6}, -\frac{1}{6}j\sqrt{6}$ **9.** 1, 9   **11.** $\frac{64}{729}, 1$

13. $-27, 125$ **15.** 256   **17.** 5   **19.** $-2, -1, 3, 4$ **21.** 18

23. $1, -2, 1 + j\sqrt{3}, 1 - j\sqrt{3}, \frac{1}{2}(-1 + j\sqrt{3}), -\frac{1}{2}(1 + j\sqrt{3})$ **25.** 2000   **27.** 24 ft by 32 ft

Exercises 13–4, p. 397

1. 12   **3.** 2   **5.** $\frac{2}{3}$ **7.** 2, 3   **9.** 32   **11.** 16   **13.** 9   **15.** 12   **17.** $\frac{1}{2}$ **19.** 7, -1 **21.** 0

23. 5   **25.** 6   **27.** 258   **29.** $\dfrac{v_0^2 - v^2}{2g}$ **31.** $\dfrac{k^2}{2n(1 - k)}$ **33.** 9 in. by 12 in.   **35.** 5.20 mi

Exercises for Chapter 13, p. 398

1. $x = -0.9, y = 3.5; x = 0.8, y = 2.6$ **3.** $x = 2.0, y = 0.0; x = 1.6, y = 0.6$
5. $x = 0.8, y = 1.6; x = -0.8, y = 1.6$ **7.** $x = -1.2, y = 2.7; x = 1.2, y = 2.7$
9. $x = 0.0, y = 0.0; x = 2.4, y = 0.9$ **11.** $x = 0, y = 0; x = 2, y = 16$
13. $x = \sqrt{2}, y = 1; x = -\sqrt{2}, y = 1$
15. $x = \frac{1}{12}(1 + \sqrt{97}), y = \frac{1}{18}(5 - \sqrt{97}); x = \frac{1}{12}(1 - \sqrt{97}), y = \frac{1}{18}(5 + \sqrt{97})$
17. $x = 7, y = 5; x = 7, y = -5; x = -7, y = 5; x = -7, y = -5$ **19.** $x = -2, y = 2; x = \frac{2}{3}, y = \frac{10}{9}$
21. $-4, -2, 2, 4$ **23.** $1, 16$ **25.** $\frac{1}{3}, -\frac{1}{7}$ **27.** $\frac{25}{4}$ **29.** $-\frac{1}{2}, -2$ **31.** 11 **33.** $1, 4$ **35.** 8

37. $\frac{9}{16}$ **39.** 7 **41.** $r = \sqrt{\dfrac{\sqrt{\pi^2 h^4 + 4S^2} - \pi h^2}{2\pi}}$ **43.** 0.75 s, 1.5 s   **45.** $x = 16, y = 15$

47. 0.25 cm^2 **49.** 493 ft

Exercises 14–1, p. 403

1. 0 **3.** 0 **5.** -40 **7.** 46 **9.** 8 **11.** 183 **13.** -28 **15.** 14 **17.** Yes   **19.** Yes
21. No   **23.** No   **25.** Yes   **27.** Yes   **29.** $(4x^3 + 8x^2 - x - 2) \div (2x - 1) = 2x^2 + 5x + 2$; No
31. -7

Exercises 14–2, p. 408

1. $x^2 + 3x + 2, R = 0$ **3.** $x^2 - x + 3, R = 0$ **5.** $2x^4 - 4x^3 + 8x^2 - 17x + 42, R = -40$
7. $2x^3 + 3x^2 + 5x + 17, R = 46$ **9.** $x^2 + x - 4, R = 8$ **11.** $x^3 - 3x^2 + 10x - 45, R = 183$
13. $2x^3 - x^2 - 4x - 12, R = -28$ **15.** $x^4 + 2x^3 + x^2 + 7x + 4, R = 14$
17. $x^5 + 2x^4 + 4x^3 + 8x^2 + 18x + 36, R = 66$ **19.** $x^6 + 2x^5 + 4x^4 + 8x^3 + 16x^2 + 32x + 64, R = 0$
21. Yes   **23.** No   **25.** No   **27.** No   **29.** Yes   **31.** No   **33.** Yes   **35.** No

Exercises 14–3, p. 412

(Note: Unknown roots listed)
1. $-2, -1$ **3.** $-2, 3$ **5.** $-2, -2$ **7.** $-j, -\frac{2}{3}$ **9.** $2j, -2j$ **11.** $3, -1$ **13.** $-2, 1$
15. $1 - j, \frac{1}{2}, -2$ **17.** $\frac{1}{4}(1 + \sqrt{17}), \frac{1}{4}(1 - \sqrt{17})$ **19.** $2, -3$ **21.** $-j, -1, 1$ **23.** $-2j, 3, -3$

Exercises 14–4, p. 418

1. $1, -1, -2$ **3.** $2, -1, -3$ **5.** $\frac{1}{2}, 5, -3$ **7.** $\frac{1}{3}, -3, -1$ **9.** $-2, -2, 2 \pm \sqrt{3}$
11. $2, 4, -1, -3$ **13.** $1, -\frac{1}{2}, 1 \pm \sqrt{3}$ **15.** $\frac{1}{2}, -\frac{2}{3}, -3, -\frac{1}{2}$ **17.** $2, 2, -1, -1, -3$
19. $\frac{1}{2}, 1, 1, j, -j$ **21.** $\frac{15}{2}$ **23.** a **25.** 20 in.   **27.** 2 in. or 1.17 in.   **29.** 0.59
31. $-1.86, 0.68, 3.18$

Exercises for Chapter 14, p. 420

1. 1 **3.** -107 **5.** Yes   **7.** No   **9.** $x^2 + 4x + 10, R = 11$ **11.** $2x^2 - 7x + 10, R = -17$
13. $x^3 - 3x^2 - 4, R = -4$ **15.** $2x^4 + 10x^3 + 4x^2 + 21x + 105, R = 516$ **17.** No   **19.** Yes
21. (Unlisted roots) $\frac{1}{2}(-5 \pm \sqrt{17})$ **23.** (Unlisted roots) $\frac{1}{3}(-1 \pm j\sqrt{14})$ **25.** (Unlisted roots) $4, -3$
27. (Unlisted roots) $-j, \frac{1}{2}, -\frac{3}{2}$ **29.** (Unlisted roots) $2, -2$
31. (Unlisted roots) $-2 - j, \frac{1}{2}(-1 + j\sqrt{3}), -\frac{1}{2}(1 + j\sqrt{3})$ **33.** $1, 2, -4$ **35.** $-1, -1, \frac{5}{2}$
37. $\frac{5}{3}, -\frac{1}{2}, -1$ **39.** $\frac{1}{2}, -1, j\sqrt{2}, -j\sqrt{2}$ **41.** 4 **43.** $1, 0.4, 1.5, -0.6$ (last three are irrational)
45. 5 **47.** 8 ft   **49.** 7 ft, 7 ft, 11 ft   **51.** 1.1 ft

Exercises 15–1, p. 428

1. 39 **3.** 30 **5.** 50 **7.** -40 **9.** -6 **11.** -86 **13.** 118 **15.** -2
17. $x = 2, y = -1, z = 3$ **19.** $x = -1, y = \frac{1}{3}, z = -\frac{1}{2}$ **21.** $x = -1, y = 0, z = 2, t = 1$
23. $x = 1, y = 2, z = -1, t = 3$ **25.** $\frac{2}{7}$ A, $\frac{18}{7}$ A, $-\frac{8}{7}$ A, $-\frac{12}{7}$ A   **27.** 100 mL, 300 mL, 200 mL

Exercises 15–2, p. 435

1. -60 **3.** -56 **5.** 0 **7.** 0 **9.** 57 **11.** -124 **13.** -13 **15.** -118 **17.** -72 **19.** 0
21. $x = 0, y = -1, z = 4$ **23.** $x = \frac{1}{3}, y = -\frac{1}{2}, z = 1$ **25.** $x = 2, y = -1, z = -1, t = 3$
27. $x = 1, y = 2, z = -1, t = -2$ **29.** $\frac{33}{16}$ A, $\frac{11}{8}$ A, $-\frac{5}{8}$ A, $-\frac{15}{8}$ A, $-\frac{15}{16}$ A
31. PPM of SO_2, NO, NO_2, CO: $0.5, 0.3, 0.2, 5.0$

Exercises 15–3, p. 441

1. $a = 1$, $b = -3$, $c = 4$, $d = 7$ **3.** $x = 2$, $y = 3$

5. Elements cannot be equated; different number of rows.   **7.** $x = 4$, $y = 6$, $z = 9$ **9.** $\begin{pmatrix} 1 & 10 \\ 0 & 2 \end{pmatrix}$

11. $\begin{pmatrix} 0 & 0 \\ 1 & 6 \\ 1 & 0 \end{pmatrix}$ **13.** $\begin{pmatrix} 0 & 9 & -13 & 3 \\ 6 & -7 & 7 & 0 \end{pmatrix}$ **15.** Cannot be added.   **17.** $\begin{pmatrix} -1 & 13 & -20 & 3 \\ 8 & -13 & 6 & 2 \end{pmatrix}$

19. $\begin{pmatrix} -3 & -6 & 5 & -6 \\ -6 & -4 & -17 & 6 \end{pmatrix}$ **21.** $A + B = B + A = \begin{pmatrix} 3 & 1 & 0 & 7 \\ 5 & -3 & -2 & 5 \\ 10 & 10 & 8 & 0 \end{pmatrix}$

23. $-(A - B) = B - A = \begin{pmatrix} 5 & -3 & -6 & -7 \\ 5 & 3 & 0 & -3 \\ -8 & 12 & 8 & 4 \end{pmatrix}$ **25.** $F_1 = 34.2$ N, $F_2 = 24.1$ N

27. $\begin{pmatrix} 24 & 18 & 0 & 0 \\ 15 & 12 & 9 & 0 \\ 0 & 9 & 15 & 18 \end{pmatrix}$

Exercises 15–4, p. 447

1. $(-8 \ -12)$ **3.** $\begin{pmatrix} -15 & 15 & -26 \\ 8 & 5 & -13 \end{pmatrix}$ **5.** $\begin{pmatrix} 29 \\ -29 \end{pmatrix}$ **7.** $\begin{pmatrix} -7 & -5 \\ 8 & 0 \\ 14 & 10 \end{pmatrix}$ **9.** $\begin{pmatrix} 33 & -22 & 7 \\ 31 & -12 & 5 \\ 15 & 13 & -1 \\ 50 & -41 & 12 \end{pmatrix}$

11. $\begin{pmatrix} -62 & 68 \\ 73 & -27 \end{pmatrix}$ **13.** $AB = (40)$, $BA = \begin{pmatrix} -1 & 3 & -8 \\ 5 & -15 & 40 \\ 7 & -21 & 56 \end{pmatrix}$ **15.** $AB = \begin{pmatrix} -5 \\ 10 \end{pmatrix}$, BA not defined.

17. $AI = IA = A$ **19.** $AI = IA = A$ **21.** $B = A^{-1}$ **23.** $B = A^{-1}$ **25.** Yes   **27.** No

29. $\begin{pmatrix} -1 & 0 \\ 0 & -1 \end{pmatrix}\begin{pmatrix} -1 & 0 \\ 0 & -1 \end{pmatrix} = \begin{pmatrix} 1 & 0 \\ 0 & 1 \end{pmatrix}$

31. 800 ft of brass pipe, 1000 ft of steel pipe; 1620 ft of brass pipe, 2240 ft of steel pipe.

Exercises 15–5, p. 454

1. $\begin{pmatrix} -2 & -\frac{5}{2} \\ -1 & -1 \end{pmatrix}$ **3.** $\begin{pmatrix} -\frac{1}{3} & \frac{1}{6} \\ \frac{2}{15} & \frac{1}{30} \end{pmatrix}$ **5.** $\begin{pmatrix} \frac{3}{4} & \frac{1}{2} \\ -\frac{1}{4} & 0 \end{pmatrix}$ **7.** $\begin{pmatrix} -\frac{1}{4} & -\frac{1}{8} \\ \frac{1}{16} & \frac{5}{32} \end{pmatrix}$ **9.** $\begin{pmatrix} -3 & 2 \\ 2 & -1 \end{pmatrix}$

11. $\begin{pmatrix} -\frac{1}{2} & -2 \\ \frac{1}{2} & 1 \end{pmatrix}$ **13.** $\begin{pmatrix} \frac{2}{9} & -\frac{5}{9} \\ \frac{1}{9} & \frac{2}{9} \end{pmatrix}$ **15.** $\begin{pmatrix} \frac{3}{8} & \frac{1}{16} \\ -\frac{1}{4} & \frac{1}{8} \end{pmatrix}$ **17.** $\begin{pmatrix} -18 & -7 & 5 \\ -3 & -1 & 1 \\ -5 & -2 & 1 \end{pmatrix}$ **19.** $\begin{pmatrix} 3 & -4 & -1 \\ -4 & 5 & 2 \\ 2 & -3 & -1 \end{pmatrix}$

21. $\begin{pmatrix} 2 & 4 & \frac{7}{2} \\ -1 & -2 & -\frac{3}{2} \\ 1 & 1 & \frac{1}{2} \end{pmatrix}$ **23.** $\begin{pmatrix} \frac{5}{2} & -2 & -2 \\ -1 & 1 & 1 \\ \frac{7}{4} & -\frac{3}{2} & -1 \end{pmatrix}$ **25.** $\begin{pmatrix} 2 & 4 & \frac{7}{2} \\ -1 & -2 & -\frac{3}{2} \\ 1 & 1 & \frac{1}{2} \end{pmatrix}$ **27.** $\begin{pmatrix} \frac{5}{2} & -2 & -2 \\ -1 & 1 & 1 \\ \frac{7}{4} & -\frac{3}{2} & -1 \end{pmatrix}$

29. $\dfrac{1}{ad-bc}\begin{pmatrix} ad-bc & -ba+ab \\ cd-dc & -bc+ad \end{pmatrix} = \begin{pmatrix} 1 & 0 \\ 0 & 1 \end{pmatrix}$

Exercises 15–6, p. 459
1. $x = \frac{1}{2}, y = 3$ **3.** $x = \frac{1}{2}, y = -\frac{5}{2}$ **5.** $x = -4, y = 2, z = -1$ **7.** $x = -1, y = 0, z = 3$
9. $x = 1, y = 2$ **11.** $x = -\frac{3}{2}, y = -2$ **13.** $x = -3, y = -\frac{1}{2}$ **15.** $x = \frac{4}{5}, y = -\frac{22}{5}$
17. $x = 2, y = -4, z = 1$ **19.** $x = 2, y = -\frac{1}{2}, z = 3$ **21.** 10 lb, 8 lb, 12 lb
23. 8 of type A, 10 of type B

Exercises for Chapter 15, p. 461
1. 6   **3.** 186   **5.** -438 **7.** 44   **9.** 6   **11.** 186   **13.** -438 **15.** 44   **17.** -9 **19.** -44
21. $a = 3, b = -1$
23. $x = 2, y = -3, z = \frac{5}{2}, a = -1, b = -\frac{7}{2}, c = \frac{1}{2}$ **25.** $\begin{pmatrix} 1 & -3 \\ 8 & -5 \\ -8 & -2 \\ 3 & -10 \end{pmatrix}$ **27.** $\begin{pmatrix} 3 & 0 \\ -12 & 18 \\ 9 & 6 \\ -3 & 21 \end{pmatrix}$

29. Cannot be subtracted.   **31.** $\begin{pmatrix} 7 & -6 \\ -4 & 20 \\ -1 & 6 \\ 1 & 15 \end{pmatrix}$ **33.** $\begin{pmatrix} 13 \\ -13 \end{pmatrix}$ **35.** $\begin{pmatrix} 34 & 11 & -5 \\ 2 & -8 & 10 \\ -1 & -17 & 20 \end{pmatrix}$ **37.** $\begin{pmatrix} -2 & \frac{5}{2} \\ -1 & 1 \end{pmatrix}$

39. $\begin{pmatrix} \frac{2}{15} & \frac{1}{60} \\ -\frac{1}{15} & \frac{7}{60} \end{pmatrix}$ **41.** $\begin{pmatrix} 11 & 10 & 3 \\ -4 & -4 & -1 \\ 3 & 3 & 1 \end{pmatrix}$ **43.** $\begin{pmatrix} \frac{1}{2} & -\frac{1}{2} & -1 \\ -3 & 2 & 1 \\ -4 & 3 & 2 \end{pmatrix}$ **45.** $x = -3, y = 1$

47. $x = -2, y = 7$ **49.** $x = -1, y = -3, z = 0$ **51.** $x = 1, y = \frac{1}{2}, x = -\frac{1}{3}$
53. $x = 3, y = 1, z = -1$ **55.** $x = 1, y = 2, z = -3, t = 1$ **57.** $J^{-1} = -J = \begin{pmatrix} 0 & 1 \\ -1 & 0 \end{pmatrix}$

59. $(A+B)(A-B) = \begin{pmatrix} -6 & 2 \\ 4 & 2 \end{pmatrix}$, $A^2 - B^2 = \begin{pmatrix} -10 & -4 \\ 8 & 6 \end{pmatrix}$ **61.** $F_1 = 12$ lb, $F_2 = 9$ lb

63. $p_1 = \$60, p_2 = \$50, p_3 = \$90$ **65.** 30 g, 50 g, 20 g   **67.** 70°, 80°, 100°, 110°

69. $\begin{pmatrix} 15{,}000 & 10{,}000 \\ 20{,}000 & 18{,}000 \\ 8{,}000 & 30{,}000 \end{pmatrix} + \begin{pmatrix} 18{,}000 & 12{,}000 \\ 30{,}000 & 22{,}000 \\ 12{,}000 & 40{,}000 \end{pmatrix} = \begin{pmatrix} 33{,}000 & 22{,}000 \\ 50{,}000 & 40{,}000 \\ 20{,}000 & 70{,}000 \end{pmatrix}$

Exercises 16–1, p. 470
1. $7 < 12$ **3.** $20 < 45$ **5.** $-4 > -9$ **7.** $16 < 81$ **9.** $x \leq 12$ **11.** $-x \leq -3$ **13.** $x > -2$
15. $x \leq 4$ **17.** $1 < x < 7$ **19.** $x < -9, x \geq -4$ **21.** $x > 0$ **23.** $x \leq 0$ **25.** $x < -1, x > 1$

27. $2 \le x < 6$ **29.** $x > 0, y > 0$ **31.** $y > x$ **33.** Multiply both members by x.   **35.** Multiply by y and take square roots.   **37.** $h > 200$ m   **39.** $T < 36°$F or $T > 40°$F   **41.** 25,000 years $< t <$ 40,000 years

43. $V = \dfrac{k}{a}$ if $r < a$, $V = \dfrac{k}{r}$ if $r \ge a$

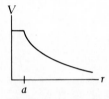

Exercises 16–2, p. 474

1.

$x > 2$

3.

$x > \dfrac{1}{2}$

5.

$x < \dfrac{8}{3}$

7.

$x \ge 1$

9.

$x < 0$ $x > 2$

11.
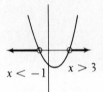
$x < -1$ $x > 3$

13.

$1 \le x \le 4$

15.

$x < -\dfrac{4}{3}$ $x > 2$

17.

$x > 1$

19.
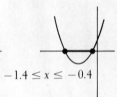
$-1.4 \le x \le -0.4$

21.

All x

23.
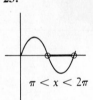
$\pi < x < 2\pi$

25.
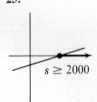
$s \ge 2000$

27.

$0 \le d < 2.3$

29.
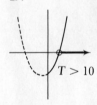
$T > 10$

31.
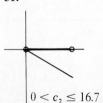
$0 < c_2 \le 16.7$

Exercises 16–3, p. 479

1. $x > 5$ **3.** $x \le -5$ **5.** $x \le 3$ **7.** $x < \frac{11}{2}$ **9.** $x \le \frac{13}{2}$ **11.** $x > -\frac{7}{9}$ **13.** $x < \frac{12}{7}$
15. $x \ge \frac{9}{2}$ **17.** $-1 < x < 1$ **19.** $x \le -2, x \ge \frac{1}{3}$ **21.** $\frac{1}{3} < x < \frac{1}{2}$ **23.** All x **25.** $-2 < x < 0, x > 1$
27. $-2 < x < -1, x > 1$ **29.** $x < 3, x > 8$ **31.** $-6 < x \le \frac{3}{2}$ **33.** $-1 < x < 2$
35. $-5 < x < -1, x > 7$ **37.** $-1 < x < \frac{3}{4}, x > 6$ **39.** $2 < x < 4, 5 < x < 9$ **41.** $x \le -2, x \ge 1$
43. $-1 \le x \le 0$ **45.** $t < \frac{15}{4}$ s   **47.** $-20°C < T < -16.2°C, 6.2°C < T < 10°C$ **49.** $0 \le n_2 \le 35$

Exercises 16–4, p. 482

1. $3 < x < 5$ **3.** $x < 1, x > \frac{7}{3}$ **5.** $\frac{1}{6} \le x \le \frac{3}{2}$ **7.** $x < -\frac{3}{2}, x > 0$ **9.** $-10 < x < 2$
11. $x \le -1, x \ge \frac{7}{3}$ **13.** $-4 < x < -2, -1 < x < 1$ **15.** $x < -4, -2 < x < -1, x > 1$
17. $y < \frac{1}{4}$ ft if $|x - 6| < 2$ ft   **19.** 5 in., 11 in.

Exercises 16–5, p. 487

1.
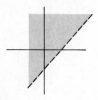

3.

5.

7.
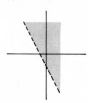

9.

11.
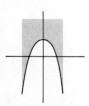

13.
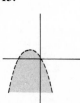

15.
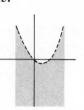

17.
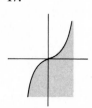

19.
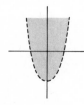

21.

23.
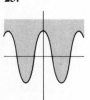

25.
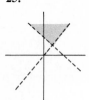

27.
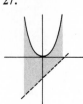

29.
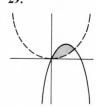

31.

33.
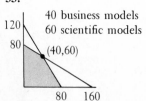

40 business models
60 scientific models
120
80
(40,60)
80   160

35.
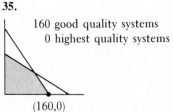

160 good quality systems
0 highest quality systems
(160,0)

Exercises for Chapter 16, p. 489

1. $x > 6$ **3.** $x \leq -\frac{5}{3}$ **5.** $x \leq -\frac{29}{2}$ **7.** $-2 < x < \frac{1}{5}$ **9.** $x < -9, x > 7$ **11.** $-4 < x < -1, x > 1$
13. $-\frac{1}{2} < x \leq 8$ **15.** $x < -4, \frac{1}{2} < x < 3$ **17.** $x = 0$ **19.** $x < 0, x > \frac{1}{2}$ **21.** $x < -1, x > 5$
23. $-2 \leq x \leq \frac{2}{3}$

25. **27.** **29.** **31.** **33.** $x \leq 3$

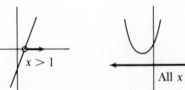

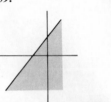

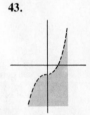

35. $x \leq -4, x \geq 0$ **37.** **39.** **41.** **43.**

45. **47.** **49.** $\frac{1}{x} > 0, \frac{1}{y} < 0, \frac{1}{x} > \frac{1}{y}$

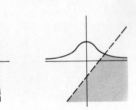

51. Divide both members by y; then use $x - 1 > y$.
53. $T \geq 4000°C$ **55.** $x \leq 80$ ft **57.** $x > 16$ in.
59. $2\ \Omega \leq R_1 \leq 6\ \Omega, 2\ \Omega \leq R_2 \leq 6\ \Omega$
61.

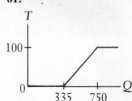

Exercises 17–1, p. 493

1. 6 **3.** $\frac{4}{3}$ **5.** $\frac{1}{5}$ **7.** 40 **9.** 40,000 to 1 **11.** 19.2 **13.** 900 lb/in.^{2} **15.** 24 calculations/min
17. 1.44 Ω **19.** 32.5 r/min **21.** 9080 g **23.** 0.335 hp **25.** 8.06 km **27.** 12.5 m/s **29.** 27.0 in.
31. 72,000 ft-lb **33.** 100 mg, 120 mg **35.** 4.0 ft, 6.0 ft

Exercises 17–2, p. 498

1. $y = kz$ **3.** $s = \frac{k}{t^2}$ **5.** $f = k\sqrt{x}$ **7.** $w = kxy^3$ **9.** $r = \frac{16}{y}$ **11.** $y = \frac{1}{4}\sqrt{x}$ **13.** $s = \frac{7}{2\sqrt{t}}$

15. $p = \frac{16q}{r^3}$ **17.** 25 **19.** 50 **21.** 180 **23.** 2.56×10^5 **25.** 2.40 in. **27.** $L_d = 0.018T_d$

29. $p = 30,000t$ **31.** 7680 BTU/h **33.** $I = \frac{10^6}{r^2}$ **35.** 28 N **37.** $R = \frac{5.00 \times 10^{-5} l}{A}$ **39.** 482 m/s

41. 135 ft **43.** 23.3 cm/s **45.** $P = \frac{k}{V^{3/2}}$, 2400 kPa **47.** 0 ft/s^2

Exercises for Chapter 17, p. 501

1. 10^6 **3.** $\frac{5}{18}$ **5.** $\frac{5}{7}$ **7.** $\frac{50}{7}$ **9.** 3.25 in.   **11.** 1.72 L   **13.** 36,000 characters   **15.** 492 g

17. 71.9 ft   **19.** 73.5 cm, 31.5 cm   **21.** $y = 3x^2$ **23.** $v = \dfrac{128x}{y^3}$ **25.** 11.7 in.   **27.** 300 N·m

29. 22.5 hp   **31.** $R = \frac{5}{18}A$ **33.** 64.0π unit^2 **35.** 2810 K   **37.** 0.536 N   **39.** 551 ft/s

41. 40.0 ft/s   **43.** $\dfrac{1.28 \times 10^5 r^4}{l^2}$ **45.** 50.0 Hz   **47.** 67.3×10^6 mi   **49.** $E = 0.139 \,(\tan\theta + \cot\theta)$

Exercises 18–1, p. 509

1. 4, 6, 8, 10, 12   **3.** 13, 9, 5, 1, -3 **5.** 22   **7.** -62 **9.** 37   **11.** $49b$ **13.** 440   **15.** $-\frac{85}{2}$
17. $n = 6$, $s = 150$ **19.** $d = -\frac{2}{19}$, $l = -\frac{1}{3}$ **21.** $a = 19$, $l = 106$ **23.** $n = 62$, $s = -4867$
25. $n = 23$, $l = 6$ **27.** $n = 8$, $d = \frac{1}{14}(b + 2c)$ **29.** $a = 36$, $d = 4$, $s = 540$
31. $a = 3$, $d = -\frac{1}{3}$, $s = 15$ **33.** 5050   **35.** 100,500   **37.** 9th   **39.** 6400 ft   **41.** 8 years
43. $a = 1$, $l = n$

Exercises 18–2, p. 513

1. 45, 15, 5, $\frac{5}{3}$, $\frac{5}{9}$ **3.** 2, 6, 18, 54, 162   **5.** 16   **7.** $\frac{1}{125}$ **9.** $\frac{1}{9}$ **11.** 2×10^6 **13.** 248   **15.** 255
17. $l = 64$, $s = \frac{1365}{16}$ **19.** $a = 16$, $l = 81$ **21.** $a = 1$, $r = 3$ **23.** $n = 5$, $s = \frac{2343}{25}$ **25.** 32
27. \$46,769.21   **29.** \$597.53   **31.** 5.68 h   **33.** 22.0 ft   **35.** 0.0319 in.   **37.** 64.6%

39. $s = \dfrac{a - rl}{1 - r}$

Exercises 18–3, p. 516

1. 8   **3.** $\dfrac{25}{4}$ **5.** $\dfrac{400}{21}$ **7.** 8   **9.** $\dfrac{10,000}{9999}$ **11.** $\frac{1}{2}(5 + 3\sqrt{3})$ **13.** $\dfrac{1}{3}$ **15.** $\dfrac{40}{99}$ **17.** $\dfrac{2}{11}$

19. $\dfrac{91}{333}$ **21.** $\dfrac{11}{30}$ **23.** $\dfrac{100,741}{999,000}$ **25.** 24.0 ft   **27.** 2.5 r

Exercises 18–4, p. 521

1. $t^3 + 3t^2 + 3t + 1$ **3.** $16x^4 - 32x^3 + 24x^2 - 8x + 1$ **5.** $x^5 + 10x^4 + 40x^3 + 80x^2 + 80x + 32$
7. $64a^6 - 192a^5b^2 + 240a^4b^4 - 160a^3b^6 + 60a^2b^8 - 12ab^{10} + b^{12}$
9. $625x^4 - 1500x^3 + 1350x^2 - 540x + 81$ **11.** $64a^6 + 192a^5 + 240a^4 + 160a^3 + 60a^2 + 12a + 1$
13. $x^{10} + 20x^9 + 180x^8 + 960x^7 + \cdots$ **15.** $128a^7 - 448a^6 + 672a^5 - 560a^4 + \cdots$
17. $x^{24} - 6x^{22}y + \frac{33}{2}x^{20}y^2 - \frac{55}{2}x^{18}y^3 + \cdots$ **19.** $a^{40} + 10\dfrac{a^{38}}{b} + \dfrac{95}{2}\dfrac{a^{36}}{b^2} + \dfrac{285}{2}\dfrac{a^{34}}{b^3} + \cdots$

21. $1 + 8x + 28x^2 + 56x^3 + \cdots$ **23.** $1 + 2x + 3x^2 + 4x^3 + \cdots$ **25.** $1 + \frac{1}{2}x - \frac{1}{8}x^2 + \frac{1}{16}x^3 - \cdots$
27. $\frac{1}{3}[1 + \frac{1}{2}x + \frac{3}{8}x^2 + \frac{5}{16}x^3 + \cdots]$ **29.** 1.049   **31.** 9.980   **33.** 3.009   **35.** 0.924   **37.** $56a^3b^5$
39. $10,264,320x^8b^4$ **41.** $T^4 + 4T^3h + 6T^2h^2 + 4Th^3 + h^4$ **43.** $1 - \dfrac{x}{a} + \dfrac{x^3}{2a^3} - \cdots$

Exercises for Chapter 18, p. 522

1. 81   **3.** 1.28×10^{-6} **5.** $-\frac{119}{2}$ **7.** $\frac{16}{243}$ **9.** $\frac{195}{2}$ **11.** $\frac{1023}{96}$ **13.** 81   **15.** $\frac{9}{16}$ **17.** -1.5
19. -0.25 **21.** 186   **23.** $\frac{455}{2}$ (AP), 127 (GP) or 43 (GP)   **25.** $\frac{7}{9}$ **27.** $\frac{75}{99}$ **29.** $\frac{41}{333}$ **31.** $\frac{1}{6}$
33. $x^4 - 8x^3 + 24x^2 - 32x + 16$ **35.** $x^{10} + 5x^8 + 10x^6 + 10x^4 + 5x^2 + 1$
37. $a^{10} + 20a^9b^2 + 180a^8b^4 + 960a^7b^6 + \cdots$ **39.** $p^{18} - \frac{3}{2}p^{16}q + p^{14}q^2 - \frac{7}{18}p^{12}q^3 + \cdots$
41. $1 + 12x + 66x^2 + 220x^3 + \cdots$ **43.** $1 + \frac{1}{2}x^2 - \frac{1}{8}x^4 + \frac{1}{16}x^6 - \cdots$
45. $1 - \frac{1}{2}a^2 - \frac{1}{8}a^4 - \frac{1}{16}a^6 - \cdots$ **47.** $1 + 6x + 24x^2 + 80x^3 + \cdots$ **49.** 30.133   **51.** 0.124
53. 1,001,000   **55.** 195   **57.** \$6633.24   **59.** \$8590   **61.** 34.9°C
63. 4.40×10^9 in. = 69,400 mi   **65.** 20.0 s   **67.** 10^{-20} atm

Exercises 19–1, p. 532

(*Note:* "Answers" to trigonometric identities are intermediate steps of suggested reductions of the left member.)

1. $1.483 = \dfrac{1}{0.6745}$ **3.** $(\frac{1}{2}\sqrt{3})^2 + (-\frac{1}{2})^2 = \frac{3}{4} + \frac{1}{4} = 1$ **5.** $\dfrac{\cos\theta}{\sin\theta} \cdot \dfrac{1}{\cos\theta} = \dfrac{1}{\sin\theta}$

7. $\dfrac{\sin x}{\dfrac{\sin x}{\cos x}} = \dfrac{\sin x}{1}\cdot\dfrac{\cos x}{\sin x}$ **9.** $\sin y\left(\dfrac{\cos y}{\sin y}\right)$ **11.** $\sin x\left(\dfrac{1}{\cos x}\right)$ **13.** $\csc^2 x(\sin^2 x)$

15. $\sin x(\csc^2 x) = (\sin x)(\csc x)(\csc x) = \sin x\left(\dfrac{1}{\sin x}\right)\csc x$ **17.** $\sin x\csc x - \sin^2 x = 1 - \sin^2 x$

19. $\tan y\cot y + \tan^2 y = 1 + \tan^2 y$ **21.** $\sin x\left(\dfrac{\sin x}{\cos x}\right) + \cos x = \dfrac{\sin^2 x + \cos^2 x}{\cos x} = \dfrac{1}{\cos x}$

23. $\cos\theta\left(\dfrac{\cos\theta}{\sin\theta}\right) + \sin\theta = \dfrac{\cos^2\theta + \sin^2\theta}{\sin\theta} = \dfrac{1}{\sin\theta}$

25. $\sec\theta\left(\dfrac{\sin\theta}{\cos\theta}\right)\csc\theta = \sec\theta\left(\dfrac{1}{\cos\theta}\right)(\sin\theta\csc\theta) = \sec\theta(\sec\theta)(1)$

27. $\cot\theta(\sec^2\theta - 1) = \cot\theta\tan^2\theta = (\cot\theta\tan\theta)\tan\theta$ **29.** $\dfrac{\sin x}{\cos x} + \dfrac{\cos x}{\sin x} = \dfrac{\sin^2 x + \cos^2 x}{\cos x\sin x} = \dfrac{1}{\cos x\sin x}$

31. $(1 - \sin^2 x) - \sin^2 x$ **33.** $\dfrac{\sin x(1 + \cos x)}{1 - \cos^2 x} = \dfrac{1 + \cos x}{\sin x}$

35. $\dfrac{(1/\cos x) + (1/\sin x)}{1 + (\sin x/\cos x)} = \dfrac{(\sin x + \cos x)/\cos x\sin x}{(\cos x + \sin x)/\cos x} = \dfrac{\cos x}{\cos x\sin x}$

37. $\dfrac{\sin^2 x}{\cos^2 x}\cos^2 x + \dfrac{\cos^2 x}{\sin^2 x}\sin^2 x = \sin^2 x + \cos^2 x$ **39.** $\dfrac{\sec\theta}{\dfrac{1}{\sec\theta}} - \dfrac{\tan\theta}{\dfrac{1}{\tan\theta}} = \sec^2\theta - \tan^2\theta$

41. $\dfrac{\sin^2 x + \cos^2 x - 2\cos^2 x}{\sin x\cos x} = \dfrac{\sin^2 x - \cos^2 x}{\sin x\cos x} = \dfrac{\sin x}{\cos x} - \dfrac{\cos x}{\sin x}$

43. $4\sin x + \dfrac{\sin x}{\cos x} = \sin x\left(4 + \dfrac{1}{\cos x}\right)$ **45.** $\dfrac{1}{\cos x} + \dfrac{\sin x}{\cos x} + \dfrac{\cos x}{\sin x} = \dfrac{\sin x + \cos^2 x + \sin^2 x}{\sin x\cos x}$

47. $\dfrac{\cos\theta + \sin\theta}{1 + \dfrac{\sin\theta}{\cos\theta}} = \dfrac{\cos\theta + \sin\theta}{\dfrac{\cos\theta + \sin\theta}{\cos\theta}}$

49. $\left(\dfrac{\sin x}{\cos x} + \dfrac{\cos x}{\sin x}\right)\sin x\cos x = \dfrac{\sin^2 x\cos x}{\cos x} + \dfrac{\sin x\cos^2 x}{\sin x}$ **51.** $(2\sin^2 x - 1)(\sin^2 x - 1)$

53. $\dfrac{\cot 2y(\sec 2y + \tan 2y) - \cos 2y(\sec 2y - \tan 2y)}{\sec^2 2y - \tan^2 2y} = \dfrac{\cot 2y\sec 2y + 1 - 1 + \cos 2y\tan 2y}{1}$

55. Infinite GP: $\dfrac{1}{1 - \sin^2 x} = \dfrac{1}{\cos^2 x}$ **57.** $\mu = \dfrac{w\sin\theta}{w\cos\theta} = \dfrac{\sin\theta}{\cos\theta} = \tan\theta$

59. $\sin\theta(\sin^2\theta) = \sin\theta(1 - \cos^2\theta)$ **61.** $\sqrt{4 + 4\tan^2\theta} = 2\sqrt{1 + \tan^2\theta}$

Exercises 19–2, p. 537

1. $\sin 105° = \sin 60°\cos 45° + \cos 60°\sin 45° = \dfrac{\sqrt{3}}{2}\cdot\dfrac{\sqrt{2}}{2} + \dfrac{1}{2}\cdot\dfrac{\sqrt{2}}{2} = 0.9659$

3. $\cos 15° = \cos(60° - 45°) = \cos 60°\cos 45° + \sin 60°\sin 45°$
$\qquad = (\tfrac{1}{2})(\tfrac{1}{2}\sqrt{2}) + (\tfrac{1}{2}\sqrt{3})(\tfrac{1}{2}\sqrt{2}) = \tfrac{1}{4}\sqrt{2} + \tfrac{1}{4}\sqrt{6} = \tfrac{1}{4}(\sqrt{2} + \sqrt{6}) = 0.9659$

5. $-\tfrac{33}{65}$ **7.** $-\tfrac{56}{65}$ **9.** $\sin 3x$ **11.** $\cos x$ **13.** 1 **15.** 1

17. $\sin(270° - x) = \sin 270°\cos x - \cos 270°\sin x = (-1)\cos x - 0(\sin x)$

19. $\cos(\tfrac{1}{2}\pi - x) = \cos\tfrac{1}{2}\pi\cos x + \sin\tfrac{1}{2}\pi\sin x = 0(\cos x) + 1(\sin x)$

21. $\cos(30° + x) = \cos 30°\cos x - \sin 30°\sin x = \tfrac{1}{2}\sqrt{3}\cos x - \tfrac{1}{2}\sin x = \tfrac{1}{2}(\sqrt{3}\cos x - \sin x)$

23. $\sin\left(\dfrac{\pi}{4} + x\right) = \sin\dfrac{\pi}{4}\cos x + \cos\dfrac{\pi}{4}\sin x = \tfrac{1}{2}\sqrt{2}\cos x + \tfrac{1}{2}\sqrt{2}\sin x$

25. $(\sin x \cos y + \cos x \sin y)(\sin x \cos y - \cos x \sin y) = \sin^2 x \cos^2 y - \cos^2 x \sin^2 y$
$$= \sin^2 x (1 - \sin^2 y) - (1 - \sin^2 x)\sin^2 y$$

27. $(\cos\alpha\cos\beta - \sin\alpha\sin\beta) + (\cos\alpha\cos\beta + \sin\alpha\sin\beta)$ **29, 31, 33, 35.** Use indicated method.

37. $\sin(\omega t + \phi - \alpha)$ **39.** $2A \sin\dfrac{2\pi t}{T}\cos\dfrac{2\pi x}{\lambda}$

Exercises 19–3, p. 541

1. $\sin 60° = \sin 2(30°) = 2\sin 30°\cos 30° = 2(\tfrac{1}{2})(\tfrac{1}{2}\sqrt{3}) = \tfrac{1}{2}\sqrt{3}$
3. $\cos 120° = \cos 2(60°) = \cos^2 60° - \sin^2 60° = (\tfrac{1}{2})^2 - (\tfrac{1}{2}\sqrt{3})^2 = -\tfrac{1}{2}$
5. $\sin 158° = 2\sin 79°\cos 79° = 0.3746066$ **7.** $\cos 96° = \cos^2 48° - \sin^2 48° = -0.1045285$
9. $\tfrac{24}{25}$ **11.** $\tfrac{3}{5}$ **13.** $2\sin 8x$ **15.** $\cos 8x$ **17.** $\cos^2\alpha - (1 - \cos^2\alpha)$
19. $(\cos^2 x - \sin^2 x)(\cos^2 x + \sin^2 x) = (\cos^2 x - \sin^2 x)(1)$ **21.** $\dfrac{2\sin\theta\cos\theta}{1 + 2\cos^2\theta - 1} = \dfrac{\sin\theta}{\cos\theta}$
23. $\dfrac{1}{\sec^2 x} - \dfrac{\tan^2 x}{\sec^2 x} = \cos^2 x - \dfrac{\sin^2 x}{\cos^2 x \sec^2 x}$ **25.** $\dfrac{2\tan x}{\sin 2x} = \dfrac{2(\sin x/\cos x)}{2\sin x\cos x} = \dfrac{1}{\cos^2 x}$
27. $\dfrac{\sin 3x\cos x - \cos 3x\sin x}{\sin x\cos x} = \dfrac{\sin 2x}{\tfrac{1}{2}\sin 2x}$
29. $\sin(2x + x) = \sin 2x\cos x + \cos 2x\sin x = (2\sin x\cos x)(\cos x) + (\cos^2 x - \sin^2 x)\sin x$
31. Use indicated method.
33. $\cos\alpha = \dfrac{A}{C}$, $\sin\alpha = \dfrac{B}{C}$:

$$A\sin 2t + B\cos 2t = C\left(\frac{A}{C}\sin 2t + \frac{B}{C}\cos 2t\right)$$
$$= C(\cos\alpha\sin 2t + \sin\alpha\cos 2t)$$

Exercises 19–4, p. 544

1. $\cos 15° = \cos\dfrac{1}{2}(30°) = \sqrt{\dfrac{1 + \cos 30°}{2}} = \sqrt{\dfrac{1.8660}{2}} = 0.9659$
3. $\sin 75° = \sin\dfrac{1}{2}(150°) = \sqrt{\dfrac{1 - \cos 150°}{2}} = \sqrt{\dfrac{1.8660}{2}} = 0.9659$ **5.** $\sin 118° = 0.8829476$
7. $\sqrt{2\left(\dfrac{1 + \cos 164°}{2}\right)} = \sqrt{2}\cos 82° = 0.1968205$
9. $\sin 3\alpha$ **11.** $4\cos 2x$ **13.** $\tfrac{1}{26}\sqrt{26}$ **15.** $\tfrac{1}{10}\sqrt{2}$ **17.** $\pm\sqrt{\dfrac{2}{1 - \cos\alpha}}$
19. $\tan\dfrac{1}{2}\alpha = \dfrac{1 - \cos\alpha}{\sin\alpha} = \dfrac{\sin\alpha}{1 + \cos\alpha}$ **21.** $\dfrac{1 - \cos\alpha}{2\sin\tfrac{1}{2}\alpha} = \dfrac{1 - \cos\alpha}{2\sqrt{\tfrac{1}{2}(1 - \cos\alpha)}} = \sqrt{\dfrac{1 - \cos\alpha}{2}}$
23. $2\left(\dfrac{1 - \cos x}{2}\right) + \cos x$ **25.** $\sqrt{\dfrac{(1 + \cos\theta)(1 - \cos\theta)}{2(1 - \cos\theta)}} = \dfrac{\sin\theta}{\sqrt{4\left(\dfrac{1 - \cos\theta}{2}\right)}}$
27. $2\left(\dfrac{1 - \cos\alpha}{2}\right) - \left(\dfrac{1 + \cos\alpha}{2}\right)$ **29.** $\sqrt{1 - 2\cos\alpha + \cos^2\alpha + \sin^2\alpha(\cos^2\beta + \sin^2\beta)} = \sqrt{2 - 2\cos\alpha}$

Exercises 19–5, p. 549

1. $\dfrac{\pi}{2}$ **3.** $\dfrac{3\pi}{4}, \dfrac{7\pi}{4}$ **5.** π **7.** $0.9273, 2.214$ **9.** $\dfrac{\pi}{3}, \dfrac{2\pi}{3}, \dfrac{4\pi}{3}, \dfrac{5\pi}{3}$ **11.** $\dfrac{\pi}{3}, \dfrac{2\pi}{3}, \dfrac{4\pi}{3}, \dfrac{5\pi}{3}$
13. $0, \dfrac{\pi}{6}, \dfrac{5\pi}{6}, \pi$ **15.** $\dfrac{\pi}{12}, \dfrac{\pi}{4}, \dfrac{5\pi}{12}, \dfrac{3\pi}{4}, \dfrac{13\pi}{12}, \dfrac{5\pi}{4}, \dfrac{17\pi}{12}, \dfrac{7\pi}{4}$ **17.** $\dfrac{\pi}{2}, \dfrac{3\pi}{2}$ **19.** $0, \dfrac{\pi}{3}, \pi, \dfrac{5\pi}{3}$

21. $\dfrac{\pi}{4}, \dfrac{3\pi}{4}, \dfrac{5\pi}{4}, \dfrac{7\pi}{4}$ **23.** 3.569, 5.856   **25.** 0.2618, 1.309, 3.403, 4.451   **27.** 0, π **29.** $\dfrac{3\pi}{8}, \dfrac{7\pi}{8}, \dfrac{11\pi}{8}, \dfrac{15\pi}{8}$

31. 0.7854, 1.249, 3.927, 4.391   **33.** 0.9553   **35.** 1.41 units of time   **37.** 0.3398, 2.802

39. 0.8861, 2.256, 4.028, 5.397   **41.** -0.95, 0.00, 0.95   **43.** 0, 4.49

Exercises 19–6, p. 552

1. y is the angle whose tangent is x.   **3.** y is the angle whose cotangent is $3x$.

5. y is twice the angle whose sine is x.   **7.** y is 5 times the angle whose cosine is $2x$.

9. $\dfrac{\pi}{3}$ **11.** $\dfrac{3\pi}{4}$ **13.** $\dfrac{3\pi}{4}$ **15.** $\dfrac{\pi}{3}$ **17.** $\dfrac{5\pi}{6}$ **19.** $\dfrac{\pi}{4}$ **21.** 1.3754

23. -1.0554, 2.0862   **25.** -0.0875, 3.2291   **27.** 0.2213   **29.** $x = \frac{1}{3}\arcsin y$

31. $x = 4\tan y$ **33.** $x = \frac{1}{3}\operatorname{arcsec}(y-1)$ **35.** $x = 1 - \cos(1-y)$ **37.** I, II

39. II, III   **41.** $k = \dfrac{m\left(\arccos \dfrac{x}{A}\right)^2}{t^2}$ **43.** $\theta = \dfrac{1}{2}\arcsin\dfrac{2E}{E_0}$

Exercises 19–7, p. 557

1. $\dfrac{\pi}{3}$ **3.** 0   **5.** $-\dfrac{\pi}{3}$ **7.** $\dfrac{\pi}{3}$ **9.** $\dfrac{\pi}{6}$ **11.** $-\dfrac{\pi}{4}$ **13.** $\dfrac{\pi}{4}$ **15.** $\frac{1}{2}\sqrt{3}$ **17.** $\frac{1}{2}\sqrt{2}$ **19.** -1

21. -1.3090 **23.** -0.9838 **25.** 1.4413   **27.** 1.4503   **29.** -1.2389

31. -0.2239 **33.** $\dfrac{x}{\sqrt{1-x^2}}$ **35.** $\dfrac{1}{x}$ **37.** $\dfrac{3x}{\sqrt{9x^2-1}}$ **39.** $2x\sqrt{1-x^2}$

41. $\sin(\operatorname{Arcsin}\frac{3}{5} + \operatorname{Arcsin}\frac{5}{13}) = \frac{3}{5}\cdot\frac{12}{13} + \frac{4}{5}\cdot\frac{5}{13} = \frac{56}{65}$ **43.** $\dfrac{\pi}{6} + \dfrac{\pi}{3} = \dfrac{\pi}{2}$ **45.** $\operatorname{Arcsin}\left(\dfrac{a}{c}\right)$

47. let b = side opp. β; $\tan \beta = \dfrac{b}{1} = b$; $\tan \alpha = a + b = a + \tan \beta$; $\alpha = \operatorname{Arctan}(a + \tan \beta)$

Exercises for Chapter 19, p. 559

1. $\sin(90° + 30°) = \sin 90° \cos 30° + \cos 90° \sin 30° = (1)(\frac{1}{2}\sqrt{3}) + (0)(\frac{1}{2}) = \frac{1}{2}\sqrt{3}$

3. $\sin(180° - 45°) = \sin 180° \cos 45° - \cos 180° \sin 45° = 0(\frac{1}{2}\sqrt{2}) - (-1)(\frac{1}{2}\sqrt{2}) = \frac{1}{2}\sqrt{2}$

5. $\cos 2(90°) = \cos^2 90° - \sin^2 90° = 0 - 1 = -1$

7. $\sin\frac{1}{2}(90°) = \sqrt{\frac{1}{2}(1 - \cos 90°)} = \sqrt{\frac{1}{2}(1 - 0)} = \frac{1}{2}\sqrt{2}$ **9.** $\sin 52° = 0.7880108$

11. $\sin 92° = 0.9993908$ **13.** $\cos 6° = 0.9945219$ **15.** $\cos 164° = -0.9612617$

17. $\sin 5x$ **19.** $4 \sin 12x$ **21.** $2 \cos 12x$ **23.** $2 \cos x$ **25.** $-\dfrac{\pi}{2}$ **27.** 0.2619

29. $-\frac{1}{3}\sqrt{3}$ **31.** 0   **33.** $\dfrac{\dfrac{1}{\cos y}}{\dfrac{1}{\sin y}} = \dfrac{1}{\cos y}\cdot\dfrac{\sin y}{1}$ **35.** $\sin x \csc x - \sin^2 x = 1 - \sin^2 x$

37. $\dfrac{1 - \sin^2\theta}{\sin\theta} = \dfrac{\cos^2\theta}{\sin\theta}$ **39.** $\cos\theta\left(\dfrac{\cos\theta}{\sin\theta}\right) + \sin\theta = \dfrac{\cos^2\theta + \sin^2\theta}{\sin\theta}$

41. $\dfrac{(\sec^2 x - 1)(\sec^2 x + 1)}{\tan^2 x} = \sec^2 x + 1$ **43.** $2\left(\dfrac{1}{\sin 2x}\right)\left(\dfrac{\cos x}{\sin x}\right) = 2\left(\dfrac{1}{2\sin x \cos x}\right)\left(\dfrac{\cos x}{\sin x}\right) = \dfrac{1}{\sin^2 x}$

45. $\dfrac{\cos^2\theta}{\sin^2\theta}$ **47.** $\dfrac{(\cos^2\theta - \sin^2\theta)}{\cos^2\theta} = 1 - \dfrac{\sin^2\theta}{\cos^2\theta}$ **49.** $\dfrac{1}{2}\left(2\sin\dfrac{\theta}{2}\cos\dfrac{\theta}{2}\right)$

51. $\dfrac{1}{\cos x} + \dfrac{\sin x}{\cos x} = \dfrac{(1 + \sin x)(1 - \sin x)}{\cos x(1 - \sin x)} = \dfrac{1 - \sin^2 x}{\cos x(1 - \sin x)}$ **53.** $\cos[(x - y) + y]$

55. $\sin 4x(\cos 4x)$ **57.** $\dfrac{\sin x}{\dfrac{1}{\sin x} - \dfrac{\cos x}{\sin x}} = \dfrac{\sin^2 x}{1 - \cos x} = \dfrac{1 - \cos^2 x}{1 - \cos x}$

59. $\dfrac{\sin x \cos y + \cos x \sin y + \sin x \cos y - \cos x \sin y}{\cos x \cos y - \sin x \sin y + \cos x \cos y + \sin x \sin y} = \dfrac{2 \sin x \cos y}{2 \cos x \cos y}$ **61.** $x = \frac{1}{2} \arccos \frac{1}{2}y$

63. $x = \frac{1}{5} \sin \frac{1}{3}(\frac{1}{4}\pi - y)$ **65.** 1.2925, 4.4341   **67.** $\dfrac{\pi}{6}, \dfrac{5\pi}{6}, \dfrac{7\pi}{6}, \dfrac{11\pi}{6}$ **69.** $0, \dfrac{\pi}{2}, \pi, \dfrac{3\pi}{2}$

71. $\dfrac{\pi}{6}, \dfrac{5\pi}{6}, \dfrac{7\pi}{6}, \dfrac{11\pi}{6}$ **73.** $0, \pi$ **75.** 0   **77.** $\dfrac{1}{x}$ **79.** $2x\sqrt{1 - x^2}$ **81.** $\sqrt{1 - \cos^2\theta}$

83. $\dfrac{\tan \theta}{\sqrt{1 + \tan^2\theta}} = \dfrac{\tan \theta}{\sec \theta}$ **85.** $\cos x(\cos^2 x) = \cos x(1 - \sin^2 x)$

87. $2 \sin^2 x = 1 - \cos 2x$; $\sin^2 x = \frac{1}{2}(1 - \cos 2x)$ **89.** $t = 3 \arcsin \dfrac{y}{20}$ **91.** $\cos^2\alpha - 3 \sin^2\alpha$

93. $1 - 2r^2 + r^4 + 2r^2 - 2r^2 \cos \beta = (1 - r^2)^2 + 4r^2\left(\dfrac{1 - \cos \beta}{2}\right)$ **95.** 54.7°

Exercises 20–1, p. 568

1. $2\sqrt{29}$ **3.** 3   **5.** $\sqrt{85}$ **7.** 7   **9.** 2.86   **11.** $\frac{5}{2}$ **13.** Undefined   **15.** $-\frac{7}{6}$ **17.** 0
19. 0.747   **21.** $\frac{1}{3}\sqrt{3}$ **23.** $-\frac{1}{3}\sqrt{3}$ **25.** 20.0°   **27.** 98.5°   **29.** Parallel   **31.** Perpendicular
33. $8, -2$ **35.** -3 **37.** Two sides equal $2\sqrt{10}$ **39.** $m_1 = \frac{1}{4}, m_2 = \frac{5}{4}$ **41.** 10   **43.** $(1, 5)$

Exercises 20–2, p. 573

1. $4x - y + 20 = 0$ **3.** $7x - 6y - 16 = 0$ **5.** $x - y + 2 = 0$ **7.** $y = -3$ **9.** $x = -3$
11. $3x - 2y - 12 = 0$ **13.** $x + 3y + 5 = 0$ **15.** $x + 2y - 4 = 0$ **17.** $4x + 3y + 6 = 0$
19. $2x + 5y + 14 = 0$

21.

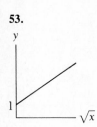

23.

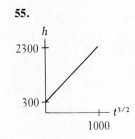

25. $y = \frac{3}{2}x - \frac{1}{2}; m = \frac{3}{2}, b = -\frac{1}{2}$ **27.** $y = \frac{5}{2}x + \frac{5}{2}; m = \frac{5}{2}, b = \frac{5}{2}$
29. -2 **31.** 1   **33.** $m_1 = m_2 = \frac{3}{2}$ **35.** $m_1 = 2, m_2 = -\frac{1}{2}$
37. $x + 2y - 4 = 0$ **39.** $3x + y - 18 = 0$ **41.** $l = 2c + 8$
43. $50x + 60y = 12{,}200$ **45.** $s = 50t + 10$ **47.** $L = \frac{2}{3}F + 15$
49. 3.35 kJ   **51.** $n = \frac{7}{6}t + 10$; at 6:30, $n = 10$; at 8:30, $n = 150$

53.

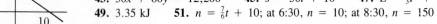

55.

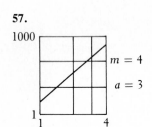

57.

59.

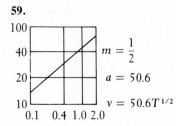

Exercises 20-3, p. 580

1. $(2, 1), r = 5$ **3.** $(-1, 0), r = 2$ **5.** $x^2 + y^2 = 9$

7. $(x - 2)^2 + (y - 2)^2 = 16$, or $x^2 + y^2 - 4x - 4y - 8 = 0$

9. $(x + 2)^2 + (y - 5)^2 = 5$, or $x^2 + y^2 + 4x - 10y + 24 = 0$

11. $(x + 3)^2 + y^2 = \frac{1}{4}$, or $4x^2 + 4y^2 + 24x + 35 = 0$

13. $(x - 2)^2 + (y - 1)^2 = 8$, or $x^2 + y^2 - 4x - 2y - 3 = 0$

15. $(x + 3)^2 + (y - 5)^2 = 25$, or $x^2 + y^2 + 6x - 10y + 9 = 0$

17. $(x - 2)^2 + (y - 2)^2 = 4$, or $x^2 + y^2 - 4x - 4y + 4 = 0$

19. $(x - 2)^2 + (y - 5)^2 = 25$, or $x^2 + y^2 - 4x - 10y + 4 = 0$; and

$(x + 2)^2 + (y + 5)^2 = 25$, or $x^2 + y^2 + 4x + 10y + 4 = 0$

21. $(0, 3)$
$r = 2$

23. $(-1, 5)$
$r = \frac{9}{2}$

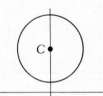

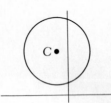

25. $(0, 0)$
$r = 5$

27. $(1, 0)$
$r = 3$

29. $(-4, 5)$
$r = 7$

31. $(1, 2)$
$r = \frac{1}{2}\sqrt{22}$

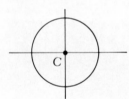

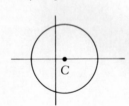

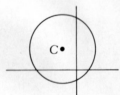

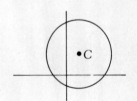

33. Symmetrical to both axes and origin.   **35.** Symmetrical to y-axis.   **37.** $(7, 0), (-1, 0)$

39. $3x^2 + 3y^2 + 4x + 8y - 20 = 0$, circle

41. $x^2 + y^2 = 2.79$ (Assume center of circle at center of coordinate system.)

43. $x^2 + y^2 = 576$, $x^2 + y^2 + 36y + 315 = 0$ (in inches)

Exercises 20-4, p. 585

1. $F(1, 0), x = -1$ **3.** $F(-1, 0), x = 1$ **5.** $F(0, 2), y = -2$ **7.** $F(0, -1), y = 1$

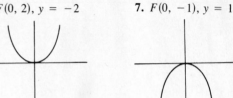

9. $F(\frac{1}{2}, 0), x = -\frac{1}{2}$ **11.** $F(0, \frac{1}{4}), y = -\frac{1}{4}$ **13.** $y^2 = 12x$ **15.** $x^2 = 16y$ **17.** $x^2 = 4y$

19. $x^2 = \frac{1}{8}y$

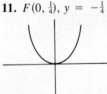

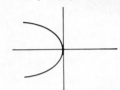

21. $y^2 - 2y - 12x + 37 = 0$　　**23.** $x^2 - 2x + 8y - 23 = 0$　**25.** $4x + 3y = 16, y = 4$

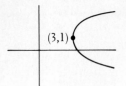

(3,1)

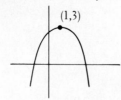

(1,3)

27. 3.68 m　　　　　　**31.**

29. $x^2 = 100y$

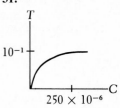

T

10^{-1}

250×10^{-6}

C

33. $y^2 = 8x$ or $x^2 = 8y$ with vertex midway between island and shore.

Exercises 20–5, p. 591

1. $V(2, 0), V(-2, 0)$
　$F(\sqrt{3}, 0), F(-\sqrt{3}, 0)$

3. $V(0, 6), V(0, -6)$
　$F(0, \sqrt{11}), F(0, -\sqrt{11})$

5. $V(3, 0), V(-3, 0)$
　$F(\sqrt{5}, 0), F(-\sqrt{5}, 0)$

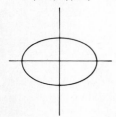

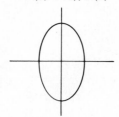

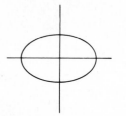

7. $V(0, 7), V(0, -7)$
　$F(0, \sqrt{45}), F(0, -\sqrt{45})$

9. $V(0, 4), V(0, -4)$
　$F(0, \sqrt{14})$
　$F(0, -\sqrt{14})$

11. $V\left(\frac{5}{2}, 0\right), V\left(-\frac{5}{2}, 0\right)$

　$F\left(\frac{\sqrt{21}}{2}, 0\right), F\left(\frac{-\sqrt{21}}{2}, 0\right)$

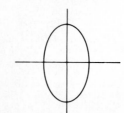

13. $\dfrac{x^2}{225} + \dfrac{y^2}{144} = 1$, or $144x^2 + 225y^2 = 32{,}400$　　**15.** $\dfrac{y^2}{9} + \dfrac{x^2}{5} = 1$, or $9x^2 + 5y^2 = 45$

17. $\dfrac{x^2}{64} + \dfrac{15y^2}{144} = 1$, or $3x^2 + 20y^2 = 192$　　**19.** $\dfrac{x^2}{5} + \dfrac{y^2}{20} = 1$, or $4x^2 + y^2 = 20$

21. $16x^2 + 25y^2 - 32x - 50y - 359 = 0$　　**23.** $9x^2 + 5y^2 - 18x - 20y - 16 = 0$
25. $2x^2 + 3y^2 - 8x - 4 = 2x^2 + 3(-y)^2 - 8x - 4$　　**27.** $9x^2 + 25y^2 = 22{,}500$
29. $2.70x^2 + 2.76y^2 = 7.45 \times 10^7$　　**31.** Major axis = 8.5 in., minor axis = 6.0 in.
33. $7x^2 + 16y^2 = 112$

Exercises 20–6, p. 597

1. $V(5, 0), V(-5, 0)$
$F(13, 0), F(-13, 0)$

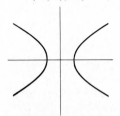

3. $V(0, 3), V(0, -3)$
$F(0, \sqrt{10}), F(0, -\sqrt{10})$

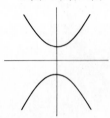

5. $V(\sqrt{2}, 0), V(-\sqrt{2}, 0)$
$F(\sqrt{6}, 0), F(-\sqrt{6}, 0)$

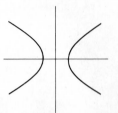

7. $V(0, \sqrt{5}), V(0, -\sqrt{5})$
$F(0, \sqrt{7}), F(0, -\sqrt{7})$

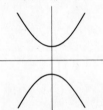

9. $V(0, 2), V(0, -2)$
$F(0, \sqrt{5}), F(0, -\sqrt{5})$

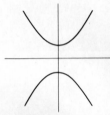

11. $V(2, 0), V(-2, 0)$
$F(\tfrac{2}{3}\sqrt{13}, 0), F(-\tfrac{2}{3}\sqrt{13}, 0)$

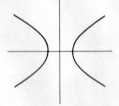

13. $\dfrac{x^2}{9} - \dfrac{y^2}{16} = 1$, or $16x^2 - 9y^2 = 144$

15. $\dfrac{y^2}{100} - \dfrac{x^2}{36} = 1$, or $9y^2 - 25x^2 = 900$

17. $\dfrac{x^2}{1} - \dfrac{y^2}{3} = 1$, or $3x^2 - y^2 = 3$

19. $\dfrac{x^2}{5} - \dfrac{y^2}{4} = 1$, or $4x^2 - 5y^2 = 20$

21.

23.

25. $9x^2 - 16y^2 - 108x + 64y + 116 = 0$

27. $9x^2 - y^2 - 36x + 27 = 0$

29.
R

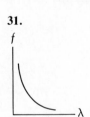

Z

31.
f

λ

33. dist(rifle to P) $-$ dist(target to P) $=$ constant
(related to distance from rifle to target).

Exercises 20-7, p. 602

1. Parabola, $(-1, 2)$ **3.** Hyperbola, $(1, 2)$ **5.** Ellipse, $(-1, 0)$ **7.** Parabola, $(-3, 1)$

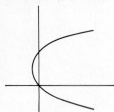 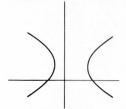 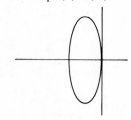

9. $(y - 3)^2 = 16(x + 1)$, or $y^2 - 6y - 16x - 7 = 0$
11. $(x + 3)^2 = 4(y - 2)$, or $x^2 + 6x - 4y + 17 = 0$

13. $\dfrac{(x + 2)^2}{25} + \dfrac{(y - 2)^2}{16} = 1$, or $16x^2 + 25y^2 + 64x - 100y - 236 = 0$

15. $\dfrac{(y - 1)^2}{16} + \dfrac{(x + 2)^2}{4} = 1$, or $4x^2 + y^2 + 16x - 2y + 1 = 0$

17. $\dfrac{(y - 2)^2}{1} - \dfrac{(x + 1)^2}{3} = 1$, or $x^2 - 3y^2 + 2x + 12y - 8 = 0$

19. $\dfrac{(x + 1)^2}{9} - \dfrac{(y - 1)^2}{16} = 1$, or $16x^2 - 9y^2 + 32x + 18y - 137 = 0$

21. Parabola $(-1, -1)$ **23.** Ellipse, $(-3, 0)$ **25.** Hyperbola, $(0, 4)$ **27.** Parabola, $(1, 0)$

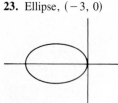 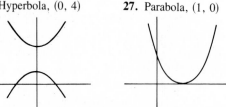

29. $x^2 - y^2 + 4x - 2y - 22 = 0$ **31.** $y^2 + 4x - 4 = 0$ **33.**

35. $\dfrac{(x - 0.9)^2}{3.7^2} + \dfrac{y^2}{3.6^2} = 1$

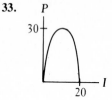

Exercises 20-8, p. 607

1. Ellipse   **3.** Hyperbola   **5.** Circle   **7.** Parabola   **9.** Hyperbola   **11.** Circle   **13.** Ellipse
15. Hyperbola   **17.** Ellipse   **19.** Parabola

21. Parabola; $V(-4, 0)$; $F(-4, 2)$ **23.** Hyperbola; $C(+1, -2)$;   **25.** Ellipse; $C(5, 0)$; $V(5, \pm 2\sqrt{2})$
$V(+1, -2, \pm\sqrt{2})$

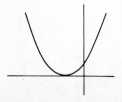 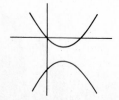 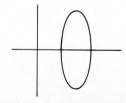

27. Parabola; $V(-\frac{1}{2}, \frac{5}{2})$; $F(\frac{1}{2}, \frac{5}{2})$ **29.** (a) Circle, (b) hyperbola, (c) ellipse

31. The origin   **33.** Hyperbola   **35.** Ellipse

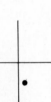

Exercises 20–9, p. 610

1.   **3.**   **5.**   **7.**   **9.**   **11.**

  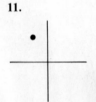

13. $\left(2, \dfrac{\pi}{6}\right)$ **15.** $\left(1, \dfrac{7\pi}{6}\right)$ **17.** $(-4, -4\sqrt{3})$ **19.** $(2.77, -1.15)$ **21.** $r = 3 \sec \theta$ **23.** $r = a$

25. $r = 4 \cot \theta \csc \theta$ **27.** $r^2 = \dfrac{4}{1 + 3\sin^2\theta}$ **29.** $x^2 + y^2 - y = 0$ **31.** $x = 4$

33. $x^4 + y^4 - 4x^3 + 2x^2y^2 - 4xy^2 - 4y^2 = 0$ **35.** $(x^2 + y^2)^2 = 2xy$

37. $s = \sqrt{x^2 + y^2} \arctan\left(\dfrac{y}{x}\right)$, $A = \dfrac{1}{2}(x^2 + y^2)\arctan\left(\dfrac{y}{x}\right)$ **39.** $r = 100 \tan \theta \sec \theta$

Exercises 20–10, p. 613

1.   **3.**   **5.**   **7.**   **9.**

   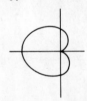

11.   **13.**   **15.**   **17.**   **19.**

21.

23.

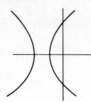

25.

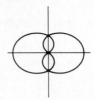

27.

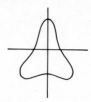

29.

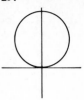

31.

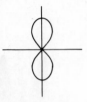

Exercises for Chapter 20, p. 614

1. $4x - y - 11 = 0$

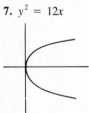

3. $2x + 3y + 3 = 0$

5. $x^2 + y^2 - 2x + 4y - 5 = 0$

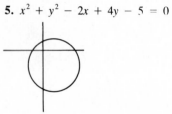

7. $y^2 = 12x$

9. $9x^2 + 25y^2 = 900$

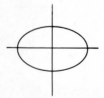

11. $144y^2 - 169x^2 = 24{,}336$

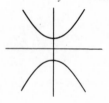

13. $(-3, 0)$, $r = 4$

15. $(0, -5)$, $y = 5$

17. $V(0, 4)$, $V(0, -4)$
$F(0, \sqrt{15})$, $F(0, -\sqrt{15})$

19. $V(2, 0)$, $V(-2, 0)$
$F(\frac{2}{5}\sqrt{35}, 0)$, $F(-\frac{2}{5}\sqrt{35}, 0)$

21. $V(4, -8)$, $F(4, -7)$

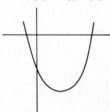

23. $(2, -1)$

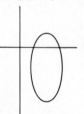

25.

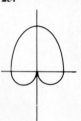

27.

29.

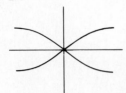

31.

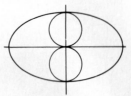

33. $\theta = \text{Arctan } 2 = 1.11$ **35.** $r^2 \cos 2\theta = 16$ **37.** $(x^2 + y^2)^3 = 16x^2y^2$
39. $3x^2 + 4y^2 - 8x - 16 = 0$ **41.** 4   **43.** 2
45. $(1.90, 1.55)$, $(1.90, -1.55)$, $(-1.90, 1.55)$, $(-1.90, -1.55)$
47. $m_1 = -\frac{12}{5}$, $m_2 = \frac{5}{12}$; $d_1^2 = 169$, $d_2^2 = 169$, $d_3^2 = 338$ **49.** $x^2 - 6x - 8y + 1 = 0$
51.   **53.**   **55.**   **57.** $x^2 + y^2 = 0.513$

59. $x^2 = -80y$ **61.**   **63.** $225y = 450x - x^2$

65. (a) 5.8 m, (b) 5.0 m   **67.** $\dfrac{x^2}{1.464 \times 10^6} + \dfrac{y^2}{1.461 \times 10^6} = 1$ **69.**   **71.** 13.5 ft

73. $r^2 = \dfrac{7.45 \times 10^7}{2.70 + 0.06 \sin^2 \theta}$

75. $(a^2 - b^2)x^2 + a^2y^2 + 2bx - 1 = 0$; $a = b$, parabola; $a^2 > b^2$, ellipse; $a^2 < b^2$, hyperbola

Exercises 21–1, p. 623

1.

No.	2	3	4	5	6	7
Freq.	1	3	4	2	3	2

3.

No.	45	46	47	48	49	50	51	52	53	54	55	56	57
Freq.	1	1	1	2	2	0	1	0	1	0	2	0	1

5.

Int.	2–3	4–5	6–7
Freq.	4	6	5

7.

Int.	43–45	46–48	49–51	52–54	55–57
Freq.	1	4	3	1	3

9.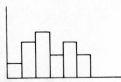

11.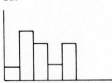

13.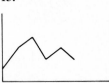

15.

17.

Vol.	746	747	748	749	750	751	752	753	754
No.	1	1	2	1	2	3	1	3	1

19.

21.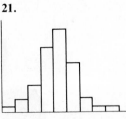

23.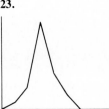

25.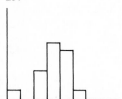

27.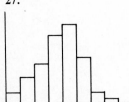

29.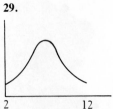

Exercises 21–2, p. 627

1. 4 **3.** 49 **5.** 4.6 **7.** 50.25 **9.** 4 **11.** 48, 49, 55 **13.** 751 mL **15.** 751 mL, 753 mL
17. 2.248 s **19.** 3.435 A **21.** 3.41 mA, 3.44 mA, 3.46 mA, 3.47 mA **23.** 20.9 mi/gal
25. $275, $300 **27.** 862 kW·h **29.** 4.5 **31.** 51

Exercises 21–3, p. 633

1. 1.50 **3.** 3.74 **5.** 1.50 **7.** 3.74 **9.** $60 **11.** 8.4 db **13.** 7 lb **15.** 147 kW·h
17. 60% **19.** 58% **21.** 64% **23.** 75%

Exercises 21–4, p. 638

1. $y = 1.0x - 2.6$

3. $y = -1.77x + 191$

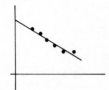

5. $V = -0.590i + 11.3$

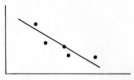

7. $V = 4.32 \times 10^{-15}f - 2.03$
   $f_0 = 0.470$ PHz

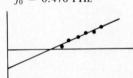

9. 0.985   **11.** -0.901

Exercises 21–5, p. 642

1. $y = 1.97x^2 + 4.8$

3. $y = 10.9/x$

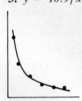

5. $y = 5.97t^2 + 0.38$

7. $y = 35.6(10^z) - 39.9$

9. $f = \dfrac{492}{\sqrt{L}} + 4$

Exercises for Chapter 21, p. 643

1. 3.6   **3.** 0.77   **5.**

Int.	101–103	104–106	107–109	110–112	113–115
Freq.	5	4	3	3	5

7. 106   **9.** 4.6   **11.**

13. 0.264 Pa·s   **15.** 0.014 Pa·s

17.

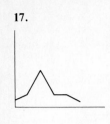

19. 0.927 in.

21. 0.93 in.

23. 0.012 in.

25. 4

27.

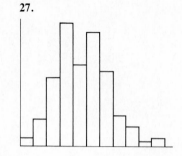

29. $R = 0.0983T + 25.0$

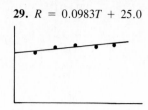

31. $y = 1.04 \tan x - 0.02$ **33.** $y = 1.10\sqrt{x}$

Exercises B–1, p. A-11

1. MHz, 1 MHz = 10^6 Hz   **3.** mm, 1 mm = 10^{-3} m   **5.** kilovolt, 1 kV = 10^3 V
7. milliampere, 1 mA = 10^{-3} A   **9.** 10^5 cm   **11.** 63,360 in.   **13.** 13.3 cm   **15.** 14.9 L
17. 144 in.^{2} **19.** 2.50×10^{-4} m^2 **21.** 2040 g   **23.** 13.6 L   **25.** 7600 kg·cm/s   **27.** 0.101 hp
29. 130 tons   **31.** 37 mi/h   **33.** 770 mi/h   **35.** 345 lb/ft^3 **37.** 101,000 Pa

Exercises B–2, p. A-14

1. 24 is exact.   **3.** 3 is exact, 74.6 is approx.   **5.** 1063 is approx.   **7.** 100 and 200 are approx.,
3200 is exact.   **9.** 3, 4   **11.** 3, 4   **13.** 3, 3   **15.** 1, 6   **17.** (a) 3.764 (b) 3.764
19. (a) 0.01 (b) 30.8   **21.** (a) same (b) 78.0   **23.** (a) 0.004 (b) same   **25.** (a) 4.93 (b) 4.9
27. (a) 57,900 (b) 58,000   **29.** (a) 861 (b) 860   **31.** (a) 0.305 (b) 0.31

Exercises B–3, p. A-17

1. 51.2   **3.** 1.69   **5.** 431.4   **7.** 30.8   **9.** 62.1   **11.** 270   **13.** 160   **15.** 27,000   **17.** 5.7
19. 4.39   **21.** 10.2   **23.** 22   **25.** 17.62   **27.** 18.85   **29.** 21.0 lb   **31.** First plane, 70 mi/h
33. 62.1 lb/ft^3 **35.** 850,000 lb   **37.** 115 ft/s

Exercises for Appendix C, p. A-24

1. 85°   **3.** 140°   **5.** 52°   **7.** 50°   **9.** 70°   **11.** 56°   **13.** 25°   **15.** 25°   **17.** 120°   **19.** 40°
21. 5   **23.** 17   **25.** 8   **27.** 5.120   **29.** 10.44   **31.** 28.89   **33.** 21   **35.** 20
37. 3.3 in., 4.0 in.   **39.** 16   **41.** 25 ft   **43.** 10 yd   **45.** 18.46 in.   **47.** 32.7 cm   **49.** 214.6 cm
51. 62.8 ft   **53.** 28 in.^{2} **55.** 13 yd^2 **57.** 24 m^2 **59.** 154 in.^{2} **61.** 216 ft^3 **63.** 20 mm^3
65. 924 in.^{3} **67.** 153 in.^{3} **69.** 27.1 ft^3 **71.** 689 m^3 **73.** 208 in.^{2} **75.** 537 m^2 **77.** 3220 ft^2
79. 343 cm^2

Exercises for Appendix D, p. A-34

(Most answers have been rounded off to four significant digits.)
1. 56.02   **3.** 4162.1   **5.** 18.65   **7.** 0.3954   **9.** 14.14   **11.** 0.5251   **13.** 13.35   **15.** 944.6
17. 0.7349   **19.** −0.7594   **21.** −1.337   **23.** 1.015   **25.** 41.35°   **27.** −1.182   **29.** 0.5862
31. 6.695   **33.** 3.508   **35.** 0.005685   **37.** 2.053   **39.** 5.765   **41.** 4.501×10^{10} **43.** 497.2
45. 6.648   **47.** 401.2   **49.** 8.841   **51.** 2.523   **53.** 10.08   **55.** 22.36   **57.** 20.3°   **59.** 4729
61. 3.301×10^4 **63.** 1.056   **65.** 55.5°   **67.** 3.277   **69.** 8.125   **71.** 1.000   **73.** 1.000
75. 12.90   **77.** 8.001   **79.** 8.053   **81.** 0.04259   **83.** 0.4219   **85.** 0.7822   **87.** 2.0736465
89. 124.3

Index of Applications

Index

Algebra

Exponents and Radicals

$$a^m \cdot a^n = a^{m+n}$$

$$\frac{a^m}{a^n} = a^{m-n} \qquad a \neq 0$$

$$(a^m)^n = a^{mn}$$

$$(ab)^n = a^n b^n$$

$$\left(\frac{a}{b}\right)^n = \frac{a^n}{b^n} \qquad b \neq 0$$

$$a^0 = 1 \qquad a \neq 0$$

$$a^{-n} = \frac{1}{a^n} \qquad a \neq 0$$

$$a^{m/n} = \sqrt[n]{a^m} = (\sqrt[n]{a})^m$$

$$\sqrt{ab} = \sqrt{a}\,\sqrt{b}$$

Special Products

$$a(x + y) = ax + ay$$

$$(x + y)(x - y) = x^2 - y^2$$

$$(x + y)^2 = x^2 + 2xy + y^2$$

$$(x - y)^2 = x^2 - 2xy + y^2$$

Quadratic Equation and Formula

$$ax^2 + bx + c = 0$$

$$x = \frac{-b \pm \sqrt{b^2 - 4ac}}{2a}$$

Properties of Logarithms

$$\log_b x + \log_b y = \log_b xy$$

$$\log_b x - \log_b y = \log_b\left(\frac{x}{y}\right)$$

$$n \log_b x = \log_b (x^n)$$

Trigonometry

$$\sin\theta = \frac{y}{r} \quad \cos\theta = \frac{x}{r} \quad \tan\theta = \frac{y}{x} \quad \cot\theta = \frac{x}{y} \quad \sec\theta = \frac{r}{x} \quad \csc\theta = \frac{r}{y}$$

Law of Sines: $\dfrac{a}{\sin A} = \dfrac{b}{\sin B} = \dfrac{c}{\sin C}$

Law of Cosines: $a^2 = b^2 + c^2 - 2bc \cos A$

Basic Identities

$$\sin\theta = \frac{1}{\csc\theta} \qquad \cos\theta = \frac{1}{\sec\theta} \qquad \tan\theta = \frac{1}{\cot\theta} \qquad \tan\theta = \frac{\sin\theta}{\cos\theta} \qquad \cot\theta = \frac{\cos\theta}{\sin\theta}$$

$$\sin^2\theta + \cos^2\theta = 1 \qquad 1 + \tan^2\theta = \sec^2\theta \qquad 1 + \cot^2\theta = \csc^2\theta$$

$$\sin 2\alpha = 2 \sin\alpha \cos\alpha \qquad \cos 2\alpha = \cos^2\alpha - \sin^2\alpha = 2\cos^2\alpha - 1 = 1 - 2\sin^2\alpha$$

$$\sin\frac{\alpha}{2} = \pm\sqrt{\frac{1 - \cos\alpha}{2}} \qquad \cos\frac{\alpha}{2} = \pm\sqrt{\frac{1 + \cos\alpha}{2}}$$